THE SPECTRAL ENERGY DISTRIBUTION OF GALAXIES

IAU SYMPOSIUM No. 284

COVER ILLUSTRATION:

M82 as viewed by Chandra in X-rays (blue), HST in Hα (orange) and visual continuum (green), and by Spitzer (red).

Picture credits: X-ray: NASA/CXC/JHU/D.Strickland; Optical: NASA/ESA/STScI/AURA/The Hubble Heritage team and IR: NASA/JPL-Caltech/Univ. of AZ/C. Engelbracht.

INTERNATIONAL ASTRONOMICAL UNION

UNION ASTRONOMIQUE INTERNATIONALE

International Astronomical Union

THE SPECTRAL ENERGY DISTRIBUTION OF GALAXIES

PROCEEDINGS OF THE 284th SYMPOSIUM OF THE INTERNATIONAL ASTRONOMICAL UNION HELD AT THE UNIVERSITY OF CENTRAL LANCASHIRE, PRESTON, UNITED KINGDOM SEPTEMBER 5–9, 2011

Edited by

RICHARD J. TUFFS

Astrophysics Dept, Max-Plank Institut für Kernphysik, Heidelberg, GERMANY

and

CRISTINA C. POPESCU

Jeremiah Horrocks Institute for Astrophysics and Supercomputing, University of Central Lancashire, Preston, UNITED KINGDOM

CAMBRIDGE UNIVERSITY PRESS
The Edinburgh Building, Cambridge CB2 2RU, UnitedKingdom
40 West 20th Street, New York, NY 10011–4211, USA
10 Stamford Road, Oakleigh, Melbourne 3166, Australia

First published 2012

Printed in the UK by MPG Books Ltd

Typeset in System $\LaTeX\ 2_\varepsilon$

A catalogue record for this book is available from the British Library

Library of Congress Cataloguing in Publication data

This journal issue has been printed on FSC-certified paper and cover board. FSC is an independent, non-governmental, not-for-profit organization established to promote the responsible management of the worlds forests. Please see www.fsc.org for information.

ISBN 9781107019843 hardback
ISSN 1743-9213

Table of Contents

Introduction

Section 1: Population Synthesis

Section 2: Understanding the Emergent SEDs of Local Universe Galaxies

Section 2-1: Understanding the Emergent SEDs of Local Universe Galaxies: Spiral and Dwarf Galaxies

Section 2-2: Understanding the Emergent SEDs of Local Universe Galaxies: Starburst Galaxies and Active Galactic Nuclei

Section 2-3: Understanding the Emergent SEDs of Local Universe Galaxies: Early Type Galaxies

Section 2-4: Understanding the Emergent SEDs of Local Universe Galaxies: Multiwavelength Surveys

Section 3: Star-Formation in Galaxies

Section 4: The Panchromatic View of the Milky Way

Section 5: Linking Low- and High-Energy Properties of Galaxies

Section 6: Understanding the Cosmological Evolution of Emergent SEDs

Preface

IAU Symposium 284 "The Spectral Energy Distribution of Galaxies" was held in Preston (UK), between 5th and 9th of September 2011. Jointly organised by the University of Central Lancashire (Preston) and the Max-Plank Institut für Kernphysik (Heidelberg), the symposium was attended by participants from 33 different countries from Europe, North America, South America, Asia, Africa and Australia/Oceania, with a balanced representation between professors, senior scientists and young career scientists, postdoctoral fellows and PhD students.

The symposium brought together developers and users of self-consistent physical or semi-empirical models for the emergent panchromatic spectral energy distributions (SEDs) of galaxies ranging over the complete accessible spectral range from gamma-rays to radio. Motivated by the rapid development in the corresponding observational capabilities in the last decade, the main goal of the symposium was to provide a forum for the interaction of modellers with both observers assembling multiwavelength datasets on galaxies and theoreticians considering fundamental physical processes in galaxies.

The program was fashioned to reflect the interconnections between the very broad range of physical processes responsible for the panchromatic photon output of galaxies. This embraced the formation, evolution and emission of stars; accretion-driven sources of photons; the chemical and physical properties of the interstellar medium, including both the gaseous and solid-state components and their interactions with ambient photon fields; and high energy processes involving cosmic rays. On the last day a final session was dedicated to models for the evolution of the panchromatic SEDs of galaxies over cosmological time, thus linking the detailed physical processes in individual galaxies discussed earlier in the week with the photon output of the Universe.

All of these topics have, of course, been the subjects of many dedicated individual symposia in the past, attended largely by their own specialized communities. However, IAU symposium 284 was unique in its concept of connecting the topics and bringing together the communities (or at least making a significant step towards achieving this). A particular challenge of this concept was to avoid the symposium becoming a sequence of self-contained mini-workshops or tutorials on each of the constituent topics, addressing selected topical issues directed at, and attended by, one particular community. However, in this respect any prior concerns proved to be completely unfounded. All the delegates, representing a mix of theoreticians, observers, and specialists in the many technical and astrophysical subfields needed to build SED models, proved to be enthusiastic and proactive participants throughout the week, generating many perceptive, and sometimes unexpected interdisciplinary discussions following both the oral presentations (many of these discussions could be documented for inclusion in the proceedings) and in the poster sessions. There were no parallel sessions.

In general, the intellectual atmosphere was open and constructive, with several known examples of new collaborations arising from discussions initiated at the symposium. For this, particular thanks must go to the authors for their careful preparation of the presentations. These were generally well directed to the broad audience, while very effectively answering the call for papers to highlight techniques and results combining measurements made from across the electromagnetic spectrum. Indeed, this response to the symposium confirms the genuine need and demand for more effective quantitative analysis techniques to exploit the already copious amounts of multwavelength data now available for galaxies near and far.

Cristina C. Popescu and Richard J. Tuffs May, 2012

Acknowledgements

The broad and challenging scope of this symposium was particularly well suited to the framework and scale of an IAU Symposium. We would therefore like to thank first and foremost the IAU for entrusting us with this challenge. Our particular gratitude is addressed to the organising committee of the coordinating division for our symposium, IAU Division VIII "Galaxies & the Universe". We are also very grateful to the broad support addressed by the other divisions and commissions of the IAU who recognised the interdisciplinary scope of our symposium, namely Division VII "Galactic System", Division VI "Interstellar Matter", Division X "Radio Astronomy", Division XI "Space & High Energy Astrophysics" and Division IX Commission 21 "Galactic and Extragalactic Background Radiation". Special thanks are addressed to the IAU's assistant general secretary Thierry Montmerle for his untiring role in supervising the link between the IAU and our symposium. The large diversity of scientists attending the symposium would not have been possible without the generous grants offered by the IAU, which we gratefully acknowledge.

Much valuable advice on the science content of the meeting was provided by the Scientific Organising Committee which was comprised of Gustavo Bruzual (CIDA, Venezuela), Françoise Combes (LERMA, Paris, France), Andrew Fabian (University of Cambridge, UK), Jay Gallagher (University of Wisconsin, Madison, USA), Yu Gao (Purple Mountain Observatory, China), Hidehiro Kaneda (Nagoya University, Japan), Nikolaos D. Kylafis (University of Crete, Heraklion, Greece), Renée Kraan-Korteweg (University of Cape Town, South Africa), Carol Lonsdale (NRAO, USA), Cristina C. Popescu (co-chair, Jeremiah Horrocks Institute, University of Central Lancashire, Preston, UK), Vladimir Ptuskin (IZMIRAN, Moscow, Russian Federation), Elaine Sadler (University of Sydney, Australia), Laura Silva (INAF, Trieste, Italy), Richard J. Tuffs (co-chair, Max Planck Institut für Kernphysik, Heidelberg, Germany), Jacqueline van Gorkom (Columbia University, New York, USA), Barbara Whitney (Space Science Institute, Boulder, USA). Their input was instrumental in bringing together such an exciting mix of diverse communities.

Many thanks are addressed to Bogdan Păstrăv from the Jeremiah Horrocks Institute (JHI) of the University of Central Lancashire (UCLan) for the continuous, careful and tireless work he did to help with the logistics of the conference, from the early stages of the conference organisation to the days when the conference took place. We would also like to thank Giovanni Natale from JHI/UCLan for taking on responsibility for the conference organisation during the critical time of the countdown days before the event and during the symposium. Meiert Grootes and Ellen Simmat from the Max Planck Institut für Kerphysik are acknowledged for their invaluable work in helping with the design of the conference poster and with the webpage of the conference. Gabi Wiese of MPI-Kernphysik was instrumental in putting together the abstract booklet of the conference and, during the conference, for marshalling the gathering of discussions for this proceedings volume. Finally we would like to thank Emma Kelly, our conference officer, for ensuring the smooth running of the conference.

We were very fortunate to have benefited from the enthusiastic support of the PhD and master students from UClan and MPI-Kernphysik during the busy days of the conference. They assisted with the IT, worked with the microphones, helped with the registration desk and were there in place if their presence was needed. Thanks to all of you: Nicky Agius, Ellen Andrae, Adam Clarke, Gareth Few, Meiert Grootes, Kelly Hambleton, James

Kelly, Michael Maxwell and Simon Murphy. Thanks also to the postdoctoral researchers and staff from the JHI/UClan who helped in various ways with the conference organisation.

For the general public in Lancashire the days of the IAU Symposium were highlighted by the fascinating public talk given by the JHI/UCLan professor Don Kurtz on "The Beauty of Galaxies: from the Milky Way to the Beginning of Time". We could not have been more proud of having such a distinguished speaker contributing to the other activities organised with the occasion of the IAU Symposium.

Our conference benefited from the generous financial support of the Great Britain Sasakawa (GBSF) foundation, which was used to support Japanese participants to attend the IAU Symposium. There is no doubt that the substantial numbers of Japanese astrophysicists attending the symposium would not have been possible without the Sasakawa grants. This, in turn, added significant value to the symposium, and, reciprocally, we believe, contributed to the propagation of the impressive results presented by our Japanese colleagues to the world wide community working in the field.

Our acknowledgements would not be complete without our thanks to the Dean of UCLan's School of Computing, Engineering and Physical Sciences, Robert Wallace, for his full support of the conference and for his welcoming address to the participants of the IAU 284, at the reception taken place at the Harris Museum in Preston. Many thanks also to the UCLan's Director of Research, Robert Walsh, for his enthusiastic speech addressed at the Harris Museum on the ocassion of the conference reception.

Finally, our special thanks go to Professor Gordon Bromage, Head of UCLan's Jeremiah Horrocks Institute, for his continuous encouragement and support of the conference and for his invaluable role in chairing the Local Organising Committee of the symposium.

THE ORGANIZING COMMITTEE

The Scientific Organising Commitee

- Gustavo Bruzual (CIDA, Venezuela)
- Françoise Combes (LERMA, Paris, France)
- Andrew Fabian (University of Cambridge, UK)
- Jay Gallagher (University of Wisconsin, Madison, USA)
- Yu Gao (Purple Mountain Observatory, China)
- Hidehiro Kaneda (Nagoya University, Japan)
- Nikolaos D. Kylafis (University of Crete, Heraklion, Greece)
- Renée Kraan-Korteweg (University of Cape Town, South Africa)
- Carol Lonsdale (National Radio Astronomy Observatory, USA)
- Cristina C. Popescu (co-chair, University of Central Lancashire, Preston, UK)
- Vladimir Ptuskin (IZMIRAN, Moscow, Russian Federation)
- Elaine Sadler (University of Sydney, Australia)
- Laura Silva (INAF, Trieste, Italy)
- Richard J. Tuffs (co-chair, Max Planck Institut für Kernphysik, Germany)
- Jacqueline van Gorkom (Columbia University, New York, USA)
- Barbara Whitney (Space Science Institute, Boulder, USA)

The Local Organising Committee

- Ellen Andrae (Max Planck Institut für Kernphysik)
- Gordon Bromage (co-chair, University of Central Lancashire, Preston, UK)
- Meiert Grootes (Max Planck Institut für Kernphysik)
- Emma Kelly (University of Central Lancashire, Preston, UK)
- Giovanni Natale (University of Central Lancashire, Preston, UK)
- Bogdan Păstrăv (University of Central Lancashire, Preston, UK)
- Cristina C. Popescu (co-chair, University of Central Lancashire, Preston, UK)
- Richard J Tuffs (co-chair, Max Planck Institut für Kernphysik)
- Gabriele Weese (Max Planck Institut für Kernphysik)
- Christopher van Eldik (Max Planck Institut für Kernphysik)

Sponsors

- The International Astronomical Union

- The Great Britain Sasakawa foundation

- The University of Central Lancashire

- The Max Planck Institut für Kernphysik

THE PARTICIPANTS OF IAU SYMPOSIUM 284 GATHER

FOR THE CONFERENCE PHOTOGRAPH.

Participants

- Viviana Acquaviva, USA
- Nicola Agius, UK
- Ellen Andrae, Germany
- Ko Arimatsu, Japan
- Steven Bamford, UK
- George Bendo, UK
- Francois Boulanger, France
- Gordon Bromage, UK
- Gustavo Bruzual, Venezuela
- Veronique Buat, France
- Martin Bureau, UK
- Denis Burgarella, France
- Sukanya Chakrabarti, USA
- Igor Chilingarian, Russia
- Adam Clarke, UK
- David Clements, UK
- Roger Clowes, UK
- Asantha Cooray, USA
- Luigi Costamante, Italy
- Roland Crocker, Germany
- Elisabete da Cunha, Germany
- Andre De Castro Milone, Brazil
- Ilse De Looze, Belgium
- Christoph Deil, Germany
- Brenda Dingus, USA
- Alberto Dominguez, USA
- Simon Driver, Australia
- Guillaume Drouart, ESO
- Loretta Dunne, New Zealand
- Andreas Efstathiou, Cyprus
- Vladimir Elkin, UK
- Bruce Elmegreen, USA
- Anant Eungwanichayapant, Thailand
- Ignacio Ferreras, UK
- Mercedes Filho, Portugal
- Nahiely Flores-Fajardo, Mexico
- John (Jay) Gallagher, USA
- Anna Gallazzi, Denmark
- Frederic Galliano, France
- Yu Gao, China
- Brad Gibson, UK
- Jean Michel Gomes, Portugal
- Meiert Willem Grootes, Germany
- Brent Groves, Germany
- Lucia Guaita, Sweden
- Philip Günster, Germany

- Kelly Hambleton, UK
- Paul Harvey, USA
- Barbara Hassall, UK
- Sara Heap, USA
- Christian Henkel, Germany
- Israel Hermelo, Germany
- Jim Hinton, UK
- Benne Holwerda, The Netherlands
- Andrew Hopkins, Australia
- Hyunjin Jeong, South Korea
- Noelia Jiménez, Argentina
- Benjamin Johnson, France
- David Jones, Germany
- Hidehiro Kaneda, Japan
- Oskar Karczewski, UK
- Ivan Katkov, Russia
- Sugata Kaviraj, UK
- Ciska Kemper, Taiwan
- Mina Koleva, Belgium
- Ralf Kotulla, USA
- Renée C., Kraan-Korteweg, South Africa
- Don Kurtz, UK
- Brian Lacki, USA
- Mark Lacy, USA
- Man I Lam, China
- Lauranne Lanz, USA
- Jaehyun Lee, South Korea
- Claus Leitherer, USA
- Ute Lisenfeld, Spain
- Lijie Liu, China
- Carol Lonsdale, USA
- Nidia Lugo, Venezuela
- Walter Maciel, Brazil
- John MacLachlan, UK
- Suzanne Madden, France
- Barry Madore, USA
- Gladis Magris Crestini, Venezuela
- Minni Mao, Australia
- Christopher Martin, USA
- Yoshiki Matsuoka, Japan
- Kalevi Mattila, Finland
- Michael Maxwell, UK
- Richard McDermid, USA
- Sharon Meidt, Germany
- Sofia Meneses-Goytia, The Netherlands
- Erin Mentuch, Canada
- Martin Meyer, Australia
- Areg Mickaelian, Armenia
- Norikazu Mizuno, Japan

- John Moustakas, USA
- Simon Murphy, UK
- Giovanni Natale, UK
- Danielle Nielsen, USA
- Ray Norris, Australia
- Kyuseok Oh, South Korea
- Rene A. Ortega-Minakata, Mexico
- Bogdan Adrian Păstrăv, UK
- Kate Pilkington, UK
- Cristina C. Popescu, UK
- Troy Porter, USA
- Almudena Prieto, Spain
- Philippe Prugniel, France
- Ando Ratsimbazafy, South Africa
- Monica Relano Pastor, Spain
- Aurelie Remy, France
- Gustavo E. Romero, Argentina
- Michael Rowan-Robinson, UK
- Kate Rowlands, UK
- Paul Ruffle, UK
- David Sanchez, Germany
- Patricia Sanchez-Blazquez, Spain
- Anne E. Sansom, UK
- Daniel Schaerer, Switzerland
- Andrew Schechtman-Rook, USA
- Chris Sedgwick, UK
- Kwang-il Seon, South Korea
- Stephen Serjeant, UK
- Zhengyi Shao, China
- Ralf Siebenmorgen, ESO
- Olga Silchenko, Russia
- Daniel Smith, UK
- Gregory Snyder, USA
- João Rodrigo Souza Leão, Brazil
- Laura Sturch, USA
- Toyoaki Suzuki, Japan
- Ryan Swindle, USA
- Fatemeh Tabatabaei, Germany
- Toshinobu Takagi, Japan
- Qinghua Tan, China
- Tomislav Terzić, Croatia
- Todd Thompson, USA
- Yoshiki Toba, Japan
- Kazufumi Torii, Japan
- Richard J. Tuffs, Germany
- Jing Wang, Germany
- Kenny Wood, UK
- Stijn Wuyts, Germany
- Tova Yoast-Hull, USA

- Fang-Ting Yuan, Japan
- Olga Zakhozhay, Ukraine
- Andreas Zezas, Greece
- Hong-Xin Zhang, USA
- Zhiyu Zhang, Germany
- Jiali Zhu, China
- Stefano Zibetti, Denmark

The Spectral Energy Distribution of Galaxies
Proceedings IAU Symposium No. 284, 2011
R.J. Tuffs & C.C. Popescu, eds.

doi:10.1017/S1743921312008599

Reification of galaxies: cognitive astrophysics and the multiwavelength inverse problem

Barry F. Madore
Carnegie Observatories, Pasadena, California, USA

Lessons learned in the history and philosophy of science have generally had little immediate impact on how we as individual astronomers conduct our research. And yet we do share many common views on how we undertake basic research, and how we translate observations and theory into communicable knowledge. In this introductory talk I will illustrate how we as extragalactic astronomers have already violated some of the basic tenets of what constitutes "science" as seen from a philosophical point of view, and I will predict what the future of astronomy as a science may soon look like. Simple examples of how we are already "cognitively closed" to many immediate and tangible aspects of the Universe will be given and some solutions to this dilemma will be proposed. We may be at a point in time where more data is not necessarily the best solution to our problems. Discovering that familiar concepts and even certain objects may not exist in the traditional sense of the word could provide a motivation for broadening our way of conceptualizing the extragalactic Universe, more as a continuum of processes and phase transitions rather than an assembly of discrete objects. Once again the Universe may be "forcing us to think".

The Spectral Energy Distribution of Galaxies
Proceedings IAU Symposium No. 284, 2011
R.J. Tuffs & C.C. Popescu, eds.

doi:10.1017/S1743921312008605

Population synthesis at the crossroads

Claus Leitherer[1] **and Sylvia Ekström**[2]

[1]Space Telescope Science Institute, 3700 San Martin Dr., Baltimore, MD 21218, USA
email: leitherer@stsci.edu

[2]Observatoire de Genève, 51 chemin des Maillettes, CH-1290 Sauverny, Switzerland
email: sylvia.ekstrom@unige.ch

Abstract. The current state-of-the-art of population synthesis is reviewed. The field is currently undergoing major revisions with the recognition of several key processes as new critical ingredients. Stochastic effects can artificially enhance or suppress certain evolutionary phases and/or stellar mass regimes and introduce systematic biases in, e.g., the determination of the stellar initial mass function. Post-main-sequence evolution is often associated with irregular variations of stellar properties on ultra-short time-scales. Examples are asymptotic giant branch stars and luminous blue variables, both of which are poorly treated in the models. Stars rarely form in isolation, and the fraction of truly single stars may be very small. Therefore, stellar multiplicity must be accounted for since many systems will develop tidal interaction over the course of their evolution. Last but not least, stellar rotation can drastically increase stellar temperatures and luminosities, which in turn leads to revised mass-to-light ratios in population synthesis models.

Keywords. binaries: close, stars: evolution, HII regions, galaxies: fundamental parameters, galaxies: photometry, galaxies: starburst, galaxies: stellar content

1. From Fundamental Data to Cosmology

Stellar population synthesis is at the heart of spectral energy distribution (SED) studies of galaxies. The very nature of population synthesis makes it a rather multi-disciplinary research area. It is built on our knowledge of stellar astrophysics, which in turn relies on fundamental data derived from laboratory physics, such as atomic line data and nuclear reaction rates. On the other hand, those studying galaxy evolution trust population synthesis models for the interpretation of galaxy colors and their evolution with cosmic time. Ultimately, cosmological models often rest on our faith in population synthesis. The chain of assumptions that must be made is usually — subconsciously — assigned errors which hierarchically decrease from the fields of cosmology over stellar astrophysics to laboratory data. One goal of this review is to raise the awareness of radically new developments and uncertainties in hitherto considered "well-known" areas and to foster feedback between the stellar and cosmological communities.

2. The Basics of Population Synthesis

Population synthesis aims to reproduce the observed galaxy SED from a set of prespecified input parameters. Evolutionary spectral synthesis is a special, widely used case of population synthesis, whose goal is to reproduce observed spectra self-consistently from the star-formation history of a galaxy and from stellar evolution models. This method was pioneered by Tinsley (1980) after the advent of modern stellar evolutionary tracks in the Hertzsprung-Russell diagram (HRD). Evolutionary synthesis has relatively few free parameters but its success depends on the reliability of the adopted stellar models. Therefore, assessing these assumptions becomes critical.

The astrophysical ingredients entering evolutionary synthesis are as follows: (i) Quantities related to the star-formation process, i.e., the star-formation rate and its evolution with time and the stellar mass spectrum at birth, also known as the initial mass function (IMF). See, e.g., the conference proceedings of Treyer *et al.* (2011). (ii) Once stars have formed, stellar evolution models provide a prescription for the variation of the stellar luminosity (L), effective temperature ($T_{\rm eff}$), and mass (M) as a function of chemical composition, initial mass, and time. (iii) Spectral libraries describe the spectrum of each star for any given (L, $T_{\rm eff}$, M). (iv) Second-order effects, such as dust attenuation or geometric effects are often accounted for in a very approximate way or even neglected altogether.

Several widely used synthesis models are made available to the community via dedicated web sites. Among the most popular packages are GALEV (Schulz *et al.* 2002; Kotulla *et al.* 2009), GALEXEV (Bruzual & Charlot 1993; 2003), population synthesis based on MILES (Vazdekis *et al.* 2010), PEGASE (Fioc & Rocca-Volmerange 1997; Le Borgne *et al.* 2004), and Starburst99 (Leitherer *et al.* 1999; Leitherer & Chen 2009). In general, the predictions made by different synthesis codes are in good agreement. When discrepancies are found, they can usually be understood in terms different intrinsic input parameters and/or treatment of peculiar stellar evolutionary phases, such as the asymptotic giant branch (AGB) phase (Vázquez & Leitherer 2005; Conroy & Gunn 2010).

The mentioned synthesis packages restrict their output to models for the stellar luminosities, with some generic contribution by nebular emission. More specialized applications generate panchromatic galaxy SEDs accounting for the stellar as well as the gas emission using full photo-ionization modeling (Dopita *et al.* 2005; 2006a; 2006b; Groves *et al.* 2008). SEDs with self-consistent inclusion of dust were first introduced by Dwek & Scalo (1980) and refined most recently by Dwek & Cherchneff (2011). Chemical evolution models for the full life-cycle of the major elements over the cosmic evolution of galaxies were discussed by Matteucci (2009; 2010).

Rather than reviewing these "traditional" synthesis models I will focus on some recent, new developments which may challenge well-established results: stochasticity, peculiar evolutionary phases, stellar multiplicity, and stellar rotation. The importance of these mechanisms and concepts had been recognized for quite some time but only in recent years have they been implemented in population synthesis codes for *quantitative* comparison with observations.

3. Stochasticity

Traditional population synthesis models scale all population properties by the star-formation rate. This approach is strictly valid only in the limit of a stellar population containing an infinite number of stars. Real star clusters or galaxies, however, contain a finite number of stars. Furthermore, most of the light is sometimes provided by very few bright stars, in particular in the near-infrared (IR). The so-called stochastic fluctuations in the integrated population properties are the result of the random presence of these luminous stars. Depending on population age and wavelength domain, stochastic effects can become important for masses as high as $\sim 10^5\ M_\odot$ (Bruzual 2002; Cerviño *et al.* 2002; Fouesneau & Lançon 2010).

da Silva *et al.* (2011) developed and released SLUG, a new code to "Stochastically Light Up Galaxies". SLUG synthesizes stellar populations with a Monte Carlo technique to treat stochastic sampling properly. Included are the effects of clustering, the stellar IMF, star-formation history, stellar evolution, and cluster disruption. Their code has been built using the same ingredients as in Starburst99, thereby allowing the immediate

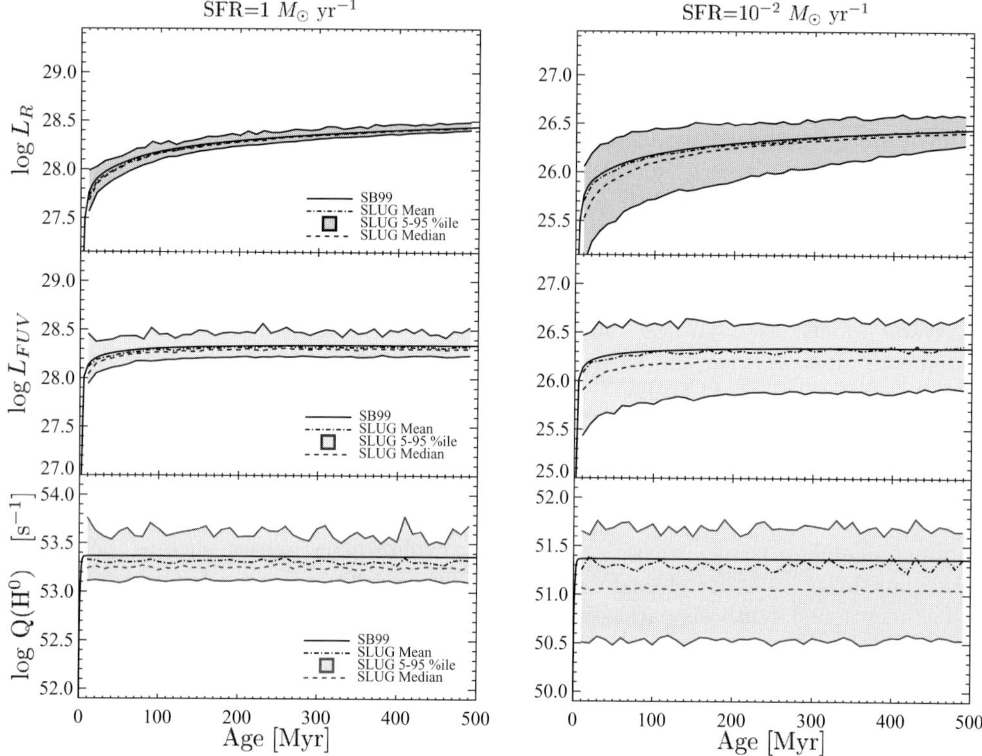

Figure 1. R-band, far-ultraviolet (UV), and ionizing luminosities vs. time for galaxies with constant star-formation rates of 1 (left) and 0.01 (right) $M_\odot$ yr^{-1} assuming clustered star formation. Starburst99 models (solid black lines) are compared with simulations from SLUG. The SLUG models are represented by their mean (dash-dotted line), median (dashed line) and 5 – 95 percentile range (shaded region). From da Silva *et al.* (2011).

separation of stochastic effects. An example is shown in Fig. 1. This figure highlights the increased scatter and systematic offsets due to stochasticity and clustering at low star-formation rates.

GALEX UV imagery suggests that star-formation rates inferred from Hα in galactic environments characterized by low stellar and gas densities tend to be less than those based on the UV luminosity (Lee *et al.* 2009). The origin of the discrepancy is still under debate. One possible explanation is that the stellar IMF is systematically deficient in high-mass stars in such environments. However, effects of uncertainties in the stellar evolutionary tracks and model atmospheres, uncertain metallicities, non-constant star-formation histories, leakage of ionizing photons, departures from Case B recombination, dust attenuation, and finally stochasticity in the formation of high-mass stars need to be considered.

Eldridge (2011) found that the scatter and variation of the Hα- and UV derived star-formation rates are less dependent on the IMF but more on the star-formation history of each individual galaxy. The general trend is that those systems with weaker star-formation in the last 10 Myr have relatively lower Hα rates, while those with most of the star formation in the last 10 Myr display the opposite behavior even at low mean star-formation rates. Therefore any simulation attempting to predict the properties of a sample of galaxy must take into account the stochastic nature of star-formation. If there

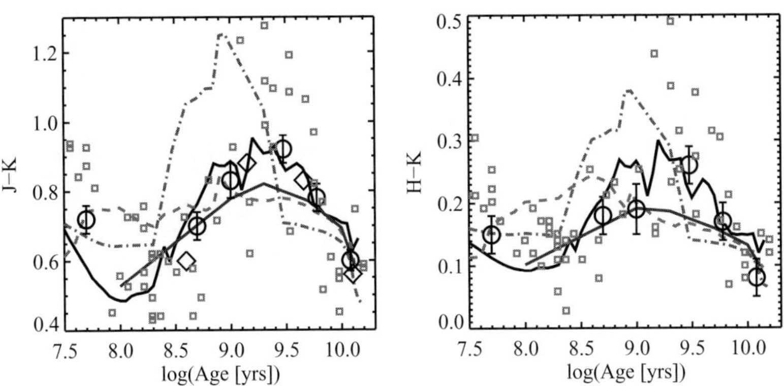

Figure 2. Comparison of observed and modeled near-IR colors as a function of age. The data are derived from star clusters in the Magellanic Clouds. The models include the predictions from Maraston (2005; dot-dashed lines), Bruzual & Charlot (2003; triple-dot-dashed lines) and two FSPS models (dashed and solid lines). The observed data are individual measurements (squares) and averages over many star clusters (circles). From Conroy & Gunn (2010).

are only a few clusters, the appearance of the galaxy-wide stellar population may be very different from what might be expected for a simple stellar population with a smooth star-formation history.

4. Rapid Evolutionary Phases

Rapid evolutionary phases pose particular challenges for population synthesis. Thermally pulsing AGB stars are especially notorious (Maraston 2011). Their uncertainties in stellar evolution modeling arise from short pulsation time-scales ($\sim 10^4$ yr), the double-shell burning, and strong mass loss in this phase. The spectral modeling during the carbon-rich phase adds to the complexity. Yet AGB stars can often make a significant contribution to the galaxy SED in the optical and near-IR because the AGB phase is bright and occurs around $\sim$1 Gyr when the overall stellar population has already faded. Different population synthesis models disagree quite severely in their predicted luminosity contribution from thermally pulsing AGB stars. The Flexible Stellar Population Synthesis (FSPS) technique of Conroy *et al.* (2009; 2010) and Conroy & Gunn (2010) is particularly well suited for addressing this issue. FSPS is a novel Monte-Carlo-type approach that is capable of flexibly handling various uncertain aspects of stellar evolution such as, e.g., thermally pulsing AGB stars, the horizontal branch, or blue stragglers. The technique allows one to marginalize over uncertain evolutionary aspects in order to understand the full uncertainties associated with the physical properties of galaxies including stellar masses, ages, metallicities, and star-formation rates. An example is in Fig. 2 where the AGB phase as implemented in different evolutionary tracks is tested. The conclusion is that models generally overestimate this phase, which is consistent with the result of Kriek *et al.* (2010) who used a sample of $\sim$60 post-starburst galaxies at $z \approx 1.6$ in the NEWFIRM medium-band survey to assess different models. The AGB contribution to the integrated SED is a factor of $\sim$3 lower than predicted. This could be due to lower bolometric luminosities, shorter lifetimes, and/or heavy dust obscuration of AGB stars.

AGB stars affect population synthesis models both directly (via their luminosity contribution) and indirectly (via their strong mass loss influencing stellar evolution). Luminous

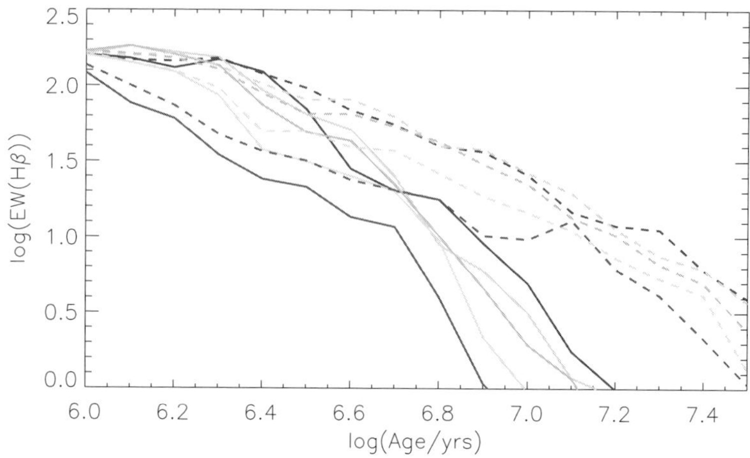

Figure 3. Hβ equivalent width vs. time for a series of instantaneous burst models generated with the BPASS code. Solid and dashed lines indicate populations without and with close binary stars, respectively. Both model families with and without binaries are shown for five different metallicities, higher metallicities generally having higher equivalent widths. From Eldridge & Stanway (2009).

blue variables (LBVs) mirror AGB stars in their eruptions and their high mass-loss rates. They introduce uncertainties in population synthesis similar to those of AGB stars. They generally do not contribute significantly to the galaxy SED but their mass ejections are decisive for the final evolution of massive stars. It has recently been recognized that most mass loss in massive stars prior to the Wolf-Rayet phase occurs in LBVs, whereas steady stellar winds close to the main-sequence are less important (Smith 2010). LBVs belong to a diverse class of transients in the upper HRD spanning a wide range of outburst luminosity which ranges from that of classical novae to that of core-collapse supernovae (Smith *et al.* 2011). The ubiquity of eruptive phases and highly time-variable evolutionary phase introduce formidable challenges to population synthesis.

5. Stellar Multiplicity

Massive stars are known for their high-degree of multiplicity. Most OB stars are found in binaries and multiple systems. Combining information from these various surveys, Mason *et al.* (2009) found a minimum multiplicity fraction close to 70%. Even bona fide single field stars may have been part of a multiple system in the past, then ejected by a supernova explosion or by dynamical interaction. While multiplicity deserves studies by its own right, neglecting this fundamental property can be a serious omission in population synthesis (Vanbeveren 2010). In the following we will not focus on the well-known vertical main-sequence displacement in the color-magnitude diagram caused by binaries but rather on the modified stellar evolution of close binary system and their consequences for population synthesis. Relevant in this context is the fraction of *close* binaries, i.e., those binaries whose evolution differs from that of single stars. Sana & Evans (2011) estimate a spectroscopic binary fraction of close to 50%, with a significant fraction likely to be interacting.

de Mink *et al.* (2009) demonstrated that the evolution of massive, close binaries still on the main-sequence sharply differs from single-star evolution. Tidal effects can lead to significant spin-up and high rotational velocities. Mixing processes induced by rotation may be so efficient that helium produced in the center is mixed throughout the envelope. Such stars evolve almost chemically homogeneously. If the metallicity is low, they remain blue and compact, while they gradually evolve into Wolf-Rayet stars and possibly into progenitors of long γ-ray bursts.

Alternatively, at low rotation velocities, the distinguishing property of binary evolution will be Roche-lobe overflow, which typically occurs off the main-sequence. Depending on the system properties, the primary will transfer material to the secondary, which will accrete material and gain mass. The crucial point for population synthesis is the "rejuvenation effect" (van Bever & Vanbeveren 1998; Eldridge & Stanway 2009): the previously less massive secondary will become more massive and more luminous and will therefore appear younger. This is illustrated in Fig. 3, which shows the evolution of the nebular Hβ emission-line equivalent width of single stellar populations. The models were obtained with the BPASS (Binary Population and Spectral Synthesis) code of Eldridge *et al.* (2008), which accounts for the evolutionary effects of massive binaries. Compared with single-star models, binary populations are predicted to have larger Hβ equivalent width at older ages due to the rejuvenation effect. Hβ is commonly used as an age indicator in H II regions (Stasińska *et al.* 2001) and would require recalibration. For instance, in single-star models, a large Hβ value at older ages would be incompatible with an instantaneous burst and suggest ongoing star formation. Binary models for single stellar populations, however, might still produce large enough Hβ emission at older ages.

6. Stellar Rotation

Massive stars exhibit significant helium and nitrogen enrichment on the surface while still on the main-sequence (Hunter *et al.* 2009). These telltale signs of nuclear processing require an efficient transportation mechanism to expose materials from the convective core. While mass removal by stellar winds had previously been thought to assume this role, the main-sequence mass-loss rates are now considered too low (see Section 4) to make this a viable option. Alternatively, mixing, mass transfer, and mass loss in close binaries might cause the enrichment. However, it seems unlikely that the evolution of *all* massive stars is driven by binarity. Many stars with surface enrichment are also fast rotators. While the relation between nitrogen overabundance and rotation velocity is complex and far from unique, there is consensus that rotation is an important, if not the prime driver of mixing in massive stars (Brott *et al.* 2011a). Stellar evolution with rotation has been modeled over the past decade, but it is only now that *quantitative* evolutionary tracks with rotation have become available for implementation of population synthesis (Brott *et al.* 2011a; Ekström *et al.* 2011).

Fig. 4 illustrates the principal effects caused by rotation. At high masses ($M > 20 M_\odot$), the dominant mechanism is the increase of the convective core and the reduced surface opacity due to the He enhancement, resulting in a higher L and higher $T_{\rm eff}$, respectively. (Recall that H is the major opacity source and any decrease of its relative abundance by mixing lowers the opacity and therefore increases $T_{\rm eff}$.) Therefore tracks with rotation tend to fall above and to the left of non-rotating tracks. At lower masses (e.g., the track for $M = 15\ M_\odot$ in Fig. 4), these effects become less important and are surpassed by the impact of the centrifugally induced radius increase and the associated lower $T_{\rm eff}$. At masses below $\sim$2 $M_\odot$ no rotational effects are expected since low-mass stars arrive on the zero-age main-sequence with essentially no velocity because of magnetic braking.

We implemented the new suite of 48 different stellar evolutionary tracks, both rotating and non-rotating, by Ekström *et al.* (2011) in Starburst99. The grid has solar chemical composition and covers the full mass range from 0.8 to 120 $M_\odot$. The rotating models

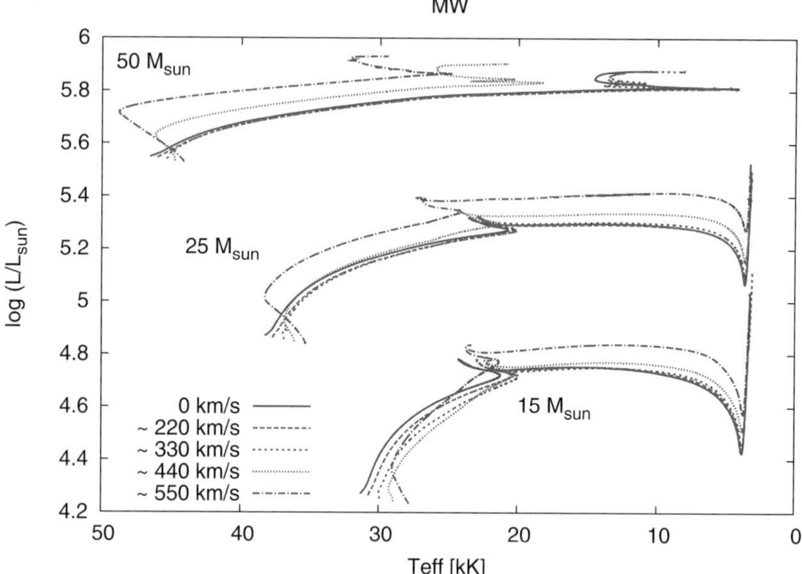

Figure 4. Evolutionary tracks for rotating stars with initial masses of 15, 25, and 50 $M_\odot$ having solar chemical composition. Line types indicate different initial rotation velocities on the zero-age main-sequence. From Brott *et al.* (2011b).

start on the zero-age main-sequence with a rotation velocity of 40% of the equatorial break-up velocity. Guidance on the choice of the initial rotation velocity is provided by measurements in B stars but data for O stars are still scarce. Clearly this parameter must be considered with care. The evolution is computed until the end of the central carbon-burning phase, the early AGB phase, or the core helium-flash for the massive, intermediate, and low-mass stars, respectively. Vázquez *et al.* (2007) used an earlier, preliminary version of these models to test the impact of stellar rotation on synthetic population properties. The new model suite will allow us to verify and extend the previous work (Levesque *et al.* 2012, in preparation).

Rotation increases both L and $T_{\rm eff}$ for stars more massive than $\sim$20 $M_\odot$. This effect sets in very early on the main-sequence. As a result, a stellar population containing hot, massive stars is predicted to be more luminous for a given mass, and its SED is shifted to shorter wavelengths. Therefore the most dramatic changes with respect to prior models occur at the short-wavelength end of the SED. Examples are reproduced in Fig. 5, which compares $M_{\rm Bol}$ and the luminosity of the nebular Brγ) resulting from models with and without rotation. The single stellar population models with rotation for $M_{\rm Bol}$ become more luminous by $\sim$0.4 mag (left part of Fig. 5). The right part of Fig. 5 shows the Brγ luminosity for an equilibrium population. Brγ is often used to determine star-formation rates in obscured galaxies, such as ultra-luminous IR galaxies. For this specific example, the rates following from the models with and without rotation would be 100 and 175 $M_\odot$ yr^{-1}, respectively. The extreme UV increases by a factor of a few in the hydrogen ionizing continuum and by several orders of magnitude in the neutral and ionized helium continua with the new models. The predictions for the latter need careful testing, as the photon escape fraction crucially depends on the interplay between the stellar parameters supplied by the evolution models and the radiation-hydrodynamics of the atmospheres. In contrast, the escape of the hydrogen ionizing photons has little

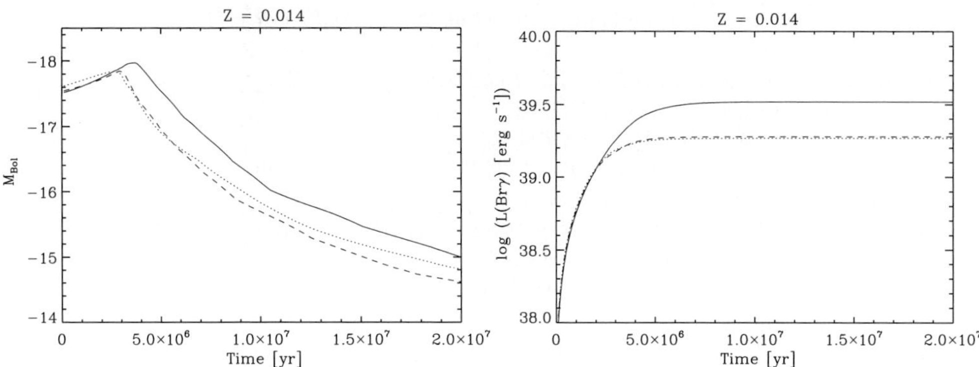

Figure 5. Comparison of stellar population properties resulting from evolutionary tracks with and without rotation. Left: bolometric luminosity vs. time for a single stellar population with a mass of 10^6 $M_\odot$. Right: Brγ luminosity for continuous star formation at a rate of 100 $M_\odot$ yr^{-1}. Solid lines: new tracks with rotation; dashed: new tracks with zero rotation velocity; dotted: previous generation of Geneva tracks released 1992 – 1994. The models are for solar chemical composition and use a standard Kroupa IMF between 0.1 and 100 $M_\odot$.

dependence on the particulars of the atmospheres and consequently is a relatively safe prediction.

7. The New Standard

Stochasticity, exotic stellar evolutionary phases, multiplicity, and rotation are now considered essential ingredients in population synthesis models. A new suite of population synthesis models which account for these effects and provide quantitative predictions have become available in the very recent past. Somewhat frustratingly, implementation of these processes cannot rely on first principles, which essentially adds new free parameters requiring empirical calibration. Care is therefore required when the first model generation is used to analyze galaxy SEDs. The community is encouraged to perform extensive testing and help making these new synthesis models the new standard.

Acknowledgements

Claus Leitherer acknowledges travel support from the Director's Discretionary Research Fund at STScI.

References

Brott, I., de Mink, S. E., Cantiello, M., *et al.* 2011b, *A&A*, 530, A115
Brott, I., Evans, C. J., Hunter, I., *et al.* 2011a, *A&A*, 530, A116
Bruzual A. G. 2002, in IAU Symp. 207, Extragalactic Star Clusters, eds. D. Geisler, E. K. Grebel, & D. Minniti (San Francisco: ASP), 616
Bruzual A. G. & Charlot, S. 1993, *ApJ*, 405, 538
——. 2003, *MNRAS*, 344, 1000
Cerviño, M., Valls-Gabaud, D., Luridiana, V., & Mas-Hesse, J. M. 2002, *A&A*, 381, 51
Conroy, C. & Gunn, J. E. 2010, *ApJ*, 712, 833
Conroy, C., Gunn, J. E., & White, M. 2009, *ApJ*, 699, 486
Conroy, C., White, M., & Gunn, J. E. 2010, *ApJ*, 708, 58
da Silva, R. L., Fumagalli, M., & Krumholz, M. 2011, arXiv:1106.3072
de Mink, S. E., Cantiello, M., Langer, N., *et al.* 2009, *A&A*, 497, 243

Dopita, M. A., Fischera, J., Sutherland, R. S., *et al.* 2006b, *ApJS*, 167, 177
Dopita, M. A., Fischera, J., Sutherland, R. S., *et al.* 2006a, *ApJ*, 647, 244
Dopita, M. A., Groves, B. A., Fischera, J., *et al.* 2005, *ApJ*, 619, 755
Dwek, E. & Cherchneff, I. 2011, *ApJ*, 727, 63
Dwek, E. & Scalo, J. M. 1980, *ApJ*, 239, 193
Ekström, S., Georgy, C., Eggenberger, P., *et al.* 2011, arXiv:1110.5049
Eldridge, J. J. 2011, arXiv:1106.4311
Eldridge, J. J., Izzard, R. G., & Tout, C. A. 2008, *MNRAS*, 384, 1109
Eldridge, J. J. & Stanway, E. R. 2009, *MNRAS*, 400, 1019
Fioc, M. & Rocca-Volmerange, B. 1997, *A&A*, 326, 950
Fouesneau, M. & Lançon, A. 2010, *A&A*, 521, A22
Groves, B., Dopita, M. A., Sutherland, R. S., *et al.* 2008, *ApJS*, 176, 438
Hunter, I., Brott, I., Langer, N., *et al.* 2009, *A&A*, 496, 841
Kotulla, R., Fritze, U., Weilbacher, P., & Anders, P. 2009, *MNRAS*, 396, 462
Kriek, M., Labbé, I., Conroy, C., *et al.* 2010, *ApJ*, 722, L64
Le Borgne, D., Rocca-Volmerange, B., Prugniel, P., *et al.* 2004, *A&A*, 425, 881
Lee, J. C., Gil de Paz, A., Tremonti, C., *et al.* 2009, *ApJ*, 706, 599
Leitherer, C. & Chen, J. 2009, *New Astronomy*, 14, 356
Leitherer, C., Schaerer, D., Goldader, J. D., *et al.* 1999, *ApJS*, 123, 3
Maraston, C. 2005, *MNRAS*, 362, 799
——. 2011, in Why AGB Stars care about Galaxies II, eds. F. Kerschbaum, T. Lebzelter, & B. Wing (San Francisco: ASP), in press (arXiv:1104.0022)
Mason, B. D., Hartkopf, W. I., Gies, D. R., Henry, T. J., & Helsel, J. W. 2009, *AJ*, 137, 3358
Matteucci, F. 2009, in Probing Stellar Populations out to the Distant Universe, ed. L. A. Antonelli *et al.* (Melville: AIP), 143
——. 2010, in Galaxies and their Masks, eds. L. David, K. C. Freeman & I. Puerari (New York: Springer), 261
Sana, H. & Evans, C. J. 2011, in IAU Symp. 272, Active OB Stars, eds. C. Neiner, G. Wade, G. Meynet, & G. Peters (Cambridge: CUP), 474
Schulz, J., Fritze-v. Alvensleben, U., Möller, C. S., & Fricke, K. J. 2002, *A&A*, 392, 1
Smith, N. 2010, in Hot and Cool: Bridging Gaps in Massive Star Evolution, eds. C. Leitherer, P. Bennett, P. Morris & J. van Loon (San Francisco: ASP), 63
Smith, N., Li, W., Silverman, J. M., Ganeshalingam, M., & Filippenko, A. V. 2011, *MNRAS*, 415, 773
Stasińska, G., Schaerer, D., & Leitherer, C. 2001, *A&A*, 370, 1
Tinsley, B. M. 1980, *Fund. Cosm. Phys.*, 5, 287
Treyer, M., Wyder, T., Neill, J., Seibert, M., & Lee, J. 2011, UP2010: Have Observations Revealed a Variable Upper End of the Initial Mass Function? (San Francisco: ASP)
van Bever, J. & Vanbeveren, D. 1998, *A&A*, 334, 21
Vanbeveren, D. 2010, in IAU Symp. 266, Star Clusters: Basic Galactic Building Blocks throughout Time and Space, eds R. de Grijs & J. R. D. Lépine (Cambridge: CUP), 293
Vazdekis, A., Sánchez-Blázquez, P., Falcón-Barroso, J., *et al.* 2010, *MNRAS*, 404, 1639
Vázquez, G. A. & Leitherer, C. 2005, *ApJ*, 621, 695
Vázquez, G. A., Leitherer, C., Schaerer, D., Meynet, G., & Maeder, A. 2007, *ApJ*, 663, 995

Discussion

CHAKRABARTI: Do the models of star cluster evolution include IR emission that are produced by young, dust-enshrouded star clusters?

LEITHERER: No – they don't extend to very young star clusters.

THOMPSON: You showed that with rotation both the lifetime and total luminosity of the most massive stars in a ZAMS stellar population are longer (larger)? Why?

LEITHERER: Rotation increases the size of the convective core by adding extra-nuclear hydrogen from the radiative envelope. This increases the stellar luminosity L because L is determined by the luminosity of the core. Since fresh fuel is added to the nuclear burning zone, the lifetimes are higher as well.

ELMEGREEN: For the stochastic models as low star formation rates, where H_α/FUV drops does that drop depend on the star formation history - i.e. whether it is bursty with a complete IMF for each burst, or continuous?

LEITHERER: Lee *et al.* – the authors of the original observational paper - did in fact simulations of different SF histories and concluded that this could not be the sole explanation of the decrease of SFR(H_α)/SFR(UV) at low SF rates.

The Spectral Energy Distribution of Galaxies
Proceedings IAU Symposium No. 284, 2011
R.J. Tuffs & C.C. Popescu, eds.

doi:10.1017/S1743921312008617

Effects of Non-Solar Abundance Ratios on Star Spectra: Observations versus Models

Anne E. Sansom[1], André de Castro Milone[2], Alexandre Vazdekis[3] and Patricia Sánchez-Blázquez[4]

[1]Jeremiah Horrocks Institute, University of Central Lancashire, Preston, Lancs PR1 2HE, UK
[2]Divisão de Astrofísica, Instituto Nacional de Pesquisas Espaciais, Av. dos Astronautas 1758, São José dos Campos, SP 12227-010, Brazil
[3]Instituto de Astrofísica de Canarias, La Laguna, E-38200 Tenerife, Spain
[4]Departamento de Física Teórica, Universidad Autónoma de Madrid, 28409, Spain

email: aesansom@uclan.ac.uk

Abstract. Element abundance ratios hold important clues to understanding the evolution of stellar populations, through the varying timescales of different nucleosynthetic contributors. Newly measured and compiled [Mg/Fe] ratios in the MILES stellar library are used to confront models of star spectra. Such models have been used in recent years to provide estimates of differential changes in spectral line strengths, due to enhancements in [α/Fe]. In this paper we test one widely used set of theoretical element response functions. Using magnesium as a proxy for all alpha elements we test the reliability of these theoretical response functions against empirical observations, and thus the reliability of current methods of measuring element abundance ratios in the stellar populations.

Keywords. stars: atmospheres, stars: abundances, techniques: spectroscopic.

1. Why are Abundance Ratios Important?

In individual stars element abundance patterns are useful for understanding the initial composition, evolution and photospheric structural and dynamical processes. In samples of stars abundance patterns are used to determine which components the stars are associated with and to constrain the history of how those populations formed. In unresolved stellar populations, the ratio of α-capture elements to iron-peak elements [α/Fe] can tell us about the timescale of star formation, since different enrichment sources contribute to the inter-stellar medium at different rates and with different patterns of heavy element abundances. Therefore it is important to be able to accurately measure abundance ratios so that a coherent picture of the data on stellar populations can be achieved, against which models of galaxy evolution and chemical feedback can be tested.

The importance of abundance ratios in understanding stellar populations is highlighted by the large number of publications in which attempts are made to measure abundance ratios in integrated stellar populations (recent examples include: Annibali *et al.* 2011, Thomas *et al.* 2011). Many of these studies are based on the use of element response functions. We use such response functions in studying different types of galaxies (e.g. Sansom & Northeast 2008). Response functions are tables of how line strengths change in theoretical spectra, due to different atmospheric abundances of elements. They can be used to make differential corrections to theoretically or empirically derived simple stellar populations (SSPs), to account for effects of non-solar abundance ratios on predicted line-strength, such as the widely used Lick indices. The most extensively used response functions are those of Korn *et al.* (2005) (hereafter K05) and it is those that we test in this current work.

2. New Measurements

The empirical data that we use to test the K05 response functions became available from our recent careful study of stars in the MILES spectral library (Sánchez-Blázquez *et al.* 2006). Our catalogue of [Mg/Fe] measurements for MILES stars is published in Milone, Sansom & Sánchez-Blázquez (2011). See also Milone *et al.* in this volume. We use [Mg/Fe] as a proxy for the α-elements in general, therefore we can assign [α/Fe] and [Fe/H] values for many of the stars in the MILES library. Response functions are applied in a differential way. Therefore we can use MILES stars that have the same effective temperature and surface gravity as the models of K05 (within observational errors) and form the ratio of line-strengths at specific abundances ([α/Fe],[Fe/H]) to line-strengths of MILES stars at solar abundances ([α/Fe] and [Fe/H] = 0.0). This relies on having a suitable star with the base abundance (0,0), which we have, within observational errors. For the corresponding predictions from theoretical response functions this is done in a two-stage process; first to correct to a specific metallicity, then to a specific [α/Fe]. This latter step is done by taking into account changes due to the α-elements only. Both MILES observations and K05 models have Lick indices measured at the Lick resolutions.

3. Comparisons with Theoretical Models

3.1. *Line-Strength Indices*

For strong lines, ratios of line-strengths are compared, however, some of the Lick indices take positive and negative equivalent width values (e.g. for H_γ and H_δ features), therefore differences rather than ratios are compared for those cases. In these comparisons of normalised observations versus normalised theoretical predictions, the data points will lie on a one-to-one line if the predictions using response functions match the observations. Below we show an example for some of the Lick indices.

Fig. 1 shows that the Lick $H_{\delta A}$ and $H_{\gamma A}$ indices are not well matched between observations and models, whereas Fe5270 and Fe5335 are. Other Fe sensitive features also behave as predicted from the response functions. The lack of agreement that we find for higher order Balmer indices may be due to a problem with our assumption of [Mg/Fe] tracking [α/Fe] in the observed stars; an anomaly in the normalising stars used; or a problem with the predictions from response functions for specific indices. The higher order Balmer lines are located in a region of the spectrum that shows a rapid change with wavelength and a stronger dependence on abundance ratios. Features in the side bands may systematically affect response function predictions in such rapidly changing regions in the blue part of the spectrum, where there are a lot of atomic and molecular line contributions.

New and alternative band definitions need to be explored, together with tests of the theoretical model spectra, in order to find more robust measures for abundance ratios in unresolved stellar populations. This also affects measurements of ages and metallicities since all indices have some sensitivity to each of these parameters.

3.2. *Spectra*

To test for the effect of abundance ratios across a stellar spectrum we form the ratio of observed spectra at enhanced versus solar [α/Fe] ratios. An example is illustrated in Fig. 2, which shows the ratio of two spectra of stars in the MILES library. The stars were chosen to have the same overall metallicity, based on inferred [O/H] abundances via

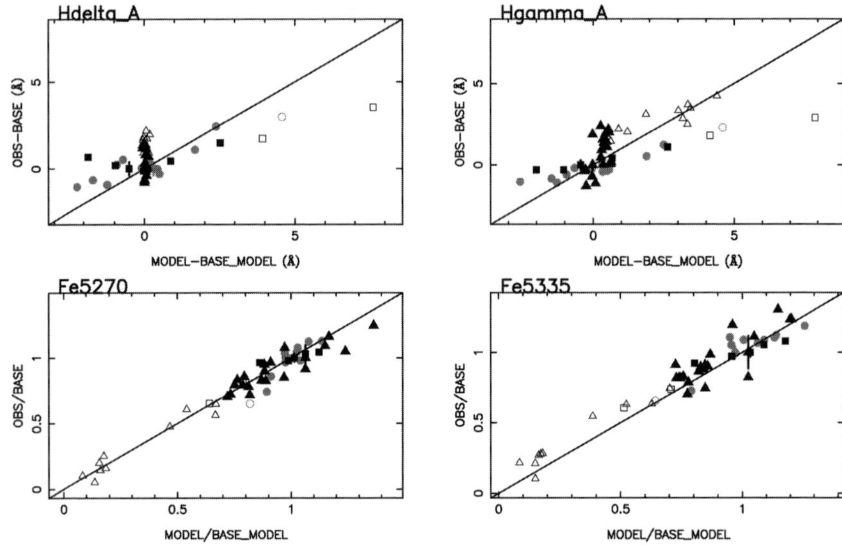

Figure 1. Comparison of empirical observations for MILES stars with theoretical predictions obtained from the K05 response functions, for $H_{\delta A}$, $H_{\gamma A}$, Fe5270 and Fe5335 Lick indices. Model values are normalised in the same way as the observations in these plots. BASE refers to indices for stars with solar abundances. Data points are stars covering abundance ranges: $-2.86<$[Fe/H]$<+0.41$ and $-0.40<[\alpha/\mathrm{Fe}]<+0.53$. The diagonal line shows the one-to-one relation. Circles are cool giant stars (T_{eff} ~4255K, log(g)~1.9), triangles are turn-off stars for a 5 Gyr old population (T_{eff} ~6200K, log(g)~4.1) and squares are cool dwarf stars (T_{eff} ~4575K, log(g)~4.6) matching the T_{eff} and log(g) values of three theoretical models with tabulated response functions in K05. Open symbols correspond to larger extrapolations from the base models at solar abundance (i.e. stars with [Fe/H]<-0.4).

the correlation of Bensby, Feltzing & Lundström (2004). This figure shows that the blue to UV part of the spectrum, below about 450nm, is increasingly affected by changes in $[\alpha/\mathrm{Fe}]$. Qualitatively similar patterns are found for ratios of theoretical spectra, as shown at the top of Fig. 2. This is similar to the earlier work of Cassisi *et al.* (2004) (their figure 2), based on ATLAS 9 models. There are several identifiable features in the blue region of the spectrum that may be worth further study for their sensitivity to abundance patterns. This very blue part of the spectrum has been little observed or modelled to date. This is because there are many uncertain predictions of lines particularly affecting the blue region of the spectrum, therefore attempts to model the flux levels there have required corrections for these uncertain contributions.

4. Interpretation and Future Work

We find that, whilst some of the Lick indices are correctly modelled for element abundance ratios, through the use of the Korn *et al.* 2005 response functions, other indices are not. In particular the iron sensitive Lick indices are quite well modelled, whereas the higher order Balmer line indices show discrepancies, when observations are compared with models of star spectra. The sense of these discrepancies is different in cool and warm stars. Therefore it is not easy to predict how these discrepancies will affect SSPs. We are currently extending the MILES empirical spectral database to encompass a wider range of stars with known abundance ratios.

Further tests of spectra, indices and response functions are ongoing and will be presented in a future paper. From the initial spectral ratios given here, future work on the

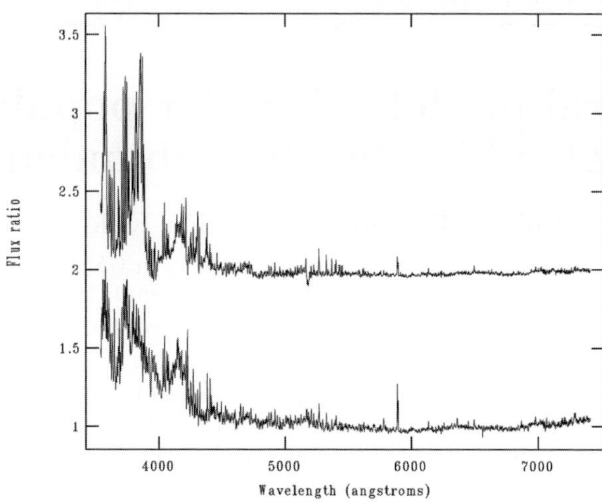

Figure 2. The lower curve shows the ratio between an observed (MILES) spectrum of a giant star with $[\alpha/\mathrm{Fe}] = +0.4$ to that of a star with $[\alpha/\mathrm{Fe}] = 0.0$, both at solar [Z/H], $T_{eff} = 4500$ and $\log(g) = 2.0$. The upper curve shows this same ratio for theoretical spectra from Coelho *et al.* (2007). The curves are separated by one in the vertical axis of relative flux ratio, to avoid overlapping. The effects of abundance ratios are strongest in the U and B bands, at <4500Å.

sensitivity of features in the blue part of the spectrum shows promise for highlighting features sensitive to varying abundance ratios. This will be an important new direction to explore, since accurate measurements of abundance ratios are vital for interpreting star formation histories in galaxies. Uncovering and accurately calibrating sensitive new $[\alpha/\mathrm{Fe}]$ indicators would help with such measurements.

References

Annibali, F., Grützbauch R., Rampazzo R., *et al.* 2011, *A&A*, 528, 19
Bensby, T., Feltzing, S., & Lundström, I. 2004, *A&A*, 415, 155
Cassisi, S., Salaris, M., Castelli, F., & Pietrinferni, A. 2004, *ApJ*, 616, 498
Coelho, P., Bruzual, G., Charlot, S., *et al.* 2007, *MNRAS*, 382, 498
Korn, A. J., Maraston, C., & Thomas, D. 2005, *A&A*, 438, 685
Milone, A. de C., Sansom, A. E., & Sánchez-Blázquez, P. 2011, *MNRAS*, 414, 1227
Sánchez-Blázquez, P., Peletier, R. F., Jiménez-Vicente, J., *et al.* 2006, *MNRAS*, 371, 703
Sansom, A. E. & Northeast, M. S. 2008, *MNRAS*, 387, 331
Thomas, D., Johansson, J., & Maraston, C. 2011, *MNRAS*, 412, 2199

Discussion

Ferreras: Concerning the "blue excess", would it appear if treating both [Fe/H] & $[\alpha/\mathrm{Fe}]$ as free parameters? Concerning the H_γ and H_δ mismatch, could this be related to "nearby" features like CN in H_δ or G-band in H_γ? Are there any better indices?

Sansom: The spectral ratios plotted show specific examples of stars with known [Fe/H] & $[\alpha/\mathrm{Fe}]$. These are not fitted parameters. The stars chosen have the same overall metallicity, and different $[\alpha/\mathrm{Fe}]$ ratios. Concerning the indices, indeed band and side band definitions are important and we will explore further definitions in the literature for the H_γ and H_δ features.

The Spectral Energy Distribution of Galaxies
Proceedings IAU Symposium No. 284, 2011
R.J. Tuffs & C.C. Popescu, eds.

doi:10.1017/S1743921312008629

Spectral models of stellar populations resolved in chemical abundances

Philippe Prugniel[1] **and Mina Koleva**[2,1]

[1]Université de Lyon, Lyon I, CRAL-Observatoire de Lyon UMR5574, CNRS, France
email: philippe.prugniel@univ-lyon1.fr

[2]Sterrenkundig Observatorium, Ghent University, Belgium

Abstract. We present model spectra of stellar populations with variable chemical composition. We derived the [α/Fe] abundance ratio of the stars of the most important libraries (ELODIE, CFLIB and MILES) using full spectrum fitting and we generated PEGASE.HR models resolved in [α/Fe]. We used a semi-empirical approach that combines the observed spectra with synthetic stellar spectra. We tested the models using them to derive [α/Fe] in galaxies and star clusters using full spectrum fitting. The present models are available from http://ulyss.univ-lyon1.fr

Keywords. galaxies: stellar content, galaxies: abundances, stars: fundamental parameters

---

1. Introduction

The comparison between models and observations is the key to derive the metallicity, age and star formation history of galaxies and star clusters. The evolutionary synthesis of galaxies consists in choosing an initial mass function (IMF) and following the evolution of the population over the time according to some theoretical evolutionary tracks. This approach predicts the distribution of the stars in the effective temperature ($T_{\rm eff}$) vs. surface gravity (log g) plan at any epoch, and the final step is to draw spectra from a library to assemble the integrated spectrum.

The modelling of stellar populations steadily progressed since the first attempts in the 80s (e.g. Bruzual 1983). Beside the metallicity, age and history, the detailed chemical composition became an important parameter to measure. This allows one to trace the source of metal enrichment, because of the different yields of type Ia and II supernovae. In turn, since both SN types correspond to different time-scales, this informs about the duration of the enrichment process. A population that produced its metal and formed its stars before the SNIa onset will be deficient in Fe (or as more commonly said, over-abundant in α-elements, like O and Mg). To vary the yield in the models, both the stellar library and the evolutionary tracks have in principle to be changed. However, Dotter *et al.* (2007) have shown that the tracks essentially depend on the total metallicity, rather than of the actual mixture. Therefore, the scaled-solar tracks can reasonably be used for any composition, provided they are associated with stellar spectra of the same total metallicity.

The models predict colours, used for SED fitting, spectra, and line-strength indices. Indices are equivalent-widths defined for a number of prominent spectral features and their study brings an approximate handle on the physical parameters (age, metallicity and/or detailed composition).

Several authors modeled indices with variable abundances of α-elements relative to Fe (Weiss *et al.* 1995; Trager *et al.* 2000; Thomas *et al.* 2003). The approach consisted in evaluating the response of individual indices to changes of composition using theoretical spectra (Tripicco & Bell 1995). Then, assuming that the abundance pattern in the library is matching that of the solar neighborhood (justified by the fact that the observed libraries consist of nearby stars), the indices could be estimated for any T_{eff} log g, [Fe/H] and [α/Fe]. In the visible region of the spectrum, the variation of [α/Fe] results in a prominent change of the Mg_b index near 5175 Å, and therefore a combination of this index with some Fe indices could be used to derive the [α/Fe] in galaxies and clusters.

Models of indices were developing faster and further than the spectral predictions, because of the lack of stellar libraries calibrated in relative flux. However, the quality of the libraries also progressed. Libraries of about 1000 stars covering a wide range of the parameter space, with an extended coverage of the optical spectrum at mediun spectral resolution and a fair flux calibration became available. The three most important are ELODIE (Prugniel *et al.* 2007b, R=$\lambda/\Delta\lambda$ $\approx$10000, $\lambda\lambda = 3900 - 6800$ Å), CFLIB (Valdes *et al.* 2004, R$\approx$4000, $\lambda\lambda = 3460 - 9400$ Å) and MILES (Sánchez-Blázquez *et al.* 2006, R$\approx$2000, $\lambda\lambda = 3536 - 7410$ Å).

In this paper, we present the first spectral models with variable [α/Fe] based on these libraries and on precise measurements of [α/Fe] in their stars.

2. Method

Previous attempts to produce semi-empirical spectral models with variable [α/Fe] were made in Prugniel *et al.* (2007a), and Walcher *et al.* (2009). The latter one uses population models based on synthetic spectra to predict the differential change of each spectral element when [α/Fe] is varied. This differential effect is then used to 'correct' empirical models to the desired α-elements enrichment. The former method determines the differential changes of the stellar spectra and 'correct' the library before computing models at different [α/Fe]. Both approachs are equivalent, and are based on the assumption that the libraries follow the abundance pattern of the solar neighborhood (i.e. they adopt a functional dependence of [α/Fe] as a function of [Fe/H]). Models of Lick indices also make the same assumption.

Although the assumption of a relation between [α/Fe] and [Fe/H] is statistically verified, [α/Fe] present a significant dispersion at a given [Fe/H] (e.g. Milone *et al.* 2011). When the library is interpolated to extract the spectra entering in the synthesis, the stochastic effects can become significant. For example, for a local interpolation only a

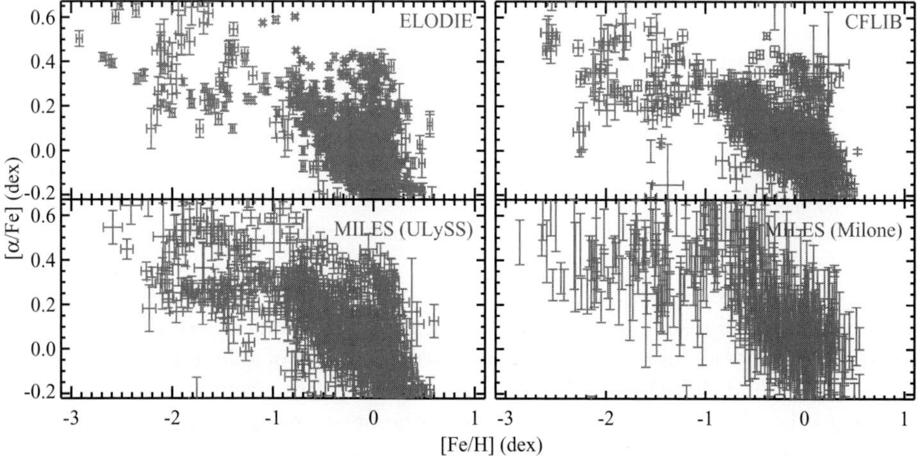

Figure 1. Distribution of [α/Fe] in the three libraries, compared to the distribution in Milone *et al.* (2011).

small number of library stars are used near a given location in the $T_{\rm eff}$, log g and [Fe/H] space, and their mean [α/Fe] may be biased. In order to avoid this effect, *the improvement introduced in the present work is to determine [α/Fe] in the individual stars of the library.* Each stellar spectrum is then corrected to a set of given [α/Fe] applying differential effects inferred from a theoretical library. An alternative may have been to measure [α/Fe] and directly interpolate the spectra in the four-dimensional space. This would have avoided the dependency toward a theoretical library, but required some extrapolations, because of the non-uniform distribution of the stars in the [Fe/H] vs. [α/Fe] plan,

The approach is finally the following:

– Measure [α/Fe] in the spectra of the libraries.

– Correct the library to some given values of [α/Fe] using differential effects computed in a theoretical library.

– Build interpolators for these corrected libraries. (these are functions returning a spectrum for given $T_{\rm eff}$, log g, and [Fe/H].

– Make model predictions at the chosen [α/Fe].

3. Determination of [α/Fe] in the stars

We used full-spectrum fitting with the ULySS package (Koleva *et al.* 2009) to measure [α/Fe]. Full-spectrum fitting was successful to measure the other atmospheric parameters in the various libraries (Wu *et al.* 2011; Prugniel *et al.* 2011; Koleva & Vazdekis 2011), and the precision and reliability was found consistent with detailed analysis using high-resolution spectroscopy.

We measured [α/Fe] by comparing the ELODIE spectra to the grid of Coelho *et al.* (2005). We minimized the χ^2 residuals to derive the four atmospheric parameters. We measured [α/Fe] only for $3500 < T_{eff} < 7000$, because of the limits of the grid of synthetic spectra. We assumed that the remaining stars follow the solar neighborhood abundance pattern. Then, we used the ELODIE library as a reference to measure [α/Fe] in the other libraries, using the same full-spectrum fitting method.

The measurements compare well with the most precise measurements from the compilation of Milone *et al.* (2011). The distribution of [α/Fe] vs. [Fe/H] is presented in Fig. 1 for the three libraries and compared to Milone *et al.* (2011).

4. Population models resolved in [α/Fe]

We generated pairs of interpopators for [α/Fe] $\in \{0, 0.4\}$ dex, and computed grids of single stellar populations (SSPs) models with PEGASE.HR (Le Borgne *et al.* 2004). These pairs of grids were then grouped together to make four dimensional arrays (wavelength, age, [Fe/H] and [α/Fe]) that can be used with ULySS to determine the three SSP-equivalent parameters. Figure 2 is a snapshot of models made with the three libraries.

We used these models to analyze Galactic globular clusters spectra as in Koleva *et al.* (2008), and we found results consistent with the determinations based on spectroscopy of resolved stars. We also applied the method to study spectra of galaxies from Koleva *et al.* (2011) and we found that the results are consistent with those from Lick indices, but they are more precise. The gain in precision reflects the better usage of the information. Mg_b accounts for less than 20% of the total opacity change in the optical range ($\lambda > 4500$ Å) when [α/Fe] varies, and using the whole spectral range with full-spectrum fitting is a clear advantage.

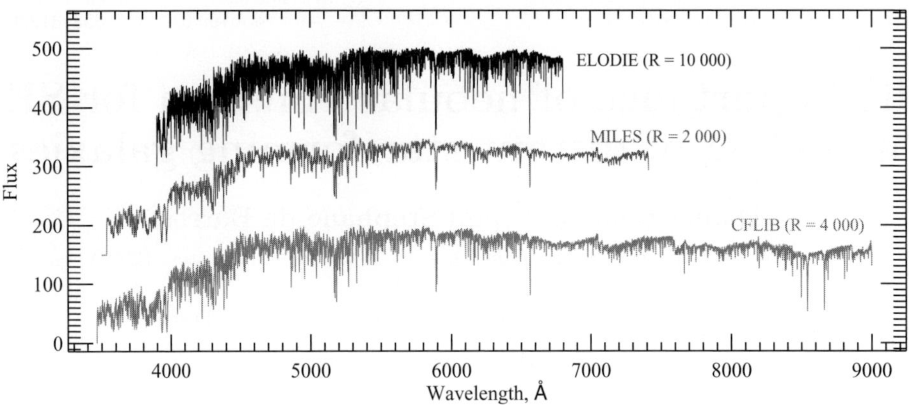

Figure 2. Single stellar population models with PEGASE.HR and ELODIE, MILES and CFLIB. The age is eight Gyr and the metallicity is solar.

5. Prospective

This semi-empirical method, used to build models of stellar populations resolved in [α/Fe], can straightforwardly be applied to resolve other individual elements. The key point is to have reliable synthetic grids of spectra to determine the differential effects. The Coelho *et al.* (2005) grid already resolve Ca independently of the α-elements, and other grids, varying other elements will be computed in the future.

The models presented here are available at `http://ulyss.uly-lyon1.fr`. They can be used with the ULySS package to measure [α/Fe] in galaxies and clusters.

References

Bruzual, G. 1983, *ApJ*, 273, 105

Coelho, P., Barbuy, B., Meléndez, J., Schiavon, R. P., & Castilho, B. V. 2005, *A&A*, 443, 735

Dotter, A., Chaboyer, B., Ferguson, J. W., *et al.* 2007, *ApJ*, 666, 403

Koleva, M., Gupta, R., Prugniel, P., & Singh, H. 2008, in Astronomical Society of the Pacific Conference Series, Vol. 390, Pathways Through an Eclectic Universe, ed. J.H. Knapen, T.J. Mahoney & A. Vazdekis, 302

Koleva, M., Prugniel, P., Bouchard, A., & Wu, Y. 2009, *A&A*, 501, 1269

Koleva, M., Prugniel, P., de Rijcke, S., & Zeilinger, W. W. 2011, *MNRAS*, 417, 1643

Koleva, M. & Vazdekis, A. 2011, *ArXiv* e-prints, 1111.5449

Le Borgne, D., Rocca-Volmerange, B., Prugniel, P., *et al.* 2004, *A&A*, 425, 881

Milone, A. D. C., Sansom, A. E., & Sánchez-Blázquez, P. 2011, *MNRAS*, 414, 1227

Prugniel, P., Koleva, M., Ocvirk, P., Le Borgne, D., & Soubiran, C. 2007a, in IAU Symposium, Vol. 241, IAU Symposium, ed. A. Vazdekis & R.F. Peletier, 68–72

Prugniel, P., Soubiran, C., Koleva, M., & Le Borgne, D. 2007b, *arXiv*:astro-ph/0703658

Prugniel, P., Vauglin, I., & Koleva, M. 2011, *A&A*, 531, A165

Sánchez-Blázquez, P., Peletier, R. F., Jiménez-Vicente, J., *et al.* 2006, *MNRAS*, 371, 703

Thomas, D., Maraston, C., & Bender, R. 2003, *MNRAS*, 339, 897

Trager, S. C., Faber, S. M., Worthey, G., & González, J. J. 2000, *AJ*, 119, 1645

Tripicco, M. J. & Bell, R. A. 1995, *AJ*, 110, 3035

Valdes, F., Gupta, R., Rose, J. A., Singh, H. P., & Bell, D. J. 2004, *ApJS*, 152, 251

Walcher, C. J., Coelho, P., Gallazzi, A., & Charlot, S. 2009, *MNRAS*, 398, L44

Weiss, A., Peletier, R. F., & Matteucci, F. 1995, *A&A*, 296, 73

Wu, Y., Singh, H. P., Prugniel, P., Gupta, R., & Koleva, M. 2011, *A&A*, 525, A71

The Spectral Energy Distribution of Galaxies
Proceedings IAU Symposium No. 284, 2011
R.J. Tuffs & C.C. Popescu, eds.

doi:10.1017/S1743921312008630

The importance of nebular emission for SED modeling of distant star-forming galaxies

Daniel Schaerer[1,2] and Stephane de Barros[1]

[1]Observatoire de Genève, Université de Genève, 51 Ch. des Maillettes, 1290 Versoix, Switzerland

[2]CNRS, IRAP, 14 Avenue E. Belin, 31400 Toulouse, France

email: daniel.schaerer@unige.ch

Abstract. We highlight and discuss the importance of accounting for nebular emission in the SEDs of high redshift galaxies, as lines and continuum emission can contribute significantly or subtly to broad-band photometry. Physical parameters such as the galaxy age, mass, star-formation rate, dust attenuation and others inferred from SED fits can be affected to different extent by the treatment of nebular emission.

We analyse a large sample of Lyman break galaxies from $z \sim 3$–6, and show some main results illustrating e.g. the importance of nebular emission for determinations of the mass–SFR relation, attenuation and age. We suggest that a fairly large scatter in such relations could be intrinsic. We find that the majority of objects ($\sim$ 60–70%) is better fit with SEDs accounting for nebular emission; the remaining galaxies are found to show relatively weak or no emission lines. Our modeling, and supporting empirical evidence, suggests the existence of two categories of galaxies, "starbursts" and "post-starbursts" (lower SFR and older galaxies) among the LBG population, and relatively short star-formation timescales.

Keywords. galaxies: starburst, galaxies: high-redshift, galaxies: evolution, galaxies: formation

1. Introduction

Until recently the importance of nebular emission on the SED of high redshift galaxies and hence on on our understanding of the physical parameters of these objects was overlooked. For example, when Spitzer started to detect the rest-frame optical part of the spectrum of high-redshift galaxies redshifted into broad-band filters of its IRAC camera at 3.6 and 4.5 μm, it was soon noted that some of these galaxies showed fairly large Balmer breaks (e.g. Egami *et al.* 2005, Eyles *et al.* 2005). Several authors interpreted the large break as being due to the presence of "old", evolved stellar populations with ages of several hundred Myr (Eyles *et al.* 2005, Labbe *et al.* 2010). If true this would imply the presence of very early galaxy populations, and very large formation redshifts ($z_f \sim 10 - 30$).

However, since emission lines are ubiquituous in star-forming galaxies and since atomic physics causes most of the strong observed lines to be in the (rest-frame) optical part of the spectrum, longward of the Balmer break, nebular emission affects broad-band photometry differentially, and can cause an apparent Balmer break. The importance of this effect for high-z galaxies was pointed out by Schaerer & de Barros (2009), who showed that significantly younger ages could be obtained when accounting for nebular lines.

It has now become clear (Schaerer & de Barros 2010, Ono *et al.* 2010, Lidman *et al.* 2011) that nebular emission (both lines and continuum emission) must be taken into account for the interpretation of photometric measurements of the SEDs of star-forming galaxies such as Lyman-alpha emitters (LAE) and Lyman break galaxies (LBGs) – the dominant galaxy populations at high-z.

Furthermore, as testified by the presence of Lyα emission, a large (and possibly growing) fraction of the currently know population of star-forming galaxies at high redshift shows emission lines (at least Lyα!; cf.Ouchi *et al.* 2008, Stark *et al.* 2010 and others).

In parallel a lot of diverse evidence for galaxies with strong emission lines and/or strong contributions of nebular emission to broad-band filters has been found at different redshifts, e.g. by Shim *et al.* (2011), McLinden *et al.* (2011), Atek *et al.* (2011), Trump *et al.* (2011), van der Wel *et al.* (2011).

Here we briefly summarise our method to account for nebular emission in a SED plus photometric-redshift fitting code and present some results obtained from a systematic study of a large sample of $z \sim 3$–6 LBGs.

2. SED modeling

Several evolutionary synthesis models, such as PÉGASE, GALEV and others include nebular emission (cf. Fioc *et al.* 1999, Charlot & Longhetti 2001, Anders & Fritze 2003, Zackrisson *et al.* 2008). However, they were not used to analyse observations of high redshift galaxies.

Our SED fitting tool, described in Schaerer & de Barros (2009, 2010), is based on a version of the *Hyperz* photometric redshift code of Bolzonella *et al.* (2000), modified to taking into account nebular emission. In de Barros *et al.* (2011, hereafter dB11) we consider a large set of spectral templates based on the GALAXEV synthesis models of Bruzual & Charlot (2003), covering different metallicities and a wide range of star formation histories. A Salpeter IMF is adopted. Nebular emission from continuum processes and numerous emission lines is added to the spectra predicted from the GALAXEV models as described in Schaerer& de Barros (2009), proportionally to the Lyman continuum photon production. The intergalactic medium (IGM) is treated with the prescription of Madau (1995).

The free parameters of our SED fits are: redshift z, metallicity Z (of stars and gas), star formation (SF) history described by the timescale τ (i.e. SFR $\propto \exp(-t/\tau)$), the age $t_\star$ defined since the onset of star-formation, and attenuation A_V described by the Calzetti law. In some models we also vary the strength of the Lyα line from zero to its maximum predicted by case B (see Schaerer *et al.* 2011). In dB11 we consider 3 sets of SF histories, exponentially declining SF with variable timescales, SFR = const, and rising star-formation following Finlator *et al.* (2011). We use Monte-Carlo simulations to determine probability distribution functions for each parameter/quantity.

3. Impact of nebular emission on physical parameters of star-forming galaxies from $z \sim 3$ to 8

In Schaerer & de Barros (2009) we have demonstrated the effect nebular emission may have on age determinations of high-z galaxies. In Schaerer & de Barros (2010) we have applied our SED fitting tool to $z \sim 6$–8 galaxies recently observed with the new WFC3 camera onboard HST and other, more luminous galaxies. We have shown an overall trend of the star-formation rate (SFR) increasing with stellar mass, albeit with a large scatter. Also, we have shown that the specific SFR may be higher than previously thought, depending on model assumptions. We have also found indications for dust in some galaxies at this redshift, and a possible trend of increasing dust attenuation with galaxy mass. However, since only few of these galaxies are detected beyond $\gtrsim$ 1.6–2 μm, the uncertainties of the physical parameters are fairly large.

To improve on this aspect, and to examine systematically the effects of nebular emission on the physical parameters of distant galaxies, we have examined a large sample of $z \sim$ 3–6 galaxies. Some results from this study (dB11) are highlighted here.

3.1. *$z \sim$ 3–6 galaxies: photometric data and selection*

We have used the GOODS-MUSIC catalogue of Santini *et al.* (2009) providing photometry in the U, B_{435}, V_{606}, i_{776}, z_{850LP}, J, H, K, bands mostly from the VLT and HST, and the 3.6, 4.5, 5.8, and 8.0 μm bands from the IRAC camera onboard *Spitzer*. Using standard criteria we have selected U, B, V, and i-drop galaxies. To reduce the contamination rate (typically $\sim$ 10–20 %) we have only retained the objects whose median photometric redshifts agree with the targetted redshift range. This leaves us with a sample of 389, 705, 199, and 60 galaxies with median photometric redshifts of $z_{\rm phot}$ = 3.3, 3.9, 4.9, and 6.0. See dB11 for more details.

3.2. *Results – Two categories of high-z galaxies*

One of the first results obtained when applying SED models with/without nebular emission to our large galaxy sample is that the majority of the objects ($\sim$ 60–70%) are better fit with nebular emission, irrespectively of the SF history. For the rest, the difference between fits with or without nebular emission is generally minor. The difference between these two categories corresponds overall to objects whose best-fit SED clearly shows nebular lines (for the majority), and weak or no lines for the rest. Empirically this can be verified for $z \sim$ 3.8–5 galaxies, where Hα falls in the 3.6 μm filter, whereas the 4.5 μm filter remains essentially free of strong emission lines. Indeed, for this redshift range we find that the objects best fit with nebular lines show a systematically bluer (3.6-4.5) color than the other category (see de Barros *et al.* 2011b). Our first category strongly resembles the "Hα emitters" identified by Shim *et al.* (2011) through their 3.6 μm excess.

From our modeling we suggest that the two categories correspond to "starburst" and "post-starburst" phases. This is supported by their difference in (instantaneous) SFR, where the objects better fit with emission lines ("starbursts") show on average a higher SFR than the "post-starburst" category (see Fig. 1), and also by an age difference. Interestingly we find a quite similar distribution of galaxies between these categories for all redshifts between $z \sim$ 3 and 6.

3.3. *Results – Mass–SFR relations and the specific SFR at high-z*

The mass-SFR relation of galaxies and its evolution with redshift has drawn of lot of attention during the last few years. Some of the issues debated include the scatter in this relation and its behaviour at $z \gg 3$ (e.g. Daddi *et al.* 2007, Bouche *et al.* 2010).

In Fig. 1 we show the derived SFR from our $z \sim$ 4 sample as a function of UV magnitude and stellar mass respectively, and how these relations depend on assumptions on the star-formation histories and on the treatment of nebular emission. The reference model (left panels) assumes SFR=const and ages $\geqslant$ 50 Myr, as often assumed in other studies. The derived SFR values are found on and above the SFR(UV) calibration of Kennicutt (1998) indicated by the dashed line, deviations from this line being due to dust attenuation. When nebular emission is treated and the SF history is kept free (among the declining SF histories) both effects contribute to increasing the scatter around the Kennicutt relation (right panels). Models with nebular emission naturally reveal

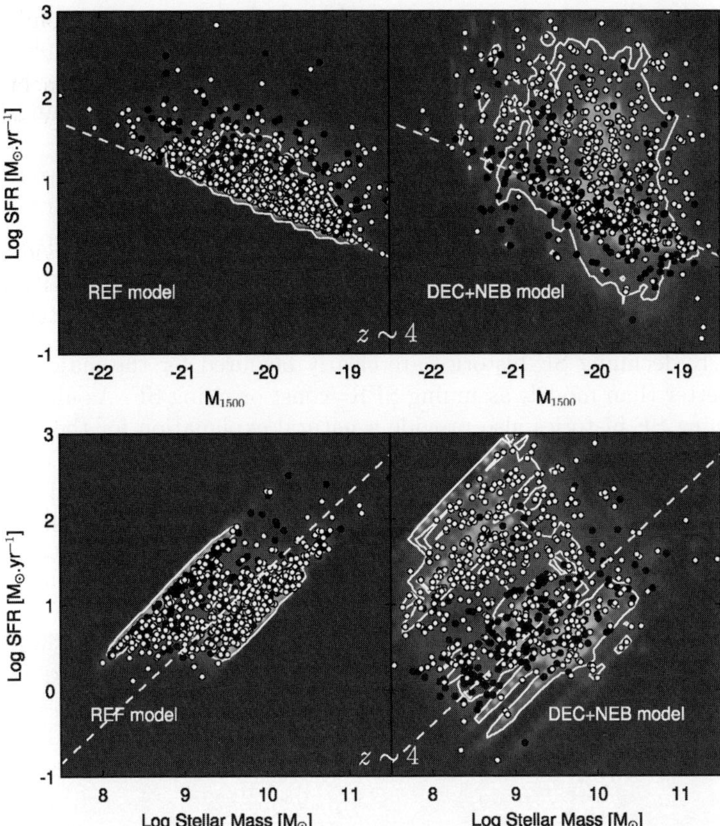

Figure 1. SFR versus absolute UV magnitude (top) and SFR versus stellar mass (bottom) for the B-drop sample ($z \sim 4$) obtained from models with constant SFR without nebular emission (left panels) and for declining SFH and nebular emission (right panels). Points show the median values for each galaxy and the white solid line encloses the 68% confidence limit of the 2D probability distribution function. White points show the "starburst" objects, black "post-starburst" galaxies. The dashed lines show the SFR(UV) relation from Kennicutt (1998) and the mean SFR-$M_\star$ relation of Daddi *et al.* (2007).

a difference in SFR between the two categories discussed above, consistent with the observed color in the IRAC bands (3.6-4.5 μm).

The derived stellar masses do not strongly depend on the treatment of nebular emission. On average we obtain a reduction of $\sim$ 30–40% for $z \sim 3$–4 and somewhat more ($\sim$ 70–80%) at higher redshift.

As shown in Fig. 1 the existence of a tight mass–SFR relation relies on the assumption of constant SFR over long enough timescales ($\gg$ 100 Myr). Again for variable SF histories and including nebular emission we find a large scatter (e.g. around the mean relation obtained by Daddi *et al.* (2007) at $z \sim 2$ shown in this Fig.). The median specific SFR (=SFR/$M_\star$) of $z \sim 3$–5 galaxies is typically a factor $\sim$ 2–3 higher when nebular emission is accounted for both for SFR=const and for variable SFHs (cf. de Barros *et al.* 2011b).

3.4. *Results – Dust attenuation, galaxy ages and SF histories*

Our analysis also yields new information on dust attenuation and on trends with UV magnitude, galaxy mass, redshift etc. For illustration, we show the attenuation A_V versus

M_{1500} for the U-, B-, and i-drop samples in Fig. 2. A clear evolution of decreasing A_V with redshift is found, both at a given UV magnitude and for the sample average. The exact value of the attenuation depends on model assumptions such as the SF history and of course also on the attenuation law. Taking nebular emission into account does not lead to a systematic shift of the inferred attenuation.

At $z \sim 3$–4 the median age of galaxies shows a wide range from very young to close to the maximum age allowed by cosmology. Including the effects of nebular emission yields on average somewhat younger ages (as already found in our earlier papers), but does not preclude old ages. No strong evolution of the average galaxy age with redshift is found between $z \sim 3$ and 5 e.g. at fixed UV magnitude or given stellar mass; at higher z our sample appear significantly younger, though.

Interestingly declining SF histories are clearly favoured for the majority of galaxies, as they fit better than models assuming SFR=const or rising SF. As already mentioned above, declining SF histories also provide a natural explanation for the two galaxy categories identified ("starburst" and "post-starburst"). Other arguments, e.g. from galaxy statistics (Lee *et al.* 2009, 2011) or others (Stark *et al.* 2009), also tend to favour short times scales of SF. Timescales and SF histories of high-z galaxies are, however, subject to debate (e.g. Maraston *et al.* 2010, Finlator *et al.* 2011). For more detailed results and discussion of other implications see dB11.

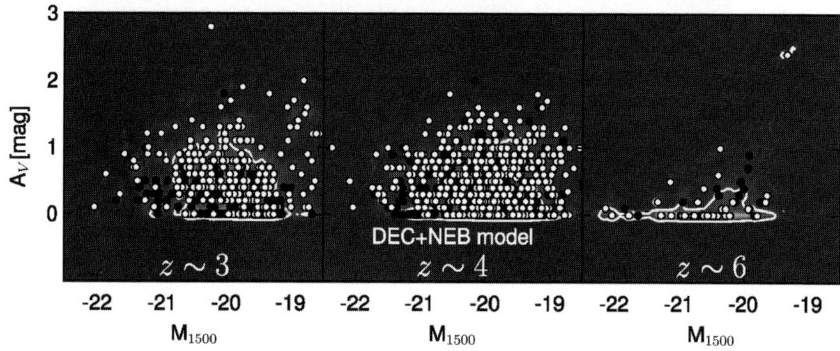

Figure 2. Attenuation A_V as a function of UV magnitude for the $z \sim 3$ (left), 4 (middle), and 6 (right) samples, as determined from models with declining SFHs and accounting for nebular emission. Symbols as in Fig. 1. Note the fairly large scatter of A_V at given M_{UV}, and the clear decrease of the attenuation at high redshift.

4. Conclusions

As our previous results on $z \sim 6$–8 galaxies (Schaerer & de Barros 2009, 2010), and the latest study of $z \sim 3$–6 objects (dB11) show, it is important to include the effects of nebular emission in SEDs models when deriving physical parameters from broad-band photometric data. SED models including nebular lines and variable Lyα can also be used to infer trends of Lyα emission from photometric data, as shown in Schaerer *et al.* (2011.

Accounting for nebular emission can lead to significant differences in the inferred age (which tend to become younger), masses (slightly lower), SFR , specific star-formation rate, attenuation and other parameters. We also find that the physical parameters depend quite significantly on the assumed star-formation histories. Our models favour relatively short (or recurrent?) star-formation timescales.

From our analysis of a large sample of high-z galaxies we find that the majority of objects ($\sim$ 60–70%) is better fit with SEDs accounting for nebular emission (with a contribution proportionaly to the Lyman continuum flux). This category of apparently strongly SF objects ("starburst") can also be distinguished empirically from the rest of the LBG population, which are probably closer to "post-starbursts". We suggest that a fairly large scatter could be intrinsic in relations between SFR–mass and others. More detailed results and implications are presented and discussed in dB11.

References

Anders, P. & Fritze-v. Alvensleben, U. 2003, *A&A*, 401, 1063
Atek, H., *et al.* 2011, *ApJ*, in press, arXiv1109.0639
Bolzonella, M., Miralles, J.-M., & Pelló, R. 2000, *A&A*, 363, 476
Bouche, N., *et al.* 2010, *ApJ*, 718, 1001
Bruzual, G. & Charlot, S. 2003, *MNRAS*, 344, 1000
Charlot, S. & Longhetti, M. 2001, *MNRAS*, 323, 887
Daddi, E., Dickinson, M., Morrison, G., *et al.* 2007, *ApJ*, 670, 156
de Barros, S., Schaerer, D., & Stark, D. 2011, *A&A*, in preparation (dB11)
de Barros, S., Schaerer, D., & Stark, D. 2011, *SF2A meeting*, arXiv:1111.***
Egami, E., Kneib, J.-P., Rieke, G. H., *et al.* 2005, *ApJ*, 618, L5
Eyles, L. P., Bunker, A. J., Stanway, E. R., *et al.* 2005, *MNRAS*, 364, 443
Fioc, M. & Rocca-Volmerange, B. 1999, *A&A*, 351, 869
Finlator, K., *et al.* 2011, *MNRAS*, 410, 1703
Kennicutt, Jr., R. C. 1998, *ARAA*, 36, 189
Labbé, I., González, V., Bouwens, R. J., *et al.* 2010, *ApJ*, 708, L26
Lee, K. S., *et al.* 2009, *ApJ*, 695, 368
Lee, K. S., *et al.* 2011, arXiv:1111.1233
Lidman., C., *et al.* 2011, *MNRAS*, in press, arXiv1109.1333
Madau, P. 1995, *ApJ*, 441, 18
Maraston, C., *et al.* 2010, *MNRAS*, 407, 830
McLinden, E. M., *et al.* 2011, *ApJ*, 730, 136
Ono, Y., *et al.* 2010, *ApJ*, 724, 1524
Ouchi, M., Shimasaku, K., Akiyama, M., *et al.* 2008, *ApJS*,, 176, 301
Santini, P., Fontana, A., Grazian, A., *et al.* 2009, *A&A*, 504, 751
Schaerer, D. & de Barros, S. 2009, *A&A*, 502, 423
Schaerer, D. & de Barros, S. 2010, *A&A*, 515, A73
Schaerer, D., & de Barros, S., & Stark, D. 2011, *A&A*, in press, arXiv1110.4398S
Shim, H., Chary, R.-R., Dickinson, M., *et al.* 2011, *ApJ*, 738, 69
Stark, D. P., Ellis, R. S., Bunker, A., *et al.* 2009, *ApJ*, 697, 1493
Stark, D. P., Ellis, R. S., Chiu, K., Ouchi, M., & Bunker, A. 2010, *MNRAS*, 408, 1628
Trump, J. R., *et al.* 2011, *ApJ*, in press, arXiv1108.6075
van der Wel, A., *et al.* 2011, *ApJ*, in press, arXiv1107.5256
Zackrisson, E., Bergvall, N., & Leitet, E. 2008, *ApJ*, 676, L9

Discussion

LEITHERER: Prior to ionizing any gas, a stellar far-UV photon can be absorbed by dust. For instance, some codes like PEGASE assume a certain fraction of photons are absorbed by dust, whereas others like starburst99 assume a fraction of zero. Isn't this an additional free parameter?

SCHAERER: Yes, in principle of course. Currently we're assuming that all Lyman continuum photons contribute to the nebular emission, which gives good fits (better than without nebular emission) in many cases. This of course maximises the contribution of nebular emission. We may explore variations of this parameter in the future.

The Spectral Energy Distribution of Galaxies
Proceedings IAU Symposium No. 284, 2011
R.J. Tuffs & C.C. Popescu, eds.

doi:10.1017/S1743921312008642

NBursts+phot: parametric recovery of galaxy star formation histories from the simultaneous fitting of spectra and broad-band spectral energy distributions

Igor V. Chilingarian† and Ivan Yu. Katkov

Sternberg Astronomical Institute, Moscow State University,
Universitetskii pr. 13, Moscow, 119992 Russia
email IC: `chil@sai.msu.ru`; IK: `katkov.ivan@gmail.com`

Abstract. We present NBURSTS+PHOT, a novel technique for the parametric inversion of spectrophotometric data for unresolved stellar populations where high-resolution spectra and broad-band SEDs are fitted simultaneously, helping to break the degeneracies between parameters of multi-component stellar population models.

Keywords. Methods: data analysis, galaxies: kinematics and dynamics, galaxies: stellar content.

---

1. Motivation

Optical spectra of galaxies contain important information about their internal kinematics and stellar content that can be extracted by fitting them using stellar population models. There are a number of approaches for fitting the full spectrum that have proved to be efficient in studying different classes of galaxies and star clusters.

On the other hand, the broad-band photometry in different spectral domains that became available for large samples of galaxies thanks to modern wide-field surveys allows one to study certain properties that cannot be derived from the spectra (e.g. internal extinction) and to break degeneracies between parameters in the case of complex star formation histories (SFH). Usually, photometric measurements and spectra are used independently. Here we propose a new approach NBURSTS+PHOT that fits in a single minimization loop spectral and photometric information, recovering both the parametric SFH, and internal kinematics of galaxies.

2. Novel Technique

NBURSTS+PHOT is the extension of the NBURSTS (Chilingarian *et al.* 2007a,b) parametric full spectrum fitting technique implemented quite similarly to the non-parametric spectrophotometric inversion proposed by Pappalardo *et al.* (2010). During the minimization, the χ^2 value is computed as a sum of the spectral and photometric contributions taken with a certain weight α:

$$\chi^2 = \sum_{N_\lambda} \frac{(F_i - P_{1p}(T_i(SFH) \otimes \mathcal{L}(v, \sigma, h_3, h_4) + P_{2q}))^2}{\Delta F_i^2} + \alpha \sum_{N_{ph}} \frac{(f_j - w_j t_j(SFH) A_j)^2}{\Delta f_j^2},$$

$$\text{where} \quad T_i(SFH) = \sum_{N_{bursts}} k_i T_i(t_n, Z_n) \qquad (2.1)$$

† Present address: CfA/SAO 60 Garden St., Cambridge, MA, 02138 USA

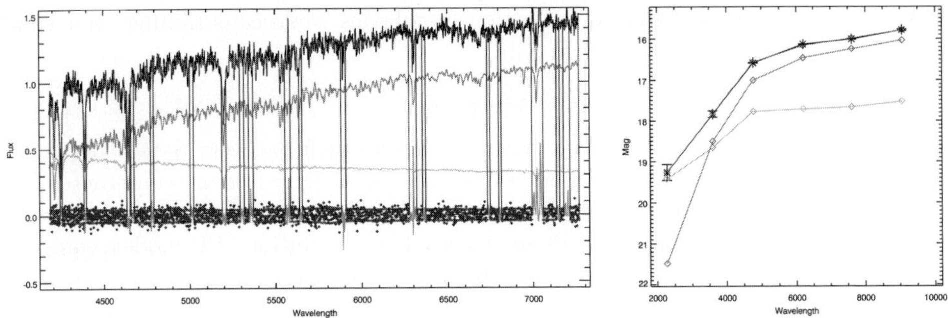

Figure 1. Example of the spectrophotometric inversion for a poststarburst E+A galaxy SDSS J230743.41+152558.4. The SDSS spectrum is shown in the left panel; the SED including GALEX NUV and 5 SDSS photometric measurements is shown in the right panel. Data (both spectroscopic and photometric) and the best-fitting model are shown in black and dark-grey, respectively, and the two components of the model are shown in light-grey.

where $\mathcal{L}$ is the line-of-sight velocity dispersion, F_i and ΔF_i are the observed spectral flux and its uncertainty, $T_i(SFH)$ is the flux from a synthetic spectrum, represented by a linear combination of N_{bursts} SSPs and convolved according to the line-spread function of the spectrograph, P_{1p} and P_{2q} are multiplicative and additive Legendre polynomials of orders p and q for correcting the continuum, t is the age, Z the metallicity, and v, σ, h_3, and h_4 are the radial velocity, velocity dispersion and Gauss-Hermite coefficients respectively (van der Marel & Franx, 1993). f_j and t_j are the observed and modelled photometric fluxes, w_j is a photometric flux normalization coefficient computed linearly, and $A_j(E(B-V))$ is a photometric extinction in a given band depending on $E(B-V)$, an additional free parameter. By adjusting the α parameter, one can choose the relative importance of photometric and spectral measurements. When dealing with galaxies at non-zero redshifts, the photometric measurements should be k-corrected.

In Fig. 1 we present an application of the NBURSTS+PHOT technique to the combined dataset for a poststarburst E+A galaxy SDSS J230743.41+152558.4 (Chilingarian *et al.* 2009). We used its SDSS DR7 (Abazajian *et al.* 2009) spectrum (left panel) and photometry from SDSS DR7 (5 optical *ugriz* bands) and GALEX GR5 (one *NUV* band; Martin *et al.* 2005) surveys. Photometric measurements were k-corrected using the analytical approximations from Chilingarian *et al.* (2010) and Chilingarian & Zolotukhin (2011) for optical and GALEX bands respectively†.

3. Pros and Cons

3.1. *Pros*

• Photometry and spectra are fitted in a single loop, so all the available information is used to constraint the SFH and internal kinematics. Therefore, certain biases originating from degeneracies between parameters of internal kinematics and stellar populations (such as metallicity – velocity dispersion) can be avoided.

• Photometry in certain bands helps to break degeneracies between parameters of multiple star formation episodes. For example, adding GALEX UV colours will help to detect small quantities of young stars and measure their mass fraction, while spectra will be helpful to recover their parameters (age and metallicity) which cannot be derived from photometric data alone.

† See the k-corrections calculator at `http://kcor.sai.msu.ru/UVtoNIR.html`

• We can estimate the internal extinction in galaxies while constraining their stellar population properties by the spectral data.

3.2. *Cons*

• The choice of α is not evident. It has to be determined for every data collection.

• The spectral and photometric models have different origins and are constructed from different ingredients. Therefore, the procedure is not fully self-consistent and some biases can be introduced. For the spectral fitting we use high-resolution SSP models computed with the PEGASE.HR (Le Borgne *et al.* 2004) code based on the ELODIE.3.1 (Prugniel *et al.* 2007) empirical stellar library, while the photometric models are computed with the PEGASE.2 (Fioc & Rocca-Volmerange 1997) code using the low-resolution BaSeL synthetic stellar library (Lejeune *et al.* 1997). Spectra of real galaxies often contain emission lines which are sometimes quite strong and may influence the corresponding photometric measurements.

• It is difficult to get homogeneous photometry in several spectral domains because the data provided by different wide-field surveys often have certain specific features different, such as apertures and photometric zero points. In addition, the photometry has to be provided for the same aperture as a spectrum (e.g. we cannot use the integrated galaxy photometry and 3-arcsec SDSS spectra).

The authors thank the IAU for the provided financial aid and RFBR grant 10-02-00062a for covering the remaining expanses.

References

Abazajian, K. N., *et al.* 2009, *ApJS*, 182, 543

Chilingarian, I. V., De Rijcke, S., & Buyle, P. 2009, *ApJL*, 697, L111

Chilingarian, I. V., Melchior, A.-L., & Zolotukhin, I. Y. 2010, *MNRAS*, 405, 1409

Chilingarian, I. V., Prugniel, P., Sil'chenko, O. K., & Afanasiev, V. 2007, *MNRAS*, 376, 1033

Chilingarian, I. V., Prugniel, P., Sil'chenko, O. K., & Koleva, M. 2007, in: Vazdekis, A. & Peletier, R.R., eds., Proc. IAU Symp. 241, "Stellar Populations as Building Blocks of Galaxies". Cambridge Univ. Press, Cambridge, p. 175

Chilingarian, I. V. & Zolotukhin, I. Y. 2011, *MNRAS* (in press)

Fioc, M. & Rocca-Volmerange, B. 1997, *A&A*, 326, 950

Le Borgne, D., Rocca-Volmerange, B., Prugniel, P., Lançon, A., Fioc, M., & Soubiran, C. 2004, *A&A*, 425, 881

Lejeune, T., Cuisinier, F., & Buser, R. 1997, *A&AS*, 125, 229

Martin, D. C., *et al.* 2005, *ApJL*, 619, L1

Pappalardo, C., Lançon, A., Vollmer, B., Ocvirk, P., Boissier, S., & Boselli, A. 2010, *A&A*, 514, A33

Prugniel, P., Soubiran, C., Koleva, M., & Le Borgne, D. 2007, astro-ph/0703658

van der Marel, R. & Franx, M. 1993, *ApJ*, 407, 525

The Spectral Energy Distribution of Galaxies
Proceedings IAU Symposium No. 284, 2011
R.J. Tuffs & C.C. Popescu, eds.

doi:10.1017/S1743921312008654

An empirical spectral library of chemically well characterized stars for stellar population modelling

André de Castro Milone[1], Anne E. Sansom[2], Patricia Sánchez-Blázquez[3], Alexandre Vazdekis[4,5], Jesus Falcón-Barroso[4,5] and Carlos Allende Prieto[4,5]

[1]Divisão de Astrofísica, Instituto Nacional de Pesquisas Espaciais, Av. dos Astronautas 1758, São José dos Campos, SP 12227-010, Brazil

[2]Jeremiah Horrocks Institute, University of Central Lancashire, Preston, Lancs PR1 2HE, UK

[3]Departamento de Física Teórica, Universidad Autónoma de Madrid, Cantoblanco 28409, Madrid, Spain

[4]Instituto de Astrofísica de Canarias, Vía Láctea s/n, E-38205 La Laguna, Tenerife, Spain

[5]Departamento de Astrofísica, Universidad de La Laguna, E-38205 La Laguna, Tenerife, Spain

email: andre.milone@inpe.br

Abstract. With the goal of assembling a new generation of more realistic single stellar population (SSP) models, we have obtained magnesium abundances for nearly 80% of the stars of the widely employed MILES empirical spectral library. Additional spectroscopic observations of carefully selected stars have recently been obtained to improve the parametric coverage of this library. Here we report on: (i) the framework of Mg abundance determination carried out at mid-resolution, (ii) the newly acquired data, and (iii) the preliminary steps towards modelling stellar populations.

Keywords. astronomical data bases: miscellaneous, stars: abundances, stars: atmospheres, techniques: spectroscopic.

1. Introduction

One limitation of current stellar population models is the use of nearby stars whose atmospheres are not completely characterized in terms of elemental abundances. Typically, these models take the iron abundance as a metallicity tracer, but stellar spectra may change considerably if the abundance ratios between iron and other metals depart from the scaled-solar mixture. Furthermore, stars from Galaxy's distinct components show different abundance patterns.

Accounting for various elemental abundances, including alpha (traced by **Mg**) and iron peak elements (traced by **Fe**), will allow us to provide an empirical stellar library that is particularly useful for modelling spectral energy distributions of evolving stellar populations. Whilst Mg is produced via C burning in massive stars and ejected into the interstellar medium mainly through Type II supernovae (SNe), Fe is mainly released in Type Ia SNe. The higher the **Mg/Fe** abundance ratio, the shorter the time scale of star formation (see, e.g., McWilliam 1997).

2. The MILES library

The MILES data base contains good-quality flux-calibrated optical spectra for 985 stars with a uniform FWHM resolution of 2.5 Å (Sánchez-Blázquez *et al.* 2006, Falcón-Barroso *et al.* 2011). This library presents an unprecedented coverage of stellar parameters. In Cenarro *et al.* (2007) the main photospheric parameters of individual stars were carefully transformed from data of several high-resolution spectroscopic analyses into a homogeneous system with great accuracy (σT_{eff} = 100 K, σlog g = 0.2, σ[Fe/H] = 0.1 dex). MILES is regarded nowadays as a standard empirical stellar library, and it is publically available through a user-friendly website (http://miles.iac.es).

3. The MILES Mg/Fe Catalogue

[Mg/Fe] abundance ratios were collated from high-resolution (*HR*) studies and transformed onto a uniform scale having the wide catalogue of Borkova & Marsakov (2005, hereafter BM2005) as reference. Milone, Sansom & Sánchez-Blázquez (2011) also carried out a spectroscopic analysis at mid-resolution (*mr*) to extend the Mg characterization by applying spectral synthesis on the MILES spectra with the MOOG LTE code (Sneden 2002, http://verdi.as.utexas.edu) through an automatic process excluding poorly matched cases. To calibrate our *mr* measurements we used a control sample with 309 library's stars (255 dwarfs, being 219 from BM2005, and 51 giants from *HR* works).

Two strong Mg I features (at λ5183.60Å and λ5528.40Å) were analysed through the pseudo-equivalent width and line profile fit methods. We linearly interpolated model atmospheres over the MARCS 2008 grid (Gustafsson *et al.* 2008) that suitably follow the general chemical pattern of our Galaxy for the α-elements. We adopted accurate atomic transition data from VALD and molecular line data compiled by Kurucz (1995).

The catalogue provides data for 752 MILES stars (411 dwarfs & 341 giants) identifying their sources. The σ[Mg/Fe] weighted average is ~0.1 dex being 0.09 and 0.12 dex for the *HR* and *mr* measurements respectively. The abundance ratio distribution is fairly flat across the MARCS models' parameter space. The *HR* data are spread along the main-sequence basically from $T_{eff} \approx$ 4500 K up to around 10000 K and on the giant branch mainly from 4000 K up to 5500 K. Our *mr* measurements have a wide distribution over the Herztsprung-Russell diagram, however there are deficiencies in the low main-sequence, red giant branch tip and hottest giants. The *mr* data statistically recover with acceptable accuracy the solar neighbourhood global pattern of [Mg/Fe] as a function of [Fe/H] traditionally defined by *HR* data.

4. New data set for MILES

We have compiled from *HR* works a new set of candidate stars to be incorporated into the MILES library. The whole set contains around one thousand stars spread over both sky hemispheres by adopting as reference the PASTEL stellar parameter catalogue (Soubiran *et al.* 2010). The motivation is to improve the sample distribution over the MILES 4-D parameter space (T_{eff}, log g, [Fe/H], [Mg/Fe]). Having excluded spectroscopic binaries, anomalous variable and peculiar stars, we have defined a sample of around 400 field stars that are observable with the Isaac Newton Telescope (INT) at Observatorio del Roque de Los Muchachos (Canary Islands, Spain).

We have observed this year 218 new stars in total using the same spectroscopic instrumental setup employed for the original MILES library (INT with the Intermediate Dispersion Spectrograph). The photospheric parameters and Mg abundances of the new sample stars have yet to be calibrated to the MILES system.

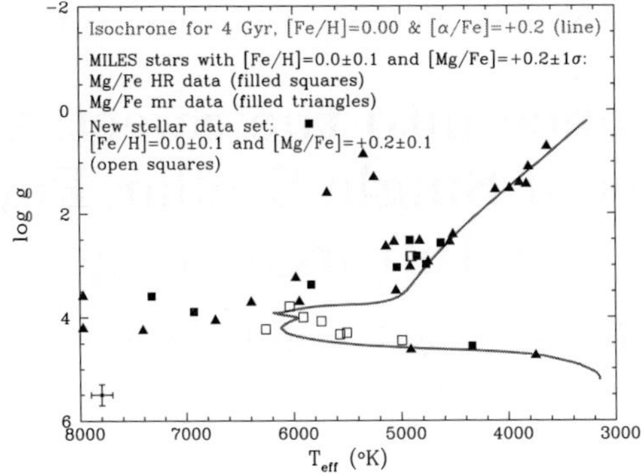

Figure 1. Selected set of MILES stars versus an isochrone from Dotter *et al.* (2008).

5. SSP modelling

Unlike the Vazdekis *et al.* (2010) models, which show a Mg/Fe scaled-solar abundance ratio at solar metallicity but increase it with decreasing metallicity following the abundance pattern of solar neighbourhood, we intend to construct α-enhanced models with variable [α/Fe] for a range in ages and metallicities. We will build up state-of-the-art SSP models with distinct α/Fe by selecting MILES stars according to their [Mg/Fe] values within a 4-D intepolation scheme, whose spectra will be integrated along isochrones of similar [α/Fe] under a self-consistent approach. Figure 1 exemplifies how the matching of MILES stars with an isochrone will be improved by including the new data set.

6. Summary

We have obtained [Mg/Fe] with a precision of ~0.1 dex for about 80% of MILES that are placed on a uniform scale (available on request or in Milone *et al.* 2011). MILES is currently being expanded in around 20% through the inclusion of new stellar data set that fills in some gaps of the library 4-D parameter space and increase the star density in other regions. We will be able to compute a set of self-consistent semi-empirical SSP models with variable α-enhancement for a range in ages and metallicities around solar, which will represent a benchmark for models based on theoretical atmospheres.

Acknowledgements

Milone thanks IAU and the Brazilian foundation Capes (AEX 3185/11-7). We thank R. Cacho, I. Martín and E. Mármol-Queraltó who helped us in one of the INT runs.

References

Borkova, T. V. & Marsakov, V. A. 2005, *AZh*, 82, 453
Cenarro, J., Peletier, R. F., Sánchez-Blázquez, P., *et al.* 2007, *MNRAS*, 374, 664
Dotter, A., Chaboyer, B., Jevremović, D., *et al.* 2008, *ApJS*, 178, 89
Falcón-Barroso, J., Sánchez-Blázquez, P., Vazdekis, A., *et al.* 2011, *A&A*, 532A, 95F
Gustafsson, B., Edvardsson, B., Eriksson, K., *et al.* 2008, *A&A*, 486, 951
Kurucz, R. 1995, *An Atomic and Molecular Data Bank for Stellar Spectroscopy*, SAO, Cambridge
McWilliam, A. 1997, *ARAA*, 35, 503
Milone, A. de C., Sansom, A. E., & Sánchez-Blázquez, P. 2011, *MNRAS*, 414, 1227
Sánchez-Blázquez, P., Peletier, R. F., Jiménez-Vicente, J., *et al.* 2006, *MNRAS*, 371, 703
Soubiran, C., Le Campion, J.-F., Cayrel de Strobel, G., & Caillo, A. 2010, *A&A*, 515, 111
Vazdekis, A., Sánchez-Blázquez, P., Falcón-Barroso, J., *et al.* 2010, *MNRAS*, 404, 1639

The Spectral Energy Distribution of Galaxies
Proceedings IAU Symposium No. 284, 2011
R.J. Tuffs & C.C. Popescu, eds.

doi:10.1017/S1743921312008666

A first glance into the Spectral Energy Distributions of Single Stellar Populations in the Infrared range

Sofia Meneses-Goytia and Reynier F. Peletier

Kapteyn Instituut, Rijksuniversiteit Groningen,
Landleven 12, 9747AD, Groningen, the Netherlands

email: s.meneses-goytia@astro.rug.nl

Abstract. The present work shows the Spectral Energy Distributions (SEDs) in the infrared using the IRTF stellar library, obtained using models based on Single Stellar population Models (SSP). We have focused on the K band in order to compare with observables of elliptical galaxies. We also present the comparisons of our models with velocity dispersions, ages and metallicities obtained with models in the optical range.

Keywords. infrared: stars, galaxies, galaxies: formation, elliptical

1. Introduction

Light is what we can actually obtain from an object and, therefore, its spectrum is one of the most important, or even the only way to gather any information about it. The spectra of individual stars can provide an idea of the behaviour of a larger object, whether they are part of it (resolved) or not (unresolved). We use unresolved Single Stellar Population (SSP) synthesis models to interpret the light of faraway galaxies. These models are based on the assumption of a single burst of stars that share the same characteristics, such as age and/or metallicity (e.g. Vazdekis *et al.* 1996). A Spectral Energy Distribution (SED) is the spectrum of a SSP model, which allows us to study the theoretical spectra as a whole, compared to an observed object (Yamada *et al.* 2008, Vazdekis *et al.* 2010). Therefore, we have to adapt the synthetic spectra to the characteristics of observed objects, in this work, elliptical galaxies, by convolving it to the observed resolution and velocity dispersion. With that, we can compare to the full spectrum allowing us to have insights into ages, metallicities, Initial Mass Functions (IMF), etc. The aim of working in the IR range is to relate this approach to early-type galaxies, since most stars in these galaxies are old (RGB) and therefore emit most of their light in the IR (Maraston 2005).

2. SSP synthesis models and fittings

The basic scheme of SSP modelling is shown in Fig. 1, based on Tinsley 1980. Using a given set of isochrones (at a certain age and metallicity), and an assumed Initial Mass Function (IMF), we interpolate the evolutionary tracks with a given stellar library in order to obtain a distribution of theoretical stars. When integrating the light of these stars, we obtain a synthetic spectrum of a galaxy with a particular set of parameters (i.e. age and metallicity). The last step is then to compare this set of spectra to observations. For this first glance, we use the Marigo *et al.* (2008) isochrones, which have a wide range of metallicities and ages. In addition, they provide the stellar flux in several bands, including in the IR. For the IMF we adopt KTG universal. As the input library, we chose the IRTF stellar library (Rayner *et al.* 2009, Cushing *et al.* 2005) for which we homogenized the stellar parameters (Teff, log G and metallicity) through a full spectrum fitting approach and interpolation, with aid of M. Koleva (IAC-Spain and University of Ghent-Belgium). For more details of the chosen ingredients of our models, see Fig. 1.

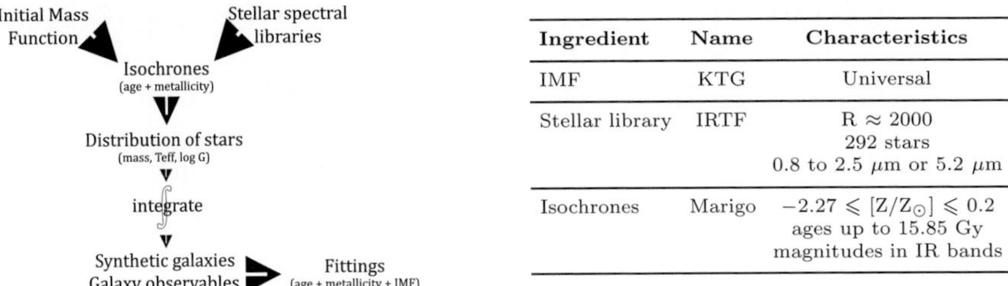

Ingredient	Name	Characteristics
IMF	KTG	Universal
Stellar library	IRTF	R ≈ 2000 292 stars 0.8 to 2.5 μm or 5.2 μm
Isochrones	Marigo	$-2.27 \leqslant [Z/Z_\odot] \leqslant 0.2$ ages up to 15.85 Gy magnitudes in IR bands

Figure 1. Basic outline and ingredients for SSP modeling in this work.

We based our approach on the Vazdekis *et al.* SSP models (Vazdekis *et al.* 1996, 1997, 2003 and 2010, Vazdekis 1999), that have been proven to work in the optical range using the CaT (Cenarro *et al.* 2001) and MILES stellar libraries (Sanchez-Blazquez *et al.* 2006). These models have also provided the spectra of these populations, i.e. the SEDs (Yamada *et al.* 2008, Vazdekis *et al.* 2003, hereafter V03 and 2010).

Here we focus on the K band, to compare with properties of observed elliptical galaxies from Marmol-Queralto *et al.* (2009, here after MQ09) selected in K band (2.19 to 2.34 μm), of which 12 are field galaxies and two more belong to Fornax. For the comparison, we use the maximum penalized likelihood approach, from Cappellari and Emsellen (2004) used in the programming package pPXF. This routine convolves the theoretical models to match the resolution and velocity dispersion of the observations and then compares them pixel by pixel to obtain the best fit.

3. Results

In Fig. 2, we show three of our models (with normalized flux) at solar metallicity, in J, H and K bands (from 0.93 to 2.41 μm). Furthermore, Fig. 3 shows fits to two observed galaxies, along with resulting age, metallicity and best fitting model as obtain using the pPXF method. The figure shows that the model (red line) provides a good representation of the observed data (black line), as can be seen by looking at the residuals (green dots). Additionally, through the fits using pPXF we were also able to obtain the kinematics of the studied galaxies, which can be compared to the ones published by MQ09 and the HyperLeda database, as shown in Fig. 4.

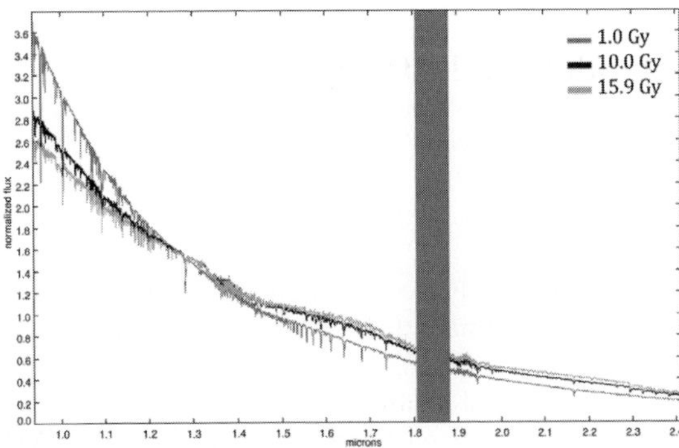

Figure 2. SSP models for a galaxy with solar metallicity at three ages

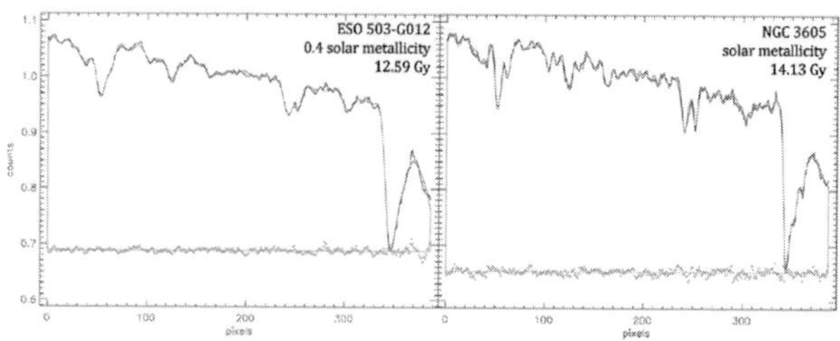

Figure 3. Comparison of the SSP models with two observed elliptical galaxies

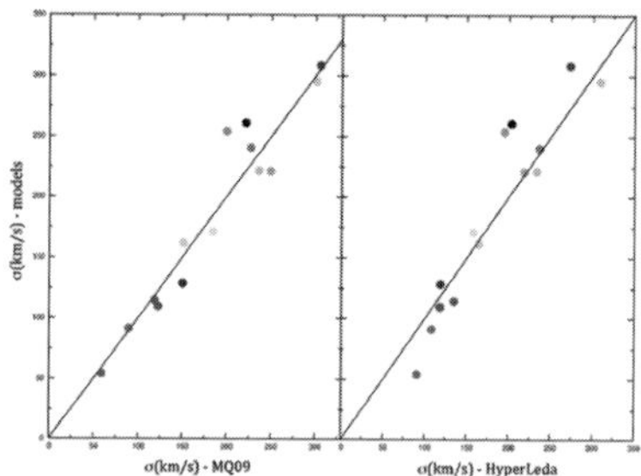

Figure 4. Comparison of the velocity dispersion obtained with models fittings and published

4. Conclusions and future work

To obtain the real accuracy of our models, some tests are being done using the IRTF stellar library and the models. Among these tests are calculating the Full Width at Half Maximum (FWHM) in collaboration with J. Falcon-Barroso (IAC-Spain); obtaining the integrated colours, (J-H), (J-K) and (H-K); and calculating the indices. Furthermore, we also need to understand the behaviour of the spectra in certain regions of the spectrum, and what is lost from the stars when they are integrated into the synthetic galaxy, etc. A future application for these models will be for the X-Shooter IR range.

We would like to thank our collaborators for their help in improving this work. We acknowledge the usage of the HyperLeda database (http://leda.univ-lyon1.fr)

References

Cappellari, M. & Emsellem, E. 2008, *PASP*, 116, 138
Cenarro, A. J., Gorgas, J., Cardiel, N., Pedraz, S., Vazdekis, A., & Peletier, R. F. 2001, *ApSSS*, 277, 319
Cushing, M. C., Rayner, J. T., & Vacca, W. D. 2005, *ApJ*, 623, 1115
Maraston, C. 2005, *MNRAS*, 362, 779
Marigo, P., Girardi, L., Bressan, A., Groenewegen, M. A. T.., Silva, L., & Granato, G. L. 2008, *A&A*, 482, 883
Marmol-Queralto, E., Cardiel, N., Sanchez-Blazquez, *et al.* 2009, *ApJ*, 705, 199
Rayner, J. T., Cushing, M. C., & Vacca, W. D. 2009, *ApJS*, 185, 289
Sanchez-Blazquez, P., Gorgas, J., Cardiel, N., & Gonzlez, J. J. 2006, *A&A*, 457, 809
Thomas D., Maraston C., & Bender R. 2003, *MNRAS*, 339, 897
Tinsley B. M. 2003, *FCPh*, 5, 287
Vazdekis, A., Casuso, E., Peletier, R. F., & Beckman, J. E. 1996, *ApJ*, 106, 307
Vazdekis, A., Peletier, R. F., & Beckman, J. E. 1996, *ApJ*, 106, 307
Vazdekis, A. 1999, *ApJ*, 111, 203
Vazdekis, A., Cenarro, A. J., Gorgas, J., Cardiel, N., & Peletier, R. F. 2003, *MNRAS*, 340, 1317
Vazdekis, A., Sánchez-Blázquez, P., Falcón-Barroso, J., *et al.* 2010, *MNRAS*, 26, 147
Yamada, Y., Arimoto, N., Vazdekis, A., & Peletier, R. F. 2008, *ApJ*, 674, 612

The Spectral Energy Distribution of Galaxies
Proceedings IAU Symposium No. 284, 2011
R.J. Tuffs & C.C. Popescu, eds.

doi:10.1017/S1743921312008678

Stellar population models in the UV: I. Characterisation of the New Generation Stellar Library

Mina Koleva[1,2] **and Alexander Vazdekis**[1]

[1]Instituto de Astrofísica de Canarias, La Laguna, E-38200 Tenerife, Spain
[2]Sterrenkundig Observatorium, Ghent University, Krijgslaan 281, S9, B-9000 Ghent

email: mina.koleva@gmail.com

Abstract. We have fully characterized the NGSL stellar spectral library, which allows us to open the UV stellar spectral range for stellar population studies. We have performed the necessary steps to prepare this library for its implementation in models synthesizing SEDs of stellar cluster and galaxy spectra. We have determined and homogenized the atmospheric parameters of the stars of this library with the aid of a full spectrum-fitting algorithm, using the MILES spectral library as a template. We also have characterized the resolution of this library and corrected systematic effects in the optical spectral range to achieve a precision of 10% of the dispersion.

Keywords. stars: abundances, stars: atmospheres, stars: fundamental parameters

1. Introduction

To study extragalactic objects we use their electro-magnetic radiation. Different physical processes and different astronomical objects emit in different wavelength domains. The ultra-violet wavelengths (UV, with $E_{typ} = h\nu \geqslant 10\,\mathrm{eV}$) are irreplaceable to characterise the metallicity and the star-formation history (SFH) of young stellar populations, to study the enhancement of α-elements or the contribution of blue horizontal branch stars to the integrated fluxes. They are also of a prime importance when studying distant galaxies whose restframe UV is observed in the optical, where the current instrumentation is most developed.

To derive relevant stellar population parameters we compare the observations to predictions from stellar population synthesis models. The quality of these models relies to a great extent on the input stellar library. The most important quantities of a stellar library is its coverage of the main atmospheric parameters (temperature, metallicity and gravity), its resolution and wavelength coverage. The theoretically computed stellar libraries would have been the ideal reference, but they cannot accurately reproduce the observations (colours, line depths etc.) of individual stars. Therefore, we rely on empirical collections of spectra. At intermediate resolutions ($R \sim 2000$, currently used in the extragalactic studies), we still lack high quality stellar spectral libraries in the UV. A good enough library, covering this spectral domain is the New Generation Stellar Library (NGSL) recently released. Here we describe the preparation of the NGSL for stellar population models.

2. Data and Analysis

The New Generation Spectral Library† (Gregg *et al.* 2006) was observed with the Hubble Space Telescope Imaging Spectrograph (STIS) and consists of 374 stars with metallicities between -2.0 dex and 0.5 dex. It contains normal stars from O to M spectral types in all luminosity classes. Its wavelength coverage from 0.2 to 1.0 μm is among the widest available observational libraries at this resolution, though it misses Lyα. Its spectral resolution is R$\sim$1000. The stars of the NGSL were rigorously chosen to have a good coverage in the space of atmospheric parameters. Unfortunately, about 200 stars were not observed owing to the failure of STIS in 2004. The released library lacks some hot- and low-metallicity stars, but is well-suited to model intermediate- and old-aged stellar populations.

We applied a full spectrum fitting approach to characterise the NGSL spectra and to infer the stellar parameters. For this purpose we employed the *ULySS* package (Koleva *et al.* 2009). We followed the approach used in Prugniel *et al.* (2011) to derive (i) the LSF to describe the intrinsic resolution and its variation with wavelength, (ii) the atmospheric parameters of the stars, and (iii) the Galactic extinction on the line-of-sight of each star.

ULySS performs a parametric minimisation of the squared differences between an observation and a linear combination of non-linear models as

$$Obs(\lambda) = P_n(\lambda) \times \left(G(v_{sys}, \sigma) \otimes \sum_{i=0}^{i=k} W_i \ \mathrm{CMP}_i \left(a_1, a_2, ..., \lambda\right) \right), \qquad (2.1)$$

where $Obs(\lambda)$ is the observed one-dimensional spectrum function of the wavelength (λ), sampled in $log\lambda$; P_n is a multiplicative polynomial of degree n; and $G(v_{res}, \sigma)$ is a Gaussian broadening function parameterised by the residual velocity v_{res}, and the dispersion σ. The CMP_i are k non-linear functions of any number of parameters, figuring the physical model. Their weights W_i can be constrained (to be positive in the present case).

Here we used three different specific cases of Eq. 2.1. First, to determine the broadening by comparing the stars in common between NGSL and a reference library, we used a single component that consists in a template spectrum (i.e. no non-linear parameter). Second, to determine the broadening with respect to a theoretical library, we used a positive linear combination of spectra taken from a grid. Finally, we measured the atmospheric parameters of the stars using a TGM component. TGM is a model spectrum, function of the effective temperature, surface gravity and metallicity, respectively, written as T_{eff}, $logg$, and [Fe/H]. The model used for the TGM component was the MILES interpolator, presented in Prugniel *et al.* (2011). This interpolator returns a spectrum for any temperature, metallicity, and gravity where each wavelength bin is computed by an interpolation over the entire reference library.

3. Results and Conclusions

The Gaussian from Eq. 2.1 encompasses the physical broadening and the relative broadening between the observation and the model. In this case the physical broadening is mostly negligible, therefore we can characterise the line spread function (LSF, analogue to the photometric PSF) by comparing the NGSL spectra to higher resolution templates. We used ELODIE (Prugniel & Soubiran 2001) and MILES (Sánchez-Blázquez *et al.* 2006) libraries to characterise the LSF in the optical domain (with the first modification of Eq. 2.1) and UVBlue (Rodríguez-Merino *et al.* 2005) and Munari (Munari et al.

† http://archive.stsci.edu/prepds/stisngsl/

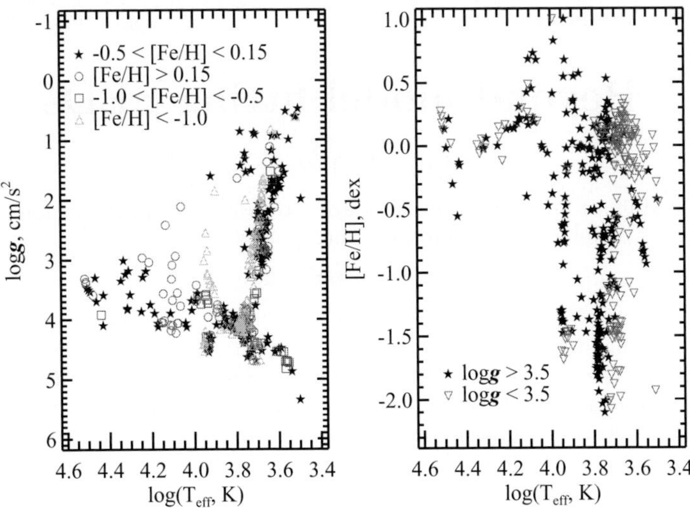

Figure 1. Resulting stellar parameter coverage for the 374 stars of the NGSL. The left panel shows their distribution in the T_{eff} - $log g$ plane. These stars were separated into four different metallicity bins as indicated in the legend. In the right panel we plot the dwarf and giant distribution in the T_{eff} -[Fe/H] plane.

2005) theoretical grids to monitor the LSF in the infra-red and UV regions. The LSFs obtained with the four reference libraries are fully consistent.

The FWHM of the LSF is varying from 3 Å at the UV end to 5 Å at 5000 Å , and to 10 Å at the NIR end. The library has a roughly constant reciprocal resolution $R = \lambda/\delta\lambda \approx 1000$ and an instrumental velocity dispersion $\sigma_{ins} \approx 130\,\mathrm{km\,s^{-1}}$. Our analysis reveals a defect of the wavelength calibration of the green grating. We used a simple linear relation to correct it: $\lambda_{cor} = \lambda - 0.7(5650 - \lambda)/(5650 - 3060)$, for $3060 < \lambda < 5650$ Å, where λ is the original wavelength in Å and λ_{cor} the corrected wavelength. We find that the wavelength calibration is precise down to 0.1 px, after correcting this systematic effect.

We determined the atmospheric parameters of the NGSL stars by fitting the spectra with the third modification of Eq. 2.1. The distribution of the stars in the parameter space is shown on Fig. 3. We use the polynomial from the fit to get the extinction A_V. We derived the atmospheric parameters homogeneously. The precision for the FGK stars is 42 K, 0.24 and 0.09 dex for T_{eff}, $log g$ and [Fe/H], respectively. The corresponding mean errors are 29 K, 0.50 and 0.48 dex for the M stars, and for the OBA stars they are 4.5 percent, 0.44 and 0.18 dex. The comparison with the literature shows that our results are not biased.

References

Gregg, M. D., Silva, D., Rayner, J., *et al.* 2006, in *The 2005 HST Calibration Workshop: Hubble After the Transition to Two-Gyro Mode*, ed. A. M. Koekemoer, P. Goudfrooij, & L. L. Dressel, 209–215

Koleva, M., Prugniel, P., Bouchard, A., & Wu, Y. 2009, *A&A*, 501, 1269

Prugniel, P. & Soubiran, C. 2001, *A&A*, 369, 1048

Prugniel, P., Vauglin, I., & Koleva, M. 2011, *A&A*, 531, A165+

Rodríguez-Merino, L. H., Chavez, M., Bertone, E., & Buzzoni, A. 2005, *ApJ*, 626, 411

Sánchez-Blázquez, P., Peletier, R. F., Jiménez-Vicente, J., *et al.* 2006, *MNRAS*, 371, 703

Munari, U., Sordo, R., Castelli, F., & Zwitter, T. 2005, *A&A*, 442, 1127

The Spectral Energy Distribution of Galaxies
Proceedings IAU Symposium No. 284, 2011
R.J. Tuffs & C.C. Popescu, eds.

doi:10.1017/S174392131200868X

Beyond model fitting SEDs

Ignacio Ferreras

Mullard Space Science Laboratory, University College London
Holmbury St Mary, Dorking, Surrey RH5 6NT, UK

email: ferreras@star.ucl.ac.uk

Abstract. Extracting star formation histories from spectra is a process plagued by numerous degeneracies among the parameters that contribute to the definition of the underlying stellar populations. Traditional approaches to overcome such degeneracies involve carefully defined line strength or spectral fitting procedures. However, all these methods rely on comparisons with population synthesis models. This paper illustrates alternative approaches based on the statistical properties of the information that can be extracted from uniformly selected samples of observed spectra, without any prior reference to modelling. Such methods are more useful with large datasets, such as surveys, where the information from thousands of spectra can be exploited to classify galaxies. An illustrative example is presented on the classification of early-type galaxies with optical spectra from the Sloan Digital Sky Survey.

Keywords. methods: statistical, techniques: spectroscopic, galaxies: stellar content

---

1. The Goal

In order to understand the process of structure formation and evolution in the Universe, it is essential to solve the problem of galaxy formation. *Ab initio* models of galaxy formation come up against the many complexities of the baryon physics transforming gas into stars. Hence, it is desirable to determine from the observations the distribution in age and composition of the stellar populations in galaxies, enabling us to backtrack their formation histories. The spectral energy distribution (SED) of galaxies encodes a treasure trove of information about the underlying stellar populations. It consists of a linear superposition of stellar spectra from a complex distribution of ages, metallicities, and dust. In principle, one could derive the star formation and chemical enrichment histories from an SED. However, the inherent degeneracies prevent us from reaching that goal using straightforward techniques.

2. The Standard Method

The most used approach to the extraction of star formation histories from spectra involves model fitting techniques, whereby the observations are compared with a grid of synthetic stellar populations, parameterised by a distribution of ages, metallicities and dust. A figure of merit is defined, often from a χ^2 function, from which one defines a likelihood, that can be combined with priors in a Bayesian way. The search for the best fit and the uncertainties in the parameters can come in many flavours, such as searches over a large library of models (Gallazzi *et al.* 2005); Metropolis-based algorithms (Cid Fernandes *et al.* 2004); search on data-compressed models (Panter *et al.* 2003), or least-squares solutions (Ockvirk *et al.* 2006). However, all these methods rely on the accuracy of the synthetic models to explain all the subtleties of galaxy spectra.

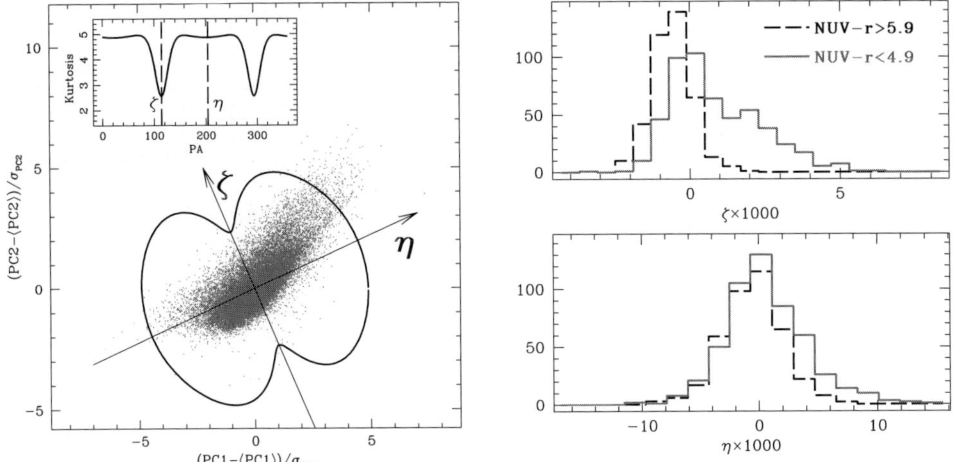

Figure 1. Independent components can be extracted from a PCA decomposition of the data. *Left:* The first two principal components from a sample of SDSS spectra of early-type galaxies is projected in different orientations, and the kurtosis of the projections (shown as a contour in the main panel and as a function of position angle in the inset) shows the preferred orientation to separate the signals from PC1 and PC2. *Right:* The distribution of the projected components (η an ζ) are shown for a subsample where GALEX photometry is available. The histograms correspond to NUV-bright (solid grey) and NUV-faint galaxies (dashed black). Notice only optical spectra are used for the definition of the components.

3. Beyond Model Fitting SEDs

Large spectroscopic surveys, such as the Sloan Digital Sky Survey (SDSS, York *et al.* 2000) have opened up the possibility of extracting information on a purely statistical basis. One could consider a sample of galaxy spectra as a set of multi-dimensional vectors, converting this problem into a multi-variate analysis method. For instance, these vectors can be expressed as linear combinations of a reduced set of "basis vectors" that constitute fundamental stellar populations from which one can disentangle the observations. Such methods have been applied to perform general classifications of spectra from surveys such as 2dFGRS (Madgwick *et al.* 2003) or SDSS (Yip *et al.* 2004). They can also be used to improve data reduction, such as in the removal of night sky lines (Wild & Hewett 2005). More relevant to this conference, these techniques can be exploited to disentangle the information from the stellar populations (see e.g. Ronen *et al.* 1999, Ferreras *et al.* 2006).

In this paper, we present and extend recent results following this approach, focused on a volume-limited sample of early-type galaxies (ETGs) from SDSS (see e.g. Rogers *et al.* 2007 and Rogers *et al.* 2010). We emphasize that the method does not rely on information from any model in order to extract information about the underlying stellar populations.

3.1. *Independent Component Analysis*

In Fig. 1, we illustrate the technique of Independent Component Analysis (Hyvärinen *et al.* 2001), where the observations are assumed to be created from superpositions of spectra that are statistically independent. One could naively relate these independent components to the populations that define the star formation histories of the galaxies under scrutiny. In the left panel of Fig. 1, we show galaxy spectra as projections on to the first and second components obtained from Principal Component Analysis (see Rogers *et al.* 2007, for details). Principal components are decorrelated signals extracted from

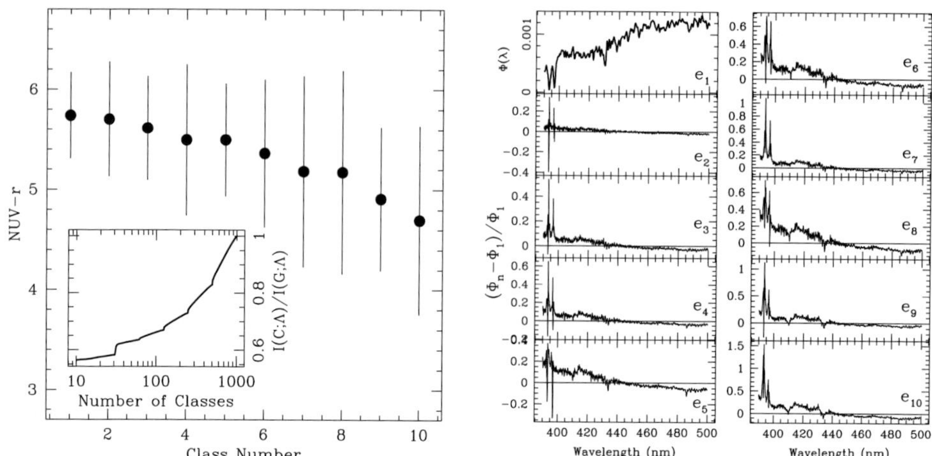

Figure 2. A sample of 1,000 galaxies from the sample of SDSS early-type galaxies, with available GALEX photometry, is presented to an Information Bottleneck sorter (see text for details). *Left:* The inset shows the decrease in information as the number of classes is reduced from the trivial set (as many classes as galaxies) to a target of 10 classes. The main panel shows the NUV–r colour of each of the 10 classes. *Right:* Spectral information of the 10 classes. The average spectrum of the first class is shown in the top-left panel. For the other 9 classes, we show the relative change with respect to the first one.

the original spectra. In order to go beyond a simple decorrelation, one can explore the non-gaussianity of the distribution: if we assume that the observations are superpositions of more fundamental spectra, according to the central limit theorem, the latter should be less gaussian. We show as a solid line the contour of the kurtosis of the (PC1,PC2) projections, as a function of the projeciton angle. There is a preferred set of directions where the kurtosis reach extrema (see inset). These directions define a new pair of "co-ordinates" for each spectra, which are the projections on to the new axes, given here as η and ζ. It is possible to interpret the meaning of these new components when we compare the distribution of η and ζ as a function of NUV–r colour, combining SDSS and GALEX photometry (rightmost panels of Fig. 1). Out of the two components, ζ is found to correlate strongly with NUV excess, a sign of recent star formation (Kaviraj *et al.* 2007).

3.2. *The Information Bottleneck*

In Fig. 2, we present another approach based on the concept of mutual information between sets of data. The Information Bottleneck technique (IB, Slonim *et al.* 2001) focuses at a classification of input data based on an algorithm that aims at minimising the complexity of the set (i.e. the number of classes), while minimising the information lost by the classification. For this example, we start with 1,000 early-type galaxies from the above mentioned sample, and classify them into ten bins, according to the IB method. In the inset of the left panel, the mutual information between spectra and classes is plotted as a function of the total number of classes in each step, from full information at the top-right corner, where we start with the trivial choice of having as many classes as galaxies, to the bottom-left corner, where only ten classes are used to describe the entire set, and 60% of the initial information retained. On the left of Fig. 2, the distributon of NUV–r colours is shown for each class, showing that there is a clear trend in the classification, with respect to NUV flux, meaning that, to first order, the most important factor driving the differences in the spectra of massive ETGs is the presence of recent star formation.

Note that the classification of the spectra uses the optical range ($\lambda = 3800$-7000Å), where the presence of recent star formation, as detected by an NUV excess, leaves a very weak trend, which is hard to detect when using traditional methods of spectral fitting. The panels on the right show the average spectra of all ten classes (from class 2 to 10, we show the fractional change with respect to the spectrum of the first class). The excess in blue light as one progresses down the class number is evident. Notice also the bumps in the higher order classes, at the positions of the Hδ ($\lambda = 4102$Å) and Hγ ($\lambda = 4340$Å) Balmer absorption lines, which is also a characteristic feature of age.

4. Epilogue

The methods described here are nothing but a tip in the iceberg of possible techniques explored in the field of multivariate analysis. Machine learning methods such as neural networks (Abdalla *et al.* 2008) or support vector machines (Tsalmantza *et al.* 2009) can be applied to galaxy data, when prior information about the classes is robust. Clustering methods have also been tentatively applied to galaxy spectra (Sánchez Almeida *et al.* 2011). However, a proper disentanglement of the underlying stellar populations still remains an open problem.

References

Abdalla, F. B., Mateus, A., Santos, W. A., Sodrè, L., Ferreras, I., & Lahav, O. 2008, *MNRAS*, 387, 945

Cid Fernandes R., Gu Q., Melnick J., Terlevich E., Terlevich R., Kunth D., Rodrigues Lacerda R., & Joguet B. 2004, *MNRAS*, 355, 273

Ferreras, I., Pasquali, A., de Carvalho, R. R., de la Rosa, I. G., & Lahav, O. 2006, *MNRAS*, 370, 828

Hyvärinen, A., Karhunen, J., & Oja, E., *Independent Component Analysis*, 2001, Wiley

Gallazzi, A., Charlot, S., Brinchmann, J., White, S. D. M., & Tremonti, C. A. 2005, *MNRAS*, 362, 41

Rogers, B., Ferreras, I., Lahav, O., Bernardi, M., Kaviraj, S., & Yi, S. K. 2007, *MNRAS*, 382, 750

Rogers, B., Ferreras, I., Pasquali, A., Bernardi, M., Lahav, O., & Kaviraj, S. 2010, *MNRAS*, 405, 329

Kaviraj, S. *et al.* 2007, *ApJ Supp. Ser.*, 173, 619

Madgwick, D., Somerville, R., Lahav, O., & Ellis, R. 2003, *MNRAS*, 343, 871

Ocvirk, P., Pichon, C., Lançon, A., & Thiébaut, E. 2006, *MNRAS*, 365, 46

Panter B., Heavens A. F., & Jimenez R. 2003, *MNRAS*, 343, 1145

Ronen, S., Aragón-Salamanca, A., & Lahav, O. 1999, *MNRAS*, 303, 284

Sánchez Almeida, J., Aguerri, J. A. L., Muñoz-Tuñón, C., & Huertas-Company, M., 2011, *ApJ*, 735, 125

Slonim, N., Somerville, R., Tishby, N., & Lahav, O. 2001, *MNRAS*, 323, 270

Tsalmantza, P. *et al.* 2009, *A&A*, 504, 1071

Wild, V. & Hewett, P. C. 2005, *MNRAS*, 358, 1083

Yip, C. W. *et al.* 2004, *AJ*, 128, 2603

York, D. G. *et al.* 2000, *AJ*, 120, 1579

The Spectral Energy Distribution of Galaxies
Proceedings IAU Symposium No. 284, 2011
R.J. Tuffs & C.C. Popescu, eds.

doi:10.1017/S1743921312008691

SED fitting with MCMC: methodology and application to large galaxy surveys

Viviana Acquaviva[1], Eric Gawiser[1] and Lucia Guaita[2]

[1]Department of Physics and Astronomy, Rutgers, The State University of New Jersey, Piscataway, NJ 08854

[2]Institutionen för Astronomi, Stockholms Universitet, SE-106 91 Stockholm, Sweden

email: vacquaviva@physics.rutgers.edu

Abstract. We present GalMC (Acquaviva *et al.* 2011), our publicly available Markov Chain Monte Carlo algorithm for SED fitting, show the results obtained for a stacked sample of Lyman Alpha Emitting galaxies at z $\sim$ 3, and discuss the dependence of the inferred SED parameters on the assumptions made in modeling the stellar populations. We also introduce SpeedyMC, a version of GalMC based on interpolation of pre-computed template libraries. While the flexibility and number of SED fitting parameters is reduced with respect to GalMC, the average running time decreases by a factor of 20,000, enabling SED fitting of each galaxy in about one second on a 2.2GHz MacBook Pro laptop, and making SpeedyMC the ideal instrument to analyze data from large photometric galaxy surveys.

Keywords. methods: statistical, galaxies: evolution

1. SED fitting with MCMC: a two-step process

SED fitting is the process of extracting information on the physical properties of galaxies, such as stellar population age, mass, star formation rate, dust content, metallicity, and redshift, starting from a set of templates that predict how galaxy spectra look like as a function of these properties, which are the SED fitting parameters. This process relies on the simple but powerful idea that since the properties of the models are known, if we can find models that resemble the observations we can infer the properties of the data. A more rigorous way of comparing models with observations – in other words, deciding whether a model resembles the data or not – is the χ^2 statistics. For each set of parameters, we need to compute the prediction of what the observations would be if that model was the true one. This step is conceptually simple but complicated in practice, because of the many astrophysical processes that need to be modeled. GalMC implements this process through the sequence described in Fig. 1. GalMC is based on Bayesian statistics. Therefore, a second step of the inference process requires to reconstruct the probability distribution of the SED fitting parameters, which are treated as random variables. This is done by exploring the parameter space with a random walk biased so that the frequency of visited locations is proportional to the probability density function. This path through parameter space is the Markov Chain. Once these probabilities are known, one can compute the desired credible intervals for each of the parameters; and because of how visited locations are chosen, integrating the probability distribution functions (PDFs) becomes a simple matter of summing over the points in the chains.

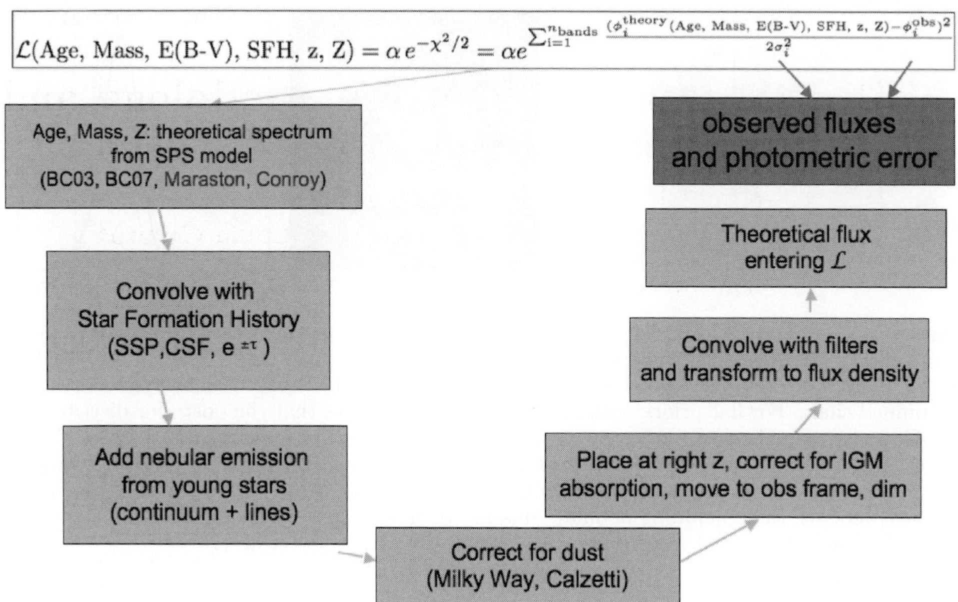

Figure 1. The series of steps performed by GalMC to obtain the predicted spectrum as a function of the SED parameters. After the convolution with filter transmission curves, this quantity can be directly compared to the data to obtain a χ^2 value.

2. Probability distributions, degeneracies, and the impact of systematics

Detailed results for two stacked samples of Lyman Alpha Emitting galaxies at $z \sim 3$ were presented in Acquaviva *et al.* (2011). Here we just show that GalMC is able to capture multi-modal probability distributions, such as the double peak in the Age vs Stellar Mass distribution, which is due to the degeneracy between these two parameters. We also want to highlight the impact of the assumptions made in modeling the stellar populations, shown for the case of Stellar Mass in the right panel of Fig. 2. The different curves all refer to models commonly used in the literature: the BC03 (Bruzual & Charlot 2003) and CB07 (Charlot & Bruzual 2011) stellar population templates, at Solar or variable metallicity, and with or without the inclusion of nebular emission. The corresponding scatter in the estimate of stellar mass (which does not include the possibility of different initial mass functions, IMFs) is a factor of ~ 2.5, significantly larger than the statistical uncertainty for the same data.

3. SpeedyMC: MCMC for large galaxy catalogs

MCMC algorithms are much more efficient ways of exploring high-dimensional parameter spaces with respect to algorithms where the probability distribution is sampled at a set of fixed locations on a grid. In fact, the "interesting" region of parameter space (the one where data and models look like each other) often occupies a small fraction of the total volume. While grid-based models need to explore all of it, Markov Chains are able to "recognize" the interesting regions and will spend most of the time visiting (sampling) those locations. Yet, the complicated process described in Fig. 1, which leads to the computation of the χ^2 value corresponding to a set of parameters, usually needs to be repeated tens of thousands of times. The computational bottlenecks in this case are the generation of a stellar population template at the right age, and the convolution with

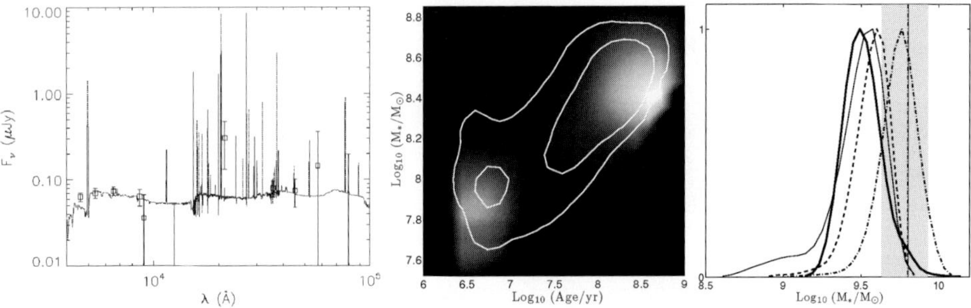

Figure 2. *Left*: Data and best-fit model of the z = 3.1 Lyman Alpha Emitters from Acquaviva *et al.* (2011). *Middle*: Marginalized constraints on age and stellar mass. The contours indicate the 68% and 95% credible regions, while the grey-scale gradient is based on average likelihood in the binned chain. For flat priors, lack of exact overlap indicates that the posterior distribution is non-Gaussian; in this case, the contours also show a bi-modal probability distribution. Markov chains are analyzed using the public software from Lewis & Bridle 2002. *Right*: Probability distribution for the Stellar Mass assuming $Z = Z_\odot$ for the BC03 (dotted-dashed) and BC07 (dashed) models, then including nebular emission (thin solid), and varying Z with a logarithmic prior (thick solid). Shaded regions show the constraints from Lai *et al.* (2008).

the filter transmission curves. To alleviate the first problem, GalMC uses our modified version of GALAXEV (Bruzual & Charlot 2011), developed in collaboration with the authors, which is $\sim$ 20 times faster than the official release. However, the typical time per iteration is still about 0.4 seconds on a 2.2GHz MacBook Pro laptop (for simplicity, all quoted running times will be referred to this machine), and therefore the typical chain per object takes a few hours to run. This becomes impractical for catalogs comprising thousands of objects. The basic idea of SpeedyMC is to find a different (faster) way to compute the χ^2 corresponding to a certain set of parameters. To achieve this objective, we take the following four steps:

(a) We compute the spectra on a grid of locations exploring the entire parameter space, saving the final product of the sequence of steps described in Fig 1 (*after* convolution with the filter transmission curves, so we retain only a handful of numbers corresponding to the flux densities in the observations' bands);

(b) We read the grid into memory;

(c) We run MCMC as usual, but to compute the χ^2 at each location we use *multi-linear interpolation* between the pre-computed spectra;

(d) We enjoy the speed up factor of 20,000, which allows us to fit the SED of each galaxy in a few seconds (even assuming to run several chains per object).

A couple of caveats are in order. First, this method doesn't have the flexibility of GalMC; because it is difficult to perform interpolation in more than three dimensions, not more than four SED fitting parameters can be used (the fourth parameter being stellar mass, which is a normalization and therefore is excluded from the interpolation process). Second, there is an "overhead" cost in computing the grid; for 50 values of age and E(B-V), and 100 values of redshift, running the initial grid takes about 24 hours, and this need to be repeated for a different survey (since the set of utilized filters change), or to use, e.g., a different star formation history or IMF. Still, for large surveys the set of filters is fixed, the number of different modeling options one might want to try is limited, and four parameters are enough to capture the general physical properties of a population of galaxies. Finally, let us observe that the resolution of the initial grid *does not* correspond to the resolution with which the PDF is sampled, as is the case in grid-based models. MCMC is still free to sample any desired location in parameter space, and the accuracy

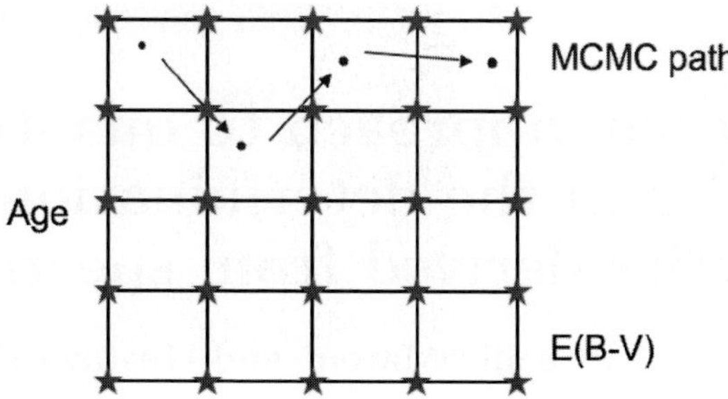

Figure 3. An example of the path of SpeedyMC for a two-dimensional grid. The visited locations do not need to lie at the locations where template spectra have been saved (indicated by stars); instead, the corresponding spectrum is obtained by very fast bi-linear interpolation between the four corner stars.

of the predicted spectrum corresponds to the accuracy of the linear interpolation between the points of the grid. This is illustrated in Fig. 3. The accuracy can be improved by increasing the number of points in the grid. A test conducted on the LAEs at $z = 3.1$ revealed that 50 points in age between 0 and the age of the Universe and 50 values of E(B-V) between 0 and 1 are enough to produce the estimate and credible intervals as the original GalMC, and that using 100 values rather than 50 does not produce any appreciable difference.

SpeedyMC is currently being used for the analysis of data from the Cosmic Assembly Near-Infrared Deep Extragalactic Legacy Survey (CANDELS), Acquaviva *et al.* (2012). The algorithm is not yet public, but you are welcome to contact the author for discussion on how to implement it, starting from GalMC.

References

Acquaviva, V., Gawiser, E., & Guaita, L. 2011, *ApJ*, 737, 47
Acquaviva, V. *et al.* 2012, *in preparation*
Bruzual, G. & Charlot, S. 2003, *MNRAS*, 344, 1000
Charlot, S. & Bruzual, G., private communication, 2011
Lai, K., *et al.* 2008, *ApJ* 674, 70
Lewis, A. M. & Bridle, S. 2002, *PRD*, D66, 103511

Discussion

CLEMENTS: Does your code output the Bayesian evidence?

AQUAVIVA: Not yet, although we plan to implement it in order to do model selection.

The Spectral Energy Distribution of Galaxies
Proceedings IAU Symposium No. 284, 2011
R.J. Tuffs & C.C. Popescu, eds.

doi:10.1017/S1743921312008708

A Bayesian approach to quantify the uncertainties in the determination of galaxy properties derived from spectral fits

Gladis Magris C.[1], **Cecilia Mateu**[1] **and Gustavo Bruzual A.**[1,2]

[1]Centro de Investigaciones de Astronomía (CIDA)
Apdo. Postal 264, Mérida 5101, Venezuela
email: magris,cmateu@cida.ve

[2]Centro de Radioastronomía y Astrofísica, UNAM, Campus Morelia
Apartado Postal 372, 58090, Morelia, Michoacán, México
email: g.bruzual@crya.unam.mx

Abstract. We use a bayesian formalism to quantify the uncertainties in the determination of the luminous mass and age of the dominant stellar population in a galaxy obtained from simple spectral fits. The analysis is performed over a sample of synthetic spectra covering a wide range of star formation histories and seen at different ages and redshifts. Using the bayesian approach we can establish quantitatively the uncertainties in the parameters derived from these fits in a straightforward manner, which is not possible using some simple algorithms, e.g. GASPEX, a non-negative least-square fitting algorithm.

Keywords. methods: data analysis, methods: statistical, galaxies: stellar content

1. Introduction

SED fitting methods have proven successful in carefully reproducing the observed SEDs with a vast variety of galaxies. It is well known that, in spite of the success in the spectral fitting process, the recovered physical parameters suffer from uncertainties that must be evaluated and made publicly available to the final users of such information. Walcher *et al.* (2010) present a wide discussion on this topic, showing the main techniques used to fit SED models to observations of galaxies, including an analysis of the limitations and uncertainties behind this processes. Bayesian fitting, and bayesian analysis of the uncertainties, is currently a common way to investigate the accuracy to which physical parameters are derived from galaxy spectra. Kauffmann *et al.* (2003), Da Cunha *et al.* (2008), Acquaviva *et al.* (2011), Benitez (2000), Pirzkal *et al.* (2011) are some of the authors who report the accuracy of their results by means of this technique.

2. Method

Our test is based on Dinbas3D, a dynamical method, successfully implemented to fit the spectra of simple and composite stellar populations, which has been proved to recover with great accuracy the mass and ages of galaxies of a variety of star formation histories (Mateu 2009; Cabrera 2011). In Dinbas3D the model galaxy's SED is assumed to be a linear combination of three simple stellar population SEDs $f_i(\lambda)$, from the library of Charlot & Bruzual (2007), of arbitrary ages t_i:

$$\sum_{i=1}^{3} a_i f_i(\lambda)$$

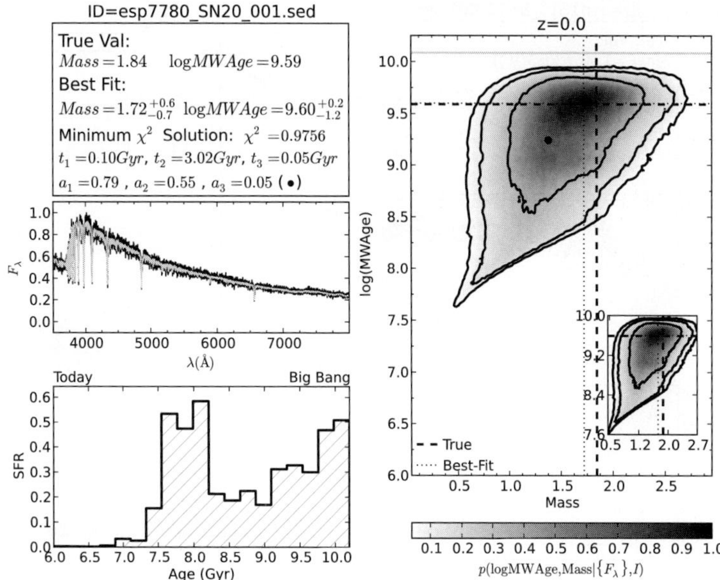

Figure 1. *Right panel*: 2D posterior pdf as a function of mass and mass-weighted age. Contours represent 1σ, 2σ and 3σ confidence intervals. The black dot corresponds to the minimun χ^2 model. *Left Bottom:* Derived star formation rate, as a function of age. *Left Middle:* Problem spectrum (black), minimun χ^2 fit (grey). *Left Top:* Input and best fit parameter summary.

and the best fitting model is the model with the three pairs a_i, t_i that minimizes the difference:

$$\chi^2 = \sum_l (F_{mod}(\lambda_l) - F_{obs}(\lambda_l))^2/\sigma_l^2$$

We use the Charlot & Bruzual (2007) stellar population synthesis models to generate a library of 'mock galaxy' spectra. For z = 0, we choose a subset of the star formation rate histories suggested by Kauffmann *et al.*(2003), and for $z > 0$, the SFR was set equal to $\exp(-t/\tau = 1)$ or SSP for $z > 1$. We add random gaussian noise corresponding to a constant S/N ratio equal to 20. In order to explore the robustness of the solutions, and to quantitatively report the uncertainties in the physical parameters (mass and age), we compute 10^8 models with a_i, t_i randomly drawn from a prior, $Prior(a,t)$, probability distribution function (pdf), which is uniform in a_i and $\log(t_i)$. We then evaluate the posterior pdf $p(a,t|F_{obs}) = Prior(a,t)exp(-\chi^2/2)$ at each of these models and marginalize it to obtain the 2-D posterior $p(Mass, MWAge|F_{obs})$, which depends only on the total mass and the mass-weighted-age. The best-fitting solution is thus given by the maximum of $p(Mass, MWAge|F_{obs})$ and the mass and age uncertainties are given by the corresponding 1σ confidence intervals.

3. Conclusions

In Figures 1 and 2 we synthesize the results of our spectral galaxy modelling. For each galaxy we show the 2D posterior pdf as a function of mass and mass-weighted age, the recovered star formation history, together with the numerical values of the best solution and the input parameters. We conclude that:

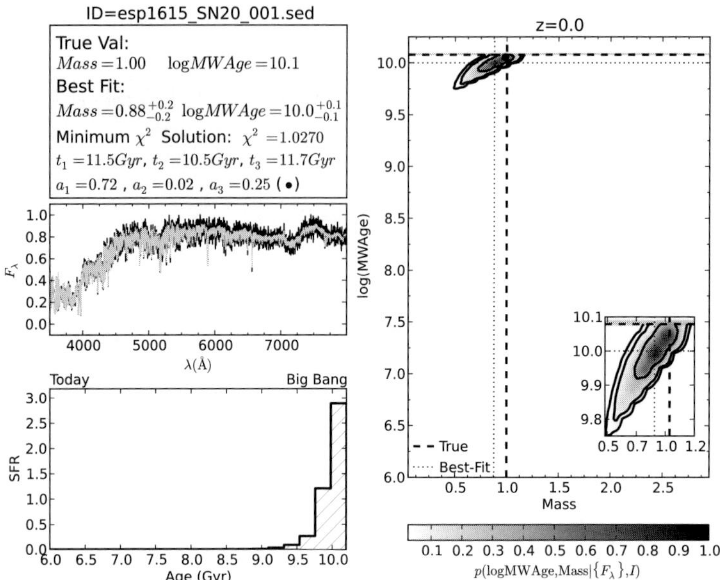

Figure 2. Same as figure 1

• For galaxies forming large amounts of stars at present or in the recent past, the uncertainty can be as large as a factor of 2 in age and a factor of 2.6 in mass. (Figure 1)

• Using a bayesian formalism we find that the mass and mean galaxy age can be estimated with a 10−15% accuracy for quiescently star forming galaxies.

• The more exact and precise solutions are obtained for galaxies with pure old population (Figure 2)

• The uncertainty in the derived mass and, to a lesser extent, in age, increases with redshift.

• In general, the same results are obtained if we use more than 3 components in the fits

4. Acknowledgements

G. Bruzual acknowledges support for this work by the National Autonomous University of Mexico, through grant IA102311.

References

Acquaviva, V., Gawiser, E., & Guaita L. 2011, *ApJ*, 737, 47

Benitez, N. 2000, *ApJ*, 536, 571

Cabrera, I. 2011, *Bachelor thesis* Universidad Central de Venezuela

Charlot, S. & Bruzual, A. G. 2007, in preparation

Da Cunha, E., Charlot, S., & Elbaz, D. 2008, *MNRAS*, 388, 1595

Kauffmann, G, Heckman, T. M., White, S. D. M. *et al.* 2003, *MNRAS*, 341, 33

Mateu, J. 2009, *Rev. Mexicana AyA* (CS) 164

Pirzkal, N., Rothberg, B., & Nilsson, K. K. *et al.* 2011, *ApJ*, (submitted) arXiv: 1104.0054v1

Walcher, J., Groves, B., Budavári, T., & Dale, D. 2010, *Ap&SS*, 331, 1

The Spectral Energy Distribution of Galaxies
Proceedings IAU Symposium No. 284, 2011
R.J. Tuffs & C.C. Popescu, eds.

doi:10.1017/S174392131200871X

What can the UV SED tell us about primitive galaxies?

Sara R. Heap

NASA's Goddard Space Flight Center, Code 667, Greenbelt Maryland, U.S.A.
email: Sally.Heap@NASA.gov

Abstract. We use the full SED of a well observed dwarf galaxy, I Zw 18, to evaluate what inferences can be made about very high-redshift galaxies from their UV SED's alone.

Keywords. galaxies: evolution, dwarf, individual (I Zw 18), high-redshift, stellar content

1. Introduction

The ultraviolet spectral energy distribution (SED) is most often described by a power-law index, β, as in $F_\lambda = \lambda^\beta$, so the "bluer" the UV continuum, the more negative the value of β. Originally defined by Calzetti *et al.* (1994), the slope of the UV continuum has been used as a measure of extinction in star-forming galaxies, particularly after Meurer *et al.* (1999) showed that β is related to the infrared excess seen in galaxies. More recently, it has been used to characterize very high-redshift galaxies discovered with Hubble's Wide Field Camera 3 (WFC3), not only to estimate dust extinction but to infer other physical properties as well (e.g. Bouwens *et al.* 2010, Finkelstein *et al.* 2010).

In fact, the UV SED has become the basis of our knowledge about very high-redshift galaxies. In a just-released study of 2000 z = 4–7 galaxies observed by Hubble's ACS and WFC3 cameras, Bouwens *et al.* (2011) found that the rest-frame UV color as given by β becomes systematically bluer towards fainter UV luminosities at all redshifts between z = 4 and z = 7. In addition, the rest-frame UV colors of galaxies at a given UV luminosity get slightly bluer with increasing redshift. They argue that dust extinction (rather than stellar age, nebular emission, etc.) is the main driver for the trends they see.

These are impressive, exciting results! Nevertheless, as illustrated in §2, a single parameter like β cannot yield three independent physical parameters: age, metallicity, and dust extinction. So, in §3, we will explore the beta method using the blue compact dwarf (BCD) galaxy, I Zw 18, as a test case. I Zw 18 is a very low metallicity dwarf galaxy ($12 + \log O/H = 7.2$). It is arguably the best local analogue to very high-redshift galaxies (Thuan 2008). I Zw 18 has been well observed from x-rays to the radio, so we can use its full SED to evaluate inferences made from the UV SED alone.

2. The UV-continuum slope parameter, β

The UV slope parameter is sensitive to many factors. Here, we describe the sensitivity of β to the stellar parameters and defer to the next section its sensitivity to gas and dust. Figure 1 shows how β changes with stellar age (i.e. the duration of star formation) and metallicity. The values of β in this figure were derived from model spectra calculated with Geneva evolutionary models (Lejeune & Schaerer 2001) and Castelli & Kurucz' (2004) grid of stellar spectra. It shows that stellar age and metallicity are degenerate: a galaxy of a given β can be young and metal-rich, or it can be older and more metal-poor.

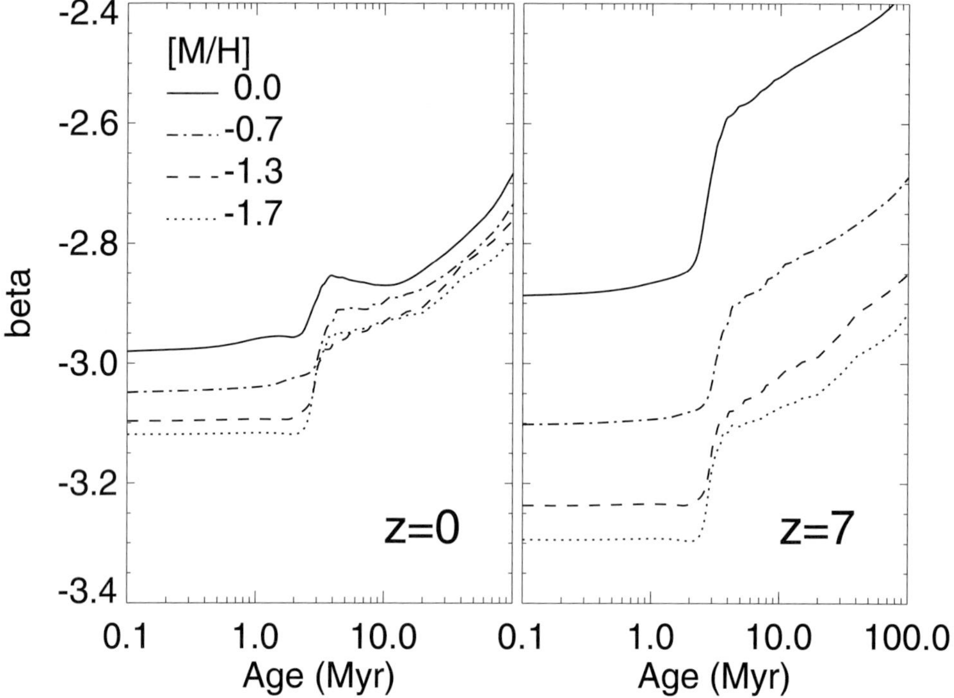

Figure 1. Variation of β with continuing star-formation (CSF) age and metallicity, [M/H], for model galaxies at z = 0 (left) measured over the wavelength range, $\lambda \sim$ 1270-1710 Å; and for z = 7 galaxies (right) observed by WFC3 at $\lambda_{\rm rest} \sim$ 1500-2000 Å.

However, we can say that a galaxy with a very blue UV continuum (large negative β) must be both very metal-poor and young (and have little to no dust extinction).

Figure 1 also shows that β depends on the rest-frame spectral region of the flux measurements. The beta method assumes that the stellar SED can be described by a power law, which is only appropriate if the UV SED follows a Rayleaigh-Jeans law. In fact, the UV SED of a young stellar population shows curvature on a log-log plot, with the slope of the SED flattening towards shorter wavelengths. Hence, it is essential to calibrate the beta method at the rest-frame spectral region where the flux measurements are made.

UV-faint galaxies at z = 7, which are the bluest galaxies in the WFC3 high-redshift sample, have $\beta = -2.7$ (Bouwens *et al.* 2011). Our studies of the northwest ionizing star cluster in I Zw 18 (Heap *et al.*, in preparation) suggest a stellar CSF age of 10-15 Myr and a metallicity, [M/H] = −1.7. Thus, the stellar component should have a $\beta = -2.9$, but its observed value (corrected for foreground reddening) is $\beta = -2.4$. The redder observed UV color may be due to our neglect of possible UV flux contributions from old stars ($\geqslant$ 1 Gyr) in the outskirts of I Zw 18 (Aloisi *et al.* 2007, Contreres Ramos *et al.* 2011). More likely, it is due to nebular continuum emission and/or dust absorption, a topic we take up in the next section.

3. UV SED vs. Full SED

The full SED of a star-forming galaxy is obviously better than a small segment, but not simply because more is better. The real advantage of a full SED is that it reveals important components of a galaxy that are not noticeable in the UV where young, massive stars are the dominant flux contributor. Figure 2 illustrates this point by showing the observed SED of I Zw 18 compared to models computed with Cloudy (Ferland *et al.* 1998). The model assumes a young massive star cluster at the center of a giant HII region, which in turn, is surrounded by a H I cloud. The figure shows how different components of a galaxy are prominent in different spectral regions: hot, massive stars are the dominant contributor to the UV flux; nebular emission becomes stronger toward longer wavelengths and exceeds the stellar emission at $\geqslant 1\mu$; and dust emission is dominant in the thermal infrared. It suggests that nebular emission and dust absorption, which both work to redden the UV spectrum, must be involved in flattening the UV continuum of I Zw 18.

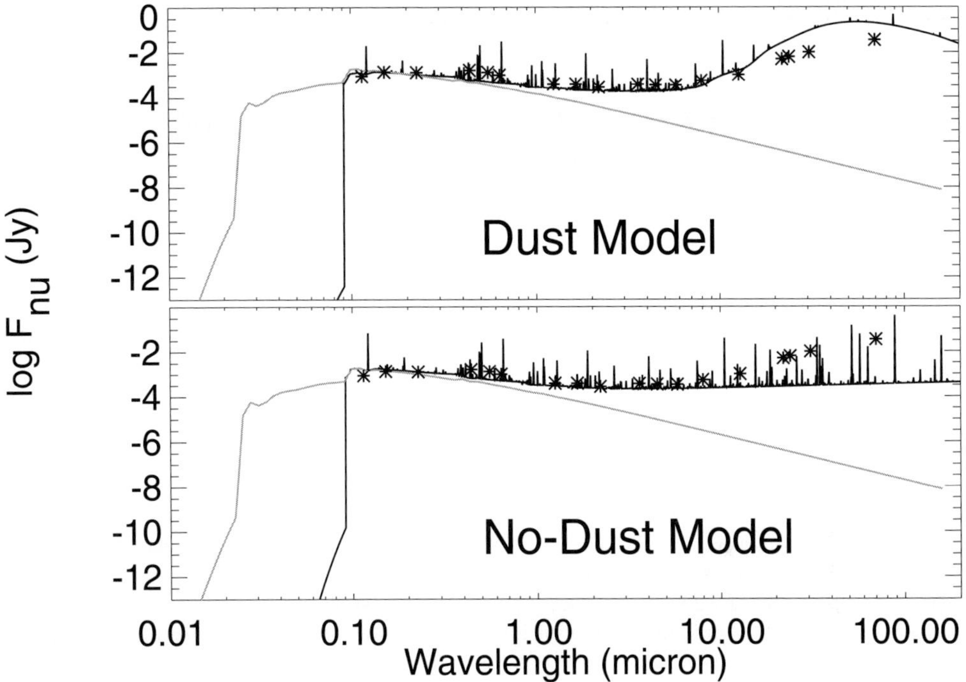

Figure 2. Observed and model SED's for I Zw 18. The observed data (asterisks) were taken from NASA's Extragalactic Database (NED). The model stellar SED is shown as a gray line; the Cloudy model SED (stars+gas+dust), as the solid black line.

Figure 3 compares a model UV SED from Cloudy to the UV spectrum of I Zw 18-NW obtained by Hubble's Cosmic Origins Spectrograph (COS). It shows that nebular emission is indeed present but at a low level. A weak nebular continuum flux is in line with other evidence that a significant fraction of stellar H-ionizing photons escapes the H II region (Dufour & Hester 1990; Pequignot's 2008). And although the surrounding H I cloud has a high H I column density, $N_{HI} = 2.2 \times 10^{21}$ cm^{-2}, it is "clumpy and irregular" (van Zee *et al.* 2006), so some ionizing photons may escape I Zw 18 altogether. Spitzer/MIPS imagery of I Zw 18 shows that 70μ dust emission follows the H I contours of 21-cm emission (Herrera *et al.* 2010), so the UV SED is likely flattened by extinction by dust produced by older stars in the extensive H I cloud.

In conclusion, the full SED of local galaxies like I Zw 18 is essential for revealing all the components and processes involved in making the UV SED of low-metallicity galaxies. The complexity of even dwarf galaxies, however, warns us to be careful in using them to interpret the radiation of very high-redshift galaxies.

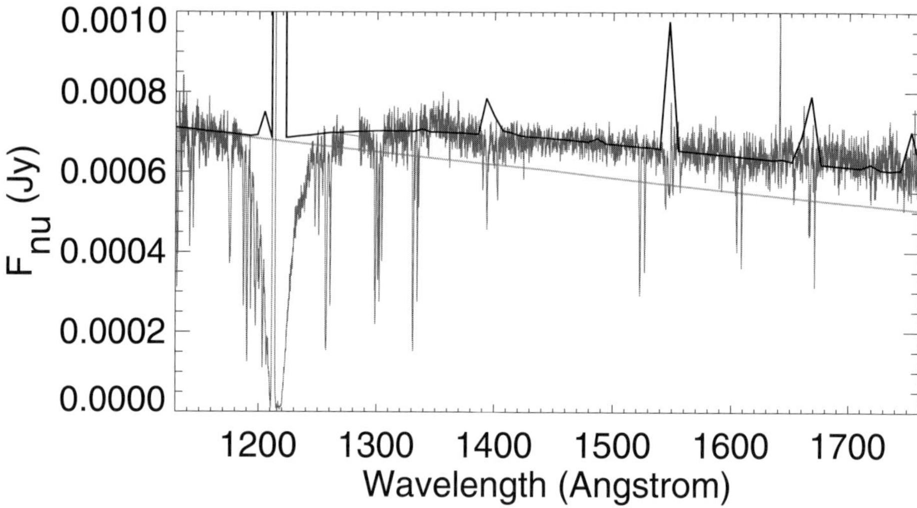

Figure 3. Comparison of the COS spectrum of I Zw 18 with our model stellar spectrum (gray) and Cloudy spectrum with coarse sampling (black).

This research was supported by NASA grant NNX08AC146 to the University of Colorado at Boulder. It has made use of NASA's Extragalactic Database (NED) and NASA Astrophysics Data System.

References

Aloisi, A., Clementini, G., Tosi, M., *et al.* 2007, *ApJ*, 667, L151
Bouwens, R. J., Illingworth, G. D., Oesch, P. A., *et al.* 2010, *ApJ (Letters)*, 708, L69
Bouwens, R. J., Illingworth, G. D., Oesch, P. A., *et al.* 2011, astro-ph 1109.0994v1
Calzetti, D., Kinney, A., & Storchi-Bergmann, T. 1994, *ApJ*, 429, 582
Castelli, F. & Kurucz, R. 2004, wwwuser.oat.ts.astro.it/castelli/grids.html
Contreras Ramos, R., Annibali, F., Fiorentino, G., Tosi, M., *et al.* 2011, astro-ph 1106.5613v1
Dufour, R. & Hester, J. 1990, *ApJ*, 350, 149
Ferland, G., Korista, K., Verner, D., *et al.* 1998, *PASP*, 110, 761
Finkelstein, S., Papovich, S., Giavalisco, M. *et al.* 2010, *ApJ*, 719, 1250
Herrara-Camus, R., Bolatto, A., Leroy, A. *et al.* 2010, www.astro.umd.edu/~rhc/Herrera_IZw18.pdf
Lejeune, T. & Schaerer, D. 2001, *A&A*, 366, 538
Meurer, G., Heckman, T., & Calzetti, D. 1999, *ApJ*, 521, 64
Reines, A., Nidever, D., Whelan, D., & Johnson, K. 2010, *ApJ*, 708, 26
Schaerer, D. 2002, *A&A*, 382, 28
Thuan, T. 2008, in: L. Hunt, S. Madden & R. Schneider (eds.), *Low-Metalllicity Star Formation*, IAU Symp. 255, (CUP), p. 348
van Zee, L., Westpfahl, D., Haynes, M., & Salzer, J. 1998, *AJ*, 115, 1000
van Zee, L., Cannon, J., & Skillman, E. 2006, *BAAS*, Vol. 38, p. 1136

The Spectral Energy Distribution of Galaxies
Proceedings IAU Symposium No. 284, 2011
R.J. Tuffs & C.C. Popescu, eds.

doi:10.1017/S1743921312008721

Hα and FUV luminosities from a stochastically formed stellar population

Nidia Lugo Lopez L.[1,2], **Gladis Magris C.**[2] **and Antonio Parravano**[1]

[1]Universidad de los Andes
email: nlugo@cida.ve, email: parravan@ula.ve

[2]Centro de Investigaciones de Astronomía (CIDA),
Apdo. Postal 264, 5101-A, Mérida, Venezuela
email: magris@cida.ve

Abstract. It has been observed that the ratio of Hα to FUV luminosity ($L_{H\alpha}/L_{FUV}$) is lower in low surface brightness galaxies. This behaviour has been attributed to systematic variations of the upper mass end and/or the slope of the Initial Mass Function (IMF) (Meurer *et al.* (2009) and Lee *et al.* (2009)). However these hypotheses do not explain the observed scatter in luminosity ratio ($L_{H\alpha}/L_{FUV}$). We present a model for the total $L_{H\alpha}$ and L_{FUV} luminosity arising from a randomly populated IMF following the Salpeter power law and the clustering law of Oey & Clarke (2007).

Keywords. methods: numerical, stars: mass function, galaxies: star clusters.

1. Method Description

We assume that the star formation process occurs in individual associations of variable size, following an universal power-law distribution (Lamb *et al.* 2010) and that in each association of stars, the stellar population is distributed with a Salpeter IMF. In order to allow stochastic fluctuacions around distributions, we assume that each association forms stars on five generations or bursts (Parravano *et al.* 2003). We use the following empirical clustering law to calculate the size by association:

$$n(N)dN \propto N^{-\beta}dN \tag{1.1}$$

where N is the number of high-mass stars (m > $8M_{\odot}$), n(N) dN is the number of associations in the range N to N + dN and β is the power law index. With this equation we calculate the number of high-mass stars per association Parravano *et al.* (2003).

$$N^{\beta-1} = \frac{N_{h,l}^{\beta-1}}{1 - x\left(1 - \left(\frac{N_{h,l}}{N_{h,u}}\right)^{\beta-1}\right)} \tag{1.2}$$

where $N_{h,l}$ and $N_{h,u}$ are the minimum and maximum number of stars of an association respectively. In this work, x is a random number between 0 and 1, $N_{h,l} = 1$, $N_{h,l} = 6\times10^4$ and $\beta = 2$. The mass of each star in the association is calculated as:

$$m_i = \left[\mu_i^{-\Gamma} - x_i\left(\mu_i^{-\Gamma} - \mu_{i-1}^{-\Gamma}\right)\right]^{-1/\Gamma} \tag{1.3}$$

with

$$\mu_i = \frac{m_u}{\left[\frac{i\phi_u}{N/5}\left(\frac{m_u}{m_h}\right)^{\Gamma} + 1\right]^{1/\Gamma}} \tag{1.4}$$

where $\Gamma = 1.35$ for the IMF Salpeter, m_u is the upper mass end ($m_u = 120M_\odot$) of the IMF, m_h is the limit for high-mass stars ($m_h = 8M_\odot$) and again x is a random number between 0 and 1.

We use the number of ionizing photons (Nio) and FUV luminosity provided by Gustavo Bruzual (private communication) corresponding to the Charlot & Bruzual (2007) models with the Padova tracks at solar metallicity. We calculate the corresponding $L_{H\alpha}$ using $L_{H\alpha}=1.36\times10^{-12}$ Nio (erg/s) (Osterbrock 1989).

To calculate the total $H\alpha$ and FUV luminosities of a galaxy, we use, as free parameter, the number of associations formed per million year (Association Formation Rate, AFR) and add the contribution of each star through its lifetime. We follow the evolution of these quantities during a simulation time of $4\times10^9 yr$ in order to assure that results are statistically reliable.

2. Results

The results of our simulation are shown in Fig. 1. In this figure, each panel corresponds to a different AFR (labelled in the figure). In the left panel AFR = 0.1 Myr^{-1} and in the right panel AFR = $4000 Myr^{-1}$, that corresponding to a star formation rate of $1.7 \times10^{-4} M_\odot yr^{-1}$ and $5.7 M_\odot yr^{-1}$, respectively.

We can observe that:

- L_{FUV} shows a smoother distribution than $L_{H\alpha}$. This is due to stars with $8M_\odot < m < 40M_\odot$ that contribute a longer time to L_{FUV} than to $L_{H\alpha}$.
- The ratio $L_{H\alpha}/L_{FUV}$ is greater in the higher AFR regime (bottom panels) and the corresponding fluctuations, measured as the standard deviation of the $L_{H\alpha}/L_{FUV}$, are lower for higher AFR.

In figure 2 we compare the results of our model with a sampled of observed galaxies. We are able to explain both, the scatter and the change of slope observed in the $L_{H\alpha}$-L_{FUV} relation for $L_{H\alpha} < 1 \times10^{38}$(erg/s). These results are similar to those published by

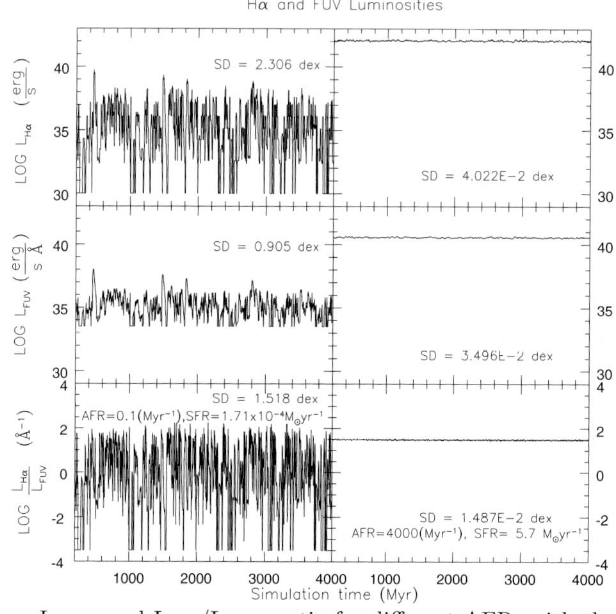

Figure 1. $L_{H\alpha}$, L_{FUV} and $L_{H\alpha}/L_{FUV}$ ratio for different AFR, with their fluctuations.

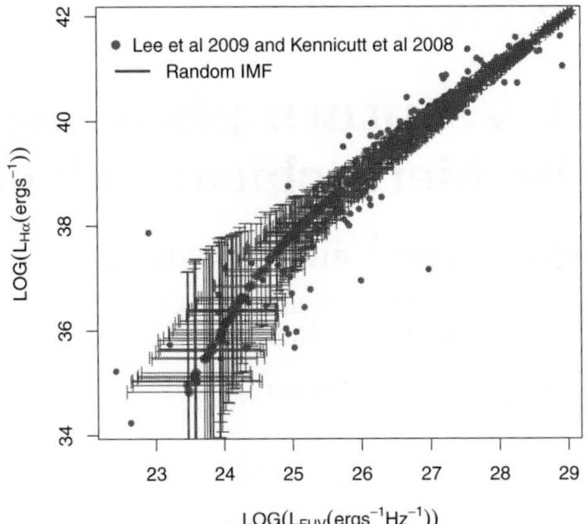

Figure 2. The squares shows the observed galaxies: We have compiled 330 galaxies from Lee *et al.* (2009) in the bands FUV and Hα luminosity from Kennicutt *et al.* (2008). The corrections for internal absorption were made as indicated in Lee *et al.* (2009). The points and errors bars respectively show average values and their fluctuations, measured as the standard deviation in our simulation.

Fumagalli *et al.* (2011), who use a slightly different model, where a distinction between the mass of the stars that contribute to the ionizing and non ionizing luminosity is assumed.

3. Conclusion

Our model, in which the IMF and clustering law for high mass stars are stochastically sampled, is able to reproduce the observed $L_{H\alpha}$-L_{FUV} trend in star forming galaxies of different brightness. It is not necessary to invoke a non-universal IMF in order to explain the $L_{H\alpha}$-L_{FUV} observed.

References

Calzetti, D., Chandar, R., Lee, J., Elmegreen, B., Kennicutt, R., & Whitmore B. 2010, *ApJ*, 719, 158

Fumagalli, M., da Silva, R. L., & Krumholz, M. R. 2011, *ApJL*, 741, 264

Kennicutt, R. C., Lee, J. C., Funes, S. J., Jose, G., Sakai, S., & Akiyama, S. 2008, *ApJS*, 178, 274

Lamb, J. B., Oey, M. S., Werk, J. K., & Ingleby, L. D. 2011, *ApJ*, 725, 1886

Lee, J. C., Gil de Paz, A., Tremonti, C., Kennicutt, J., Salim, S., Bothwell, M., Calzetti, D., Dalcanton, J., Dale, D., Engelbracht, C., Funes, S., Johnson, B., Sakai, S., Skillman, E., van Zee, L., Walter, F., & Weisz, D. 2009, *ApJ*, 706, 599

Meurer, G. R., Wong, O. I., Kim, J. H., Hanish, D. J., Heckman, T. M., Werk, J., Bland-Hawthorn, J., Dopita, M. A., Zwaan, M. A., Koribalski, B., Seibert, M., Thilker, D. A., Ferguson, H. C., Webster, R. L., Putman, M. E., Knezek, P. M., Doyle, M. T., Drinkwater, M. J., Hoopes, C. G., Kilborn, V. A., Meyer, M., Ryan-Weber, E. V., Smith, R. C., & Staveley-Smith, L. 2009, *ApJ*, 695, 765

Oey, M. & Clarke, C. 2007, *ApJ*, 719, 158

Parravano, A., Hollenbach, D., & McKee, C. 2003, *ApJ*, 584, 797

Parravano, A., McKee, C., & Hollenbach, D. 2011, *ApJ*, 726, 27

Weisz, D. R., Johnson, B., Johnson, L., Skillman, E. D., Lee, J. C., Kennicutt, R. C., Calzetti, D., van Zee, L., Bothwell, M.,Dalcanton, J. J., Dale, D. A., & Williams, B. F. 2011, *ArXiv:1109.2905v1*

The Spectral Energy Distribution of Galaxies
Proceedings IAU Symposium No. 284, 2011
R.J. Tuffs & C.C. Popescu, eds.

doi:10.1017/S1743921312008733

The UV upturn phenomenon in the hierarchical universe

Jaehyun Lee[1,2] and Sukyoung K. Yi[1]

[1]Department of Astronomy & Yonsei University Observatory, Yonsei University, Seoul 120-749, Republic of Korea

email: syncphy@galaxy.yonsei.ac.kr

Abstract. Recent studies show that an old stellar population with high metallicity in the monolithic paradigm can explain the UV upturn. Numerical simulations and empirical studies however point out that massive early-type galaxies have evolved hierarchically with an extended star formation history. This obviously has an impact on our traditional understanding on the UV upturn and requires a new investigation on its origin. We report on our investigation on the evolutionary history of model galaxy SEDs in the hierarchical scenario. The use of conventional population models (calibrated to the monolithic picture) in combination with merger trees and extended star formation fails to reproduce the observed UV upturn. If a hierarchical picture is thought to be more realistic than a monolithic one, new calibration on the population models is required.

Keywords. galaxies: evolution, ultraviolet: galaxies

1. Introduction

The ultraviolet (UV) upturn phenomenon is one of the interesting features in elliptical galaxies. It has been believed that most of elliptical galaxies are composed of old stellar populations which are dim in the UV, and thus the phenomenon has attracted much attention. However, recent studies revealed that a considerable fraction of elliptical galaxies show residual star formation (RSF) (Yi *et al.* 2005), and that even old horizontal branch stars can produce UV light (e.g., Greggio & Renzini 1990; Buzzoni 2007). Since the UV excess from young stellar populations shows a different SED morphology, it is likely that the UV upturn has the origin of extended horizontal-branch stars.

Numerical simulations and empirical studies show that elliptical galaxies are formed in a hierarchical buildup (e.g., Toomre & Toomre 1972; Negroponte & White 1983) involving mergers even recently (van Dokkum 2005). In this picture elliptical galaxies have a complex evolutionary history and various stellar sub-populations.

The UV upturn phenomenon provides a good constraint for stellar population and galaxy formation models. Yi, Demarque, & Oemler (1998) shows that the SED of an old stellar population model with high metallicity in the monolithic paradigm shows a good agreement with the empirical data from the UV upturn of early-type galaxies. Going a step further, in this study, we look into the SEDs of model galaxies in the hierarchical universe.

2. Model

Since the semi-analytic approach was first introduced by White & Frenk (1991), the semi-analytic model has become a good way to investigate galaxy formation and evolution. In this study, SEDs of galaxies are derived, based on our semi-analytic model.

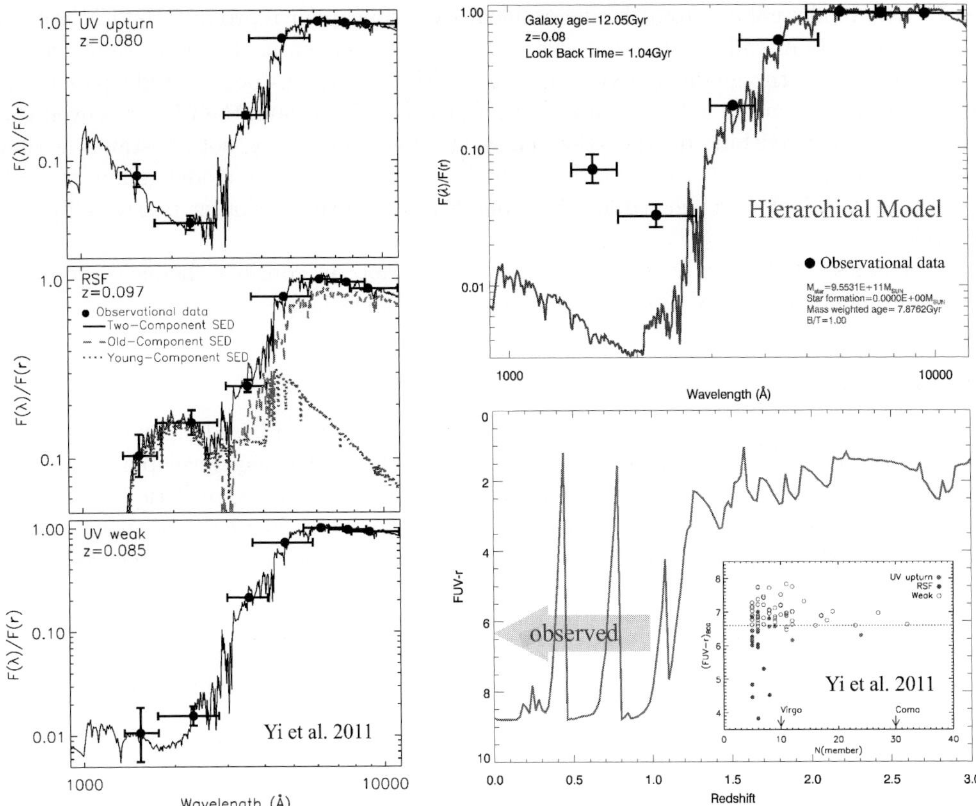

Figure 1. Left: Spectral fits for UV upturn, RSF, and UV weak galaxies. Upper right: SED of a model galaxy in the hierarchical scenario. Bottom right: FUV-r evolutionary history of the model galaxy.

We have developed a new semi-analytic model by considering accurate merging time scales, ram pressure and tidal stripping, merger-induced starburst, stellar mass loss and its recycling, and hot gas components of satellite galaxies.

The model SED library in this study is based on an aggregation of Yi (2003) and Bruzual & Charlot (2003)(hereafter BC03). The Yi models are calibrated to trace the UV spectral evolution of old populations while the BC03 models cover a wide age range.

The evolutionary histories of galaxies in a $10^{14}M_\odot$ model cluster are calculated by our semi-analytic model based on the extended Press-Schechter formalism. Since our semi-analytic model cannot consider chemical evolution yet and it is suggested that the UV upturn is mainly caused by metal-rich stellar populations, we assume super-solar metallicity($Z = 0.04$) for all stellar populations in the model galaxies.

After calculating the evolutionary histories of model galaxies, the model SEDs of galaxies are synthesized by combining star formation histories and the model SED library.

3. Results

In Yi *et al.* (2011), a metal-composite stellar population with a simple old-age model can reproduce the UV upturn phenomenon while RSF and passive early-type galaxies show different SED morphology (Fig. 1). On the other hand, the SED of a typical massive early-type model galaxy from the hierarchical scenario cannot reproduce the UV upturn even though its metallicity is super solar (Z = 0.04), as shown in the upper right panel.

The FUV-r of the model galaxy evolves as redshift decreases(bottom right panel). At high redshift, when the galaxy evolves with active star formation, the FUV-r is even less than 0. At low redshift, on the other hand, it rapidly increases notwithstanding that several starbursts change the color occasionally. The FUV-r of the model galaxy at z = 0 is roughly 9, which is much redder than found in UV upturn or other passive elliptical galaxies.

A weakness of this study is that our model does not yet consider chemical evolution consistently while the UV upturn is sensitive to the metallicity.

4. Discussion

Until today, population studies on the UV upturn have been based on the monolithic formation scenario. In this simplistic picture, elliptical galaxies may develop a strong UV flux from hot HB stars. This has been a classical view. In the more realistic hierarchical scenario, however, giant elliptical galaxies grow with many mergers and hence likely have extended star formation histories and thus have smaller mass-weighted ages. For this reason, our attempt to reproduce the observed UV upturn using the traditional population models (calibrated for the monolithic "old"ellipticals) in combination with hierarchical halo merger trees bluntly failed. Thus, population models should be re-calibrated to be effective and meaningful for UV upturn studies in the hierarchical merger picture.

References

Bruzual, G. & Charlot, S. 2003, *MNRAS*, 344, 1000
Buzzoni, A. 2007, *ASPC* , 374, 311
Greggio, L. & Renzini, A. 1990, *ApJ* , 364, 35
Negroponte, J. & White, S. D. M. 1983, *MNRAS*, 205, 1009
Toomre, A. & Toomre, J. 1972, *BAAS* , 4, 214
van Dokkum, P. G. 2005, *AJ*, 130, 2647
White, S. D. M. & Frenk, C. S. 1991, *ApJ*, 379, 52
Yi, S. K., Demarque, P., & Oemler, A. Jr. 1998, *ApJ*, 492, 480
Yi, S. K. 2003, *ApJ*, 572, 202
Yi, S. K. *et al.* 2005, *AJ*, 619, 111
Yi, S. K., Lee, J., Sheen, Y.-K., Jeong H., Suh, H., & Oh, K. 2011, *ApJS*, 195, 22

The Spectral Energy Distribution of Galaxies
Proceedings IAU Symposium No. 284, 2011
R.J. Tuffs & C.C. Popescu, eds.

doi:10.1017/S1743921312008745

Star formation history and the SED of galaxies: insights from resolved stars

Benjamin D. Johnson[1] **and Daniel R. Weisz**[2]

[1]UPMC-CNRS, UMR7095, Institut d'Astrophysique de Paris, F-75014, Paris, France
[2]Astronomy Department, University of Washington, Box 351580, Seattle, WA 98195-1580

email: `johnson@iap.fr`

Abstract. The star formation history (SFH) of galaxies is a principle uncertainty in SED modeling, and simple parameterizations of the SFH in typical SED fitting techniques may introduce biases in the resulting derived parameters. It is possible to constrain the SFH of galaxies more tightly through the observations of resolved stellar colour-magnitude diagrams with HST. This work is a first attempt to combine constraints on galaxy SFH from resolved stars with broadband SED modeling from the UV to the IR. This combination allows for the effects of different realistic SFHs on the SED to be quantified.

Keywords. galaxies:dwarf, galaxies:fundamental parameters

1. Introduction

Estimates of the stellar mass and star formation rate of galaxies from broadband SEDs rely on population synthesis models. An important ingredient of these models is the star formation history (SFH). The simplest models assume a constant SFR, for example to derive linear scalings between SFR and luminosity (e.g., Madau *et al.* 1998, Kennicutt 1998). More sophisticated modeling involves allowing the SFR to vary with time, though it is usually parameterized to be smoothly varying (e.g. τ-models with exponentially declining SFR). The parameterized SFH is then constrained by the SED of individual galaxies. Difficulties in this approach arise from 1) degenerecies between SFH and dust attenuation and stellar metallicity and 2) biases due to SFH that deviate from the assumed parameterization (e.g. with a different long-term SFR evolution or variable SFR on short timescales).

Stronger constraints on the SFH of nearby galaxies come from analysis of the colour-magnitude diagram (CMD) of resolved stars. Using HST these can be obtained for galaxies outside the local group. With such data it is possible to determine the effects of observationally constrained SFHs on measures of SFR and stellar mass that typically assume a smooth SFH. It is also possible to compare the SED expected given the SFH to the observed SED, both to constrain the effects of other galaxy properties on the SED and to assess the performance of SED models in the ultraviolet (UV) and near-infrared (NIR).

2. The Effects of Realistic SFH on the SED

The sample of galaxies that we consider is drawn from the ANGST survey (Dalcanton *et al.* 2009). Galaxies in this survey have been observed in multiple bands with either the WFPC2 or ACS instrument onboard HST. Their small distances (< 4 Mpc) insure that individual stars can be detected to several magnitudes below the tip of the red giant branch. The distance limit also results in a sample that is dominated by dwarf galaxies (median $M_* \sim 10^{7.4}$ $M_\odot$), mostly of later morphological type, but including some dwarf spheroidals. The sample generally has low metallicity and low dust attenuation.

The construction of CMDs and the method for determining the SFH from the CMD is described in Weisz *et al.* 2011 and references therein. Briefly, linear combinations of colour magnitude diagrams for SSPs with different ages are combined to fit the observed CMD, and the combination coefficients are used to construct the SFH. The time binning of the recovered SFH that we use is $\Delta \log t = 0.1$. An example SFH is shown in Figure 1.

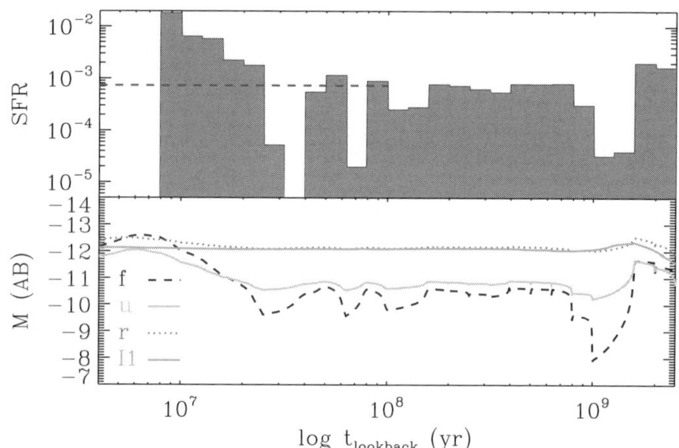

Figure 1. *Top:* Example SFH for UGC8091, inferred from the colour-magnitude diagram of resolved stars. The dashed line indicates $\langle SFR \rangle_8$ *Bottom:* Resulting evolution of the luminosity in several broad bands.

We can use the SFH derived from resolved stars as input for population synthesis codes. This allows the prediction of the SED in any bands at any lookback time. For this study we have used the FSPS v2.1 code of Conroy *et al.* (2009) with $0.2Z_\odot$ metallicity. This population synthesis code uses stellar evolutionary tracks that are consistent with those used for the derivation of the SFH from the CMDs, and we have been careful to use consistent stellar initial mass functions (IMF). We obtain very similar results using the population synthesis code of Bruzual (2007). The input SFH is also used to derive the SFR averaged over 100 Myr.

With the input SFH and the predicted FUV luminosity, we then construct the modeled SFR conversion factor

$$\epsilon_{FUV,8} = \nu L_{\nu,FUV} / \langle SFR \rangle_8$$

where $\nu L_{\nu,FUV}$ is the predicted far-UV luminosity (here in the *GALEX* FUV band) and $\langle SFR \rangle_8$ is the input SFR averaged over the last 100 Myr. This timescale is typically assumed to be the timescale probed by the far-UV ($\lambda \sim 1300 - 1800 \mathring{A}$), as this is the approximate lifetime of stars that dominate the far-UV luminosity of a population. This conversion factor is shown in Figure 2 as a function of several colours. We also show the conversion factor of Kennicutt (1998) (for the IMF used here) and that obtained for several exponentially declining models with different decline rates. We have ignored the effects of dust attenuation and metallicity variations. The important thing to note is the significant scatter in the SFR conversion factor at a given colour, much larger than predicted for different τ-models. The rms dispersion is a factor of two for this sample, and the range of $\epsilon_{FUV,8}$ is more than an order of magnitude. Furthermore, the scatter is not well correlated with UV/optical colour. For some redder galaxies a large fraction of the UV luminosity appears to be contributed by stars older than 100 Myr.

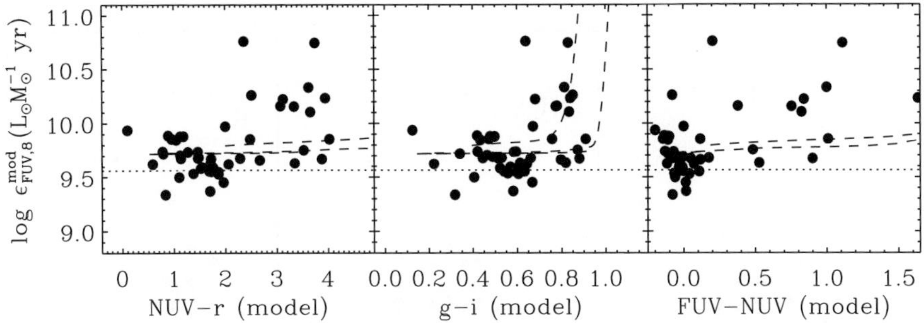

Figure 2. Conversion between SFR and FUV luminosity, as inferred from the SFH. The conversion factors are shown as a function of several colours. The dotted line shows the conversion factor of Kennicutt (1998), while the dashed lines show the expectations for several exponentially declining models.

3. Comparison to Observations

Observations of the sample include broadband photometry from *GALEX*, SDSS, and *Spitzer* Dale *et al.* (2009), Lee *et al.* (2011). Comparison of the predicted luminosties described above to the observed luminosities yields excellent agreement in the g or B band, expected since the resolved stars constitute a significant fraction of the total optical flux. In Figure 3 we compare observed UV colours to the colours predicted given the SFH. This shows that the observed UV and UV/optical colours are much redder than the prediction; observed galaxies are fainter in the UV (and have redder UV colours) than expected. The average offset is ~ 1 mag in the FUV and ~ 0.6 mag in the NUV.

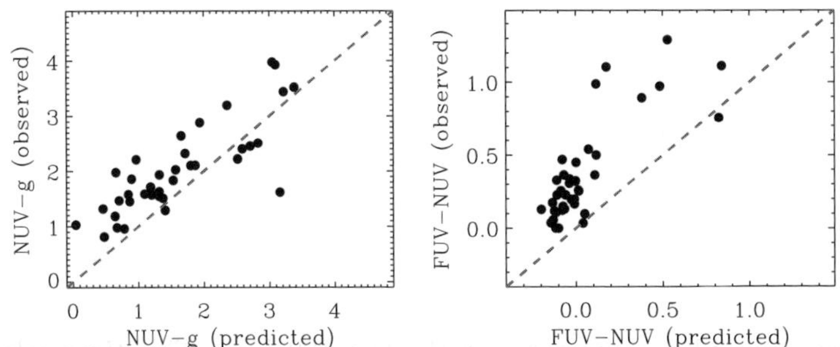

Figure 3. Comparison of observed UV/optical colours to the colours predicted given the SFH. The dashed line indicates perfect agreement. The observed colours are systematically redder in both NUV-g (*left*) and FUV-NUV (*right*).

One obvious possible cause for such dimming and reddening is dust attenuation. Estimates of dust attenuation are available from the ratio of 24μm *Spitzer* fluxes to the FUV fluxes, following Hao *et al.* (2011). As the sample consists largely of low mass galaxies, the metallicity and dust content is typically very small ($A_{FUV} < 0.35$ mag). Figure 4 shows that the inferred dust attenuation is lower than required to explain the observed

UV faintness of the sample galaxies relative to the predictions. Similar tests suggest that metallicity effects and stochastic star formation cannot explain the differences in the UV. Other possibilities include systematic errors in the determination of the recent SFH, or deficiencies in the modeling of the UV in the population synthesis models.

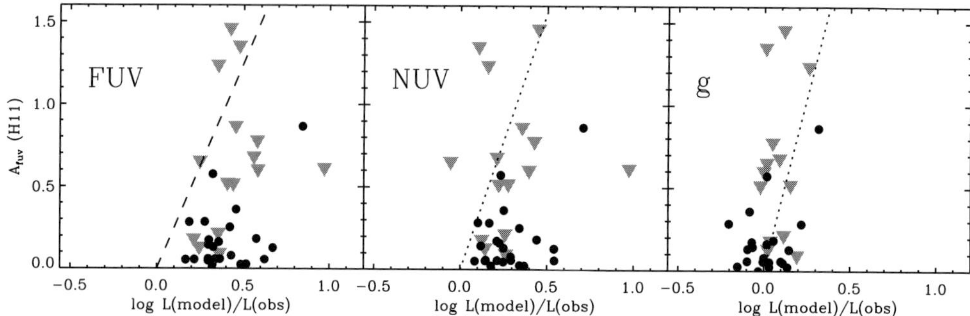

Figure 4. Dust attenuation. The FUV dust attenuation inferred from the 24μm to UV ratio is plotted against the ratio of the predicted luminosity to the observed luminosity. Gray triangles indicate upper limits to A_{FUV}. The attenuation required to explain any difference is shown as the dashed line. In the FUV and NUV bands the measured dust attenuation is always below this line and there is no correlation.

The model predictions in the *Spitzer* IRAC bands are sensitive to the treatment of TP-AGB stars. The predictions of models that include the unmodified isochrones of Marigo *et al.* (2008) are brighter than the observed 3.6 μm and 4.5 μm luminosities by 0.5 mag on average. A modification of the TP-AGB isochrones as suggested by Conroy *et al.* (2009) results in no average offset.

4. Conclusions

We have made a first step towards combining constraints from resolved stars with the broadband SED of galaxies. We have shown that it is possible to gain insight into the effects of observationally constrained SFHs on the conversion between SFR and UV luminosity. Effects on measures of stellar mass will be explored in a future work. We highlight an apparent discrepency between the observed UV luminosities and those expected given the SFH. We suggest that future work to simultaneously fit both the resolved stellar populations and the UV through NIR broadband SED may improve constraints on the SFH of nearby galaxies and provide a test of population synthesis models.

References

Bruzual, A. G. 2007, in *IAU Symposium*, Vol. 241, IAU Symposium, ed. A. Vazdekis & R. F. Peletier, 125

Conroy, C., Gunn, J. E., & White, M. 2009, *ApJ*, 699, 486

Dalcanton, J. J., Williams, B. F., Seth, A. C., *et al.* 2009, *ApJS*, 183, 67

Dale, D. A., Cohen, S. A., Johnson, L. C., *et al.* 2009, *ApJ*, 703, 517

Hao, C.-N., Kennicutt, R. C., Johnson, B. D., *et al.* 2011, *ApJ*, 741, 124

Kennicutt, R. C. 1998, *ARAA*, 36, 189

Lee, J. C., Gil de Paz, A., Kennicutt, Jr., R. C., *et al.* 2011, *ApJS*, 192, 6

Madau, P., Pozzetti, L., & Dickinson, M. 1998, *ApJ*, 498, 106

Marigo, P., Girardi, L., Bressan, A., *et al.* 2008, *A&A*, 482, 883

Weisz, D. R., Dalcanton, J. J., Williams, B. F., *et al.* 2011, *ApJ*, 739, 5

The Spectral Energy Distribution of Galaxies
Proceedings IAU Symposium No. 284, 2011
R.J. Tuffs & C.C. Popescu, eds.

doi:10.1017/S1743921312008757

Direct constraints on the impact of TP-AGB stars on the SED of galaxies from near-infrared spectroscopy

Stefano Zibetti[1], Anna Gallazzi[1], Stéphane Charlot[2], Anna Pasquali[3] and Daniele Pierini†

[1]Dark Cosmology Centre, Niels Bohr Institute, University of Copenhagen, Juliane Maries Vej 30, DK-2100 Copenhagen, Denmark

[2]Institut d'Astrophysique de Paris, France

[3]Astronomisches Rechen-Institut, Heidelberg, Germany

email: `zibetti, gallazzi@dark-cosmology.dk`

Abstract. We present new spectro-photometric NIR observations of 16 post-starburst galaxies especially designed to test for the presence of strong carbon features of thermally pulsing AGB (TP-AGB) stars, as predicted by recent models of stellar population synthesis. Selection based on clear spectroscopic optical features indicating the strong predominance of stellar populations with ages between 0.5 and 1.5 Gyr and redshift around 0.2 allows us to probe the spectral region that is most affected by the carbon features of TP-AGB stars (unaccessible from the ground for $z \sim 0$ galaxies) in the evolutionary phase when their impact on the IR luminosity is maximum. Nevertheless, none of the observed galaxies display such features. Moreover the NIR fluxes relative to optical are consistent with those predicted by the original Bruzual & Charlot (2003) models, where the impact of TP-AGB stars is much lower than has been recently advocated.

Keywords. galaxies: general - galaxies: galaxies: stellar content galaxies: fundamental parameters - infrared: galaxies, stars - stars: AGB and post-AGB.

---

1. TP-AGB Stars: the big unknown

The modeling and interpretation of the spectral energy distribution (SED) of stellar populations via stellar population synthesis (SPS) is a fundamental tool to understand galaxy properties and their evolution. Yet we are far from a complete comprehension and a reliable modeling of some stellar evolutionary phases which strongly affect the energy output of stellar populations. Among them, the so-called thermally pulsing - asymptotic giant branch (TP-AGB for short) phase has been the focus of debate among modelists for several years (e.g. Maraston 2005; Bruzual 2007; Marigo *et al.* 2008). TP-AGB stars are believed to be major contributors to the NIR flux of stellar populations with age between 0.5 and 1.5 Gyr. However, quantitative predictions of the lifetimes, luminosities and emission continuum shapes and spectral features of these stars and of their influence on the integrated spectra of even "simple" stellar populations, are still very uncertain and strongly model-dependent. Different treatments of TP-AGB stars (i.e. prescriptions to model their lifetimes and luminosities, different stellar spectral libraries) in popular SPS codes can yield significant differences, in particular concerning *i)* the flux ratios in different NIR bands and in NIR vs optical bands and *ii)* the presence (or absence) of sharp spectral features due to molecules in the atmospheres of those stars. The age and spectral ranges affected by TP-AGB stars are of particular interest for two main applications. *a)*

† Visitor astronomer at the Max-Planck-Institut für extraterrestrische Physik, Garching

At low redshift, the NIR is often used as tracer of stellar mass in galaxies, both unresolved (e.g. Bell *et al.* 2003) and resolved ones (e.g. Zibetti, Charlot & Rix 2009), because of the (supposed) lower sensitivity to the young stellar components that often dominate the visible part of the spectrum and because of lower dust extinction. However, different treatments of TP-AGB stars in SPS models result in systematic differences in global stellar mass estimates of several tens of percent and in substantially different estimates of the stellar mass density contrast of structures like spiral arms within galaxies (see Zibetti, Charlot & Rix, 2009). *b)* At $z \gtrsim 2.5$, the bulk of stellar populations has mean ages in the range 0.5–1.5 Gyr that is (allegedly) mostly affected by TP-AGB stars. As demonstrated by, e.g., Maraston *et al.* (2006) and Kannappan & Gawiser (2007), uncertainties on TP-AGB models severely affect stellar mass and age determinations for galaxies at those redshifts (see also Conroy, Gunn & White 2009).

2. Testing SPS model predictions: NIR spectrophotometric observations of post-starburst galaxies at $z \sim 0.2$

We directly test the impact of TP-AGB stars on the integrated SED of galaxies by obtaining spectra in H and K bands, which we combine with available optical spectroscopy and photometry. Such an SED can be confronted with SPS predictions and used to constrain model templates for galaxies at low and high redshifts. In order to maximize the impact of TP-AGB stars, we target "post-starburst" galaxies, whose star formation an intense burst 0.5-1.5 Gyr back in the past, followed by negligible activity. They can be relatively easily selected from the large SDSS spectral database by requiring no emission in Hα or in [OII] and equivalent width of H$\delta > 5$Å in absorption (following Goto 2005). In addition we require light-weighted mean stellar age between 0.5 and 1.5 Gyr as determined from optical spectral indexes (little sensitive to TP-AGB modeling) according to the bayesian method of Gallazzi *et al.* (2005). As the most prominent NIR features of TP-AGB stars are located at 1.41 and 1.77μm and, thus, fall in the atmospheric gaps between J and H, and H and K bands for $z \sim 0$ galaxies, we selected galaxies at $z \sim 0.2$, so that those two features move to the middle of the H and K atmospheric windows, respectively. The resulting sample of 16 galaxies provides a good coverage of the age-metallicity domain that is most strongly affected by TP-AGB stars.

NIR spectra in H and K bands have been obtained with ISAAC at the ESO-VLT during P86 and assembled with the SDSS spectra to provide flux-calibrated SEDs ranging from ~3000 to ~21000Å with a rest-frame resolution of 3Å in the optical and 40Å in the NIR and typical S/N>30 per resolution element (see examples in Fig. 1). Spectra have been rescaled to match the broad-band "total" fluxes derived from SDSS, UKIDSS and ISAAC imaging in matched apertures corresponding to the SDSS Petrosian aperture.

Where are the TP-AGB NIR spectral features? The H- and K-band spectra of all 16 galaxies in the sample appear smooth and featureless. In particular we do not detect either of the two sharp features at 1.41μm and 1.77μm (C_2) rest-frame, in clear contrast with the Maraston (2005) models predicting extremely strong and abrupt flux drops of 10 to 30% at those wavelengths. Our observations are the first ones to prove this *directly* thanks to the NIR spectroscopy and the choice of a suitable redshift range. Nine examples out of our 16 observed galaxies are reported in Fig. 1 to demonstrate the lack of such NIR spectral features. Spanning the full range of metallicities over the age range where TP-AGB stars are expected to peak, there is no sign of sharp NIR spectral features.

"TP or not TP? that is the question" Over the last few years a general consensus has grown among SPS modelists about the need for a boosted NIR flux due to TP-AGB stars with respect to earlier generation models (e.g. BC03), although with strong differences

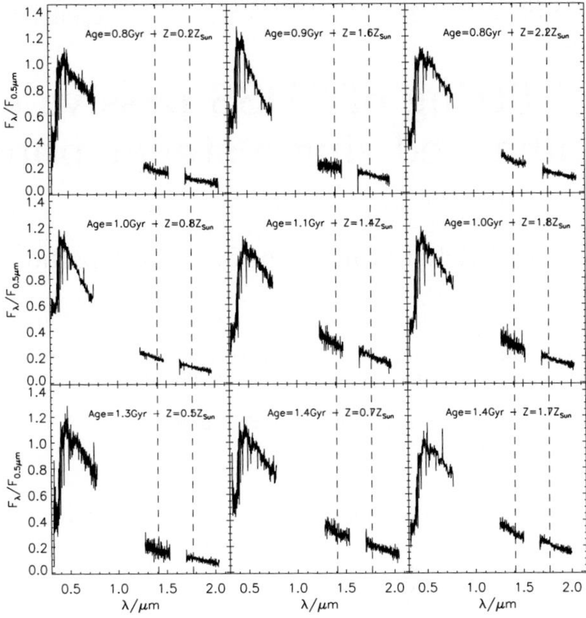

Figure 1. Nine examples of combined optical-NIR spectra (ISAAC for the NIR, SDSS for the optical) of the post-starburst galaxies in our sample. The three rows show galaxies of increasing mean light-weighted age, from ~0.8 Gyr (top), to ~1 Gyr (middle) to ~1.4 Gyr (bottom). In each row light-weighted mean metallicity increases from left to right, from sub-solar to super-solar values. The spectra in each panel are cross-calibrated based on SDSS, UKIDSS and ISAAC photometry normalized to the flux at 0.5 μm (rest) and shown as a function of rest-frame wavelength. The vertical dashed lines mark 1.41 and 1.77 μm, where sharp spectral features are expected based on the Maraston (2005) models (cf. their Fig. 14, 15 and 18). However neither is observed. Each panel is labeled with the light-weighted stellar age and metallicity derived using the bayesian method of Gallazzi *et al.* (2005).

concerning its exact amount. Kriek *et al.* (2010) have recently challenged this, by showing that the composite (medium-band NIR) SED of 62 post-starburst galaxies at $0.7 < z < 2.0$ is fully consistent with BC03 and inconsistent with Maraston (2005). Although very careful analysis is needed to compare our new observations with different models (in prep.), preliminary analysis shows that the majority of our spectra is extremely well reproduced by BC03 simple stellar populations (SSP) with no need for further tweaking/fitting or inclusion of dust. This suggests that predictions from early models might not require very substantial correction in terms of NIR output.

References

Bell, E. F., McIntosh, D. H., Katz N., & Weinberg, M. D. 2003, *ApJS*, 149, 289
Bruzual, G. 2007, arXiv:astro-ph/0703052
Bruzual, G. & Charlot, S. 2003, *MNRAS*, 344, 1000
Conroy, C., Gunn, J. E., & White, M. 2009, *ApJ*, 699, 486
Gallazzi, A., *et al.* 2005, *MNRAS*, 362, 41
Goto, T. 2005, *MNRAS*, 357, 937
Kannappan, S. J. & Gawiser, E. 2007, *ApJ*, 657, L5
Kriek, M., *et al.* 2010, *ApJL*, 722, 64
Maraston, C. 2005, *MNRAS*, 362, 799
Maraston, C. *et al.* 2006, *ApJ*, 652, 85
Marigo, P., *et al.* 2008, *A&A*, 482, 883
Zibetti, S., Charlot, S., & Rix, H.-W. 2009, *MNRAS*, 400, 1181

The Spectral Energy Distribution of Galaxies
Proceedings IAU Symposium No. 284, 2011
R.J. Tuffs & C.C. Popescu, eds.

doi:10.1017/S1743921312008769

Spectral fitting of SDSS passive galaxies with α-enhanced single stellar populations

Jean Michel Gomes[1] **and Paula Coelho**[2]

[1]Centro de Astrofísica, Universidade do Porto, Rua das Estrelas, 4150-762 Porto, Portugal
[2]Núcleo de Astrofísica Teórica, Universidade Cruzeiro do Sul,
R. Galvão Bueno 868, Liberdade, 01506-000, São Paulo, Brasil

email: `jean@astro.up.pt`

Abstract. The power of population synthesis as a mean to estimate the star-formation and chemical histories of galaxies has been well established in the last decade. The major developments were due to a huge avalanche of methods, codes and high-quality galaxy data sets, such as the 2dF, 6dF and SDSS surveys. Semi-empirical spectral synthesis allows for the decomposition of a galaxy spectrum in terms of linear combinations of base elements, i.e. Single Stellar Populations (SSPs) of different ages and metallicities, which are computed from evolutionary synthesis codes (BPASS, GALEV, GALAXEV, MILES, PÉGASE, etc...), containing distinct ingredients like: stellar library, evolutionary tracks, metallicities and Initial Mass Function. In general, they have solar-scaled relative abundances, but this is about to change with the unfolding of new α-enhanced SSP models (Coelho *et al.* 2007). However, passive galaxies have some spectral features corresponding to "enhanced-ratios" ([E/Fe]), like O, Ne, Si, S, Mg, Na, C and N over Fe that are not well modeled using solar-scaled SSPs (Trager *et al.* 2000), leading to residuals between observed and modeled spectra, which also correlate with the velocity dispersion ($\sigma_\star$) and stellar mass ($M_\star$): Massive galaxies exhibit a larger [E/Fe] discrepancy than less massive ones. This result can be interpreted as a signature of distinct previous star-formation efficiencies in passive galaxies, leading to distinctive ratios of type Ia and II SNe.

We have applied the STARLIGHT spectral synthesis code (Cid Fernandes *et al.* 2005) to a sample of $\sim$ 1000 passive galaxies from the SDSS DR7 with a S/N at the continuum $\geqslant$ 20 to investigate possible enhancements in the derived [E/Fe] ratios. Three sets of SSPs based on Coelho *et al.* (2007) theoretical models and Walcher *et al.* (2009) prescriptions were computed for [α/Fe]=0.0, [α/Fe]=0.2 and [α/Fe]=0.4. Our aim is to determine: (1) the quality of the fits, (2) the mean stellar age and metallicity distributions, and (3) the star-formation history of passive galaxies.

Using [α/Fe]=0.0 SSPs, we have identified the strongest residuals in the CN (4142.125 – 4177.125 Å), Na D (5876.875 – 5909.375 Å) and Mg (5069.125 – 5196.625 Å) bands. On the other hand, [α/Fe]=0.2 and [α/Fe]=0.4 SSP models tend to reproduce better the Mg band, as compared to solar-scaled SSPs ([α/Fe]=0.0). The residuals are decreased by 1.77 Å ([α/Fe]=0.2) and 2.92 Å ([α/Fe]=0.4). However, as expected, these α–enhanced models lead to worse fits for the CN and Na D bands. These residuals may even reach up to 2.08 Å (CN) and 4.20 Å (Na D), using [α/Fe]=0.2 SSPs and 2.28 Å (CN) and 7.94 Å (Na D), using [α/Fe]=0.4 SSPs.

In terms of mean stellar ages and metallicities, we obtain non-negligible biases in both quantities when we compare the solar-scaled SSPs with α-enhanced ones, which tend to have mean stellar ages by 0.12 dex ([α/Fe]=0.2) and 0.14 dex ([α/Fe]=0.4) higher and mean stellar metallicities by 0.1 dex ([α/Fe]=0.2) and 0.2 dex ([α/Fe]=0.4) lower.

Keywords. galaxies: evolution, galaxies: formation, galaxies: stellar content

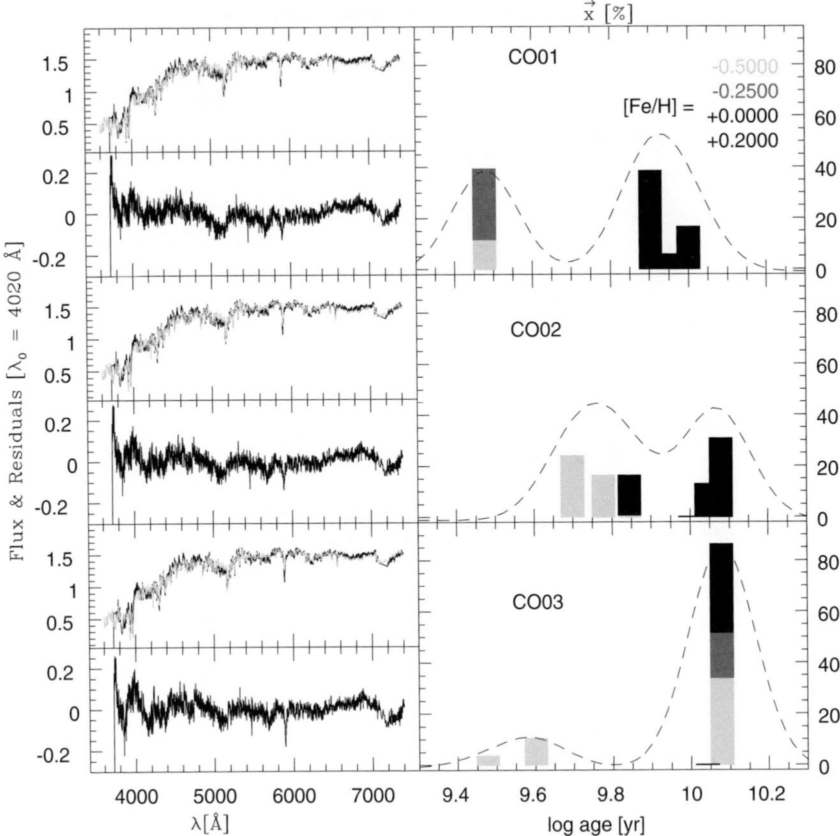

Figure 1. Spectral fit to the SDSS passive galaxy J100329.97+372734.6 using three different sets of SSPs with the following α-enhancement ratios: CO01 ([α/Fe]=0.0), CO02 ([α/Fe]=0.2) and CO03 ([α/Fe]=0.4), spanning 10 distinct ages from 3 to 12 Gyr and 4 different metallicities ([Fe/H]=-0.50, -0.25, +0.00 & +0.20). Left panels: Observed (black) and modeled (grey) spectrum with the corresponding residuals for CO01, CO02 and CO03 (from top to bottom). The fits are equally good in terms of their reduced χ^2. The corresponding Star Formation Histories (SFHs) are displayed in the right panels. The smoothed SFHs (re-scaled) are shown (black dashed line) with a Gaussian FWHM of 0.2 dex, further illustrating the tendency for higher mean stellar ages with increasing [α/Fe] SSP abundances.

1. Sample of passive galaxies

We have applied STARLIGHT to $\sim$ 1000 passive galaxies, lacking nebular emission, from the SDSS DR7, using three different libraries with distinct α-enhanced SSP models: CO01 ([α/Fe]=0.0), CO02 ([α/Fe]=0.2) and CO03 ([α/Fe]=0.4), spanning a range in ages from 3 to 12 Gyr, and 4 metallicities ([Fe/H]=-0.50, -0.25, +0.00 & +0.20). In Fig. 1 we show an example for the spectral fits and the corresponding Star Formation Histories (SFH) to the SDSS passive galaxy J100329.97+372734.6 using CO01, CO02 and CO03 SSPs.

2. Mean stellar age and metallicity

Fig. 2 shows the degeneracy diagram of age versus metallicity, showing a clear trend for a higher mean stellar age and a lower mean stellar metallicity with increasing α-element enhancement in model SSPs.

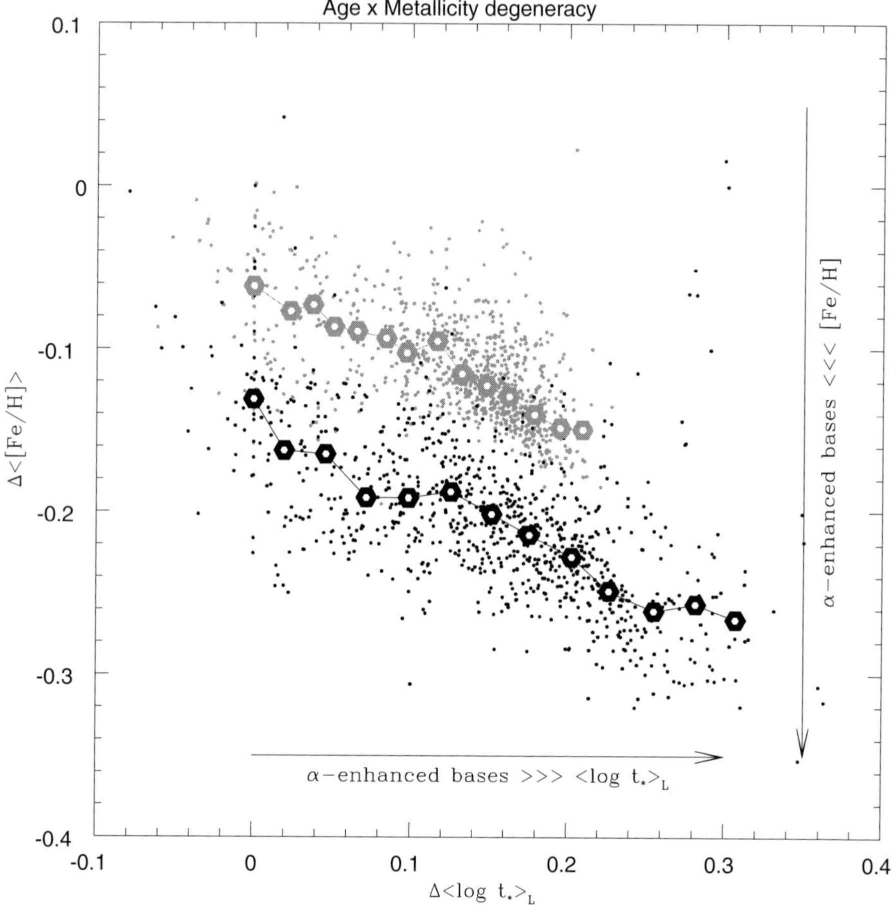

Figure 2. The degeneracy diagram of Age vs Metallicity. We show the difference in mean stellar age and mean stellar metallicity for CO02 [α/Fe=0.2] (grey) / CO03 [α/Fe=0.3] (black) compared to CO01 [α/Fe=0.0]. The hexagons correspond to the median values: top for CO02–CO01 and bottom for CO03–CO01. A clear trend for a higher mean stellar age and a lower mean stellar metallicity with increasing α-element enhancement is apparent.

Acknowledgements

J. M. Gomes is supported by a Post-Doctoral grant, funded by FCT/MCTES (Portugal) and POPH/FSE (EC). Special thanks also to Privatdozent Dr. P. Papaderos for making suggestions and revising the manuscript.

References

Cid Fernandes R. *et al.* 2005, *MNRAS*, 358, 363
Coelho P. *et al.* 2007, *MNRAS*, 382, 498
Lee H.-C. *et al.* 2009, *AJ*, 138, 1442
Trager S. C. *et al.* 2000, *AJ*, 120, 165
Walcher C. J. *et al.* 2009, *MNRAS*, 398, 44
Worthey G., Faber S. M., & Gonzalez J. J. 1992, *ApJ*, 398, 69
Worthey G. 1994, *ApJ*, 95, 107

The Spectral Energy Distribution of Galaxies
Proceedings IAU Symposium No. 284, 2011
R.J. Tuffs & C.C. Popescu, eds.

doi:10.1017/S1743921312008770

Multi-component parametric inversion of galaxy kinematics and stellar populations using full spectral fitting

Ivan Yu. Katkov and Igor V. Chilingarian

Sternberg Astronomical Institute, Moscow State University,
Universitetskii pr. 13, Moscow, 119992 Russia
email IK: katkov.ivan@gmail.com; IC: chil@sai.msu.ru

Abstract. The stellar line-of-sight velocity distribution (LOSVD) can be strongly asymmetric in regions where the light contributions of both disc and bulge in spiral and lenticular galaxies are comparable. Existing techniques for the stellar kinematics analysis do not take into account the difference of disc and bulge stellar populations. Here we present a novel approach to the analysis of stellar kinematics and stellar populations. We use a two-component model of spectra where different stellar population components are convolved with pure Gaussian LOSVDs. For this model we present Monte-Carlo simulations demonstrating degeneracies between the parameters.

Keywords. Methods: data analysis, galaxies: kinematics and dynamics, galaxies: stellar content.

1. Introduction

A flat rotating stellar disc and a slowly rotating spheroidal bulge in spiral and lenticular galaxies usually possess very different stellar population properties. Consequently, the resulting stellar line-of-sight velocity distribution (LOSVD) can be strongly asymmetric in regions where the light contributions of both disc and bulge are comparable. At the first approximation, this can be accounted by the Gauss-Hermite parametrization of the LOSVD (van der Marel & Franx, 1993), however, different absorption features (e.g. sensitive to age and metallicity) will have different effective LOSVDs. The first attempt to recover parametrically the multi-component dynamics connected to multi-component stellar populations was done by De Bryuine *et al.* (2004), who used stellar spectra to model different stellar populations. Here we present a more realistic approach based on the full spectral fitting, where multiple dynamical components are represented by multiple simple stellar population (SSP) models. More sophisticated stellar population models can be used instead of SSPs.

2. Two-component parametric LOSVD recovery

As an example, we present a two-component model having two pure Gaussian LOSVD components with different stellar populations characterised by their ages and metallicities. An optimal template is represented by the linear combination of two SSPs each convolved with its own LOSVD; hence the χ^2 value is computed as:

$$\chi^2 = \sum_{N_\lambda} \frac{[F_i - P_p \cdot \sum_j k_j \cdot S(T_j, Z_j) \otimes \mathcal{L}(v_j, \sigma_j)]^2}{\delta F_i^2}, \tag{2.1}$$

where $\mathcal{L}(v, \sigma)$ - pure Gaussian LOSVD; F_i and δF_i are the observed flux and its uncertainty; $S(T_j, Z_j)$ is the flux from the j-th synthetic spectrum of SSP with given age T_j

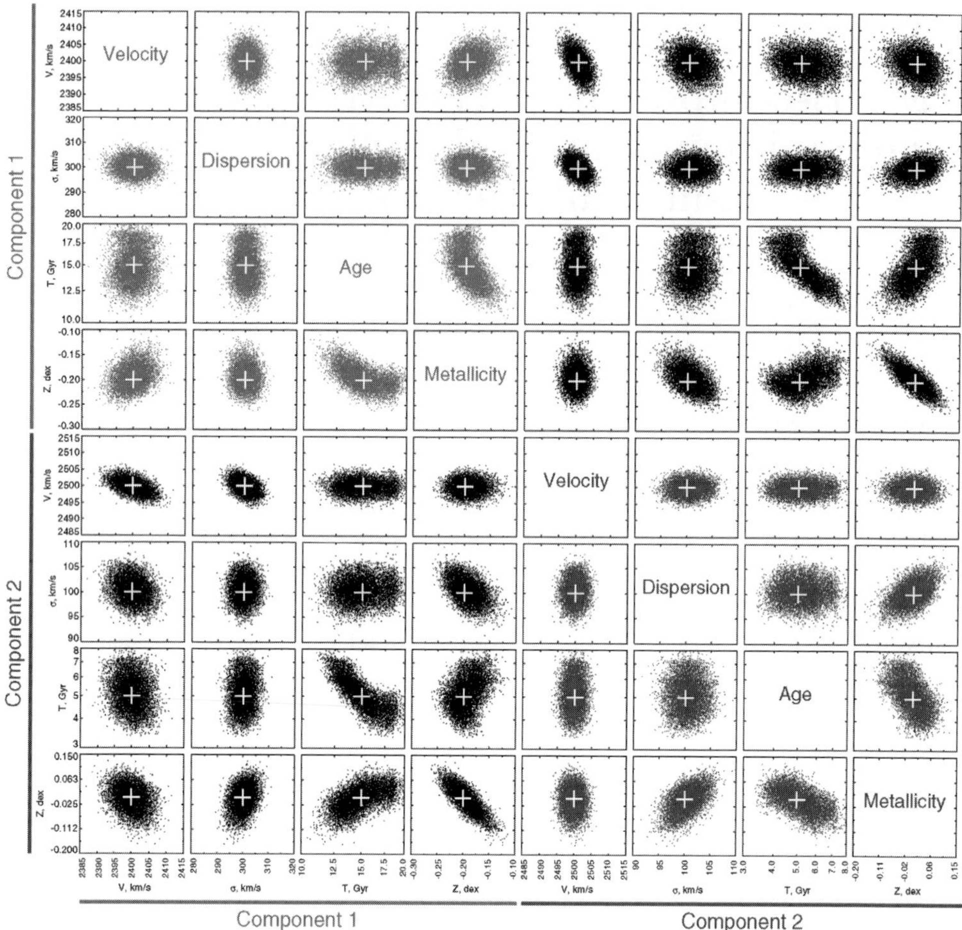

Figure 1. Covariance matrix of the model parameters. Left-top quadrant and right-bottom quadrant present a self-correlation of parameters of bulge and disc components, correspondingly. Left-bottom and right-top quadrants correspond to mutual correlation of stellar components. White crosses show the input values.

and metallicity Z_j; P_p is a multiplicative Legendre polynomial of order p for correcting the continuum, determined at each step of minimization loop by solving the linear least-square problem; k_j is the j-th component weight (normally found by the linear minimization). The important point in this study is that we fixed the relative SSP contributions k_j to the values derived from the photometric light profile decomposition. This approach was implemented on top of the NBursts full spectral fitting technique (Chilingarian *et al.* 2007a, Chilingarian *et al.* 2007b).

Our new approach was applied to the real spectra of the luminous early-type galaxy NGC 524 (see Katkov *et al.* 2011). It was shown that the LOSVD of NGC 524 exhibits

Table 1. Parameter of stellar components.

Component	v, km/s	σ, km/s	Age, Gyr	Z, dex	Weight K_j
1 (Bulge)	0	300	15	−0.2	0.7
2 (Disc)	100	100	5	0.0	0.3

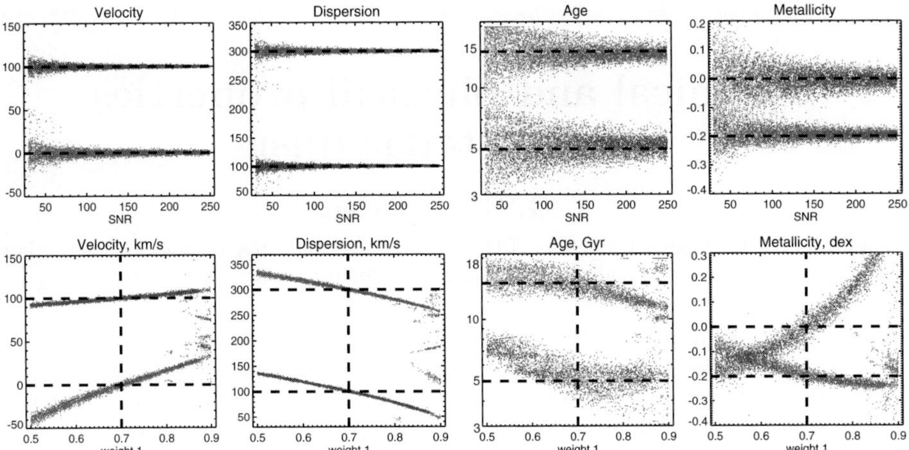

Figure 2. Top line: The dependence on signal-to-noise ratio of output parameter dispersion. Bottom line: Output parameters with varied relative contribution of components.

strong asymmetry and a stellar component of galaxy can be described as a bulge, inner disc and a conterrotating outer disc.

3. Monte-Carlo simulation

In order to explore the parameter space, we constructed and analysed using our technique a sample of 8000 realizations of synthetical spectra created as a linear combination of two components (see Table 1), adding random noise corresponding to the signal-noise ratio (SNR) 100. The relative contributions $k_{1,2}$ were fixed. The degeneracies between all the parameters are shown in Fig. 1 as a covariance matrix. Here and after red and blue points correspond to the bulge and disc components respectively. We can see that the most degenerate pair of parameters are $T_1 - T_2$ and $Z_1 - Z_2$.

In order to demonstrate how the parameters are recovered depending on the SNR we fitted a sample of 8000 models varying a SNR between 30 and 250. Fig. 2(top line) shows how the output parameter uncertainty decreases when increasing SNR. The dashed lines show the input parameters of synthetic spectrum.

We used the same approach to demonstrate the importance of knowledge of the relative contributions k_j. Fig. 2 (bottom line) presents 4000 MC realizations where we varied the relative SSP contribution of bulge component k_1 in a range between 0.5 and 0.9, while the real value was equal to 0.7. SNR = 150 was adopted in these realisations. It is obvious that even a small offset in k_j can lead to important biases in the recovered parameters.

The authors thank the IAU for the provided financial aid and RFBR grant 10-02-00062 for covering the remaining expanses.

References

Chilingarian, I. V., Prugniel, P., Sil'chenko, O. K., & Afanasiev, V. 2007, *MNRAS*, 376, 1033

Chilingarian, I. V., Prugniel, P., Sil'chenko, O. K., & Koleva, M. 2007, In: *Stellar Populations as Building Blocks of Galaxies*, Proc. of the IAU Symp. no 241, eds. A. Vazdekis and R. F. Peletier, p. 175, arXiv:0709.3047

De Bryuine, V., De Rijcke, S., Dejonghe, H., & Zeilinger, W. W. 2004, *MNRAS*, 349, 461

Katkov, I., Chilingarian, I., Sil'chenko, O., Zasov, A., & Afanasiev V. 2011, *arXiv:1106.2527*

van der Marel, R. & Franx, M. 1993, *ApJ*, 407, 525

The Spectral Energy Distribution of Galaxies
Proceedings IAU Symposium No. 284, 2011
R.J. Tuffs & C.C. Popescu, eds.

doi:10.1017/S1743921312008782

Chemical and physical properties of interstellar dust

A. G. G. M. Tielens

Leiden Observatory, Leiden University, PO Box 9513, NL-2300 RA Leiden, the Netherlands
email: tielens@strw.leidenuniv.nl

Abstract. The characteristics of interstellar dust reflect a complex interplay between stellar injection of stardust, destruction in the ISM, and regrowth in clouds. Astronomical observations and analysis of stardust isolated from meteorites have revealed a highly diverse interstellar and circumstellar grain inventory, including both amorphous materials and highly crystalline compounds (silicates and carbon). This review summarizes this dust budget and inventory. Interstellar dust is highly processed during its sojourn from its birthsite (stellar ejecta) to its incorporation into protoplanetary systems. Processing by strong shocks due to supernova explosions is particularly important. Sputtering by impacting gas ions in shocks in the intercloud medium of the ISM is counteracted by accretion in cloud phases and their balance sets the observed, interstellar, elemental depletion patterns. Astronomical and meteoritical-stardust evidence for these processes is reviewed and it is concluded that dust formation in the ISM is very rapid. Not surprisingly, the characteristics of interstellar dust are expected to vary widely reflecting local stellar sources, the effects of SNe processing, and the interstellar accretion process.

---

1. Introduction

The presence of small dust grains in the interstellar medium of galaxies near and far is very apparent through extinction of stellar and nebular photons, through scattered light, through optical and infrared polarization, and through infrared emission. Polycyclic Aromatic Hydrocarbon (PAHs) molecules, the extension of the interstellar grain size distribution into the molecular domain, are equally prominent in images and spectra of galaxies at mid-IR wavelengths. In addition, many of the key processes that drive the evolution of galaxies – including star and planet formation as well as the accretion onto central black holes – occur deeply inside dust enshrouded regions. Hence, understanding the characteristics of interstellar dust is of key importance for analyzing observations of galaxies and their evolutionary processes: star formation and black hole interaction.

In addition, small dust grains play an important role in the physical and chemical evolution of the interstellar medium of galaxies. Dust grains shield the gas from the destructive effects of FUV photons. In addition, dust grains provide surfaces for the formation of H_2 and other molecules. Such species play a key role in the star formation process because they control the degree of ionization and therefore the coupling of the gas to magnetic fields. Moreover, molecules dominate the cooling of dense clouds, facilitating gravity to overcome thermal supporting forces. Small grains and PAHs also dominate the heating of neutral gas through the photo-electric effect and thereby the physical conditions and the phase structure of the ISM. Last but not least, dust grains are the building blocks of cometesimals and planetesimals in the first steps towards the formation of planetary systems.

Studies of interstellar dust are therefore of key importance for our understanding of the evolution of galaxies. The recent observations that relatively large amounts of dust are present in galaxies as early as redshifts of 6 (Bertoldi *et al.* 2003) drive a number of key

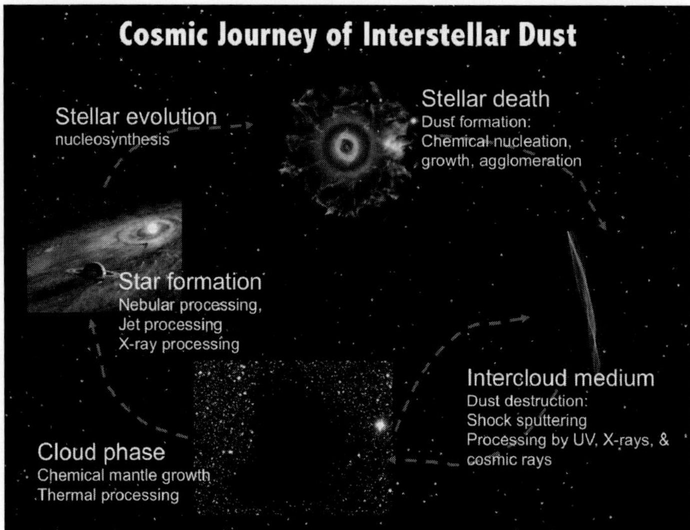

Figure 1. The lifecycle of large molecules and grains in space. At the end of their life, stars return most of their material back to the interstellar medium enriched by their nucleosynthetic products. In the interstellar medium, the ejected material mixes with material left over from the big bang and with material from other stars. Eventually, new stars and planetary systems are born from this material. During this cosmic cycle, materials are heavily processed. Some of these processes are indicated in this figure. (Figure adapted from Tielens 2005).

questions: "What are the sources of dust and how did their contribution vary over the age of the Universe ?", "What processes modify the dust characteristics in the interstellar medium ?", and "How did this dust influence the evolution of galaxies ?". In this review, I will address a number of these issues and indicate future observational opportunities that may help us address the characteristics and role of dust in the Universe.

2. The lifecycle of interstellar dust

The lifecycle of dust in space starts with the nucleation and growth of high temperature condensates such as silicates, oxides, graphite, and carbides at high densities and temperatures in the ejecta from stars (Fig. 1). This ejected material is rapidly mixed with other gas and dust in the interstellar medium of galaxies. In the interstellar medium, dust cycles many times between the intercloud and cloud phases on a very fast timescale ($\simeq 3 \times 10^7$ yr). In the low density, warm neutral and ionized intercloud media, material is processed by strong shocks driven by supernova explosions. The hot gasses in the shock sputter atoms from the dust while high velocity collisions among grains lead to vaporization, melting, phase transformation, and shattering of the projectile and target. In the denser media – diffuse and dense clouds –, gas phase species accrete onto grains forming a mantle. Coagulation may also play a role in increasing the grain size inside clouds. If the grain survives the onslaught of interstellar shocks, eventually, during one of these cycles, a grain may find itself in a dense cloud core when this core becomes gravitationally unstable against collapse. The grain may then wind up in the star or in the surrounding planet-forming disk and eventually become part of a planet-forming body. The complete cycle from injection by a star until formation of a new star and any associated planets typically takes some 3×10^9 yr.

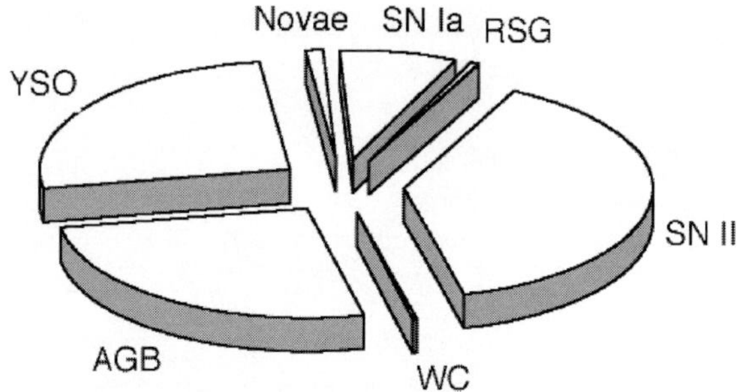

Figure 2. The contribution of different stellar sources to the interstellar dust budget. For details, see § 3. Note that all these values are very uncertain. (Taken from Tielens 2001).

3. Sources of stardust

Stars in the last stages of their evolution return much of their material to the ISM. For most stars, as these ejecta expand and cool, small dust grains nucleate and grow when the temperature drops to about 1500 – 1000 K, depending on the material and the chemical kinetics (Cherchneff 2000, Cherchneff & Dwek 2010). The gas mass return rate to the ISM – mostly H and He – is dominated by Asymptotic Giant Branch stars (AGB), as expected since these objects are the descendants of low mass stars that dominate the stellar mass budget and the Milky Way is approximately in steady state. In terms of condensible elements, the mass return rate is about equally due to AGB stars and SNe (Tielens 2005). Observations show that AGB stars are important contributors to the stardust budget of the Milky Way, returning some 8 $M_\odot$ kpc^{-2} Myr^{-1} in solid form. By comparison, red supergiants, the descendants of massive stars, return only a paltry 0.2 $M_\odot$ kpc^{-2} Myr^{-1} of dust. The contribution of SNe to the dust mass budget is unclear. If all the condensible material were turned into dust, type II SNe might contribute some 12 $M_\odot$ kpc^{-2} Myr^{-1}. Observationally, dust is known to form typically between 300 and 1000 days after the SN explosion as revealed by a sudden drop in optical light, a concommittant increase in IR emission, and the development of a pronounced blue-red asymmetry in the emission line profiles of the ejecta (Wooden *et al.* 1993, Lucy *et al.* 1989). However, IR studies of core collapse SNe implied less than 5×10^{-3} $M_\odot$ of dust (somewhat dependent on the adopted grain material properties, very sensitive to the adopted clumpy distribution of the ejecta, and widely varying between SNe), corresponding to an efficiency of $\simeq 10^{-3} - 10^{-1}$ (Wooden *et al.* 1993, Sugarman *et al.* 2006, Ercolano *et al.* 2007, Meikle *et al.* 2007; see Barlow 2009 for a recent review). Young supernova remnants (SNR) provide another view of the dust formation efficiency of SNe and one that indicates much higher dust formation efficiencies. Spitzer studies of the young ($\simeq 330$ yr) supernova remnant, Cas A, revealed a warm dust mass of 0.025 $M_\odot$ in the volume processed by the reverse shock (Rho *et al.* 2008). A Herschel study adds to that 0.075 $M_\odot$ of cold dust mass interior to the reverse shock (Barlow *et al.* 2010). The total dust mass ($\simeq 0.1$ $M_\odot$) should be compared to the estimated total ejected mass of 2-4 $M_\odot$, much of this in the form of oxygen ($\simeq 2$ $M_\odot$) which will not condense (Willingale *et al.* 2002, Vink *et al.* 1996). In analogy to the well-studied type IIb SN 1993J, some 0.6 $M_\odot$ of condensible elements (C, Mg, Si, S & Fe) were ejected during the SN explosion of Cas A (Thielemann *et al.* 1996), corresponding to a very high dust formation efficiency of ~ 0.2. Likewise, Herschel observations of the extremely young SNR associated with SN 1987A – which has just entered the reverse shock phase as the ejecta slammed into the previous stellar wind remnant some 10 years after the explosion – reveal ~ 0.5 $M_\odot$

Table 1. Inventory of dust in space

Oxide dust	Carbonaceous dust	Other
amorphous silicates (I,C,S)	PAHs (I,C,S)	silicon nitride Si_3N_4 (S)
crystalline forsterite, Mg_2SiO_4 (C,S)	Fullerene, C_{60} (C,I)	magnesium sulfide, MgS (C)
crystalline enstatite ($MgSiO_3$) (C,S)	Amorphous Carbon (C,I,S)	Carbonate (C,I)
Silica, SiO_2 (C)	Graphite (C,I,S)	Ice (C,I)
aluminum oxide, Al_2O_3 (C,S)	Diamond (C,S)	
spinel, $MgAl_2O_4$ (C,S)	silicon carbide, SiC (C,I ?,S)	
titanium oxide, TiO_2 (S)	other carbides (C ?, S)	
hibonite, $CaAl_{12}O_{19}$ (S)		
Magnesium iron oxide, $Mg_{0.1}Fe_{0.9}O$ (C)		

Legend: I: Spectroscopic evidence for presence in interstellar dust. C: Spectroscopic evidence for presence in circumstellar dust. S: Present as stardust in meteoritic or cometary material (For a a discussion, see Tielens 2001, Zinner 2003).

of cold dust† (Matsuura *et al.* 2011); many orders of magnitude larger than estimated from observations during the dust condensation period of $\sim 500 - 1000$ days (Wooden *et al.* 1993). Given the bewildering zoo of SNe types and the uncertainties and conflicting observational results on dust formation in these environments, the contribution of SNe to the dust budget can presently only be guessed at but this anecdotal evidence suggests that it is high. Figure 2 assumes that all of the condensibles form dust and type II SNe are then more important than AGB stars for the dust budget.

Figure 2 also includes a contribution of dust formed in the inner regions of protoplanetary disks and entrained and ejected by the protostellar wind. This estimate is also at the high end as it is based upon the assumption that 1/3 of the accreting mass is ejected as a wind (Shu & Shang 1997) and that all condensibles form dust. Other estimates of protostellar wind characteristics typically result in a ~ 3 times smaller wind mass loss rate (Hartmann 1995). It should also be kept in mind that, for the dust budget, protostellar winds are only a pseudo source of dust. The wind likely originates from the inner region where all preexisting dust has sublimated and recondensed. The net addition of dust may then actually be negative (carbon will not condense as dust in these environments). Certainly, in terms of the *stardust* budget, protostars present a sink.

In summary, while the details are heatedly debated, it is clear that many different sources contribute to the dust budget of the ISM of galaxies.

4. Composition of stardust

The composition of interstellar dust can be studied through IR spectroscopy and through the analysis of genuine stardust grains retrieved from meteorites (Tielens 2001, Zinner 2003). These records show that the composition of circumstellar dust is rich and varied (Table 1) with silicates and other oxides, various forms of pure carbon-based materials and carbides, and a variety of other materials. In the astronomical record, much of this heterogeneity refers to circumstellar dust either in late type stars returning much of their material back to the ISM or in (the disks of) young stellar objects such as T Tauri stars and Herbig AeBe stars. Interstellar dust mainly consists of amorphous silicates and amorphous carbon. Limits on other materials are at the percent level (Kemper

† As this was observed after the SN ejecta reached the blue supergiant ejecta and the reverse shock has started to process the SN material, contamination by line emission may have affected this estimate.

et al. 2004). In addition, a substantial fraction (~ 0.05) of the elemental carbon in the Universe is locked up in PAHs and a much smaller fraction ($10^{-5} - 10^{-4}$) in the fullerene, C_{60}. (Tielens 2005, Sellgren 2010, Berne & Tielens 2011).

5. Processing of dust in the interstellar medium

5.1. *Shock destruction*

Interstellar shocks destroy interstellar dust (Jones *et al.* 1994, Jones *et al.* 1996) mainly through sputtering by impinging energetic ions. Supernova explosions drive strong shock waves into the surrounding interstellar medium. As the supernova remnant expands, the shock velocity drops. Initially, this expansion is adiabatic (the Sedov-Taylor phase) because very hot gas cools slowly, but when the shock velocity drops below $\simeq 250$ km/s, cooling becomes important (the radiative phase). Eventually, the remnant will merge with the ISM. Near threshold, the sputtering yield of materials rises steeply reaching a broad maximum of $\sim 10^{-2}$ for H atoms impacting with ~ 1 keV. Because a much larger volume of the ISM is processed by low velocity shocks than by fast shocks, most of the destruction is done by those radiative shocks where impacting H-atoms are near the threshold of the sputtering yield (e.g., $v_s \simeq 10^2$ km/s). Because, for a given SNR pressure, the shock velocity in a phase scales with $\rho^{1/2}$, most of the destruction will occur by 100 km/s shocks in the tenuous intercloud medium (rather than by shocks propagating in the diffuse cloud phase). In the plane of the Milky Way, such shocks occur typically once every sojourn of the grain into the warm intercloud medium and scrapes off some 30 Å from a grain surface. The destruction timescale is then set by the number of shocks required to erode a grain "fully" which depends on the interaction of supernovae with the ISM. Estimated timescales are $\simeq 500$ Myr and are not very sensitive to the details of the model for the ISM (e.g., 2-phase, 3-phase; Jones *et al.* 1994).

5.2. *Interstellar depletions*

This timescale for dust destruction, τ_{des}, is much shorter than the timescale for replenishment of dust material by old, dying stars ($\tau_{inj} \simeq 3$ Byr). Hence, in a steady state between dust injection by stars and destruction by shocks in the ISM, the expected depletion of condensible elements is $\delta/\delta_o = (1 + \tau_{inj}/\tau_{des})^{-1} \simeq 0.15$ where δ_o is the fraction of the element injected in the form of dust by stars. This is much less than the observed depletion of such dust-forming elements as Mg, Si, and Fe ($0.9 - 0.99$; Savage & Sembach 1996). Indeed, in such a model, the high observed depletions would require/imply a destruction timescale $\tau_{des} >> \tau_{inj}$ and that is incompatable with sputtering studies of materials and the "violent" nature of the ISM. In fact, some 10% of the condensible elements are injected into the ISM by stars (OB & WR stars) which do not form dust at all and hence such a steady state model can never explain depletions in excess of 0.9 (e.g., depletions of elements such as Fe, Ti, Ca; Jura 1987).

It is clear that dust has to regrow very quickly in the ISM itself and there is direct observational evidence for this. Observations have revealed a large and systematic difference in the elemental depletions in the different phases of the interstellar medium (Fig. 5.2; Savage & Sembach 1996, Cartledge *et al.* 2006) and these variations can be used to get an "observational" handle on the rate at which dust is destroyed in the intercloud medium and reformed through accretion in the cloud phase. The simple model for interstellar depletions, outlined above, can be extended to include accretion of elements in the cloud phase of the ISM; so, there is now a balance between dust injection by stars, dust destruction in the Warm Intercloud Medium, and accretion in the (diffuse) cloud phase of the ISM (Tielens 1998). The observed large depletion variations demonstrate

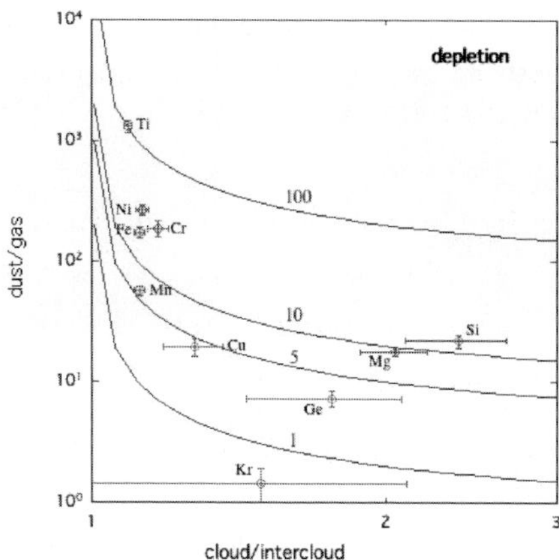

Figure 3. Observed depletions of elements in the ISM (Savage & Sembach 1996, Jenkins 2009). The y-axis is the ratio of the abundance of an element in the dust phase to its abundance in the gas phase, both measured in diffuse clouds. The x-axis is the ratio of the depletion of these elements in the diffuse cloud medium to that in the intercloud medium. The solid lines are the results of a simple model balancing destruction in the intercloud medium with accretion in the cloud medium (Tielens 1998). The labels indicate the adopted values for the accretion rate relative to the cloud-to-intercloud mixing ratio. As these observations demonstrate, the rates for destruction in the intercloud phase, accretion in the cloud phase, and mixing between these phases have to be within a factor of a few of each other and destruction and accretion are thus very rapid compared to the injection rate of dust by stars into the ISM.

rather directly that the processes involved – shock destruction and accretion – operate on a timescale similar to the timescale at which material is mixed from the cloud to the intercloud medium and back (Fig. 5.2). This mixing timescale is much less ($\simeq$ 30 million years) than the timescale at which new dust is injected into the ISM. Thus, specifically, some 10% of the iron, 15% of the magnesium, and some 30% of the silicon is returned to the gas phase upon each sojourn into the intercloud medium and then rapidly reaccreted once the material is cycled back to the (diffuse) cloud phase. Again, as for the material injected into the ISM by non-dust-producing stellar sources, this points towards the importance of the formation of a thin, outer layer in the ISM. This thin coating is readily sputtered in the warm intercloud medium and in that way protects underlying (stardust) grains against the destructive effects of shocks.

5.3. *Noble gasses in stardust grains*

Genuine stardust grains have been isolated from carbonaceous meteorites. Analysis of these stardust grains has revealed an isotopic composition which is distinctly non-Solar and derives directly from the stellar birthsites of these dust grains (Anders & Zinner 1993). Apparently, these dust grains formed in stellar ejecta – enriched by the nucleosynthetic products of processes taking place in the deep interiors of these stars and then mixed to the surface – were injected into the ISM, processed by shocks and other energetic events, became part of a region of star formation that collapsed to form the Solar system, saw the hot gases swirl in the Solar nebula, experienced possibly the shocks and lightning processes rampant in this environment, were incorporated into the planetary body from which the meteorite was derived, survived the violent break-up of this body, crashed on Earth and then was taken apart in the laboratory; and through all this arduous and torturous history, these stardust grains never equilibrated fully with the gas and managed to preserve their stellar heritage. Detailed studies of these stardust grains have opened a new window on the dusty universe, providing new insights in the composition

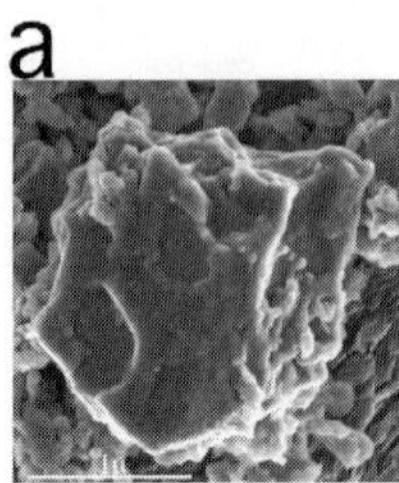

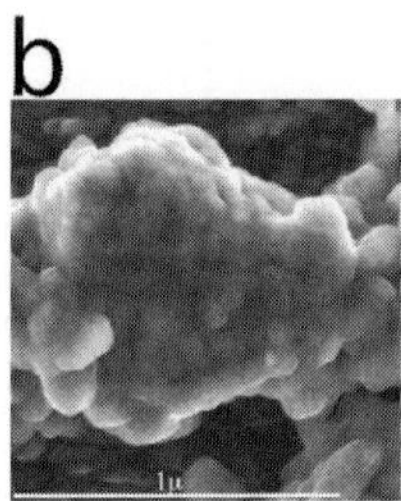

Figure 4. Field emission scanning electron microscope images of pristine presolar SiC grains from the Murchison meteorite (Bernatowicz *et al.* 2003): (a) exhibiting primary growth crystal faces and polygonal depressions; (b) coated with an apparently amorphous, possibly organic phase. Scale bars are 1 μm.

of dust and the processes that play a role in their formation (Bernatowicz *et al.* 2003). In addition, recent studies have started to explore the impact of interstellar processes on these grains (Henkel *et al.* 2010, King *et al.* 2010). Here, I will focus on the latter aspect. SiC stardust contains appreciable amounts of noble gasses and analysis of the isotopic composition of these noble gasses reveals the presence of two distinct components: The G-component characteristic for the isotopic composition of AGB stars (e.g., s-process) and the N-component with a solar system/ISM isotopic composition (Lewis *et al.* 1994). The abundances of each of these two components show fractionation with the lighter elements less abundant than expected from the abundance of Xenon. This fractionation is size dependent with smaller grains showing more severe fractionations. Noble gasses trapped in SiC grains represent the effects of ion implantation and the implantation depth is very energy dependent. For a constant grain-gas relative velocity, heavier species are more energetic and therefore implant deeper. The G-component is then thought to result from implantation of AGB-material during the PN nebula phase when radiation pressure rapidly accelerates grains to velocities of $\simeq$ 250 km/s relative to the gas. The N-component is the result of implantation during $\sim$ 100 km/s shock processing in the ISM (Guillard *et al.* 2011). The implantation of the N-component is accompanied by sputtering of the surface layer and it is this sputtering that preferentially removes the shallowier implanted, lighter noble gasses and leads to fractionation. The results for a simple model are compared to the measurements in Figure 5. Thus, the measured fractionation of the noble gases preserved in SiC stardust grains provide rather direct evidence for the shock processing of dust in the interstellar medium. Additional evidence for the importance of ion implantation is provided by depth-dependent studies of the concentration of trace elements in a few individual stardust grains. These reveal a concentration towards the surface as expected for ion-implantation at velocities of $\sim$ 500 km/s (Henkel *et al.* 2010, King *et al.* 2010).

6. Dust formation in the ISM

This review emphasizes the importance of dust growth in the ISM. For some elements (the heavily depleted ones: Mg, Si, Fe, Ca, Al, Ti, ...), this is an efficient process. But other elements are hardly affected. Sodium and potasium are obvious examples of elements that are never highly depleted. Oxygen seems to be involved to some extent but sulfur not. Most telling perhaps, carbon does not seem to show a difference in depletion between the warm intercloud medium and the diffuse cloud phase of the ISM and oxide rather than carbonaceous mantles seem indicated. From shock models, the average binding energy of atoms in this mantle is $\simeq$ 2 eV (Tielens 1998) and they are quite refractory. Past models

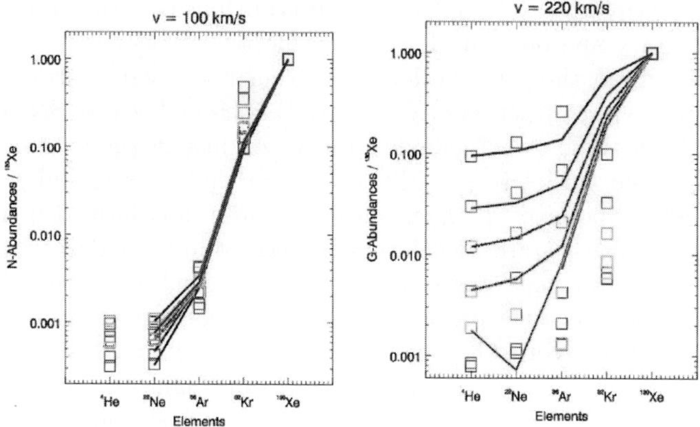

Figure 5. Comparison between measured noble gas abundances (symbols) in SiC grains extracted from the Murchison meteorite and model predictions (lines) for the N- (Left) and G- (Right) components. Elemental abundances have been normalized to ^{130}Xe and to Solar abundances for N or AGB abundances for G. Each line refers to a different size fraction (top to bottom: big to small). Error bars are smaller than the symbols (Guillard *et al.* 2011).

have invariably focused on carbonaceous mantles formed by energetic processing (e.g., UV photolysis or ion bombardment) of simple molecular ice mixtures in dense molecular clouds. However, while organic "goop" is readily formed this way in the laboratory, there is no observational evidence for the importance of this process in the ISM.

The chemistry of mantle formation seem to be very different. Likely, this is because this oxide mantle is predominantly formed in the diffuse ISM. A number of effects may then play a role (Tielens & Allamandola 1987). The preponderance of atomic hydrogen may favor the formation of simple hydrides and the more volatile ones (H_2O, CH_4, ...) are readily photodesorbed. Other atoms may preferentially form coordination complexes or salts, which are not so susceptible to photo-erosion. Moreover, the high UV photon flux may promote chemical interlinking of the growing surface layer and the underlying material. Of course, the high degree of ionization may also have some influence on the accretion/sticking rate; particularly for elements with ionization potentials below the grain work function (Na, K). At this point, these are mere speculations and a systematic study of the chemical routes and kinetics of grain surface chemistry is called for. The large body of data on interstellar elemental depletions may well provide a good guide for such endeavors.

7. Future

A good understanding of the characteristics of interstellar dust is a prerequisite for the interpretation of the spectral energy distribution of galaxies. We have a global understanding of the processes involved and some of these processes have been studied in great detail. It is clear that interstellar dust is very diverse, has many sources, and is heavily processed in the ISM. Amorphous silicates and some form of aromatic carbon dust are key components of interstellar dust. The properties of the dust will reflect local conditions, including the relative importance of different stardust sources, the evolution of SNR and the rate at which they process the ISM, and the detailed chemistry involved in mantle formation in the ISM.

The characteristics of dust at high z is of particular interest. At present, this is highly uncertain and the role of dust in the early evolution of the Universe can only be guessed at.

From studies in the (relatively) local Universe, it is obvious that the properties of dust are very dependent on its environment. The UV extinction curves in the Magellanic Clouds are quite different from those in the local solar neighborhood (Cartledge *et al.* 2005), either because of the large fraction of C-rich AGB stars or because SNe can process the ISM more readily in these dwarf galaxies. Likewise, dust in the extreme environments of ULIRG and quasar nuclei is very different from that in the Milky Way (Marwick Kemper *et al.* 2005, Spoon *et al.* 2006) either because it is formed in massive star or quasar outflows or because stardust is heavily processed in these energetic environments. Much has been made in recent studies on whether (the high mass end of) the low mass stars can evolve rapidly enough to produce the observed dust at high redshifts in their AGB phase or whether SN are the dominant producers of this dust. It is important to realize that this may actually be very moot point. In the high star formation environment of the early Universe, SNR may rapidly process any stardust and the dust we see could largely be produced by the chemical accretion processes discussed above for the Milky Way.

Despite our general ignorance in all matters dusty, the future looks bright. Recent studies of the stardust budget have focuses on the Magellanic clouds where the SAGE Spitzer Legacy programs have been very successful (Sargent *et al.* 2010, Mikako *et al.* 2010). With the launch of Gaia, accurate distances of stars in the Milky Way can be determined, removing the largest uncertainty in studies of the dust mass budget of the Milky Way. With SOFIA operational, well-determined mass loss rates for various classes of AGB stars and supergiants are then in sight for our own galaxy. The James Webb Space Telescope (JWST) can extend this to galaxies in the local group. Most importantly, JWST will be able to survey IR emission of all SNe & LBV within 50 Mpc and determine masses and composition of dust formed in the ejecta. This volume is large enough that all denizens of the SN-zoo can be reasonably expected to be probed. JWST can also systematically probe characteristics of dust in regions of extreme star formation. The long wavelength capabilities of Herschel, SOFIA and SPICA will be essential to probe cold dust characteristics in young SNR, while warm dust (in the reverse shock) can be studied by JWST and SPICA. Linking these to the characteristics of the progenitor and the conditions (eg., elemental composition) of the dusty knots may then shed much light on the interaction of SN with the dust they produce. The contours of the dusty Universe, observed through a looking glass darkly in this review, may then well focus in full glory.

References

Anders, E. & Zinner, E. 1993, *Meteoritics*, 28, 490

Barlow, M. J., Krause, O., Swinyard, B. M., *et al.* 2010, *A&A*, 518, L138

Barlow, M. J. 2009, *Astrophysics in the Next Decade*, (Berlin: Springer Verlag), 247

Bernatowicz, T. J. *et al.* 2003, *Geochim. Cosmochim. Acta*, 67, 4679

Berné, O. & Tielens, A. G. G. M. 2011, *PNAS*, in press

Bertoldi, F., Carilli, C. L., Cox, P., *et al.* 2003, *A&A*, 406, L55

Cartledge, S. I. B., Clayton, G. C., Gordon, K. D., *et al.* 2005, *ApJ*, 630, 355

Cartledge, S. I. B., Lauroesch, J. T., Meyer, D. M., & Sofia, U. J. 2006, *ApJ*, 641, 327

Cherchneff, I. & Dwek, E. 2010, *ApJ*, 713, 1

Cherchneff, I. 2000, *The Carbon Star Phenomenon*, 177, 331

Ercolano, B., Barlow, M. J., & Sugerman, B. E. K. 2007, *MNRAS*, 375, 753

Guillard, P., Jones, A. P., & Tielens, A. G. G. M. 2011, *A&A*, submitted

Hartmann, L. 1995, *Revista Mexicana de Astronomia y Astrofisica Conference Series*, 1, 285

Henkel, T., King, A., & Lyon, I. 2007, *Lunar and Planetary Institute Science Conference Abstracts*, 38, 2351

Jenkins, E. B. 2009, *ApJ*, 700, 1299
Jones A. P., Tielens, A. G. G. M., Hollenbach, D. J., & McKee, C. F. 1994, *ApJ*, 433, 797
Jones, A. P., Tielens, A. G. G. M., & Hollenbach, D. J. 1996, *ApJ*, 469, 740
Jura, M. 1987, in "Interstellar Processes", *ASSL*, 134, p3
Kemper, F., Vriend, W. J., & Tielens, A. G. G. M. 2004, *ApJ*, 609, 826
King, A. *et al.* 2010, *Lunar and Planetary Institute Science Conference Abstracts*, 41, 1976
Lewis, R. S., Amari, S., & Anders, E. 1994, *GeCoA*, 58, 471
Lucy, L. B. *et al.* 1989, in "structure and dynamics of the interstellar medium", eds. G. Tenorio-Tagle, M. Moles, J. Melnick (Berlin: Springer Verlag), p164
Markwick-Kemper, F., Gallagher, S. C., Hines, D. C., & Bouwman, J. 2007, *ApJ*, 668, L107
Matsuura, M., *et al.* 2011, *Science*, 333, 6047
Meikle, W. P. S., Mattila, S., Pastorello, A., *et al.* 2007, *ApJ*, 665, 608
Matsuura, M., Barlow, M. J., Zijlstra, A. A., *et al.* 2009, *MNRAS*, 396, 918
Rho, J., Reach, W. T., Tappe, A., *et al.* 2009, *ApJ*, 700, 579
Rho, J., Kozasa, T., Reach, W. T., *et al.* 2008, *ApJ*, 673, 271
Sargent, B. A., Srinivasan, S., Meixner, M., *et al.* 2010, *ApJ*, 716,
Savage, B. D. & Sembach, K. R. 1996, *Ann Rev Astron Astrophys*, 34, 279
Sellgren, K., *et al.* 2010, *ApJ*, 722, L54
Shu, F. H. & Shang, H. 1997, *Herbig-Haro Flows and the Birth of Stars*, 182, 225
Spoon, H. W. W., Tielens, A. G. G. M., Armus, L., *et al.* 2006, *ApJ*, 638, 759
Sugerman, B. E. K. *et al.* 2006, *Science*, 313, 196
Thielemann F. K., Nomoto, K., & Hashimoto, M-A. 1996, *ApJ*, 460, 408
Tielens, A. G. G. M. & Allamandola, L. J. 1987, in "Interstellar Processes", *ASSL*, 134, p333
Tielens, A. G. G. M. 1998, *ApJ* 499, 267
Tielens, A. G. G. M. 2001, in Tetons 4: Galactic Structure, Stars and the Interstellar Medium, eds. C. E. Woodward, M. D. Bicay, and J. M. Shull (San Francisco, ASP) 231, 92
Tielens, A. G. G. M. 2005, *Physics and Chemistry of the Interstellar Medium*, (Cambridge: University of Cambridge Press)
Vink, J. *et al.* 1996, *A&A*, 307, L41
Willingale, R., *et al.* 2002, *A&A*, 381, 1039
Wooden, D., *et al.* 1993, *ApJS*, 88, 477
Zinner, E. 1998, *Ann. Rev. Earth Planet. Sci.*, 26, 147
Zinner, E. K. 2003, In Treatise on Geochemistry Vol 1, ed. K. K. Turekian, H. D. Holland, A. M. Davis), (Amsterdam: Elsevier), p17

Discussion

HENKEL: Traditionally, the production of the rare nitrogen isotope 15N was tentatively attributed to novae. In your plot of carbon and nitrogen isotope ratios from individual dust grains, however, grains related to novae show particularly large 14N/15N ratios. Does this imply that 15N production by novae can be discarded?

TIELENS: Nucleosynthesis is beyond my expertise. However, the indicated nova grain in the stardust isotope data are characterized by low N14/N15 and low C12/C13 ratios. They also show excess Si30, and high 26Al/27Al. This indicates explosive hydrogen burning and a likely origin in nova.

TUFFS: Do you think it is feasible to condense hydrogen onto grain cores - can you get cold enough?

TIELENS: The key point is indeed the dust temperature. The sublimation temperature of solid H2 is $\sim$4K, which is only slightly higher than the background temperature of 2.7K. So, dust really does not get that cold, observationally. Dust in the diffuse ISM is at $\sim$16-20K. Dust in dense clouds may get as cold as 6K.

The Spectral Energy Distribution of Galaxies
Proceedings IAU Symposium No. 284, 2011
R.J. Tuffs & C.C. Popescu, eds.

doi:10.1017/S1743921312008794

The processing of radiation by dust in galaxies

Ralf Siebenmorgen[1] **and Frank Heymann**[1,2]

[1]European Southern Observatory, Karl-Schwarzschild-Str. 2, D-85748 Garching b. München, Germany

[2]Department of Physics and Astronomy, University of Kentucky, Lexington, KY 40506-0055, USA

email: `Ralf.Siebenmorgen@eso.org`

Abstract. Optical/UV photons and even harder radiation components in galaxies are absorbed and scattered by dust and re-emitted at infrared wavelengths. For a better understanding of the obscured regions of the galaxies detailed models of the interaction of photons with dust grains and the propagation of light are required. A problem which can only be solved by means of numerical solution of the radiative transfer equation. As a prologue we present high angular mid IR observations of galactic nuclei in the spirit of future ELT instrumentation. Dust models are discussed, which are suited to fit the extinction curves and relevant to compute the emission of external galaxies. Self-consistent radiative transfer models have been presented in spherical symmetry for starburst nuclei, in two dimensions for disk galaxies (spirals) and, more recently, in three dimensional configuration of the dust density distribution. For the latter, a highlighting example is the clumpy dust tori around AGN. Modern advances in the field are reviewed which are either based on a more detailed physical picture or progress in computational sciences.

Keywords. radiative transfer,scattering,instrumentation: high angular resolution, galaxies: ISM, galaxies: nuclei, (galaxies:) quasars: general,galaxies: Seyfert,galaxies: spiral, galaxies: starburst, infrared: galaxies

---

1. Observations

Dust enshrouded activity of a galaxy can be studied ideally by mid–infrared (MIR) observations. To explore the origin of the nuclear MIR emission of galaxies as being due to either active galactic nuclei (AGN) or star formation, observations of high spatial resolution are required.

The nuclear MIR surface brightness is introduced as a quantitative measurement for AGN and starburst activity. However, one is unable to distinguish between both activity types using the nuclear MIR surface brightness derived from 4m class telescopes, even when adopting the theoretical diffraction limit of $0.7''$ (FWHM) of such telescopes (cmp. small gray symbols in Fig. 1). Since the PSF width is twice as large as for a 8m class telescope and the point source sensitivity is a factor 16 lower, it becomes more difficult for a 4m to resolve starburst and the surface brightness of unresolved sources is reduced. Data recently obtained at 8m class telescopes (Siebenmorgen *et al.* 2008) show that, out to a distance of 100Mpc, the MIR surface brightness acquired clearly differentiate AGN from SB behavior (Fig. 1). Utilizing VISIR at the VLT the AGN still appear point like whereas most starburst are resolved in the MIR. This discrimination was made possible by an increase in spatial resolution by a factor 2. Therefore it provides a clue to what will be possible by increasing the spatial resolution by another factor 5 when going from the VLT to the proposed extreme large telescope such as the E-ELT which will be 40m

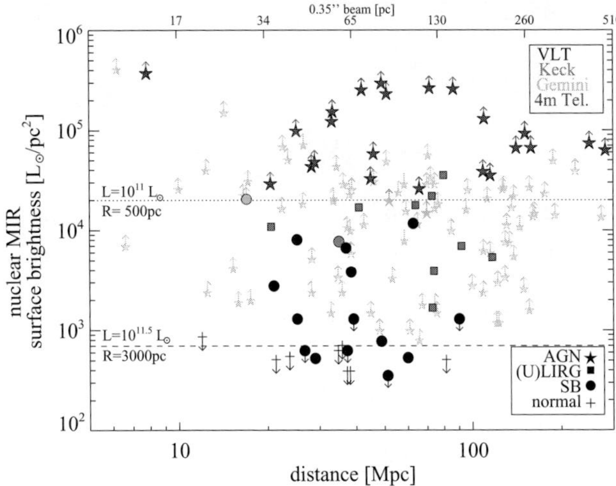

Figure 1. Nuclear mid IR surface brightness versus distance (as plotted by Siebenmorgen *et al.* 2008). Large symbols mark data from VLT: Haas *et al.* (2007), Horst *et al.* (2006, 2008), *this work*; Keck: Soifer *et al.* (1999, 2000, 2001) ; Gemini: Alonso-Herrero *et al.* (2006), Mason *et al.* (2007); small symbols from 4 m class telescopes (gray): Haas *et al.* (2007). AGN (stars) have $S > 20000\ L_\odot/pc^2$ when observed with 8 m class telescopes, starburst and (U)LIRGs are below (with the exception of VV114A which may probably harbor an AGN). Normal galaxies (+) are not detected. The horizontal dotted and long-dashed lines give the surface brightness computed with starburst models by Siebenmorgen & Krügel (2007) for given total luminosity L and where stars and dust are distributed in a $A_V = 18$ mag nucleus of radius R.

class. For the E-ELT a mid infrared instrument is included in the instrumentation plan and it has beside imaging also high resolution spectroscopic and polarimetric observing capabilities (Brandl *et al.*, 2010).

2. Dust model

Teams interested in modelling the processing of radiation by dust in galaxies often apply a dust model as derived for the diffuse ISM of the Milky Way. Dust cross sections are computed using similar optical constants and temperature fluctuating particles such as PAHs are included. We developed one of such dust models in which *large* ($60\,\text{Å} < a < 0.2 - 0.3\,\mu$m) silicate (Draine 2003) and amorphous carbon (Zubko *et al.* 2004) grains and *small* graphite ($5\,\text{Å} < a < 80\,\text{Å}$) grains are considered. We apply a power law size distribution: $n(a) \propto a^{-3.5}$ and absorption and scattering cross-sections are computed with Mie theory. In addition there are *PAHs* with 30 and 200 C atoms with absorption cross section as given by Schutte *et al.* (1993). By computing cross sections above 100 eV, we consider an approximation of kinetic energy losses (Dwek& Smith 1996) and apply it to all particles. The choice of parameters is set up to achieve a fit of the mean extinction curve of the ISM (Fitzpatrick & Massa 2007), as shown in Fig.2.

With the advent of *ISO* and *Spitzer* more PAH emission features and more details of their band structures are detected (Tielens 2008). We consider 17 emission bands and take Lorentzian profiles (Siebenmorgen *et al.* 1998). Parameters of what we call astronomical PAH are calibrated using mid-IR spectra of starburst nuclei and the RT

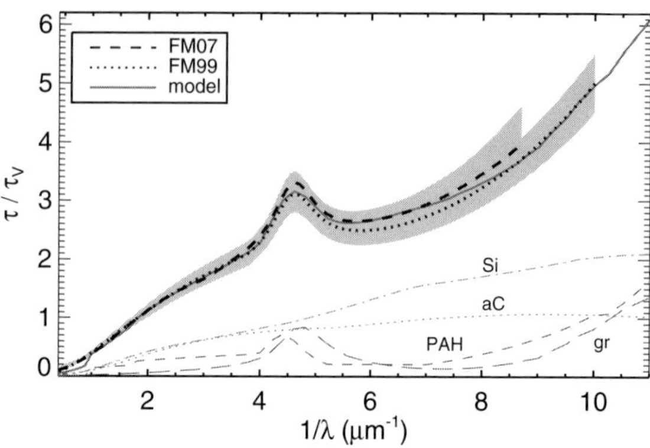

Figure 2. Mean extinction curve of the ISM by Fritzpatrick&Massa (2007, black) and a fit by the dust model of Sect. 2. Individual dust components are shown as labeled. The grey areas indicate the 1σ deviation as of the samples.

model as of Sect. 3. PAH cross sections of the emission bands are listed by Siebenmorgen & Krügel (2001). In the model, we use dust abundances of [X]/[H] (ppm) of: 31 for [Si], 150 [amorphous C], 50 [graphite] and 30 [PAH], respectively; which is in agreement with cosmic abundance constraints (Asplund *et al.* 2009). We are in the process of upgrading the model to be consistent with the polarization of the ISM (Voshchinnikov 2004). Besides extinction the model accounts for the diffuse emission of solar neighborhood when the dust is heated by the interstellar radiation field (Mathis *et al.* 1983).

3. Starbursts

There are three different ways in the literature trying to reproduce the SED of extra-galactic nuclei. A first one uses a SED of a well-known galaxy as a template to match other objects (Lutz *et al.* 2002). A second group reproduce the shape of the SED by optical thin dust emission using a scaled up version of the interstellar radiation field (Draine & Lee 2007). A third and more ambitious method is to solve the radiative transfer (RT) problem using assumptions about the galaxy. The latter is done at various levels of sophistication and it may be instructive to point out technical differences. Teams solving the RT problem evaluate the emission from a dusty medium of spheroidal shape filled with stars and dust. At first glance, the model results appear to agree, but upon closer inspection one finds that deviations of derived parameters are substantial. For example for Arp220 Grooves *et al.* finds an optical depth of a few whereas we derive values between 70 – 120. We admit that we did not always find it easy to pin down exactly which approximations our colleagues used, still we try to summarize the main features of some RT models which are in widespread use. Monte Carlo techniques are discussed in Sect. 5. We use the term *dust self–absorption* when photons which are emitted by a dust particles may be absorbed by other dust particles within the model sphere and we call *exact RT* when the RT equation is solved accurately including multiple scattering and dust self–absorption.

– Groves (2004, this volume): Shock, photoionization and dust radiative transfer code called MAPPINGIII. The dust is distributed in a screen and the RT is solved in one dimension ignoring dust self-absorption and treating scattering in forward direction only.

– Efstathiou & Rowan–Robinson (2003), Efstathiou (this volume): Dust is distributed in spherical symmetry. There are two components: molecular clouds with exact RT computation and cirrus which is added as a foreground screen. Both components are uncoupled in the RT equation. The code includes a free parameter to scope with the observed optical and UV spectrum.

– Takagi *et al.* (2003, this volume): Exact RT in spherical symmetry where dust and stars are homogeneously distributed. Molecular clouds (clumps) are not treated. The code includes a free parameter to scope with the observed optical and UV spectrum.

– Siebenmorgen *et al.* (2007): Exact RT in spherical symmetry where dust and a young and old population of stars are distributed in the galaxy. A fraction of the young stars are in molecular clouds for which a second exact RT computation is solved. The coupling of both RT solutions, that of the galaxy and the embedded sources, is treated and this is a particular feature of the model.

– Silva *et al.* (1998): Present a code called GRASIL in which dust is distributed in axial-symmetry. There is a molecular cloud component with exact RT. A cirrus component is added in which the RT is solved by ignoring dust self-absorption and in which dust scattering is simplified by altering the dust absorption cross section. In addition the code includes a free parameter to scope with the observed optical and UV spectrum. All three components are uncoupled in the RT equation.

– Popescu *et al.* (2002, this volume): Dust is distributed in axial-symmetry. The model is fine tuned to Spiral galaxies (Sect. 4) and includes a bulge, two galactic disk components and molecular clouds (clumps). The RT is solved by ignoring dust self-absorption and only the first scattering event is treated. Clumps are added by a pre-computed template spectrum. There is a free parameter to scope with the observed optical and UV spectrum. All five components are uncoupled in the RT equation.

As the coupling between the RT calculations of the galaxy and the embedded sources is a computer intensive problem we provide a library of some 7000 SEDs† for the nuclei of starburst and ultra–luminous galaxies (Siebenmorgen *et al.*, 2007). Its purpose is to quickly obtain estimates of the basic parameters of the object, such as luminosity, size and dust or gas mass and to predict the flux at yet unobserved wavelengths using a physical model. Unfortunately, for faint or red-shifted objects photometry is sometimes only provided at two MIR bands, for example at 8 and 24μm from Spitzer. In Fig. 3 we demonstrate for the ULIRG NGC6240 that the SED library can be used to estimate the total luminosity to within a factor ~ 2 in case only two such MIR fluxes are known. We also see (Fig. 3) that the SED will be quite well constrained by an additional submm data point. The model is applied to high red shifts ($z \approx 3$, Efstathiou & Siebenmorgen 2009) where PAH have been detected.

4. Spirals

The geometrical distribution of dust in spiral galaxies has been investigated by fitting images in the optical/NIR. This is done by means of 2d radiative transfer codes which include a bulge component and an old stellar population of stars; originally only dust absorption and scattering was treated (Kylafis & Bahcall 1987, Xilouris *et al.* 1999, Misiriotis *et al.*, 2000). Building into the models typical dust emission properties gives a puzzle when compared to observations in the FIR. Typically the models underestimate

† SED library available at: `http://www.eso.org/~rsiebenm/sb_models/`

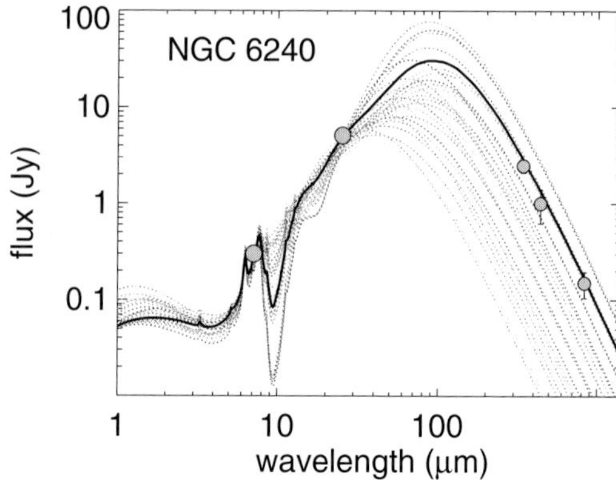

Figure 3. Elements of the SED starburst library by Siebenmorgen *et al.* (2007, dotted) which fit photometry (circles) of NGC6240 at 8μm (Siebenmorgen *et al.* 2004) and 24μm (Klaas *et al.* 2001) to within 30%; submm data by Benford (1999) and Klaas *et al.* (2001), best fit is indicated as thick line.

the FIR luminosities by a factor of 3. The FIR excess could be explained by Popescu *et al.* (2000, 2010, this volume) introducing more components. In their models there is a distribution of diffuse dust associated with the old and young stellar disk populations as well as a clumpy component arising from dust in the parent molecular clouds in star forming regions. Basic parameters of their models are: angular size and inclination of the disk, the central face-on dust opacity in the B-band, a clumpiness factor for the star-forming regions, the star-formation rate, the normalized luminosity of the stellar population and the bulge-to-disk ratio and a wavelength dependent escape probability of stellar radiation (which is sometimes treated as model output). A schematic view of the geometry of the RT model is presented in Fig. 4. The model is successfully applied to a large sample of galaxies from the Millennium Galaxy Catalog Survey (Driver *et al.* 2007). The observed attenuation–inclination relation is fit when a two dust disks are considered whereas data are not fit in a single disk scenario.

Over the past decade the edge on spiral NGC891 become a benchmark test for RT models of spiral galaxies. Bianchi (2008) uses a clumpy disks model and applied MC techniques which are of advantage when dealing with such complicated geometries (see Sect. 5). His results give support to the idea that the diffuse dust disk is more extended than the stellar disks. De Looze *et al.* (20011, this volume) present a MC radiative transfer model of the Sombrero galaxy which is able to fit the SED and optical/NIR images and extinction profiles. Using only an old stellar population the dust luminosity in the FIR is again underestimated by a factor 3. The discrepancy is solved by assuming a star forming stellar component both in the ring and inner disk to account for emission at 24 and 70μm. In the submm an additional dust component is used, accounting for 75% of the total dust content and distributed in quiescent compact clumps. A possibility that part of the dust could be composed of grains with a higher submm emissivity, of for example large fluffy grains (Krügel & Siebenmorgen 1994), could not be ruled out. On the other hand a similar approach introducing a clumpy medium in a single disk model is applied to the edge-on spiral galaxy UGC4754 (Baes *et al.* 2011). The model has a deficit in the FIR luminosity and the authors propose higher submm dust emissivities to solve the energy ballance at such long wavelengths.

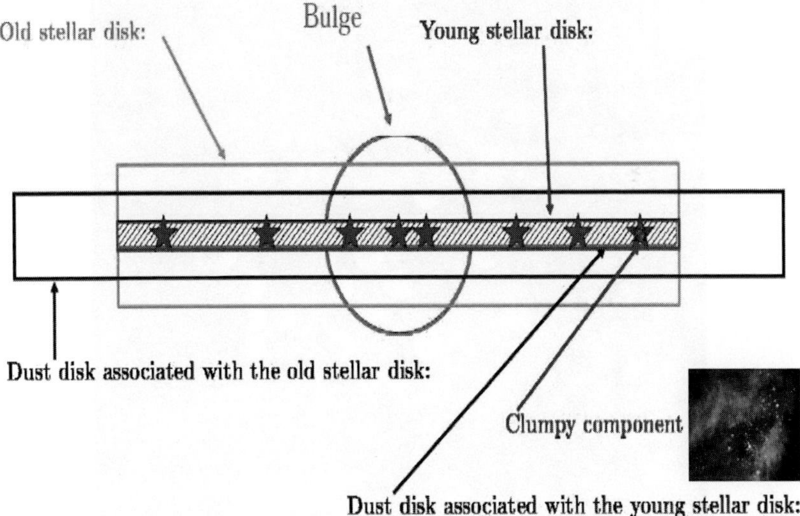

Figure 4. Schematic view of the geometry applied in the models of Spirals by Popescu *et al.* (2011)

5. AGN and Monte Carlo

Dust is detected in the majority of active galactic nuclei (Haas *et al.*, 2008). According to the unified scheme (Antonucci & Miller, 1985), AGN are surrounded by a dust obscuring torus. However, observations are not able to resolve the inner parts of AGNs so that the geometrical distribution of the dust is a matter of debate. Theoretical considerations favor a clumpy structure in a torus like configuration of optical thick dust clouds surrounding the black hole accretion disk (Pier & Krolik (1992). Some evidence of a clumpy or filamentary structure of the torus is given by VLTI observations of the nearby active galactic nuclei in Circinus (Tristram *et al.* 2007). Radiative transfer models of homogeneous toroidal structures over-predict the silicate absorption and emission band at around 10μm when compared to observations. The silicate emission feature in type I AGN is rather shallow (Siebenmorgen *et al.* 2005). A statistical attempt to describe the radiation from a clumpy AGN torus is given by Nenkova *et al.* (2002) and more detailed radiative transfer computations using the Monte Carlo technique are presented by Hönig *et al.* (2006) and Schartmann *et al.* (2008).

We develop a vectorized three dimensional Monte Carlo (MC) technique to solve the radiative transfer problem in such a fairly complicated geometry of a clumpy dust torus. Originally the MC radiative transfer method for scattered light in astronomical sources is presented by Witt (1977) and important improvements are introduced by Lucy (1999). A first version of the code which we are using is developed by Krügel (2006). To reduce the computational effort of MC dust radiative transfer methods different optimization strategies are developed (Lucy 1999, Bjorkman & Wood 2001, Gordon *et al.* 2001, Misselt *et al.* 2001, Baes 2008, Bianchi 2008, Baes *et al.* 2011). A numerical solution of the radiative transfer equation which is specifically developed to be vectorized and to run on graphical computer units (GPU) is presented by Heymann (2010). A GPU version including stochastic heated grains is presented (Siebenmorgen *et al.* 2010), in which some detailed physics of the destruction of PAHs by soft (photospheric) and hard (X–ray) radiation components is treated, (see Siebenmorgen & Krügel 2010 for a discussion of PAH

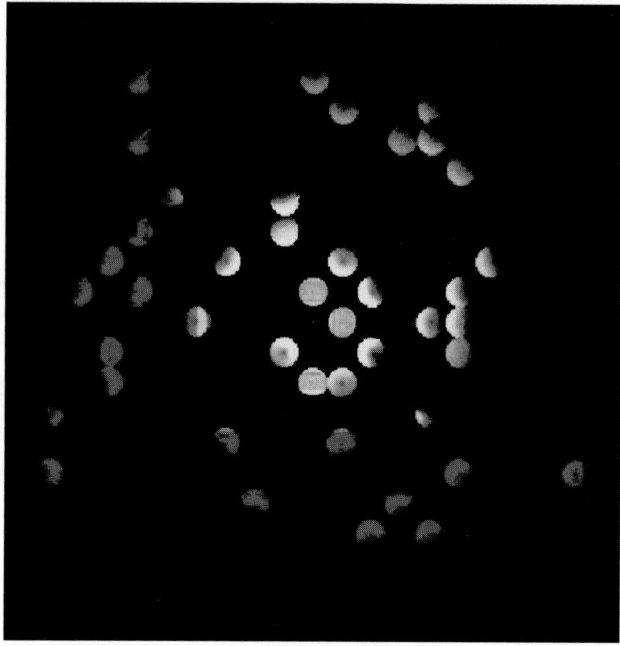

Figure 5. Zoom into the clumpy AGN torus displaying the temperature structure of individual clouds.

dissociation). The speed up factor of the GPU method is proportional to the number of processors available. Therefore the GPU scheme is on our standard PC with a conventional graphic card about 100 times faster when compared to the original scalar version of the MC program. The numerical solution provides the temperature of the dust, spectra and images within acceptable timescales. The code is tested against existing benchmark results. The method handles arbitrary dust distributions in a three dimensional Cartesian model space at various optical depths. Therefore it is well adopted to be applied to a clumpy torus structure around an AGN.

The primary heating source of the dust in the torus emerges from the accretion disc around the massive black hole. For the spectral shape we apply a broken power law as suggested by Rowan-Robinson (1995). As example we use a total AGN luminosity of $L_{\rm AGN} = 10^{45}$ erg/s. This sets the inner radius $R_{\rm in} \sim 0.4\sqrt{L_{\rm AGN}/10^{45}}$ [pc] where the dust evaporation zone is located. We treat the torus to an outer radius of $R_{\rm out} = 50 R_{\rm in}$ and consider a torus opening angle of $\sim 20^{\rm o}$. Clouds are randomly distributed in the otherwise optical thin dust torus. Individual clouds are approximated with a sphere of constant density and radius of 3 pc. The optical depth through the center of the clouds is $\tau_{\rm V} = 30$. In Fig. 5 we zoom into a part of the clumpy torus and show the temperature structure of the clouds. The individual hemispheres which are oriented in direction of the central heating source are warmer as compared to their dark sides. Shadowed clumps become also colder than those directly heated by the central engine. Intensity maps of the emission at 10μm and scattered light at 0.5μm of the clumpy AGN dust torus model are displayed in Fig. 6. The maps show that in a clumpy medium even in the edge on case some radiation from the central region is visible. A more detailed description of the clumpy AGN torus model and a further application to the silicate emission in Quasars is given in Heymann & Siebenmorgen (2011).

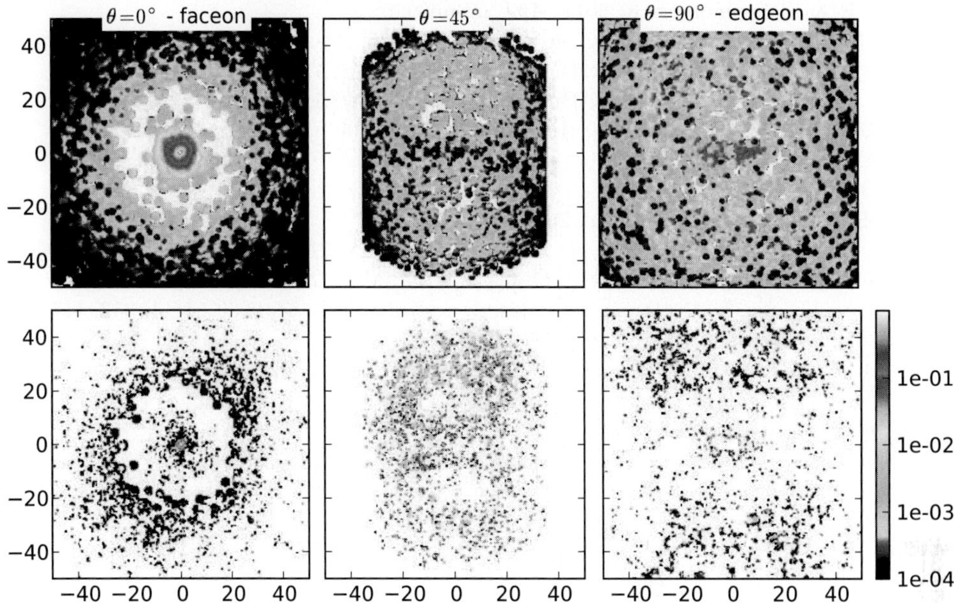

Figure 6. Intensity maps of the clumpy AGN dust torus model (color-code: cgs units normalized to the peak intensity). The model is made up of 1000 clumps and each clump has a total optical depth of $\tau_V = 30$. Mid IR images at $10\,\mu$m are dominated by dust emission and displayed in the top panel. The optical light at $0.55\,\mu$m represents the scattering intensities and are shown in the bottom panel. The AGN torus is viewed face-on ($\theta = 0°$, left), at $\theta = 45°$ (middle) and edge-on ($\theta = 90°$, right).

References

Alonso-Herrero, A., Colina, L., Packham, C. *et al.* 2006, *ApJ*, 652, L83

Antonucci, R. R. J. & Miller, J. S. 1985, *ApJ*, 297, 621

Asplund, M., Grevesse, N., Sauval, A. J., & Scott, P. 2009, *ARA&A*, 47, 481

Baes, M. 2008, *MNRAS*, 391, 617

Baes, M., Verstappen, J., De Looze, I., *et al.* 2011, *ApJS*, 196, 22

Bianchi, S. 2008, *A&A*, 490, 461

Bjorkman, J. E. & Wood, K. 2001, *ApJ*, 554, 615

Brandl, B. R., Lenzen, R., Pantin, E., *et al.* 2010, *SPIE*, 7735, 83

De Looze, I., Baes, M., Fritz, J., & Verstappen, J. 2011, *MNRAS* subm

Dopita, M. A., Groves, B. A., Fishera, J., *et al.* 2005, *ApJ*, 619, 755

Draine, B. T. 2003, *ApJ*, 598, 1026.

Driver, S. P., Popescu, C. C., Tuffs, R. J., *et al.* 2007, *MNRAS*, 379, 1022

Dwek, E. & Smith, R. K. 1996, *ApJ*, 459, 686

Efstathiou, A. & Rowan–Robinson, M. 2003, *MNRAS*, 343, 322

Efstathiou, A. & Siebenmorgen, R. 2009, *A&A*, 502, 541

Fitzpatrick, E. L. & Massa, D. 2007, *ApJ*, 663, 320

Gordon, K. D., Misselt, K. A., Witt, A. N., & Clayton, G. C. 2001, *ApJ*, 551, 269

Grooves, B. 2004, Austalian National Univ., PhD,
`http://www.mpia-hd.mpg.de/ brent/documents/BrentThesis.pdf`

Haas, M., Siebenmorgen, R., Pantin, E., *et al.* 2007, *A&A*, 473, 369

Heymann, F. 2010, PhD, University of Bochum,
`http://www-brs.ub.ruhr-uni-bochum.de/netahtml/HSS/Diss/HeymannFrank/diss.pdf`

Heymann, F. & Siebenmorgen, R. 2011, *ApJ*, submitted
Hönig, S. F., Beckert, T., Ohnaka, K., & Weigelt, G. 2006, *A&A*, 452, 459
Horst, H., Smette, A., Poshak, G., & Duschl, W. J. 2006, *A&A*, 457, 17
Horst, H., Poshak, G., Smette, A., & Duschl, W. J. 2008, *A&A*, 479, 389
Krügel, E. & Siebenmorgen, R. 1994, *A&A*, 288, 929
Kylafis, N. D. & Bahcall, J. N. 1987, *ApJ*, 317, 637
Lucy, L. B. 1999, *A&A*, 344, 282
Lutz, D., Sturm, E., Genzel, R., *et al.* 2003, *A&A*, 409, 867
Mathis, J. S., Mezger, P. G., & Panagia, N. 1983, *A&A*, 128, 212
Misselt, K. A., Gordon, K. D., Clayton, G. C., & Wolff, M. J. 2001, *ApJ*, 551, 277
Misiriotis, A., Kylafis, N. D., Papamastorakis, J., & Xilouris, E. M. 2000, *A&A*, 353, 117
Nenkova, M., Ivezic, Z., & Elitzur, M. 2002, *ApJ*, 570, L9
Popescu, C. C., Misiriotis, A., Kylafis, N. D., *et al.* 2000, *A&A*, 362, 138
Popescu, C. C., Tuffs, R. J., Dopita, M. A., *et al.* 2010, *A&A*, 527, 109
Rowan-Robinson, M. 1995, *MNRAS*, 272, 737
Schartmann, M., Meisenheimer, K., Camenzind, M., *et al.* 2008, *A&A*, 482, 67
Schutte, W. A., Tielens, A. G. G. M., & Allamandola, L. J. 1993, *ApJ*, 415, 397.
Siebenmorgen, R., Natta, A., Krügel, E., & Prusti, T. 1998, *A&A*, 339, 134
Siebenmorgen, R., Krügel, E., & Laureijs, L. 2001, *A&A*, 377, 735
Siebenmorgen, R., Haas, M., Krügel, E., & Schulz, B. 2005, *A&A*, 436, L5
Siebenmorgen, R. & Krügel, E. 2007, *A&A*, 461, 445
Siebenmorgen, R. & Krügel, E., 2010, *A&A*, 511, A6
Siebenmorgen, R., Heymann, F., & Krügel, E. 2011, *EAS*, 46, 285
Silva, L., Granato, G. L., & Bressan, A. 1998, *ApJ*, 506, 600
Soifer, B. T., Neugebauer, G., Matthews, K., *et al.* 1999, *ApJ*, 513, 207
Soifer, B. T., Neugebauer, G., Matthews, K., *et al.* 2000, *AJ*, 119, 509
Soifer, B. T., Neugebauer, G., & Matthews, K. 2001, *AJ*, 122, 1213
Takagi, T., Arimoto, N., & Hanami, H. 2003, *MNRAS*, 340, 813
Tielens, A. G. G. M. 2008, *ARA&A*, 46, 289
Tristram, K. R. W., Meisenheimer, K., Jaffe, W., *et al.* 2007, *A&A*, 474, 837
Voshchinnikov, N. V. 2004, Optics of cosmic dust I, *ASPR*, 12, 1
Witt, A. N. 1977, *ApJS*, 35, 1
Wood, K., Whitney, B. A., Robitaille, T., & Draine, B. T. 2008, *ApJ*, 688, 1118
Xilouris, E. M., Byun, Y. I., Kylafis, N. D., *et al.* 1999, *A&A*, 344, 868
Zubko, V., Dwek, E., & Arendt, R. G. 2004, *ApJS*, 152, 211

Discussion

KEMPER: In the Monte Carlo fit of the AGN you showed, the silicate feature seems to be shifted to $\sim 11\ \mu$m both in the observations and the model. Did you have to invoke any special dust properties, such as porosity to obtain this result?

SIEBENMORGEN: The peak of the silicate emission band at $\sim 11\ \mu$m is obtained using dust of the ISM or by a dust model using fluffy dust aggregates of composite grains (Kruegel & Siebenmorgen, 1994). It requires to set a different number for the clumps in the torus as a function of radius Nclump(r).

POPESCU: I'd like to comment that fluffy grains can explain the excess of FIR/submm seen in galaxies, but this cannot explain the steep slope in the attenuation/inclination curve.

THOMPSON: Some have claimed that the radiation pressure force on dusty clumps on parsec scales around AGNs is very important. You have an exquisite calculation of the dust transport in your models. Can you comment on the effect on the dynamics of clumps due to this force in your models?

SIEBENMORGEN: A first application of the vectorized Monte Carlo was done very recently without being able to investigate the pressure force. The Monte Carlo code, however, seems fast enough to combine hydro-dynamical codes with radiative transfer models.

The Spectral Energy Distribution of Galaxies
Proceedings IAU Symposium No. 284, 2011
R.J. Tuffs & C.C. Popescu, eds.

doi:10.1017/S1743921312008800

A detailed dust energy balance study of the Sombrero galaxy

Ilse De Looze, Maarten Baes, Jacopo Fritz, Gianfranco Gentile and Joris Verstappen

Universiteit Gent, Sterrenkundig Observatorium
Krijgslaan 281, S9, 9000 Gent

email: ilse.delooze@ugent.be

Abstract. The Sombrero galaxy (M104) is an interesting object for a dust energy balance study due to its very symmetric dust lane, its proximity and its (nearly edge-on) inclination of 84°. From a panchromatic radiative transfer analysis, including scattering, absorption and thermal dust re-emission, we construct a standard model for M104 accounting for observations in the optical wave bands (stellar SED, images and extinction profiles in the V and R_C band). This standard model underestimates the observed dust emission at infrared wavelengths by a factor of ~ 3, similar to the discrepancy found in other energy balance studies of edge-on spirals. Supplementing this standard model with a young stellar component of low star formation activity in both the inner disk (SFR ~ 0.21 $M_{\odot}$ yr^{-1}) and dust ring (SFR ~ 0.05 $M_{\odot}$ yr^{-1}), we are capable of solving the discrepancy in the dust energy budget of the Sombrero galaxy at wavelengths shortwards of 100 μm. To account for the remaining discrepancy at longer wavelengths, we propose a secondary dust component distributed in quiescent clumps. This model with a clumpy dust structure predicts three-quarters of the total dust content to reside in compact dust clouds with no associated embedded sources.

Keywords. radiative transfer – ISM: dust, extinction – galaxies: M104

1. Introduction

Although dust grains make up only a small part of a galaxy's interstellar medium (ISM), they play an important role in several astrophysical processes. From optical studies, the extinguishing effect of dust grains on stellar light is evident. Alternatively, dust can be probed in emission at FIR/submm wavelengths. Ideally, a complementary study of the dust component from optical/UV images and its thermal re-emission at FIR/submm wavelengths allows to put strong constraints on the properties of dust in a galaxy. Edge-on spiral galaxies are particularly suited for detailed studies of their dust energy budget (Popescu *et al.* 2000, 2011, Misiriotis *et al.* 2001, Popescu & Tuffs 2002, Alton *et al.* 2004, Dasyra *et al.* 2005, Baes *et al.* 2010). The edge-on view offers the opportunity to trace the dust out to large radii and to resolve the dust structure vertically, but also implies that all details of dust distribution vanish when integrating along the line-of-sight.

With an inclination of $\sim 84°$, the Sombrero galaxy (M104), an early-type spiral (Sa), combines the advantages of edge-on galaxies with an inside view of the distribution of stars and dust in the inner disk. Along with its proximity ($D = 9.2$ Mpc), the large dataset of available FIR/submm observations (IRAS: Rice *et al.* 1988; MIPS: Bendo *et al.* 2006; LABOCA & MAMBO: Vlahakis *et al.* 2008) and the very regular and symmetric shape of its dust lane, a spatially resolved analysis of the dust energy balance in M104 is possible.

2. Modeling approach and results

To perform the radiative transfer modeling, we used the SKIRT code (Baes *et al.* 2003, 2011). SKIRT is a 3D Monte Carlo radiative transfer code that can model the dust extinction, including both absorption and scattering, and the thermal re-emission of dust, under the assumption of non local thermal equilibrium. Several advanced optimization techniques (stellar foam to generate random positions, forced scattering and continuous absorption) were implemented to improve the efficiency of the code.

To study the energy balance, we construct a model for the stars and dust in M104 based on optical data: SED fluxes from Dale *et al.* (2007), V and R_C band images from the Spitzer Infrared Nearby Galaxies Survey (SINGS) archive (Kennicutt *et al.* 2003) and V and R_C extinction curves from Emsellem (1995). Based on this optical model, the radiative transfer code SKIRT simulates the re-emission from dust at FIR/submm wavelengths. From a direct comparison of the predicted dust emissivity with infrared observations, one can study the energy budget in the Sombrero galaxy. In case of discrepancies, the optical model must be refined to balance the absorption of stellar light by dust grains and dust emission in this galaxy.

2.1. *Standard model*

For the stellar component in the bulge and disk, we adopt the axi-symmetric geometry from Emsellem (1995), who modeled the stars in M104 as a linear combination of 15 Multi-Gaussian Expansion (MGE) components. The luminosities of the MGE components were scaled linearly to fit the stellar SED. The shape of the stellar SED is parametrized by a Maraston *et al.* (2005) single stellar population (SSP) with an age of 10 Gyr and a close to solar metallicity $Z = 0.016$ (Vazdekis *et al.* 1997). For the dust component, a least-square fitting procedure to the observed V and R_C band images and extinction curves results in a dust ring modeled as a linear combination of 4 Gaussian rings:

$$\rho_d(R, z) = \sum_{i=1}^{4} \rho_{0,i} \, \exp\left[-\frac{(R - R_i)^2}{2\sigma_i^2}\right], \quad -\frac{h_z}{2} \leqslant z \leqslant \frac{h_z}{2}. \tag{2.1}$$

where $\rho_{0,i}$, R_i and σ_i correspond to the relative mass contribution, radial distance to the maximum value and width of the Gaussian function, respectively, and $h_z = 20$ pc. The dust in the inner disk is distributed in an exponential disk ($h_R = 2000$ pc, $h_z = 20$ pc), in correspondence to the model by Bendo *et al.* (2006). The total amount of dust is found to be $M_d = 7.9 \times 10^6$ $M_\odot$, of which $M_d = 7.5 \times 10^6$ $M_\odot$ is distributed in the dust ring and the remaining $M_d = 4.1 \times 10^5$ $M_\odot$ resides in the inner disk. The nucleus, famous for its strange FIR/submm SED (Bendo *et al.* 2006, Vlahakis *et al.* 2008), is modeled as a point source, with an intensity field obtained through interpolation between the observed fluxes from the nucleus at IR/submm wavelengths.

When comparing the FIR/submm emission predicted from radiative transfer calculations for this standard model to the observed quantities, our model is found to underestimate the true FIR/submm emission by a factor of ~ 3 (see Figure 1, left panel). This "dust energy balance problem" is similar to what has been found in dust energy balance studies for several edge-on spirals (Popescu *et al.* 2000, Misiriotis *et al.* 2001, Popescu & Tuffs 2002, Alton *et al.* 2004, Dasyra *et al.* 2005, Baes *et al.* 2010).

2.2. *Model with embedded star formation*

In an attempt to solve the discrepancy in the energy budget for the Sombrero galaxy, we adjust the standard model from the previous section by adding a young stellar population with a constant, continuous star formation during the past 100 Myr, distributed in the inner disk and the inner part of the dust ring. The shape of the SED is parametrized as a Starburst SED, generated from data of the Starburst99 library (Leitherer *et al.* 1999), with the same metallicity (Z = 0.016) as the old stellar population (10 Gyr) in the Sombrero galaxy. A least-square fitting procedure to the observations at 24 and 70 μm result in a young stellar component with a modest SFR $\sim$ 0.05 $M_\odot$ yr^{-1} in the dust ring and a more active component (SFR $\sim$ 0.21 $M_\odot$ yr^{-1}) in the inner disk. The total star formation activity (SFR $\sim$ 0.26 $M_\odot$ yr^{-1}) in our model agrees well with the SFR estimate based on Hα and radio observations (Kennicutt *et al.* 2009). The dissimilarity between model and observations in the FUV and NUV wavebands (see Figure 1, middle panel) is likely a consequence of the highly obscured nature of the young stellar population, which should be more embedded in our model. The dust mass in the ring was increased by a small amount (M_d = 7.8 $\times$ 10^6 $M_\odot$) to maintain the resemblance between the modeled and observed V band extinction curve. While this model solves the discrepancy in the dust energy balance at infrared wavelengths shortwards of 100 μm, the SED and MIPS 160 μm image (see Figure 1, middle panel, and 2) clearly show that this model with embedded star formation can not account for the IR emission at longer wavelengths.

3. Discussion

Keeping in mind the agreement of the SFR activity in our model with 24 and 70 μm observations and SFR calibrations, we argue that an additional component of heavily extinguished star formation will not solve the remaining discrepancy at longer wavelengths. It seems that either an additional dust component or dust grains with optical properties diverging from Galactic values (e.g. Alton *et al.* 2004 and Dasyra *et al.* 2005) need to be invoked to explain the inconsistency in the energy balance. While high emissivity dust grains can influence the shape of the SED at some wavelengths, they are incapable of solving the difference in the integrated dust energy balance. With a view on the inner regions of M104, we can rule out an additional dust component associated to a second dust disk, similar to the works Popescu *et al.* (2000, 2011) and Misiriotis *et al.* (2001). The assumption of a dust component distributed in dense clumps, which are not associated to localized embedded sources, seems more likely and is supported by high resolution optical imaging (HST, Christian *et al.* 2003), revealing a clumpy dust structure. When fitting a single-component modified blackbody to the residual SED fluxes, we require a dust mass M_d = 2.0 $\times$ 10^7 $M_\odot$ at a temperature T_d $\sim$ 16.2 K (see Figure 1, right panel). This result is in contrast to energy balance studies of the late-type galaxies NGC891 (Bianchi 2008) and UGC4754 (Baes *et al.* 2010), which require an additional clumpy dust component associated to localized embedded sources to explain the discrepancy in their energy balance. This might be a first indication for a difference in the dust distribution for spiral galaxies along the Hubble sequence. While early-type spirals tend to

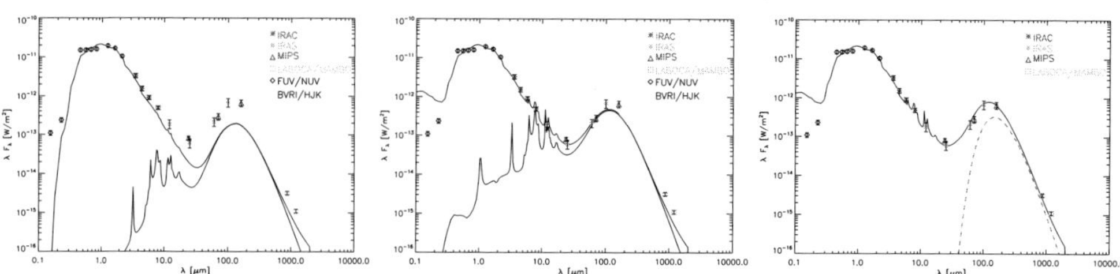

Figure 1. The total SED as obtained from SKIRT calculations (solid black line) for the standard model (left), model with embedded star formation (middle) and clumpy dust component (right).

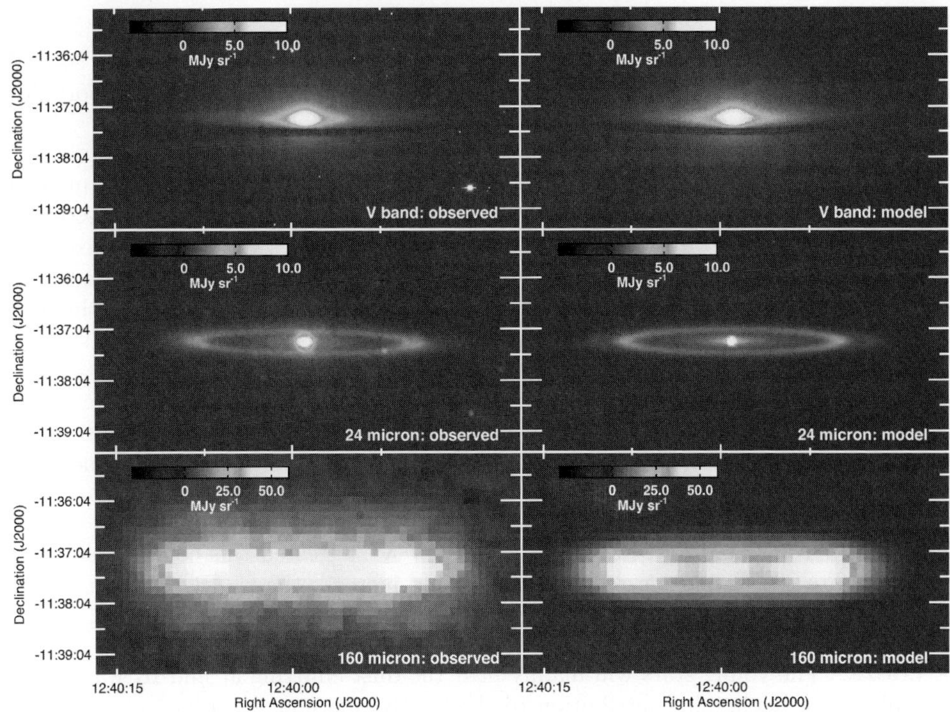

Figure 2. Results for the model with embedded star formation.

have the majority of dust distributed in quiescent clumps, most of the dust in late-type galaxies is associated to embedded localized sources.

References

Alton, P. B., Xilouris, E. M., Misiriotis, A., Dasyra, K. M., & Dumke, M. 2004, *A&A*, 425, 109

Baes, M., *et al.* 2010, *MNRAS*, 343, 1081

Baes, M., *et al.* 2010, *A&A*, 518, L39

Baes, M., Verstappen, J., De Looze, I. *et al.* 2011, *ApJS*, 196, 22

Bendo, G. *et al.* 2006, *AJ*, 645, 134

Bianchi, S. 2008, *A&A*, 490, 461

Christian, C. A., Bond, H. E., & Frattare, L. M. *et al.* 2003, *Bulletin of the AAS*, Vol. 35, p. 1398

Dale, D. A. *et al.* 2007, *AJ*, 655, 863

Dasyra K. M., Xilouris E. M., Misiriotis A. *et al.* 2005, in *AIP* Conf. Proc., Vol. 761, pp. 197–202

Emsellem, E. 1995, *A&A*, 303, 673

Kennicutt, R. C., *et al.* 2003, *PASP*, 115, 928

Kennicutt, R. C., *et al.* 2009, *ApJ*, 703, 1672

Leitherer, C., *et al.* 1999, *ApJS*, 123, 3

Maraston, C. 2005, *MNRAS*, 362, 799

Misiriotis, A., Popescu, C. C., Tuffs, R., & Kylafis, N. D. 2001, *A&A*, 372, 775

Popescu, C. C., Misiriotis, A., Kylafis, N. D., Tuffs, R. J., & Fischera J. 2000, *A&A*, 362, 138

Popescu, C. C. & Tuffs, R. J. 2002, *MNRAS*, 335, L41

Popescu, C. C., Tuffs, R. J., Dopita, M. A. *et al.* 2011, *A&A*, 527, A109

Rice, W., Lonsdale, C. J., Soifer, B. T. *et al.* 1988, *ApJS*, 68, 91

Vazdekis, A., Peletier, R. F., Beckman, J. E., & Casuso, E. 1997, *AJ*, 111, 203

Vlahakis, C., Baes, M., Bendo, G., & Lundgren, A. 2008 *A&A*, 485, L25

Discussion

WANG: I am wondering why you don't include a disk of old stars, which would solve the discrepancy at in near NIR without over-producing UV as you did when star formation was included instead.

DE LOOZE: A combination of a young and more evolved stellar population might be an option, but we don't think an old stellar population alone will solve the discrepancy at MIR wavelengths. Only young stars can provide enough UV emission to excite the PAHs. Furthermore, the amount of young stars we put in our model is consistent with several SFR tracers. Therefore, we argue that the young stars in our model should be more embedded, to solve the over-estimation of the UV emission in our model.

BENDO: I'd like to comment upon the use of a disk of evolved stars in radiative transfer models: In my follow-up work to Bendo *et al.* (2006), I found evidence for a near-infrared exponential disk that would suggest that a second evolved star component is needed in the model.

CLEMENTS: You noted an over prediction of UV & suggested this could be solved by adding dust to star-formation regions. This would increase the far-IR. Would this solve the under-prediction of far-IR flux?.

DE LOOZE: The young stars will mainly heat the dust emitting at mid-IR wavelengths and will have a negligible contribution to the far-IR emission. In order to solve for the FIR/submm energy balance problem, the only solution seems to be an extra cold dust component distributed in quiescent clumps.

TUFFS: Following up on the comment from Dave Clements, I would also like to comment that a geometrically correlated clumpy distribution of stars and dust might also help to resolve the discrepancy between the model-predictions and observations in the UV, due to the stronger local coupling between the UV & dust. Also, a potential difficulty of invoking "Passive" clumps, without stars, is that they would lead to high dust-to- gas ratios. What dust-to-gas ratio is implied by your model?

DE LOOZE: When scaling the HI mass (3.2 x $10^8 M_\odot$; Bejeje *et al.* 1984) and H_2 mass (4.4 x $10^8 M_\odot$; Bejeje *et al.* 1991) with a factor of 1.4 to include Helium, we get a total gas content of 1.06 x $10^9 M_\odot$ at a distance of D=9.2 Mpc. Our radiative transfer model, including the additional dust distribution in quiescent dust clumps, accounts for a total dust mass of $M_d =$ 2.8 x $10^7 M_\odot$, implying a gas-to-dust ratio of $\sim$38. Considering that the molecular gas has only been observed towards some directions in the ring and the total molecular gas mass was derived under the assumption of a constant ratio of the H_2/HI surface mass density of 1.3 throughout the HI emission regions, we need to take into account a large uncertainty factor for this estimate of the gas-to-dust ratio in M104.

The Spectral Energy Distribution of Galaxies
Proceedings IAU Symposium No. 284, 2011
R.J. Tuffs & C.C. Popescu, eds.

doi:10.1017/S1743921312008812

Investigations of dust heating in M81, M83 and NGC 2403 with Herschel and Spitzer

George J. Bendo[1] and the Herschel-SPIRE Local Galaxies Guaranteed Time Programs

[1]UK ALMA Regional Centre Node, Jodrell Bank Centre for Astrophysics
School of Physics and Astronomy, University of Manchester
Oxford Road, Manchester M13 9PL, United Kingdom
email: george.bendo@manchester.ac.uk

Abstract. We use Herschel Space Observatory and Spitzer Space Telescope 70-500 μm data along with ground-based optical and near-infrared data to understand how dust heating in the nearby face-on spiral galaxies M81, M83, and NGC 2403 is affected by the starlight from all stars and by the radiation from star-forming regions. We find that 70/160 μm flux density ratios tend to be more strongly influenced by star-forming regions. However, the 250/350 and 350/500 μm micron flux density ratios are more strongly affected by the light from the total stellar populations, suggesting that the dust emission at >250 μm originates predominantly from a component that is colder than the dust seen at <160 μm and that is relatively unaffected by star formation activity. We conclude by discussing the implications of this for modelling the spectral energy distributions of both nearby and more distant galaxies and for using far-infrared dust emission to trace star formation.

Keywords. galaxies: ISM, infrared: general

1. Introduction

After the completion of the all-sky surveys by the Infrared Astronomical Satellite, astronomers have found conflicting evidence for the heating sources of the dust producing far-infrared emission in nearby galaxies. Some authors claimed that the dust was heated primarily by star formation (e.g. Devereux & Young 1990; Buat & Xu 1996) while others indicated that evolved stellar populations could heat the dust (e.g. Sauvage & Thuan 1992; Walterbos & Greenawalt 1996). This issue has become more important since the launch of the *Herschel* Space Observatory (Pilbratt *et al.* 2010). *Herschel* has been able to produce high signal-to-noise >200 μm images of both nearby and more distant galaxies, so it will be highly sensitive to colder dust that may have been missed by telescopes primarily observing at shorter wavelengths.

Several papers have been published on the sources of the heating for the dust emitting at *Herschel* wavelengths (e.g. Bendo *et al.* 2010, Rowan-Robinson *et al.* 2010, Boquien *et al.* 2011). We will focus on the results from Bendo *et al.* (2011) on the spiral galaxies M81, M83, and NGC 2403. Their analysis was based on comparing the infrared surface brightness ratios to Hα emission (used as a tracer of star formation) and 1.6 μm emission (used as a tracer of the emission from the total stellar population). The infrared surface brightness ratios depend on dust heating, so they will appear correlated with the emission tracing the dust heating sources.

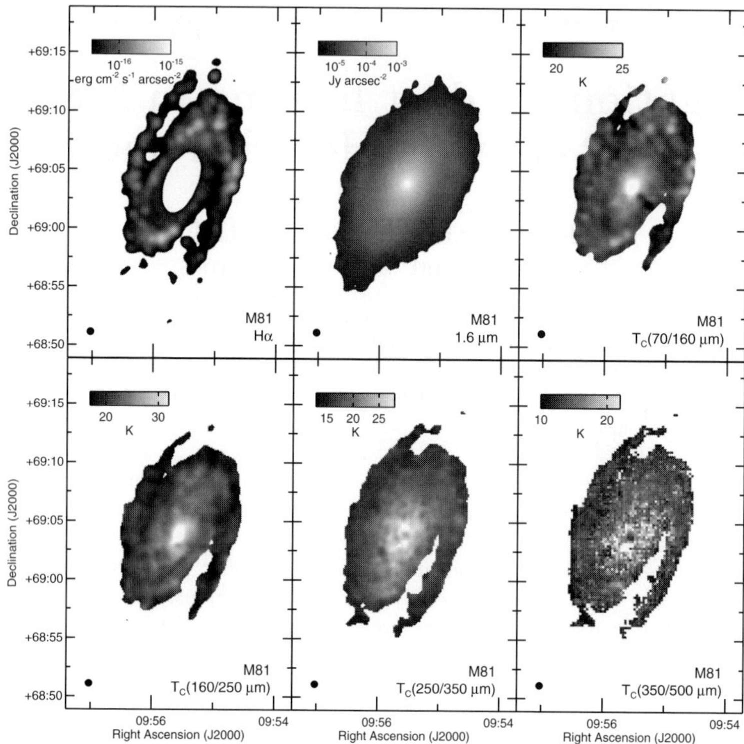

Figure 1. The colour temperature maps of M81 and Hα and 1.6 μm images at the same resolution. The 70 μm data are from *Spitzer*, the 160-500 μm data are from *Herschel*, the Hα data are from Boselli & Gavazzi (2002), and the 1.6 μm data are from Garrett *et al.* (2003). The colour temperatures are only shown for visualisation purposes; the quantitative analysis was performed on surface brightness ratios. All of these maps were created from data where the point spread function (PSF) was matched to the PSF of the 500 μm data, which has a FWHM of 36$''$ (shown by the circles in the lower left corner of each panel). Pixels where the signal was not detected at the 5σ level in either the single band shown (in the case of the Hα and 1.6 μm images) or in both bands (in the case of the colour temperature maps) were left blank (white). Each map is 30$'$ $\times$ 20$'$, and north is up and east is left in each panel. See Bendo *et al.* (2011) for information on the creation of the colour temperature maps.

2. Results

Figures 1–3 show the data that were used in this comparison. Each galaxy yields similar but slightly different results. Details on the analysis, including quantitative analyses based on rebinned versions of these maps, are given by Bendo *et al.* (2011); we present a summarized version of the results here.

M81 has infrared surface brightness ratios that vary mostly as a function of the 1.6 μm emission, which mainly depends on galactocentric radius. Except in the outer disc in the 70/160 μm ratio map, very little local dust heating is seen. These results suggest that most of the dust seen at 160-500 μm as well as some of the dust seen at 70 μm is primarily heated by the total stellar population, including the evolved stellar population in the bulge.

In M83, both the star formation and the total stellar population trace the grand design spiral pattern of the galaxy, and so it is difficult to determine whether the infrared surface brightness ratios depend more on either the Hα or the 1.6 μm emission. To disentangle the relative contributions of star-forming regions and the total stellar populations to dust

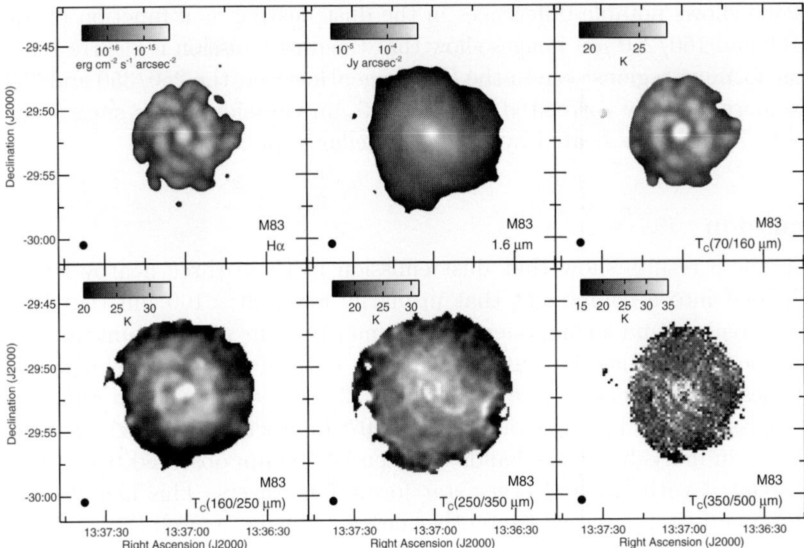

Figure 2. The colour temperature maps of M83 and Hα and 1.6 μm images at the same resolution. The data sources are the same as for Figure 1 except that the 70 μm data are from *Herschel* and the Hα data are from Meurer *et al.* (2006). Each map is $20' \times 20'$. See the caption of Figure 1 for additional information about the layout.

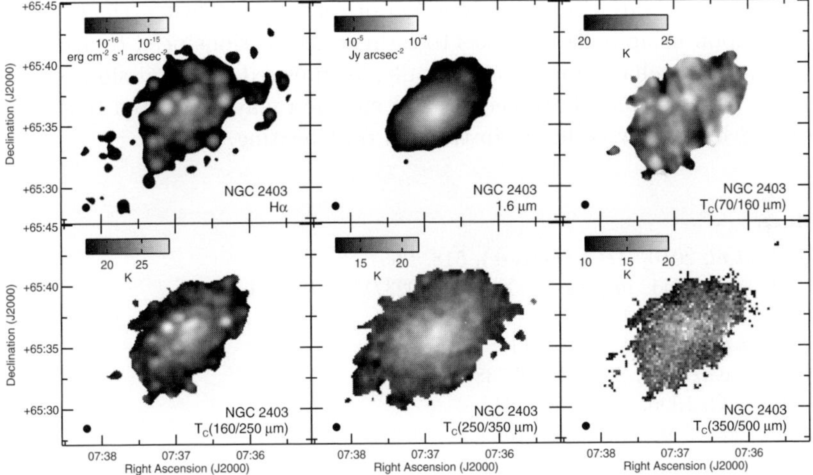

Figure 3. The colour temperature maps of NGC 2403 and Hα and 1.6 μm images at the same resolution. The data sources are the same as for Figure 1. Each map is $21' \times 18'$. See the caption of Figure 1 for additional information about the layout.

heating as traced by any pair of wavelengths λ_1 and λ_2, we fit the data with

$$\ln\left(\frac{I_\nu(\lambda_1)}{I_\nu(\lambda_2)}\right) = \alpha \ln(I(\mathrm{H}\alpha) + A_1 I_\nu(1.6\,\mu\mathrm{m})) + A_2 \tag{2.1}$$

and then examined the ratio of the $I(\mathrm{H}\alpha)$ and $A_1 I_\nu(1.6\ \mu\mathrm{m})$ terms. We found that the the 70/160 and 160/250 μm ratios primarily depend upon the Hα intensity while the 250/350 and 350/500 μm ratios primarily depend upon the 1.6 μm surface brightness. This suggests that the emission from M83 at >250 μm originates from dust heated by the total stellar population.

NGC 2403 shows notable differences in the dust heating as a function of wavelength. The 70/160 and 160/250 μm images show that the dust emission is clearly heated locally by the star-forming regions seen in the Hα image. However, the 250/350 and 350/500 μm ratios are more strongly correlated with the 1.6 μm emission, again suggesting that the dust seen at >250 μm is heated by the total stellar population.

3. Discussion

Overall, these results show that dust emission in these three nearby spiral galaxies can be divided into a component that primarily emits at <160 μm that is heated by star-forming regions and a component that primarily emits at >250 μm that is primarily heated by the total stellar population, including evolved stars in the bulge and discs of these galaxies. This is generally consistent with the expectations from radiative transfer and dust emission models (e.g. Draine *et al.* 2007; Popescu *et al.* 2011). Despite this, the dust emission in individual wave bands between 70-500 μm observed from these galaxies is still correlated with Hα and other star formation tracers. This has also been found for >100 μm emission observed from other nearby galaxies as well (Boquien *et al.* 2010, 2011; Calzetti *et al.* 2010; Verley *et al.* 2010). These relations could arise even if the dust is not directly heated by the star-forming regions if the dust mass is correlated with star formation through the Schmidt law (e.g. Kennicutt 1998).

This has multiple implications. First of all, dust models and templates need to be adjusted to not only account for this very cold dust component but also to replicate the change in the colour variations as a function of wavlength. Dust extinction corrections that depend upon using infrared fluxes to estimate corrections need to account for dust heating by evolved stellar populations. Finally, far-infrared dust emission should be used cautiously as a star formation tracer, as dust emission may be related to star formation through the Schmidt law rather than through dust heating.

References

Bendo, G. J. *et al.* 2010, *A&A* (Letters), 518, L65
Bendo, G. J. *et al.* 2011, in press (arXiv:1109.0237V1 [astro-ph])
Boquien, M. *et al.* 2010, *A&A* (Letters), 518, L70
Boquien, M. *et al.* 2011, *AJ*, 142, 111
Boselli, A. & Gavazzi, G. 2002, *A&A*, 386, 124
Buat, V. & Xu, C. 1996, *A&A*, 306, 61
Calzetti, D. *et al.* 2010, *ApJ*, 714, 1256
Devereux, N. A. & Young, J. S. 1990, *ApJ* (Letters), 350, L25
Draine, B. T. *et al.* 2007, *ApJ*, 663, 866
Kennicutt, R. C., Jr. 1998, *ApJ*, 498, 541
Jarrett, T. H., Chester, T., Cutri, R., Schneider, S. E., & Huchra, J. P. 2003, *AJ*, 125, 525
Meurer, G. R. *et al.* 2006, *ApJS*, 165, 307
Pilbratt, G. *et al.* 2010, *A&A* (Letters), 518, L1
Popescu, C. C., Tuffs, R. J., Dopita, M. A., Fischera, J., Kylafis, N. D., & Madore, B. F. 2011, *A&A*, 527, A109
Rowan-Robinson, M. *et al.* 2010, *MNRAS*, 409, 2
Sauvage, M. & Thuan, T. X. 1992, *ApJ* (Letters), 396, L69
Verley, S. *et al.* 2010, *A&A* (Letters), 518, L68
Walterbos, R. A. M. & Greenawalt B. 1996, *ApJ*, 460, 696

The Spectral Energy Distribution of Galaxies
Proceedings IAU Symposium No. 284, 2011
R.J. Tuffs & C.C. Popescu, eds.

doi:10.1017/S1743921312008824

The dust distribution in late-type low surface brightness disks

John MacLachlan[1], Lynn Matthews[2]†, Kenny Wood[1] and Jay Gallagher[3]

[1]School of Physics and Astronomy, University of St Andrews,
North Haugh, St Andrews, Fife, KY16 9SS, Scotland

[2]Harvard-Smithsonian Center for Astrophysics,
60 Garden Street, MS-42, Cambridge, MA 02138, USA

[3]Department of Astronomy, University of Wisconsin,
475 N. Charter Street, Madison, WI, 53706, USA

email: jmm55@st-andrews.ac.uk

Abstract. Late-type low surface brightness (LSB) disk galaxies are common in the local universe and appear dynamically and chemically under evolved compared to their high surface brightness (HSB) counterparts. We have utilized multi-wavelength imaging and photometry of three edge-on, low-mass LSB disk galaxies to investigate the dust distribution in such systems. Through the use of Monte Carlo radiation transfer models to interpret the data, we find that the dust disk appears to have a vertical scale height similar to the stellar disk. This is in contrast to previous findings from HSB galaxies, where the dust is believed to be more concentrated in the galactic mid-plane. We believe the change in the relative scale heights of the dust and stellar disks is likely associated with the increased stability of the ISM against vertical collapse and the thin nature of the stellar disks.

Keywords. galaxies: ISM, galaxies: spiral, radiative transfer

1. Introduction

Late type LSB disk galaxies, with *B* band central, face-on surface brightnesses below 23.0 mag arcsec^{-2} are a common class of galaxy in the local universe. Often their total HI masses are comparable to or larger than their HSB counterparts, while their HI surface densities are lower by a factor ~ 2 (van der Hulst *et al.* 1993). This fact may may lead to a lower than expected star formation efficiency causing LSB galaxies to evolve more slowly than HSB galaxies. These galaxies frequently show signs of being dark matter dominated at nearly all radii (de Blok & Bosma 2002). Due to the difficulties of directly detecting molecular gas in low mass LSB spirals (cf. Matthews *et al.* 2005), an alternative method to investigate the composition and structure of the cool phases of the ISM is to observe far-infrared (FIR) emission produced by dust.

2. Method

In order to investigate the dust content of such galaxies we have gathered multi-wavelength data (*GALEX*, SDSS, 2MASS and *Spitzer*) for a sample of three edge-on examples. By using Monte Carlo radiation transfer codes that include the effects of transiently heated grains and polycyclic aromatic hydrocarbon molecules (Wood *et al.*

† Current address: Massachusetts Institute of Technology, Haystack Observatory, Off Route 40, Westford, MA 01886, USA

2008) in conjunction with GALEV (Kotulla *et al.* 2009) stellar population models we have attempted to model the dust distributions. As well as the diffuse dust component we also include a template for emission from small star forming regions that are below the current resolution of our models. This emission from compact regions, powered by young stars, is necessary to reproduce the observed far-infrared (FIR) emission

3. Results

We are able to reproduce the global, optical appearance of all three galaxies using smooth emissivity and dust distributions. Figure 1 shows optical, near and mid-infrared imaging of UGC 7321 along with our best fitting model images. The SEDs for the three galaxies studied can be seen in Figures 2, 3 and 4. Our smooth, axisymmetric models provide a good fit to the total FIR emission in all three cases.

Figure 1. Observed (left) and synthetic (right) images of UGC 7321 at, from top to bottom: B band, 3.6μm, 8μm.

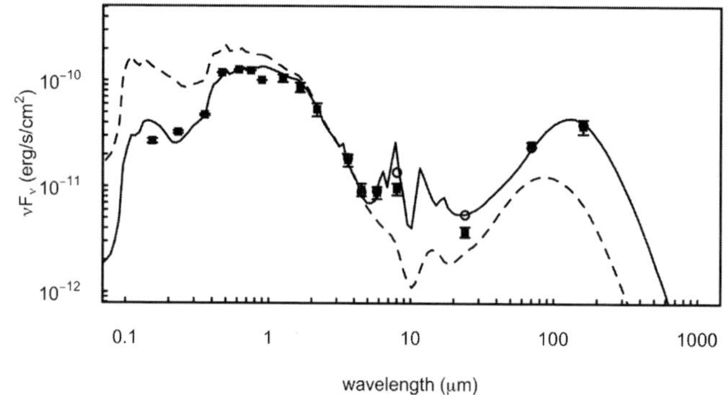

Figure 2. The observed and modeled SED of UGC 7321. Square points indicate an observed flux from one of our data sources either *GALEX*, SDSS, 2MASS or *Spitzer* IRAC/MIPS along with their associated errors. The solid line shows the model output SED. Open circles show the predicted fluxes for *Spitzer* IRAC/MIPS bands. The long dashed line indicates the input, unattenuated, stellar template plus star forming region dust emission template.

4. Implications

In our analysis we find that unlike high surface brightness galaxies previously modeled (Xilouris *et al.* 1999), the dust disks in edge-on LSB galaxies appear to have scale heights equal to those of the stellar disks. The comparable scale heights in the dust and stellar disks are likely associated with the increased stability of the ISM in low mass disks against vertical collapse (Dalcanton *et al.* 2004) and the thin nature of the stellar disks, which suggests minimal dynamical heating. Dust masses are found in the range 1.16

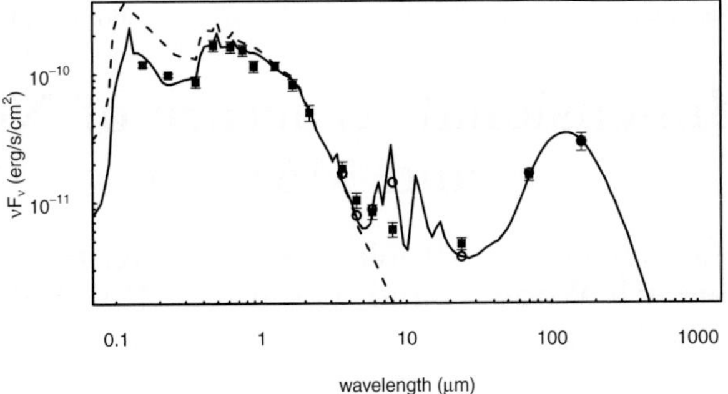

Figure 3. Same as Figure 2 but for IC 2233.

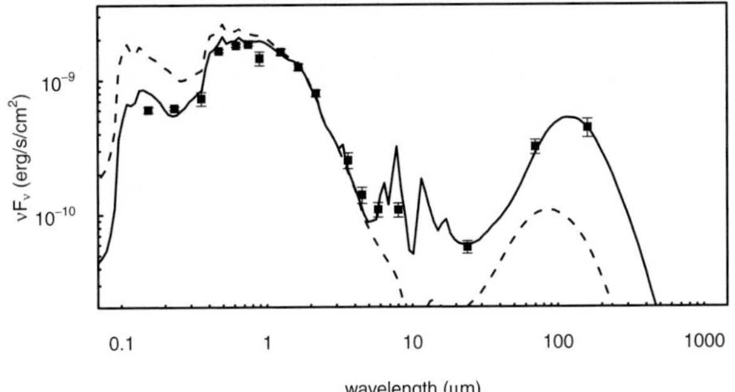

Figure 4. Same as Figure 2 but for NGC 4244.

to $2.38 \times 10^6 M_\odot$, corresponding to face-on (edge-on), V-band, optical depths $0.034 \lesssim \tau_{face} \lesssim 0.106$ ($0.69 \lesssim \tau_{eq} \lesssim 1.99$).

In future work we hope to develop our radiation transfer models to include small scale non-axisymmetric structures which may shed further light on the structure of the ISM and star formation processes in LSB disk galaxies. The inclusion of radial variations in the dust disk scale heights, associated with a flaring gas disk, could also prove important. Additional sub-mm observations may also allow us to uncover cold dust that is associated with the extended HI, as has been found in some HSB galaxies (Popescu & Tuffs 2003).

References

Dalcanton, J. J., Yoachim, P., & Bernstein, R. A. 2004, *ApJ*, 608, 189

de Blok, W. J. G. & Bosma, A. 2002, *A&A*, 385, 816

Kotulla, R., Fritze, U., Weilbacher, P., Anders, P., & the galev team. 2009, *MNRAS*, 396, 462

Matthews, L. D., Gao, Y., Uson, J. M., & Combes, F. 2005, *ApJ*, 129, 1849

Popescu, C. C. & Tuffs, R. J. 2003, *A&A*, 410, L21

van der Hulst, J. M., Skillman, E. D., Smith, T. R., *et al.* 1993, *AJ*, 106, 548

Wood, K., Whitney, B. A., Robitaille, T., & Draine, B. T. 2008, *ApJ*, 688, 1118

Xilouris, E. M., Byun, Y. I., Kylafis, N. D., Paleologou, E. V., & Papamastorakis, J. 1999, *A&A*, 344, 868

The Spectral Energy Distribution of Galaxies
Proceedings IAU Symposium No. 284, 2011
R.J. Tuffs & C.C. Popescu, eds.

doi:10.1017/S1743921312008836

The 3-dimensional structure of NGC 891 and M51

Andrew Schechtman-Rook[1], Matthew A. Bershady[1], Kenneth Wood[2,1], and Thomas P. Robitaille[3]

[1]Department of Astronomy, University of Wisconsin-Madison, 425 N. Charter St., Madison, WI 53706
email: andrew@astro.wisc.edu
mab@astro.wisc.edu

[2]School of Physics & Astronomy, University of St Andrews, North Haugh, StAndrews, Fife, Scotland, KY16 9SS
email: kw25@st-andrews.ac.uk

[3]Harvard-Smithsonian Center for Astrophysics, 60 Garden St., Cambridge, MA 02138
email: trobitaille@cfa.harvard.edu

Abstract. We investigate the three-dimensional structure of the nearby edge-on spiral galaxy NGC 891 using 3D Monte Carlo radiative transfer models, with realistic spiral structure and fractally clumped dust. Using the spiral and clumpiness parameters found from recently completed scattered light models we produce lower resolution SED models which reproduce the spatially-integrated UV-to-FIR SED of NGC 891. Our models contain a color gradient across the major axis of the galaxy - similar to what is seen in images of the NGC 891. With minor adjustment our SED models are able to match the SED of M51, a similar galaxy at a near face-on inclination.

Keywords. radiative transfer, galaxies: ISM, galaxies: spiral, galaxies: structure

1. Introduction

The interplay between starlight and dust in spiral galaxies is enormously complex. In the last decade, space-based observatories have allowed astronomers to measure spectral energy distributions (SEDs) from the ultraviolet (UV) through the far-infrared (FIR). These SEDs show that radiative transfer (RT) models based only on optical data (e.g. Xilouris *et al.* 1999) significantly overpredict the amount of escaping UV radiation and underpredict the amount of dust present in edge-on spirals (Popescu *et al.* 2000; Bianchi 2008, hereafter B08). To combat this problem astronomers have created more complex models, including additional stellar disk components and semi-analytic approximations for the effects of enshrouded star formation with clumpy dust and light.

Most of these advanced models have focused exclusively on fitting galaxy SEDs at the expense of creating models with observationally accurate images, due to the difficulty in fitting non-axisymmetric parameters like clumpy dust substructures or spiral arms.

We have recently completed a preliminary investigation into optimizing non-axisymmetric RT models fit to high-resolution *Hubble Space Telescope* optical images of the central region of the edge-on spiral galaxy NGC 891 (Schechtman-Rook *et al.* 2011). These models included fractally clumped dust and logarithmic spiral arms specially designed to match observations of the arm/interarm ratios of face-on spiral galaxies. The best-fitting models show a strong preference for spirality and clumpiness. Here we investigate the effect of the spirality and clumping parameters determined from these fits to

Table 1. Model Parameters

Model	Spirality	Clumpy Dust	Clumpy Disk-2	M_d (10^8 $M_\odot$)	L_1 (10^{10} $L_\odot$)	$L_{1+Bulge}/L_2$	T (g cm^{-3})
1	No	No	No	1.0	4.2	60	—
2	No	Yes	No	1.0	4.2	7.5	—
3	Yes	Yes	Yes	1.5	4.2	7.5	0
4	Yes	Yes	Yes	1.0	4.2	2.1	3.42×10^{-27}
5	Yes	Yes	Yes	1.75	2.0	1.4	3.42×10^{-27}

optical images on fitting the SEDs of NGC 891 and M51. We use a new 3D Monte Carlo RT code (HYPERION, Robitaille *et al.* 2011), which accepts any dust and emissivity grids, and includes multiple dust grain sizes (including PAHs).

2. Model

Our $100 \times 100 \times 50$ pixel SED models are mapped to a 40x40x20 kpc volume. Stellar emission is set to the Sb spectrum of Fioc & Rocca-Volmerange (1997), split into a old population of stars with masses <3 $M_\odot$(bulge and Disk-1) and a young population ($M > 3$ $M_\odot$; Disk-2). Model parameters are shown in Table 1. M_d is the total dust mass, L_1 and L_2 are the total bolometric *disk* luminosities, and T denotes the density threshold for young stellar emission (where appropriate). All models have a Sersic n = 4 bulge with $L_{Bulge} = 1.8 \times 10^{10} L_\odot$ and structural parameters identical to those used in Bianchi (2008). The model ratios of old-to-young stellar populations span unredenned, spatially-integrated SEDs corresponding to a wide range of Hubble types (for reference template E,Sa,Sc,Im galaxies from Fioc & Rocca-Volmerange (1997) have old-to-young bolometric luminosity ratios of 35,9.8,2.3,1.1, respectively).

Disk-1 has half the scale-length and twice the scale-height of the dust. Disk-2 has the same scales as the dust, and when it is clumpy it follows the distribution of the clumpy dust; higher dust density cells are weighted to have more stellar emission. Clumpiness and spirality are consistent with what we found in Schechtman-Rook *et al.* (2011), i.e., 50% of the dust mass is put in clumps distributed in 130 parent cells; spiral arms have a pitch angle of 20° and a relative strength of 50%. For models 4 and 5 there is a minimum dust density required to place Disk-2 emission in a cell; this threshold mimics the effects of embedded star formation.

3. Results

Single disk models are unable to simultaneously match the UV and FIR SED of NGC 891, even when the dust mass and stellar luminosity are increased or clumpiness and spirality are added. The poor fit from single disk models was also noted by B08. Adding spirality, regardless of the amplitude of the spiral perturbations, has almost no effect on the spatially-integrated SED morphology, a result found also by Popescu *et al.* (2011).

Shown in Figure 1, restricting the young stellar emission to areas of higher dust density (Model 4) allows us to dramatically increase the luminosity of the second disk without increasing the dust mass. This density threshold appears to be unique for any combination of disk luminosities and dust masses — higher thresholds push the dust peak too far into the blue, while at lower thresholds the model cannot simultaneously reproduce the UV, FIR, and 12 μm PAH feature. While this model still predicts a low 24 μm flux, high-resolution simulations of individual dust clouds indicate that this discrepancy is due to

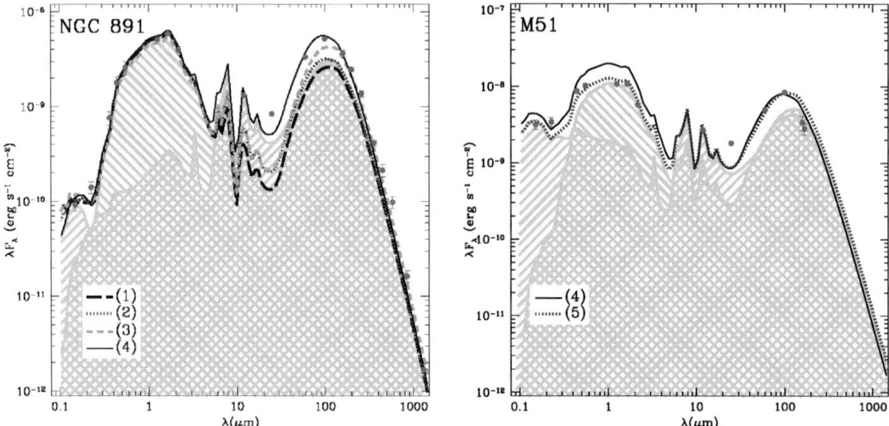

Figure 1. Spatially integrated SEDs. Left: Two-disk models (lines) compared to data from NGC 891 (points, from sources in Bianchi 2008). Decreasing (left to right) and increasing line shading denote contributions to Model 4's SED from old and young stars, respectively. Right: Face-on two-disk models compared to data for M51 (from NED).

the unresolved nature of the dense clumps in our simulations (B08). We investigated the viability of our models at other projected inclinations by comparing them to M51 ($i \sim 20$ deg; Figure 1). For simplicity we fix all geometric parameters (including clumping and spirality) of the dust and stellar distributions as well as L_2, and allow only M_d and L_1 to vary. A good match to the observed, spatially integrated SED (except at 24μm, as explained above) is found by increasing the dust mass by 75% and halving the old-disk luminosity (model 5). The FIR flux distributions indicate that M51 has a dearth of cold dust compared to N891 despite M51's increased dust mass, consistent with elevated star-formation induced by M51's ongoing interaction.

While the addition of spirality has little effect on the shape of the spatially-integrated SED, it does produce a gradient in the UV/blue light across the major axis of the model. This gradient is present in NGC 891 and is commonly considered to be due to the presence of a a grand design spiral pattern Kamphuis *et al.* (2007). This gradient presents an opportunity to better constrain the otherwise hidden thin disk of edge-ons, and can only be fit by models with non-axisymmetric structure like ours.

Acknowledgements

Research was supported by NSF/AST1009471. With data obtained from MAST/STScI, and the NASA/IPAC Extragalactic Database (NED). We thank M. Wolff for help installing HYPERION; C. Howk, B. Benjamin, and B. Whitney for useful discussions.

References

Bianchi, S. 2008, *A&A*, 490, 461
Fioc, M. & Rocca-Volmerange, B. 1997, *A&A*, 326, 950
Kamphuis, P., *et al.* 2007, *A&A*, 471, L1
Popescu, C. C., Misiriotis, A., Kylafis, N. D., Tuffs, R. J., & Fischera, J. 2000, *A&A*, 362, 138
Popescu, C. C., Tuffs, R. J., Dopita, M. A., *et al.* 2011, *A&A*, 527, A109
Robitaille, T. P. *et al.* 2011, *A&A*, Submitted
Schechtman-Rook, A., Bershady, M. A., & Wood, K. *ApJ*, Submitted
Xilouris, E. M., *et al.* 1999, *A&A*, 344, 868

The Spectral Energy Distribution of Galaxies
Proceedings IAU Symposium No. 284, 2011
R.J. Tuffs & C.C. Popescu, eds.

doi:10.1017/S1743921312008848

Stellar and dust SED modelling of the Whirlpool interacting galaxy system

Erin Mentuch and Christine Wilson
Dept. of Physics & Astronomy, McMaster University, Hamilton, Ontario, L8S 4M1, Canada
email: mentuch@physics.mcmaster.ca

Abstract. Some 300-500 Myr ago, the Whirlpool galaxy (NGC 5194/M51a) and its nearby post-starburst galaxy neighbour, NGC 5195/M51b closely interacted, resulting in significant changes to their star formation activity. Both galaxies display colors indicative of enhanced star formation during closest passage, but since then, star formation has ceased in NGC 5195 yet remained ongoing in the spiral NGC 5194. With a wealth of multi-wavelength (0.2–500 μm for this study) observations available, this nearby (10 Mpc) system, whose star formation history is well constrained through optical colors of individual stars and its dynamical history, provides the optimal laboratory to test the relation between dust emission and stellar emission within the fundamental framework of today's stellar population synthesis and dust emission models.

Keywords. galaxies: individual (NGC 5194/M51a, NGC 5195/M51b), galaxies: general, galaxies: fundamental parameters, galaxies: stellar content, infrared: galaxies, techniques: photometric

1. Introduction

Standard stellar population techniques can tell us how much stellar mass is in a galaxy, its metallicity, the amount of dust extinction and can even shed light on the star formation history within a galaxy. Most dust emission models (Dale & Helou 2002, Draine *et al.* 2007) probe the total dust mass and some (Désert *et al.* 1990, Compiègne *et al.* 2011) the relative abundance of various dust emission components (eg. polycyclic aromatic hydrocarbons (PAHs), very small grains, big grains). The amount of dust emission observed in a galaxy is tied to both the interstellar radiation field (provided by stellar radiation, in addition to non-stellar sources such as active galactic nuclei) and the abundance of dust grains. How these two processes relate to the stellar populations which both provide the heating and create the dust is not yet well constrained by observations. Fortunately, observations of nearby galaxies in the infrared with the *Herschel Space Observatory* offer the next step towards the development of fully consistent galaxy population models that can account for dust created and heated by stellar populations, in the context of a galaxy's star formation history.

Observed as part of the *Herschel Very Nearby Galaxies Survey* (VNGS, P.I. Christine Wilson), we have performed a spatially resolved spectral energy distribution (SED) fitting analysis on the well studied grand design spiral, the Whirlpool galaxy (NGC 5194/M51a) and its post-starburst nearby neighbour, NGC 5195/M51b. While star formation is ubiquitous and ongoing across NGC 5194 as indicated by its ~ 19 600 HII regions (Lee *et al.* 2011) and optical colors (Tikhonov *et al.* 2009), the smaller (in angular extent) neighbour NGC 5195 shows no signs of recent star formation through a deficiency of ionized gas (Thronson *et al.* 1991). This lack of star formation is supported by evidence for dynamically stable (and thus, unable to form stars) molecular gas (Kohno *et al.* 2002). An encounter between the two galaxies 300–500 Myr ago is inferred from both kinematic (Salo & Laurikainen 2000) and hydrodynamic modelling (Dobbs *et al.* 2010)

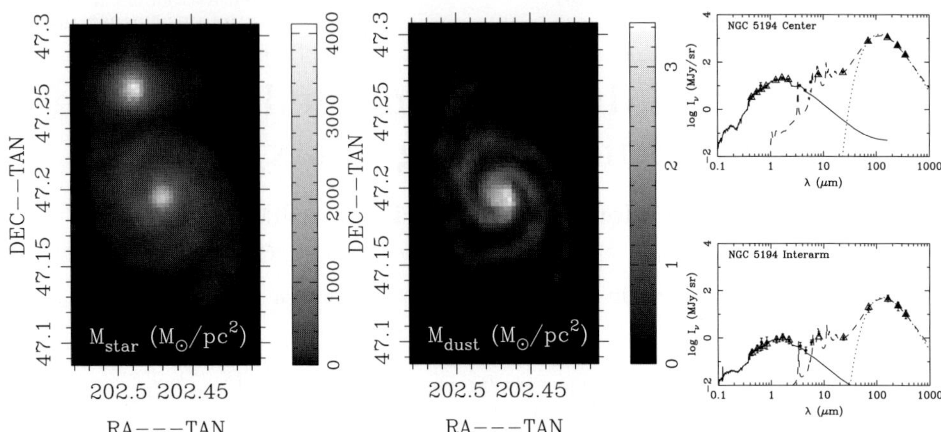

Figure 1. Stellar mass (*left*) and dust mass (*middle*) maps of the Whirlpool system. *Far right:* We plot two example SED fits from the nuclear region of NGC 5194 (top) and the inter arm region of NGC 5194 (bottom). The solid curve is the best fit stellar emission model (PEGASE.2, Fioc & Rocca-Volmerange 1997) and the dashed curve is the best fit dust emission model from Draine & Li (2007). For comparison, we also show the best fitting modified greybody model (dotted curve) for both examples.

of the Whirlpool system. In agreement, a recently constructed color-magnitude diagram of its individual stars, measured using *HST* images, suggests a burst of star formation occurred 390–450 Myr ago in NGC 5195 (Tikhonov *et al.* 2009). In contrast, this same work posits that NGC 5194 is better described by two episodes of star formation, one at 265-325 Myr and another at 380-450 Myr. Not unexpectedly, this actively star forming disk galaxy cannot be easily described by just a simple stellar population.

As both galaxies display peaks in their star formation during the epoch of closest encounter, how well do SED models retrieve this information? What about the dust emission? How much can be attributed to more evolved stars and how much to more recent star formation? In these proceedings, we highlight the results (to be published in detail in Mentuch *et al.* 2012, in prep) of our spatially resolved SED fitting analysis of the Whirlpool system in the context of these fundamental questions.

2. Data and models

Our analysis includes twenty multi-wavelength images of the Whirlpool system, exploiting the great efforts of a number of large legacy surveys. Our optical (B, V, R, I & $H\alpha$) and mid-infrared (IRAC 3.6, 4.5, 5.6, 8.0 and MIPS 24 μm) observations come from the SINGS survey (Kennicutt *et al.* 2003). Near-infrared J, H, K imaging is from the 2MASS LGA survey (Jarrett *et al.* 2003). VNGS's observations with *Herschel* cover 70, 160, 250, 350 and 500 μm. In addition, imaging of the neutral gas is provided by the BIMA SONG survey (Helfer *et al.*2003) and we have molecular gas observations at CO(1-0) observations from the Nobeyama 45 m telescope provided by Koda *et al.* (2011). All optical and near-IR images are corrected for galactic foreground extinction and matched to a common resolution of FWHM = 28" using either gaussian convolutions or in the case of our *Herschel* observations, kernels from Aniano *et al.* (2011). At this common resolution and a distance of 8.4 Mpc (Feldmeier *et al.* 1997), we resolve the galaxies on physical scales of ~1 kpc.

We chose to model stellar and dust emission as separate components using template SED libraries that have both been extensively tested and the parameter space well constrain by observations. Spectral SED libraries were created using standard code from PEGASE.2 (Fioc & Rocca-Volmerange 1997). Details can be found in Mentuch *et al.* (2012, in prep), but in brief, our spectral library spans metallicities from 0.005 to 0.04 and the star formation history is modelled as a sum of a continuous star forming population described by a tau exponential with an additional burst of star formation whose amplitude and age are both free parameters. Dust extinction, whose amplitude is an additional parameter, is modelled using a Milky Way extinction law (Pei 1992).

A least-squares fitting algorithm selects the best fit model from our suite of 350 000 stellar models to our $BVIRJHK$ images. We use the modelled stellar maps of the galaxy to subtract the contribution of stellar emission from our mid-infrared images in the IRAC 5.6 and 8.0 μm bands. These resulting images and our other infrared (24–500 μm) images are fit to the two-component emission model of Draine & Li (2007). In their model, the mid- through far-infrared emission due to PAHs, very small dust grains and carbonaceous dust grains is expressed as a function of the underlying interstellar radiation field (ISRF) comprised of a high energy component surrounding massive star forming regions and then a large scale interstellar medium component with a constant ISRF. The bulk of dust mass is contained in this latter component. In our least-squares fitting procedure, free parameters included the minimum strength (U_{min}, see Draine *et al.* 2007 for details) of the radiation field, the fractional contributions of both ISRF components and the fraction of PAH molecules.

3. Maps of SED fitted parameters

A large number of output parameters are yielded from each pixel fit. We generate maps of each parameter using the average of the best fits from a Monte Carlo fitting simulation (see Mentuch *et al.* 2012, in prep). Here, we highlight some of the results of our fitting which confirm the encounter model and some of the surprises. Figure 1 shows the stellar and dust mass maps in the left and middle panels with examples of our SED fits on the right. The stellar mass map displays a radially smooth distribution with NGC 5194 displaying a typical profile for a spiral galaxy and NGC 5195 showing a steeper profile as an S0. There are some enhancements in stellar mass along the spiral arms, but they are minimal. Coincidentally, both galaxies have very similar stellar masses of $M_\star=(3.62\pm0.03)\times10^{10}\ M_\odot$ and $M_\star=(2.01\pm0.04)\times10^{10}\ M_\odot$ for NGC 5194 and NGC 5195, respectively, despite their different morphologies and angular sizes. On the other hand, their dust masses differ by an order of magnitude; $M_{dust}=(7.75\pm0.02)\times10^7\ M_\odot$ and $M_{dust}=(7.67\pm0.04)\times10^6\ M_\odot$ for NGC 5194 and NGC 5195, respectively. It is difficult to speculate whether the difference in the dust-to-stellar mass ratio is primarily due to morphology, the presence of active star formation, or a consequence of the gas fraction in each galaxy prior to interaction. Currently, the gas-to-dust ratio across both galaxies is fairly constant with values in the range of 80–120, typical for spiral galaxies. Figure 2, left panel, shows a mass map of the total gas, traced well by the dust mass surface density shown in black contours.

Reassuringly, the SED fits are consistent with an increased epoch of star formation 300–500 Myr ago. Both galaxies have old stellar populations that have been forming stars continuously (assuming an exponentially declining SFH) for about 8.6 billion years and both had about 10$\pm$6% of their stellar mass formed in a burst of star formation which occurred 360$\pm$10 Myr ago. We note here that because of the simplicity of our chosen

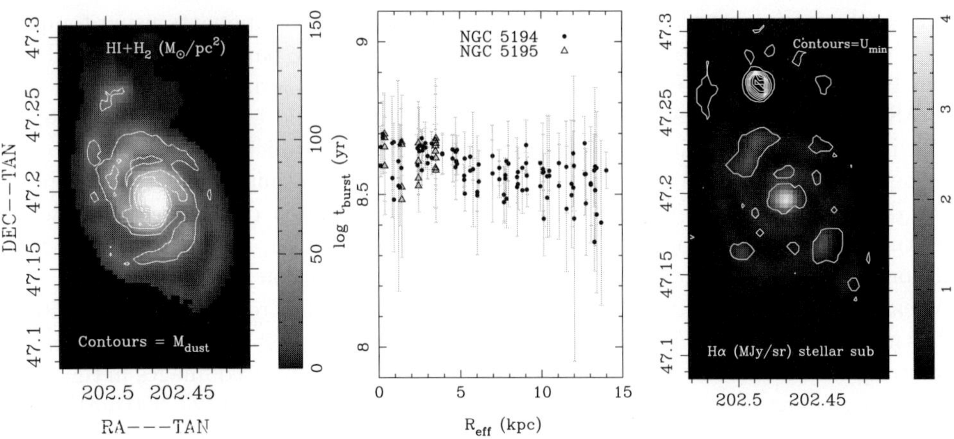

Figure 2. *Left:* Total gas (molecular H_2 from CO(1-0) plus neutral HI gas) mass map with dust mass overlaid in contours. *Middle:* Time since an instantaneous burst of star formation from our SED fitting analysis matches the time of closest approach between the galaxy pair derived from dynamical simulation and colors of individual stars. Here we plot a radial profile of this parameter for each galaxy, NGC 5194 in solid dots and NGC 5195 in open triangles. *Right:* Map of the stellar subtracted Hα emission with the derived ISRF, quantified as U_{min} from Draine & Li(2007), shown as contours. The highest ISRF is found in the post-starburst galaxy, NGC 5195, where a clear lack of Hα emission is found.

SFH, we do not pick out all of the current star formation occurring in the spiral arms of NGC 5194 in which relatively little stellar mass contributes to the total SED.

One surprising result to come out of our study is the presence of large amounts of dust in the post-starburst galaxy NGC 5195 as is seen in the middle panel of Figure 1. Dust extinction is seen at visible wavelengths extending from one of the spiral arms of NGC 5194 and Tikhonov *et al.* (2009) confirm that a large population of post-main-sequence stars make up this region. The dust in NGC 5195 is heated by a strong radiation field as indicated by large U_{min} values (and high dust temperatures), despite a lack of Hα emission (ie. active star formation). Thus, in agreement with Kohno *et al.* (2002) who found a lack of cold unstable dense gas, but large amounts of stable molecular gas, NGC 5195 appears to be deficient in cold dust but contains a good amount of warm dust.

Acknowledgements

E. Mentuch would like to thank the IAU organizing committee for the travel grant and acknowledge the continued support for this research provided by the National Science and Engineering Research Council (NSERC) of Canada through her PDF fellowship.

References

Aniano, G., Draine, B. T., Gordon, K. D., & Sandstrom, K. 2011, *PASP*, 123, 1218
Compiègne, M., Verstraete, L., Jones, A., *et al.* 2011, *A&A*, 525, A103
Dale, D. A. & Helou, G. 2002, *ApJ*, 576, 159
Désert, F.-X., Boulanger, F., & Puget, J. L. 1990, *A&A*, 237, 215
Dobbs, C. L., Theis, C., Pringle, J. E., & Bate, M. R. 2010, *MNRAS*, 403, 625
Draine, B. T. & Li, A. 2007, *ApJ*, 657, 810
Feldmeier, J. J., Ciardullo, R., & Jacoby, G. H. 1997, *ApJ*, 479, 231
Fioc, M. & Rocca-Volmerange, B. 1997, *A&A*, 326, 950
Helfer, T. T., Thornley, M. D., Regan, M. W., *et al.* 2003, *ApJS*, 145, 259
Jarrett, T. H., Chester, T., Cutri, R., Schneider, S. E., & Huchra, J. P. 2003, *AJ*, 125, 525

Kennicutt, R. C., Jr., Armus, L., Bendo, G., *et al.* 2003, *PASP*, 115, 928
Koda, J., Sawada, T., Wright, M. C. H., *et al.* 2011, *ApJS*, 193, 19
Kohno, K., Tosaki, T., Matsushita, S., *et al.* 2002, *PASJ*, 54, 541
Lee, J. H., Hwang, N., & Lee, M. G. 2011, *ApJ*, 735, 75
Pei, Y. C. 1992, *ApJ*, 395, 130
Salo, H. & Laurikainen, E. 2000, *MNRAS*, 319, 377
Thronson, H. A., Jr., Rubin, H., & Ksir, A. 1991, *MNRAS*, 252, 550
Tikhonov, N. A., Galazutdinova, O. A., & Tikhonov, E. N. 2009, *Astron. Lett.*, 35, 599

Discussion

GALLAGHER: Do you have any idea about what is going on in the nucleus of NGC 5195? It is the most intense source in the system in Spitzer 24 μm images, a stronger point source than the M51 nucleus.

MENTUCH: The bright emission at both 24 and 10 μm is consistent with the emission by dust predicted by the Draine & Lee (2007) model corresponding to a very high interstellar radiation field and a relatively low PAH fraction. It appears not to be related to star formation.

CHACKRABARTI: Following up on Jay Gallagher's question - you see M51's companion in the B-band (which traces gas and new stars) and at 24 μm (which traces hot dust). Could there be some star formation that's not apparent in H-alpha because it's obscured by dust?

MENTUCH: The B-band emission is well fit by the SED with SFH parameters for a tau model + burst (implying an old stellar population plus a starburst 400 Myr ago), so there is no additional emission in the band associated with recent (within 10 Myr) star-formation. The hot dust is due to the high interstellar radiation field in the galaxy, whose origin is uncertain, although M51b is very bright in X-ray emission, which may be a clue.

The Spectral Energy Distribution of Galaxies
Proceedings IAU Symposium No. 284, 2011
R.J. Tuffs & C.C. Popescu, eds.

doi:10.1017/S174392131200885X

Hot & cold dust in M31: the resolved SED of Andromeda

Brent Groves[1], **Oliver Krause**[1] **and the MPIA Herschel Andromeda Team**

[1]Max Planck Institute for Astronomy, Königstuhl 17, D-69117 Heidelberg, Germany

email: brent@mpia.de, krause@mpia.de

Abstract. Due to its proximity, the Andromeda galaxy (M31, NGC 224) offers a unique insight into how the spectra of stars, dust, and gas combine to form the integrated Spectral Energy Distribution (SED) of galaxies. We introduce here *Herschel Space Observatory* PACS and SPIRE photometric observations of M31 which cover the far-infrared to sub-mm wavelengths (70-500 μm). These new observations reveal that the total IR luminosity of M31 is relatively weak, with $L_{\rm IR} = 10^{9.65} L_{\odot}$, only 10% of the total luminosity of M31. However, as seen in the previous studies of M31, the IR luminosity is dominated by a 10 kpc ring in all *Herschel* bands. This is distinct from the optical, where the bulge in the central 2kpc, dominates the luminosity, clearly demonstrating how different components at distinct positions in a galaxy contribute to make the integrated SED.

Keywords. Galaxies:individual:M31,ISM:dust

1. Introduction

The Andromeda galaxy (M31) has helped forge our understanding of the association of dust with gas and stars on galactic scales. The seminal work of Baade & Gaposchkin (1963) noted that in Andromeda the dust (as seen through extinction) was predominantly associated with young stars and hence star formation. Yet it was with far-infrared space telescopes that the dust revealed itself. The IRAS observations of M31 (Habing *et al.* 1984, Soifer *et al.* 1986, Walterbos & Schwering 1987, Xu & Helou 1996a,b) found the IR luminosity of M31 was relatively weak compared to many other galaxies found in the IRAS surveys. The IRAS survey found the IR maps dominated by the 10 kpc ring, as also found at 175 μm in the ISO survey of M31 (Haas *et al.* 1998).

The *Spitzer Space Telescope* view of M31 took this IR image to the next level of resolution, with the 10 kpc ring clearly visible in the IRAC 8 μm band (Barmby *et al.* 2006), and in the longer wavelength MIPS bands (Gordon *et al.* 2006). In addition to the larger 10 kpc ring, the MIPS imagery also revealed the central peak, and inner spirals and outer loops. Using this *Spitzer* data, in association with GALEX and SDSS imaging, Montalto *et al.* (2009) modelled the dust of M31, determining the dust mass distribution and importantly the heating of the dust across the disk. They found that over the majority of the disk the dust was heated predominantly by optical photons and stars older than 1 Gyr, as suggested in the earlier work by Xu & Helou (1996b).

In this paper we continue this investigation into the far-infrared properties of M31 with *Herschel Space Observatory* (Pilbratt *et al.* 2010) observations of the Andromeda galaxy. This data has given us an unprecedented look into the long wavelength emission from M31 and demonstrates that, overall, M31 has a low total IR luminosity and relatively low total dust mass. However, even with cold dust (long wavelengths), the 10kpc ring

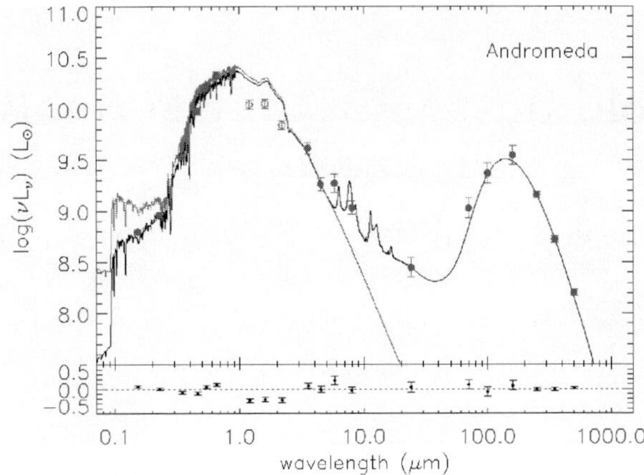

Figure 1. The spectral energy distribution of Andromeda (filled points), and the best fit model (black curve) from the SED fitting code MAGPHYS (da Cunha *et al.* 2008). Note that the 2MASS data (empty circles) were not used in determining the best fit due to the smaller aperture used. The solid curve shows the best model fit, while the grey line shows the unattenuated stellar spectrum. In the lower plot we show the offset of the data from the best fit model.

dominates the emission, but dust emission can be found out to radii even further than this. For full details of this work, we refer the reader to Groves *et al.* (in prep.).

2. The Integrated SED of Andromeda

The Andromeda galaxy was observed in all 6 *Herschel* photometric bands (PACS 70, 100, and 160 μm and SPIRE 250, 350, and 500 μm as a Guaranteed Time (GT1) program. A $\sim 3^\circ \times 1^\circ$ region centred on M31 was observed in slow parallel mode ($20''$/s) for a total time of $\sim$ 24 hours (86412 seconds). The full observations and reduction are described in detail in Krause *et al.* (in prep.).

Using literature data (UV from Gil de Paz *et al.* (2007), optical from Walterbos & Kennicutt (1987), near-IR from Jarrett *et al.* (2003), and the mid-IR from Barmby *et al.* (2006) and Gordon *et al.* (2006)) and the new *Herschel* data, it is now possible to obtain the full UV–IR integrated spectral energy distribution (SED) of the Andromeda galaxy, enabling a comparison of our nearest massive neighbour with more distant galaxies. We obtained the full SED of M31 within a $r_{\rm maj} \sim 95'$ ellipse, with an inclination of 75° and a P.A. of 37.7°, shown by the filled points in Figure 1.

Figure 1 also shows the best fit model SED from the SED fitting code MAGPHYS (da Cunha *et al.* 2008). Due to the significantly smaller aperture used to determine the 2MASS fluxes and the observable offset between these fluxes and the R and 3.6 μm bands, we have not included these in the determination of the best fit model. The offset between the best fit model and the observations can be seen in the lower part of Figure 1, including the offset of the 2MASS points.

It is immediately clear from the integrated SED that M31 is an early type spiral, with a relatively low UV and IR luminosity (c.f. the work of Mutch *et al.* 2011, who compare Andromeda with SDSS galaxies, and found that it fell in the "green valley"). The total luminosity of M31 from the best-fit model is $\log(L_*/L_\odot) \sim 10.61$. Only ~10% of the radiation in M31 has been processed by dust, with $L_{\rm dust} = 4.44^{+0.15}_{-0.10} \times 10^9$ ($L_\odot$), and the subsequent radiation from the dust is predominantly cool, peaking at $\sim 160\mu$m. The total mass of dust returned by the model is $\log M_{\rm d} = 7.43 \pm 0.06 (M_\odot)$, supporting the idea that M31 is a dust poor galaxy. The mean cold dust temperature in M31 is $16.8^{+0.5}_{-0.7}$K,

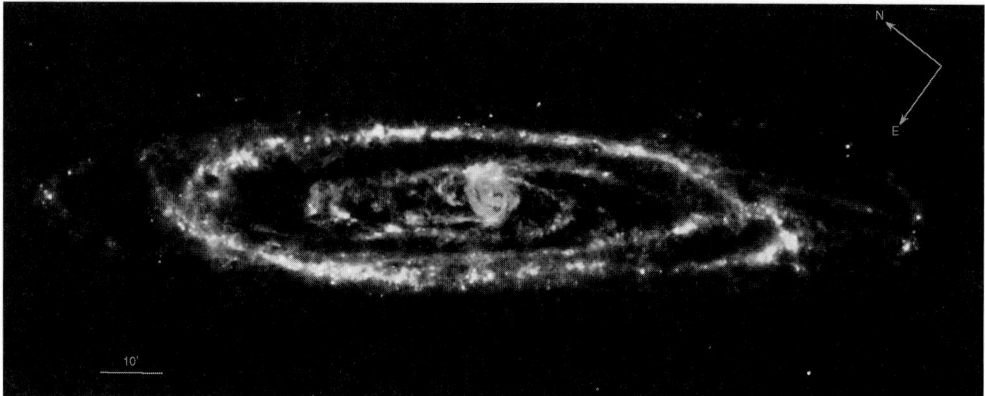

Figure 2. Far-infrared image of Andromeda using the Herschel bands showing PACS 70 μm, PACS 100 μm, and SPIRE 250 μm, all convolved to SPIRE 250 μm resolution. All three bands have the same square root scaling from 10 MJy/sr to 150 MJy/sr The angular scale is shown by the 10' bar in the lower right.

as expected with the peak of the IR bump at such long wavelengths. Altogether, M31 is a relatively dust-poor spiral with relatively cool dust, mostly because it is dominated by the emission from the diffuse ISM.

3. The resolved SED of Andromeda

The unprecedented spatial resolution of *Herschel* ($6-8''$, or 20–60pc at $70-250\mu$m), enables us to go further and determine the heating of the dust at the scales of HII regions and giant molecular clouds, and obtain the SED of Andromeda on small scales. In Figure 2 we show a three-colour far-infrared image of M31, using the PACS 70 and 100 μm and SPIRE 250 μm bands. All three bands have the same square root scaling from 15 MJy/ to 150 MJy/sr.

Clearly visible in the image is the 10 kpc starforming ring of M31 and its spiral structure, with bright white points indicating positions of HII regions. Outside the 10kpc ring, little dust emission is seen at these levels of surface brightness, except for the extended loops along the major axis. Inside the ring, the inner spiral arms are visible, linking the ring to both the nuclear and bulge emission at the centre. These features have been visible in both the earlier IRAS and *Spitzer* MIPS far-IR maps of M31, yet never before at such spatial resolution at these wavelengths. This image clearly demonstrates which regions contribute at which wavelengths, with the 10kpc ring dominating long wavelengths but the central region dominated by the shortest wavelengths.

This is more clear when we fit a simple modified blackbody, $F_\nu \propto B_\nu(\mathrm{T_d})(\frac{\nu}{\nu_0})^\beta$, to the 100–500 μm data. By minimizing over the other parameters, we determined the dust temperature T_d as a function of radius, shown in Figure 3. The radial variation quantitatively displays what was clear in Figure 2, with a high temperature in the central 2 kpc where the bulge of M31 is, while the disk of M31 has an approximately constant temperature ($T_\mathrm{d} \approx 17 \pm 1$K). This even includes the 10kpc ring, where little variation is seen with the mean T_d, though clearly there are individual hot HII regions within the ring. To the right of the radial temperature we show the integrated IR SEDs of the central 2 kpc (the bulge), and between 9 and 11 kpc (the ring) and the best fitting simple modified blackbody. While these two SEDs demonstrate the difference in dust temperatures between the two regions, they also reveal the relative contribution of these

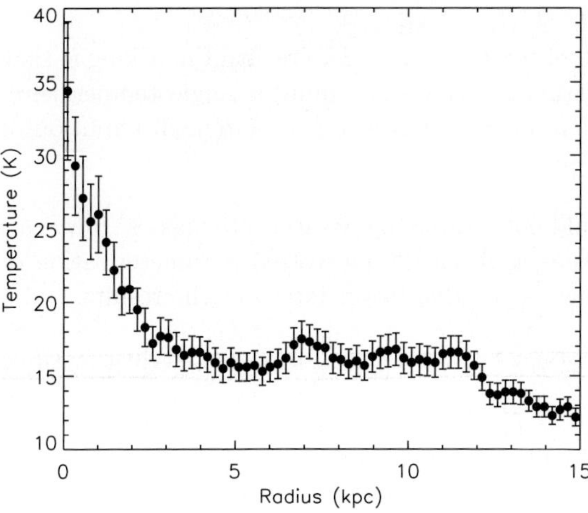

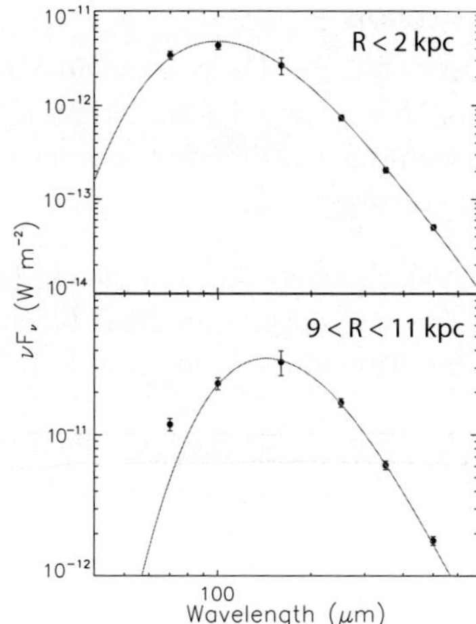

Figure 3. *Left*: Radial variation of mean temperature in M31. The temperature is determined by fitting the median flux in all bands in radial bins of 1′ (0.23kpc) width using elliptical annuli. The error bars show the 16-84 percentiles of marginalized PDFs. *Right*: The *Herschel* IR SED from the central 2kpc (*top*) and from the 10kpc ring (*bottom*) as labelled.

regions to each band, with the overall bulge IR emission an order of magnitude weaker than the ring, but having a relatively greater (though still weak) contribution at the shortest wavelengths.

As a final note, we point out that this work presents only a small fraction of the available information with these new *Herschel* images. With existing optical, GALEX, *Spitzer*, HI, radio, and CO data, and the currently ongoing Pan-chromatic Hubble Andromeda Treasury survey which will cover roughly a third of the star forming disk, using HST 6 filters from the UV–NIR, in addition to the *Herschel* images presented here, there exists a seemingly-overwhelming wealth of data on the Andromeda galaxy that should prove to be a heritage in understanding the connections between stars, gas, and dust and how they combine to make a galaxy.

References

Baade, W. & Gaposchkin, C. H. P. 1963, *Evolution of stars and galaxies.*, by Baade, W; Gaposchkin, C. Cambridge, Harvard University Press

Barmby, P. *et al.* 2006, *ApJL*, 650, L45

da Cunha, E., Charlot, S., & Elbaz, D. 2008, *MNRAS*, 388, 1595

Gil de Paz, A., Boissier, S., Madore, B. F., *et al.* 2007, *ApJS*, 173, 185

Gordon, K. D., *et al.* 2006, *ApJL*, 638, L87

Haas, M., Lemke, D., Stickel, M., *et al.* 1998, *A&A*, 338, L33

Habing, H. J., Miley, G., Young, E., *et al.* 1984, *ApJL*, 278, L59

Jarrett, T. H., Chester, T., Cutri, R., Schneider, S. E., & Huchra, J. P. 2003, *AJ*, 125, 525

Montalto, M., Seitz, S., Riffeser, A., *et al.* 2009, *A&A*, 507, 283

Mutch, S. J., Croton, D. J., & Poole, G. B. 2011, *ApJ*, 736, 84

Pilbratt, G. L., Riedinger, J. R., Passvogel, T., *et al.* 2010, *A&A*, 518, L1

Soifer, B. T., Rice, W. L., Mould, J. R., *et al.* 1986, *ApJ*, 304, 651

Xu, C. & Helou, G. 1996a, *ApJ*, 456, 152

Xu, C. & Helou, G. 1996b, *ApJ*, 456, 163

Walterbos, R. A. M. & Kennicutt, R. C., Jr. 1987, *A&AS*, 69, 311

Walterbos, R. A. M. & Schwering, P. B. W. 1987, *A&A*, 180, 27

Discussion

CHAKRABARTI: Do you consider a range of temperatures? The reason I'm asking is that Chris McKee and I found that if you estimate masses assuming a single temperature, you would get a different answer than if you considered a self-consistent distribution of temperatures.

GROVES: I agree. The simple models would not be able to give accurate masses. However, you can use simple modified blackbodies as a physically-motivated parameterization of the far IR colours. They are useful so long as you don't over-interpret the results.

WANG: Since M31 is highly inclined, do you see any evidence for projected dust features from the front side of galactic disk.

GROVES: While most of the disk is optically thin we can see on the 3-colour Herschel image of M31 a portion where we have the disk dust emission projected onto the bulge.

The Spectral Energy Distribution of Galaxies
Proceedings IAU Symposium No. 284, 2011
R.J. Tuffs & C.C. Popescu, eds.

doi:10.1017/S1743921312008861

Resolved optical-infrared SEDs of galaxies: universal relations and their break-down on local scales

Stefano Zibetti[1] and Brent Groves[2]

[1]Dark Cosmology Centre, Niels Bohr Institute, University of Copenhagen, Juliane Maries Vej 30, DK-2100 Copenhagen, Denmark

[2]Max-Planck-Institut für Astronomie, Königstuhl 17, D-69117 Heidelberg, Germany

email: `zibetti@dark-cosmology.dk` `brent@mpia.de`

Abstract. A large body of evidence has demonstrated that the global rest-frame optical and IR colours of galaxies correlate well with each other, which can be readily interpreted as a sign of typically smooth star formation histories. However the processes that lead to the observed correlations are contrary: the stellar light that contributes to the optical is readily absorbed by dust which emits in the IR. Thus on small scales we expect these correlations to break down. In this contribution we present our recent results (Zibetti & Groves 2011) from a pixel-by-pixel multi-wavelength (u-band to 8μm) analysis of seven nearby galaxies ranging from early- to late-types. We show that such a break-down occurs already on scales on few 100 pc, as a result of the different physical conditions in spatially distinct regions inside the galaxy, as we demonstrate by means of a Principal Component Analysis. Despite the lack of internal correlation between optical and IR within individual galaxies, when the pixels of all galaxies are compared the well known optical-IR colour correlations return, demonstrating that the variance observed within galaxies is limited around a mean which follows the well-known trends. We also examine the extremely strong correlations between the mid IR (*Spitzer*-IRAC)-NIR colours which extend continuously across all galaxies. These correlations arise from the differing contribution of stellar light and dust to the IRAC bands, enabling us to determine pure stellar colours for these bands, but still demonstrating the need for dust (or stellar) corrections in these bands when being used as stellar (dust) tracers.

Keywords. galaxies: general - galaxies: individual: NGC4254 galaxies: photometry galaxies: stellar content infrared: ISM.

---

1. Introduction

The optical to mid IR SED of a galaxy (350 nm to $\sim$ 10 μm for the scope of this paper) is characterised by the presence of two distinct regimes, short-ward and long-ward of $\sim 1-2$ μm, respectively. At optical wavelengths we observe the radiation of stars, possibly extincted and reddened by dust. Moving into the mid IR we see a combination of stellar radiation and dust glowing as it re-radiates the energy that has been absorbed from stars, either via thermal mechanisms or molecular transitions (e.g. the PAH bands); dust emission becomes more and more dominant as one moves to longer wavelengths. Nowadays it is an obvious and well established fact that the integrated optical colours of galaxies are well correlated to each other (e.g. Blanton *et al.* 2003 for an illustration of it with the large statistical sample of the SDSS): this reflects the simple fact that the shape of the SED at these wavelengths is mainly dictated by the mean stellar population age (to first order) and metallicity and modulated by dust extinction. Mid IR colours are also correlated with each other as they reflect, to first order, the balance between dust and emission from (old) stars. What is much less obvious is the correlation between integrated optical and mid IR colours, in the sense that optically bluer galaxies are normally relatively brighter in the mid IR and, vice versa, red galaxies (unless heavily dust obscured) are dim in mid IR (e.g. Hogg *et al.* 2005). This correlation indicates that *on galaxy scale* the hot dust and PAH emission which trace the most recent star formation (on $\sim 10^7$ yr timescale, Calzetti *et al.* 2007) is associated with the diffuse presence of a young stellar component whose age spans up to 1 Gyr, as reflected in the optical colours. Such a correlation between integrated optical and mid IR colours is basically an indication of the typical smoothness of the star formation history (SFH) of galaxies.

There is no simple physical reason why such correlations may apply also *within* galaxies on sub-kpc scales. In fact, on a purely physical ground the interplay between stars and dust can be very complex; in particular, the hot optical-UV radiation needed to power the (hot) dust emission has to be promptly absorbed by the dust itself, with the side effect of reddening the optical spectrum. In fact only a fine tuned balance between the local density of young stars and dust can result in tight optical-mid IR colour correlations.

In order to explore this issue we have performed pixel-by-pixel SED analysis of a small sample of seven nearby ($D \lesssim 20$Mpc) galaxies with available imaging in the optical from the SDSS (York *et al.* 2000, u, g, r, i, z), in the near IR H-band (either from GOLDMine, Gavazzi *et al.* 2003, or UKIDSS Lawrence *et al.* 2007) and mid IR in the Spitzer-IRAC bands (from SINGS, Kennicutt *et al.* 2003), plus narrow-band Hα. These galaxies span the full range of mass and morphologies of normal non-dwarf galaxies. We refer the reader to Zibetti & Groves (2011) for a fully detailed description of the analysis and results; in this contribution we briefly summarise them and highlight the most important points.

2. Main Results

The images of each galaxy in the ten broad bands (plus Hα) are sampled with pixels of approximately 200 pc on a side. For each of these pixels the local SED is constructed by normalising the monochromatic flux νf_ν in each band to the flux in H band. This not only allows to study all SEDs on a similar scale but also is optimal in separating the two wavelength regimes dominated by stellar and dust emission respectively. For each galaxy individually we study the distribution of pixels in all possible combinations of "colour-colour" plots (where by "colour" we actually mean H-band normalised flux). As expected we find that optical colours correlate with each other and the same holds true for the mid IR/IRAC colours. However mixed optical-mid IR colour-colour diagrams display very weak correlations if any at all. Figure 1 shows an example of such diagrams for NGC 4254, the most actively star forming galaxy in our sample. This galaxy has been chosen because it spans the largest range in colours with its pixels and as such it best illustrates the result. Similar lack of correlation is obtained for all other galaxies. It is worth noting that, although these diagrams indicate that optical colours are independent of mid IR colours for the pixels inside each individual galaxy, the pixels are typically clustered around the barycentre roughly corresponding to the integral colours of the galaxy and do not cover the full parameter space. We find that indeed when all galaxies are put together on these diagrams they form much more correlated sequences, thus lifting the seeming contradiction between the known global colour-colour correlations and the lack thereof that we see on local scales.

The study of local colour-colour correlations *within* individual galaxies indicates that local SEDs are essentially a two-parameter family: one parameter controls the relative emission of hot dust and PAHs in the mid IR and the other controls the shape of the

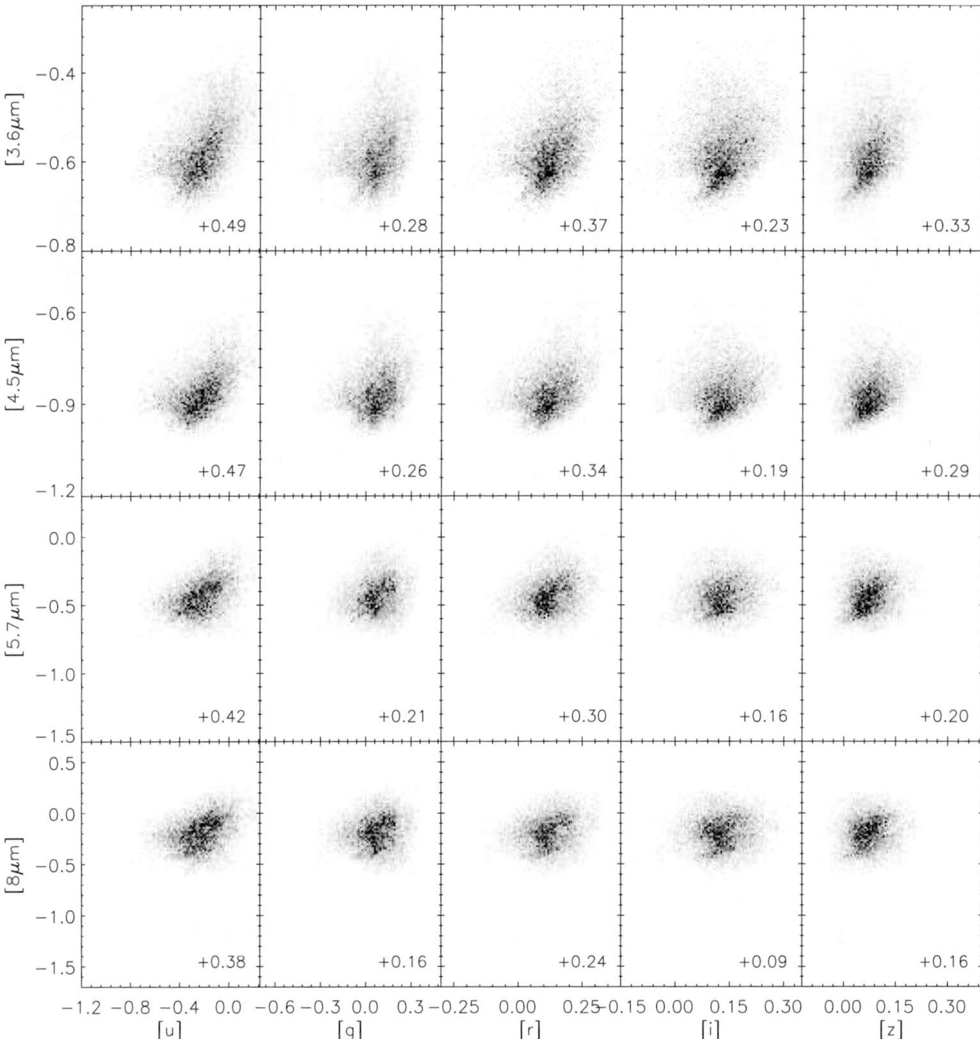

Figure 1. Mid IR–NIR colours vs optical–NIR colours for individual pixels in NGC 4254. The very low degree of correlation between the two spectral regions is also quantified by the low Spearman rank correlation coefficients, in the bottom right corner of each panel. Axis units are $\log(\nu_X f_\nu(X)/\nu_H f_\nu(H))$, X being the band in the axis title.

optical continuum. We have explored this hypothesis by means of a Principal Component Analysis (PCA) of the pixels SEDs in each galaxy independently. We find indeed that most of the variance is accounted for by the first two principal components (PC), one (PC1) mainly connected to the relative mid IR emission and the other (PC2) to the opical-NIR colours. By mapping the coefficients of PC1 and PC2 we see that they feature distinct galaxy components, such as the bulge, spiral arms, inter-arm regions, young stellar associations and star forming regions. This shows that the optical-IR colour correlations break down already on the scale of this structure and are not just due to stochastic effects nor noise. In order to gain physical insight into the meaning of these PCs, we construct diagrams like those presented in Fig. 2, relative to NGC 4254, for all galaxies. As a function of the surface brightness (SB) in H-band ($\approx M^*$) and in Hα ($\approx$ SFR) we colour code the median coefficient for PC1 and PC2 in the left and right panels, respectively.

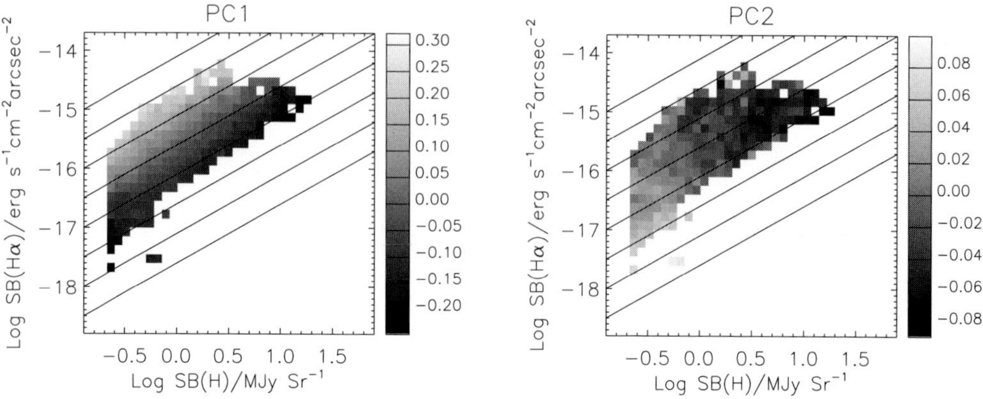

Figure 2. Median coefficients of the two principal components of the local SEDs in NGC 4254 as a function of local surface brightness in H-band($\approx M^*$) and in Hα($\approx$ SFR). Diagonal lines are lines of constant SB(Hα)/SB(H)$\approx$specific SFR.

The strength of PC1 (left panel) is tightly correlated to the specific SFR, in that curves of equal PC1 strength are roughly parallel to the lines of constant ratio of Hα over H-band intensity. PC2 strength (right panel) is essentially a measure of blue vs red optical-NIR colours and displays only a mild dependence on SB(H), in line with the well know age gradients in spiral galaxies (e.g. MacArthur *et al.* 2004) which produce bluer colours in the outer parts of lower SB. This picture is similar for most galaxies with some ongoing star formation, while completely passive galaxies do not have much variation in the mid IR flux. Galaxies like NGC 3521 however show that dust scattering in inclined discs can play a significant role and destroy the PCA properties described so far.

In summary, we find that locally optical-NIR colours are largely uncorrelated from the relative mid IR emission, thus showing that a high surface density of specific star formation does not necessarily imply blue optical colours typical of young stellar populations. In fact, on the one hand a substantial part of the (blue) radiation from young stars can be reddened by dust in correspondence to star forming regions, while on the other hand blue stellar populations may well be present in regions where the star formation has terminated since several hundreds Myr. However, globally these effects average to a relation between current specific SFR and optical-NIR colours. We argue that this is a consequence of the global SF history of a galaxy being relatively smooth on the timescales for such colours to evolve, i.e. $\lesssim 1$ Gyr, such that relatively young stars of age $\gtrsim 100$ Myr are always accompanied by current SF and also the SF episodes extend longer than the times it takes to liberate the young stars from the obscuring birth cloud.

Tight universal mid IR colour correlations. We show that pixels from *all* galaxies form extremely tight sequences in the mid IR (IRAC) colour-colour plots with typical scatter of a few percent only around the median relations. In practice, at (almost) each pixel of any of our galaxies, given the near IR flux (H band) and the mid IR flux in one band (e.g. 8 μm) it is possible to predict the flux in the other three IRAC bands to percent accuracy using a universal fitting formula (see Zibetti & Groves 2011). These relations also highlight the fact that not even at 3.6 or 4.5 μm is the stellar emission pure from any dust contaminant. The reason for such a small scatter is currently unclear, although it might be partly related to the small spread in metallicity of our sample. Moreover, by extrapolating these relations to very low mid IR flux relative to H, we are able to estimate pure stellar colours in the mid IR, which can be extremely useful to separate stellar from dust emission at these wavelengths.

References

Blanton, M. R., *et al.* 2003, *ApJ*, 594, 186
Calzetti, D. *et al.* 2007, *ApJ*, 666, 870
Gavazzi, G., Boselli, A., Donati, A., Franzetti, P., & Scodeggio, M. 2003, *A&A*, 400, 451
Hogg, D. W., Tremonti, C. A., Blanton, M. R., Finkbeiner, D. P., Padmanabhan, N., Quintero, A. D., Schlegel, D. J., & Wherry, N. 2005, *ApJ*, 624, 162
Kennicutt, R. C., *et al.* 2003, *PASP*, 115, 928
Lawrence, A., *et al.* 2007, *MNRAS*, 379, 1599
MacArthur, L. A., Courteau, S., Bell, E., & Holtzman, J. A. 2004, *ApJS*, 152, 175
York, D. G., *et al.* 2000, *AJ*, 120, 1579
Zibetti, S. & Groves, B. 2011, *MNRAS*, 417, 812

Discussion

DA CUNHA: Is the very tight correlation you find between the 8 and 5.8 μm fluxes an indication of a constant ratio between PAH bands?

ZIBETTI: The relation between the fluxes at 8 and 5.8 μm is indeed very tight, with an rms scatter of only a few percent, much less than the $\sim 30\%$ PAH variation expected from, for example, the Draine & Li models. It should be noted that galaxies along the sequence contribute with a very small overlap, such that the small scatter could be a consequence of a roughly homogeneous state inside individual galaxies (especially metallicity). However, all galaxies are very well alligned along the sequence and this supports further the idea of a constant PAH ratio among the galaxies of this small sample.

The Spectral Energy Distribution of Galaxies
Proceedings IAU Symposium No. 284, 2011
R.J. Tuffs & C.C. Popescu, eds.

doi:10.1017/S1743921312008873

Spectral Energy Distributions of a set of H II regions in M33 (HerM33es)

Monica Relaño[1], Simon Verley[1], Isabel Pérez[1], Carsten Kramer[2], Manolis Xilouris[3], Médéric Boquien[4], Jonathan Braine[5], Daniela Calzetti[6], Christian Henkel[7], and the HerM33es Team

[1]Dept. de Física Teórica y del Cosmos, Universidad de Granada, Spain, [2]Instituto de Radioastronomía Milimétrica, Granada, Spain, [3]Institute of Astronomy and Astrophysics, NOA, Athens, Greece, [4]Laboratoire d'Astrophysique de Marseille, Marseille, France, [5]Laboratoire d'Astrophysique de Bordeaux, France, [6]Deparment of Astronomy, University of Massachusetts, USA, [7]MPI für Radioastronomie, Bonn, Germany

email: `mrelano@ugr.es`

Abstract. Within the framework of the HerM33es Key Project for Herschel and in combination with multi-wavelength data, we study the Spectral Energy Distribution (SED) of a set of H II regions in the Local Group Galaxy M33. Using the Hα emission, we perform a classification of a selected H II region sample in terms of morphology, separating the objects in *filled*, *mixed*, *shell* and *clear shell* objects. We obtain the SED for each H II region as well as a representative SED for each class of objects. We also study the emission distribution of each band within the regions. We find different trends in the SEDs for each morphological type that are related to properties of the dust and their associated stellar cluster. The emission distribution of each band within the region is different for each morphological type of object.

Keywords. (ISM:) dust, ISM: evolution, (ISM:) H II regions, galaxies: M33.

1. Introduction

The study of the star formation rate (SFR) in galaxies of different types has been lately improved due to the new available data. Remarkably, the focus has been turned to the closest galaxies as well as to star-forming regions within our Galaxy. For these nearby objects, the resolution of the data offers us an opportunity to test whether the proposed SFR calibrators trace indeed the location of the stellar births (Churchwell *et al.* 2006; Relaño & Kennicutt 2009, among others). Recently, a new study on the star-forming regions in the Magellanic Clouds has analysed the relation of the amount of flux at the different wavelengths via the SEDs of these objects (Lawton *et al.* 2010).

Within the HerM33es Key Project (Kramer *et al.* 2010) we are obtaining maps of the entire galaxy M33 at wavelengths between 100 μm and 500 μm using PACS and SPIRE instruments on Herschel. Verley *et al.* (2010) identified a set of H II regions in the north of M33 showing a shell-like morphology in these infrared bands, and also traced by the Hα emission. In order to further study this phenomenon we have analysed the SEDs of a set of H II regions in M33 covering different morphologies. The resolution of our data (from $\sim 2''$ to $\sim 20''$) is good enough to perform such as study.

2. SED of H II regions

We use an Hα image of M33 (Hoopes & Walterbos 2000) to identify H II regions for which clear morphology can be recognised. A morphological classification was obtained as follows: *filled* regions are objects showing a compact knot, *mixed* regions are

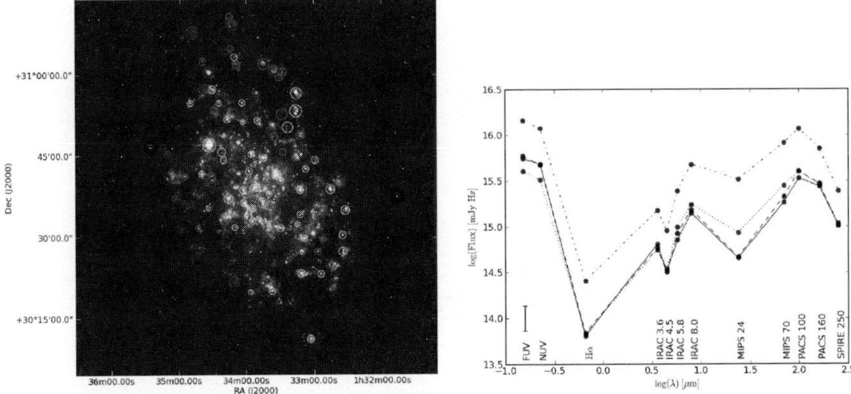

Figure 1. Left: Location of the H II region sample on the continuum-subtracted Hα image of M33 (Hoopes & Walterbos 2000). Circles correspond to the apertures used to perform the photometry. Right: SEDs for our set of H II regions: filled (dotted), mixed (dashed-dotted), shells (solid), and clear shells (dashed). Typical errors are shown in the lower left corner of the figure.

those presenting several compact knots and filamentary structures, and *shells* are regions showing arcs. We add another classification for the most spherical, closed shells called *clear shells*. Out of the 120 selected H II regions, 9 are filled, 47 mixed, 37 shell and 27 clear shells. Our sample is distributed over the whole disk of M33 (see Fig. 1, left).

We use multi-wavelength data from FUV (GALEX) to 250 μm (Herschel) smoothed to a common 20″ resolution and regridded to the 6″ pixel size of the 250 μm Herschel image. Photometry was performed using individual apertures for each object and local background was subtracted to eliminate the contribution of the diffuse medium to the H II region fluxes. In Fig. 1 (right) we show the SEDs for our objects together with a *characteristic* SED for each classification obtained with the mean values in each band for all the H II regions in the corresponding classification. From Fig. 1 (right) we observe that: (i) mixed regions are more luminous in all bands as they normally have several knots of star formation, (ii) the slope of the SED between the FUV-NUV wavelength range and Hα is steeper for shells and clear shells than for filled and mixed, (iii) filled and mixed objects have more 24 μm relative to 8 μm than the shells and clear shells do.

3. Dust Temperature

The 100 μm/70 μm, 160 μm/70 μm or 160 μm/100 μm ratios normally trace the temperature of the warm dust emitting from 24 μm to 160 μm. In Fig. 2 (left) we plot the 100 μm/70 μm ratio versus the Hα surface brightness for our sample. At high Hα surface brightness the 100 μm/70 μm flux density ratio decreases showing that highly luminous H II regions tend to have warmer dust. This agrees with the correlation observed by Boquien *et al.* (2010, 2011) in M33, and in a similar study by Bendo *et al.* (2011) who observed M81, M83 and NGC 2403 with Herschel.

However, the filled regions seem to have a constant 100 μm/70 μm ratio, independent of the Hα surface brightness. For shells and clear shells there is a dispersion in the 100 μm/70 μm ratio showing that these regions present a range of dust temperatures. For the shells, the relative location between stars and dust can affect more the temperature of the dust than the intensity of the stellar radiation field and therefore they tend to present a wider dust temperature range.

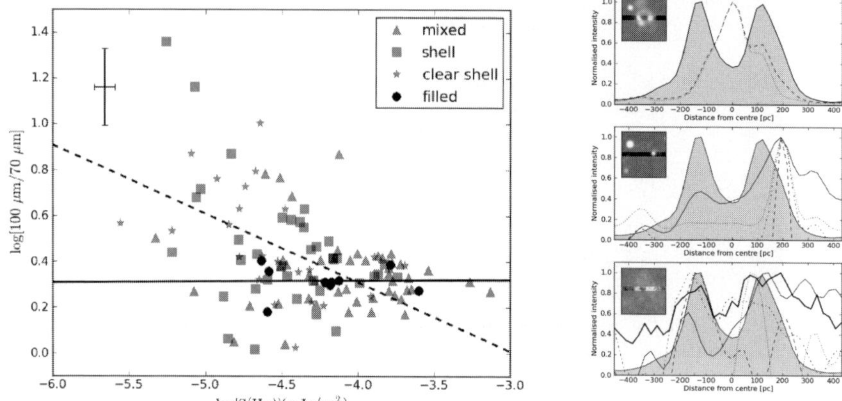

Figure 2. Left: 100 μm/70 μm ratio versus the Hα surface brightness for our sample. Right: Normalised emission line profiles for a shell region in the horizontal direction. Top: Hα (solid line), FUV (dotted), and NUV (dashed), middle: 3.6 μm (dotted line), 4.5 μm (dashed-dotted), 5.8 μm (dashed), and 8.0 μm (solid) and bottom: 24 μm (dotted), 70 μm (dashed-dotted), 100 μm (dashed), 160 μm (solid), and 250 μm (thick solid)). At top-left corner in each panel we show the location of the profile in the region in the Hα, 4.5 μm, and 250 μm images for the top, middle and bottom panels, respectively. The Hα profile is depicted in grey in all the panels for reference.

4. Emission line profiles for clear shells

We have performed a multi-wavelength study of the emission distribution in the interior of the clear shells. From FUV to 250 μm we have obtained profiles in the horizontal (East-West) direction crossing the centre of the H II regions. Each profile corresponds to the integration of a line of 4 pixels ($\sim 24''$) width perpendicular to the direction of the profile.

In Fig. 2 (right) we show the emission line profile for one of the clear shells of our sample. The Hα profile shows the characteristic double peak of the shell emission, and there is a displacement between the Hα and FUV/NUV emission, with the FUV/NUV emission located in the inner part of the region. The ratio Hα/FUV is lower in the centre than in the boundaries of the shell, which could be due to: (i) the existence of a young stellar population within the shell, or (ii) the ionising photons from the central cluster reaching the shell and ionising the gas within the rim. The emission at all IR bands follows clearly the Hα shape of the shell: at 24 μm and 250 μm the emission decays in the centre of the shell and they are enhanced at the boundaries. The same trend, though not so clear, is seen at 70 μm, 100 μm, and 160 μm. The emission of the Polycyclic Aromatic Hydrocarbon molecules (PAH) at 8 μm is marked by the location of the shell boundaries.

References

Bendo, G. J., Boselli, A., Dariush, A. *et al.* 2011, *astro-ph*, 1109.0237
Boquien, M., Calzetti, D., Kramer, C. *et al.* 2010, *A&A*, 518, 70
Boquien, M., Calzetti, D., Combes, F. *et al.* 2011, *AJ*, 142, 111
Churchwell, E., Povich, M. S., Allen, D. *et al.* 2006, *ApJ*, 649, 759
Hoopes, C. G. & Walterbos, R. A. M. 2000, *ApJ*, 541, 597
Kramer, C., Buchbender, C., Xilouris, E. M. *et al.* 2010, *A&A*, 518, 67
Lawton, B., Gordon, K. D., Babler, B. *et al.* 2009, *ApJ*, 716, 453
Relaño, M. & Kennicutt, R. C. Jr. 2009, *ApJ*, 699, 1125
Verley, S., Relaño, M., Kramer, C. *et al.* 2010, *A&A*, 518, 68

The Spectral Energy Distribution of Galaxies
Proceedings IAU Symposium No. 284, 2011
R.J. Tuffs & C.C. Popescu, eds.

doi:10.1017/S1743921312008885

Variation in the dust spectral index across M33

Fatemeh S. Tabatabaei[1], Jonathan Braine[2], Carsten Kramer[3], Manolis Xilouris[4], Mederic Boquien[5], Simon Verley[6], Eva Schinnerer[1], Daniela Calzetti[7], Francoise Combes[8], Frank Israel[9], Christian Henkel[10] and The HerM33es Team

[1]Max-Planck-Institut für Astronomie, Königstuhl 17, 69117-Heidelberg, Germany
[2]Laboratoire d'Astrophysique de Bordeaux, Université de Bordeaux, 33271 Floirac, France
[3]Instituto Radioastronomia Milimetrica, 18012 Granada, Spain
[4]Institute of Astronomy and Astrophysics, National Observatory of Athens, GR-15236 Athens, Greece
[5]Laboratoire d'Astrophysique de Marseille, UMR 6110 CNRS, Marseille, France
[6]Department Física Teórica y del Cosmos, Universidad de Granada, Spain
[7]Department of Astronomy, University of Massachusetts, Amherst, MA 01003, USA
[8]Observatoire de Paris, 75014 Paris, France
[9]Sterrewacht Leiden, Leiden University, PO Box 9513, 2300 RA, Leiden, The Netherlands
[10]Max-Planck-Institut für Radioastronomie, Auf dem Hügel 69, 53121-Bonn, Germany

email: taba@mpia.de

Abstract. Using the Herschel PACS and SPIRE FIR/submm data, we investigate variations in the dust spectral index β in the nearby spiral galaxy M33 at a linear resolution of 160 pc. We use an iteration method in two different approaches, single and two-component modified black body models. In both approaches, β is higher in the central disk than in the outer disk similar to the dust temperature. There is a positive correlation between β and Hα as well as with the molecular gas traced by CO(2-1). A Monte-Carlo simulation shows that the physical parameters are better constrained when using the two-component model.

Keywords. Interstellar Medium, Star formation, Dust, Galaxies: M33

---

1. Introduction

Inferring the nature of dust and dust heating requires knowledge about the dust emissivity, and its power law index. There is some evidence supporting variation of dust emissivity with environmental conditions (e.g. Lisenfeld *et al.* 2000, Paradis *et al.* 2009). The dust spectral index could change depending on grain properties, e.g. geometry, size distribution, and chemical composition. Dust grains might be affected by different physical processes like shattering, sputtering, grain - grain collision, condensation and coagulation. Hence, depending on recent star formation activities, dust/gas content and metallicity changes across galaxies, variations in the dust spectral index in inner/outer disks or arm/inter-arm regions may occur. Herschel observations of the nearest Scd galaxy M33, (HerM33es, Kramer *et al.* 2010, Boquien *et al.* 2010) ideally enable us to derive distribution of β across M33.

We use the Herschel PACS & SPIRE data at 160, 250, 350, and 500μm together with the Spitzer MIPS data at 70μm to investigate the physical properties of a standard dust model (Big grains+vsg+PAH, Krügel 2003 and references therein). These wavelengths

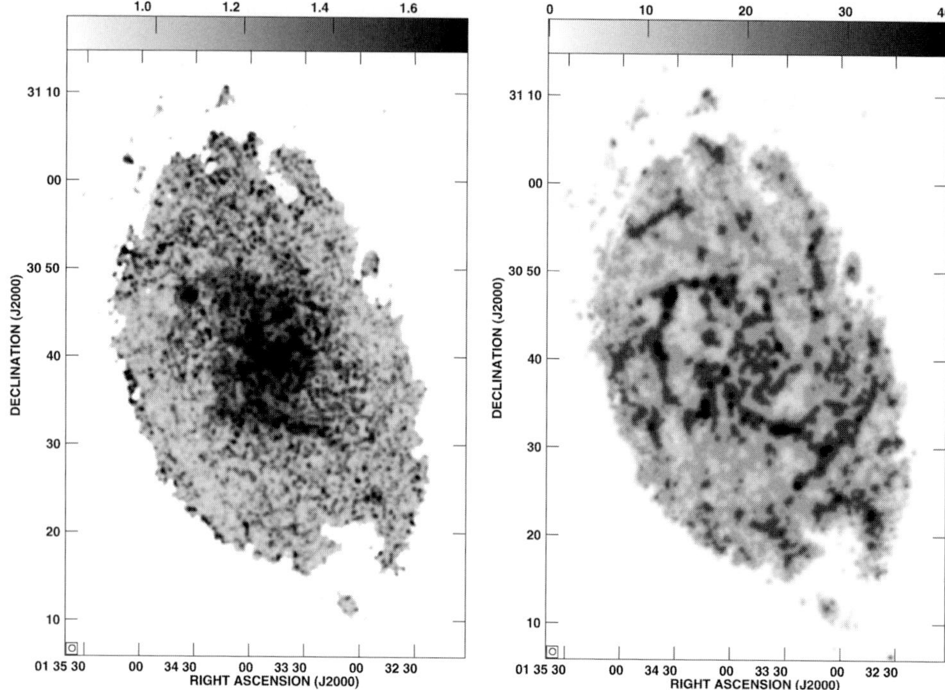

Figure 1. Distribution of dust spectral index (right) and column density (in μg cm^{-2}, right) across M33, based on the single-component approach.

spread over a range where the dust emission mainly emerges from the big grains, which are believed to be in thermal equilibrium with the interstellar radiation field.

2. Analysis and Results

We use the Newton-Raphson iteration method to derive T, β and column density, pixel by pixel, in two different approaches:
1) We assume that dust emitting at the SPIRE bands is mainly heated by the interstellar radiation field (ISRF) and obeys single - component gray body radiation. In this case we use the data at longer wavelengths (250, 350 and 500μm).
2) We consider another dust component heated by young massive stars with higher temperature (warm dust) besides the dust heated by the ISRF (cold dust), hence, dust emits as a two-component gray body. Here, we use the data from 70μm to 500μm wavelengths. Using a Monte-Carlo simulation, we further investigate the reliability of each approach and estimate the errors and uncertainties. We derive the following results:

Single Component Approach As shown in Fig. 1, in the central disk (R < 4 kpc), the dust spectral index is higher in the central disk (β >1.4) than in the outer disk. Similarly, both T and column density are higher in the central disk. The flocculent spiral structure of M33 is evident in the column density map. The average T and β for the entire galaxy are 22±9 K and 1.3±0.5, respectively.
Two Component Approach The cold and warm dust components are separated assuming a variable β for the dominant colder component. Although we have to use a fixed value of $\beta_{\rm warm}$ for the warm dust, we repeat the analysis for different possibilities of $\beta_{\rm warm}$ = 1, 1.5, and 2. Interestingly, we find that β is again higher in the central

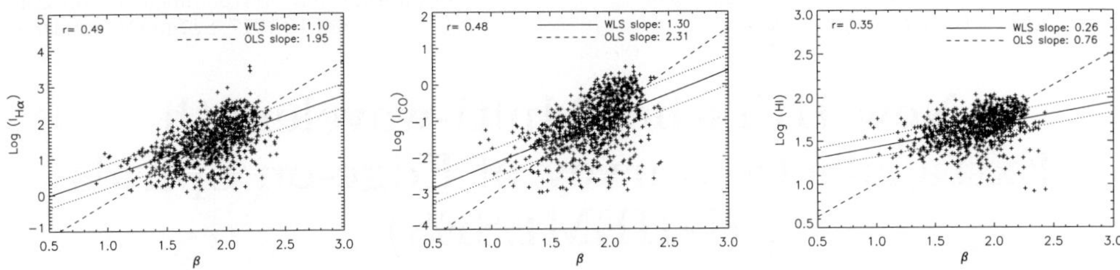

Figure 2. Correlations between the dust spectral index and Hα, CO(2-1), and HI integrated intensities. The WLS weighted least square fit (solid line) and its 5-sigma confidence levels (dotted lines) are shown. The dashed lines show the ordinary least square fit (OLS).

disk and even shows the arm/inter-arm contrast, irrespective of $\beta_{\rm warm}$ assumed. Also in all three cases, the cold dust follows spiral arms and dust lanes and is denser where the temperature is lower. The warm dust (both temperature and column density) follows the distribution of Hα and star forming regions. For the entire galaxy, we derive $T_{\rm cold} = 16 \pm 3$ K, $T_{\rm warm} = 40 \pm 13$ K, and $\beta = 1.7 \pm 0.4$ assuming $\beta_{\rm warm} = 2$. Our simulations show that the two-component approach, generally, constrains the parameters better than the single-component approach, particularly the temperature.

3. Correlation with ISM Tracers

The M33's outer/inner disk and even the spiral arms are visible in the β-maps. Having similar distribution as of the molecular gas and radiation field traced by Hα, β seems to have a meaningful change across the galaxy. Figure 2 shows correlations between β and Hα, CO(2-1) and HI intensities (Gratier *et al.* 2010) as tracers of the ionized, molecular, and atomic phases of the ISM, respectively. In both approaches, β is found to be better correlated with both Hα and CO than with HI. This could indicate that β is not the same in star forming and non-starforming regions in a galaxy. Star formation could in principle influence the grain composition (via a change in metallicity) and size distribution (via shattering of dust grains due to strong Shocks), which could modify the dust emissivity index i.e. β.

References

Boquien, M., Calzetti, D., Kramer, C. *et al.* 2010, *A&A*, 518, 70
Gratier P., Braine, J., Rodriguez-Fernandez, N. J. *et al.* 2010, *A&A*, 522, 3
Kramer, C., Buchbender, C., Xilouris, E. M. *et al.* 2010, *A&A*, 518, 67
Krügel, E. 2003, *The Physics of Interstellar Dust* (IoP Series in astronomy and astrophysics, ISBN 0750308613)
Lisenfeld, U., Isaak, K. G., & Hills, R. 2000, *MNRAS*, 312, 433
Paradis, D., Bernard, J.-Ph., & Mény, C. 2009, *A&A*, 506, 745

The Spectral Energy Distribution of Galaxies
Proceedings IAU Symposium No. 284, 2011
R.J. Tuffs & C.C. Popescu, eds.

doi:10.1017/S1743921312008897

New HErschel Multi-wavelength Extragalactic Survey of Edge-on Spirals (NHEMESES)

B. W. Holwerda[1], S. Bianchi[2], M. Baes[3], R. S. de Jong[4], J. J. Dalcanton[5], D. Radburn-Smith[5], K. Gordon[6] and M. Xilouris[7]

[1]European Space Agency (ESTEC), Keplerlaan 1, 2200 AV Noordwijk, The Netherlands
[2]INAF-Arcetri Astrophysical Observatory, Florence, [3]University of Gent, [4]Astronomisch Insitüt Potsdam, [5]University of Washington, [6]Space Telescope Science Institute, [7]National Observatory of Athens

email: `benne.holwerda@esa.int`

Abstract. Edge-on spiral galaxies offer a unique perspective on the vertical structure of spiral disks, both stars and the iconic dark dustlanes. The thickness of these dustlanes can now be resolved for the first time with *Herschel* in far-infrared and sub-mm emission. We present NHEMESES, an ongoing project that targets 12 edge-on spiral galaxies with the PACS and SPIRE instruments on *Herschel*. These vertically resolved observations of edge-on spirals will impact on several current topics.

First and foremost, these *Herschel* observations will settle whether or not there is a phase change in the vertical structure of the ISM with disk mass. Previously, a dramatic change in dustlane morphology was observed as in massive disks the dust collapses into a thin lane. If this is the case, the vertical balance between turbulence and gravity dictates the ISM structure and consequently star-formation and related phenomena (spiral arms, bars etc.). We specifically target lower mass nearby edge-ons to complement existing *Herschel* observations of high-mass edge-on spirals (the HEROES project).

Secondly, the combined data-set, together with existing *Spitzer* observations, will drive a new generation of spiral disk Spectral Energy Distribution models. These model how dust reprocesses starlight to thermal emission but the dust geometry remains the critical unknown.

And thirdly, the observations will provide an accurate and unbiased census of the cold dusty structures occasionally seen extending out of the plane of the disk, when backlit by the stellar disk. To illustrate the NHEMESES project, we present early results on NGC 4244 and NGC 891, two well studies examples of a low and high-mass edge-on spiral.

Keywords. (ISM:) dust, extinction ISM: structure galaxies: fundamental parameters (typical scales, dust mass) galaxies: individual (NGC 4244, NGC 891) galaxies: ISM galaxies: spiral galaxies: structure infrared: ISM submillimeter

---

Edge-on spiral galaxies offer a unique perspective on spiral disks. An observer can explore the vertical structure of both stars and ISM using different wavelengths. Dust is mechanically linked to the cold ISM (Allen *et al.* 1986; Weingartner & Draine 2001). Dalcanton *et al.* (2004) used the appearance of dust lanes as a probe of vertical stability of spiral disks. They found that in massive spiral disks, the ISM collapses into a thin dustlane, while the less massive disks show more flocculant dust morphology. In addition, observations of edge-ons often reveal dark structures extending out of the plane of massive disks (e.g., Howk 1999), but the quantity of extra-planar dust has been impossible to constrain from extinction measures.

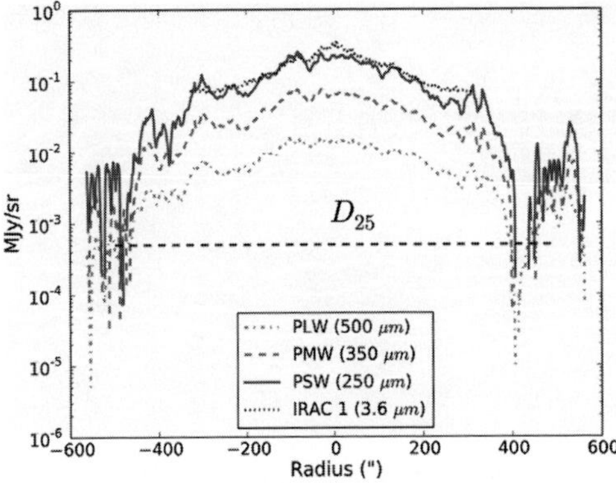

Figure 1. The radial profiles of NGC4244; *Spitzer* 3.6 (dotted line with closely spaced dots), SPIRE 250 (dotted), 350 (dashed) and 500 (solid) μm.

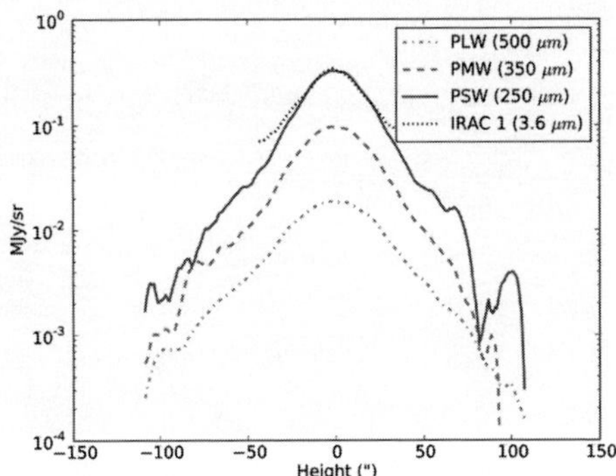

Figure 2. The vertical profile of NGC4244, same legend as Fig. 1. The SPIRE profile stays within the 3.6 μm height.

A comprehensive approach is to model multi-wavelength images of edge-on spirals with a bulge, a stellar and a dust disk to constrain stellar and ISM structure (e.g., Xilouris *et al.* 1999; Bianchi 2008; Baes *et al.* 2010; Popescu *et al.* 2011). However, until now, these models are often degenerate in the vertical distributions of stellar light and ISM due to lack of resolution and wavelength coverage. With the advent of *Herschel*, the vertical structure of some nearby edge-on disks can be now resolved. Our *Herschel* program NHEMESES† (PI B.W. Holwerda) aims to (1) measure the vertical scale of dust in spiral disks (an unknown in the above models), (2) the extra-planar dust component, and (3) whether or not the strong break seen in dust lanes by Dalcanton *et al.* (2004) is indeed a real phase-change in the cold ISM with disk mass.

We show some first NHEMESES results with the SPIRE morphology of NGC 4244, the prototypical disk-dominated low-mass spiral galaxy ($v_{rot} = 95$ kms^{-1}).

The radial profile (Fig. 1) appears to truncate beyond a peak in flux from a small star-formation region on both sides of the disks (see also Fig. 3). The radius of truncation is similar to those initially found by van der Kruit & Searle (1981) and more recently by de Jong *et al.* (2007) for different stellar populations.

The vertical sub-mm profiles (Fig. 2) are similar in width to the *Spitzer* 3.6 μm emission, a good tracer of the stellar mass (e.g., Meidt *et al.*, *this volume*). Comerón *et al.* (2011) find evidence for a thick and thin stellar disk in this galaxy and the dusty ISM appears associated with the inner (thin) disk. In contrast, a second, thicker vertical component in NGC 891 has been reported (Kamphuis *et al.* 2007, Seon *et al.*, this volume).

Fig. 3 shows the colour image based on the 250, 350 and 500 μm SPIRE images with the HI contours from Zschaechner *et al.* (2011); Heald *et al.* (2011). The dust emission is restricted to the highest HI contour; it is a single, concentrated disk. This is in contrast to the more massive NGC 891, where Popescu & Tuffs (2003); Bianchi & Xilouris (2011) find evidence for dust throughout the HI disk and extended envelope (Oosterloo *et al.* 2007).

The fit to the Spectral Energy Distribution (SED) using the model of MacLachlan *et al.* (2011) (Fig. 4) finds an optically thin disk with a similar scale-height for both dust and stars (as did Seth *et al.* 2005, based on stellar populations). MacLachlan *et al.* report

† New **HE**rshel **M**ulti-wavelength **E**xtragalactic **S**urvey of **E**dge-on **S**pirals, a 10.3 hour, OT1, priority 2 program to supplement the GT program on large disks, **HER**shel **O**bservations **E**dge-on **S**pirals, PI M. Baes.

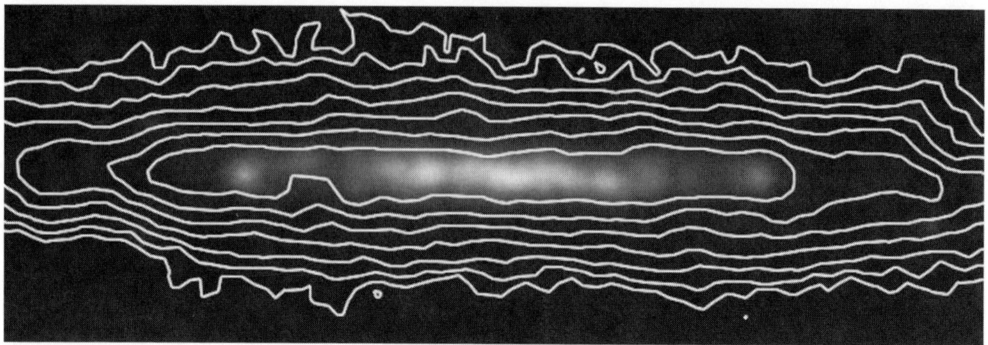

Figure 3. A three-colour image of the SPIRE observations of NGC 4244 with 250, 350 and 500 μm and the HI contours from Zschaechner *et al.* (2011) (contours at 6.4×10^{19} cm^{-2} increasing by factors of 2). The SPIRE flux is contained mostly within the 8.1×10^{19} cm^{-2} contour.

a dust mass of $2.38 \times 10^6 M_\odot$ and a dust scale-length 1.8 times its stellar scale length. We find that their predicted sub-mm fluxes are lower than what we observe, possibly because the dusty ISM is much clumpier than the smooth diffuse ISM they assume for the SED model. However, we find that their high value of the scale-length of the dusty disk is comparable with the radial profile of sub-mm emission we observe (Fig. 1).

We find that dust in NGC 4244 is distributed vertically throughout the (thin) stellar disk, in accordance with the prediction from Dalcanton *et al.* (2004), but that, unlike the much more massive NGC 891, there is little evidence for a second dust component associated with the outer HI envelope. SED models for NGC 4244 are converging to a solution in which the large-scale distribution (in height and length) for the dust is comparable to the stellar ones, but distributed almost exclusively in clumps. This different distribution of the dusty geometry for lower-mass spiral disks may help explain the colder dust temperature and higher relative dust mass observed by H-ATLAS (Dunne *et al.* 2011, Dunne *et al. this volume*).

The NHEMESES project aims to observe 12 low-mass spirals with PACS and SPIRE on board *Herschel*. Combined with existing *Herschel* observations of massive edge-on

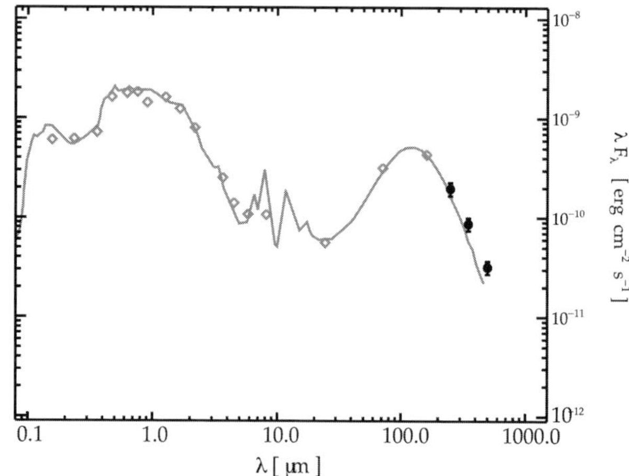

Figure 4. The Spectral Energy Distribution model from MacLachlan *et al.* (2011) (see also MacLachlan *et al.* (*this volume*) and the SPIRE fluxes for NGC 4244. The SED under-predicts the sub-mm fluxes by only a factor 2.

spiral galaxies, we aim to (1) ascertain if the ISM indeed goes through a phase-change at 120 km/s, (2) take a census of dusty outflows in nearby, quiescent, spiral disks and (3) provide a suite of multi-wavelength data (in combination with SDSS and *Spitzer* observations) to serve as a benchmark for SED models of edge-on spirals galaxies.

References

Allen, R. J. *et al.* 1986, *Nature*, 319, 296
Baes, M. *et al.* 2010, *A&A* 518, L39+
Bianchi, S. 2008, *A&A*, 490, 461
Bianchi, S. & Xilouris, E. M. 2011, *A&A*, 531, L11+
Comerón, S. *et al.* 2011, *apj*, 729, 18
Dalcanton, J. J. *et al.* 2004, *ApJ*, 608, 189
de Jong, R. S. *et al.* 2007, *ApJ*, 667, L49
Dunne, L. *et al.* 2011, *MNRAS*, 1395
Heald, G. *et al.* 2011, *ArXiv e-prints*
Howk, J. C. 1999, *Ap&SS*, 269, 293
Kamphuis, P. *et al.* 2007, *A&A*, 471, L1
MacLachlan, J. M. *et al.* 2011, *ArXiv e-prints*
Oosterloo, T. *et al.* 2007, *AJ*, 134, 1019
Popescu, C. C. & Tuffs, R. J. 2003, *A&A*, 410, L21
Popescu, C. C. *et al.* 2011, *A&A*, 527, A109+
Seth, A. C. *et al.* 2005, *AJ*, 129, 1331
van der Kruit, P. C. & Searle, L. 1981, *A&A*, 95, 105
Weingartner, J. C. & Draine, B. T. 2001, *ApJ*, 553, 581
Xilouris, E. M. *et al.* 1999, *A&A*, 344, 868
Zschaechner, L. K. *et al.* 2011, *ArXiv e-prints*

The Spectral Energy Distribution of Galaxies
Proceedings IAU Symposium No. 284, 2011
R.J. Tuffs & C.C. Popescu, eds.

doi:10.1017/S1743921312008903

Ionization of the diffuse gas in galaxies: hot low-mass evolved stars at work

N. Flores-Fajardo[1], C. Morisset[2,3], G. Stasińska[4] and L. Binette[2]

[1]Centro de Radioastronomía y Astrofísica, Universidad Nacional Autónoma de México; Antigua Carretera a Pátzcuaro no. 8701 Col. Ex Hacienda San José de la Huerta, Morelia, Michoacán C.P. 58089, México

[2]Instituto de Astronomía, Universidad Nacional Autónoma de México; Apdo. postal 70–264; Ciudad Universitaria; México D.F. 04510; México.

[3]Instituto de Astrofísica de Canarias; C/ Vía Láctea, s/n, E38205, La Laguna (Tenerife), España

[4]LUTH, Observatoire de Paris, CNRS, Université Paris Diderot; Place Jules Janssen 92190 Meudon, France

email: nahieflores@gmail.com

Abstract. We revisit the question of the ionization of the diffuse medium in late type galaxies, by studying NGC 891. The most important challenge for the models considered so far was the observed increase of [O III]/Hβ, [O II]/Hβ, and [N II]/Hα with increasing distance to the galactic plane. We propose a scenario based on the *expected* population of massive OB stars and hot low-mass evolved stars (HOLMES) in this galaxy to explain this observational fact. In the framework of this scenario we construct a finely meshed grid of photoionization models. For each value of the galactic latitude z we look for the models which simultaneously fit the observed values of the [O III]/Hβ, [O II]/Hβ, and [N II]/Hα ratios. For each value of z we find a range of solutions which depends on the value of the oxygen abundance. The models which fit the observations indicate a systematic decrease of the electron density with increasing z. They become dominated by the HOLMES with increasing z only when restricting to solar oxygen abundance models, which argues that the metallicity above the galactic plane should be close to solar. They also indicate that N/O increases with increasing z.

Keywords. galaxies: individual (NGC 891) — galaxies: ISM — galaxies: abundances — stars: AGB and post-AGB

---

1. Introduction

The Diffuse Ionized Medium (DIG) was detected through its optical line emission outside the classical H II regions (Reynolds 1971) and turns out to be a major component of the interstellar medium in galaxies (Reynolds 1991). Most specialist agree that OB stars in galaxies likely represent the main source of ionizing photons for the DIG (see Haffner *et al.* 2009). However, an additional ionizing source is suggested by reported increase of such emission line ratios as [N II]/Hα, [S II]/Hα with galactic height. Several sources of additional ionization/heating have been suggested without complete success.

2. Proposed Scenario

The extraplanar DIG is destributed in clouds (for simplicity represented by rectangles) that are ionized by two star populations:1) OB stars located in the thin disk and whose ionizing radiation escape from the disk through small "holes", which ionize the "bottom" part of the gas clouds. They constitute only a small fraction of the entire population of

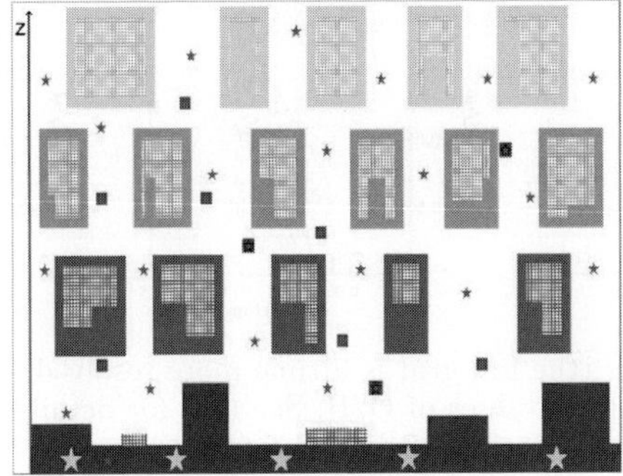

Figure 1. Schematic representation of the extraplanar gas and ionizing stars. Ionized gas is represented in plain shade and neutral gas in a pattern shade. For both components, a darker shade indicates gas of higher density. The big stars in the plot represent OB stars. The small stars represent the HOLMES, which can be either central stars of present-day planetary nebulae, or hot pre-white dwarfs.

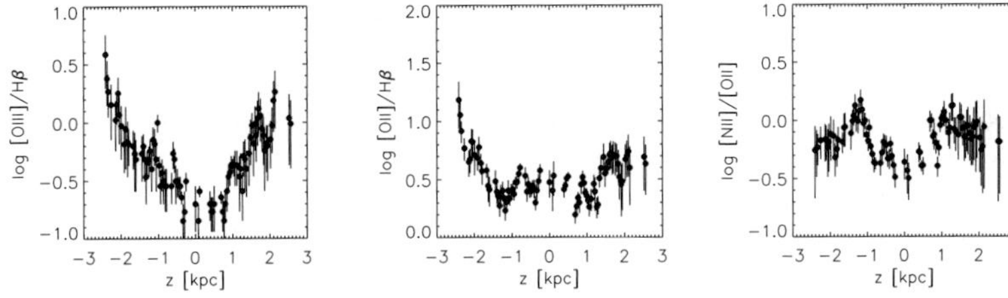

Figure 2. Observed values of [O III]/Hβ, [O II]/Hβ and [N II]/[O II]in NGC 891, as a function of the distance to the galactic plane. The data are from Otte *et al.* (2001).

the OB stars in the disk. 2) HOLMES, which are distributed in the galaxy thick disk and halo. Their influence with respect to that of OB stars increases away from the galacti plane. Figure 1 shows the datails of the proposed scenario.

3. Observational Data

We chose to focus on the edge on spiral galaxy NGC 891, a galaxy that has been extensively observed, especially in optical emission lines (Figure 2), providing the best diagnostics for our scenario.

4. Modeling

The ionizing spectral energy distribution (SED) from OB stars is obtained using the code Starburst99 (Leitherer *et al.* 1999), considering a continuous star formation. The SED of HOLMES is obtained using the code PEGASE (Fioc & Rocca-Volmerange 1997) considering a instantaneous starburst at look back after 10 Gyr. The photoionization models for the DIG were computed with Cloudy (Ferland *et al.* 1998). Each one is defined by the ratio of the surface fluxes: $\Phi_{\rm HOLMES}/\ (\Phi_{\rm OB} + \Phi_{\rm HOLMES})$, the ionization parameter U, O/H and N/O.

Figure 3 shows the location of the observaional points in the [O III]/Hβ vs [O II]/Hβ diagram, with respect to our "reduced" grid of models. The dashes lines join models with same values of log $\Phi_{\rm OB}$ (equal to 3.5, 4, 4.5, 5, 5.5, 6, 6.5), while the continuous line join models with the same values of U (equal to -4, -3.5, -3). Form left to right, each panel corresponds to a different value of log O/H (-3.9, -3.3, -2.9). The value of N/O is the

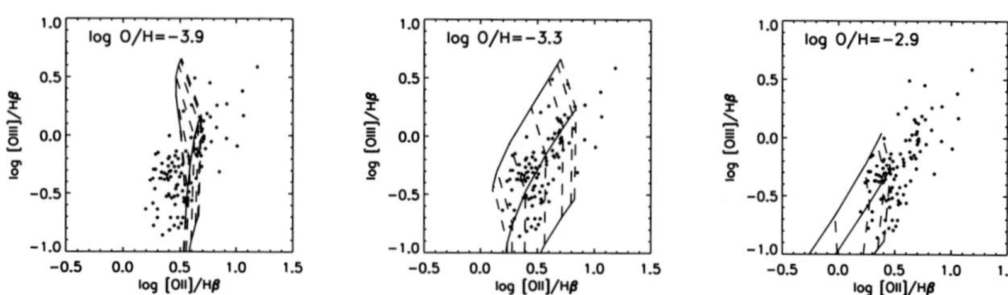

Figure 3. "Reduced" grids (the full grid is 5 time more resolved) of models in the [O III]/Hβ vs. [O II]/Hβ plane, for various values of O/H. See text for details on the varying parameters. The observational points are superimposed on the grid.

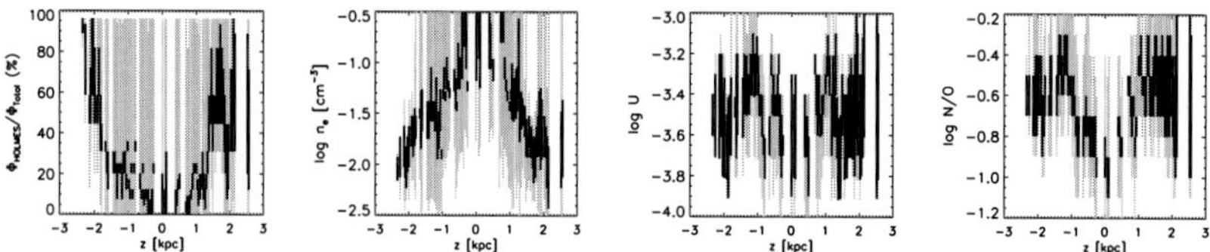

Figure 4. $\Phi_{\rm HOLMES}/\Phi_{\rm OB}$, $n_{\rm e}$, U and N/O vs z for the models that fit the observational [O III]/Hβ, [O II]/Hβ, and [N II]/Hα simultaneously. The light shade bars indicate the range of the values for which the models in the grid fit the observations. The dark bars show the same, but restricting to solar abundances models.

same in all the panels, namely -0.5 dex. The observational points that have the highest values of [O III]/Hβ and [O II]/Hβ correspond to the largest values of $\Phi_{\rm HOLMES}/\Phi_{\rm total}$.

5. Results

For each observational point in the DIG of NGC 891 we select the models of our finely meshed grid that reproduce simultaneously the values of [O III]/Hβ, [O II]/Hβ and [N II]/Hα within the observational uncertainties. We find solutions for each value of z. In Figure 4 we show the acceptable ranges of $\Phi_{\rm HOLMES}/\Phi_{\rm total}$, U, $n_{\rm e}$ and N/O as a function of z. The models clearly indicate a systematic decrease of the electron density with increasing z. If we restrict to solar metallicity (dark bars in Fig. 4), we find that the models which fit the observations become dominated by the HOLMES as z increases. Turning the argument around, this might be an indication that the metallicity of the DIG is roughly solar. Our models also indicate that N/O increases with increasing destance to the galactic plane, at least until $|z| \sim 1.5$ kpc.

6. Conclusions

Our scenario, which considers both the population of OB stars and that of HOLMES, is able to explain the long standing problem of the ionization of the DIG (see Flores-Fajardo *et al.* 2011).

References

Ferland, G. J., Korista, K. T., Verner, D. A., *et al.* 1998, *PASP*, 110, 761

Fioc, M. & Rocca-Volmerange, B. 1997, *A&A*, 326, 950

Flores-Fajardo, N., Morisset, C., Stasińska, G., & Binette, L. 2011, *MNRAS*, 415, 2182

Haffner, L. M., Dettmar, R.-J., Beckman, J. E., *et al.* 2009, *Reviews of Modern Physics*, 81, 969

Leitherer, C., Schaerer, D., Goldader, J. D., *et al.* 1999, *ApJS*, 123, 3

Otte, B., Reynolds, R. J., Gallagher, J. S., III, & Ferguson, A. M. N. 2001, *ApJ*, 560, 207

Reynolds, R. J. 1971, Ph.D. Thesis

Reynolds, R. J. 1991, *The Interstellar Disk-Halo Connection in Galaxies*, 144, 67

The Spectral Energy Distribution of Galaxies
Proceedings IAU Symposium No. 284, 2011
R.J. Tuffs & C.C. Popescu, eds.

doi:10.1017/S1743921312008915

Detection of a large amount of diffuse extraplanar dust in NGC 891

Kwang-Il Seon[1] and Adolf N. Witt[2]

[1]Korea Astronomy and Space Science Institute, Republic of Korea
email: kiseon@kasi.re.kr

[2]Ritter Astrophysical Research Center, University of Toledo, USA
email: awitt@utnet.utoledo.edu

Abstract. Significant discrepancies have been found between the dust masses derived from various tracers (optical/near-IR, far-IR/sub-millimeter observations, and the variation of dust attenuation with viewing angle). Here we report the first detection of the extended far-UV (FUV) and near-UV (NUV) haloes perpendicular to the galactic plane of NGC 891, which can be interpreted as scattered stellar light from the galactic plane. An additional "geometrically thick" dust disk, which contains about the same mass as the standard thin dust disk, is needed to reproduce the vertically extended UV profiles

Keywords. radiative transfer, galaxies: spiral, galaxies: individual (NGC 891)

---

1. Energy balance problem in spiral galaxies

Internal dust attenuation of the photon flux from galaxies reduces the total emergent luminosity of galaxies, introducing an inclination dependence of the observed flux. The amount of dust may then be inferred from the inclination dependence of the dust attenuation. The classical test to examine the opacity of spirals, dust attenuation versus the inclination angle, however, caused debates over whether face-on spiral galaxies are optically thin or thick (Valentijn 1990; Burnstein *et al.* 1991). Motivated by these arguments, realistic radiative transfer models for dusty galaxies were developed and compared with optical/near-IR (NIR) images (e.g., Kylafis & Bahcall 1987; Xilouris *et al.* 1997; Xilouris *et al.* 1998). The result seemed to confirm the conventional viewpoint that spiral galaxies are optically thin. However, comparisons between the optical/NIR and far-IR (FIR)/submm observations have been puzzling. The SED in FIR/submm requires at least a dust mass twice as large as estimated from the radiative transfer model of optical/NIR images (Popescu *et al.* 2000; Bianchi 2008). To resolve this discrepancy, an extra dust mass in the form of a secondary thin (even thinner than the thin disk found from optical/NIR observations) disk + clumpy dust clouds associated with molecular clouds that was supposed to be hidden in optical images was introduced (Popescu *et al.* 2000; Tuffs *et al.* 2004; Bianchi 2008; Popescu *et al.* 2011). However, the model produces a dust lane in K-band that is much stronger than that observed (Dasyara *et al.* 2005; Bianchi 2008).

We note that the previous radiative models assumed that the geometrically thin dust disk and/or the dust clouds are concentrated in the galactic midplane. We may thus infer that the conventional thin-disk picture of dust may have a limitation in explaining the observational facts consistently. In other words, an additional dust component that is existing in a form different from the standard thin disk and containing a significant amount of mass may be needed to overcome the discrepancies.

In fact, there have been various attempts to determine whether large amounts of dust reside outside the galactic plane. In addition, the possibility of dust being repelled away from the galactic plane by radiation pressure of starlight in the disk has also been theoretically investigated (Greenberg *et al.* 1987; Ferrara *et al.* 1990). Filamentary dust structures above the galactic plane have been observed up to $|z| \lesssim 2$ kpc in nearby edge-on spiral galaxies using high-resolution optical images (e.g., Howk & Savage 1997; Thompson *et al.* 2004). The extraplanar dust filaments were found to contain too small an amount of dust ($\sim 1-2\%$ of the total dust mass) to be considered as the additional dust component (Popescu *et al.* 2000). The filamentary features, however, were traced in absorption against the background starlight, thereby implying preferentially "dense" dust features visible only to heights limited by the vertical extent of the background starlight. Therefore, any "diffuse" dust component above the galactic plane was not traceable in the studies and the full vertical extent and overall amount of the diffuse extraplanar dust remained unknown.

2. Search for the vertically extended dust layer

If there is a significant amount of diffuse dust above the galactic plane, the extraplanar dust may provide some, if not all, of the extra dust mass required to explain FIR/submm observations. This possibility was investigated by searching the vertically extended submm (450 and 850 μm) emissions in SCUBA images of NGC 891 (Alton *et al.* 2000a; Alton *et al.* 2000b). However, detection of a significant amount of extraplanar dust was not successful because the instrumental point spread function (PSF) was too broad, made worse by the presence of a wide wing (Alton *et al.* 2000a). Recently, Bianchi & Xilouris (2011) analyzed *Herschel*/SPIRE images of NGC 891 at 250, 350, and 500 μm and found no excess emission above the galactic plane beyond that of the thin, unresolved, disk. However, we should note that the spatial resolution of *Herschel*/SPIRE (18.1″, 24.9″, and 36.4″ FWHM for 250, 350, and 500 μm, respectively) is even worse than that of SCUBA (16″ FWHM).

Here we report the results of a search for the diffuse extraplanar dust based on the fact that the dust should appear as a faint extended reflection nebula illuminated by starlight. The scattered light would not be easily distinguished from direct starlight when the scaleheight of the light source is greater than or comparable to that of the extraplanar dust. Therefore, UV observations of edge-on galaxies can provide the means for detecting scattered light from the diffuse extraplanar dust, because OB stars, the main source of the UV continuum, have a scaleheight < 0.2 kpc and have no bulge or halo component. We thus examined the FUV data of the edge-on galaxy NGC 891 obtained from the*GALEX* mission (Gil de Paz *et al.* 2007), and indeed found vertically extended FUV and NUV emission in the galaxy.

Radiative transfer models of the dust-scattered UV radiation were calculated using a Monte-Carlo simulation code (Seon 2009). We assumed that stars and dust are exponentially distributed in both vertical and radial directions. The dust scattering properties were assumed to be identical to those of Milky Way dust. The models with a single exponential dust disk were obviously not able to produce the vertical profile of the data. Two exponential (geometrically thin + thick) dust disks were needed to reproduce the UV data (Fig. 1). The result indicates that an extraplanar dust disk with a mass as large as the standard thin dust disk, but with a 10 times higher scaleheight, is required to explain the UV profiles.

We note that the thick disk exhibits very little concentration toward the central plane compared to the thin dust disk. The vertical opacity profile of the two dust disks is in

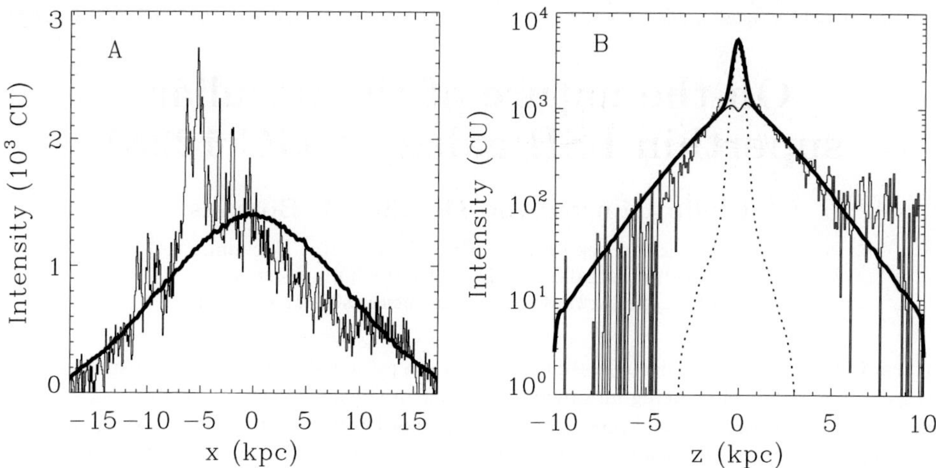

Figure 1. Observed FUV intensity profiles of NGC 891 plotted together with the corresponding profiles from the radiative transfer model (smooth thick solid line). (A) The radial intensity profile and (B) the vertical intensity profile. Smooth thin dotted and solid lines in (B) are the model profiles of directly escaped and scattered radiations, respectively. The broad wing features of the direct starlight in (B) are due to the Galex PSF. Obviously, the effect of the wing of the PSF is negligible. Here, 1 continuum unit (CU) = 1 photons cm^{-2} s^{-1} Å^{-1} sr^{-1}.

fact well represented by a single exponential with a scaleheight of 0.2 kpc, which are consistent with optical/NIR observations, in the central extinction-lane region ($|z| <$ 0.5 kpc) where models of optical/NIR images are mainly fitted. Therefore, the secondary thick disk can be completely hidden from the radiative transfer models of optical/NIR images when only a single dust disk is assumed.

References

Alton, P. B., *et al.* 2000, *A&AS*, 145, 83

Alton, P. B., *et al.* 2000, *A&A*, 356, 795

Bianchi, S., Davies, J. I., & Alton, P. B. 2000, *A&A*, 359, 65

Bianchi, S. 2008, *A&A*, 490, 461

Bianchi, S. & Xilouris, E. M. 2011, *A&A*, 531, L11

Burnstein, D. Haynes, M., & Faber, S. M. 1991, *Nature*, 353, 515

Davies, J. I., Phillipps, S., Boyce, P. J., & Disney, M. J. 1993, *MNRAS*, 260, 491

Dasyra, K. M., *et al.* 2005, *A&A*, 437, 447

Ferrara, A., Ferrini, F., Barsella, B., & Aiello, S. 1990, *A&A*, 240, 259

Gil de Paz, A., *et al.* 2007, *ApJS*, 173, 185

Greenberg, J. M., Ferrini, F., Barsella, B., & Aiello, S. 1982, *Nature*, 327, 214

Howk, J. C. & Savage, B. D. 1997, *AJ*, 114, 2463

Kylafis, N. D. & Bahcall, J. N. 1987, *ApJ*, 317, 637

Popescu, C. C., *et al.* 2000, *A&A*, 362, 138

Popescu, C. C., *et al.* 2011, *A&A*, 527, A109

Seon, K.-I. 2009, *ApJ*, 703, 1159

Thompson, T. W. J., Howk, J. C., & Savage, B. D. 2004, *AJ*, 128, 662

Tuffs, R. J., *et al.* 2004, *A&A*, 419, 821

Valentijn, E. A. 1990, *Nature*, 346, 153

Xilouris, E. M., Kylafis, N. D., Papamastorakis, J., Paleologou, E. V., & Haerendel, G. 1997, *A&A*, 325, 135

Xilouris, E. M., Alton, P. B., Davies, J. I., Kylafis, N. D., Papamastorakis, J., & Trewhella, M. 1998, *A&A*, 331, 894

The Spectral Energy Distribution of Galaxies
Proceedings IAU Symposium No. 284, 2011
R.J. Tuffs & C.C. Popescu, eds.

doi:10.1017/S1743921312008927

On the nature of the peculiar superthin LSB galaxy UGC 12281

Philip Günster and Dominik J. Bomans

Astronomisches Institut, Ruhr-Universität Bochum,
Universitässtraße 150, D-44801 Bochum, Germany
email: guenster@astro.rub.de, bomans@astro.rub.de

Abstract. UGC 12281 has been classified as having a pure disk and being a low surface brightness galaxy (LSBG), thus being an obvious member of the so-called superthin galaxies. At the same time it represents an extremely untypical type of LSBG due to its remarkable amount of current star formation and evidence for extraplanar ionized gas. This makes it become a perfect tool to investigate the triggering of star formation in LSB galaxies, located in an alleged isolated area. By means of deep photometry and long-slit spectroscopy we analyse the Hα halo and verify the existence of a potential dwarf companion which we found on processed SDSS images.

Keywords. galaxies: evolution, galaxies: dwarf, galaxies: halos

1. Introduction

Galaxies in their variety reveal an interesting subset, very late-type spirals without bulge viewed edge-on. They were first mentioned by Goad & Roberts (1981) and called superthin galaxies from then on. The superthins can be distinguished easily as having very thin disks causing a measured axis ratio of more than 9. The modest gradients of their rotation curves to the center imply a small central mass concentration. This is in good agreement with their tendency to not show any central bulge. Based on the edge-on disk galaxy catalog by Kautsch *et al.* (2006) about 38% among the superthins are LSB galaxies showing a blue central surface brightness of more than 23 mag$\cdot$arcsec^{-2} (Impey & Bothun 1997). This implies they must have a relatively small star formation rate (SFR) over large time scales. To address the question why they never formed enough stars to appear significantly brighter, Rosenbaum *et al.* (2009) looked at the large-scale environment in which they are embedded. They conclude that LSBGs must have evolved in low density areas without any tidal interaction with companions to effectively trigger star formation. Therefore superthin LSB galaxies with high SFR promise to be an excellent tool to understand LSBG evolution. We investigate here the superthin UGC 12281. This galaxy has an axis ratio of almost 12. However, Rossa & Dettmar (2003) revealed it as Hα bright implying a high star formation activity. Indeed they determined a radial extent of strong star formation to 20 kpc which accounts for 2/3 of the total disk. Hints for extraplanar diffuse ionized gas (eDIG) were detected, too.

2. Data

We observed UGC 12281 with the 2.2 m telescope at the Calar Alto Observatory (CAHA). The CAFOS instrument provided deep imaging data in the bands Roeser BV, R and Hα. We also did long-slit spectroscopy at two slit positions: first we aim at measuring an integral spectrum of the diffuse halo gas. For that the slit is placed parallel to the disk with a 2 kpc offset from the midplane. Secondly, a slit is set perpendicular to the galaxy's major axis hitting the spot of the suspected dwarf companion (position flagged with an arrow in Fig. 1) and crossing the disk.

3. UGC 12281

Our structural analysis includes disk profile fits: We trace its shape first using the Source Extractor software by Bertin & Arnouts (1996). The derived astrometric and photometric parameters of the Roeser BV image are fed into the Galaxy Fitting code GALFIT (Peng *et al.* 2010). We apply a two-dimensional single profile to the disk according to Sérsic (1968), namely of the kind $I(R) = I_0 \cdot \exp[-\left(\frac{R}{a}\right)^{1/n}]$ where I_0 is the central intensity and a the scale length. The Sérsic index n is found to 1.25 which indicates pure-disk structure. GALFIT also provides the residual of the original and the model fit which shows no sign of a bulge, but spiral structure.

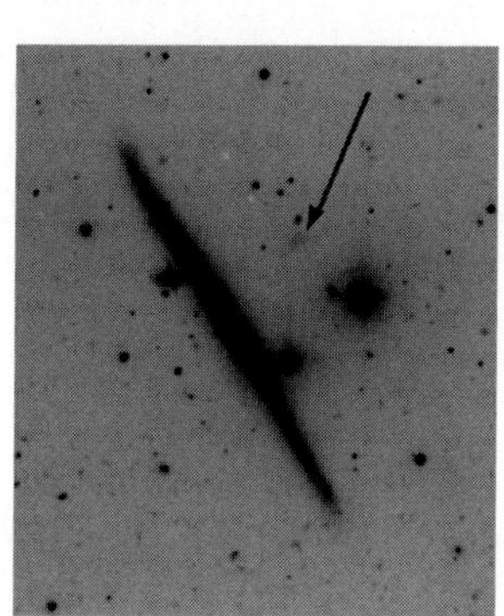

Figure 1. Roeser BV image of UGC 12281 taken with the CAHA 2.2 m telescope. The arrow marks the position of the dwarf companion.

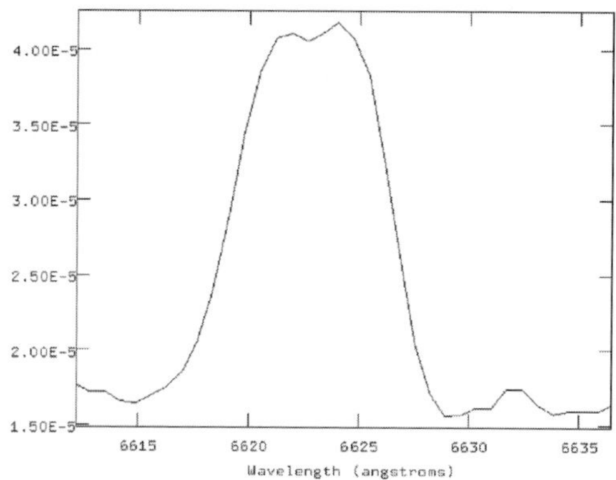

Figure 2. Closeup view of the Hα line in the disk spectrum of UGC 12281. The line obviously has a double peak. The split was measured to 170 km$\cdot$s^{-1}.

4. A close-by dwarf companion

From archival data of the Sloan Digital Sky Survey (SDSS) we were able to derive the following parameters of the possible dwarf companion: Its absolute magnitude is $M_{g'}$ = -12.01 and the average surface brightness $\mu'_g = 25.73$ mag$\cdot$arcsec^{-2}. The size could be estimated to approx. 2.0 x 1.0 kpc. The dwarf is elongated towards the host galaxy hinting at gravitational interaction with UGC 12281. If truely a satellite galaxy, this may be part of the explanation for UGC 12281's relatively high star formation providing stirring of the disk. Compared to the Milky Way satellite Fornax (M_V = -13.3, major axis = 2.8 kpc, $\mu'_g = 23.4$ mag$\cdot$arcsec^{-2}; Lokas 2009), the companion is similar in absolute magnitude, but more compact and with lower surface brightness, again supporting the idea of gravitational disturbance. Its B-V color is measured to 1.13 which is redder than Fornax (B-V = 0.63; Mateo 1998) and implies a higher metallicity or an intermediate-age stellar population.

5. A nearly transparent disk?

The slit orientation parallel to the galaxy's disk but slightly above it should reveal the diffuse halo gas. Unfortunately, in our initial spectral analysis no emission line features are detected while deep Hα imaging shows diffuse Hα halo emission. When we put the slit going through the galaxy's disk the Hα and the [N II] line appear to have double peaks (visible in Fig. 2) and account for a split of 170 km·s^{-1}, similar to the HI $v_{rot,max} = 146$ km·s^{-1} (Warmels 1988). This raises the question if we detected both the fore and the back side of the rotating disk. Since we look at an extreme flat edge-on object this assumption would require UGC 12281 to be nearly complete transparent. A multi-color analysis shows neither in the SDSS nor in our deep data an obvious dust lane. Such a smooth distribution of dust components is in agreement with results for other LSBGs of MacLachlan *et al.* (this conference).

6. Chain galaxies

Superthin galaxies with strong star formation may be in close connection to objects called chain galaxies. They are high axis ratio, clumpy systems at high redshifts. Although edge-on galaxies have obviously more extinction than the ones viewed face-on, their average surface brightnesses can be brighter (Elmegreen *et al.* 2004). This is due to their extinction path length which is usually larger than the actual disk thickness. Holmberg (1958) showed that the inclination-corrected face-on surface brightness is brighter than the actual average surface brightness by the extinction effect. This means that this distant galaxies can compensate for the effect of the cosmological Tolman dimming. The high SFR superthins, in the end, may be the low-redshift counterparts of chain galaxies.

7. Conclusions & Outlook

UGC 12281 appears to be an untypical LSB galaxy. Its recent star formation is significant and we found strong hints for a diffuse Hα halo. The disk is well fitted by a Sérsic index of 1.25 with no strong disturbances. The Hα line shows a split which is about in the order of the rotational velocity. This implies a nearly transparent disk. Our data verifies the existence of a very low surface brightness dwarf companion which is a candidate for interacting with its host and thus triggering its star formation. Star forming LSB superthins like UGC 12281 could allow us to overcome the Tolman dimming and extend our study of LSBG evolution to much higher redshifts.

References

Bertin, E. & Arnouts, S. 1996, *A&AS*, 117, 393
Goad, J. W. & Roberts, M. S. 1981, *ApJ*, 250, 79
Elmegreen, D. M., Elmegreen, B. G., & Sheets, C. M. 2004, *ApJ*, 603, 74
Holmberg, E. 1958, *Medd. Lunds. Astron. Obs., Ser II*, 136, 1
Impey, C. & Bothun, G. 1997, *ARAA*, 35, 267
Kautsch, S. J., Grebel, E. K., Barazza, F. D., & Gallagher, III, J. S. 2006, *A&A*, 445, 765
Lokas, E. L. 2009, *MNRAS*, 394, L102
Mateo, M. L. 1998, *ARAA*, 36, 435
Peng, C. Y., Ho, L. C., Impey, C. D., & Rix, H.-W. 2010, *AJ*, 139, 2097
Rosenbaum, S. D., Krusch, E., Bomans, D. J., & Dettmar, R.-J. 2009, *A&A*, 504, 807
Rossa, J. & Dettmar, R.-J. 2003, *A&A*, 406, 505
Sérsic, J. L. 1968, *Atlas de galaxias australes*
Warmels, R.-H. 1988, *A&AS*, 73, 453

The Spectral Energy Distribution of Galaxies
Proceedings IAU Symposium No. 284, 2011
R.J. Tuffs & C.C. Popescu, eds.

doi:10.1017/S1743921312008939

Low Metallicity ISM: excess submillimetre emission and CO-free H_2 gas

Suzanne C. Madden[1], Aurélie Rémy[1], Frédéric Galliano[1], Maud Galametz[2], George Bendo[3], Diane Cormier[1], Vianney Lebouteiller[1], Sacha Hony[1] and the *Herschel* SAG 2 consortium

[1]CEA Saclay, DSM, AIM, Service d'Astrophysique, Gif-sur-Yvette 91911, France
[2]Institute of Astronomy, University of Cambridge, Madingly Rd., Cambridge, UK
[3]Alma Regional Center, University of Manchester, Oxford Rd., Manchester, UK

email: suzanne.madden@cea.fr

Abstract. The low metallicity interstellar medium of dwarf galaxies gives a different picture in the far infrared(FIR)/submillimetre(submm)wavelengths than the more metal-rich galaxies. Excess emission is often found in the submm beginning at or beyond 500 μm. Even without taking this excess emission into account as a possible dust component, higher dust-to-gas mass ratios (DGR) are often observed compared to that expected from their metallicity for moderately metal-poor galaxies. The Spectral Energy Distributions (SEDs) of the lowest metallicity galaxies, however, give very low dust masses and excessively low values of DGR, inconsistent with the amount of metals expected to be captured into dust if we presume the usual linear relationship holding for all metallicities, including the more metal-rich galaxies. This transition seems to appear near metalllicities of 12 + log(O/H) 8.0 - 8.2. These results rely on accurately quantifying the total molecular gas reservoir, which is uncertain in low metallicity galaxies due to the difficulty in detecting CO(1-0) emission. Dwarf galaxies show an exceptionally high [CII] 158 μm/CO (1-0) ratio which may be indicative of a significant reservoir of 'CO-free' molecular gas residing in the photodissociated envelope, and not traced by the small CO cores.

Keywords. galaxies: dwarf, galaxies: ISM, ISM: molecules

1. Introduction

Understanding how galaxies evolve requires fundamental comprehension of the process of star formation and the subsequent effects on the interstellar medium (ISM) under conditions in the early universe - an epoch when the ISM had not yet endured many cycles of chemical enrichment. We can attempt to extrapolate to such conditions, by studying the interplay between star formation and dust and gas in low metallicity dwarf galaxies, of which our local universe hosts a veritable zoo.

The first systematic study of dust in dwarf galaxies observed with wavelengths as long as 100 μm with *IRAS*, showed lower infrared luminosity (L_{IR}) compared to Hα than spirals. $L_{12\mu m}/L_{25\mu m}$ values were lower than spirals, while the higher $L_{60\mu m}/L_{100\mu m}$ highlighted the presence of warmer dust and enhancement of very small grains (e.g. Hunter *et al.* 1989; Melisse & Israel 1994). The *ISO* and *Spitzer* missions opened up the window onto MIR spectroscopy, demonstrating the dearth of PAHs in dwarf galaxies, a consequence of the prevailing hard radiation field, shocks or delayed injection by AGB stars (e.g. Madden *et al.* 2006; Wu *et al.* 2006; O'Halloran *et al.* 2008; Galliano *et al.* 2008) as well as cooler dust revealed by longer wavelengths, extending to 160 μm (e.g. Popescu *et al.* 2002). SCUBA/JCMT and Laboca/APEX observations exposed excess emission in dwarf galaxies at 850/870 μm which, if interpreted as cold dust, could comprise a large amount of dust in low metallicity galaxies compared to their measured gas reservoirs and their low metallicities (e.g. Galliano *et al.* 2003; Galliano *et al.* 2005; Zhu *et al.* 2009; Galametz *et al.* 2011). One constraint on the dust mass should be the measured dust-to-gas mass ratio (DGR), as this parameter should be effected by chemical evolution. Atomic gas is often widespread in dwarf galaxies, in contrast to molecular gas, if the CO(1-0) is a reliable tracer of the molecular reservoir in these galaxies. On the other hand, the 158 μm [CII], often assumed to arise from the photodissociation regions (PDRs) around molecular clouds, is excessively bright in dwarf galaxies compared to the CO, in contrast to more metal-rich galaxies (Poglitsch *et al.* 1995; Israel *et al.* 1996; Madden *et al.* 1997; Madden 2000). What is this telling us about the total molecular gas reservoir, the galaxy morphology and the distribution of the various gas phases in dwarf galaxies?

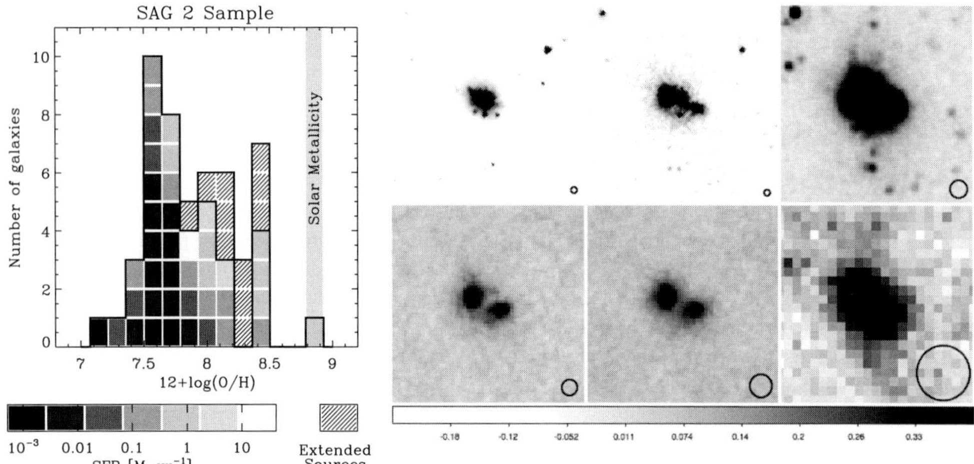

Figure 1. (left) Metallicity and star formation properties of the Herschel Dwarf Galaxy Survey. (right) *Spitzer* and *Herschel* images of the dwarf galaxy, NGC 1705. Images from left to right and top to bottom are: 3.6 μm *Spitzer*/IRAC, 8 μm *Spitzer*/IRAC, 24 μm *Spitzer*/MIPS, 70 μm *Herschel*/PACS, 100 μm *Herschel*/PACS, 250 μm *Herschel*/SPIRE. Image sizes are 130"x130" and beam sizes are indicated in lower right corner of images.

Herschel has opened up the submm wavelength window beyond 160 μm, with observations covering the 50 to 500 μm window. The Dwarf Galaxy Survey (DGS) is a *Herschel* key program targeting 48 local universe dwarf galaxies with a wide range of star formation properties and metallicity values, as low as 1/50 $Z_\odot$ (Fig. 1) and as nearby as the Magellanic Clouds to study the multiphase components of the stars, gas and dust under low metallicity environments.

2. SED models of dwarf galaxies

PACS observations cover 3 photometric bands: 70, 100 and 160 μm (FWHM $\sim$ 10"; Poglitsch *et al.* 2010) while SPIRE observes at 250, 350 and 500 μm (FMHM=18" to 38"; Griffin *et al.* 2010). Comparison of the *Spitzer* 24 μm images (Bendo *et al.* 2012) with the PACS 70 μm and SPIRE 250 μm (Fig. 1) highlights the extent of the cooler dust traced by the 250 μm emission in contrast to the warmer dust emitting at 24 μm, favoring the more compact HII regions. Coverage at *Spitzer* and *Herschel* wavelengths provides well-sampled SEDs for wide-ranging studies of the dust properties in dwarf galaxies.

Metallicity and β and T effects. The FIR-submm behavior of the DGS sample is investigated via *Spitzer* and *Herschel* colour-colour diagrams, to obtain an overview of the

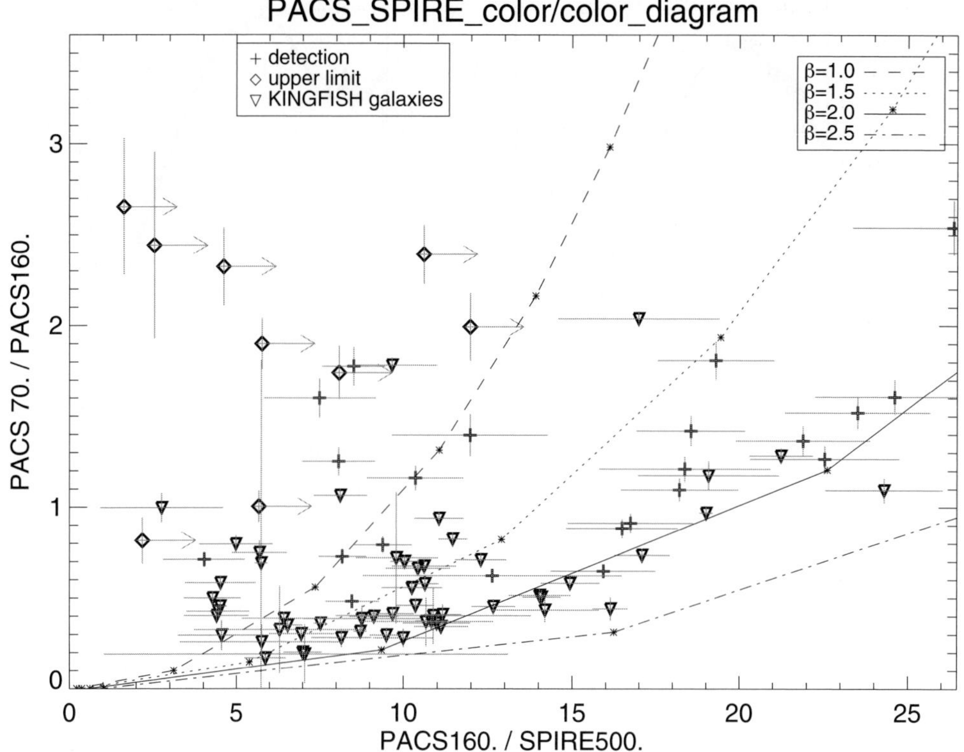

Figure 2. Herschel colour-colour diagram of DGS galaxies detected in at least all PACS bands: PACS70/160 and PACS160/SPIRE500. For comparison, KINGFISH galaxies (Dale *et al.* 2012) are also shown (downward triangles). Modeled modified black body fits with β = 1.0, 1.5, 2.0, 2.5 are shown in curves, with temperature values noted as dots on the curves, increasing from 10 K in steps of 10 K, starting from the lower left of the curves to the upper right.

total sample (Rémy *et al.* in preparation; also this volume). Fig. 2 shows the effect of varying emissivity indices (β) and average dust temperatures (T) presuming a single modified black-body to explain the FIR-submm emission. For comparison the KINGFISH galaxies (Dale *et al.* 2012) are also included. The effect of the overall higher metallicity clusters the bulk of the KINGFISH sample between β =1.5 and 2.0. The most active dwarf galaxies, particularly the blue compact dwarfs, peak at wavelengths less than 70 μm and some between 35 and 60 μm, much shorter wavelengths compared to the more metal-rich starburst galaxies. The difference in the shape of the SEDs of dwarf galaxies can be noted via the *Herschel* colour-colour diagram: as a consequence of their overall hotter dust peaking at MIR wavelengths (higher PACS70/PACS160 values for the lower metallicity bins), their Rayleigh-Jeans slope drops off at FIR and submm wavelengths. Consequently, these galaxies are often not detected at all SPIRE wavelengths for the lowest metallicity galaxies, particularly at 500 μm. A prominent example of this effect is illustrated in one of the lowest metallicity galaxies of our sample, SBS0335-052E (12+1og(O/H) = 7.29; Fig. 3). The integrated SED shows the flatter MIR-FIR emission peaking between 20 and 30 μm (Houck *et al.* 2004; Galliano *et al.* 2008; Sauvage *et al.* in preparation). The very dense super star clusters dominate the warm overall dust emission, with the observed SED leaving little evidence for cold submm-emitting dust.

For those dwarf galaxies detected at both PACS and SPIRE wavelengths, modified black-body fits (omitting 70 μm in the fits) give a wide range of β and T solutions with

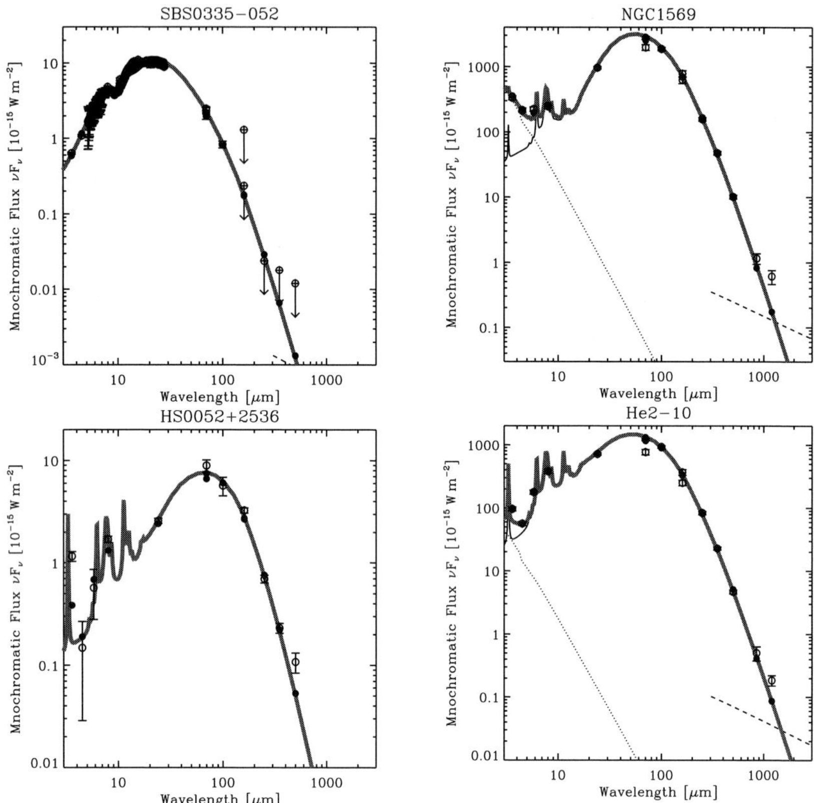

Figure 3. SEDs of dwarf galaxies. SBS0335-052 has on average very hot dust, with the SED peaking between 20 and 30 μm(Galliano *et al.* 2008; Sauvage *et al.* in preparation]). HS0052+2536 shows a very low β in the *Herschel* colour-colour diagram (Fig. 2) and does indeed show a 500 μm excess. NGC 1569 and He2-10 both fall near $\beta \sim 2$ in the colour-colour diagram, but their full SED model unveils an excess *beyond* 500 μm. *Spitzer* and *Herschel* data are used to constrain the SED models. *Herschel* observations are used for the model constraints in the cases of data redundancy (70 and 160 μm). Stellar contribution is shown as dotted lines, the dashed lines are the free-free continuum, extrapolated from observed radio observations when available and the solid grey curve is the total modeled SED. The open squares are the observations while the black points show the predicted model fluxes, often overlapping completely with the opens squares.

a mean β of 1.6 and mean T of 32 K (Fig. 4). Since the most metal-poor galaxies of the DGS sample do not have 500 μm detections, this distribution of modeled β and T is primarily representing galaxies with metallicity values greater than 12 +log (O/H) $\sim$ 8.0 - 8.2. For comparison, the KINGFISH galaxies show a similar mean β but the T distributions are cooler, peaking between $\sim$ 20 and 25 K (Dale *et al.* 2012). Galaxies requiring an exceptionally low β solution, may be indicative of the presence of a submm excess which could start to be detectable at wavelengths as low as 500 μm.

Submm excess examples. Taking into account the optical to submm and radio wavelengths, full SED models (Galliano *et al.* 2008) can help to further interpret the Herschel colour-colour diagrams. For example, from the *Herschel* colour-colour diagram (Fig. 2), HS0052+2536 has an exceptionally low β value $\sim$ 1. Inspection of the fully modeled SEDs (Fig. 3) shows that a submm excess is indeed present. About 50% of the DGS galaxies detected at 500 μm show a submm excess of $\sim$ 7% to 100% above the SED model (Rémy *et al.*, in preparation). A relationship between metallicity and submm excess within the

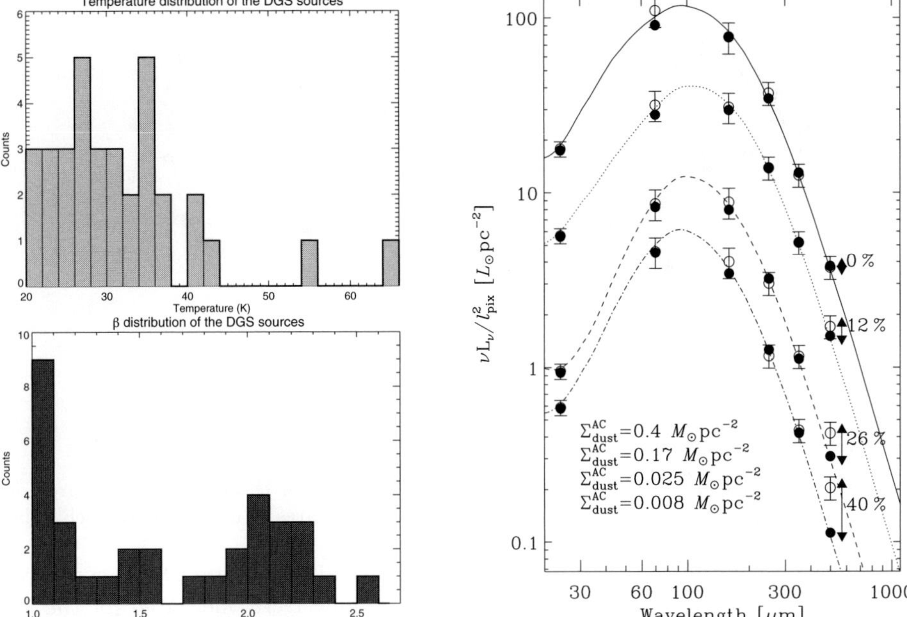

Figure 4. (left) T and β distribution of the DGS sample with 500 μm detections. (right) Selected SEDs in the LMC showing submm excess ranging up to 40% compared to that expected from the SED models (Galliano *et al.* 2011). SEDs are for a range of dust mass surface densities, ranging from the lowest to the highest values, corresponding to lower to higher profiles, respectively.

DGS sample is not yet obvious. This may be somewhat due to the requirement for 500 μm detections which often omits the lowest metallicity galaxies. While this *Herschel* colour-colour diagram highlights potential galaxies with submm excess, galaxies for which the excess begins beyond 500 μm will be missed without observations at longer wavelengths. For example, NGC 1569 and He2-10, show $\beta \sim 2$ in the Herschel colour-colour diagram but do indeed show a submm excess when including their ground-based 850 μm observations (Fig. 3; see also Galliano *et al.* 2003; Galliano *et al.* 2005; Galametz *et al.* 2011).

DGRs and possible origins of the submm excess. Understanding how the DGR varies as a function of metallicity is important to have an accurate picture of the gas and dust life cycle in galaxies. How are the heavy metals incorporated into dust and how do metallicity or other local or global parameters control this process? Understanding the behavior of the DGR is also important since many studies determining the dust mass in galaxies then quantify the gas reservoir in galaxies by assuming a DGR. Numerous studies have noted a proportionality of DGR with metallicity, but results can vary depending on the long wavelength data used. Engelbracht *et al.* (2008) noted a decrease in DGR from the more metal rich galaxies to the moderately metal-poor galaxies, until 12 + log (O/H) $\sim$ 8, beyond which the DGR appears to be constant. Adding 850 μm observations beyond *Spitzer* wavelengths, however, can increase dust masses (Galametz *et al.* 2011). The low metallicity end of the DGR relationship has been particularly ambiguous, since dust mass estimates of dwarf galaxies are sometimes hampered by a submm excess.

For the lowest metallicity galaxies (12 + log (O/H) $<$ 8.0) where the overall SED peaks at short wavelengths and where we have mostly only upper limits in the submm, we find low upper limits to dust masses and low DGRs compared to the expected value. The upper limit dust mass from SBS0335-052, for example, is very low (Galliano *et al.*

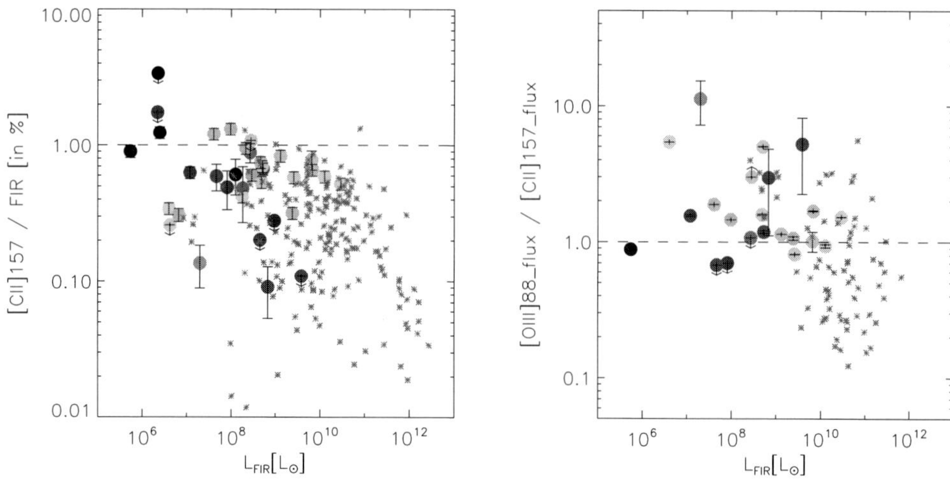

Figure 5. [CII]/FIR and [OIII]/FIR as a function of $L_{[FIR]}$ for the DGS sample (large dots) compared to that of Brauher *et al.* 2008 ISO data (small squares) - mostly metal-rich galaxies.

2008; Sauvage *et al.* in preparation) giving an unusually low total DGR, compared to that expected for its extremely low metallicity. The low DGR measured for the lowest metallicity galaxies may be telling us that perhaps metals are not necessarily incorporated into dust in the same way for the lowest metallicity galaxies as for the more metal rich galaxies. On the other hand, perhaps the total gas reservoir is underestimated.

The submm excess has been noted in dwarf galaxies for almost 10 years now - since the first SCUBA observations of dwarf galaxies at 850 μm. Yet the origin remains uncertain. In the LMC, where Herschel brings 10 pc resolution at 500 μm, Galliano *et al.* (2011) (see also Galliano, this volume) have highlighted locations of submm excess ranging from 15% to 40% compared to that expected from the SED models (Fig. 4). The excess seems to be anticorrelated with the dust mass surface density. Possible explanations for the submm excess include: 1) very cold dust component (e.g. Galliano *et al.* 2005; Galametz *et al.* 2011; 2) excessive free-free emission; 3) unusual dust emissivity properties (e.g. Lisenfeld *et al.* 2002; Mény *et al.* 2002; Galliano *et al.* 2011; Paradis *et al.* 2012); 4) anomalous spinning dust (e.g. Draine & Lazarian 1998; Ysard *et al.* 2010). With more sensitive submm wavelength coverage, *Herschel* is confirming the flater submm slope in more dwarf galaxies. Problems exist with these explanations put forth for the origin of the excess. For example, invoking a very cold dust component to account for the submm excess can augment the dust mass excessively, resulting in a *very high DGR* compared to that expected for the metallicity. But, are we correctly quantifying the total molecular gas mass in low metallicity environments?

3. The molecular gas in dwarf galaxies

While CO is the widely used means to access M_{H_2} in galaxies, it is by now a well-established fact that the Galactic X factor, which converts CO to M_{H_2}, underestimates the mass of molecular gas in low metallicity galaxies. What is uncertain, however, is how to accurately correct up this factor as a function of metallicity to obtain the total gas reservoir. During the last 2 decades valiant effort has been invested in detecting and interpreting CO emission in low luminosity dwarf galaxies to quantify the molecular gas reservoir (e.g. Leroy *et al.* 2009 and references within). With the first detections of the 158 μm [CII] line in dwarf galaxies on the KAO, the surprisingly high [CII]/CO values highlighted the fact that CO could be missing a large reservoir of molecular gas (factors of 10 to 100) due to the lower dust abundance. The consequently deeper penetration of UV photons further photodissociate CO, reducing the CO cores - hence, the dearth of detected CO. Due to the self-shielding of H_2, the larger C^+-emitting envelope can harbor H_2 which is not accounted for via CO observations (e.g. Poglitsch *et al.* 1995; Madden *et al.* 1997; Wolfire *et al.* 2010). The observed [CII]/CO, thought to be a useful tracer of star formation in galaxies, can sometimes be a factor of a 2 to 5 (or more) times higher in dwarf galaxies than in metal-rich galaxies (e.g. Madden 2000; Stacey *et al.* 2010).

Dwarf galaxies generally emit a larger fraction of their FIR in the [CII] line, $\sim$ 0.5 to 2 %, in contrast to the more metal-rich spirals and starbursts which usually show [CII]/FIR less than $\sim$ 0.5% (Fig. 5). If [CII] is the dominant coolant in galaxies and the photoelectric effect is responsible for the primary heating of the ISM, then [CII]/FIR can be thought of as a proxy for the grain photoelectric heating efficiency (e.g. Rubin *et al.* 2009) – the fraction of the power being absorbed by the grains that goes into heating of the ISM. The high [CII]/FIR values observed in dwarf galaxies thus, imply high photoelectric heating efficiencies which may be attributed to properties of the low-metallicity ISM, including lower dust abundance, clumpy ISM, etc. Traversing the galaxy, UV photons suffers less attenuation in the metal-poor ISM, and as a consequence of the longer photon mean-free path, the dust on galaxy-wide scales is subject to lower UV flux, effectively decreasing the overall FIR flux and increasing the observed [CII]/FIR. The 88 μm [OIII] line is typically the brightest FIR line in dwarf galaxies - usually brighter than the [CII] line (Fig. 5), normally considered to be the brightest FIR line in galaxies. This suggests a substantial filling factor of ionised gas in dwarf galaxies and places an important constraint on quantifying the origin of the [CII] line, which also can be excited by electrons in the diffuse ionized gas, not only PDRs (Lebouteiller *et al.* 2012). The [CII] and [OIII] together are important calibrators of star formation activity and characterize the diffuse and molecular phases in galaxies. With ALMA, these lines are valuable accessible diagnostics of the ISM of high red-shift galaxies offering new insight on the evolution of the star formation and ISM throughout earlier epochs.

References

Bendo, G., Galliano, G., & Madden, S. C. 2012, *MNRAS* submitted
Brauher, J. R. Dale, D. A., & Helou, G. 2008, *ApJS*, 178, 280
Dale, D. A., Aniano, G., Engelbracht, C. W., Hinz, J. L., Krause, O. *et al.* 2012, *ApJ*, 745, 95
Draine, B. T. & Lazarian, A. 1998, *ApJ*, 508, 157
Engelbracht, C. W., Rieke, G. H., Gordon, K. D., Smith, J.-D. T. *et al.* 2008, *ApJ*, 678, 804
Galametz, M., Madden, S. C., Galliano, F., Hony, S., Bendo, G. J. *et al.* 2011, *A&A*, 532, 56
Galliano, F., Madden, S. C., Jones, A., Wilson, C., *et al.* 2003, *A&A*, 407, 159
Galliano, F., Madden, S. C., Jones, A. P., Wilson, C. D., & Bernard, J.-P. 2005, *A&A*, 434, 867
Galliano, F., Dwek, E., & Chanial, P. 2008, *A&A*, 672, 214
Galliano, F., Hony, S., Bernard, J.-P., Bot, C., Madden, S. C. *et al.* 2011, *A&A*, 536, 88
Griffin, M., Abergel, A., Abreu, A., Ade, P. A. R. *et al.* 2010, *A&A*, 518, 3
Houck, J. R., Charmandaris, V., Brandl, B. R., Weedman, D. *et al.* 2004, *ApJS*, 154, 211
Hunter, D. A., Gallagher, J. S., III, Rice, W. L., *et al.* 1989, *A&A*, 336, 152
Israel, F. P., Maloney, P. R., Geis, N., Herrmann, F., Madden, S. C. *et al.* 1996, *ApJ*, 465, 738
Lebouteiller, V., Cormier, D., Madden, S. C., *et al.* 2012, *A&A*, submitted
Leroy, A. K., Walter, F., Bigiel, F., Usero, A., Weiss, A. *et al.* 2009, *AJ*, 137, 4670
Lisenfeld, U., Israel, F. P., Stil, J. M., & Sievers, A.l 2002, *A&A*, 382, 860
Madden, S. C., Poglitsch, A., Geis, N., Stacey, G. J., & Townes, C. H. 1997, *ApJ*, 483, 200
Madden, S. C. 2000, *NewAR*, 44, 249

Madden, S. C., Galliano, F., Jones, A. P., & Sauvage, M. 2006, *A&A*, 446, 877
Melisse, J. P. M. & Israel, F. P. 1994, *A&A*, 285, 51
Mény, C., Gromov, V., Boudet, N., Bernard, J.-Ph., *et al.* 2007, *A&A*, 468, 171
O'Halloran, B., Madden, S. C., & Abel, N. P. 2008, *ApJ*, 681, 1205
Paradis, D., Paladini, R., Noriega-Crespo, A., Mény, C. *et al.* 2012, *A&A*, 537, 113
Poglitsch, A., Krabbe, A., Madden, S. C., Nikola, T., Geis, N. *et al.* 1995, *ApJ*, 454, 293
Poglitsch, A., Waelkens, C., Geis, N., Feuchtgruber, H. *et al.* 2010, *A&A*, 518, 1
Popescu, C. C., Tuffs, R. J., Völk, H. J. *et al.* 2002, *ApJ*, 567, 221
Rubin, D., Hony, S., Madden, S. C., Tielens, A. G. G. M. *et al.* 2009, *A&A*, 494, 647
Stacey, G. J., Hailey-Dunsheath, S., Ferkinhoff, C., Nikola, T. *et al.* 2010, *ApJ*, 724, 957
Wolfire, M. G., Hollenbach, D., & McKee, C. F. 2010, *ApJ*, 716, 1191
Wu, Y., Charmandaris, V., Hao, L., Brandl, B. R., *et al.* 2006, *ApJ*, 639, 157
Ysard, N., Miville-Deschnes, M. A., & Verstraete, L. 2010, *A&A* 509, 1
Zhu, M., Papadopoulos, P., Xilouris, *et al.* 2009, *ApJ*, 706, 941

Discussion

CHAKRABARTI: The low β values you are finding would suggest considerable grain growth. This would happen only in regions of very high density. If you used a range of temperatures (instead of a single temperature), I suspect you would find that you could fit the SEDs with $\beta = 2$.

MADDEN: The Herschel colour-colour plots I presented with overlays based on modified Black-Body fits are only suggestive in locating the potential flatter submm slopes we are seeing in low metallicity galaxies. The actual modelled SEDs I showed incorporate the sophisticated SED models of Galliano *et al.* and do indeed solve for the wide range of dust temperatures.

GALLAGHER: Dwarfs contain cold HI clouds, e.g. the Young & Lo studies of yore as well more recent SMC studies. Do these clouds show up as [CII] sources as one might expect/hope?

MADDEN: These cold HI clouds of narrow velocity are tracing the cold Neutral medium and would be good candidates for [CII] detection. It would be a good experiment for Herschel.

The Spectral Energy Distribution of Galaxies
Proceedings IAU Symposium No. 284, 2011
R.J. Tuffs & C.C. Popescu, eds.

doi:10.1017/S1743921312008940

Characterisation of the submillimeter excess in dwarf galaxies: Presentation of the *Herschel* Dwarf Galaxies Survey

Aurélie Rémy[1], Suzanne C. Madden[1], Frederic Galliano[1], Maud Galametz[2], Sacha Hony[1] and the *Herschel* SAG2 Consortium

[1]CEA- Service d'Astrophysique, CEA Saclay
Orme des Merisiers Bat. 709, 91191 Gif-sur Yvette, France

[2]Institute of Astronomy , University of Cambridge
Madingley Road, Cambridge, CB3 0HA, UK

email: aurelie.remy@cea.fr

Abstract. The *Herschel* Space Observatory is revolutionizing our view of dust in galaxies with high sensitivity observations in the far infrared(FIR)/submillimeter (submm) regime from 70 to 500 μm. *Herschel* is confirming the submm excess that has been noted previously in low metallicity dwarfs. We present here the Dwarf Galaxies Survey sample through a *Herschel* colour-colour diagram. We will then focus on two galaxies, Haro11 and NGC4449 presenting different interesting behaviours in the FIR as revealed by *Herschel*.

Keywords. galaxies: dwarf, galaxies: ISM, ISM: dust, infrared: ISM, submillimeter,

1. The *Herschel* Dwarf Galaxies Survey

The Interstellar Medium (ISM) plays a key role in the evolution of a galaxy, being the repository of stellar ejecta and the site of stellar birth. The spectral energy distribution (SED) is a current snapshot with this historical information integrated over the lifetime of the galaxy. To understand star formation and its feedback on the ISM in conditions that may be representative of different stages of early universe environments, the *Herschel* Dwarf Galaxy Survey (DGS; P.I. Madden) is studying the gas and dust properties of a wide variety of 48 low-metallicity (Z) galaxies at far infrared (FIR) and submillimeter (submm) wavelengths. Our sample has been observed with both *Herschel* photometers : PACS (Poglitsch *et al.* 2010) at 70, 100 and 160 μm and SPIRE (Griffin *et al.* 2010) at 250, 350 and 500 μm spanning a wide range in metallicity : 12+log(O/H) = 7.2 to 8.5.

2. Far Infrared Colours

To obtain a better overall view of the DGS and its mid infrared (MIR) to submm behaviour, we constructed several *Herschel* colour-colour diagrams, and present one of them here (example : Fig. 1). Total fluxes were extracted from the maps to compute the various ratios (Remy *et al.* 2011, in prep). We eliminated galaxies with more than one upper limit in the considered bands. We then computed the theoretical *Herschel* flux ratios of simulated modified black bodies spanning a range in temperatures (from 10 to 100K) and emissivity index (from 1.0 to 2.5). The spread of galaxies in the diagram in Fig. 1 reflects variations in metallicity, T, and β in our survey. The submm SED for local and distant galaxies is commonly described by $\beta = 2$. The flattening of the submm slope ($\beta<2$) may signal possible excess emission appearing at or beyond 500 μm. Therefore, the $PACS_{70}/PACS_{160}$ vs $PACS_{160}/SPIRE_{500}$ diagram can be considered a

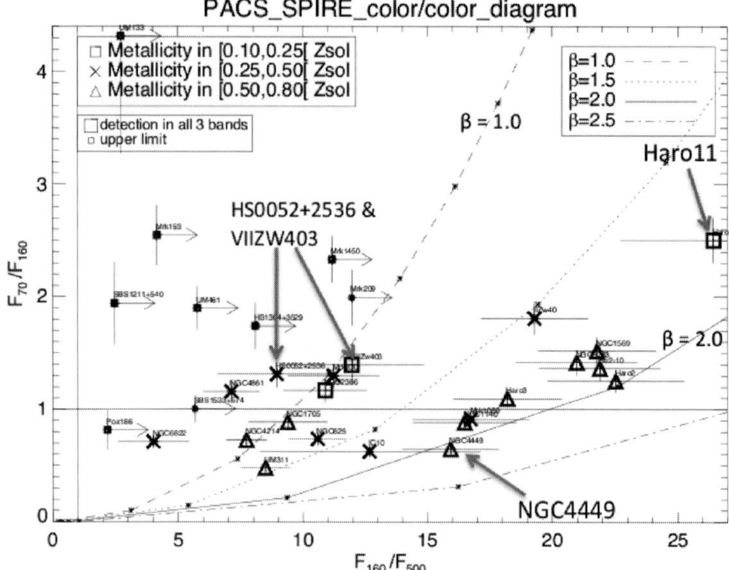

Figure 1. DGS colour colour diagram : $PACS_{70}/PACS_{160}$ vs $PACS_{160}/SPIRE_{500}$. The symbols delineate the metallicity bins for the galaxies. The large symbols are for galaxies with detections in all 3 bands; smaller ones are sources that are not detected at 500 μm. The four curves give theoretical *Herschel* flux ratios for simulated modified black bodies for β=1.0 to 2.5 and T=10 to 100K, increasing in 10K bins from left to right. Sources mentionned in the text are indicated by arrows.

good diagnostic tool, for example, to spot some potential "submm excess" galaxies (eg. Haro11, HS0052+2536 or VIIZw403).

3. Zoom on particular cases

From our diagnostic diagram, we select, for illustration, Haro11 (D$\sim$90 Mpc, Z=$0.2Z_\odot$) for its apparent submm excess, and NGC4449 (D$\sim$4.1 Mpc, Z=$0.5Z_\odot$) for its excess starting after 500 μm.

The Haro 11 MIR to submm SED was modelled by Galametz *et al.* (2009) with an additional Very Cold Dust (VCD) component (β=1, T=10 K) to explain the submm excess, and NGC 4449 has been modelled here with the Galliano *et al.* (2011) model.

Note how Haro11 SED peaks at $\sim$ 40 μm, shorter than metal rich starbursting galaxies, and typical of the very active starbursting dwarf galaxies (Fig. 2). Also note the submm excess seen in both galaxies : Haro 11 begins to show submm excess at 500 μm, as highlighted by the *Herschel* colour-colour diagram (Fig. 1). For NGC4449 however the excess is not apparent at 500 μm : NGC4449 falls on the $\beta = 2$ line when considering only the Herschel bands (Fig. 1). This is confirmed by the modelled SED (Fig. 2). The excess in dwarf galaxies is often only appearing at longer submm wavelengths and is not always detected with *Herschel.* This highlights the necessity of having submm constraints beyond *Herschel* for SED modelling.

Assuming a VCD component for Haro 11, the derived dust mass is $1.7\times10^7 M_\odot$. Haro11 only has an HI upper limit of $\sim10^8 M_\odot$ (Bergvall *et al.* 2000), and only an upper limit of CO(1-0) (D. Cormier, priv. com.), which, if we assume this traces the molecular gas, leaves little H_2. Therefore, the measured gas-to-dust mass ratio (G/D) is $\sim$5.8, much too low compared to that expected from chemical evolution models. Assuming a Galactic G/D $\sim$158 (Zubko *et al.* 2004) we should find G/D$\sim$790 for Haro11, considering its metallicity. Assuming that the dust mass determination is reliable, a large fraction of the gas mass seems to be missing and could exist as a large reservoir of CO-dark molecular gas (as suggested by Madden *et al.* 1997 for IC10, another dwarf irregular galaxy) or as a large reservoir of ionized gas. For NGC4449, we derive a dust mass of $1.9 \times 10^6 M_\odot$ from our modelled SED, which is comparable to that found by Böttner *et al.* (2003) using SCUBA 450 and 850 μm observations. Hunter *et al.* (1998, 2000) determined a total gas mass (HI and H_2) of $1.1 \times 10^9 M_\odot$, giving us a G/D of ~578. The scaling from the Galactic value would be ~316, considering NGC4449 metallicity. As we did not yet model the potential excess seen in NGC 4449, some cold dust could have been missed by our present dust mass estimate.

Other explanations have been proposed to explain the origin of the submm excess in addition to an extra VCD component. For example, the use of amorphous carbons instead of graphites, with a lower emissivity index, requires less dust mass to account for the same luminosity (Galliano *et al.* 2011). The electric dipole emission from fast rotating grains ("spinning dust", e.g. Bot *et al.* 2010) has also been suggested as a possible source of the submm excess. To date, no one explanation suffices. Further studies on the DGS may bring new answers on this present issue.

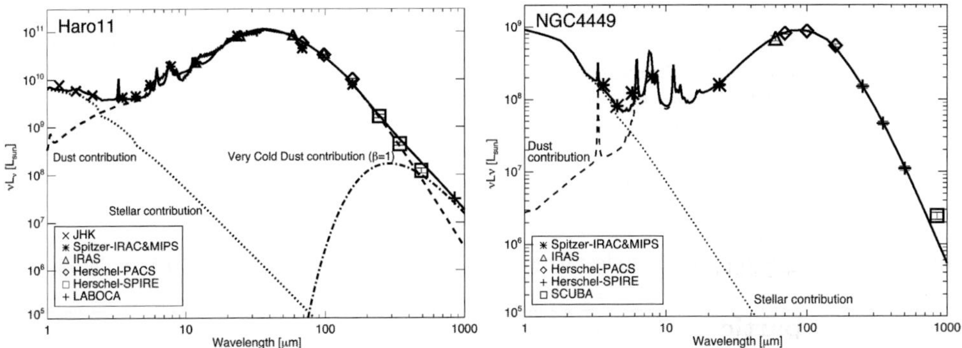

Figure 2. (left :) SED of Haro11 from Galametz *et al.* 2009, with *Herschel* points overlayed as diamonds and squares. (right :) The SED of NGC4449 has been obtained with the Galliano *et al.* 2011 model, including the new *Herschel* points. Other symbols are observational constraints from previous missions such as 2MASS, ISO, IRAS, Spitzer and IRS spectrum between 5-40 μm for Haro11. The total SEDs are the sum of the different contributions displayed here : stellar, dust and VCD for Haro11. The VCD hypothesis have been tested on Haro11 to explain the submm excess (Galametz *et al.* 2009) but led to extremely low G/D. NGC4449 however begins to show potential excess after 500 μm which is not detected with *Herschel*.

References

Bergvall, N., Masegosa, J., Östlin, G., & Cernicharo, J. 2000, *A&A*, 359, 41
Bot, C., Ysard, N., Paradis, D., *et al.* 2010, *A&A*, 523, A20
Böttner, C., Klein, U., & Heithausen, A. 2003, *A&A*, 408, 493
Galametz, M., Madden, S. C., Galliano, F. *et al.* 2009, *A&A*, 508, 645
Galliano, F., Hony, S., Engelbracht, C. *et al.* 2011, arXiv1110.1260
Griffin, M. J., Abergel, A., Abreu, A., *et al.* 2010, *A&A*, 518, L3
Hunter, D. A., Wilcots, E. M., van Woerden, H. *et al.* 1998, *ApJ*, 495, 47
Hunter, D. A., Walker C. E. & Wilcots, E. M. 2000, *AJ*, 119, 668
Madden, S. C., Poglitsch, A., Geis, N. *et al.* 1997, *ApJ*, 483, 200
Poglitsch, A., Waelkens, C., Geis, N., *et al.* 2010, *A&A*, 518, L2
Zubko, V., Dwek, E., & Arendt, R. 2004, *ApJS* 152, 211

The Spectral Energy Distribution of Galaxies
Proceedings IAU Symposium No. 284, 2011
R.J. Tuffs & C.C. Popescu, eds.

doi:10.1017/S1743921312008952

Dust in dwarf galaxies: The case of NGC 4214

U. Lisenfeld[1], I. Hermelo[1], M. Relaño[1], R. J. Tuffs[2], C. C. Popescu[3], J. Fischera[4], and B. Groves[5]

[1]Universidad Granada, Granada, Spain, email: ute@ugr.es, mrelano@ugr.es,israelhermelo@ugr.es

[2]Max-Planck-Institut für Kernphysik, Heidelberg, Germany, email: Richard.Tuffs@mpi-hd.mpg.de

[3]Jeremiah Horrocks Institute for Astrophysics and Supercomputing, University of Central Lancashire, Prestion, U.K., email: cpopescu@uclan.ac.uk

[4]Canadian Institute for Theoretical Astrophysics (CITA), University of Toronto, email: fischera@cita.utoronto.ca

[5]Max-Planck-Institut für Astronomie, Heidelberg, Germany, email: brent@mpia.de

Abstract. We have carried out a detailed modelling of the dust heating and emission in the nearby, starbursting dwarf galaxy NGC 4214. Due to its proximity and the great wealth of data from the UV to the millimeter range (from GALEX, HST, *Spitzer*, Herschel, Planck and IRAM) it is possible to separately model the emission from HII regions and their associated photodissociation regions (PDRs) and the emission from diffuse dust. Furthermore, most model parameters can be directly determined from the data leaving very few free parameters. We can fit both the emission from HII+PDR regions and the diffuse emission in NGC 4214 with these models with "normal" dust properties and realistic parameters.

Keywords. dust, extinction, galaxies: irregular, galaxies: individual (NGC 4214), galaxies: ISM

1. Introduction

The dust spectral energy distributions (SEDs) of dwarf galaxies frequently show differences to those of spiral galaxies. The two main differences are: (i) a relatively low emission at 8 μm, most likely due to a lower PAH content at low metallicities (e.g. Draine *et al.* 2007, Engelbracht *et al.* 2008), and (ii) a submillimeter (submm) "excess" which has been found in the SED of many starbursting, low-metallicity galaxies (Lisenfeld *et al.* 2002, Galliano *et al.* 2003, 2005, Bendo *et al.* 2006, Galametz *et al.* 2009, 2011, Israel *et al.* 2010, Bot *et al.* 2010). Different reasons have been suggested to explain this excess:

(1) A large amount of cold ($<$ 10 K) dust (Galliano *et al.* 2003, 2005, Galametz 2009, 2011). However, extraordinarily large dust masses are needed for this explanation and it is unclear, how these large amounts of cold dust can be shielded efficiently from the interstellar radiation field (ISRF).

(2) A low dust emissivity spectral index of $\beta = 1$ in the submm. Different dust grains have been suggested to be responsible for this, from very small grains (Lisenfeld *et al.* 2002), fractal grains (Reach *et al.* 1995) to amorphous grains (Meny *et al.* 2007).

(3) Spinning grains (Ferrara & Dettmar 1994; Draine & Lazarian 1998). Bot *et al.* (2010) showed that this grain type could explain the submm and mm excess in the Large and Small Magellanic Cloud.

In order to interpret the dust SED of a galaxy a physical model is needed, based on realistic dust properties and taking into account the heating and emission of dust immersed in the wide range of ISRFs. Ideally, radiation transport in a realistic geometry should be done, but it is often difficult due to the complex geometry and large number of parameters. Models can generally be classified into three broad groups: (1) modified blackbody fits, which are too simple to describe reality correctly but give a first idea of the range of dust temperatures, (2) semiempirical models that try, in a simplified way, to describe dust immersed in a range of different radiation fields (e.g. Dale *et al.* 2001, Draine *et al.* 2007, Galametz *et al.* 2009, 2011, da Cunha *et al.* 2008) and (3) models that include full radiation transfer (e.g. Popescu *et al.* 2000, 2011 for spiral galaxies, Siebenmorgen & Krügel 2007 for starburst galaxies) which are the most precise description of a galaxies if all parameters, including the geometry, is known.

We model the dust emission of the nearby ($D = 2.9$ Mpc) starbursting dwarf galaxy NGC 4214. Due to its proximity and the large amount of ancillary data it is possible to apply physical models that take into account the full radiation transfer and constrain their input parameters. Here, we describe the general outline of our work and the results. A more detailed description of the available data, the data reduction and the determination of the model input parameters is presented in Hermelo *et al.* in this volume.

2. The observed SED of NGC 4214

NGC 4214 is an irregular galaxy, dominated by two bright, young star-forming (SF) regions in its center (called NW and SE in this work). A great wealth of data is available for this object, ranging from GALEX ultraviolet (UV), Hubble Space Telescope (HST) UV to infrared (IR) images, Spitzer IRAC and MIPS, Herschel SPIRE, IRAM MAMBO at 1.2mm, as well as Planck detections at 350, 550 and 850 μm. Furthermore the galaxy has been mapped in HI as part of the THINGS project (Walter *et al.* 2008) and has been observed with OVRO in CO(1-0) (Walter *et al.* 2001).

The large amount of data, most of them at a high spatial resolution, allows to determine the dust SED separately for the emission from the SF regions SE and NW, where the dust is heated by nearby massive stars, and the diffuse dust heated by the general interstellar radiation field. Figure 1 shows the observed dust SED of the individual SF regions (top) and of the diffuse medium (bottom), determined by subtracting the emission of both SF regions from the total dust emission.

3. The models

We separately modelled the emission from the massive SF regions and the diffuse dust emission. We use the model of Groves *et al.* (2008) that describes the emission from an HII region together with its surrounding PDR. For the diffuse emission we used the library of model SEDs from Popescu *et al.* (2011).

The model by Groves *et al.* describes the luminosity evolution of a star cluster of mass $M_{\rm cl}$ from stellar population synthesis, and incorporates the expansion of the HII region and PDR due to the mechanical energy input of stars and SNe. The dust emission from the HII region and surrounding PDRs is calculated from radiation transfer. The main parameters in this model are: (1) metallicity, (2) age of the cluster, (4) external pressure, (3) compactness parameter, C, which parametrizes the heating capacity of the stellar cluster and depends on $M_{\rm cl}$ and the external pressure of the ambient medium, (5) column density of the PDR, $N_{\rm H}$, and (6) covering factor, $f_{\rm cov}$, defining which fraction of the HII region is surrounded by the PDR. The large amount of data and results from previous studies allow us to constrain most parameters (all except $N_{\rm H}$) very tightly from observations (see Hermelo *et al.* in this volume).

The library of models from Popescu *et al.* (2011) is calculated including the full radiation transfer for a disk galaxy. The model galaxy consists of two exponential stellar disks, describing old and young stars, respectively, together with their associated dust disks. Two additional components of the model can be neglected in our case: a bulge which is absent in NGC 4214 and individual dusty SF regions, which we subtracted from the SED and modelled separately with the model of Groves *et al.* (2010). The main parameters of the diffuse model are: (1) the total central face-on opacity in the B-band, τ_B^f, (2) the star formation rate, SFR, (3) the clumpiness factor F which is the same as $f_{\rm cov}$ of Groves *et al.*, (3) the normalized luminosity of the old stellar population, *old*, and (4) the exponential scale-length of the stellar emission in the blue band, h_s. From the primary parameters SFR and F, Popescu *et al.* define the star formation rate powering the diffuse dust emission, $SFR' = SFR \times (1-F)$. We observationally determined $h_s = 873^{+172}_{-123}$ pc and $F = 0.55$, the average of the values of NW and SE. We left SFR', τ_B^f and *old* as free parameters.

4. Results and conclusions

In Fig. 1 we show the fits of the models to the data, separately for both SF regions and for the diffuse emission. The parameters were chosen within the tight constraints given by the observations. Note that for the diffuse component the flux level of the dust emission is not free but is fixed by the observationally measured SFR.

The Groves *et al.* (2008) model fits all data points of the region NW longwords of 10 μm within the errors. The fit for the SE region is good for the dust SED but overestimates the thermal radio emission. Here, a better fit could be achieved for ages higher than those estimated for the central clusters (4.5 instead of 3.5 Myr). This could indicate that the

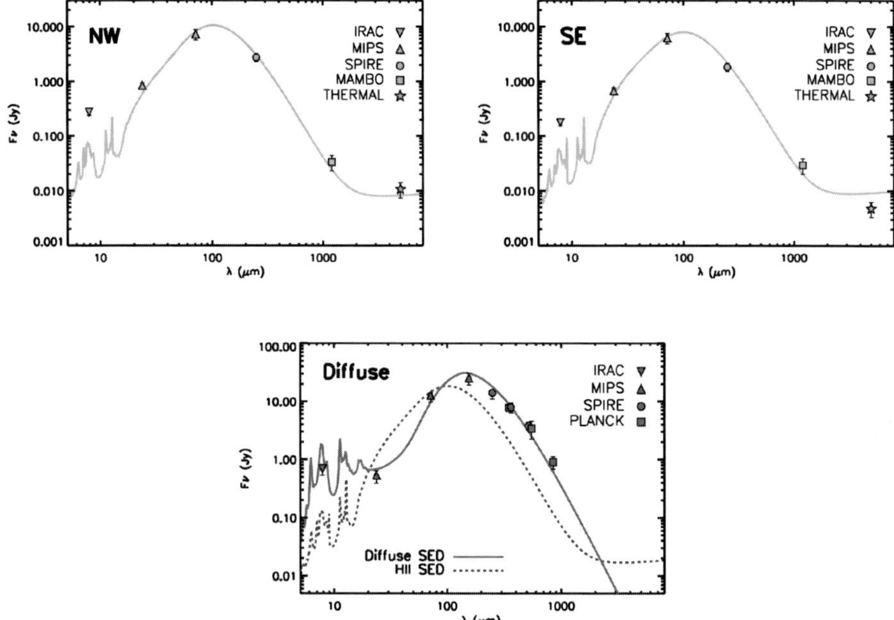

Figure 1. *Top:* The dust emission of the HII regions NW (left), SE (right) and the best-fit model for the parameter range determined by the observations. *Bottom:* The same for the diffuse emission. For illustration, the sum of the fit to NW+SE is also included (dashed line).

dust within our apertures is not only heated by the central clusters but also by an older stellar population. The model underestimates the emission at 8 μm, the reasons for this still need to be investigated. We achieve an excellent fit for the diffuse dust emission for the best-fit values of $\tau_B^f = 2$, $SFR' = 0.0071\, M_\odot\ \mathrm{yr}^{-1}$ and $old = 0$. However, the best-fit SFR derived from SFR' and F is a factor of about 1.7 below the corresponding value derived from the observed, attenuation corrected integrated UV-to-optical luminosity of NGC 4214. A possible reason for this discrepancy is that part of the stellar UV emission leaves the galaxy before contributing to the diffuse dust heating. This is not an unlikely scenario for a starbursting dwarf galaxy where massive stars can create holes in the ISM around them. Alternatively, differences in the geometry of the model galaxy and NGC 4214 (e.g. a higher ratio of the vertical scale-height to the radial scale-length in NGC 4214) could also have an effect.

In conclusion, we obtain an in general good agreement between model predictions, based on standard dust properties, and the data. There are no indications for a submm excess in this galaxy. The fact that we achieve a good agreement between models and data shows that this approach is a fruitful way to better understand the SED of nearby dwarf galaxies.

References

Bendo, G. J., Dale, D. A., Draine, B. *et al.* 2006, *A&A*, 523, 20

Bot, C., Ysard, N., Paradis, D., Bernard, J. P., Lagache, G., Israel, F. P., & Wall, W. F. 2010, *ApJ*, 652, 283

Dale, D. A., Helou, G., Contursi, A., Silbermann, N. A., & Kolhatkar, S. 2001, *ApJ*, 549, 215

da Cunha, E., Charlot, S., & Elbaz, D. 2008, *MNRAS*, 388, 1595

Draine, B. T. & Lazarian, A. 1998, *ApJ*, 508, 157

Draine, B. T., Dale, D. A., Bendo, G., Gordon, K. *et al.* 2007, *ApJ*, 663, 866

Engelbracht, C. W., Rieke, G. H., Gordon, K. D., Smith, J.-D. T., Werner, M. W., Moustakas, J., Willmer, C. N. A., & Vanzi, L. 2008, *ApJ*, 678, 804

Ferrara, A. & Dettmar, R.-J. 1994, *ApJ*, 427, 155

Israel, F. P., Wall, W. F., Raban, D., Reach, W. T., Bot, C., Oonk, J. B. R., Ysard, N., & Bernard, J. P. 2010, *A&A*, 519, 67

Galametz, M., Madden, S., Galliano, F., *et al.*, 2009, *A&A*, 508, 645

Galametz, M., Madden, S. C., Galliano, F., Hony, S., Bendo, G. J., & Sauvage, M. 2011, *A&A*, 532, 56

Galliano, F., Madden, S. C., Jones, A. P., Wilson, C. D., & Bernard, J.-P., & Le Peintre, F. 2003, *A&A*, 407, 159

Galliano, F., Madden, S. C., Jones, A. P., Wilson, C. D., & Bernard, J.-P. 2005, *A&A*, 434, 867

Groves, B., Dopita, M. A., Sutherland, R. S., Kewley, L. J., Fischera, J., Leitherer, C., Brandl, B., & van Breugel, W. 2008, *A&A*, 176, 438

Lisenfeld, U., Israel, F. P., Stil, J. M., & Sievers, A. 2002, *A&A*, 382, 860

Meny, C., Gromov, V., Boudet, N. Bernard, J.-P., Paradis, D., & Nayral, C. 2007, *A&A*, 468, 171

Reach, W. T., Dwek, E. Fixsen *et al.* 2007, *ApJ*, 451, 188

Popescu, C. C. and Misiriotis, A. and Kylafis, N. D., Tuffs, R. J., & Fischera, J. 2000, *A&A*, 363, 138

Popescu, C. C., Tuffs, R. J., Dopita, M. A., Fischera, J., Kylafis, N. D., & Madore, B. F. 2011, *A&A*, 527, 109

Siebenmorgen, R. & Krügel, E. 2007, *A&A*, 461, 445

Walter, F., Taylor, C. L., Hüttemeister, S., Scoville, N., & McIntyre, V. 2001, *AJ*, 121, 727

Walter, F., Brinks, E., de Blok, W. J. G., Bigiel, F., Kennicutt, Jr., R. C., Thornley, M. D., & Leroy, A. 2008, *AJ*, 136, 2563

The Spectral Energy Distribution of Galaxies
Proceedings IAU Symposium No. 284, 2011
R.J. Tuffs & C.C. Popescu, eds.

doi:10.1017/S1743921312008964

Modeling the dust Spectral Energy Distribution of NGC 4214

Israel Hermelo[1], Ute Lisenfeld[1], Mónica Relaño[1], Richard J. Tuffs[2], Cristina C. Popescu[3], Jörg Fischera[4] and Brent Groves[5]

[1]Departamento de Física Teórica y del Cosmos, Universidad de Granada, Spain
email: israelhermelo@ugr.es, ute@ugr.es, mrelano@ugr.es

[2]Max Planck Institut für Kernphysik, Heidelberg, Germany
email: Richard.Tuffs@mpi-hd.mpg.de

[3]Jeremiah Horrocks Institute for Astrophysics and Supercomputing, University of Central Lancashire, Preston, U.K.
email: cpopescu@uclan.ac.uk

[4]Canadian Institute for Theoretical Astrophysics (CITA), University of Toronto
email: fischera@cita.utoronto.ca

[5]Max Planck Institut für Astronomie, Heidelberg, Germany
email: brent@mpia.de

Abstract. We have carried out a detailed modeling of the dust Spectral Energy Distribution (SED) of the nearby, starbursting dwarf galaxy NGC 4214. A key point of our modeling is that we distinguish the emission from (i) HII regions and their associated photodissociation regions (PDRs) and (ii) diffuse dust. For both components we apply templates from the literature calculated with a realistic geometry and including radiation transfer. The large amount of existing data from the ultraviolet (UV) to the radio allows the direct measurement of most of the input parameters of the models. We achieve a good fit for the total dust SED of NGC 4214. In the present contribution we describe the available data, the data reduction and the determination of the model parameters, whereas a description of the general outline of our work is presented in the contribution of Lisenfeld *et al.* in this volume.

Keywords. dust, extinction, galaxies: irregular, galaxies: individual (NGC 4214), galaxies: ISM

1. Introduction

NGC 4214 is a nearby (2.94 Mpc; Maíz-Apellániz *et al.* 2002) barred Magellanic irregular galaxy consisting of a smooth extended disk (Figure 1a) and a centrally concentrated young star forming (SF) region dominated by the two SF complexes NW and SE (Figure 1b). The metallicity of both complexes has been measured by Kobulnicky & Skillman (1996), who found $Z \sim 0.3 Z_{\odot}$. The main stellar cluster in the NW complex, referred here as NW-A, is located at the center of the galaxy and has removed most of the gas in our line of sight. Using stellar synthesis models, Úbeda *et al.* (2007) determined the age, the mass, the radius and the extinction of the stellar clusters within the complexes NW and SE.

We have modeled the dust SED of NGC 4214 separately for the dust emission from SF regions and from the diffuse component. Due to its proximity, the large amount and good quality of the available data and previous studies of this object, most of the model parameters can be directly calculated, making this galaxy an excellent laboratory to improve our knowledge about dust in dwarf galaxies.

2. Photometry

In order to determine the observational dust SED of NGC 4214 we used data from *Spitzer*, Herschel, and Planck archives, as well as our own observations at 1200 μm using the MAMBO bolometer at the IRAM 30m telescope on Pico Veleta (Spain). For all the photometric measurements (i) color corrections, (ii) aperture corrections, and (iii) line decontamination were done. With the exception of the atomic line [CII] 158 μm, which contributes 4.26% to the flux of MIPS 160 μm, the contamination of the other considered lines ([OI] 63, [OIII] 88, [NII] 122, [OI] 146 and [NII] 205 μm as well as CO rotational lines) was found to be negligible. Besides, the stellar contribution in the IRAC 8 μm image was subtracted. Figures 1a and 1b show the apertures used to measure the total emission and the emission from the SF complexes, respectively. In order to constrain the thermal radio emission we made use of the extinction corrected Hα flux reported by MacKenty *et al.* (2000) and eq. (3) and (4a) in Condon (1992). The diffuse dust SED was determined by subtracting the combined emission of NW and SE complexes from the total emission of the galaxy.

3. Determination of the parameters

To model the **dust emission from the HII regions** we used the model from Groves *et al.* (2008) based on radiation transfer calculations. This model computes the whole emergent spectrum including the attenuated stellar spectrum, the line emission spectrum, as well as a dust SED. The input parameters of this model are (i) the age of the stellar cluster, (ii) the metallicity, Z, (iii) the hydrogen column density of the surrounding PDR, $N_{\rm H}$, (iv) the external pressure, p_0, (v) the compactness, C, which parametrizes the dust grain temperature distribution, and (vi) the covering factor, $f_{\rm cov}$, which represents the fraction of the surface of the HII region which is covered by the PDR. With exception of $N_{\rm H}$, all the other parameters were observationally determined:

• Following Úbeda *et al.* (2007), we adopted 5 and 3.5 Myr for the age of the NW and SE complex, respectively.

• We used a metallicity of Z$= 0.2$ Z$_\odot$ (results shown here) and also tested Z$= 0.4$ Z$_\odot$ which yielded very similar results (no template for Z$= 0.3$ Z$_\odot$ is available).

• We determined p_0 by comparing the expected and the observed radii of the HII regions as a function of the age. We found values of $\log\left[(p_0/k)/\mathrm{cm}^{-3}\mathrm{K}\right] \sim 6-7$.

• C was directly determined from the external pressure and the mass of the clusters. We found values of 5.0-6.0 and 4.5-5.5 for the NW and SE complexes, respectively.

• In order to calculate f_{cov} we assumed an optically thick, uniformly fragmented PDR surrounding the ionized gas. With these assumptions, the intrinsic luminosity of the central stellar cluster can be approximated by the sum of the observed luminosities of the stars, L_s, and the luminosity re-emitted by the dust, L_d. In this picture, the covering factor is $f_{cov} = \frac{L_d}{L_d+L_s}$. We obtained $f_{\rm cov} = 0.44$ and 0.65 for NW and SE, respectively.

To model the **diffuse dust emission** we used the model from Popescu *et al.* (2011) based on full radiation transfer calculations of the propagation of starlight in galaxy disks. The relevant parameters of this model in the case of NGC 4214 are (i) the radial B-band scale-length of the old stellar disk, h_s, (ii) the central B-band face-on opacity, τ_B^f, (iii) the star formation rate, SFR, (iv) the luminosity of the old stellar population, *old*, and (v) the clumpiness factor, F, which is equivalent to the covering factor of the Groves *et al.* (2008) model. From the primary parameters SFR and F, Popescu *et al.* (2011) define the star formation rate powering the diffuse dust emission as $SFR' = SFR \times (1-F)$. The diffuse SED, as well as the parameters SFR' and *old*, which control the radiation field, must be scaled by comparing the scale-length of NGC 4214 with the reference scale-length $h_s^{ref} = 5670\,pc$. In our case, we observationally determined:

• The scale-length, for which we obtained h_s=873^{+172}_{-123} pc by fitting elliptical isophotes.

• The clumpiness factor, which was calculated as the average covering factor of the NW and SE regions ($< f_{\rm cov} >= 0.55$).

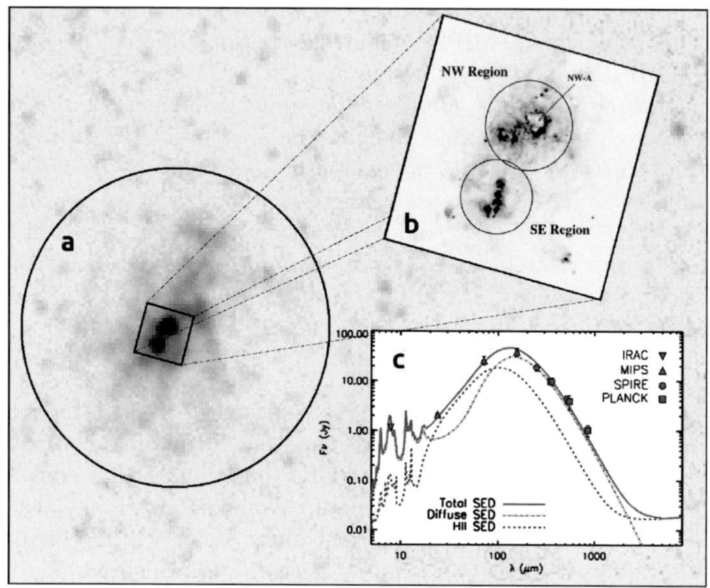

Figure 1. a) SPIRE 250 μm image of the disk of NGC 4214. The circle corresponds to the aperture we used to measure the global emission and it has a radius of ~4 kpc. **b)** HST-WFC3 Hα map showing the complexes NW and SE and the location of the SSC NW-A. **c)** Best-fit model (solid line) to the global SED for the parameters described in section 3. The global SED was obtained as the sum of the best-fits for the complexes NW and SE (dashed line) and the best-fit for the diffuse dust (dashed-dotted line).

4. Results

Figure 1c shows the data points of the total observed dust SED, together with the best-fit model (solid line) calculated as the sum of the models of the SF regions (dashed line) and the model of the diffuse dust (dashed-dotted line). The fits to the emission of the SF regions were obtained individually using the parameters described in Sec. 3, whereas the fit to the diffuse emission was obtained leaving τ_B^f, SFR', and *old* as free parameters. We achieved a good agreement between the observed and the modelled FIR/submm SED of NGC 4214. However, from our SED modelling we inferred $SFR = 0.16\,M_{\odot}{\rm yr}^{-1}$ which is a factor 1.7 smaller than the value derived from the observed, integrated UV-to-optical luminosity, indicating that part of the stellar radiation does not contribute to the diffuse dust heating (see Lisenfeld *et al.* in this volume for more details).

References

Condon, J. J. 1992, *ARAA*, 30, 575

Groves, B., Dopita, M. A., Sutherland, R. S., *et al.* 2008, *ApJS*, 176, 438

Kobulnicky, H. A. & Skillman, E. D. 1996, *ApJ*, 471, 211

Maíz-Apellániz, J., Cieza, L., & MacKenty, J. W. 2002, *AJ*, 123, 1307

MacKenty, J. W., Maíz-Apellániz, J., Pickens, C. E., *et al.* 2000, *AJ*, 120, 3007

Popescu, C. C., Tuffs, R. J., Dopita, M. A. *et al.* 2011, *A&A*, 527, A109

Úbeda, L., Maíz-Apellániz, J., & MacKenty, J. W. 2007, *AJ*, 133, 932

The Spectral Energy Distribution of Galaxies
Proceedings IAU Symposium No. 284, 2011
R.J. Tuffs & C.C. Popescu, eds.

doi:10.1017/S1743921312008976

A Multi-wavelength MOCASSIN model of the Magellanic-type galaxy NGC 4449

O. Ł. Karczewski[1], M. J. Barlow[1], M. J. Page[2] and S. C. Madden[3]

[1]Department of Physics and Astronomy, University College London, Gower Street, London, WC1E 6BT, UK

[2]Mullard Space Science Laboratory, University College London, Holmbury St. Mary, Dorking, RH5 6NT, UK

[3]CEA Saclay Service d'Astrophysique, L'Orme des Merisiers Bat 709, 91191 Gif-sur-Yvette Cedex, France

e-mail: olk@star.ucl.ac.uk

Abstract. We use the photoionisation and dust radiative transfer code MOCASSIN to create a model of the dwarf irregular galaxy NGC 4449. The best-matching model reproduces the global optical emission line fluxes and the observed spectral energy distribution (SED) spanning wavelengths from the UV to sub-mm, and requires the bolometric luminosity of 6.25×10^9 $L_\odot$ for the underlying stellar component, $M_{\rm dust}/M_{\rm gas}$ of 1/680 and $M_{\rm dust}$ of 2.2×10^6 $M_\odot$.

Keywords. dust, extinction, galaxies: dwarf, galaxies: general, galaxies: individual (NGC 4449), galaxies: stellar content, methods: numerical

---

1. Introduction

The spectral energy distribution (SED) of a galaxy emerges as a combination of the underlying stellar emission, the degree of reprocessing taking place in the interstellar medium (ISM) and thermal or non-thermal emission due to dynamical or evolutionary processes. As a consequence, galaxies emit across the entire range of the electromagnetic spectrum and the distribution of the energy output as a function of wavelength provide clues to the galaxy's past and present. Numerical models that reproduce the observed global properties of a galaxy over a wide range of wavelengths can be useful in understanding the physical and chemical nature of the underlying stellar and dust components. Here we present a MOCASSIN model of the dwarf irregular galaxy NGC 4449.

MOCASSIN (Ercolano *et al.* 2003, 2005, 2008) is a fully 3-dimensional photoionisation and dust radiative transfer code. Recent applications of the code include modelling the planetary nebula NGC 6302 (Wright *et al.* 2011) and studying dust formation by SN 2008S (Wesson *et al.* 2010). Modelling entire galaxies is technically challenging due to their large physical sizes and the high level of complexity of the galactic environments. Dwarf galaxies are ideal candidates due to their intrinsically small sizes, relatively simple star formation histories and the low degree of processing of their interstellar material.

NGC 4449 (Fig. 1) is a metal-poor magellanic-type irregular galaxy. It is not a typical dwarf irregular: at the distance of 4.21 Mpc its H I envelope extends to a radius of 14.2 kpc (11′.6), equal to approximately $6R_{25}$ (Swaters *et al.* 2002), and exhibits a counter-rotating core. It was selected for this study for its extensive coverage of photometric measurements for wavelengths ranging from the UV to sub-mm. Recently, new FIR observations were acquired by the *Herschel Space Observatory* (Pilbratt *et al.* 2010) as part of the Guaranteed Time Key Project: Dwarf Galaxy Survey (PI: Suzanne Madden).

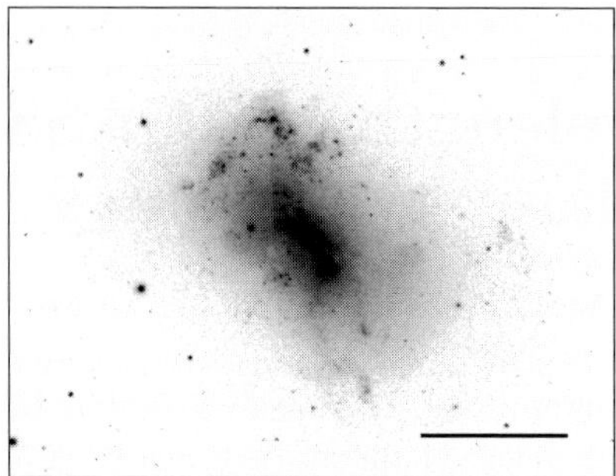

Figure 1. SDSS r-band image of NGC 4449. North is up, east is to the left. The image is centred at $12^h 28^m 11^s.1$, $+44°05'37''$ (J2000). The bar is 2′ in length.

The emission line fluxes, used to constrain the models, were derived from global spectra, acquired with the drift scanning technique, and are representative of the entire galaxy (Kobulnicky *et al.* 1999).

2. Methods and Results

Our MOCASSIN simulations use 0.5 million grid cells to establish the observed H I profile of Swaters *et al.* (2002) within the inner 3.3 kpc (2′.7), enclosing the visible part of NGC 4449 and about a half of its total H I mass. The geometry, gas density distribution and the observed abundances are fixed inputs. Trial families of simulations are run and a range of $M_{\rm dust}/M_{\rm gas}$, bolometric stellar luminosities $L_\star$ and input stellar radiation fields are tested. A parameter set providing the best match to the shape of the low-resolution SED, the global emission line fluxes and the integrated Hα luminosity (Hunter *et al.* 1999) is expanded into a new, higher resolution parameter space and the procedure is iterated until a satisfactory match with all observables is obtained. Fig. 2 gives a schematic summary of this method.

The stellar radiation field is the most crucial input, as it results from the assumed star formation history. In this work, two continuous episodes of constant star formation activity were assumed and an initial (and subsequently expanded, according to Fig. 2) set of 45 templates generated with STARBURST99 (Leitherer *et al.* 1999; version 2011-01-17) was tested. A small subset of the resulting SEDs is shown in Fig. 3.

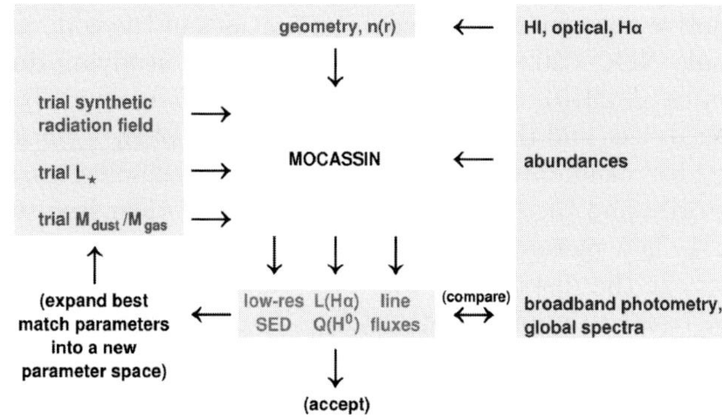

Figure 2. A schematic summary of the modelling method used. The shaded regions group together the inputs to MOCASSIN (left) the outputs (middle) and the observational constrains (right).

To break the degeneracy it is helpful to ask if both *(i)* the shape of the observed SED and *(ii)* the total emergent $L(\mathrm{H}\alpha)$ for a particular star formation scenario *can* be matched by varying $M_{\mathrm{dust}}/M_{\mathrm{gas}}$ and/or $L_{\star}$, whose effect on the resulting SED is relatively simple to predict. This condition allows for more than 90 per cent of the initial templates to be rejected at an early stage. The final best-matching model is selected by inspecting the predicted emission line fluxes. For NGC 4449 the best-matching SED is presented in Fig. 4 and the derived physical quantities are given in Table 1.

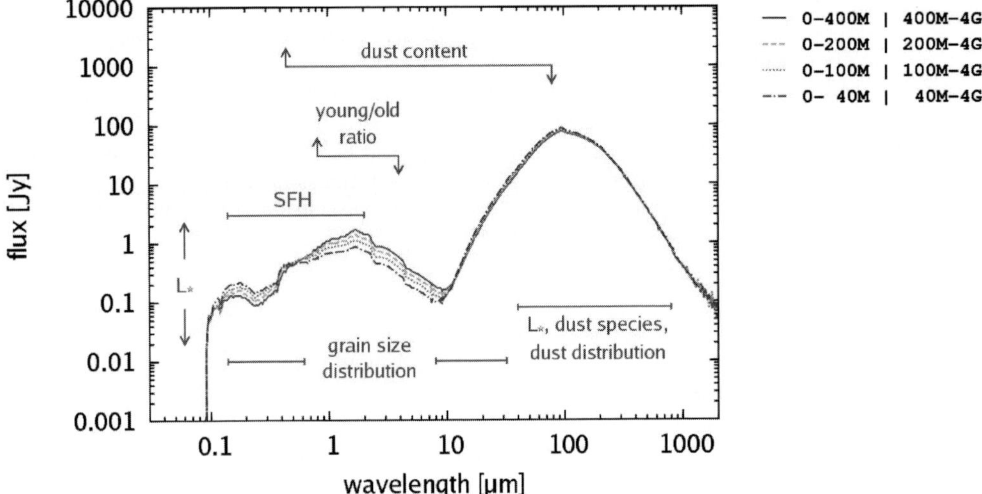

Figure 3. A family of predicted SEDs for four trial star formation scenarios assuming two episodes of continuous star formation with two different star formation rates. All input parameters are identical except for the length of the two star formation episodes (see legend for details). The approximate wavelength ranges affected by the various free parameters are indicated.

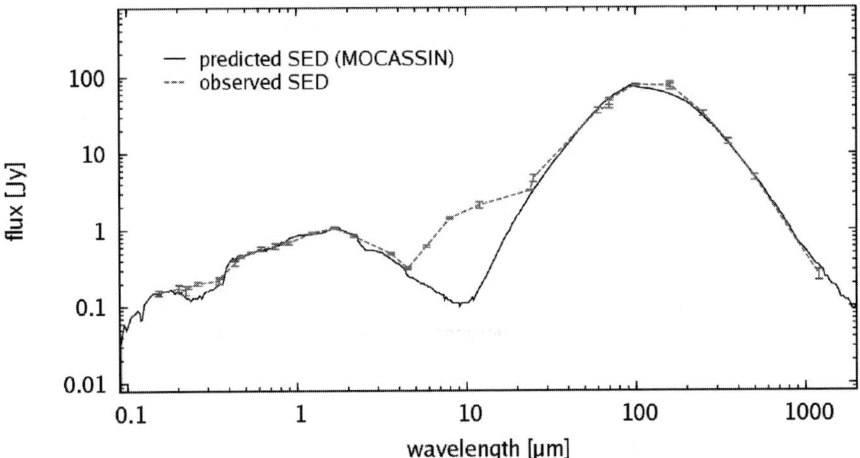

Figure 4. The predicted SED of the best-matching MOCASSIN model of NGC 4449. PAHs are not included in this model. The observed SED is based on the archive data (*GALEX* in the UV, *Swift* in the UV/optical, SDSS in the optical, 2MASS in the NIR) and the published photometry (Engelbracht *et al.* 2008, Hunter *et al.* 1986, Böttner *et al.* 2003; at longer wavelengths). In the FIR new *Herschel* PACS and SPIRE measurements are included (Remy *et al.*, in prep.).

parameter	MOCASSIN model of NGC 4449	
$L_\star$	$6.25 \pm 0.25 \times 10^9\ L_\odot$	
$M_\star$	$1.2 \pm 0.1 \times 10^9\ M_\odot$	
$M_{\rm gas}$	$1.5 \pm 0.2 \times 10^9\ M_\odot$	
$M_{\rm dust}$	$2.2 \pm 0.2 \times 10^6\ M_\odot$	
$M_{\rm dust}/M_{\rm gas}$	1/680 (1/850 to 1/540)	
dust composition	100% amorphous carbon	
dust grain sizes	0.005–1 μm, $n \propto a^{-3.5}$	
SF episodes	6 ± 2 Gyr to 120 ± 40 Myr	120 ± 40 Myr to present
relative $M_\star$	120	1
<SFR>	0.2 to 0.3 $M_\odot\,{\rm yr}^{-1}$	0.06 to 0.12 $M_\odot\,{\rm yr}^{-1}$

Table 1. Summary of global parameters derived from the best-matching MOCASSIN model of NGC 4449 assuming two continuous episodes of star formation. The global emission line fluxes are matched to within 20 per cent or better, with the exception of the sulphur lines [S II] $\lambda\lambda$6717,31, which are overpredicted by a factor of four.

3. Discussion

The gas and dust masses derived from our MOCASSIN model of NGC 4449 are consistent with earlier estimates (Hunter *et al.* 1999; Engelbracht *et al.* 2008). Our best-matching star formation scenario broadly agrees with the presence of an old (3–5 Gyr) and a young population of stars, and continuous star formation over the last 1 Gyr (Bothun 1986; Martin & Kennicutt 1997; Annibali *et al.* 2008). However, the predicted present-day star formation rate is only one-fourth of the estimate based on the Hα luminosity (Hunter *et al.* 1999). The discrepancy between these two independent estimates highlights the differences in the adopted initial mass function (IMF) as well as the limitations of the two-episode model.

The presented technique assumes spherical symmetry for NGC 4449 and a two-episode star formation history, both of which are great simplifications. Also, at present PAHs are not included in the models and, as a consequence, the 4–20 μm range is not correctly predicted. Nevertheless, most of the global parameters are predicted to within the observational uncertainties, and therefore this technique may prove useful in studying the characteristics of dust and the underlying stellar components in other dwarf galaxies.

References

Annibali, F., Aloisi, A., Mack, J., *et al.* 2008, *AJ*, 135, 1900
Bothun, G. D. 1986, *AJ*, 91, 507
Ercolano, B., Barlow, M. J., Storey, P. J., & Liu, X.-W. 2003, *MNRAS*, 340, 1136
Ercolano, B., Barlow, M. J., & Storey, P. J. 2005, *MNRAS*, 362, 1038
Ercolano, B., Young, P. R., Drake, J. J., & Raymond, J. C. 2008, *ApJS*, 175, 534
Böttner, C., Klein, U., & Heithausen, A. 2003, *A&A*, 408, 493
Engelbracht, C. W., Rieke, G. H., Gordon, K. D., *et al.* 2008, *ApJ*, 678, 804
Hunter, D. A., Gillett, F. C., Gallagher, III, J. S., Rice, W. L., & Low, F. J. 1986, *ApJ*, 303, 171
Hunter, D. A., van Woerden, H., & Gallagher, J. S. 1999, *AJ*, 118, 2184
Kobulnicky, H. A., Kennicutt, Jr., R. C., & Pizagno, J. L. 1999, *ApJ*, 514, 544
Leitherer, C., Schaerer, D., Goldader, J. D., *et al.* 1999, *ApJS*, 123, 3
Martin, C. L. & Kennicutt, Jr., R. C. 1997, *ApJ*, 483, 698
Pilbratt, G. L., Riedinger, J. R., Passvogel, T., *et al.* 2010, *A&A* (Letters), 518, L1
Swaters, R. A., van Albada, T. S., van der Hulst, J. M., & Sancisi, R. 2002, *A&A*, 390, 829
Wesson, R., Barlow, M. J., Ercolano, B., *et al.* 2010, *MNRAS*, 403, 474
Wright, N. J., Barlow, M. J., Ercolano, B., & Rauch, T. 2011, *MNRAS*, 418, 370

The Spectral Energy Distribution of Galaxies
Proceedings IAU Symposium No. 284, 2011
R.J. Tuffs & C.C. Popescu, eds.

doi:10.1017/S1743921312008988

Identification of Spitzer-IRS staring mode targets in the Magellanic Clouds

Paul M. E. Ruffle[1], Paul M. Woods[2] and Francisca Kemper[3]

[1]Jodrell Bank Centre for Astrophysics, Alan Turing Building, School of Physics and Astronomy, The University of Manchester, Oxford Road, Manchester M13 9PL, UK

[2]Department of Physics and Astronomy, University College London, Gower Street, London WC1E 6BT, UK

[3]Academia Sinica, Institute of Astronomy and Astrophysics, Taipei 10617, Taiwan

email: `paul.ruffle@manchester.ac.uk`

Abstract. The SAGE-LMC, SAGE-SMC and HERITAGE surveys have mapped the Magellanic Clouds in the infrared using the Spitzer and Herschel Space Telescopes. Over 8.5 million point sources were detected and catalogued in the LMC alone. Staring mode observations using the InfraRed Spectrograph (IRS) on board Spitzer have been obtained for 1,000 positions in the LMC and ~250 in the SMC. From the infrared spectroscopy we have identified the nature of the sources for which spectroscopy is available. These IRS staring mode targets represent an important contribution to the SED of these dwarf galaxies. Here we report on our latest results.

Keywords. galaxies: Magellanic Clouds – line: identification, profiles – stars: early-type, supergiants, AGB, post-AGB, carbon, late-type – ISM: dust, HII regions, planetary nebulae

---

The SAGE survey (Meixner *et al.* 2006) detected and catalogued 8.5 million point sources in the Large Magellanic Cloud (LMC) with Spitzer's IRAC and MIPS instruments. 2.5 million point sources have been detected and catalogued (Gordon *et al.* 2011) in the Small Magellanic Cloud (SMC). The Spitzer archive contains 1,000 IRS staring mode observations within the SAGE-LMC footprint (including 197 from the SAGE-Spec legacy

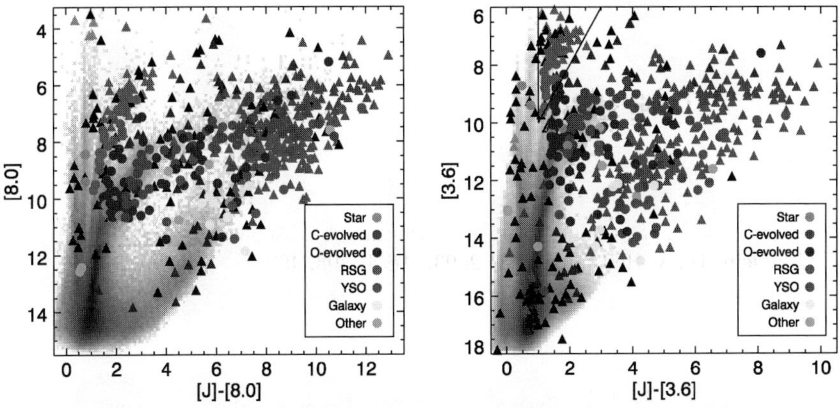

Figure 1. Example CMD distributions for 1,000 LMC point sources. Left: The initial 197 SAGE-Spec point sources from Woods *et al.* (2011) are shown as filled circles, with additional archival sources shown as filled triangles. Black points are sources that still need to be classified. O-AGB, O-PAGB, O-PN, C-AGB, C-PAGB and C-PN classifications are merged into two broad groups. Right: The cut for RSGs can be clearly distinguished from other oxygen-rich evolved stars. The greyscale backgrounds show a Hess diagram of the entire SAGE-LMC sample.

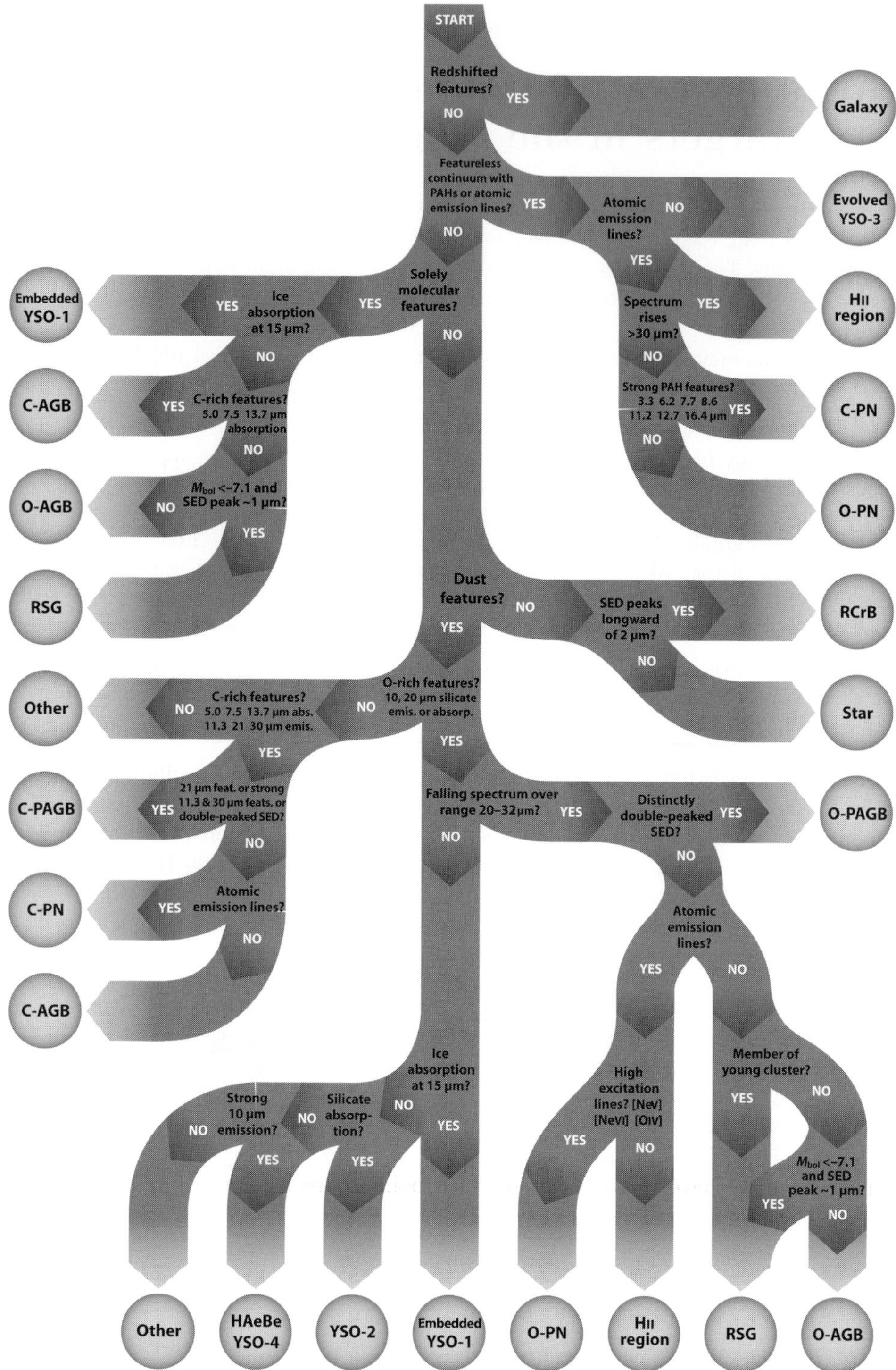

Figure 2. The logical steps of the redesigned classification decision tree of Woods *et al.* (2011), where Spitzer IRS spectra ($\lambda = 5.2$–$38\ \mu$m), associated UBVIJHK, IRAC and MIPS photometry, luminosity, variability, age and other information are used to classify LMC and SMC infrared point sources. See Table 1 for key to classification group codes.

Table 1. Classification groups used in the decision tree shown in Fig. 2.

Code	Object type	Code	Object type
YSO-1	Embedded Young Stellar Objects	RCrB	R Coronae Borealis stars
YSO-2	Young Stellar Objects	C-PAGB	Carbon-rich post-AGB stars
YSO-3	Evolved Young Stellar Objects	O-PAGB	Oxygen-rich post-AGB stars
YSO-4	HAeBe Young Stellar Objects	C-PN	Carbon-rich planetary nebulae
Star	Stellar photospheres	O-PN	Oxygen-rich planetary nebulae
C-AGB	Carbon-rich AGB stars	Other	Other known or unknown types
O-AGB	Oxygen-rich AGB stars	HII	HII regions
RSG	Red SuperGiants	Gal	Galaxies

Table 2. Summary of interim classifications for 1,000 LMC point sources.

Type	Count	Type	Count	Type	Count	Type	Count
YSO	321	RSG	67	C-PN	29	Unknown	8
Star	35	C-PAGB	26	O-PN	32	Unclassified	78
C-AGB	152	O-PAGB*	42	HII	105		
O-AGB	98	(*inc. RVTau	9)	Galaxies	7		

program) and ~250 in the SMC. These targets are the brightest infrared point sources in both clouds, and as such represent an important contribution to the SED of these dwarf galaxies. We are now extending the initial classification of 197 sources in the LMC (Woods *et al.* 2011) to ~1,250 Spitzer-IRS staring mode targets in the Magellanic Clouds.

This spectral classification will allow us to verify the larger photometric classifications that have come out of recent studies of the Magellanic Clouds (e.g. Boyer *et al.* 2011). By relating the spectral identification of objects to regions in colour-magnitude space, a grand picture of the stellar populations of both clouds can be obtained. Also, our spectroscopic classifications can be used to test photometric classification methods, e.g. those by Boyer *et al.* (2011). Moreover, since spectral features provide information on dust composition, we can investigate the physical and chemical environments in each cloud. Both the LMC and the SMC have sub-solar metallicities of 0.5 $Z_\odot$ and 0.2 $Z_\odot$ respectively (Boyer *et al.* 2011). This affects the chemistry in circumstellar environments, both for molecular and dust species, thus altering the spectral appearance. Our classification of the IRS staring mode targets in the SMC will enable us to compare the chemistry to similar environments in the LMC, and get a clearer picture of dust formation in a metal-poor star-forming galaxy similar to those in the early Universe.

Point sources are classified according to their Spitzer IRS spectra ($\lambda = 5.2$–$38\ \mu$m), associated UBVIJHK, IRAC and MIPS photometry, luminosity, variability, age and other information. Fig. 2 shows the redesigned classification decision tree of Woods *et al.* (2011), which we have used to classify each point source. For the LMC we find that we have 321 spectroscopically confirmed YSOs, 250 AGB stars, 105 HII regions, 68 post-AGB stars, 67 RSGs and 61 PNe. Objects are classified as per the groups in Table 1 and a summary of our interim classifications for 1,000 LMC sources is listed in Table 2 (Woods *et al. in prep*). Work on the SMC is also in progress (Ruffle *et al. in prep*).

References

Boyer, M. L., Srinivasan, S., van Loon, J. T., *et al.* 2011, *AJ*, 142, 103
Gordon, K. D., Meixner, M., Meade, M. R., *et al.* 2011, *AJ*, 142, 102
Meixner, M., Gordon, K. D., Indebetouw, R., *et al.* 2006, *AJ*, 132, 2268
Woods, P. M., Oliveira, J. M., Kemper, F., *et al.* 2011, *MNRAS*, 411, 1597

The Spectral Energy Distribution of Galaxies
Proceedings IAU Symposium No. 284, 2011
R.J. Tuffs & C.C. Popescu, eds.

doi:10.1017/S174392131200899X

The S^{4}G view of stellar mass, mid-IR dust, and evolved, intermediate-age stars in nearby galaxies

Sharon E. Meidt[1] **and S^{4}G team**

[1]Max Planck Institute für Astronomie
Königstuhl 17,
DE-69117, Heidelberg, Germany
email: meidt@mpia.de

Abstract. With deep imaging at 3.6 and 4.5 μm where the light in nearby galaxies is dominated by old stars, the Spitzer Survey of Stellar Structure in Nearby Galaxies (S^{4}G) promises to be the ultimate inventory of stellar mass and structure in the local universe. We present results from a novel technique that makes it possible to fully exploit the information contained in these images, pertaining not only to the stellar light (and, ultimately, mass distribution), but also the nature and distribution of the mid-IR dust and the properties of evolved, intermediate age stars (e.g. in AGB-dominated star clusters). We apply Independent Component Analysis (ICA) to the 3.6 and 4.5 μm bands to separate the light from the old stars from the secondary non-stellar (i.e. PAH and hot dust) sources of emission, which are identified via comparison to the non-stellar emission imaged at 8 μm. Then, within the context of age and mass estimation at high z, we extract optical-to-mid-IR SEDs for a sample of ICA-identified AGB-dominated clusters to constrain the typically uncertain fractional contribution of AGB light to the total stellar emission in (rest-frame) NIR bands.

Keywords. galaxies: structure, galaxies: stellar content, galaxies: star clusters, stars: AGB and post-AGB

---

1. Introduction

The Spitzer Survey of Stellar Structure in Galaxies (S^{4}G; roughly 2300 galaxies within 40 Mpc; Sheth *et al.* (2010)) provides an un-paralleled inventory of stellar mass and structure in nearby galaxies, with deep imaging at 3.6 μm and 4.5 μm (reaching $\mu_{3.6\mu m}(AB)\sim 27$ mag arcsec^{-2}) where the light mainly traces the oldest stars. But alongside the K and M giants that dominate the flux in these two shortest IRAC wavebands, several other sources appear in star-forming galaxies: PAH (the 3.3 μm emission feature in the 3.6 μm band and the PAH continuum at 4.5 μm; Flagey *et al.* 2006), accessory non-stellar thermal continuum emission (i.e. from hot dust; Mentuch *et al.* 2009, 2010), and lower-M/L asymptotic giant branch (AGB) and red supergiant (RSG) stars can all make significant local contributions to the emission at these wavelengths.

To obtain a contamination-free view of the old stellar disk, we have developed a technique using ICA (FastICA; Hyvärinen (1999) and Hyvärinen & Oja(2000)), which requires only the information from 3.6 and 4.5 μm bands – the only data available for a large fraction of the sample – and accounts for the fact that dust and stellar SEDs have different shapes between these two wavelengths. Cleaned maps of the old stellar light exhibit [3.6]-[4.5] colors consistent with those of K and M giants (cf. Pahre *et al.* 2004; Hunter *et al.* 2006) and retain full 2D structural information, as demonstrated below for M100 (and see Figure 1).

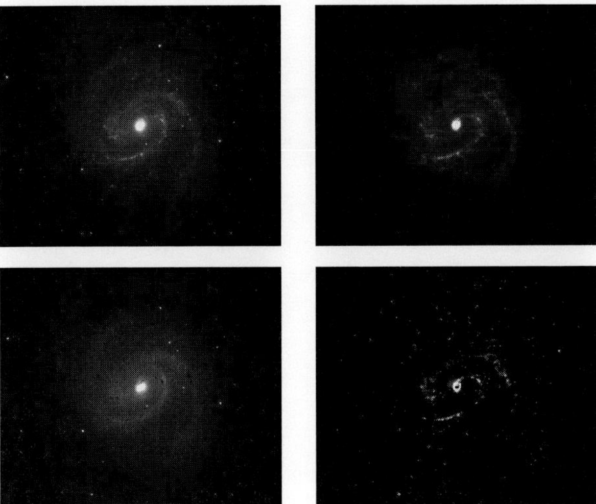

Figure 1. Clockwise from top left: S^4G pipeline-processed IRAC 3.6 μm image of M100 (Sheth *et al.* 2010); SINGS IRAC 8 μm image (Kennicutt *et al.* 2003) shown with the stellar continuum subtracted; ICA s2 map of non-stellar (PAH and hot dust) contaminant emission at 3.6 μm; ICA s1 map of old stellar light at 3.6 μm tracing the stellar mass. The map at the bottom left is a clean, smoothed version of the uncorrected image (top left), while the contaminant map (bottom right) contains most, if not all, of the bright and knotty features tracing star formation (cf. the PAH-dominated 8 μm emission; top right). All images are shown on a square root intensity scale. Field Dimensions: 10.6'×8.75'.

2. ICA separation between old stars and dust

The old stellar light Contaminant-free maps at 3.6 μm compare well with other standard NIR tracers of the old stellar light, as shown in the top left panel of Figure 2. The H-[3.6] color profile for the cleaned map shows better agreement with the predictions from synthesized stellar SED fits to an old, dust-free elliptical for H-[3.6]$\sim$0.4.† The bluer colors reflect a localized reduction in the amount of light relative to the original maps occuring, i.e., in the spiral arms.

Sources of dust emission The quality in the map of the old stellar light stems from the ability of ICA to detect all sources of emission apart from the oldest stars: hot dust, PAH and bright but low M/L intermediate-age stars. By comparing the ICA contaminant maps at 3.6 μm to the non-stellar emission imaged at 8 μm all three of these sources can be identified (see the top right panel of Figure 2). Non-stellar 3.6 μm fluxes in M100 show a clear linear relation to 8 μm fluxes over almost the full range in $F_{8,ns}$, with a slope that is consistent with the expected scaling α between the PAH emission at 3.3μm and at 8 μm ($0.03 \lesssim \alpha \lesssim 0.09$, as measured in the Milky Way; see Flagey *et al.* 2006 and references therein). Other galaxies exhibit a second, steeper relation $F_{3.6}/F_{8,ns}\sim0.3$ characteristic of hot dust (Meidt *et al.* 2011a; less prominent in the particular case of M100).

A third source of emission can be identified as the thin, almost vertical concentration of points at the far left ($F_{8,ns}\lesssim2$ MJy/sr) in the top right plot in Figure 2. This emission arise from 'red knots' that are spatially distinct from the majority of the dust (see the bottom left panel of Figure 2) and populate a different region in the color-magnitude space at the bottom right in Figure 2: knots with $F_{s2,3.6}>0.3F_8$ extend to fainter magnitudes and have redder [3.6]-[4.5] colors (at a given magnitude) than dusty regions with $F_{s2,3.6}<0.3F_8$. The red colors of the former set are consistent AGB stars, in particular (the most luminous

† see http://web.ipac.caltech.edu/staff/jarrett/irac/calibration/galaxies.html

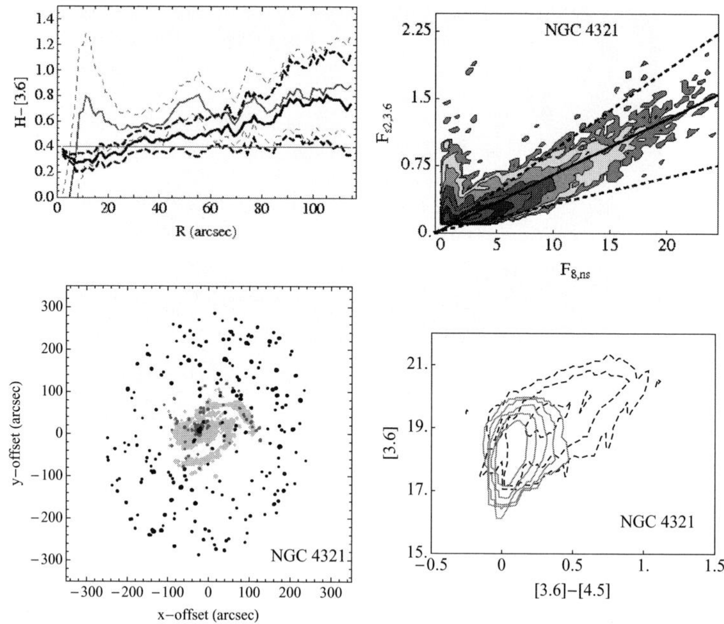

Figure 2. *(Top left)* H-[3.6] color profiles in M100 for the original and corrected maps shown in thin and thick black solid lines, respectively. Dashed errors show the range of values at each radius. The H-[3.6] color expected for an old, dust free population, H-[3.6]=0.4 (see text), is shown as a gray horizontal line. *(Top right)* Contour plot of non-stellar 3.6 μm fluxes $F_{s2,3.6}$ vs. non-stellar 8 μm fluxes $F_{8,ns}$. Overlaid dashed lines represent the expected range in $F_{s2,3.6}=\alpha F_{8,ns}$ for PAH emission, with α=0.09 or 0.03 (see text). A third solid line shows the best-fit linear relation with slope α=0.062. *(Bottom left)* Map of pixels representative of the 'dust' and 'red knot' components in the ICA contaminant distribution. Emission from PAH (light gray), hot dust (dark gray) and 'red knots' (black) is selected from within the range $F_{s2,3.6}/F_{8,ns}<0.09$, $0.1<F_{s2,3.6}/F_{8,ns}<0.3$, and $F_{s2,3.6}/F_{8,ns}>0.3$, respectively, in the plot at the top right. *(Bottom right)* CMD for pixels representative of the primary 'dust' ($F_{s2,3.6}<0.3F_{8,ns}$; in gray solid line, including emission from both PAH and hot dust) and secondary 'red knot' ($F_{s2,3.6}>0.3F_{8,ns}$; in black dashed line) components in the ICA contaminant distribution.

with 0.5<[3.6]-[4.5]<1; Groenewegen (2006)), which dominate the light from all other stars at rest-frame wavelengths $\lambda \gtrsim 1$ μm between ages 0.2-1 Gyr.

3. Circumstellar dust extinction of AGBs in M100

The isolation of AGB-dominated clusters with ICA provides us with a novel opportunity to explore variation in the typically uncertain fractional contribution of AGB light to intermediate-age populations in (rest-frame) NIR bands. For a sample of 17 AGB-dominated clusters chosen from regions least susceptible to the effects of dust extinction/reddening, we extract optical-to-mid-IR SEDs using archival SINGS data and new HAWK-I JHK imaging (i.e. Grosbol & Dottori 2011). We find that NIR brightness is coupled to the mid-IR dust emission such that a significant reduction of AGB light of up to 1 mag in K-band follows from extinction by the dust shell formed during this stage (Meidt *et al.* 2011b). Our sample of clusters–each the analogue of a ~1 Gyr old post-starburst galaxy–has implications within the context of mass and age estimation via SED modelling at high z: we find that the average ~0.5 mag extinction estimated here may be sufficient to reduce the AGB contribution in (rest-frame) K-band from ~70%, as predicted in the latest generation of synthesis models, to ~35% (see Figure 3).

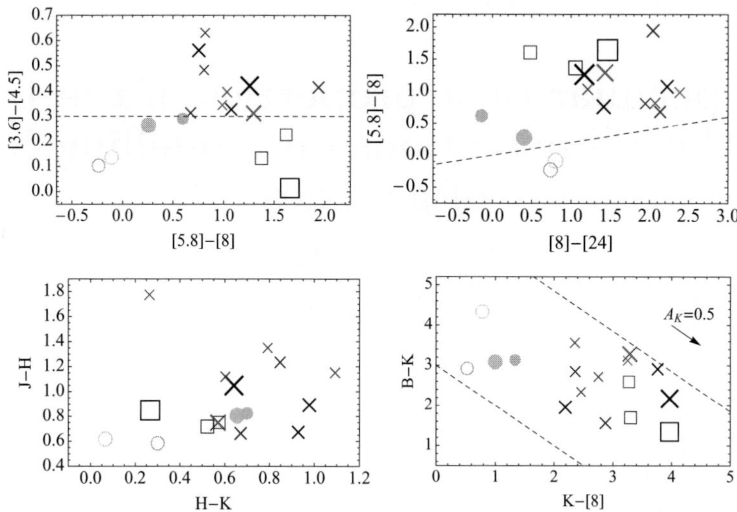

Figure 3. *(Top left, right)* Separation between clusters hosting C-rich (closed), O-rich (open) or extreme (crosses) AGB stars, as motivated by Boyer *et al.* 2011 (and see, e.g., Groenewegen 2006). Square symbols mark a fourth subset of 'mid-IR bright' objects that could potentially be dusty young clusters or HII regions (see Corbelli *et al.* 2011). The dashed black lines show our adopted criteria for distinguishing between types. The gray scaling shows variation in B-V color, from B-V=0.5 (black) to B-V=1.5 (light gray), used here as a proxy for age, while the symbol size varies according to metallicity, from low (small) to high (large). *(Bottom left)* J-H vs. H-K for AGB-dominated clusters, with symbols and colors as above. *(Bottom right)* B-K vs. K-[8] colors showing the impact of the different AGB dust chemistries and mass loss rates. The B-K color traces the prominence of the AGB relative to the other (less evolved) cluster members that alone contribute in the optical. The K-[8] color serves as a measure of the amount of dust produced around each AGB. Clusters dominated by extreme AGBs preferentially lie at bluer B-K and redder K-8 colors than their O-rich and C-rich counterparts. The downward shift corresponds to on average 0.5-1.0 mag extinction in K at fixed V-band brightness.

References

Boyer, M., Srinivasan, S., van Loon, J. *et al.* 2011, *AJ*, 142, 103
Bruzual, G. & Charlot, S. 2003, *MNRAS*, 344, 1000
Corbelli, E., Giovanardi, C., Palla, F., & Verley, S. 2011, *A&A*, 528, 116
Flagey, N., Boulanger, F., Verstraete, L., *et al.* 2006, *A&A*, 453, 969
Groenewegen, M. A. T. 2006, *A&A*, 448, 181
Grosbol, P. & Dottori, H. 2011, *astro-ph*/1109.4255
Hunter, D. A., Elmegreen, B. G., & Martin, E. 2006, *AJ*, 132, 801
Hyvärinen A., 1999, IEEE Signal Processing Lett., 6, 145
Hyvärinen, A. & Oja, E. 2000, *Neural Networks*, 13, 411
Kennicutt, Jr., R. C., *et al.* 2003, *PASP*, 115, 928
Meidt, S. *et al.* 2011a, *ApJ* in press
Meidt, S. *et al.* 2011b, in prep.
Mentuch, E., Abraham, R. G., Glazebrook, K., *et al.* 2009, *ApJ*, 706, 1020
Mentuch, E., Abraham R. G., Zibetti S. 2010, *ApJ*, 725, 1971
Pahre, M. A., Ashby, M. L. N., Fazio, G. G. & Willner, S. P. 2004a, *ApJS*, 154, 229
Sheth, K, Regan, M., Hinz, J., *et al.* 2010, *PASP*, 122, 1397

The Spectral Energy Distribution of Galaxies
Proceedings IAU Symposium No. 284, 2011
R.J. Tuffs & C.C. Popescu, eds.

doi:10.1017/S1743921312009003

The peculiar dust properties of the LMC and their implications on modelling SEDs of galaxies

Frédéric Galliano

Service d'Astrophysique, CEA/Saclay, L'Orme des Merisiers, 91191 Gif-sur-Yvette
email: frederic.galliano@cea.fr

Abstract. In a recent study based on the modelling of spatially resolved *Spitzer* and *Herschel* observations of the Large Magellanic Cloud (LMC), we have shown that the infrared (IR) to submillimetre (submm) grain opacities are systematically higher in the LMC than in our Galaxy. This discovery demonstrates an evolution of the dust properties with environment. Here, we discuss the consequences of this evolution on the modelling of SEDs of galaxies, emphasizing that the uncertainties on the grain properties can lead to erroneous interpretations of the structure of the interstellar medium (ISM) of galaxies, if no additional independent constraints are used.

Keywords. dust, extinction, galaxies: ISM, Magellanic Clouds

---

1. The Dilemma of dust SED modelling: emissivity versus geometry

The SED modelling of a region large enough to encompass different components of the ISM (H II region, PDR, cirrus, etc.) is degenerate. This degeneracy comes from the fact that, using sparse broadband fluxes and no additional constraints, the SED can be fit with different combinations of *macroscopic* properties (*i.e.* the spatial distribution of matter and starlight intensity) and *microscopic* properties (grain cross-sections, size distributions, etc.). Fig. 1 demonstrates this effect on the integrated SED of 1/4 of the LMC (Galliano *et al.* 2011).

Dust radiative transfer models are designed to constrain these *macroscopic* properties, assuming that the *microscopic* properties are well known. However, there are several reasons pointing toward significant variations of the intrinsic grain properties with the environment. The size distribution could be altered by grain destruction in shocks or grain growth in dense regions. Likewise, the grain composition and abundance are affected by grain coating in molecular clouds or by the evolution of the various sources responsible for the elemental enrichment (SN, AGB, etc.). The latter would imply a systematic evolution of the dust composition, and therefore of the grain cross-sections, with metallicity.

2. Single black body or 3D radiative transfer, in the *Herschel* era?

We have attempted to constrain some of these evolutionary processes by focussing on the high spatial resolution ($\simeq 10$ pc) *Spitzer* and *Herschel* observations of the LMC. Our model is based on realistic grain properties and accounts for a distribution of starlight intensity (Galliano *et al.* 2011, for more details). When fit to the observations, it provides the dust mass and mean starlight intensity of each pixel. Since we include SPIRE fluxes (250, 350, 500 μm), which are fully determined by the distribution of equilibrium temperature of the big grains, the distribution of starlight intensity is crucial. We have shown that performing single black body fits on these data leads to systematic underestimates of the dust mass (appendix C.2 of Galliano *et al.* 2011). Moreover, the discrepancy is larger

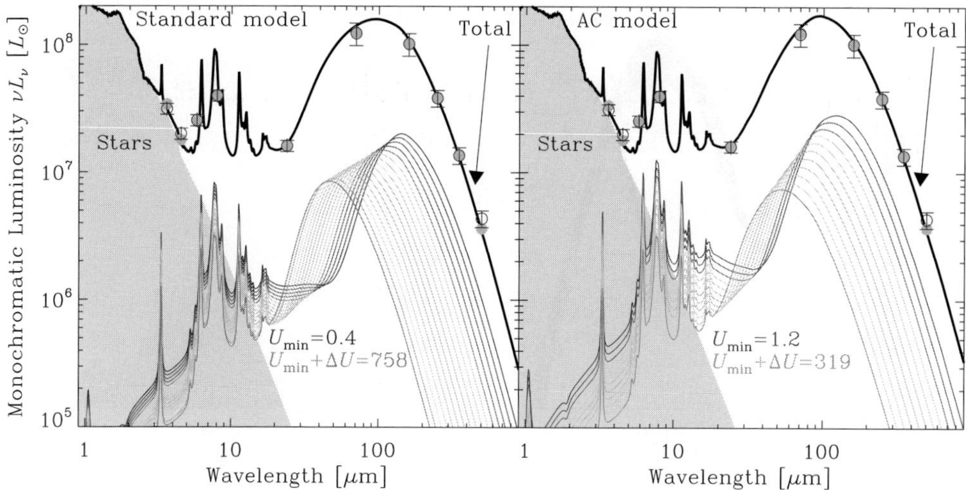

Figure 1. *SED model of 1/4 of the LMC with two different grain mixtures (Galliano et al. 2011).* For each panel: the gray filled component is a stellar continuum; the SEDs correspond to starlight intensities varying between $U_{\rm min}$ and $U_{\rm min} + \Delta U$ in units of 2.2×10^{-5} W m^{-2}; the thick black line is the total model; the filled dots are the synthetic photometry; the open circles with error bars are the observations. The *standard model* is composed of the mixture of PAHs, graphite and silicate satisfying the extinction, emission and depletion constraints for our Galaxy (Zubko *et al.* 2004). The *AC model* is the *standard model* replacing graphite with amorphous carbons. The *AC model* satisfies depletion constraints but has a higher submm opacity than the *standard model.* Both models fit equally well the LMC fluxes.

in dense regions, since the gradient of temperature is larger. It means that performing single black body fits biases the trends of dust mass surface density. In particular, it may affect the dust derived X(CO) factors.

A detailed radiative transfer is not necessary in our case, since our objective is not to derive the 3D structure of the ISM, within each pixel. Having a numerically light model allows us to rigorously propagate the observational uncertainties throughout the entire fitting process (Fig. 2). It allows to account for correlated calibration errors and the effect they have on the non-linearity of the model (Galliano *et al.* 2011).

3. The Test case of the Large Magellanic Cloud: enhanced emissivity

Assuming dust properties of the Milky Way (Zubko *et al.* 2004, *standard model)* with this approach leads to gas-to-dust mass ratios lower than the values permitted by the elemental abundances. In other words, Galactic grains applied to the LMC require more heavy elements locked up in dust than what is available in the ISM (Galliano *et al.* 2011). This model is not physical. We have explored the possbility that a large fraction of the gas could have been overlooked and concluded that, although this effect plays a role, it can not explain this discrepancy. In particular, the gas-to-dust mass ratio is unrealistic, even in the diffuse ISM, where the gas mass is more secure than in the dense ISM.

We have fit an alternate realistic dust composition, which is more emissive (replacing graphite by amorphous carbons; *AC model*). This composition is consistent with the elemental abundances. In summary, this study demonstrates that the grains in the LMC are systematically more emissive than in the Milky Way, in the submm regime. If this property is the consequence of the different evolutionary stages of the Milky and the LMC ($Z \simeq 1/2\, Z_\odot$), then we could expect such a variation of the grain opacities to be

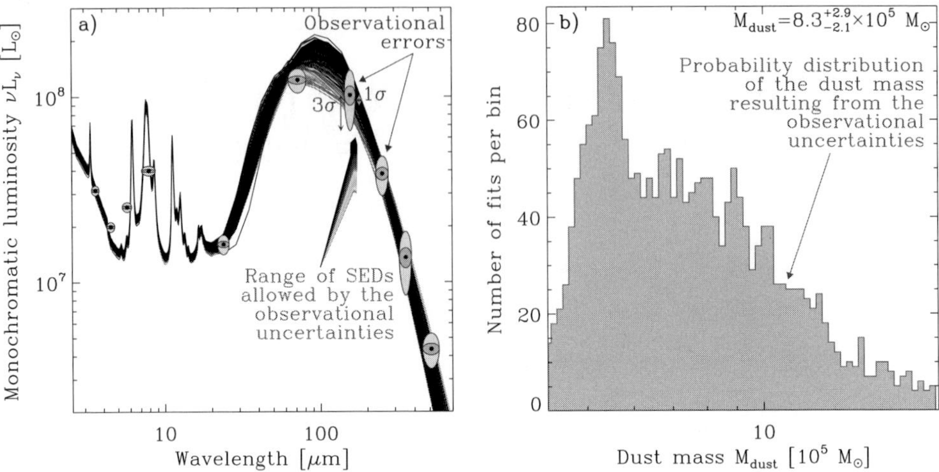

Figure 2. **a)** *Fits of the randomly perturbed SED of Fig. 1.* The grey ellipses show the observational uncertainties. The lines are models fit to the randomly perturbed values, taking into account the correlation between the systematic calibration errors. This process is described in details in Galliano *et al.* (2011). **b)** *Resulting probability distribution of the dust mass for the fits in panel a).* It allows us to estimate the uncertainty on the dust mass for several confidence intervals.

enhanced in low-metallicity galaxies. Independent studies suggest this effect is indeed more pronounced in the SMC than in the LMC (Roman-Duval *et al.* 2011).

4. Implications: degeneracies of dust SED Models of Galaxies

Our study has demonstrated that there is a systematic variation of the grain opacities between galaxies. Consequently, modelling the SED of a galaxy with the grain properties of the Milky Way, without additional constraints (gas tracers or detailed geometry, etc.), could lead to gross errors on its derived dust content and ISM structure. For instance, assuming that the grains in this particular galaxy are more emissive than Galactic dust, an excess IR emission excess could be misinterpreted as a population of inactive clumps, while it would simply be the emission from the regular diffuse ISM and PDRs.

There are several ways to tackle this degeneracy, depending on the available data. 1. Observations of the dust emission at high enough spatial resolution can allow to extract a typical SED of the diffuse ISM, assumed isothermal, and constrain its grain emissivity. 2. Direct estimate on the elemental depletion would also provide valuable information on the dust abundance and composition. 3. Extinction (not attenuation) curves can be useful to test different dust mixtures. 4. The total gas content, together with the metallicity, provides estimates of the expected gas-to-dust mass ratio.

References

Galliano, F., Hony, S., Bernard, J.-P., *et al.* 2011, *A&A*, *in press*

Roman-Duval, J., Gordon, K. D., Meixner, M., *et al.* 2011, in *AAS Meeting* 218, 129.01

Zubko, V., Dwek, E., & Arendt, R. G. 2004, *ApJS*, 152, 211

The Spectral Energy Distribution of Galaxies
Proceedings IAU Symposium No. 284, 2011
R.J. Tuffs & C.C. Popescu, eds.

doi:10.1017/S1743921312009015

Spikes in the SED and ripples in the outskirts of galaxies

Sukanya Chakrabarti

777 Glades Road, Florida Atlantic University, Boca Raton, FL 33431
email: schakra1@fau.edu

Abstract. We describe a new method that allows us to quantitatively characterize galactic satellites from analysis of disturbances in outer gas disks, without requiring knowledge of their optical light. We have demonstrated the validity of this method, which we call Tidal Analysis, by applying it to local spirals with known optical companions, including M51 and NGC 1512. These galaxies span the range from having a low mass companion ($\sim$ one-hundredth the mass of the primary galaxy) to a fairly massive companion ($\sim$ one-third the mass of the primary galaxy). This approach has broad implications for many areas of astrophysics – for the indirect detection of dark matter (or dark matter-dominated dwarf galaxies), and for galaxy evolution in its use as a decipher of the dynamical impact of satellites on galactic disks. Here, we present some preliminary results on the emergent SEDs and images, calculated along the time sequence of these dynamical simulations using the 3-D self-consistent Monte Carlo radiative transfer code RADISHE. We explore star formation prescriptions and how they affect the emergent SEDs and images. Our goal is to identify SED colors that are primarily affected by the galaxy's interaction history, and are not significantly affected by the choice of star formation prescription. If successful, we may be able to utilize the emergent UV-IR SED of the primary galaxy to understand its recent interaction history.

Keywords. galaxies: evolution, galaxies: interaction, cosmology: dark matter, hydrodynamics, radiative transfer

1. Introduction

The current paradigm of structure formation in the Universe (White & Rees 1978) builds galaxies by merging smaller units, producing a universal distribution of sub-halos, with smaller haloes embedded within larger haloes on all scales. This paradigm successfully recovers the observed large-scale distribution of galaxies (Colless *et al.* 2001) using numerical simulations with increasing fidelity (Springel *et al.* 2006). However, it is not yet clear whether it applies equally well to sub-galactic scales. The excess of dark matter-dominated dwarf galaxies in dissipationless cosmological simulations of the Milky Way relative to Local Group dwarfs (Klypin *et al.* 1999) – dubbed the "missing satellites problem" – can make one wonder about the applicability of the prevailing cold dark matter model on sub-galactic scales.

We recently developed a novel method whereby one can infer the mass, and relative position (in radius and azimuth) of satellites from analysis of observed disturbances in outer gas disks, without requiring knowledge of their optical light (Chakrabarti & Blitz 2009, henceforth CB09; Chakrabarti & Blitz 2011, henceforth CB11; Chakrabarti, Bigiel, Chang & Blitz 2011, henceforth CBCB; Chang & Chakrabarti 2011, henceforth CC11. We applied this method to M51 and inferred that its satellite has a mass one-third that of the primary galaxy, with a pericentric approach distance of 15 kpc. These estimates are corroborated by observations (Smith *et al.* 1990) and recent simulation studies (Salo & Laurikainen 2000). Moreover, at the time when our simulations achieve the best-fit to the

HI data, the azimuth of the satellite in the simulations agrees closely with its observed location. CBCB note that the derivation of these numbers is uncertain at the factor of two level due to variations in the initial conditions of the simulated M51 galaxy, orbital inclination and orbital velocity of the satellite. We call this method "Tidal Analysis" (henceforth TA). Most recently, we have found that we can build on our earlier results to infer the scale radius of the dark matter halo in the primary galaxy itself (Chakrabarti 2011), and thereby constrain the potential of the dark halo.

Our work in this sequence of papers is motivated by the question – can dark (or nearly dark) galactic satellites (and the dark matter density profile) be characterized from tidal gravitational effects on the outer gas disks of galaxies? This question and our method have far-reaching implications in many areas of astrophysics. Our method is complementary to gravitational lensing in probing mass distributions without requiring knowledge of their stellar light, although it is not subject to uncertainties in the projected mass distribution (Treu *et al.* 2002), as is lensing. It provides a means of indirect detection of dark matter-dominated objects, and may be correlated with gamma ray studies (Strigari *et al.* 2008) to hunt for dark matter dominated dwarf galaxies. TA also offers a potential route to address the missing satellites problem. Therefore, it may allow us to investigate whether the prevailing cold dark matter model applies equally well to sub-galactic scales. Finally, recent observations of disturbances in the outskirts of spiral galaxies (Levine, Blitz & Heiles 2006; Thilker *et al.* 2007; Bigiel *et al.* 2010) prompt the question whether these disturbances arise from passing galactic companions, and trigger the observed star formation in the very outskirts.

The contemporary hunt for dark matter has much in common with the hunt for planets back in the 1800s. In 1846, Urbain Le Verrier analyzed disturbances in the orbit of Uranus. He hypothesized that these perturbations were due to an yet unseen planet. He was able to calculate the azimuth of the perturbing planet, now called Neptune, to within a degree. To the best of this author's knowledge, this is the first example of the discovery of an essentially dark object from analysis of its gravitational effects on another body. The work that we are trying to do – to characterize dark matter-dominated satellites and the dark matter density profile – from gravitational effects on outer gas disks, is in a similar spirit. After the discovery of Neptune, Le Verrier became interested in understanding the perihelion precession of Mercury. He believed it was due to a planet he called Vulcan. Le Verrier's mistake in this case is due to his lack of investigation of the possible incidence of false positives in attributing orbital anomalies to other planetary bodies. However, given that there was little available data at that time that he could test his theories on, his error was perhaps understandable.

The most significant difference in the field of astronomy then and now is the explosion of data, providing theorists ample opportunity (if they are able to avail themselves of it) to test their models. We are currently exploring the effects of intrinsic processes (such as torques arising from a non-spherical halo) to see if they can mimic tidal effects. Preliminary work (Chakrabarti, Debattista & Blitz, in preparation) finds that the evolution of the halo shape (and thereby the strength of the Fourier amplitudes in the outskirts of the gas disk as determined by intrinsic processes) is significantly affected by gas cooling and angular momentum transport from the gas to the halo. The shapes of halos are considerably rounder close to present day, when the stellar to dark matter masses are comparable to local spirals ($M_{\star}/M_{\rm DM} \sim 0.03$) (Leroy *et al.* 2008), relative to $z \sim 2$. Examination of the Fourier amplitudes of these simulations close to present day indicates that the Fourier amplitudes in the outskirts are $< 10\%$, which suggests that non-spherical halos would not be the primary contributor to the observed strength of the Fourier amplitudes in local spirals. This is certainly valid for M51, which is known to be a tidally interacting system (CBCB; Salo & Laurikainen 2000). Local spirals that display small ($< 10\%$ in the Fourier amplitudes, as defined in CB09, relative to the axisymmetric mode) perturbations in the outskirts, and have no visible tidally interacting companion may require a more careful treatment of the effect of halo shapes.

We are currently in the process of building on our earlier results to explore whether the emitted SED can also be used to independently characterize galactic satellites. An extensive multi-wavelength (from the radio to the x-ray) database exists for local spirals, e.g., built around the SINGS (Spitzer Infrared Nearby Galaxy Survey) project (Kennicutt *et al.* 2003). In addition to generating simulations that match the observed HI data, calculating the SEDs of these simulated galaxies and comparing with multi-wavelength data gives us another observational handle, in addition to the HI maps, to attest to the use of these simulations as prototypes in modeling local spirals. Examining relations between commonly used star formation tracers (such as the 24 μm flux) and the emitted luminosity will also be useful in interpreting observations of high redshift galaxies.

It is now well known that major mergers have higher FIR/UV flux than secularly evolving galaxies (Chakrabarti *et al.* 2008). What we would like to understand here is if SED colors (such as the FIR/UV flux) can be a tracer of tidal interactions for minor mergers also (and down to how low a satellite mass). For this tracer to be useful, the colors would need to be robust to the choice of star formation (SF) prescription, dust model (or ISM decomposition). This work builds on our earlier work on calculating the emergent SEDs from SPH simulations (Chakrabarti *et al.* 2007; 2008; Chakrabarti & Whitney 2009). The choice of ISM decomposition (Jonsson *et al.* 2010; Narayanan *et al.* 2010) will affect the emergent SED, and we are exploring a range of prescriptions to understand the primary dependences. We have also recently implemented a model of PAH emission following Wood *et al.* (2008). Particularly massive satellites may induce some star formation in the outskirts of galaxies. Therefore, we are exploring a range of SF prescriptions, including works that have treated the effects of shock heated star formation (Barnes 2004), pressure (Blitz & Rosolowsky 2006), and molecular gas fraction (Krumholz *et al.* 2009), to understand the effect of varying star formation prescriptions on the SED. Our goal here is to identify SED colors that are primarily affected by the galaxy's interaction history, and not significantly affected by the choice of SF prescription. This study may provide a way to diagnose the strength of interactions in high-redshift galaxies (many of which are thought to be produced by minor mergers; Genzel *et al.* 2010) that are being observed by *Spitzer* and *Herschel*, but cannot be mapped adequately in HI.

2. Proof of Principle of Tidal Analysis: M51

The simulations we discuss here have the same setup as in CBCB. We carry out SPH simulations using the GADGET code (Springel 2005), of M51 interacting with its companion. CBCB carried out a simulation parameter survey and compared the resultant Fourier amplitudes of the low order modes of the gas surface density with the observed HI data of M51. They found that placing simulations on a variance vs variance plot (where the variance is with respect to the low order modes of the simulations and the data) made the best-fit simulations visually apparent. Specifically, CBCB found that the best-fit to the HI data occurred for a 1:3 mass ratio satellite with a pericentric distance of 15 kpc, parameters that are corroborated observationally and are in agreement with other simulation studies (Smith *et al.* 1990; Dobbs *et al.* 2010; Salo & Laurikainen 2000). CBCB also found that the azimuthal location of the companion of the best-fit simulation agrees very closely with the observed location of M51's companion, at the time when the

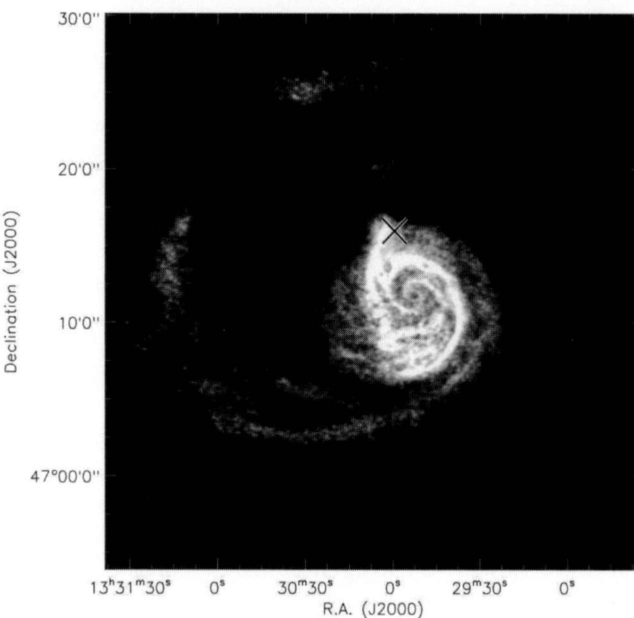

Figure 1. THINGS VLA image of M51 showing the HI distribution. Note that the companion of M51 sits at the short arm, as marked by the cross. Adapted from Chakrabarti *et al.* (2011) .

Fourier amplitudes most closely match the data. Figure 1 shows the HI map of M51, with the companion's location marked by the cross. Figure 2 shows the gas density images of the best-fit simulation. The best-fit time to the Fourier amplitudes occurs at $t \sim 0.3$ Gyr, when the satellite's location lies close to the tip of the short arm, as it does in the real galaxy. CBCB also found that the method was successful in characterizing NGC 1512's satellite ($\sim 1:100$ mass ratio), which is a much lower mass satellite than M51's relatively massive satellite. Thus, CBCB concluded that analysis of observed disturbances in the extended HI disks of galaxies can be used to infer the mass and current distance (in radius and azimuth) of galactic satellites. Earlier work in this series of papers presented the basic reasoning as to why the mass-pericentric approach degeneracy in the tidal force can be broken when the time integrated response of the primary galaxy is considered (CB09), and described the method to find the azimuth of galactic satellites from the phase of the modes (CB11). The reason why we focus our analysis on the gaseous disk is that disturbances in the gas disk dissipate on the order of a dynamical time – leaving a clean slate, and therefore allow an easier interpretation of satellite interactions than the stellar disk, where past interactions are still visible after a dynamical time.

2.1. *Exploring Star Formation Prescriptions*

We begin by contrasting the effects of two different star formation prescriptions, namely the Kennicutt-Schmidt star formation prescription and the Barnes shock heated star formation prescription. The former depends only on the density (the implementation in GADGET is described in Springel *et al.* 2005), while the latter depends on the density and the time rate of change of entropy. Barnes (2004) proposes the following form for the probability of gas particle i undergoing star formation in time Δt:

$$p_i = C_\star \rho^{n-1} \mathrm{MAX}(\dot{\mathrm{u}}, 0)^{\mathrm{m}} \Delta \mathrm{t} \tag{2.1}$$

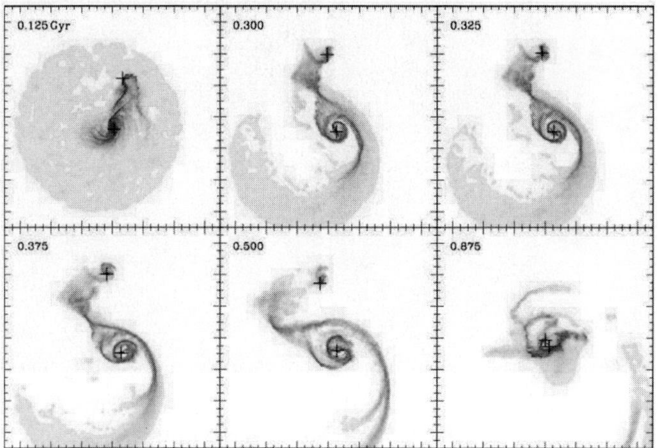

Figure 2. Gas density images of best-fit simulation of M51, with crosses marking centers of both galaxies. The best-fit time to the Fourier amplitudes occurs at $t \sim 0.3$ Gyr. The box extends from -80 kpc to 80 kpc. Adapted from Chakrabarti *et al.* (2011).

where u is the entropy and the MAX function, which returns the larger of its two arguments, is used to handle cases where $\dot{u}$ is less than zero. Shocks are identified as regions with $\dot{u}$ significantly larger than zero. Thus, setting $m = 0$ and $n > 1$ yields a density dependent prescription (with $n = 2.5$) yielding Kennicutt-Schmidt, and setting $m > 0$ yields shock heated star formation. Here, we set $m = 1$ and $n = 2.5$.

3. Emergent SEDs & Images

We use the self-consistent 3-D Monte Carlo radiative transfer code RADISHE to calculate the emergent SEDs and images through the time outputs of the SPH simulations, as in prior work (Chakrabarti *et al.* 2007; 2008; Chakrabarti & Whitney 2009). The radiative transfer methodology is described in Chakrabarti & Whitney (2009; henceforth CW09). Since dust envelopes can in principle be optically thick to their own reprocessed emission, it is necessary to calculate the temperature in a global way, i.e., the optically thin approximation is not valid in general. This is particularly true for dusty galaxies like ULIRGs. We described the implementation of the dust temperature calculation in CW09. CW09 found that the Lucy (1999) temperature calculation algorithm is very efficient for large 3-D grids, and described rules of thumb for its implementation. PAH emission is modeled following Wood *et al.* (2008), by sampling precomputed PAH emissivity files for a wide range of values of the mean intensity of the radiation field.

Figure 3 (a,b) displays simulated images of M51 close to the best-fit time in the J, H, K bands. Using the Kennicutt-Schmidt prescription (a), yields lower near-IR flux from the tidal bridge than when the shock heated star formation prescription is employed (b). This is to be expected as the Barnes (2004) prescription takes both the density (which is low in tidal tails overall) and shocks (which do occur in tidal tails) into account, while the K-S prescription only accounts for the density. Figure 4 shows the emergent SED (at the time that best fits the Fourier amplitudes of the HI data). Copious PAH emission features are present in the mid-IR, which we find commonly emerge in tidally interacting simulations. We are in the process of analyzing SEDs of these simulations to search for trends that may allow us to infer the galaxy's interaction history from the SED.

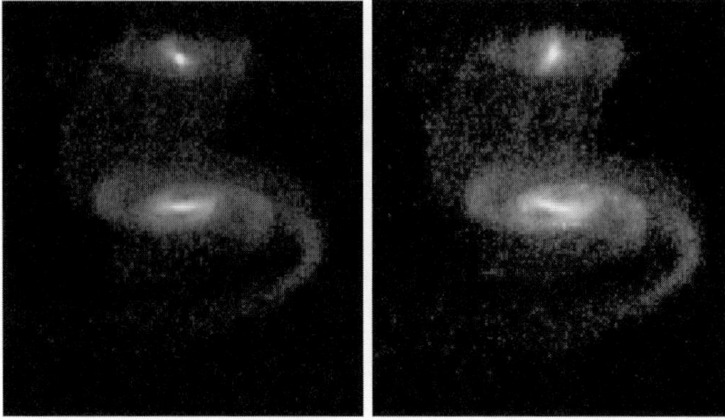

Figure 3. Simulated near-IR images of M51, using the (a) Kennicutt-Schmidt prescription, and (b) the shock heated star formation prescription.

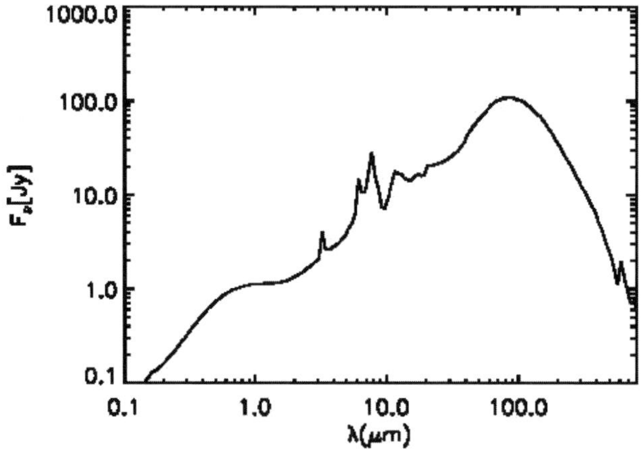

Figure 4. Simulated SED of M51 showing clear PAH features.

4. Summary & Future Work

In summary, we find that the dynamical echoes of the last generation of galactic satellites on the fragile extended HI disks of spirals can be analyzed to learn a number of things about dwarf galaxies (their mass and location), without requiring any knowledge of their optical light. This method, which we call Tidal Analysis, gives us a means of mapping the dark matter distribution. In this way it is analogous to gravitational lensing. Most recently, we have used the radiative transfer code RADISHE (Chakrabarti *et al.* 2008) to calculate emergent SEDs and images through the time outputs of these simulations of locally interacting spirals. The goal is to utilize the emergent UV-IR SED of the primary galaxy to understand its interaction history, in addition to the HI map. Future work includes studying the effects of a range of dust models, and more realistic star formation prescriptions to understand their effects on the emergent SED. The dynamical models and the Tidal Analysis method will be tested by applying it to the large THINGS sample to determine its statistical viability. Cross comparison of the SEDs and images against the data from the SINGS survey will allow us to test the radiative transfer models and implicit assumptions therein.

References

Bigiel, F., Leroy, A., Seibert, M., *et al.* 2010, *Astronomical Journal*, 140, 1194
Chakrabarti, S., Cox, T. J., Hernquist, L., *et al.* 2007, *ApJ*, 658, 840
Chakrabarti, S., Fenner, Y., Cox, T. J., Hernquist, L., & Whitney, B. A.,2008, *ApJ*, 688, 972
Chakrabarti, S. & Whitney, B. A., 2009, *ApJ*, 690, 1432 [CW09]
Chakrabarti, S. & Blitz, L., 2009, *MNRAS*, 399, L118 [CB09]
Chakrabarti, S. & Blitz L., 2011, *ApJ*, 731, 40C [CB11]
Chakrabarti, S., Bigiel, F., Chang, P., & Blitz, L. 2011, *ApJ*, 743, 35 [CBCB]
Chakrabarti, S., 2011, *ApJ*, submitted, arXiv: 1112.1416
Chang, P. & Chakrabarti, S., 2011, *MNRAS*, 416, 618C [CC11]
Colless, M., Dalton, G., Maddox, S., *et al.* 2001, *MNRAS*, 328, 1039
Debattista, V., Moore, B., Quinn, T., *et al.* 2008, *ApJ*, 681, 1076
Jonsson, P., Groves, B. A., & Cox, T. J. 2010, *MNRAS*, 403, 17
Kennicutt, R. C., Jr., Armus, L., Bendo, G., *et al.* 2003, *PASP*, 115, 928
Klypin, A., Kravtsov, A. V. *et al.* 1999, *ApJ*, 522, 82
Levine, E. S., Blitz, L. & Heiles, C., 2006, *Science*, 312, 1773L
Leroy, A., Walter, F., Brinks, E., *et al.* 2008, *ApJ*, 136, 2782
Maccio, A. *et al.* 2008, *MNRAS*, 391, 1940
Mandelbaum, R., Seljak, U., Kauffmann, G., Hirata, C. M., & Brinkmann, J. 2006, *MNRAS*, 368, 715
Salo, H. & Laurikainen, E., 2000, *MNRAS*, 319, 393
Smith, J., *et al.* 1990, *ApJ*, 362, 455S
Springel, V., 2005, MNRAS, 364, 1105
Springel, V., Di Matteo, T., & Hernquist, L. 2005, *MNRAS*, 361, 776
Springel, V., Frenk, C. S., & White, S. D. M., 2006, *Nature*, 440, 1137
Strigari, L., Koushiappas, S., Bullock, J., *et al.* 2008, *ApJ*, 678, 614S
Thilker, D., Bianchi, L., Meurer, G., *et al.* 2007, *ApJS*, 173, 538
Treu, T. & Koopmans, L., 2002, ApJ, 575, 84
Walter, F. *et al.* 2008, *Astronomical Journal*, 136, 2563
White, S. D. M. & Rees, M. J., 1978, *MNRAS*, 183, 341
Wood, K., Whitney, B. A., Robitaille, T., & Draine, B. T. 2008, *ApJ*, 688, 1118
Wright, C. O. & Brainerd, T. G. 2000, *ApJ*, 534, 34

The Spectral Energy Distribution of Galaxies
Proceedings IAU Symposium No. 284, 2011
R.J. Tuffs & C.C. Popescu, eds.

doi:10.1017/S1743921312009027

TYPHOON observations of the Lindsay-Shapley Ring

Laura K. Sturch[1] **and Barry F. Madore**[2]

[1]Institute for Astronomical Research, Dept. of Astronomy, Boston University
725 Commonwealth Ave, Boston, MA 02215, USA
email: lsturch@bu.edu

[2]Observatories of the Carnegie Institute of Washington
813 Santa Barbara St., Pasadena, CA 91101, USA
email: barry@obs.carnegiescience.edu

Abstract. We present the first results of the TYPHOON program on the ring galaxy AM 0644-741. TYPHOON is a program for producing highly resolved spectrophotometric data cubes with wavelength coverage ranging from [OII] 3727A to 7000A. Using the first results of TYPHOON we will show its efficacy in producing images and from that we will create velocity maps, one of the many uses of TYPHOON results. From this program we will deduce the motion of gas in the ring structure of AM 0644-741 to better understand how the galaxy has evolved to its present state.

Keywords. galaxies: kinematics and dynamics, galaxies: interactions, surveys, techniques: spectroscopic

---

TYPHOON is a newly developed methodology for producing highly resolved spectrophotometric data cubes using conventional spectrographic capabilities. Using TYPHOON, we are currently undertaking a survey of 100 of the closest and largest southern-hemisphere galaxies visible from the Las Campanas Observatory in Chile where observations are made on the 2.5m du Pont telescope. TYPHOON has a spectral range from 3727 to 7000 Angstroms, 3-6 Angstrom spectral resolution, and a spatial resolution that is seeing-limited at less than one arcsecond. Our program using TYPHOON has many far reaching scientific goals, but for the first foray into data analysis, we have chosen to explore its ability to study kinematic structure.

Here we show the results of our kinematic structure study of the Lindsay-Shapley Ring, also known as AM 0644-741. AM 0644-741, a classic collisional ring galaxy, is easily observed from the southern hemisphere, large and bright at redshift 0.022029 (Fisher *et al.* 1995) , and is a good candidate for TYPHOON. We have extracted those channels at and around the doppler-shifted wavelength of Hα for AM 0644-741, as shown in (Fig. 1). This image was created by imposing a narrow-band "filter" on the data to collect and collapse the doppler-shifted emission from the ring. Using the multi-dimensional capabilities of the data cube, the greyscaled velocity moment map of AM 0644-741 is shown in (Fig. 2).

The kinematic structure of AM 0644-741 has been previously studied, but not with the spectral resolution or areal completeness as with TYPHOON (e.g., Few *et al.* 1982, Higdon & Wallin 1997). As can be seen in Fig. 2, there is obvious rotation of AM 0644-741 progressing from the southwest to the northeast directions. The ring is seen rotating from -550 to 350 km/s giving a peak-to-peak velocity range of 900 km/s. Given the velocity resolution of TYPHOON, we can additionally see that the velocity changes from the inner to the outer edges of the ring on the order of 200-300 km/s depending on the position in the ring. We also note the asymmetry of the ring in velocity space. This is

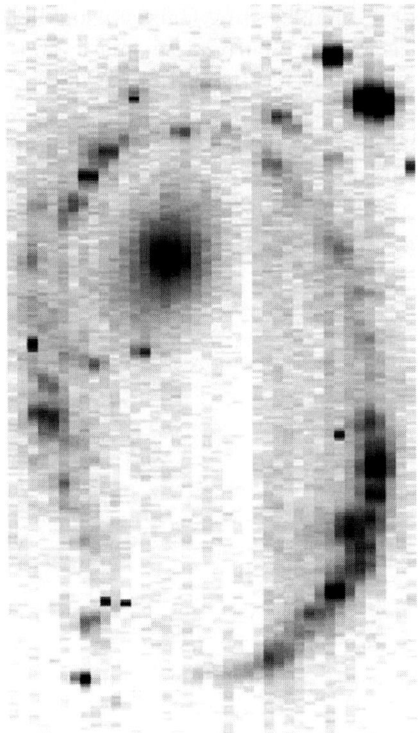

Figure 1. Data extracted from around the doppler shifted Hα to create a narrow band "filter" image of AM 0644-741.

reasonable due to the projection effects of a combined rotation and expansion of the ring, as noted by other studies (e.g., Few *et al.* 1982).

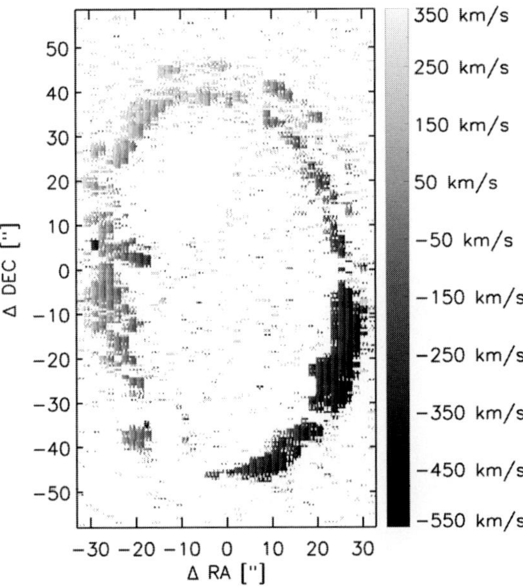

Figure 2. Greyscaled velocity moment map of AM 0644-741.

Future work on the Lindsay-Shapley Ring Galaxy using TYPHOON will be to better understand the origin of the galaxy's current kinematic and spatial structure. We will be modeling the collision and adjusting the initial parameters and masses of AM 0644-741 and the impacting galaxy to replicate our results. From there, we can deproject the ring, due to its angle on the plane of the sky, by comparing it to the model and thereby separately determine the rotation and expansion rates of the ring material. We should be able to accurately age date this collision event and place reasonable constraints on the many collision input parameters.

References

Few, J. M. A., Arp, H. C., & Madore, B. F. 1982, *MRNAS*, 199, 633
Fisher, Karl, B., *et al.* 1995, *ApJS*, 100, 69
Higdon, J. L. & Wallin, J. F. 1997, *ApJ*, 474, 686

The Spectral Energy Distribution of Galaxies
Proceedings IAU Symposium No. 284, 2011
R.J. Tuffs & C.C. Popescu, eds.

doi:10.1017/S1743921312009039

Sources of X-rays from galaxies

Q. Daniel Wang

Astronomy Department, University of Massachusetts, USA
email: wqd@astro.umass.edu

Abstract. Galactic X-ray emission is a manifestation of various high-energy phenomena and processes. The brightest X-ray sources are typically accretion-powered objects: active galactic nuclei and low- or high-mass X-ray binaries. Such objects with X-ray luminosities of $\gtrsim 10^{37}$ ergs s^{-1} can now be detected individually in nearby galaxies. The contributions from fainter discrete sources (including cataclysmic variables, active binaries, young stellar objects, and supernova remnants) are well correlated with the star formation rate or stellar mass of galaxies. The study of discrete X-ray sources is essential to our understanding of stellar evolution, dynamics, and end-products as well as accretion physics. With the subtraction of the discrete source contributions, one can further map out truly diffuse X-ray emission, which can be used to trace the feedback from active galactic nuclei, as well as from stars, both young and old, in the form of stellar winds and supernovae. The X-ray emission efficiency, however, is only about 1% of the energy input rate of the stellar feedback alone. The bulk of the feedback energy is most likely gone with outflows into large-scale galactic halos. Much is yet to be investigated to comprehend the role of such outflows in regulating the ecosystem, hence the evolution of galaxies. Even the mechanism of the diffuse X-ray emission remains quite uncertain. A substantial fraction of the emission cannot arise directly from optically-thin thermal plasma, as commonly assumed, and most likely originates in its charge exchange with neutral gas. These uncertainties underscore our poor understanding of the feedback and its interplay with the galaxy evolution.

Keywords. galaxies: general; X-rays: galaxies, binaries, ISM; stellar dynamics

---

1. Introduction

Why are X-rays from galaxies interesting? X-ray emission from a galaxy typically arises from high-energy (high temperature or relativistic) processes in extreme conditions. Such processes are important to understand in their own right, but are also believed to play a major role in shaping the structure and evolution of galaxies via various types of feedback (e.g., Tang *et al.* 2009; Oppenheimer *et al.* 2010). In particular, the X-ray emission effectively traces diffuse hot plasma, which may dominate the volume and account for the bulk of the baryon mass in and around galaxies, especially massive ones (e.g,, Crain *et al.* 2010). The X-ray wavelength range also contains major atomic transitions (e.g., all K-shell transitions of carbon through iron), providing key thermal, chemical, and kinematic diagnostics of the interstellar medium (ISM) in all phases (cold, warm, and hot) and all forms (atomic, molecular, and dust). X-rays are also energetic and penetrating, which may heat and evaporate dust grains/molecules and even companion stars in X-ray binaries, significantly affecting stellar and interstellar radiation. X-rays can often be observed through dense clouds with column densities up to $\gtrsim 10^{24}$ cm^{-2}. Therefore, X-ray observations are a powerful tool in examining the stellar and interstellar properties of galaxies.

Substantial progress in understanding the X-ray emission from nearby galaxies has been made over the past decade or so. *Chandra* and *XMM-Newton* X-ray Observatories, in particular, have provided a combination of superb spatial and spectral resolutions (reaching $\sim 1''$ and 500 km s^{-1} FWHM) as well as large collecting areas (up to

~ 3000 cm^2) over a broad energy range of 0.3-10 keV. These capabilities have enabled extensive studies of various classes of X-ray sources: their luminosity functions, spectral characteristics, and relationships to other galactic properties such as the star formation rate (SFR) and total stellar mass. These are the topics that I am going to review, focusing on recent results.

2. Discrete X-ray Sources

Active Galactic Nuclei (AGN). AGNs are powered by the accretion of matter by supermassive black holes (SMBHs) from their surroundings. This process can be very efficient, energetically, converting up to about 10% of the accreted mass into radiation and mechanical energy. For example, the energy released from the growth of a SMBH can be about 100 times the gravitational binding energy of its host galaxy. This energy feedback, if well coupled to the surrounding medium, should then have enormous impact on galaxy formation and evolution.

While much of the SMBH growth occurs at redshifts greater than 1, present AGNs typically have only moderate radiation luminosities $L_x \sim 10^{42} - 10^{45}$ ergs s^{-1} and appear in only about 10% of nearby galaxies. Such AGNs are most easily identified in X-ray, since their luminosities are still considerably higher than those of other galactic sources. AGNs typically have intrinsic X-ray spectra that can be characterized by a power law with a photon index of ~ 1.7. However, a significant fraction of AGNs are known to be severely obscured or even Compton-thick to X-rays. Such AGNs may still be detected in reflected or reprocessed light, mostly in optical, mid-IR, and/or hard X-ray (from photon-ionized gas, hot dust, and/or Fe florescence; e.g., LaMassa *et al.* 2009). Observations of the AGNs can also provide us with a better view of their feedback effects on the immediate surroundings. Their X-ray spectra typically show a mixture of collisionally excited hot thermal plasma, characterized by numerous strong resonance lines, and photo-ionized plasma, represented by strong forbidden lines and recombination continuum (e.g., Guainazzi & Bianchi 2007). Interestingly, the X-ray luminosities required to model such spectra sometimes far exceed those of the present AGNs, indicating that they may have been substantially brighter in the recent past (e.g., Wang *et al.* 2010) or that some other processes such as charge exchange (CX) between hot ions and neutral atoms may need to be considered (see later discussions; Liu *et al.* 2011b).

When a SMBH is in a low-L_x ($\lesssim 10^{42}$ ergs s^{-1}) or even moderate-L_x state, the accretion is likely in the so-called radiatively inefficient (radio) mode (e.g., Yuan 2007). Much of the gravitational energy of the accreted matter may be released in form of mechanical energy (e.g., an accretion disk wind and/or jet; Omma *et al.* 2004). Such energy release may play an important role in regulating the nuclear environment of galaxies. A SMBH may also undergo intermittent accretion episodes, resulting in cycles of heating and cooling of surrounding gas (e.g., Yuan *et al.* 2009; Pellegrini *et al.* 2011). Various studies are ongoing to explore the interplay between the SMBH accretion and the nuclear environment of very nearby galaxies ($D \lesssim 10$ Mpc; e.g., Li *et al.* 2009, 2011). For more distant galaxies, confusion with stellar sources typically becomes too severe to even identify individual low-L_x AGNs with certainty, especially in galaxies with active nuclear SF.

Sources Related to Recent Star Formation. A nearby galaxy, excluding its possible AGN, typically has $L_x \lesssim 10^{42}$ ergs s^{-1}, depending primarily on its mass and SFR. In an active SF galaxy, the luminosity is normally dominated by high-mass X-ray binaries (HMXBs), which contain neutron stars or stellar mass black holes accreting typically from stellar winds of companions more massive than $\sim 5M_\odot$. The spectrum of such a

HMXB can be characterized by a power law with a photon index of ~ 1.2, a cutoff at ~ 20 keV, plus an Fe-K emission feature centered at 6.4-6.7 keV and with an equivalent width of 0.2-0.6 keV (White *et al.* 1983). The X-ray luminosity function (XLF) of HMXBs can be characterized by a power law with a differential slope of 1.5-1.6 over a very broad luminosity range of $10^{36} - 10^{40}$ ergs s^{-1} (Grimm *et al.* 2003; Persic & Rephaeli 2007 and references therein). The XLF shows an exponential cutoff at $\sim 2 \times 10^{40}$ ergs s^{-1} (Swartz *et al.* 2011) and seems to flatten out again at $\gtrsim 10^{41}$ ergs s^{-1}.

Much attention has been placed on so-called ultra-luminous X-ray sources (ULXs) — non-nuclear, point-like, X-ray sources with apparent isotropic $L_X \gtrsim 10^{39}$ ergs s^{-1}, which is roughly the Eddington luminosity of a stellar mass BH (e.g., Swartz *et al.* 2011 and references therein). ULXs tend to have steep spectra with a mean power law photon index of ~ 1.7. There is on average ~ 1 ULX per 0.5 M$_\odot$ yr^{-1} SFR in local galaxies. ULXs are clearly a heterogeneous population of X-ray sources, including very young supernova remnants (SNRs, due to the interaction of SN ejecta with dense circumstellar materials), as well as probably mildly beamed and/or super-Eddington HMXBs. Some of these sources, especially those rare ones with $L_X \gtrsim 10^{41}$ ergs s^{-1}, may represent the so-called intermediate-mass BHs (IMBHs with $10^2 M_\odot \lesssim M_{BH} \lesssim 10^4 M_\odot$) undergoing sub-Eddington accretion. IMBHs may be formed from the collapse of Pop III stars and at centers of globular clusters (GCs) and dwarf elliptical galaxies, which may since being destroyed dynamically.

There are other young stellar objects and their remnants such as YSOs and colliding wind binaries, as well as normal SNRs. Such sources are typically quite faint individually ($\lesssim 10^{35}$ ergs s^{-1}); but collectively they can still account for a significant fraction of the X-ray emission from galaxies, especially when the contribution from brighter sources has been excised: 2-10 keV $L_x \approx 10^{38.2}$ ergs s$^{-1}[M_\odot$ yr$^{-1}]^{-1}$ (Bogdán & Gilfanov 2011).

The *total* X-ray luminosity of young stellar objects and their remnants in a galaxy is strongly correlated with the SFR: $L_x = 10^{39.4}$ ergs s$^{-1}[M_\odot$ yr$^{-1}]^{-1}$ with a scatter of 0.4 dex (e.g., Mineo *et al.* 2011). With this correlation, accounting for its possible cosmological evolution (Dijkstra *et al.* 2011) as well as the scatter, one can estimate the SFR of a distant galaxy from its X-ray luminosity. Or if we know the SFR from its other proxies, we can then compare the predicted and observed luminosities to constrain a potential AGN contribution.

Sources from Old Stars. A galaxy with little recent SF can still contain copious X-ray sources, both discrete and diffuse. In this case, the brightest off-nuclear discrete sources are low-mass X-ray binaries (LMXBs), which contain neutron stars or stellar-mass BHs accreting from their typically post-main-sequence, Roche-lobe-overflowing companions with masses $\lesssim 1 M_\odot$. Therefore, LMXBs trace stellar populations older than ~ 1 Gyr.

The average X-ray spectrum of LMXBs can be characterized by a power law with a photon index of ~ 1.6, considerably steeper than that of HMXBs. LMXBs at the high end of their 0.3-10 keV luminosity range, $\sim 10^{39}$ ergs s^{-1}, tend to have even softer spectra and are similar to Galactic black hole X-ray binary candidates in their very high state (Irwin *et al.* 2003).

The XLF of LMXBs has a convex shape. It can be represented approximately by a power law with a differential slope of ~ 2.0 in the luminosity range $L_X \sim 5 \times 10^{37} - 5 \times 10^{38}$ ergs s^{-1} (Gilfanov 2004; Kim & Fabbiano 2004). The XLF steepens at higher luminosities, which appears to be more significant with the increasing galaxy age (Kim & Fabbiano 2010); the XLF of young elliptical galaxies (a few Gyr old) is intermediate between that of typical old elliptical galaxies and that of star-forming galaxies. The overall steepening of the XLF with the luminosity can be explained by the mass transfer in binary systems with giants, which significantly shortens their stellar lifetime (Revnivtsev *et al.*

2011). In contrast, the majority of binary systems at luminosities $\lesssim 5 \times 10^{37}$ have main-sequence secondary companions (except for those with white dwarf donors), apparently responsible for the flattening of the XLF to a slope ~ 1 (Revnivtsev *et al.* 2011). The normalization of the XLF is proportional to the stellar mass, which can be estimated from the K-band luminosity using a color-based correction for the mass-to-light ratio of a galaxy (Bell & de Jong 2001). On average, the combined luminosity of LMXBs with individual $L_X \gtrsim 10^{37}$ ergs s^{-1}, which can be detected in a nearby galaxy with a reasonable exposure of *Chandra* or *XMM-Newton*, is $(8 \pm 0.5) \times 10^{39}$ ergs s^{-1} per $10^{11} M_\odot$ (Gilfanov 2004); the residual contribution down to $L_x \sim 10^{35}$ ergs s^{-1} (typically $\lesssim 20\%$, especially in the $\lesssim 2$ keV band) can be well estimated with the flat XLF.

The average properties of fainter X-ray sources, mostly cataclysmic variables (CVs) and active binaries (ABs), which are typically in the range of $10^{30} < L_x < 10^{34}$ ergs s^{-1} (2-10 keV), have also been measured. Sazonov *et al.* (2006) show that the specific luminosity of such sources (per unit stellar mass) in the solar neighborhood is $(2.0 \pm 0.8) \times 10^{27}$ and $(1.1 \pm 0.3) \times 10^{27}$ erg s^{-1} M$_\odot^{-1}$. The inferred total contribution of ABs and CVs to the 2-10 keV luminosity of the Milky Way is $\sim 2 \times 10^{38}$ ergs s^{-1}, or 3% of the integral luminosity of LMXBs. The specific luminosity has also been estimated for nearby low-mass early-type galaxies (M32 and NGC 3379; Revnivtsev *et al.* 2008) and is shown to be a factor of ~ 2 lower than the solar neighborhood value. The exact cause of this difference is yet to be identified. One possibility is that it represents the genuine difference in the binarity of stars, resulting from different stellar formation and evolution environments.

Indeed, the environment-dependence of the XLF is clearly present for LMXBs (e.g., Zhang *et al.* 2011 and references therein). The XLF shape of sources in globular clusters (GCs) is shown to be significantly different from that in the field, especially in $10^{35} - 10^{37}$ ergs s^{-1}. Over this luminosity range (relative to more luminous ones), the fraction of sources in GCs is about 4 times less than in the field. A similar result is also obtained in a recent study of the XLFs of sources over a substantially lower luminosity range (e.g., Fig. 1; Xu, Wang, & Li 2011). The relatively high luminosity range ($L_X \gtrsim 10^{31}$) should be dominated by CVs. The specific population of such sources in a GC depends on its dynamical state and can be greater than that in the Galactic bulge or in the solar

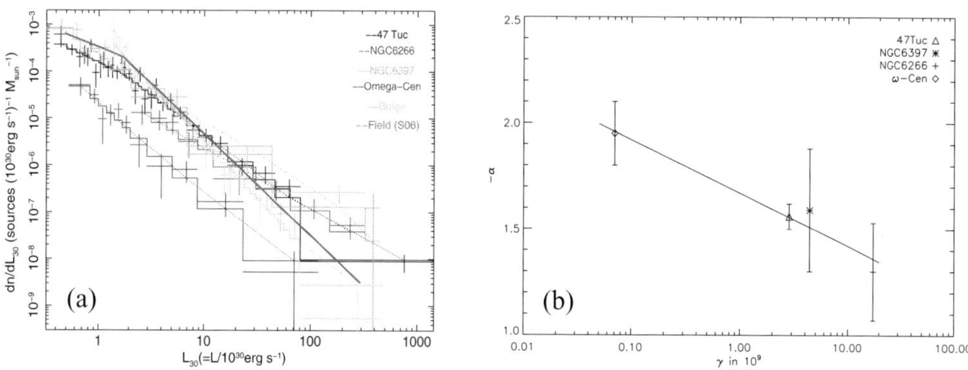

Figure 1. (a) Specific XLFs of four GCs and a Galactic bulge field, as well as the best-fit broken power law model for sources in the solar neighborhood (Field; Xu, Wang, & Li 2011). The incompleteness and bias of the source detection have been approximately corrected, while the contributions from interlopers have been subtracted statistically from the data. (b) Correlation between the power law slopes of the LFs and the mass-averaged encounter rate (γ) of the GCs, demonstrating the importance of the dynamical effect on formation and evolution of binaries.

neighborhood. Interestingly, the source population, particularly in the lower luminosity range, is generally smaller in the GCs (especially in those dynamically less evolved ones) than in the bulge and field. To explain this result, one probably needs to invoke both the dynamical formation of binaries and the assumption that the initial binary fraction of stars is very small in GCs, consistent with existing optical observations (e.g., Davis *et al.* 2008). The dynamical effect has also been shown to be important in explaining the increasing specific LMXB population in the central bulge of M31 (Zhang *et al.* 2011). In regions outside of GCs and Galactic nuclear regions, however, we can now reasonably determine and subtract the faint stellar X-ray contribution (Revnivtsev *et al.* 2008; Zhang *et al.* 2011) to facilitate the measurements of truly diffuse X-ray emission from galaxies.

3. Diffuse Soft X-ray Emission

Diffuse soft X-ray emission has commonly been used to trace various types of galactic feedback in nearby starburst and normal galaxies, as well as the cooling of hot gaseous halos or coronae resulting from the galaxy formation (accretion) processes (e.g, Crain *et al.* 2010). Assuming an origin of this emission in optically-thin thermal (collisionally-excited) hot plasma, one may estimate its mass, energy, and chemical contents and even their outflow or accretion rates from a galaxy.

Emission from Galactic Spheroids. While galactic coronae associated with giant elliptical galaxies have been studied extensively (e.g, Sun *et al.* 2007; Mulchaey & Jeltema 2010; Pellegrini 2011), the very detection of truly diffuse X-ray emission in and around low-L_x stellar spheroids (galactic bulges or low- to intermediate-mass ellipticals) becomes possible only recently, thanks largely to the calibration of the stellar L_x to mass ratio, as discussed above. A good example of such a detection is the bulge of M31 (Fig 2a; Li *et al.* 2007; Bogdán & Gilfanov 2011). The large-scale X-ray emission shows a bipolar morphology along the minor axis of the galaxy, indicating an outflow of hot plasma, which is most likely driven by Ia SNe because there is essentially no AGN and no recent SF in the bulge. But the X-ray luminosity accounts for only $\sim 2\%$ of the Ia SN mechanical energy input in the bulge. There is no bulge-wide distributed cool gas to potentially consume or convert the energy into other forms. Therefore, the outflow is actually required to explain the lack of the accumulation of the stellar feedback, not only the energy, but the mass (mainly from redgiant winds and planetary nebulae) and the iron (from Ia SNe, about 0.7 $M_\odot$ each) as well. Theoretically, it has been shown that such outflows play an important role in maintaining large-scale hot gaseous halos around galactic spheroids (Tang *et al.* 2009).

The dynamical state of a corona around a galactic spheroid depends on its gravitational mass and environment (e.g., Wang *et al.* 2004; Owen *et al.* 2006; Sun *et al.* 2007, 2010; Mulchaey & Jeltema 2010; Weźgowiec *et al.* 2011; Tang *et al.* 2009; Lu & Wang 2011). The corona of an intermediate-mass spheroid fast moving in a rich cluster, for example, most likely represents an Ia SN-driven outflow semi-confined by the ram pressure of the intracluster medium (ICM; Lu & Wang 2011). The ram pressure affects the size and lopsidedness of the coronae. Because of the semi-confinement, the outflow should also be subsonic. The hot gas density increases with the ICM thermal pressure, which may lead to the compression of cool gas clouds, if present, and hence the formation of stars (Young & Scoville 1991; Bekki & Couch 2003). Such density increase also enhances radiative cooling of the hot gas, which may fuel central supermassive black holes, explaining why the frequency of active galactic nuclei observed in clusters tends to be higher than that in the field (e.g., Hart, Stocke & Hallman 2009) and is apparently responsible for a substantially higher surface brightness of the X-ray emission detected from corona in the

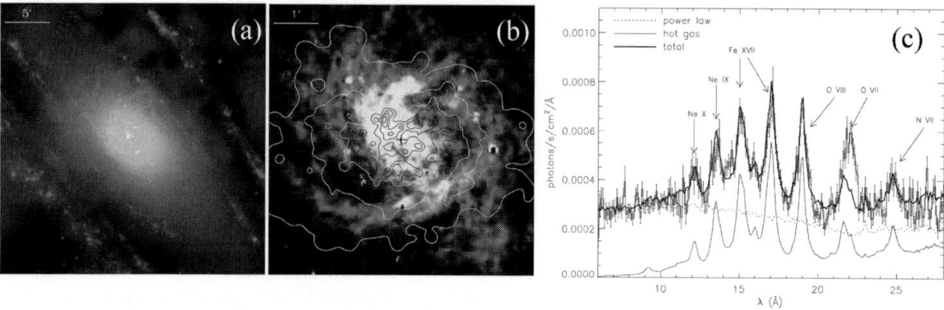

Figure 2. **(a)** Tri-color image of the spheroid region of M31: *Spitzer*/MIPS 24 μm emission, 2MASS K-band emission, and *Chandra* 0.5-2 keV emission of truly diffuse hot gas; Li & Wang 2007). **(b)** Intensity contours of the diffuse emission overlaid on an H_α image in the M31 central region. A detailed analysis of the data shows evidence for strong interaction between the hot and cool gas phases, which may provide a mechanism for starving the nucleus as marked by the plus sign (Li *et al.* 2009). **(c)** An *XMM-Newton* Reflection Grating Spectrometer (RGS) spectrum of the M31 central region (Liu *et al.* 2010). The curves show the fit with a single-temperature thermal plasma, plus a power-law that characterizes the discrete source contribution. Note the intensity excess of the O VII Kα triplet above the model, giving a strong indication for emission from the CX between hot ions and neutral atoms.

cluster environment. Furthermore, the total X-ray luminosity of a corona depends on the relative importance of the surrounding thermal and ram pressures. These environment dependencies should at least partly explain the large dispersion in the observed diffuse X-ray luminosities of spheroids with similar stellar properties. Furthermore, an outflow powered by the distributed feedback can naturally produce a positive radial gradient in the hot gas entropy, mimicking a cooling flow.

However, the soft X-ray emission in the inner regions of spheroids is still far from being well understood. For example, the low temperature and metal abundance of hot plasma, as inferred from the existing spectral studies, are still difficult to be reconciled with models and simulations (e.g., Tang & Wang 2010). Even the very assumption that the X-ray line emission originates predominantly in optically-thin thermal plasma is problematic (Liu *et al.* 2010). This becomes apparent when the so-called G [$= (f+i)/r$, forbidden+intercombination to resonance line] ratio of the Kα triplet of such a diagnostic He-like ion as O VII is analyzed. For an optically-thin thermal plasma with a temperature of a few 10^6 K and in collisional ionization equilibrium, G should be considerably less than one (e.g., Porquet *et al.* 2010). Fig. 2c presents an RGS spectrum of the M31 central region (Fig 2b), showing a clear intensity excess above a simple thermal plasma model fit at ~ 22 Å. This excess is due to the presence of the strong forbidden line (at 22.1 Å), relative to the resonant one (21.6 Å) of the O VII Kα triplet. Furthermore, the excess is shown to extend throughout much of the cross-dispersion range of the RGS observations and seems to be correlated well with the distribution of the cool gas (Fig. 2b). One natural explanation is that the excess or the strong forbidden line represents the contribution from the CX between the two gas phases.

Emission from Active SF Galaxies. Similar conclusions are also reached in studies of diffuse soft X-ray emission from active SF galaxies. Such emission has been mapped out for many face-on galaxies (e.g., Tyler *et al.* 2003; Doane *et al.* 2004; Owen & Warwick 2009; Kuntz & Snowden 2010). The emission is closely correlated with recent SF. The study of M101, for example, shows that the bulk ($\gtrsim 80\%$) of the emission is associated with SF regions with ages $\lesssim 20$ Myr, but is not due to individual SNRs, as identified in optical and radio (e.g., Kuntz & Snowden 2010). The emission traces diffuse hot plasma, which can naturally be heated by fast stellar winds and SNe of massive stars. This is most vividly demonstrated by the emission from within cool gas cavities around massive stellar clusters, as in the 30 Doradus nebula (Wang *et al.* 1999). The X-ray CCD spectra of the emission from galaxies can typically be approximately fitted with an optically-thin thermal plasma model with two characteristic temperatures of $\sim 0.2 \pm 0.1$ and 0.7 ± 0.1 keV; the low-temperature component typically accounts for the bulk of the observed flux. However, the total luminosity of the emission accounts for a very small fraction ($\lesssim 1\%$) of the expected mechanical energy input rate from fast stellar winds and SNe of massive stars. The estimated thermal energy of the plasma is also substantially smaller than the expected total energy input. Although a fraction of the energy may be carried away by cosmic rays or converted to radiation in other wavelength bands, outflows from recent SF regions (i.e., blown-out superbubbles or galactic chimneys) are expected, venting chemically-enriched hot plasma into the large-scale halos of galaxies. Indeed, extraplanar soft X-ray emission is observed around active SF galaxies (Wang *et al.* 2001, 2003; Tüllmann 2006; Wang 2010 and references therein; Li & Wang 2011; Anderson & Bregman 2011). Such emission also often appears to be correlated with extraplanar Hα-emitting and/or dusty features. This correlation has led to the conclusion that the bulk of the emission arises from the interaction between the outflowing hot plasma and entrained or pre-existing cool gas clouds, rather than from the superwinds themselves (e.g., Strickland & Heckman 2009).

To determine the nature of the interaction, we have recently examined the spectroscopic properties of the soft X-ray emission from nine nearby active SF galaxies, which show no significant AGN activities (Liu *et al.* 2011a,b), using high spectral resolution RGS data with good signal-to-noise ratios. The RGS spectra of these galaxies typically show large G ratios of the OVII Kα triplet, indicating that a substantial fraction of the emission cannot simply arise from collisionally-excited hot plasma, as commonly assumed. The bulk of the OVII Kα line emission is consistent with an origin in the CX between highly ionized ions (i.e., OVIII) and neutral atoms (e.g., HI and HeI); other possible mechanisms for producing a high G ratio, such as a collisional non-equilibrium-ionization recombining/ionizing plasma, are not favored, although they cannot be completely ruled out. In the central region of M82, for example, the putative CX contribution to the key He-like Kα triplets is on average $\gtrsim 50\%$, decreasing with the increasing ionization states (i.e., the G ratio is smaller for NeIX and Mg XI than for OVII; Liu *et al.* 2011a). To quantify the true contribution of the CX, one needs to have a spectral model of it, which is not yet available. Nevertheless, the joint analysis of multiple emission lines indicates that the thermal plasma is substantially hotter, probably consistent with the high-temperature component indicated in the 2-temperature characterization mentioned above. The decomposition of the CX and thermal contributions will also be essential to a correct estimation of the metal abundances of the plasma (Liu *et al.* 2011a).

The low X-ray emission efficiency of the galactic feedback ($\sim 1\%$) has strong implications for understanding the formation and evolution of galaxies. The efficiency shows no significant correlation with the stellar mass or gas content of a galaxy (Li & Wang 2011). There is also little evidence that the feedback energy is consumed at other wavelengths. For example, Grimes *et al.* (2009) show that O VI emission is not generally detected in their sample of starburst galaxies, suggesting that radiative cooling or turbulent mixing is not significant in draining energy from galactic winds. One may thus conclude that the bulk of the energy is gone with the winds and is released to the halos of galaxies, a topic that certainly deserves further attention.

4. Summary and Conclusions

I have reviewed some basic properties of various types of X-ray sources, particularly in relation to the spectral energy distribution (SED) of galaxies. The key points are as follows:

• AGNs are typically the brightest X-ray sources in galaxies and can be detected relatively easily even at high redshifts with little confusion. Such detection is important in modeling the SED of galaxies. Even low-L_x AGNs, as typically seen in nearby galaxies, can be important in regulating galactic nuclear environments.

• In nearby galaxies, the bulk of HMXBs and LMXBs can readily be detected. Their overall galactic populations are well correlated with the SFR and stellar mass and can also be independently estimated (e.g., with infrared observations). The comparisons of these independent estimates can then be used to constrain the presence of AGNs in distant galaxies, even if they are not well resolved.

• The X-ray source population of a stellar system is particularly sensitive to its binary fraction. Thus X-ray observations provide a powerful tool to probe the initial binary fraction and dynamical evolution of stellar clusters and even galactic spheroids.

• Diffuse X-ray emission traces the galactic feedback, primarily due to massive SF (via fast stellar winds and core-collapsed SNe) in galactic disks, to old stars (mass-loss and Ia SNe), and to possible AGNs. However, the luminosity of the X-ray emission from a galaxy typically accounts for $\lesssim 1\%$ of the feedback energy. The bulk of the energy is released in outflows of hot plasma and cosmic-rays/magnetic fields from stellar spheroids, as well as from SF regions.

• A substantial fraction of the diffuse X-ray emission may arise from the CX at the interface between hot plasma and cool gas. The CX, in principle, can be used as a new tool to probe the thermal, chemical, and kinematic properties of hot plasma tracing various high-energy feedback processes in the galaxies.

While much can still be done with the existing X-ray observatories, we will also have new tools in the near future. *NuSTAR* (scheduled to be launched in late 2012) will offer the first imaging capability in the 8-80 keV range, while *Astro-H* (2013) will provide the calorimeter-based imaging spectroscopic capability in the 0.5-10 keV range. Furthermore, *eROSITA* (2013) will conduct the first all-sky X-ray CCD imaging survey over the 0.3-10 keV range, while *GEMS* (2014) will enable polarization measurements in the 2-10 keV range. With these new capabilities, we can expect a substantial advance in our understanding of X-ray sources in galaxies.

I am grateful to the organizers of the IAU Symposium for inviting me to give a review talk on this topic, which the present article is based on, and to my students and collaborators for their contributions to some of the work described above. My research is largely supported by NASA via various grants.

References

Anderson, M. E. & Bregman, J. N. 2011, *ApJ*, 737, 22
Bekki, K. & Couch, W. J. 2003, *ApJ*, 596, 13
Bell, E. F. & De Jong, R. S. 2001, *ApJ*, 550, 212
Bogdán, A. & Gilfanov M. 2011, *MNRAS*, in press
Crain, R. A., *et al.* 2010, *MNRAS*, 407, 1403
Davis, D. S., Richer, H. B., Anderson, J., *et al.* 2008, *AJ*, 135, 2155
Dijkstra, M. *et al.* 2011, *MNRAS*, submitted (arXiv1108.4420)
Doane, N. E., *et al.* 2004, *AJ*, 128, 2712
Gilfanov, M. 2004, *MNRAS*, 349, 146

Grimes, J. P., *et al.* 2009, *ApJS*, 181, 272
Grimm, H.-J., Gilfanov M., & Sunyaev R. 2003, *MNRAS*, 339, 793
Guainazzi, M. & Bianchi, B. 2007, *MNRAS*, 374, 1290
Hart, Q. N., Stocke, J. T., & Hallman, E. J. 2009, *ApJ*, 705, 854
Irwin, J. A., Athey, A. E., & Bregman, J. N. 2003, *ApJ*, 587, 356
Kim, D.-W. & Fabbiano, G. 2004, *ApJ*, 611, 846
Kim, D.-W. & Fabbiano, G. 2010, *ApJ*, 721, 1523
Kuntz, K. D. & Snowden, S. L. 2010, *ApJS*, 188, 46
LaMassa, S. M. 2009, *ApJ*, 705, 568
Li, Z. & Wang, Q. D. 2007, *ApJL*, 668, 39
Li, Z., Wang, Q. D., & Wakker, B. P. 2009, *MNRAS*, 397, 148
Li, Z., *et al.* 2011, *ApJL*, 728, 10
Li, J. & Wang, Q. D. 2011, *MNRAS*, submitted
Liu, J., Wang, Q. D., Li Z., & Peterson J. R. 2010, *MNRAS*, 404, 1879
Liu, J., Mao, S., & Wang, Q. D. 2011a, *MNRAS*, 415, 64
Liu, J., Wang, Q. D., & Mao, S. 2011b, *MNRAS*, in press (arXiv:1111.5915)
Lu, Z. & Wang Q. D. 2011, *MNRAS*, 413, 347
Mineo, S., Gilfanov, M., & Sunyaev, R. 2011, *MNRAS*, in press (arXiv:1105.4610v2)
Mulchaey, J. S. & Jeltema, T. E. 2010, *ApJL*, 715, 1
Omma, H., Binney, J., Bryan, G., & Slyz, A. 2004, *MNRAS*, 348, 1105
Oppenheimer, B. D. 2010, *MNRAS*, 406, 2325
Owen, F., *et al.* 2006, *AJ*, 131, 1974
Owen, R. A. & Warwick, R. S. 2009, *MNRAS*, 394, 174
Pellegrini, S., 2011, *ApJ*, 738, 57
Pellegrini, S., Ciotti, L., & Ostriker, J. P. 2011, *ApJ*, submitted (arXiv1107.3675)
Persic, M. & Rephaeli, Y. 2007, *A&A*, 463, 481
Porquet, D., Dubau, J., & Grosso, N. 2010, *SSRv*, 157, 103
Ranalli, P., *et al.* 2008, MNRAS, 386, 1464
Revnivtsev, M. *et al.* 2008, *A&A*, 490, 37
Revnivtsev, M., *et al.* 2011, *A&A*, 526, 94
Sazonov, S., *et al.* 2006, *A&A*, 450, 117
Strickland, D. & Heckman, T. 2009, *ApJ*, 697, 2030
Sun, M., *et al.* 2007, *ApJ*, 657, 197
Sun, M., *et al.* 2010, *ApJ*, 708, 946
Swartz, D. A., *et al.* 2011, *ApJ*, 741, 49
Tang, S., *et al.* 2009, *MNRAS*, 392, 77
Tang, S. & Wang, Q. D. 2010, *MNRAS*, 408, 1011
Townsley, L. K., *et al.* 2011, *ApJS*, 194, 16
Tüllmann, R., *et al.* 2006, *A&A*, 457, 779
Tyler, K., *et al.* 2003, *ApJ*, 610, 213
Voss, R., *et al.* 2009, *ApJ*, 701, 474
Wang, J., *et al.* 2010, *ApJ*, 719, 208
Wang, Q. D. 1999, *ApJL*, 510, 139
Wang, Q. D., *et al.* 2001, *ApJ*, 555, 99
Wang, Q. D., Chaves, T., & Irwin, J. A. 2003, *ApJ*, 598, 969
Wang, Q. D., Owen, F., & Ledlow, M. 2004, *ApJ*, 611, 821
Wang, Q. D. 2010, *PNAS*, 107, 7168
Weźgowiec, M., *et al.* 2011, *A&A*, 531, 44
White, N. E., Swank, J. H., & Holt, S. S. 1983, *ApJ*, 270, 711
Xu, X., Wang, Q. D., & Li, X.-D. 2011, in preparation
Young, J. S. & Scoville, N. Z. 1991, *ARA&A*, 29, 581
Yuan, F., Xie, F., & Ostriker, J. P. 2009, *ApJ*, 691, 98
Yuan, F. 2007, *ASPC*, 373, 95
Zhang, Z., *et al.* 2011, *A&A*, 533, 33

Discussion

GAO: Regarding your correlation plot between soft X-ray emission and star-formation rate (SFR): I agree there is a solid correlation there. Did you compare the hard X-ray to the SFR in the same sample yet? The comparisons of these correlations might help better understand the scattering in soft X-ray vs. SFR.

WANG: Yes, the hard X-ray correlation with the SFR is relatively well known, mostly in terms of point-like sources. We are yet to do this correlation study for the total hard X-ray emission from galaxies in our sample.

ZEZAS: Do you find any difference in the diffuse X-ray emission - SFR correlation between starburst and spiral galaxies?

WANG: No. But we have not examined this issue carefully. There are other effects, such as cool gas, stellar mass, and environment.

The Spectral Energy Distribution of Galaxies
Proceedings IAU Symposium No. 284, 2011
R.J. Tuffs & C.C. Popescu, eds.

doi:10.1017/S1743921312009040

Optical SED models of galaxy mergers

Gregory F. Snyder[1], T. J. Cox[2], Christopher C. Hayward[1], Lars Hernquist[1] and Patrik Jonsson[1]

[1]Harvard-Smithsonian Center for Astrophysics
60 Garden ST, Cambridge, MA 02138

[2]Carnegie Observatories, Pasadena, CA

email: gsnyder@cfa.harvard.edu

Abstract. I discuss recent work in which we construct models of poststarburst galaxies by combining fully three-dimensional hydrodynamic simulations of galaxy mergers with radiative transfer calculations of dust attenuation. The poststarburst signatures can occur shortly after a bright starburst phase in gas-rich mergers, and thus offer a unique opportunity to study the formation of bulges and the effects of feedback. Several additional applications of spatially-resolved spectroscopic models of interacting galaxies include multi-wavelength studies of AGN/starburst diagnostics, mock integral field unit data to interpret the evolution of ULIRGs, and the 'Green Valley'.

Optical spectra of simulated major gas-rich galaxy mergers can be found at
http://www.cfa.harvard.edu/~gsnyder

Keywords. galaxies: evolution, galaxies:interactions, galaxies: starbursts, methods:n-body simulations, radiative transfer

---

1. Introduction

Poststarburst (or 'K+A') galaxies are defined by recent but not ongoing star formation (Dressler & Gunn 1983). Detailed studies (Yang *et al.* 2008) of their properties suggest that many of them will evolve into typical ($\sim L^*$) elliptical galaxies, and thus present an opportunity to study the formation of bulges. The merger hypothesis has been proposed as a formation mechanism for poststarburst galaxies by Lavery & Henry (1988).

To evaluate this hypothesis, realistic physical models of K+A galaxies are essential for two reasons: 1) the merger or event rates inferred from wide-area surveys of K+A galaxies depend sensitively on the K+A duration timescales assumed for the underlying event population, and 2) the physical processes responsible for this bulge assembly and/or star formation quenching can be inferred directly from the model components.

In Snyder *et al.* (2011), we utilized hydrodynamic simulations of galaxy mergers coupled with radiative transfer calculations of dust attenuation to reconcile the merger rate inferred from K+A galaxies with that derived from other methods. Here I present a more in-depth look at a handful of these simulations, working toward a detailed observational representation of state-of-the-art galaxy simulations.

The starting point of this analysis is $\sim$ 40 gas-rich merger simulations with a variety of initial conditions, from which we selected poststarburst galaxies based on moderate-resolution ($R \approx 1000$) optical spectroscopy, requiring strong Balmer absorption and weak [O II] emission (cf. Zabludoff *et al.* 1996). We generated these spectra with the Monte Carlo dust radiative transfer code Sunrise (Jonsson 2006; Jonsson *et al.* 2010), producing spatially-resolved spectra from multiple viewing angles. Herein I focus on 15 mergers of identical Milky Way-like disks discussed in Cox *et al.* (2006), that differ only in the initial

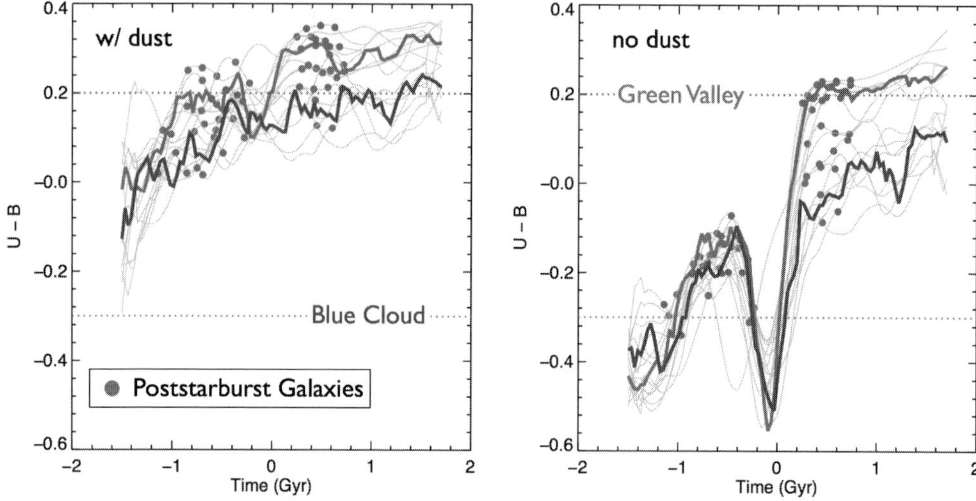

Figure 1. Time evolution of U-B colors of galaxy mergers. The dotted gray curves show the evolution of 15 major mergers of identical Milky Way-like disks. These mergers vary only in the initial relative orientation of the disk spin vectors. From Figure 8 of Snyder *et al.* (2011) we subdivide this set of mergers into two classes: "strong burst" (8) and "weak burst" (7), where the former are defined as likely forming a K+A galaxy for longer than 100 Myr occurring after the final merger ($t = 0$). The two solid curves show the median of each subclass, with the redder remnant forming from the "strong burst" class in both panels. Filled circles represent snapshots for which we identify the galaxy as a K+A from at least one of seven viewing angles. The left panel is our calculation of the colors including attenuation by dust, where for the right panel we have turned off attenuation by dust in the diffuse phase of the ISM, so that these colors reflect the *intrinsic* stellar populations. We find that K+As form redder remnants regardless of the presence of dust, but also that dust causes even the intrinsically bluest remnants to enter the 'Green Valley' of U-B color.

relative orientation of their angular momentum vectors. These simulations were run with a version of GADGET2 (Springel 2005) that includes a two-phase ISM/star-formation model Springel & Hernquist (2003) and thermal coupling of AGN feedback into the local ISM (Springel *et al.* 2005). Our goal is to determine what are the physical causes of K+A formation, and whether or not they are likely predecessors of red-and-dead galaxies.

2. Findings

We find a wide diversity of outcomes, which is not surprising given recent work showing that even major gas-rich mergers can result in a very disk-like remnant (e.g. Robertson *et al.* 2006; Hopkins *et al.* 2009b). Likewise, the merger of identical disks gives rise to a long-lived K+A galaxy only when the dynamics of the final merger lead to a short, intense starburst that is "complete" in the sense that the gas is either mostly consumed or expelled. We find that merger-induced K+A galaxies are indeed likely to evolve into redder remnants, as shown in Figure 1, whether or not dust is considered. Furthermore, the difference in their final colors arises almost exclusively to differences in the merger properties *during coalescence*, either strong AGN feedback and/or gas consumption, because at $t < 0$ (pre-merger), the intrinsic color evolution (right panel) of strong or weak K+A-inducing mergers are identical.

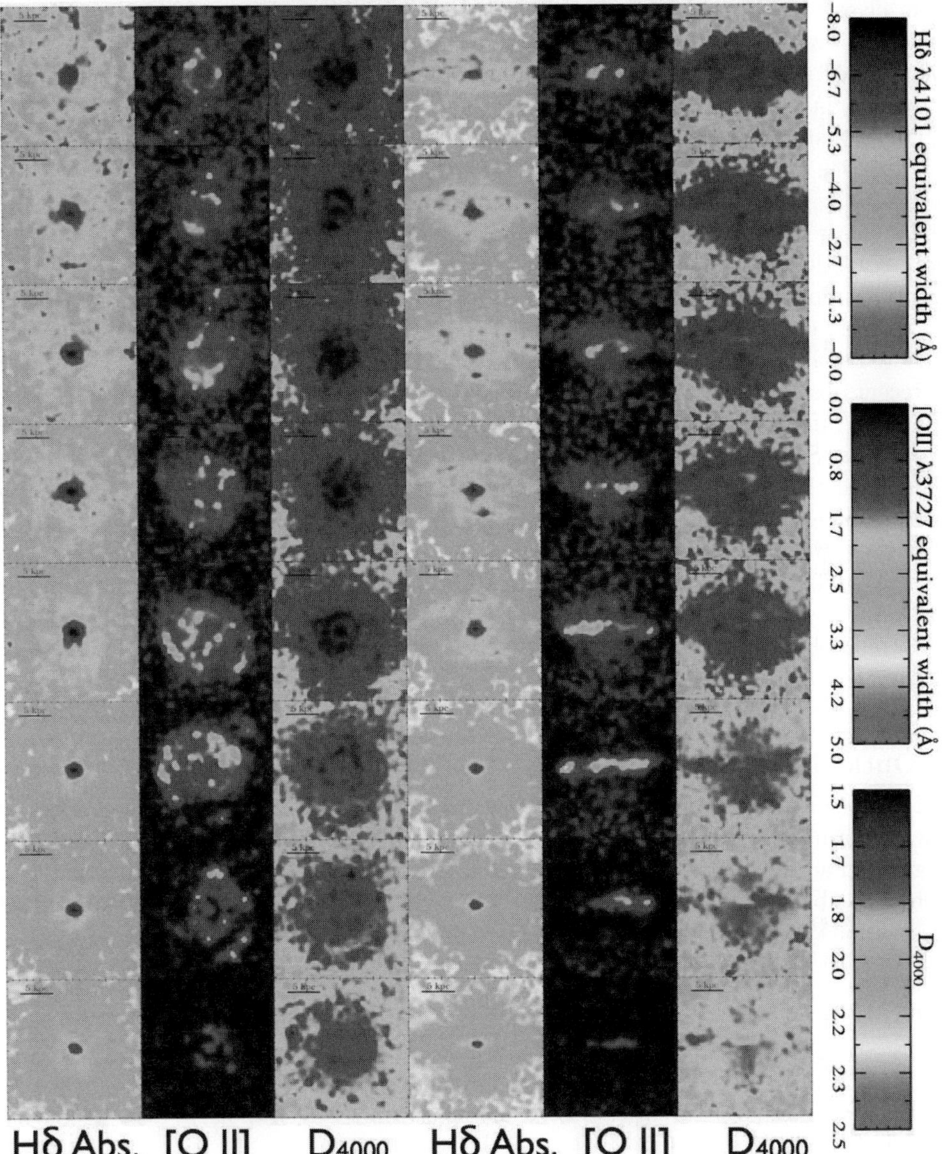

Figure 2. Time evolution of three resolved stellar population diagnostics after a simulated gas-rich major galaxy merger. We show two viewpoints: one from which a relic star-forming disk can be seen roughly face on (left three columns), and the other from which this disk is inclined (right three). The topmost row shows the remnant 100 Myr after the peak star formation rate of the starburst, and each subsequent row is 70 Myr later, ending 600 Myr post-merger. The burst relic is the clear peak in the Balmer absorption maps (as this is a K+A galaxy at these times). We see low levels of [O II] emission in an extended, rotating disk, comprised of star-forming gas that survived the merger (cf. Lotz *et al.* 2008). This demonstrates one way in which merger outcomes are diverse.

In some cases it is thought that the rapid quenching required to make a poststarburst galaxy is induced by heating from an accreting supermassive black hole, or "AGN feedback". However, Wild *et al.* (2009) found that the rapid truncation of star formation in merger-induced K+A galaxies does not require such feedback. Instead, the starbursts themselves consume most of the available cool gas, while subsequent AGN or starburst winds remove the small amount of gas and dust that happens to remain. We found that the Balmer absorption features may be strengthened by this "un-screening" process, because light from the young stars formed during the burst can become less attenuated.

This is a statement primarily about *local analogues*, and so significantly more massive and gas-rich progenitors may require powerful AGN or starburst winds to create the objects of e.g. Tremonti *et al.* (2007). Furthermore, modeling enhancements such as stellar gas recycling that allow simulations with more stable gas-rich progenitors and more realistic evolution of the gas reservoir throughout the merger, may be required to faithfully reproduce such observations. In either case, how can we determine the effect of starburst or AGN-powered outflows on the gas reservoir of galaxies? And what is the expected spatial structure of the gas, dust, and stars in recent merger remnants?

Recent observational work on the evolution of merger remnants has sought to leverage geometric information, such as the spatial extent of different stellar populations and their kinematics. SUNRISE is naturally extensible to the spectro-imaging techniques required to perform these studies, and so we demonstrate this with spatial maps of three diagnostics in Figure 2, which follows a remnant for ≈ 600 Myr or so after the merger. We clearly see the strong, spatially-concentrated central burst in the A-star tracer Balmer absorption, from either viewing angle. In addition, we see a surviving, star-forming disk. For simplicity, we have ignored dust attenuation; from the edge-on perspective only, this disk would obscure the central starburst relic almost completely (it would not necessarily be classified K+A) if dust were included.

3. Outlook

Here we have focused primarily on the optical regime of the SED. However, clearly much information can be learned by analyzing merger-induced evolution at all wavelengths. In addition, now that models including toy prescriptions of complicated physics such as feedback can be constructed and compared "apples-to-apples" with sophisticated observations, it may be possible to further constrain their parameter space.

References

Dressler, A. & Gunn, J. E. 1983, *ApJ*, 270, 7
Cox, T. J., Dutta, S. N., Di Matteo, T., *et al.* 2006, *ApJ*, 650, 791
Hopkins, P.F., Cox, T.J., Younger, J.D., & Hernquist, L. 2009b, *ApJ*, 691, 1168
Jonsson, P. 2006, *MNRAS*, 372, 2
Jonsson, P., Groves, B. A., & Cox, T. J. 2010, *MNRAS*, 403, 17
Lavery, R. J. & Henry, J. P. 1988, *ApJ*, 330, 596
Lotz, J. M., Jonsson, P., Cox, T. J., & Primack, J. R. 2008, *MNRAS*, 391, 1137
Robertson, B., Bullock, J. S., Cox, T. J., *et al.* 2006, *ApJ*, 645, 986
Snyder, G. F., Cox, T. J., Hayward, C. C., Hernquist, L., & Jonsson, P. 2011, *ApJ*, 741, 77
Springel, V. & Hernquist, L. 2003, *MNRAS*, 339, 289
Springel, V. 2005, *MNRAS*, 364, 1105
Springel, V., Di Matteo, T., & Hernquist, L. 2005b, *MNRAS*, 361, 776
Tremonti, C. A., Moustakas, J., & Diamond-Stanic, A. M. 2007, *ApJL*, 663, L77
Wild, V., Walcher, C. J., Johansson, P. H., *et al.* 2009, *MNRAS*, 395, 144
Yang, Y., Zabludoff, A. I., Zaritsky, D., & Mihos, J. C. 2008, *ApJ*, 688, 945
Zabludoff, A. I., Zaritsky, D., Lin, H., *et al.* 1996, *ApJ*, 466, 104

Discussion

CROCKER: Do I understand correctly that the photons from the stellar population you model have no dynamical effect on the gas? If yes, is that a concern?

SNYDER: Yes, and this is a concern. Ideally we would perform coupled hydro/radiative transfer simulations to understand the effect of stellar/AGN winds, but our current methods do not treat this.

CHAKRABARTI: How does the emission in the poststarbust phase depend on SFR prescriptions?

SNYDER: We haven't yet looked into this - it will be interesting to see what a more realistic treatment with molecular gas and shocks will do to the story.

MOUSTAKAS: Did I understand correctly that you don't need AGN feedback to quench the star formation because of the inclusion of radiative transfer, and do you find any evidence of high-velocity fossil winds?

SNYDER: The conclusion about AGN feedback doesn't depend on the radiative transfer, it is mostly about the fact that merger histories can give low-enough star formation without it (though the AGN does reduce it further). We do not look into winds since our simulations do not include kinetic feedback - that is definitely something to check.

The Spectral Energy Distribution of Galaxies
Proceedings IAU Symposium No. 284, 2011
R.J. Tuffs & C.C. Popescu, eds.

doi:10.1017/S1743921312009052

The SEDs of interacting galaxies

Lauranne Lanz[1], Nicola Brassington[1,2,3], Andreas Zezas[1,3], Howard A. Smith[1], Matthew L. N. Ashby[1], Elisabete da Cunha[4], Christopher Klein[5], Patrik Jonsson[1], Christopher C. Hayward[1], Lars Hernquist[1], and Giovanni Fazio[1]

[1]Harvard Smithsonian Center for Astrophysics
60 Garden Street, MS-10, Cambridge, MA 02138, USA

[2]School of Physics, Astronomy and Mathematics, University of Hertfordshire,
College Lane, Hatfield, AL10 9AB, UK

[3]Physics Department, University of Crete, P.O. Box 2208, 710 03 Heraklion, Crete, Greece

[4]Max Planck Institute for Astronomy (MPIA), Königstuhl 17, 69117, Heidelberg, Germany

[5]Department of Astronomy, University of California, Berkeley, CA 94720-3411, USA

email: llanz@cfa.harvard.edu

Abstract. The evolution of galaxies is greatly influenced by their interactions. As part of a program to study interacting galaxies, we have measured and modeled the spectral energy distributions (SEDs) from the ultraviolet (UV) to the far-infrared (FIR). We describe the constraints imposed on star formation histories by these SEDs, and the variations therein seen across the interaction sequence, and we compare the results of different star formation rate prescriptions applied to the data. The sample itself is based on the Spitzer Interacting Galaxy Survey (SIGS) of 111 galaxies in 50 systems, a project designed to probe a range of galaxy interaction parameters in the infrared. Our SEDs combine the Spitzer results with multiwavelength data from other missions, in particular GALEX and Herschel. The subset presented here is the sample for which FIR Herschel observations are currently publicly available.

Keywords. galaxies:interactions, galaxies:ISM, galaxies:photometry, infrared:galaxies, ultraviolet:galaxies

---

1. The Spitzer Interacting Galaxy Survey

1.1. *Sample Selection*

The SIGS sample is comprised of the Keel-Kennicutt (Keel *et al.* 1985) complete sample of interacting galaxies, which are chosen on the basis of association likelihood, supplemented with galaxies from their "Arp sample". Most of the systems within the complete sample are galaxies in early phases of interaction, so the inclusion of the Arp sample increases the range of interaction parameters covered. The Keel *et al.* (1985) selection criteria for the complete sample included a $| \Delta \text{ v} | < 600$ km s^{-1} requirement in order to exclude non-associated pairs. We imposed an additional distance constraint of cz $<$ 4000 km s^{-1} to retain high spatial resolution. These criteria yield a sample of 111 galaxies in 50 systems.

Previous samples tended to be chosen morphologically or based on infrared emission or optical emission line diagnostics. Due to the criteria used to select our sample, it spans a broad range of types of interactions and includes systems likely to be on their first pass as well as evolved systems. Similarly, in contrast to infrared-selected samples that tend to contain only the more active systems, our sample covers a large range of degrees of nuclear and/or starburst activity. Finally, our use of infrared diagnostics, rather than optical, significantly reduces the effects of obscuration.

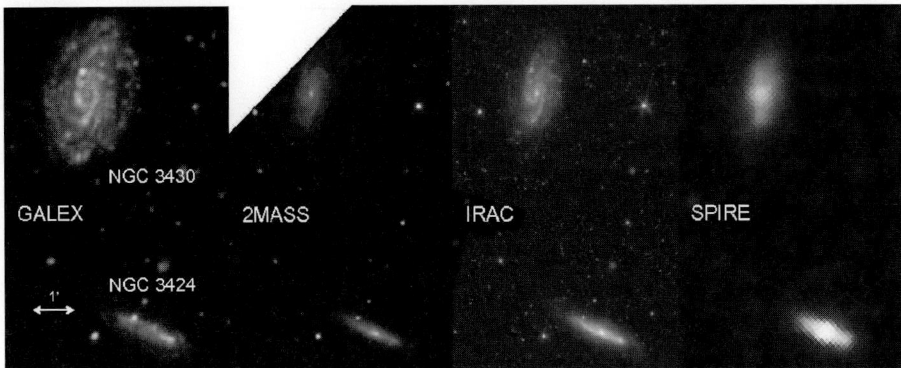

Figure 1. The interacting spiral pair of NGC 3430 (top) and NGC 3424 (below) as observed by GALEX (NUV and FUV), 2MASS (J, H and K_s), IRAC (3.6 μm, 4.5 μm, and 8.0 μm), and SPIRE 250 μm. These galaxies are in an early interaction stage and show some of SIGS's range in relative UV-to-IR morphology, exemplified by the contrast of NGC 3430's extended UV disk and NGC 3424's consistent morphology.

1.2. *Observations*

The sample has a complete set of mid-infrared (MIR) observations taken by the Spitzer Space Telescope, including photometry with both the Infrared Array Camera (IRAC; Fazio *et al.* 2004) and the Multiband Imaging Photometer (MIPS; Rieke *et al.* 2004) as well as nuclear spectroscopy with the Infrared Spectrograph (IRS; Houck *et al.* 2004). In addition, we have $> 90\%$ coverage in the near-ultraviolet (NUV) and far-ultraviolet (FUV) with Galaxy Evolution Explorer (GALEX; Martin *et al.* 2005). Presently, $\sim$20% of the galaxies also have publicly available FIR photometry from the Spectral and Photometric Imaging Receiver (SPIRE; Griffin *et al.* 2010) aboard the Herschel Space Observatory. Together, these infrared and ultraviolet observations give a view of both the obscured and unobscured star formation. We supplement these data with near-infrared observations from the Two Micron All Sky Survey (2MASS; Skrutskie *et al.* 2006), far-infrared photometry from the Infrared Astronomical Satellite (IRAS; Neugebauer *et al.* 1984), and UBV photometry from the Third Reference Catalogue (de Vaucouleurs *et al.* 1991). We are in the process of obtaining additional UV data from Swift Satellite imaging (Roming *et al.* 2005). Figure 1 shows a subset of these observations for a typical sample pair of galaxies, NGC 3430 (top) and NGC 3424 (bottom).

The apparent extent of a galaxy can vary significantly between the ultraviolet and the infrared. Therefore, we used S-Extractor to determine the optimum aperture size in both NUV and 3.6 μm images. The larger aperture was then consistently applied to all the observations when measuring photometry. Uncertainties in the measured fluxes consist of a sum in quadrature of the calibration and Poisson noise values.

2. Spectral Energy Distributions

Fig. 2 presents two typical SEDs from the subsample: NGC 3430, part of a spiral galaxy pair and M51B, a dwarf galaxy interacting with M51A. These SEDs demonstrate some of the variety observed. To extract quantitative parameters such as the star formation rate and stellar and dust masses, we used MAGPHYS (da Cunha *et al.* 2008), which uses stellar libraries from Bruzual & Charlot (2003) in combination with infrared dust spectral libraries to fit the SED such that energy absorbed in the UV and optical is re-absorbed in the infrared. This code provides posteriors of such parameters as temperatures of warm

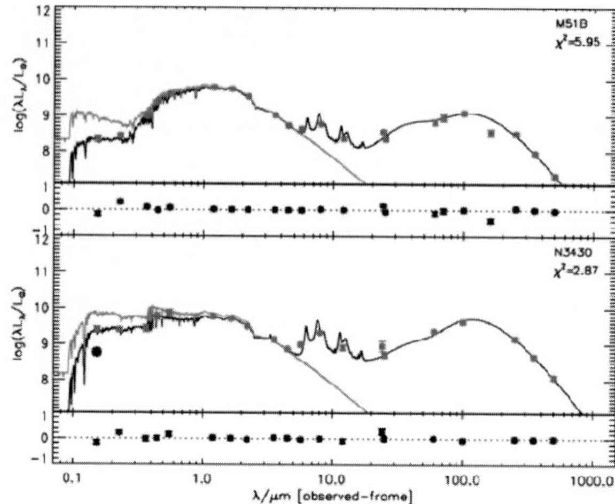

Figure 2. SEDs of M51B and NGC 3430, showing some of the range in SED shape seen, particularly in terms of the relative strengths of the MIR to the optical and the UV to the optical, overplotted with the fits from MAGPHYS (da Cunha *et al.* 2008) showing the best fit model (black) and the unattenuated stellar fitted spectrum (grey).

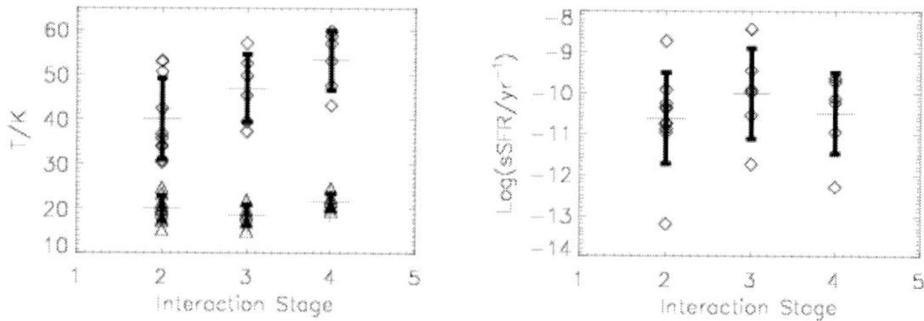

Figure 3. Specific star formation rates (right) and warm and cold dust temperatures (left) plotted against the Dopita *et al.* (2005) interaction stage, showing that based on the first 22 galaxies, the only one of these three parameters to vary significantly with interaction stage is the temperature of the 30-60 K dust associated with birth clouds and the ISM.

(30-60 K) and cold (15-25 K) dust, stellar and dust masses, and star formation rate (SFR). We find good agreement between the dust luminosity determined by this method and the integrated 8-1000 μm luminosity.

3. Results

3.1. *Variations with Interaction Stage*

Since it is notoriously difficult to precisely identify the interaction stage of any one galaxy system, we use the Dopita *et al.* (2005) five-stage classification scheme, where stages 2-4 represent mildly, moderately, and strongly interacting systems. Fig. 3 shows the warm (30-60K; blue) and cold (15-25K; green) dust temperatures (left) and specific star formation rate (sSFR; right) derived by MAGPHYS for the 22 galaxies in the 10 systems with currently available SPIRE observations with the mean and standard deviation of each set over-plotted. The mean specific star formation rate does not appear to show any significant variation as the interactions evolve in this dataset. While the cold diffuse dust likewise does not increase in temperature with interaction stage, the warm dust, associated with both interstellar medium and the birth clouds, shows a probable increase with interaction evolution.

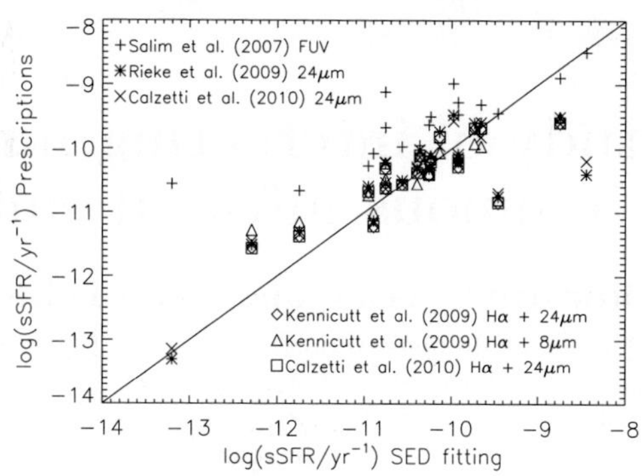

Figure 4. Comparison of specific star formation rates derived by MAGPHYS from fitting the SED to those calculated using one- or two-band prescriptions. The two objects with the highest SED-derived sSFR both have extended UV disks.

3.2. *Comparison of SED-derived sSFR with sSFR prescriptions*

Future galaxy surveys, particularly those studying objects at high redshirt, will probably not have the opportunity or the capacity to exploit the same wealth of observations we used for this sample. As such, it is important to understand which one- or two-band SFR prescriptions work best for interacting systems. In Fig. 4, we compare several indicators based either solely on UV or IR emission or a sum thereof to the sSFR derived from MAGPHYS. We generally find that the SED-derived sSFR agrees better with a prescription that includes contribution from the IR. However, NGC 3430 is a counter-example: its extended UV disk suggests that the IR-dependent prescriptions, which generally assume optical thickness, miss a significant portion of the recent star formation. We are currently extending this work to include all 111 galaxies in the SIGS sample, allowing us to thoroughly test the sSFR indicators.

Acknowledgements We are grateful to Don Neill and Susan Neff for their assistance with the GALEX data and to Volker Tolls and Joe Hora for their assistance in reducing Herschel SPIRE observations. This work was supported by the Harvard College Observatory and NASA through an award issued by JPL/Caltech via contract #1369566.

References

Bruzual, G. & Charlot, S. 2003, *MNRAS*, 344, 1000

da Cunha, E., Charlot, S., & Elbaz, D. 2008, *MNRAS*, 388, 1597

de Vaucouleurs, G., *et al.* 1991, *Third Reference Catalogue of Bright Galaxies*, (New York: Springer)

Dopita, M. A., Pereira, M., Kewley, L. J., & Capaccioli, M. 2002, *ApJS*, 143, 47

Fazio, G., *et al.* 2004, *ApJS*, 154, 10

Griffin, M. J., *et al.* 2010, *A&A*, 518, L3

Houck, J. R., *et al.* 2004, *ApJS*, 154, 18

Keel, W. C., Kennicutt Jr., R. C., Hummel, E., & Van der Hulst, J. M. 1985, *AJ*, 90, 708

Martin, D. C., *et al.* 2005, *ApJ*, 619, L1

Neugebauer, G., *et al.* 1984, *ApJ*, 498, 579

Rieke, G. H., *et al.* 2005, *ApJS*, 154, 25

Roming, P. W. A., *et al.* 2005, *Space Sci. Revs*, 120, 95

Skrutskie, M. F., *et al.* 2006, *AJ*, 131, 1163

The Spectral Energy Distribution of Galaxies
Proceedings IAU Symposium No. 284, 2011
R.J. Tuffs & C.C. Popescu, eds.

doi:10.1017/S1743921312009738

A Spitzer study of interacting luminous and ultra-luminous infrared galaxies

João Rodrigo S. Leão[1] **and Claus Leitherer**[2]

[1]Universidade Federal do Rio Grande - FURG,
Instituto de Matemática, Estatística e Fśica - IMEF
Caixa Postal 474, Campus Carreiros, Rio Grande, RS, Brazil
email: joaoleao@furg.br

[2]Space Telescope Science Institute
3700 San Martin Drive, Baltimore, MD, USA
email: leitherer@stsci.edu

Abstract. We conducted a Spitzer Space Telescope survey of 28 Luminous ($11 < log(L_{IR}/L_{\odot}) < 12$, LIRGs) and Ultra-Luminous Infrared Galaxies ($log(L_{IR}/L_{\odot}) > 12$, ULIRGs). Many of these galaxies are found in pairs or associations and are powered by either nuclear activity or star-formation (Sanders & Mirabel 1996). Our main goal is to understand the relative importance of starbursts and AGNs in interacting systems. Is the frequency of AGN and starbursts in these interacting galaxies related to their luminosities? What is the importance of the merger stage and the frequency of AGNs? We present our conclusions and diagnostic diagrams based in the observed near infrared lines and compare to studies based solely in optical data.

Keywords. techniques: spectroscopy, galaxies: starburst, galaxies: statistics, line: identification, galaxies: interactions

1. Introduction and Goals

We present the results of a Spitzer Space Telescope survey of 28 Luminous and Ultra-Luminous Infrared Galaxies (Leão & Leitherer 2005, Leão & Leitherer 2008). The observations were performed using the IRS spectrograph in low-resolution to cover the $4 - 35$ μm wavelength range. We measured the flux and equivalent widths of the several observed lines. We are particularly interested in emission lines which characterize the presence of AGNs or starbursts as will be discussed in section 2. Some of the observed SEDs are shown in figure 1. In this study the LIRGs were chosen from the study of Arribas *et al.* (2004) and images from the Nordic Optical telescope (NOT) are available for these galaxies. The ULIRGs were chosen from the study of Bushouse *et al.* (2002) and Hubble Space Telescope (HST) images are available. Many of these galaxies are found in pairs or associations (e.g., Bushouse *et al.* 2002, Arribas *et al.* 2004) and are believed to be powered by a combination of nuclear activity and star-formation (e.g., Genzel & Cesarsky 2000, Sanders & Mirabel 1996).

The study of Sanders *et al.* (1988) suggests that the fraction of AGNs inreases with the luminosity of the host galaxy. In fact, the optical study of Kim, Veilleux & Sanders (1998), hereafter KVS98, indicates that for LIRGs, AGNs are present in about 32% of their galaxies, while for ULIRGs, AGNs (Sey. 1, 2 and LINERs) dominate and are present in 70% of the systems. In many cases AGNs and starbursts coexist, making it even harder to understand the relative importance of nuclear activity and star-formation in LIRGs and ULIRGs.

The main goal of this project in to understand if diagnostics based in the near infrared spectra yield the same results as KVS98. Do we find the same relative frequency of starbursts and AGNs as a function of luminosity? What is the importance of the merger stage and the frequency of AGNs?

2. Diagnostic Diagrams

We used the observed nearIR spectra to determine the activity class of our galaxies independently of their optical classification. The presence of the PAH lines at 6.2, 7.7, and 11.3 μm are strong indications for the presence of starbursts since the PAH molecules are destroyed (or hidden) by nuclear activity (AGN). The presence of the [Ne V] line at 14.3 and 24.3μm is a strong evidence for an AGN since it takes 97 eV to ionize [NeIV] and this energy cannot come from OB stars. The AGNs also exhibit a featurefree continuum below 5μm that cannot be found in starbursts.

In figure 2 we investigate the strength of the 7.7 PAH emission line vs. the [NeV] 14.3 μm emission line and find that starbursts and AGN's are clearly separated. We designed other diagnostic diagrams using the [OIV] at 25.9 μm and [NeII] at 12.8 μm. All these diagrams successfully separate galaxies with AGN activity from those powered by starbursts.

3. Results

Our main findings are:(i) Most AGNs are also advanced mergers. This is in agreement with the Sanders *et al.* (1988) scenario to explain LIRGs and ULIRGs; (ii) We find that for $11 < log(L_{IR}/L_{\odot} < 12$ (LIRGs) starbursts dominate with 75% (25% AGNs), while

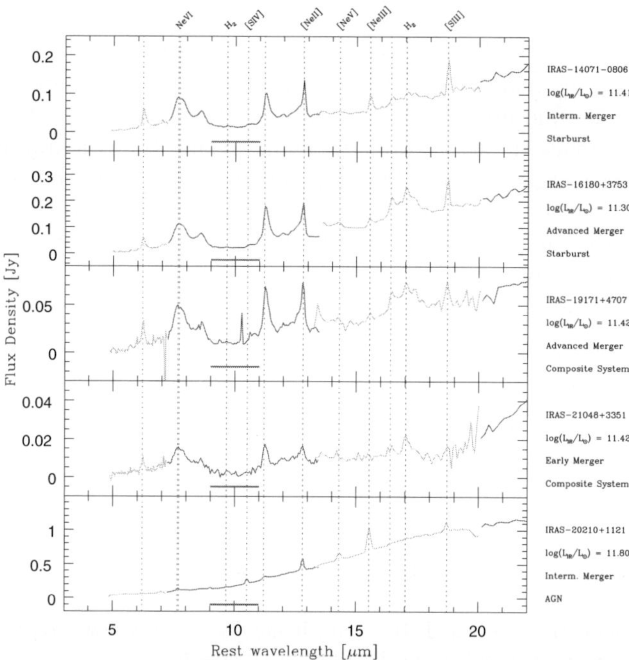

Figure 1. The IRS/SPitzer lowresolution spectra showing in detail the region between 4-22μm for the indicated galaxies, representative of the whole sample. Infrared luminosities, merger stages and activity class are also indicated. The main emission lines are indicated by vertical dotted lines: PAH lines, forbidden emission lines and molecular Hidrogen. The horizontal bar indicates the extent of the silicate absorption around 10μm.

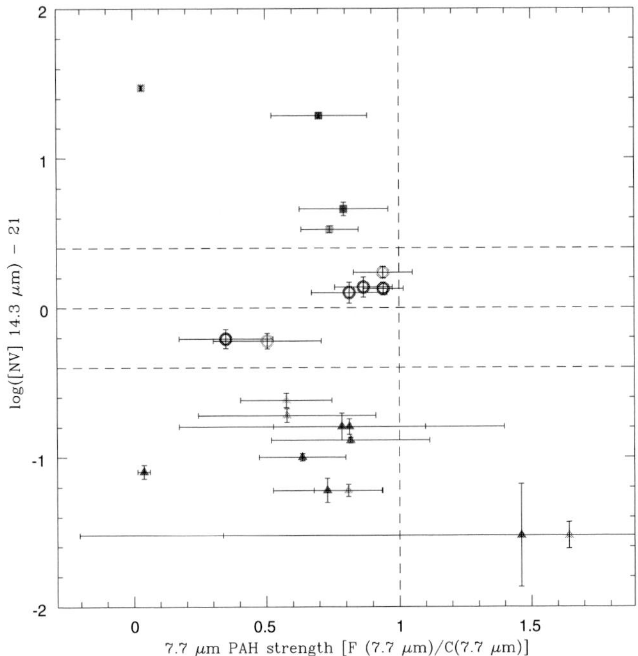

Figure 2. In this plot triangles denote starbursts, circles composite systems and squares AGNs. We see that AGNs and composite systems have stronger [N V] emission lines. Strong 7.7μm PAH lines are seen in AGN's, composites and starbursts, possibly indicating that star formation and nuclear activity might coexist in a single galaxy.

for $log(L_{IR}/L_{\odot}) \geqslant 12$ (ULIRGs), AGNs and composite systems dominate with 92%. We thus find a strong evidence for the so called luminosity dependence; (iii) Based on this small sample we find no evidence that advanced mergers are preferentially found in more luminous systems. In other words, we cannot claim that the infrared luminosity is a precise tracer of merger activity. Maybe a larger sample can shed some light on this issue. Many of these galaxies show a deeep silicate absorption and this indicates a large optical depth to the nucleus. We thus suspect that the AGN contribution for many of the galaxies in the sample may be larger. This would imply that AGNs remain hidden even at mid-IR wavelengths.

References

Sanders, D. B., Mirabel, I. F. 1996, *ARAA*, 34, 749

Leão, J. R. S. & Leitherer, C. 2005, General Obs. Proposal 20589, Spitzer Space Telescope

Leão, J. R. S. & Leitherer, C., 2008, General Obs. Proposal 20589, Final Report, Spitzer Space Telescope

Bushouse, H. *et al. ApJ*, 138, 1

Arribas, S., Bushouse, H., & Lucas, R. A. 2004, *ApJ*, 127, 2522

Genzel, R. & Cesarsky, C. 2000, *ARAA*, 38, 76

Sanders, D. B., Soifer, B. T., Elias, J. H., Neugebauer, G., & Matthews, K. 1988, *ApJ*, 328, L35

Kim, D. C., Veilleux, S., & Sanders, D. B. 1998, *ApJ*, 508, 627

The Spectral Energy Distribution of Galaxies
Proceedings IAU Symposium No. 284, 2011
R.J. Tuffs & C.C. Popescu, eds.

doi:10.1017/S1743921312009064

A new model for the infrared emission of IRAS F10214+4724

Andreas Efstathiou[1], Natalie Christopher[2], Aprajita Verma[2], and Ralf Siebenmorgen[3]

[1]School of Sciences, European University Cyprus, Engomi, 1516 Nicosia, Cyprus
[2]Oxford Astrophysics, Denys Wilkinson Building, University of Oxford, Keble Rd, Oxford OX1 3RH, United Kingdom
[3]European Southern Observatory, Karl-Schwarzschildstr. 2, 85748 Garching b. Munchen, Germany

email: a.efstathiou@euc.ac.cy

Abstract. We present a new model for the infrared emission of the high redshift hyperluminous infrared galaxy IRAS F10214+4724 which takes into account recent photometric data from *Spitzer* and *Herschel* that sample the peak of its spectral energy distribution. We first demonstrate that the combination of the AGN tapered disc and starburst models of Efstathiou and coworkers, while able to give an excellent fit to the average spectrum of type 2 AGN measured by *Spitzer*, fails to match the spectral energy distribution of IRAS F10214+4724. This is mainly due to the fact that the νS_ν distribution of the galaxy falls very steeply with increasing frequency (a characteristic of heavy absorption by dust) but shows a silicate feature in emission. We propose a model that assumes two components of emission: clouds that are associated with the narrow-line region and a highly obscured starburst. The emission from the clouds must suffer significantly stronger gravitational lensing compared to the emission from the torus to explain the observed spectral energy distribution.

Keywords. galaxies: evolution, galaxies: formation, galaxies: high-redshift, infrared: galaxies, cosmology: observations

1. Introduction

The title of the discovery paper of IRAS F10214+4724 at a redshift of 2.285 (Rowan-Robinson *et al.* 1991) was 'An IRAS galaxy with huge luminosity - hidden quasar or protogalaxy?'. It soon became clear that the reason for the enormous luminosity was that the galaxy was gravitationally lensed by a factor of 10-100 (e.g. Broadhurst & Lehar 1995). Until recently, however,it was not clear whether the dominant energy source was an active galactic nucleus (AGN) or a starburst.

Models for the infrared spectral energy distribution (SED) of IRAS F10214+4724 in the mid-90s remained inconclusive (e.g. Rowan-Robinson *et al.* 1993) mainly due to the limited wavelength coverage. Mid-infrared spectrophotometry has been shown for some time now to be able to differentiate between a starburst and AGN origin of the infrared emission (Roche *et al.* 1991 and references therein, Genzel *et al.* 1998). The mid-infrared spectrum of IRAS F10214+4724 obtained by Teplitz *et al.* (2006) using the IRS onboard *Spitzer* showed a puzzling emission feature at rest frame $10\mu m$ instead of the expected absorption feature given the type 2 nature of the AGN suggested by optical observations. Efstathiou (2006) proposed that the SED of IRAS F10214+4724 could be understood in terms of three components of emission: clouds that are associated with the narrow-line region and with a covering factor of 17%, an edge-on torus and a highly obscured starburst. It was also noted by Efstathiou (2006), however, that because the emission from the torus and the narrow-line region clouds arises from diferent regions their magnification may be different.

In this paper we develop a new model for the infrared emission of IRAS F10214+4724 which takes into account new photometric data from *Spitzer* and *Herschel* that cover the wavelength range 70-170μm and therefore complete its SED.

2. Radiative transfer models

2.1. *Starburst models*

Efstathiou, Rowan-Robinson & Siebenmorgen (2000) developed a starburst model that has three main features: The first feature of the model is that it incorporates the stellar population synthesis model of Bruzual & Charlot that gives the spectrum of the stars as a function of their age. Efstathiou, Rowan-Robinson & Siebenmorgen (2000) used the table that assumes a Salpeter IMF and stellar masses in the range 0.1-125 $M_{\odot}$. The second important feature of the model is that it involves detailed radiative transfer that takes into account multiple scattering from grains and incorporates a dust model that includes small transiently heated grains and PAH molecules as well as large classical grains. For the calculation of the emission of the small grains and PAHs the method of Siebenmorgen & Krügel (1992) was employed. The third feature of the model is that it incorporates a simple model for the evolution of the giant molecular clouds that constitute the starburst once a stellar cluster forms instantaneously at the center of the cloud. The most important characteristic of this evolutionary scheme is that by about 10 Myrs after star formation, the expansion of the HII region leads to the formation of a cold narrow shell of gas and dust. This naturally explains why the mid-infrared spectra of starburst galaxies are dominated by the PAH emission and not by the emission of hot dust. The sequence of molecular clouds at different ages can be convolved with a star formation history to give the spectrum of a starburst at different ages. Farrah *et al.* (2003) showed that the starburst models of Efstathiou, Rowan-Robinson & Siebenmorgen (2000) in combination with the tapered discs of Efstathiou & Rowan-Robinson (1995), which are described below, provided good fits to the SEDs of 41 ultraluminous infrared galaxies. One interesting prediction of the Efstathiou, Rowan-Robinson & Siebenmorgen (2000) model is that young starbursts show strong near-infrared continuum, small 6.2μm equivalent widths and deep silicate absorption features and can therefore explain the galaxies in class 3A of Spoon *et al.* (2007). The starburst model has recently been revised by Efstathiou & Siebenmorgen (2009)

2.2. *AGN torus models*

The most important constraint on early models for the torus in AGN (Pier & Krolik 1992, Granato & Danese 1994, Efstathiou & Rowan-Robinson 1995) was provided by mid-infrared spectroscopic observations from the ground in the 8-13μm window by Roche, Aitken and collaborators (Roche *et al.* 1991 and references therein). The observations showed moderate absorption features in type 2 AGN and featureless spectra in type 1 AGN. By comparison the flared discs (whose thickness increases linearly with distance from the central source) of Efstathiou & Rowan-Robinson (1995)) and Granato & Danese (1994) showed strong silicate emission features when observed face-on. Efstathiou & Rowan-Robinson (1995)) proposed that tapered discs (whose thickness increases with distance from the central source in the inner part of the disc but tapers off to a constant value in the outer disc) with a density distribution that followed r^{-1} could give flat spectra in the mid-infrared for an opening angle of around 45° and an equatorial optical depth at 1000Å of 1000. Models with a smaller opening angle predicted face-on spectra with shallow silicate absorption features whereas models with a larger opening angle showed emission features. Recent Spitzer observations showed emission features in quasars (e.g. Hao *et al.* 2005, Hao *et al.* 2007, Siebenmorgen *et al.* 2005, Spoon *et al.* 2007). The tapered discs of Efstathiou & Rowan-Robinson (1995), in combination with the starburst models of Efstathiou, Rowan-Robinson & Siebenmorgen (2000) have been very successful in fitting the spectral energy distributions of a number of AGN even in cases where mid-infrared spectroscopy was available (Alexander *et al.* 1999, Ruiz *et al.* 2001, Farrah *et al.* 2003, Efstathiou & Siebenmorgen 2005).

In this paper we use a grid of tapered disc models computed with the method of Efstathiou & Rowan-Robinson (1995). In this grid of models we consider four discrete values for the equatorial 1000Å optical depth (500, 750, 1000, 1250), three values for the ratio of outer to inner disc radii (20, 60, 100) and three values for the opening angle of the disc (30°, 45° and 60°). The spectra are computed for 37 or 74 orientations which are equally spaced in the range 0 to $\pi/2$.

The grid of tapered disc models, as well as the grid of starburst models discussed above, are available from AE on request (a.efstathiou@euc.ac.cy).

2.3. *Comparison with the average spectrum of Seyfert 2s*

We first explore whether the combination of the tapered disc and starburst models described above can fit the average spectrum of Seyfert 2s presented by Hao *et al.* 2007. The observed spectrum has been determined by first normalizing the spectra at 15μm and then averaging. The AGN models have also been averaged in an attempt to mimic the averaging of the observed spectra.

We find that the best fit to the Seyfert 2 average is provided by a model that assumes a torus opening angle of 45°, an equatorial 1000Å optical depth of 750 and a ratio of outer to inner torus radii of 20. The best fit starburst model assumes an initial optical depth of the molecular clouds τ_V of 200 and a starburst age of 72Myr.

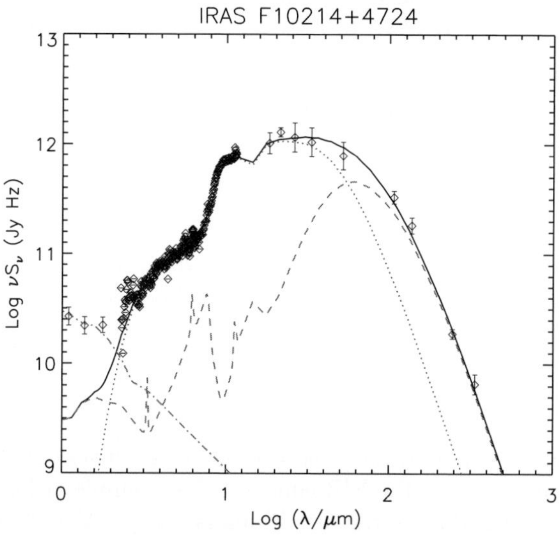

Figure 1. Fit to the spectral energy distribution of IRAS F10214+4724. The dotted and dashed lines show the contributions of the AGN and starburst respectively to the total emission (solid line). The dash dotted line also shows a fit to the IRAC data with a 1Gyr old stellar population. Data from Teplitz *et al.* (2006), IRAS, Rowan-Robinson *et al.* (1993), Sturm *et al.* (2010).

3. A new model for IRAS F10214+4724

We find that the models described above provide a very poor fit to the SED of IRAS F10214+4724. This is clearly due to the fact that the SED of F10214+4724 shows simultaneously characteristics of almost edge-on and almost face-on emission. The failure of torus emission alone to fit the mid-infrared SED of F10214+4724 motivated Efstathiou (2006) to propose a three component model: an edge-on torus, clouds associated with the narrow-line region and a highly obscured starburst. The model assumed that the narrow line region clouds have a covering factor of 17% and two discrete values for the dust temperature (610 and 200K). The new MIPS and PACS data suggest that there is about a factor of two more power in the rest frame wavelength range 20-50μm compared with the prediction of the model presented in Efstathiou (2006). Considering that the new data samples the peak of the SED of F10214+4724 this implies that the covering factor needs to increase even more. It therefore seems likely that the magnification of the emission from the narrow-line region is considerably higher than that from the torus. We find that the new data can be fitted by assuming an additional component of narrow-line region dust at a temperature of 80K. The new fit is shown in Figure 1. The torus emission is assumed to be negligible in this fit.

References

Alexander, D. M., Efstathiou, A., Hough, J. H. *et al.* 1999, *MNRAS*, 310, 78
Broadhurst, T. & Lehar, J. 1995, *ApJ* (Letters), 450, L41
Efstathiou A. & Rowan-Robinson M. 1995, *MNRAS*, 273, 649
Efstathiou A., Rowan-Robinson M. & Siebenmorgen R. 2000, *MNRAS*, 313, 734
Efstathiou, A. & Siebenmorgen, R. 2005, *A&A*, 439, 85
Efstathiou, A. & Siebenmorgen, R. 2009, *A&A*, 502, 541
Efstathiou, A. 2006, *MNRAS*, 371, L70
Farrah D., Afonso J., Efstathiou A., *et al.* 2003, *MNRAS*, 343, 585
Genzel R. *et al.* 1998, *ApJ*, 498, 579
Granato G. L. & Danese L. 1994, *MNRAS*, 268, 235
Hao, L. *et al.* 2005, *ApJ* (Letters), 625, L75
Hao, L. *et al.* 2007, *ApJ* (Letters), 655, L77
Pier E. A., & Krolik J. H. 1992, *ApJ*, 401, 99
Roche P. F., Aitken D. K., Smith C. H., & Ward M. J. 1991, *MNRAS*, 248, 606
Rowan-Robinson *et al.* 1991, *Nature*, 351, 719
Rowan-Robinson M. *et al.* 1993, *MNRAS*, 261, 513
Ruiz, M., Efstathiou, A., Alexander, D. M., & Hough, J. 2001, *MNRAS*, 325, 995.
Siebenmorgen, R. & Krügel, E. 1992, *A&A*, 259, 614
Siebenmorgen, R., Haas, M., Kruegel, E., & Schulz, B. 2005, *A&A*, 436, L5.
Spoon, H. W. W., Marshall, J. A., Houck, J. R. *et al.* 2007, *ApJ* (Letters), 654, L49
Sturm *et al.* 2010, *A&A*, 518, L36
Teplitz *et al.* 2006, *ApJ* (Letters), 638, L1

Discussion

TUFFS: AGNs are fairly randomly alligned with the dusty disks of their hosts. Do you expect measurable components of the FIR emission (for example at longer wavelengths) to be originating from diffuse dust in the disk illuminated by UV/optical from a favourable alligned AGN?

EFSTATHIOU/SIEBENMORGEN: The heating of the diffuse dust of the host galaxies at kiloparsec scales by an AGN is discussed by Siebenmorgen *et al.* (2004); basically the luminosity of the AGN may be seen as a point source heating the dust with a flux scaling as L_{AGN}/R^2.

GROVES: If the IR NLR emission in F10214 is boosted by gravitational lensing wouldn't you expect the line emission to be relatively boosted as well?

EFSTATHIOU: Lacy *et al.* (1998) also found that the emission lines in F10214 are particularly strong.

THOMPSON: A starburst of $\sim$ 100M$_\odot$/yr, on $\sim$ 100 pc scales would produce $\sim 10^{13}$L$_\odot$, implying T$_{eff}$ $\sim$90 K. Why couldn't such a starburst produce the $\sim$80 K component you require to explain the SED of IRAS F10214+4724?

EFSTATHIOU: It would be very strange that we would then have two starbursts in this object. One with a temperature of 40K and one with 80K. Such a starburst would also emit strong PAH features which would show up on the mid-infrared spectrum.

The Spectral Energy Distribution of Galaxies
Proceedings IAU Symposium No. 284, 2011
R.J. Tuffs & C.C. Popescu, eds.

doi:10.1017/S1743921312009076

Properties of mid- to far-infrared dust emission in the nearby superwind galaxy M82

Ko Arimatsu[1,2], Takashi Onaka[1], Itsuki Sakon[1], Fumi Egusa[2], and Hidehiro Kaneda[3]

[1]Department of Astronomy, Graduate School of Science, The University of Tokyo,
7-3-1 Hongo, Bunkyo-ku, Tokyo 113-0033, Japan.
email: arimatsu@astron.s.u-tokyo.ac.jp

[2]Institute of Space and Aeronautical Science, Japan Aerospace Exploration Agency,
3-1-1 Yoshinodai, Sagamihara, Kanagawa 229-8510, Japan.
email: arimatsu@ir.isas.jaxa.jp

[3]Graduate School of science, Nagoya University,
Chikusa-ku, Nagoya 464-8602, Japan.

Abstract. Using the reconstructed imaging data obtained with the Infrared Camera (IRC) on board AKARI, mid-infrared (MIR; 5-30 μm) emission characteristics of the superwind galaxy M82 are studied. The MIR images at four wavelengths (7, 11, 15, and 24 μm) show extended (out to distances of 4 kpc) emission mainly from polycyclic aromatic hydrocarbons (PAHs). The MIR SED of M82 halo is surprisingly constant. Using far-infrared imaging data obtained by Herschel/SPIRE, we reveal that the PAH abundance relative to the big (sub-micron sized) grains radially increases by about a factor of three. These results imply that PAHs may be formed in small and dense molecular clumps in the halo and efficiently supplied to the intergalactic space by the galactic superwind.

Keywords. infrared: galaxies, galaxies: starburst

1. Introduction

Galactic superwinds carry neutral and molecular gas and dust from the galactic disk into the halo and the intergalactic space (e.g. Heckman *et al.* 1990). It is valuable to investigate the mid- to far-infrared dust emission properties in the superwind to understand the chemical enrichment process of the galactic halo and the intergalactic space.

M82 (NGC 3034) is one of the most studied superwind galaxies at various wavelengths, thanks to its proximity and edge on geometry. Engelbracht *et al.* (2006) and Kaneda *et al.* (2010) found very extended mid- to far-infrared emission associated with the superwind and the tidal arms in the halo region of M82. The observed mid-infrared (MIR) emission in the halo is thought to come from PAHs, which survive in the harsh superwind environment. Further investigations are required to understand the properties of the extended infrared emission.

This report presents the latest results of the re-analysis of the imaging data obtained with *AKARI*/IRC, which is provided by artifact-subtraction processes and a new image reconstruction method developed by Arimatsu *et al.* (2011). The latest results of the comparison between the *AKARI* MIR data and the far-infrared (FIR) data obtained with *Herschel*/SPIRE (Roussel *et al.* 2010) are also presented in this report.

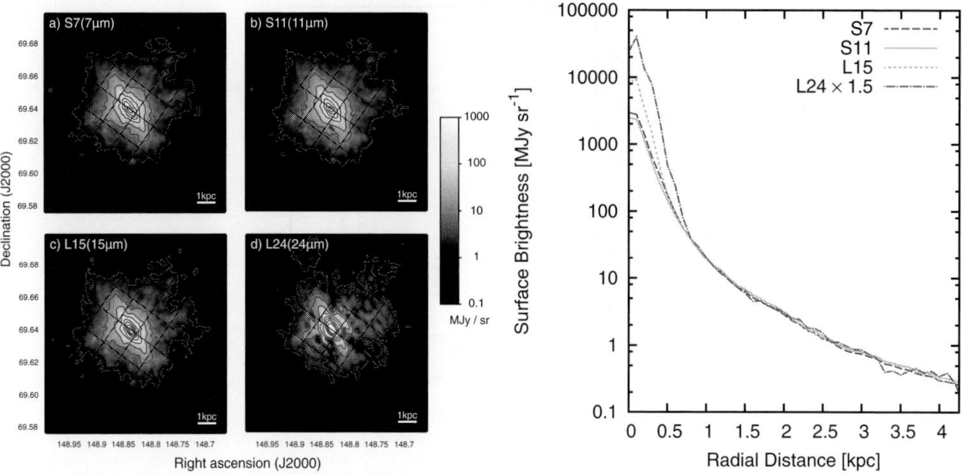

Figure 1. Left: IRC MIR band images of M82. Right: Radial surface brightness properties.

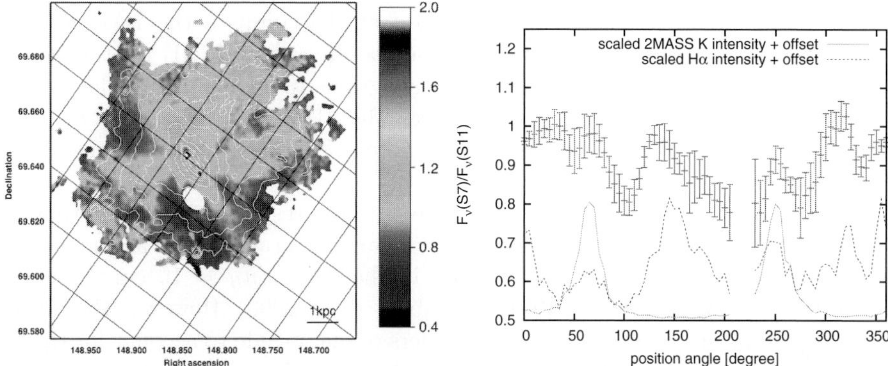

Figure 2. Left: 7μm/11μm flux ratio map of M82. Right: Position angle profile of 7μm/11μm flux ratio overlaid with the Hα and 2 MASS K band intensity profiles.

2. Mid- to far-IR Emission from M82

The reconstructed four band (7,11,15, and 24μm) MIR images of M82 taken with the *AKARI*/IRC MIR channels, which are shown in the left panel of Figure 1, reveal extended emission in the halo. The spatial distribution of the emission is surprisingly similar at the four wavelenths.

MIR radial profile. Fig. 1 (right-hand panel) shows the radial surface brightness profiles of M82 at the IRC four MIR bands. The MIR radial surface brightness profiles are surprisingly similar among the four bands at $r > 1$ kpc. These profiles can be well approximated by a power law function. The fitted power-law index is about -3, which indicates that the intensity density of the PAH emission, which reflects the PAH density distribution and the intensity of the radiation field, gently decreases with radius ($\propto r^{-4}$).

MIR intensity ratio map. The 7μm/11μm flux ratio is thought to trace the ratio between the 7$-$9 μm PAH features (enhanced when PAHs are ionized) and the PAH and very small grain emission at 10$-$13μm (Sakon *et al.* 2006). The 7μm/11μm flux ratio map (Figure 2 left) shows a relatively flat flux ratio at $r > 1$ kpc. This result suggests that the ionization fraction of PAHs is independent of galactocentric radius. 7μm/11μm position angle profile (Fig. 2 right-hand panel) shows a trend that seems to correlate

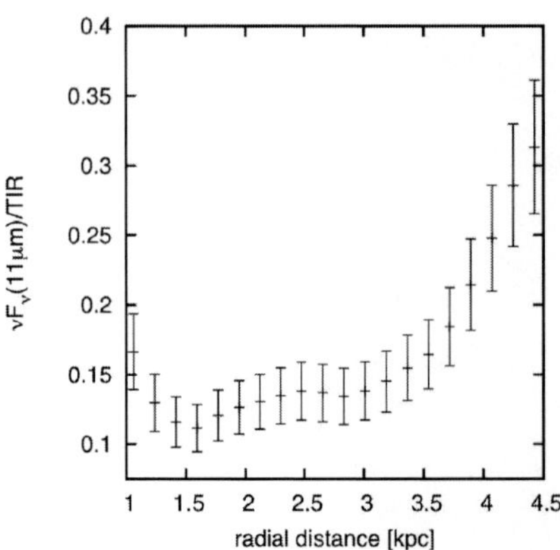

Figure 3. Radial profile of 11μm/TFIR.

with the Hα intensity profile. This colour variation suggests that PAHs are ionized near the Hα emission, with neutral PAHs located behind the ionized gas regions (which may be associated with molecular gas - e.g., Walter *et al.* 2002).

PAH abundance relative to the FIR emitting dust. To compare the PAH abundance relative to bigger, classical dust grains, we use the *Herschel*/SPIRE photometric data and estimate the mass and temperature of bigger grains to compare with the MIR data. The radial profile of the intensity ratio between the 11μm intensity and the total FIR luminosity is shown in Figure 3. This result is interpreted as a radial increase of PAHs relative to big grains.

The explanation of this trend is not yet clear. This result may suggest that PAHs are transported into the halo more efficiently than big grains. This abundance trend could also be due to the formation of PAHs in the dense ($n_H \sim 10^4$cm^{-3}) molecular clumps in the superwind (Herbst 1991, Greenberg *et al.* 2000). However, the formation efficiency is insufficient by more than three orders of magnitude to account for the trend. Alternatively, this trend could be explained by the formation of PAHs by shock-induced shattering of big grains (Jones *et al.* 1996) in the turbulent superwind environment. In any case, PAHs have to be shielded from hot ($\sim 10^6$K) plasma gas, which destroys PAHs by electron sputtering (Micelotta *et al.* 2010). The radial increase of the PAH abundance relative to big grains is thus suggested to be due to the enhancement of the emergence of PAHs from the molecular clumps in the halo, which can be a container and may be the source of the delicate materials in the harsh environment. Wider area (5 kpc $\times$ 5 kpc), and multiwavelength (CO, IR H_2 lines, etc.) observations are required to constrain a clear explanation of this result.

References

Arimatsu, K., *et al.* 2011, *PASP*, 123, 981
Engelbracht, C. W., Kundurthy, P., Gordon, K. D., *et al.* 2006, *ApJ*, 642, L127
Heckman, T. M., Armus, L., & Miley, G. K. 1990, *ApJS*, 74, 833
Herbst, E. 1991, *ApJ*, 366, 133
Jones, A. P., Tielens, A. G. G. M., & Hollenbach, D. J. 1996, *ApJ*, 469, 740
Kaneda, H., Ishihara, D., Suzuki, T., *et al.* 2010, *A&A*, 514, 14
Micelotta, E. R., Jones, A. P., & Tielens, A. G. G. M. 2010b, *A&A*, 510, A37
Roussel, H., *et al.* 2010, *A&A*, 518, L66
Sakon, I., *et al.* 2006, *ApJ*, 651, 174
Walter, F., Weiss, A., & Scoville, N. 2002, *ApJ*, 580, L21

The Spectral Energy Distribution of Galaxies
Proceedings IAU Symposium No. 284, 2011
R.J. Tuffs & C.C. Popescu, eds.

doi:10.1017/S1743921312009088

The SED of the nearby H I-massive LIRG HIZOA J0836−43: from the NIR to the radio domain

Renée C. Kraan-Korteweg[1] **and Michelle E. Cluver**[2]†,

[1]Astronomy Department, Astrophysics, Cosmology and Gravity Centre (ACGC), University of Cape Town, Rondebosch 7700, South Africa
email: kraan@ast.uct.ac.za

[2]IPAC, California Institute of Technology, Pasadena, CA 91125, USA
email: mcluver@aao.gov.au

Abstract. HIZOA J0836−43 is one of the most H I-massive galaxies in the local ($z < 0.1$) Universe. Not only are such galaxies extremely rare, but this "coelacanth" galaxy exhibits characteristics – in particular its active, inside-out stellar disk-building – that appear more typical of past ($z \sim 1$) star formation, when large gas fractions were more common. Unlike most local giant H I galaxies, it is actively star forming. Moreover, the strong infrared emission is not induced by a merger event or AGN, as is commonly found in other local LIRGs. The galaxy is suggestive of a scaled-up version of local spiral galaxies; its extended star formation activity likely being fueled by its large gas reservoir and, as such, can aid our understanding of star formation in systems expected to dominate at higher redshifts. The multi-wavelength imaging and spectroscopic observations that have led to these deductions will be presented. These include NIR (JHK) and MIR (Spitzer; $3 - 24\mu$m) imaging and photometry, MIR spectroscopy, ATCA H I-interferometry and Mopra CO line emission observations. But no optical data, as the galaxy is heavily obscured due to its location in Vela behind the Milky Way.

Keywords. galaxies: individual (HIZOA J0836−43), galaxies: evolution, infrared: galaxies

---

1. Introduction of HIZOA J0836−43

The galaxy HIZOA J0836−43, originally discovered in the Parkes deep Multibeam H I-survey of the Zone of Avoidance (ZOA; e.g. Kraan-Korteweg et al. 2005), is a rapidly rotating disk galaxy ($v_{\rm rot} = 630\,{\rm km\,s^{-1}}$; $D_{\rm HI} = 130$kpc) with an H I-mass of $M_{\rm HI} = 7.5 \cdot 10^{10}\,{\rm M_\odot}$ and a dynamical mass of $M_{\rm tot} = 1.4 \cdot 10^{12}\,{\rm M_\odot}$ (Donley *et al.* 2006). It has a prominent bulge and smallish stellar disk ($B/D \sim 0.8$) of evolved stars of $M^* = 4.4 \cdot 10^{10}\,{\rm M_\odot}$ embedded in the five times larger HI-disk, while its current hearty star forming activity makes it a luminous infrared galaxy (LIRG; next section).

HIZOA J0836−43 is the most H I-rich galaxy known. Such galaxies are extremely rare ($\sim 3{\cdot}10^{-8}/{\rm Mpc^3}$) according to the current best determined local galaxy HI-mass function (HIMF; Zwaan *et al.* 2005) and should not have been found at all in the explored volume. Interestingly, the recent, larger volume ALFALFA survey 40%-data release (Haynes *et al.* 2011) identify several, similarly extreme H I-massive galaxies in excess of the predicted HIMF number density. Independently, galaxies with such vast reservoirs of gas were more common in the past, as were LIRGs. Hence this enigmatic, relatively nearby ($v = 10\,689\,{\rm km\,s^{-1}}$) galaxy is an ideal *local* probe that enables detailed studies of evolutionary processes and the transformation of gas into stars.

† Present address: AAO, Epping NSW 1710, Australia

2. The Observational Data

The galaxy lies deep in the ZOA ($\ell = 262\overset{\circ}{.}48, b = -1\overset{\circ}{.}64$) and suffers from severe Galactic foreground extinction ($A_V = 7\overset{\text{m}}{.}3$), rendering it practically invisible in the optical. Hence our detailed follow-up observations focus on the infrared, radio and mm domain.

IRSF NIR imaging survey. The deep JHK-imaging survey of 2.2□° with the Infrared Survey Facility (IRSF) in Sutherland (SAAO) aimed at learning more about an environment that permits such a gas giant to evolve. The survey uncovered 404 galaxies to the completeness limit of $K_s < 15\overset{\text{m}}{.}8$ (Cluver 2009). It finds the volume around HIZOA J0836−43 to be overdense in sub-L^* systems, and none in the L^* range. Hence it seems to live in a tranquil, underdense environment on the edge of a void. There are no indications of a major merger that could have triggered a starburst (see also Fig. 2). This may have allowed the formation of its giant H I-disk through minor merger or accretion – and aided its survival.

Spitzer IRAC and MIPS imaging and spectroscopy. The photometry from the Spitzer IRAC and MIPS band was presented in Cluver et al. 2008. It is reproduced in the SED (Fig. 1) together with the IRSF data, the DENIS I-band (Donley *et al.* 2006) and, now available, WISE 12 & 22μm fluxes (Jarrett, priv. comm.).

The resulting total infrared luminosity $L_{\text{TIR}} = 1.2 \cdot 10^{11}\,\text{L}_\odot$ defines it as a LIRG. Despite the high star formation rate (SFR= 20.5 M$_\odot$/yr), the comparison of its SED to various galaxy model templates, and the archetypal local starburst M82 in Fig. 1, shows that HIZOA J0836−43 is neither a starburst galaxy nor – as originally assumed from its NIR morphology – a S0/Sa-galaxy. It rather resembles a Sc galaxy with strong emission in the mid-IR $5-20\mu$m range, due to the prevalence of PAH molecules, and a prominent FIR cold dust continuum. The latter is substantiated by the mid-IR spectroscopy (Cluver *et al.* 2010) which shows particularly luminous 6.2 and 7.7μm PAH features (stronger than typical starbursts; e.g. Brandl *et al.* 2006) in contrast to the continuum emission from warm dust. The latter is surprisingly weak (i.e. lower than any of the local LIRGs in GOALS), suggestive of an extended dust geometry. Cold dust is plentiful, however,

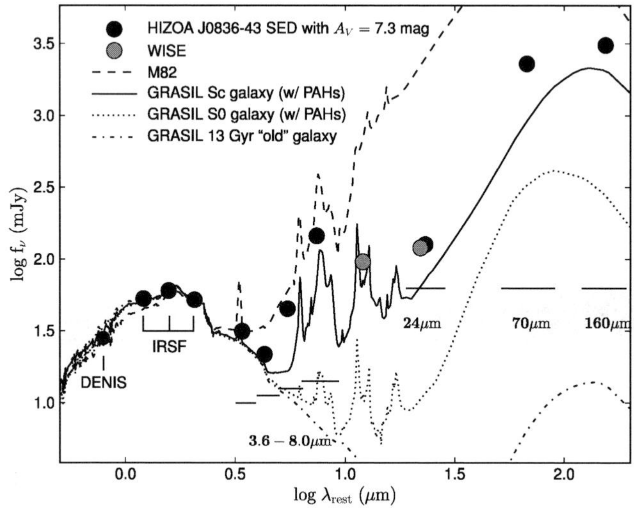

Figure 1. SED based on DENIS I, IRSF J, H, K_s, Spitzer IRAC ($3.6-8.0\mu$m) and MIPS (24, 70 & 160μm), as well as the WISE 12 & 22μm bands – all corrected for extinction ($A_V = 7\overset{\text{m}}{.}3$). Superimposed are GRASIL model templates and the M82 spectrum for comparison. While the high MIR/FIR luminosity makes this galaxy a LIRG, it clearly is not consistent with a starburst. (Adapted from Cluver *et al.* 2008).

and a dominant contributor to the total infrared luminosity with its steeply rising FIR emission for $\lambda \gtrsim 60\mu$m.

The IRS spectroscopy of the nucleus further reveals strong excited nebular lines (e.g. [Ne II], [S III], [Si II], [Ar II]). While these are typical of starburst regions, the line ratios indicate a softer radiation field (like PDR's) and AGN activity is weak or absent. This is independently confirmed by the absence of [Ne V] and [O IV] lines.

ATCA and MOPRA observations. With the surprising dearth of warm dust despite the strong star forming activity (PAH's) and the large reservoir of cold H I gas, Mopra CO line observations (1-0) were used to probe the cold molecular gas content. The spectrum shows a double horn signal, matching the H I-profile width perfectly. However, the total molecular gas mass (H_2+He) of $3.9 \cdot 10^9$ $M_\odot$ is lower by a factor of three than expected from the infrared luminosity, and only a meager 5% compared to the atomic H I (Cluver *et al.* 2010). Both the total and cold molecular gas fraction (gas/gas+stellar mass) are high (64% and 8%) compared to the mean of local galaxies (e.g. 20% total; Leroy *et al.* 2008, and 6% molecular for the CO-detected stellar-selected massive galaxies; Saintonge *et al.* 2011), but low compared to the local LIRG sample (Wang *et al.* 2006). Such high gas fraction values are the standard though at higher redshifts (e.g. Daddi *et al.* 2010 for their NIR selected BzK sample at $z \sim 1.5$).

Given this dichotomy, it seemed of importance to learn more about the interplay between star formation activity and gas physics in the galaxy itself. This required longer baseline observations with ATCA to complement the earlier Donley *et al.* (2006) data, which only had 30″ angular resolution. Of interest is the connection – if any - with the star forming region, which extends beyond the old stellar bulge, suggestive of inside-out stellar disk-building, and exhibits a warp towards the eastern side. This is evident in the composite NIR & MIR image (left panel of Fig. 2) as well as the composite spectral map (2″ resolution) of the PAH 6.2, 7.7 and 11.3μm features (right panel). Note that the latter was obtained from SL (5–14μm) IRS observations mapping across the IR disk.

A preliminary H I-map from the newly acquired data is superimposed on the right panel in Fig. 2 for comparison. We note that the improved spatial resolution reveals an enhancement in the atomic gas exactly at the location of the warp. This may be the result of a recent tidal interaction which has triggered star formation as gas flows towards the center of the galaxy. This H I-data will be explored in detail in Cluver *et al.* (in prep).

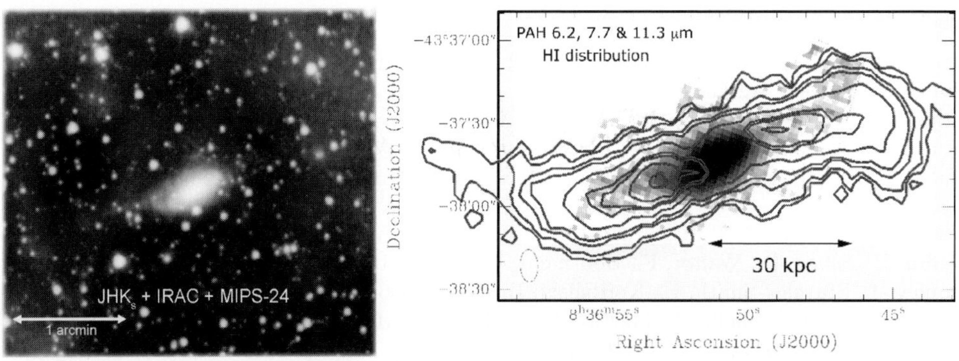

Figure 2. Left Panel: Composite of J, H, K_s and Spitzer IRAC & MIPS 24μm images. Right panel: H I distribution superimposed on composite spectral maps of the 6.2, 7.7, and 11.3μm PAH emission bands from Spitzer IRS. Note the asymmetry of PAH emission on the left (beyond the edge of the stellar disk), coinciding with an enhancement seen in the H I.

3. Connection to evolutionary processes at higher redshifts

How does this galaxy relate to systems at varying redshift/evolutionary stages? Albeit at the extreme end, HIZOA J0836−43 seems to fall exactly on the H I-mass and SFR relation defined by the nearby SINGS galaxy sample (contrary to M82 or Malin 1). This suggests it to be a scaled-up version of local disk galaxies. Comparing it to the local LIRG sample ($z \lesssim 0.1$; Wang *et al.* 2006) its specific starformation rate of 0.47 Gyr^{-1} appears to follow the overall trend. But, it again lies at an extreme stage, implying active stellar building; in an additional 2 Gyrs of time it will double its stellar mass and fit the 'current' local LIRG data. This hint of relic SF properties is also seen when comparing HIZOA J0836−43 to the higher redshift SF galaxy sample by e.g. Bell *et al.*(2005), where it fits well on sSFR-relation of $z \sim 0.7$ galaxies, but does not conform at all to the local SF galaxy sample. Idem ditto when compared to the Genzel *et al.* (2010) sample where it closely aligns with the $z \sim 1.2$ sample. Tellingly, when entering the parameters of HIZOA J0836−43 in the so-called main sequence of SFR as a function of stellar mass and redshift (Bouché *et al.*2010) a redshift (age) of $z \sim 0.95$ is returned, suggesting a tantalising link to a distant epoch of star formation.

HIZOA J0836−43 displays a combination of high gas fraction with low efficiency, but extended, star formation in a low excitation disk, also found in the sample of normal, near-IR selected (BzK) galaxies at $z \sim 1.5$ of Daddi *et al.* (2010), who consider them scaled-up local disk galaxies. Since larger gas fractions at higher redshift permit larger $L_{\rm IR}$ before invoking special events like mergers that show an offset in correlations with $L_{\rm IR}$ (Nordon *et al.* 2011), HIZOA J0836−43 can be studied as an analogue of disk building processes dominating at the peak of stellar mass growth.

The lower fraction of molecular gas in HIZOA J0836−43 compared to detected high-redshift galaxies (Tacconi *et al.* 2010) highlights the need for an accurate molecular gas inventory, in parallel to increased high z detections. Though, as pointed out by Groves at this conference, "dark" molecular gas may be a feature of the active ISM enrichment occurring in the disk. Herschel and ALMA is needed to provide a cohesive picture that can be used to understand disk building in gas-rich systems.

Acknowledgements

RKK thanks the South African National Research Foundation and the University of Cape Town for financial support.

References

Bell, E. F., Papovich, C., Wolf, C., Le Floc'h, E. *et al.* 2005, *ApJ*, 625, 23
Bouché, N., Dekel, A., Genzel, R., Genel, S. *et al.* 2010, *ApJ*, 718, 1001
Brandl, B. R., Bernard-Salas, J., Spoon, H. W. W., Devost, D. *et al.* 2006, *ApJ*, 653, 1129
Cluver, M. E. 2009, *PhD thesis, University of Cape Town*
Cluver, M. E., Jarrett, T. H., Appleton, P. N., Kraan-Korteweg, R. C., *et al.* 2008, *ApJL*, 686, L17
Cluver, M. E., Jarrett, T. H., Kraan-Korteweg, R. C., Koribalski, B. *et al.* 2010, *ApJ*, 725, 1550
Daddi, E., Elbaz, D., Walter, F., Bournaud, F. *et al.* 2010, *ApJ*, 714, L118
Donley, J., Staveley-Smith, L., Koribalski, B. *et al.*, 2006, *MNRAS*, 369, 1741
Haynes, M. P., Giovanelli, R., Martin, A. M., Hess, K. M. *et al.* 2011, *AJ*, 142, 170
Genzel, R., Tacconi, L. J., Gracia-Carpio, J., Sternberg, A. *et al.* 2010, *MNRAS*, 407, 2091
Kraan-Korteweg, R. C., Staveley-Smith, L., Donley, J. *et al.* 2004, *IAU Symp.* 216, 203
Leroy, A. K., Walter, F., Brinks, E., Bigiel, F., & de Blok, W. J. G. 2008, *AJ*, 136, 2782
Nordon, R., Lutz, D., Berta, S., Wuyts, S. *et al.* 2011, arXiv:1106.1186
Saintonge, A., Kauffmann, G., Kramer, C., Tacconi, L. J. *et al.* 2011, *MNRAS*, 415, 32

Tacconi, L. J., Genzel, R., Neri, R., Cox, P. *et al.* 2010, *Nature*, 463, 781
Wang, J. L., Xia, X. Y., Mao, S., Cao, C. *et al.* 2006, *ApJ*, 649, 722
Zwaan, M. A., Meyer, M. J., Staveley-Smith, L., & Webster, R. L. 2005, *MNRAS*, 359, L30

Discussion

CHAKRABARTI: What is your interpretation of the high IR emission? My interpretation is that this is due to a minor merger. We now have scaling relations that will allow you to derive the satellite mass from the H I-map. So I suggest that you calculate the Fourier amplitude.

KRAAN-KORTEWEG: Thank you. That is an interesting idea for the future, but we will await the reduction of our new H I-data and its analysis first. The current H I-data maps are quite smooth over most of its gas disk without any kinks in its contours. This makes the minor-merger scenario unlikely; at least, it could not have occurred over the last 2 Gyr or so to account for such a stable disk. Secondly, while a minor merger (or instability caused by the accretion of a satellite) might explain the strong IR luminosity due to triggered star formation/nuclear starburst, it does not account for the extreme mass in H I-gas in HIZOA J0836−43, nor the abnormally high gas fraction. In addition, the latter scales beautifully with normal – albeit higher redshift – starforming galaxies. What remains unresolved as yet is the question, how HIZOA J0836−43 could have acquired/accreted all this cold gas.

BUREAU: The comparison of the CO and H Iin the spectra shows a lack of CO (thus H_2) in the rising part of the rotation curve, hence central part of the galaxy. How can this be reconciled with signatures of star formation in the centre, e.g. in the nuclear infrared spectrum shown.

KRAAN-KORTEWEG: Note that the "nuclear spectrum" covers the "nuclear region" and is a $9\farcs25$ extraction, so it picks up more than just the nucleus itself. Secondly, the central low-velocity gas might well be present at a lower level compared to the H I-gas distribution (i.e. not detectable with the current sensitivity of our Mopra observations). This is seen quite often in other spiral galaxies when comparing H I and CO observations. Much of the very central CO may also have been used in a previous epoch of star formation which produced the current old stellar bulge (which has a disk scale length of 4 kpc, respectively $7''$). But interesting is that the steep rotation in the H I velocity appears matched to the CO, so the central region achieves a high velocity close in. This may be important for extended building into the disk. We really need spatial information accompanied by velocity resolution (as we now have with H I) to understand this, and ALMA will make this possible. Also, given the prominence of [Si II] in the spectrum, it is not inconceivable that some of the CO is being photodissociated, whereas the molecular hydrogen is shielded, making it "dark". Herschel (proposal submitted) in combination with ALMA (planned) will enable a detailed study of such a mechanism.

The Spectral Energy Distribution of Galaxies
Proceedings IAU Symposium No. 284, 2011
R.J. Tuffs & C.C. Popescu, eds.

doi:10.1017/S174392131200974X

Retrieving the stellar content in distant starbursts

Myriam A. Rodrigues[1,2], François Hammer[2] and Mathieu Puech[2]

[1]European Southern Observatory, Alonso de Cordova 3107
Casilla 19001 - Vitacura -Santiago, Chile

[2]GEPI , Observatoire de Paris, CNRS, University Paris Diderot
5 Place Jules Janssen, 92195 Meudon, France

email: marodrig@eso.org

Abstract. In starburst galaxies, the light emitted by the young and massive stars dominates the photon budget along most of the SED and hides the old and intermediate stellar populations. The fraction of old stars and the stellar mass are systematically underestimated by current methods (Wuyts *et al.* (2009)). We have implemented a new method to retrieve stellar masses and stellar populations in distant galaxies from photometry and spectral features. The method uses a complex SFH description and a new constraint has been introduced: the star-formation rate (SFR).

Keywords. galaxies: high-redshift, galaxies: photometry , galaxies: starburst

---

1. Introduction

Many works have emphasised the fundamental role of stellar mass in galaxy evolution. Stellar mass is found to correlate with many galaxy properties, such as luminosity, gas metallicity, colour and age of stellar populations, star-formation rate, morphology, and gas fraction, to enumerate a few of them Brinchmann & Ellis (2000), Bell *et al.* (2003), Kauffmann *et al.* (2003), Tremonti *et al.* (2004), Bell *et al.* (2007). Ironically, the stellar mass is one of the most poorly constrained quantities in distant galaxies. This difficulty arises from the fact that one of the main contributors to stellar mass, the low-mass stars on the main sequence (type G to K stars) are extremely faint and almost do not contribute to the integrated light of the stellar population as a whole. The dominance of young massive stars in any stellar population is responsible for the strong sensitivity of the SED shape to the star-formation history.

2. Complex star formation history and the SFR constraint

In starburst galaxies the principal issue in the proper recovery of the stellar populations and the stellar mass is the fact that the young stars dominate the light budget throughout the SED. The methods available in the literature suffer from systematic effects due to the simplicity of the assumed star formation history for these objects. The implementation of a complex SFH without parametrisation - such as the combination of several single stellar population templates- is not possible because of the degree of degeneracy of the problem. To break the degeneracy, we propose to add a new constraint in the optimization problem: the total star-formation rate (SFR_{UV} and SFR_{IR}). Adding the total SFR constraint allows an alleviation of the problem by limiting the amount of young stars with ages younger than 100 Myr. In particular, it induces strong constraints in the blue part of the spectra - most especially the amount of light due to young stars - and permits a

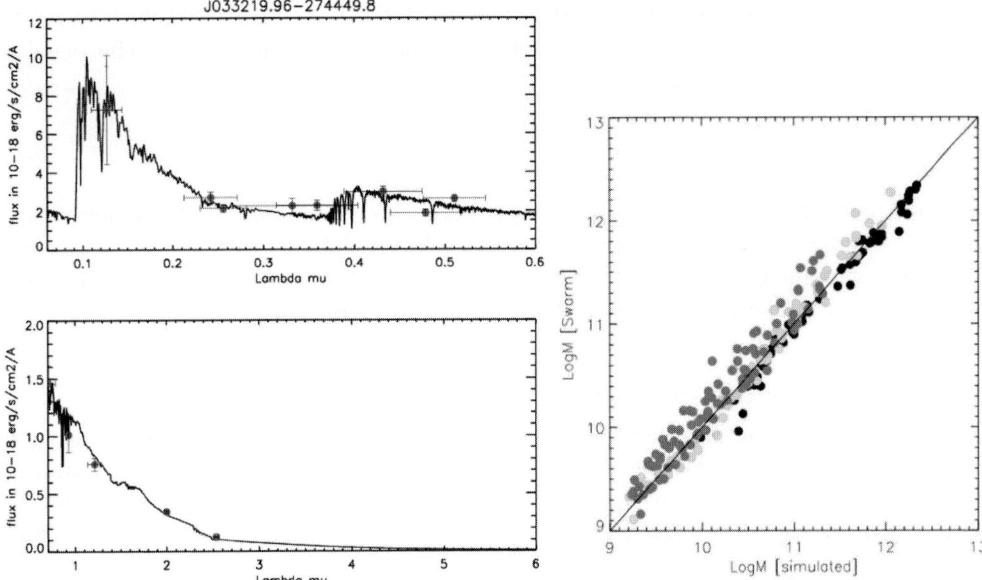

Figure 1. Left: Synthesis of a $z \sim 0.6$ starburst SED from photometry and observed SFR using a combination of 6 CSP with exponential decline timescale of 100Myrs and a two-component extinction law. The minimization of the χ^2 statistic has been realized with the swarm intelligence algorithm. **Right**: Comparison of the Mstar of a sample of simulated galaxies with the Mstar estimated by our code. The colour codes the mass fraction of old stellar population (older than 3 Gyr) in the galaxy: in black (light grey) galaxies with more than 80% (50-80%) of their Mstar locked into old stellar population. The dark grey symbols correspond to very young systems with 50% of stars younger than 3 Gyr.

better evaluation of dust obscuration from the blue slope of the SED. Moreover, adding the SFR constraint solves the problem encountered in spectrum synthesis Cid Fernandes *et al.* (2005), Mathis *et al.* (2006), i.e. the difficulty to recover the mass fraction of the intermediate age population.

3. Optimization problem

We have fitted the SED by a set of $N_\star$ single metallicity population synthesis templates $T^i(\lambda)$ (i=1,..., N). The best-fitting model SED $F(\lambda)$ is given by:

$$F(\lambda) = \sum_{i=0}^{N_\star} x_i \, T^i(\lambda) \otimes Dust(E(B-V), R_V) \tag{3.1}$$

where x_i are non-negative coefficients representing the fraction of light of the template T^i, and Dust is the two parameter extinction law of Cardelli et al. (1989). We have assumed that only the young stellar population (age $< 500\,$Myr) is affected by the dust and the same E(B-V) and R_V have been applied to these templates. The SFR constraint is imposed in the problem by a penalty function. The optimization function to minimize is thus:

$$\chi^2(x^i, E(B-V), R_V, SFR_{obs}) = \chi^2_{photo}(x^i, E(B-V), R_V) + P(SFR). \tag{3.2}$$

$$\chi^2_{photo}(x_i, E(B-V), R_V) = \sum_{j=1}^{N_{filter}} \frac{(f_j^{obs} - f_j^{model})^2}{\sigma^2 f_j^{obs}} \tag{3.3}$$

where f_j^{obs} is the rest-frame flux of the object in filter j, its associated uncertainty is $\sigma fobs_j$, and f_j^{model} is the flux of the synthetic spectrum in the j-band. The penalty function on the SFR is given by:

$$P(SFR) = \left(\frac{SFR_{obs} - SFR_{model}}{\sigma SFR_{obs}} \right)^2 \tag{3.4}$$

Adding the SFR constraint introduces a plethora of local minima to the minimization problem, rendering the localization of the best solution difficult with conventional code based on the gradient or Hessian function. The global solution of the optimization problem was found with a metaheuristic approach, the Swarm intelligence algorithm.

4. Validation of the method

We have tested the accuracy of the algorithm to retrieve the stellar mass using a library of fake galaxies. The results are shown in figure 1. The algorithm recovers the stellar mass within a random uncertainty of $\pm 0.15\,dex$. There is a small systematic of +0.1 dex in the derived stellar mass for galaxies with extremely high fractions of young stars. The preliminary results show that the constraint on the SFR improves the accuracy on the derived stellar mass and stellar population in star-forming galaxies. Work is in progress to include the spectral information in the optimization process using lick indices or principal component analysis.

References

Bell, E. F., McIntosh, D. H., Katz, N., & Weinberg, M. D. 2003, *ApJ*, 149, 289

Bell, E. F., Zheng, X. Z., Papovich, C., Borch, A., Wolf, C., & Meisenheimer, K. 2007, *ApJ*, 663, 834

Brinchmann, J. & Ellis, R. S. 2000, *ApJ*, 536, L77

Cardelli, J. A., Clayton, G. C., & Mathis, J. S 1989, *ApJ*, 345, 245

Cid Fernandes, R., Mateus, A., Sodré, L., Stasińska, G., & Gomes, J. M. 2005, *MNRAS*, 358, 363

Conroy, C., Gunn, J. E., & White, M. 2009. *ApJ*, 699, 486

Conroy, C. & Gunn, J. E. 2010, *ApJ*, 712, 833

Kauffmann, G., Heckman, T. M., White, S. D. M., Charlot, S., Tremonti, C., *et al.* 2003a, *MNRAS*, 341, 33

Mathis, H., Charlot, S., & Brinchmann, J. 2006, *MNRAS*, 365, 385

Tremonti, C. A., Heckman, T. M., Kauffmann, G., Brinchmann, J., Charlot, S., *et al.* 2004, *ApJ*, 613, 898

Wuyts, S., Franx, M., Cox, T. J., Hernquist, L., Hopkins, P. F., Robertson, B. E., & van Dokkum, P. G. 2009, *ApJ*, 696, 348

The Spectral Energy Distribution of Galaxies
Proceedings IAU Symposium No. 284, 2011
R.J. Tuffs & C.C. Popescu, eds.

doi:10.1017/S174392131200909X

What makes a galaxy radio-loud?

R. A. Ortega-Minakata[1], J. P. Torres-Papaqui[1], H. Andernach[1], R. Coziol[1], J. M. Islas-Islas[1], I. Plauchu-Frayn[2], D. M. Neri-Larios[1] and M. del C. Rojas-Granados[3]

[1]Departamento de Astronomía, Universidad de Guanajuato, C.P. 36000, Guanajuato, Mexico

[2]Instituto de Astrofísica de Andalucía (CSIC), E-18008, Granada, Spain

[3]División de Ingenierías, Universidad de Guanajuato, C.P. 36885, Salamanca, Mexico

email: rene@astro.ugto.mx

Abstract. We compare the Spectral Energy Distribution (SED) of radio-loud and radio-quiet AGNs in three different samples observed with SDSS: radio-loud AGNs (RLAGNs), Low Luminosity AGNs (LLAGNs) and AGNs in isolated galaxies (IG-AGNs). All these galaxies have similar optical spectral characteristics. The median SED of the RLAGNs is consistent with the characteristic SED of quasars, while that of the LLAGNs and IG-AGNs are consistent with the SED of LINERs, with a lower luminosity in the IG-AGNs than in the LLAGNs. We infer the masses of the black holes (BHs) from the bulge masses. These increase from the IG-AGNs to the LLAGNs and are highest for the RLAGNs. All these AGNs show accretion rates near or slightly below 10% of the Eddington limit, the differences in luminosity being solely due to different BH masses. Our results suggests there are two types of AGNs, radio quiet and radio loud, differing only by the mass of their bulges or BHs.

Keywords. galaxies: active, galaxies: evolution, galaxies: fundamental parameters, galaxies: statistics, radio continuum: galaxies

1. Introduction

The Spectral Energy Distribution (SED) of galaxies is a tool that could allow to make a physical and possibly evolutionary connection between galaxies showing different levels of AGN activity. In this study, we compare the SEDs of radio-loud and radio-quiet AGNs in three different samples with the typical SEDs of QSOs and quasars. Our samples are composed of SDSS galaxies which are Radio Loud (RLAGNs) with extended radio structures (Lin *et al.* 2010), Low Luminosity AGNs (LLAGNs), which turn out to reside mostly in groups and clusters (Torres-Papaqui *et al.* 2011), and AGNs in isolated galaxies (IG-AGNs, Coziol *et al.* 2011). The SEDs are based on radio flux densities from NVSS (1.4 GHz), FIR magnitudes from IRAS (100, 60, 25, 12 μm) and optical fluxes from SDSS (5100 Å). Except for a few detections with Chandra ($\sim$ 5 keV) in IG-AGNs (5 of 25 radio-detected IG-AGNs), no X-rays were found for the other AGNs. We used the stellar population synthesis code STARLIGHT to subtract a template from which we deduce the bulge mass of the galaxies.

2. Sample Description and Results

The RLAGNs and LLAGNs in our sample are mostly early-type galaxies in groups or clusters. Very few (142 of 4197) LLAGNs are detected in radio (NVSS at 1.4 GHz). Despite the difference in radio emission the RLAGNs and LLAGNs show similar spectra in the optical: both are narrow-line emission galaxies, frequently with some emission

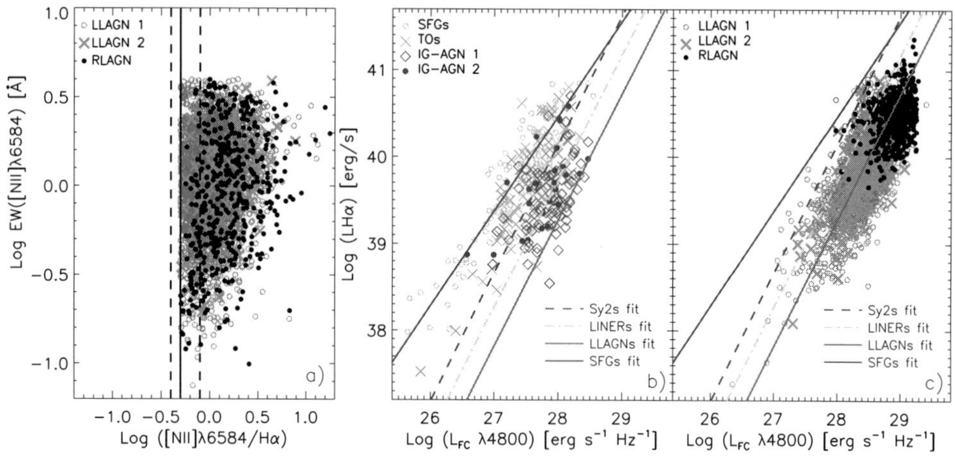

Figure 1. a) NII diagnostic diagram used to identify the LLAGNs and RLAGNs; b) and c) relations between the continuum luminosity at 4800 Å and the Hα luminosity for the AGNs in IG-AGNs (b) and the LLAGNs (c). For comparison, in (b) and (c) we show power law fits for Sy2, LINERs, LLAGNs, and Star Forming Galaxies (SFGs), based on a total of $\sim 3{\times}10^5$ SDSS galaxies (Torres-Papaqui *et al.* 2011).

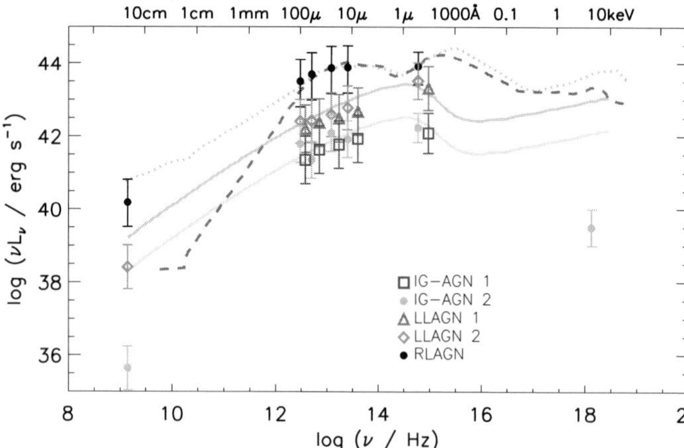

Figure 2. The SEDs of the different AGNs in our sample are compared with characteristic SEDs for quasars (dotted curve), QSOs (dashed curve) and LINERs (solid curves; one each for the LLAGNs and IG-AGNs). The symbols with error bars represent the median and quartiles. The data points for the IG-AGN 1 and LLAGN 1 samples were shifted slightly to the blue for clarity sake.

lines missing ([OIII]λ5007 and/or Hβ). The NII diagnostic diagram, Fig. 1a, was used to determine the nature of their activity (Torres-Papaqui *et al.* 2011). Galaxies with *log* [NII]/H$\alpha > -0.3$ and *log* EW[NII] < 0.6 are classified as LLAGNs.

The IG-AGNs are mostly spiral galaxies with intense narrow emission lines. A standard diagnostic diagram was used to identify 104 AGNs among the 292 IGs. Only 25 of these were detected in NVSS. In Fig. 1b we show that both the radio-undetected (IG-AGN 1) and radio-detected (IG-AGN 2) IG-AGNs are mostly LINERs. In Fig. 1c we show that the RLAGNs are generally more luminous than both the radio-undetected (LLAGN 1) and radio-detected (LLAGN 2) LLAGNs.

In Fig. 2 we show the median SEDs of the AGNs in our sample, as compared with characteristic SEDs for quasars, QSOs and LINERs (Elvis *et al.* 1994; Younes *et al.* 2010). The SEDs of the quasars and QSOs were scaled down in luminosity to fit the optical data for the RLAGNs. The LINER SED was scaled up in luminosity to fit the optical and FIR data for the LLAGNs and the IG-AGNs. The observed radio power

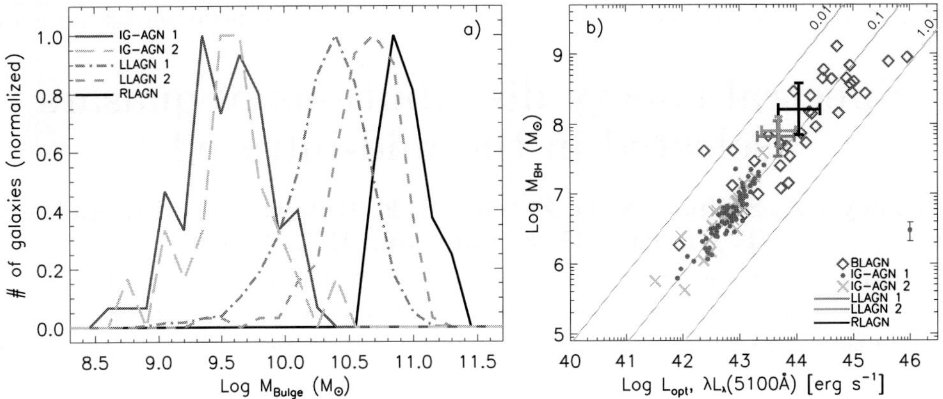

Figure 3. a) Normalized histograms for the bulge masses; b) relation between continuum luminosity at 5100Å and black hole mass as deduced from the mass of the bulge. The data for broad line AGNs (BLAGNs) are from Peterson *et al.* (2005).

is generally lower than expected from the characteristic SEDs and models. The few X-ray detections with Chandra of galaxies in the IG-AGN 2 sample also fall much below expectation.

In Fig. 3a we find that the mass of the bulge decreases in the following order: RLAGNs, LLAGN 2, LLAGN 1, IG-AGN 2 and IG-AGN 1. In each sample, the radio-loud galaxies always have the more massive bulges. In Fig. 3b we show that all the black holes (BHs) accrete at rates near or slightly below 10% of the Eddington limit. The differences in luminosity seem solely due to different BH masses, inferred from the relation by Häring & Rix (2004).

3. Discussion and Conclusions

The similarity of the optical spectral characteristics of RLAGNs and LLAGNs suggests they have a common nature, implying they are intrinsically two types of AGNs, radio loud and radio quiet. The median SEDs of LLAGNs and IG-AGNs are consistent with scaled-up versions of the characteristic SED of LINERs, while that of RLAGNs is consistent with a scaled-down SED of quasars. The difference in bulge mass between radio-loud and radio-quiet AGNs suggests the central BH mass to be higher in radio-loud AGNs. The fact that the IG-AGNs, which have formed and evolved in isolation, have smaller BH masses, suggests that the environment of a galaxy may not only determine its morphology but also its AGN type.

J.P. T-P acknowledges support grants by PROMEP (103.5-10-4684), and DAIP-UGto (65/11). I. P-F acknowledges postdoctoral fellowship #145727 by CONACyT, Mexico.

References

Coziol, R., Torres-Papaqui J. P., Plauchu-Frayn, I., *et al.* 2011, *Rev. Mexicana AyA*, 47, 361
Elvis, M., Wilkes, B. J., McDowell, J. C., *et al.* 1994, *ApJS*, 95, 1
Häring, N. & Rix, H-W. 2004, *ApJ*, 604, L89
Lin, Y.-T., Shen, Y., Strauss, M. A., Richards, G. T., & Lunnan, R. 2010, *ApJ*, 123, 1119
Peterson, B. M., Bentz, M. C., Desroches, L.-B., *et al.* 2005, *ApJ*, 632, 799
Torres-Papaqui, J. P., Coziol, R., Andernach, H., *et al.* 2011, *MNRAS*, submitted
Younes, G., Porquet, D., Sabra, B., *et al.* 2010, *A&A*, 517, A33

The Spectral Energy Distribution of Galaxies
Proceedings IAU Symposium No. 284, 2011
R.J. Tuffs & C.C. Popescu, eds.

doi:10.1017/S1743921312009106

Spectral energy distributions of quasars selected in the mid-infrared

M. Lacy[1], A. Sajina[2], A. O. Petric[3], S. E. Ridgway[4], D. M. Nielsen[5], T. Urrutia[6], D. Farrah[7], and E. L. Gates[8]

[1]National Radio Astronomy Observatory, 520 Edgemont Road, Charlottesville, VA22903, USA
[2]Dept. of Physics and Astronomy, Tufts University, Medford, MA02155
[3]Department of Astronomy, California Institute of Technology, 1200 E. California Boulevard, Pasadena, CA91125, USA
[4]National Optical Astronomy Observatory, 950 N. Cherry Avenue, Tucson, AZ85719, USA
[5]Dept. of Astronomy, University of Wisconsin, Madison, WI53706
[6]Liebnitz Institut für Astrophysik Astrophysics Potsdam, An der Sternwarte 16, 14482, Potsdam, Germany
[7]Department of Physics and Astronomy, University of Sussex, Falmer, Brighton BN1 9QH, UK
[8]UCO/Lick Observatory, University of California, 1156 High Street, Santa Cruz, CA95064

email: `mlacy@nrao.edu`

Abstract. We present preliminary results on fitting of SEDs to 142 $z > 1$ quasars selected in the mid-infrared. Our quasar selection finds objects ranging in extinction from highly obscured, type-2 quasars, through more lightly reddened type-1 quasars and normal type-1s. We find a weak tendency for the objects with the highest far-infrared emission to be obscured quasars, but no bulk systematic offset between the far-infrared properties of dusty and normal quasars as might be expected in the most naive evolutionary schemes. The hosts of the type-2 quasars have stellar masses comparable to those of radio galaxies at similar redshifts. Many of the type-1s, and possibly a one of the type-2s require a very hot dust component in addition to the normal torus emission.

Keywords. galaxies:active, quasars: general, infrared: galaxies

---

1. Introduction

Over the past three years we have been conducting an extensive campaign of follow-up spectroscopy of dust obscured quasars, using the selection technique described in Lacy *et al.* (2004; 2007). We now have over 700 spectra of AGN and quasars selected in this manner, forming the *Spitzer* Mid-InfraRed Quasar Survey (SMIRQS). SED fitting has been a key activity carried out in parallel with the spectroscopic follow-up. It allows us to separate the stellar, AGN and starburst components of the quasar and its host, it enables a consistency check on the spectroscopic redshift (which may be of low signal-to-noise in the case of highly-obscured objects), it helps with classification of $z > 1$ objects between type-2 and reddened type-2 that lack rest-frame optical (i.e. near-infrared) spectra, and it allows us to test theories for the origin and nature of dusty quasars and their various SED components.

The two most popular explanations for the existance of dust-reddened quasars are orientation, where a dusty nuclear torus provides a large amount ($A_V > 10$) of extinction to the quasar (Antonucci 1993), and evolution, where a quasar begins its life highly obscured (again $A_V >> 10$) following a galaxy merger and ensuing starburst, but then gradually expels its dusty envelope and eventually shines as an unobscured quasar

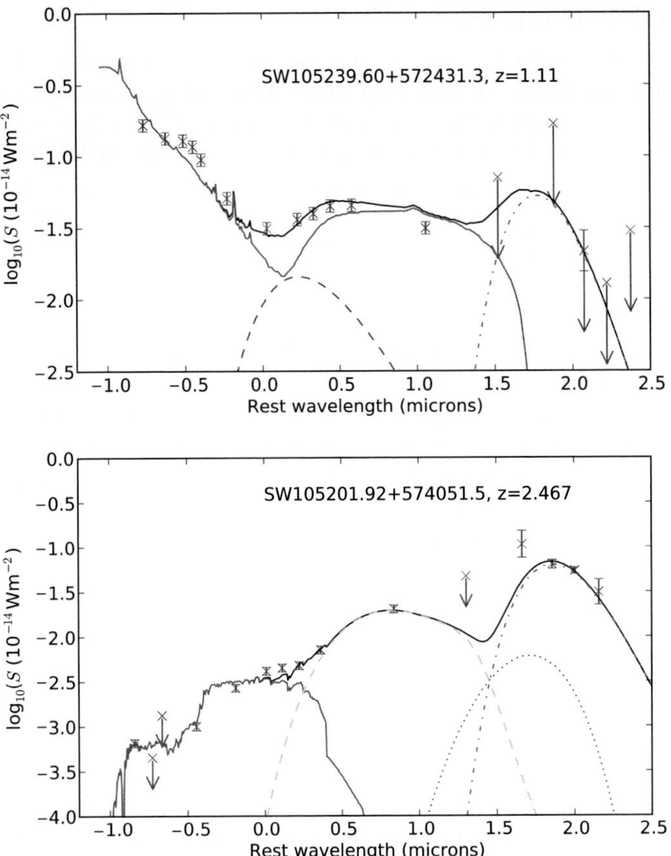

Figure 1. Example SED fits: *Top* A type-1 quasar SED fit, split into components (from left to right), an optical/IR quasar template (solid), a hot dust bump (dashed line) and a far-infrared modified black body (dot-dashed line). *Bottom* a type-2 quasar fit, components (from left to right), stellar light (solid), AGN-heated dust (dashed line), a warm dust component (dotted line) and a far-infrared modified black body (dot-dashed line).

(Sanders *et al.* 1988; Hopkins *et al.* 2008). The relative roles of orientation and evolution have been a topic of some debate, one simple, if somewhat naive, test is whether the obscured quasars show significantly more signs of ongoing star formation than their unobscured counterparts, consistent with them being at an earlier stage in the evolutionary sequence. We have also used our SED fits to study the very hot dust component in the near-infrared. The strength of this recently-recognized component appears to correlate with the bolometric luminosity of the quasar. Mor & Trakhtenbrot (2011) use WISE data to demonstrate that, because this dust is above the sublimation temperature of silicate grains, it must arise in pure graphite grains in clouds beyond the dust-free broad-line region, but inside the torus.

In this paper we present the results of a preliminary analysis of the 142 $z > 1$ objects in our mid-infrared selected quasar sample (two example fits are shown in Figure 1). Full results will be presented in Farrah *et al.* (in preparation) and Petric *et al.* (in preparation).

2. The very hot component

We confirm the need for a very hot dust component in the near-infrared spectra of many of our type-1 objects, and also marginal evidence in one of our type-2s. Extinction is probably responsible for the lack of these in most of our dusty type-1 and type-2 quasars. The apparent need for a very hot component in one of our type-2s (SW105213.39+571605.0; $z = 1.242$) is intriguing, and may help to constrain the geometry of the emission from this component.

3. Stellar masses

The objects classified as type-2 have had their stellar masses estimated by fitting stellar population models from Maraston (2005). We fit models that are very simple single or dual stellar populations (see Lacy *et al.* 2011 for details). The results (Figure 2, left) show that the mean stellar masses are around $2 \times 10^{11} M_{\odot}$, but with a wide range, $10^{10-12.4} M_{\odot}$. There is a loose correlation with quasar luminosity. These stellar masses are similar to those seen in radio galaxies (de Breuck *et al.* 2010) at these redshifts, somewhat surprising given that Kukula *et al.* (2001) claim a significant difference in the host galaxy luminosities of radio-loud and radio quiet quasars at these redshifts.

4. The cool dust component

Over 50% of our quasars show a cool dust component, visible as detections in archival *Herschel* data and/or *Spitzer* data. The detected objects have far-infrared luminosities characteristic of Ultraluminous Infrared Galaxies (ULIRGS). Dust temperatures are seen to vary widely, from ~ 30 to > 100K. We suspect many of the warmer objects may have contributions to their dust heating from AGN (see also Schumacher *et al.* 2012).

The luminosity of the cold component correlates with the AGN luminosity (Figure 2, right) with a logarithmic slope well below unity, as also seen in low-z, normal quasars

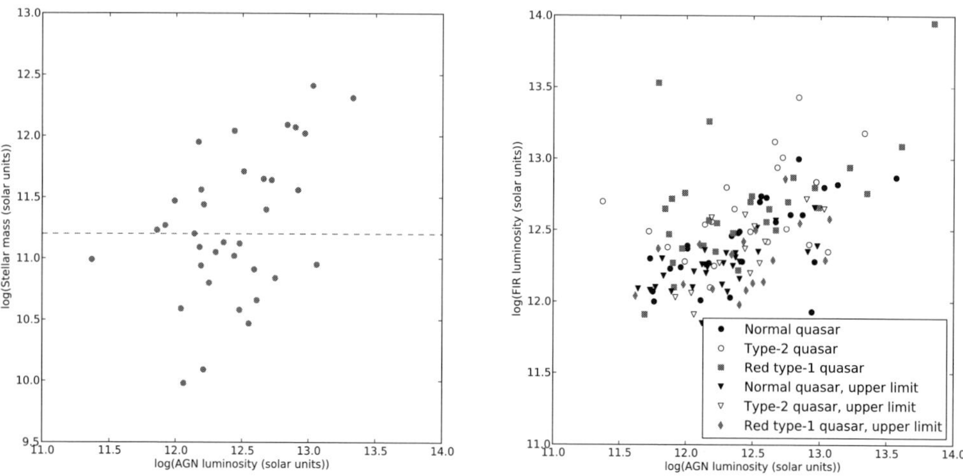

Figure 2. *Left* Stellar masses of the $z > 1$ type-2 quasars in the sample. The dashed horizontal line shows the mean stellar mass of both this sample and the radio galaxy sample of de Breuck *et al.* (2010). *Right* Far-infrared luminosity versus AGN luminosity for our quasars.

e.g. Netzer *et al.* (2007). For the most part, the dusty type-1 and type-2 quasars fall on the same correlation as the normal type-1s, however, all the objects well above the correlation are dusty or type-2 objects.

5. Discussion

Our results show that more than half of mid-infrared selected quasars, regardless of optical classification as type-1, reddened type-1 or type-2, are hosted by galaxies having the far-infrared luminosities of ULIRGs, although the amount of far-infrared emission contributed by the AGN is unclear in many cases. The most far-infrared luminous of these quasars are all obscured to some extent, however, thus evolutionary schemes, in which dusty quasars are an early stage in the evolution of quasars are certainly not ruled out by this study.

The stellar masses of the host galaxies of the type-2 quasars are high, consistent with those of radio galaxies at the same epoch, suggesting that these quasars are relatively mature systems even at $z \sim 2$. Where we have been able to estimate CO masses, these seem lower than submm-selected galaxies at these redshifts, also consistent with the idea that these are relatively mature systems (Lacy *et al.* 2011).

References

Antonucci, R. 1993, *ARA&A*, 31, 473

de Breuck, C. *et al.* 2010, *ApJ*, 725, 36

Hopkins, P. F., Hernquist, L., Cox, T. J., & Keres, D. 2008, *ApJS*, 175, 356

Kukula, M. J., Dunlop, J. S., McLure, R. J., Miller, L., Percival, W. J., Baum, S. A., & O'Dea, C. P. 2001, *MNRAS*, 326, 1533

Lacy, M. *et al.* 2004, *ApJS*, 154, 166

Lacy, M. Petric A. O., Sajina, A., Canalizo, G., Storrie-Lombardi, L. J., Armus, L., Fadda, D., & Marleau, F. R. 2007, *AJ*, 133, 186

Lacy, M., Petric A. O., Martínez-Sansigre, A., Ridgway, S. E., Sajina, A. Urrutia, T., & Farrah, D. 2011, *AJ*, 142, 196

Maraston, C. 2005, *MNRAS*, 362, 799

Mor, R. & Trakhtenbrot, B. 2011, *ApJ*, 737, L36

Netzer, H. *et al.* 2007, *ApJ*, 666, 806

Sanders, D. B., Soifer, B. T., Elias, J. H., Madore, B. F., Matthews, K., Neugebauer, G., & Scoville, N. Z. 1988, *ApJ*, 328, L35

Schumacher, H. *et al.* 2012, *MNRAS*, submitted

The Spectral Energy Distribution of Galaxies
Proceedings IAU Symposium No. 284, 2011
R.J. Tuffs & C.C. Popescu, eds.

doi:10.1017/S1743921312009118

The Mid-Infrared luminosity function of galaxies using the AKARI mid-infrared All-Sky Survey Catalogue

Y. Toba[1,2], S. Oyabu[3], H. Matsuhara[2], D. Ishihara[3], M. Malkan[4], T. Wada[2], H. Kataza[2], Y. Ohyama[5], and S. Takita[2]

[1]Department of Space and Astronautical Science, the Graduate University for Advanced Studies (Sokendai), 3-1-1 Yoshinodai, Chuo-ku, Sagamihara, 252-5210 Kanagawa, Japan

[2]Institute of Space and Astronautical Science, Japan Aerospace Exploration Agency, 3-1-1 Yoshinodai, Chuo-ku, Sagamihara, 252-5210, Kanagawa, Japan

[3]Graduate School of Science, Nagoya University, Furo-cho, Chikusa-ku, Nagoya 464-8602, Japan

[4]Department of Physics and Astronomy, University of California, Los Angeles, CA 90095-1547, USA

[5]Academia Sinica, Institute of Astronomy and Astrophysics, Taiwan

email: `toba@ir.isas.jaxa.jp`

Abstract. We present the first determination of the 18 μm luminosity function (LF) of galaxies at $0.006 < z < 0.7$ (the average redshift is ~ 0.04) using the AKARI mid-infrared All-Sky Survey catalogue. We have selected a 18 μm flux-limited sample of 243 galaxies from the catalogue in the SDSS spectroscopic region. We then classified the sample into four types; Seyfert 1 galaxies (including QSOs), Seyfert 2 galaxies, LINERs and Star-Forming galaxies using mainly [OIII]/Hβ vs. [NII]/Hα line ratios obtained from the SDSS.

As a result of constructing Seyfert 1 and Seyfert 2 LFs, we found the following results; (i) the number density ratio of Seyfert 2s to Seyfert 1s is 3.98 ± 0.41 obtained from Sy1 and Sy2 LFs; this value is larger than the results obtained from optical LFs. (ii) the fraction of Sy2s in the entire AGNs may be anti-correlated with 18 μm luminosity. These results suggest that the torus structure probably depends on the mid-infrared luminosity of AGNs and most of the AGNs in the local Universe are obscured by dust.

Keywords. catalogs, surveys, galaxies: active, galaxies: fundamental parameters,galaxies: Seyfert, galaxies: luminosity function, infrared: galaxies

1. Introduction

The luminosity function (LF) of galaxies is one of the fundamental statistics to describe the formation and evolution of galaxies in the Universe. When we focus on the infrared (IR), an infrared luminosity function is an important probe of the activity of galaxies and galaxy formation history hidden by dust since galaxies emit their entire IR through dust.

The infrared astronomical satellite AKARI, launched in 2006, is the first Japanese space mission for infrared astronomy in Japan (Murakami *et al.* 2007). One of the most important results of AKARI is a mid-infrared (9 and 18 μm) all-sky survey, the detection limits for which reach 50 mJy and 100 mJy respectively (Ishihara *et al.* 2010).

Using AKARI, we constructed 18 μm LFs of galaxies, a unique output since only AKARI only has an observing band at 18 μm. The coverage of objects from which the LFs are constructed is about 8032 deg^2, with a depth in redshift of about 0.7.

2. Data and analysis

Our sample is constructed out of objects detected by AKARI which are detected in the the Sloan Digital Sky Survey Data Release 7 (Abazajian *et al.* 2009) spectroscopic region (8032 deg^2). We achieved this by cross-matching with the SDSS and ZCAT (eq: Huchra *et al.*1995). We then selected the objects which met the following criteria as the final sample.

(*a*) $z > 0.006$ (in order to exclude stars and nearby galaxies.)

(*b*) $F_{18\mu m} > 0.181$ Jy (which corresponds to a 50 % completeness).

These processes resulted in a 18 μm flux-limited sample of 243 galaxies.

Furthermore, we classified these 243 galaxies into four catagories: Seyfert 1 galaxies, Seyfert 2 galaxies, LINERs and Star-Forming galaxies. We used the optical flux line ratio and the information in the catalogue for a type classification. Th final division of galaxies between the four types is 36 Seyfert 1 galaxies, 62 Seyfert 2 galaxies, 12 LINERs, and 68 Star-Forming galaxies.

3. Results

After adopting K-corrections and a correction for completeness, we derived LFs using the $1/V_{max}$ method (Schmidt *et al.* 1968). An advantage of the $1/V_{max}$ method is that it allows direct computation of the LF from the data, with no parameter dependence or a model assumption.

Figure 1 shows the resulting luminosity functions at 18 μm. 12 μm LFs from (Rush *et al.* 1993) are also plotted for comparison (after converting their result for a Hubble constant of 70 km s^{-1} Mpc^{-1}. The shapes of these LFs resemble each other closely.

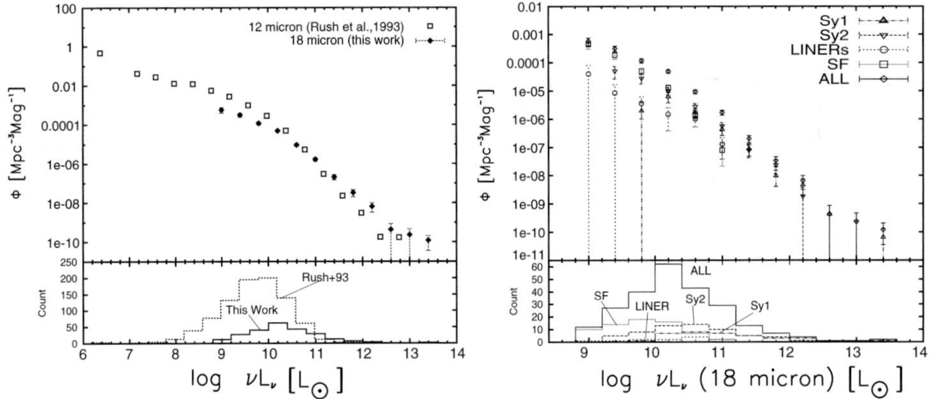

Figure 1. The 18 μm luminosity function of all samples compared to the 12 μm luminosity function after correcting for Hubble constant (Rush *et al.* 1993). The error bar is calculated considering Poisson-statistical uncertainties (left). The LFs including their types (right).

4. Discussion

4.1. *The number density ratio of Sy1s and Sy2s*

In order to establish the comparison between the number densities of Sy1s and Sy2s, we used the 18 μm luminosity, which is expected to be direct radiation from the dust torus and to be uninfluenced by dust extinction. By integrating the LFs of Seyfert 1s and Seyfert 2s (Fig.1), we obtained the number density ratio, Φ_{Sy2}/Φ_{Sy1}. In this work,

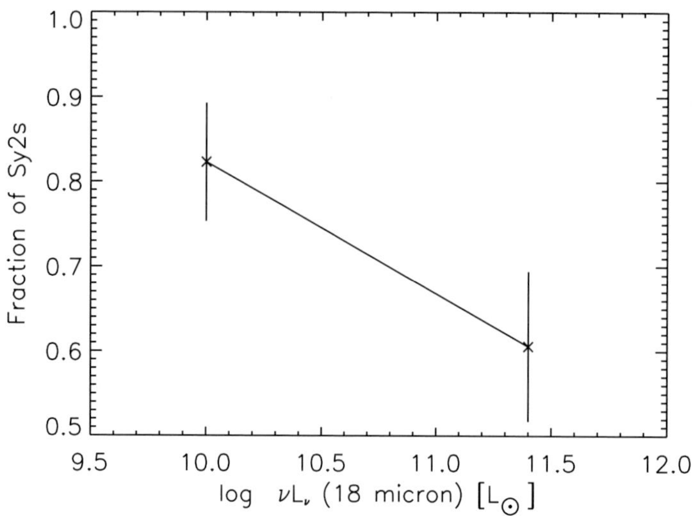

Figure 2. The fraction of Sy2s versus 18 μm luminosity.

the integral range is $\log(\nu L_\nu) > 9.6$ L$_\odot$ for both AGNs. Our results yield a value for $\Phi_{\rm Sy2}/\Phi_{\rm Sy1}$ of 3.98 ± 0.41, which is greater than the value of 0.96 obtained from the corresponding optical LFs (Huchra *et al.* 1992).

This results indicate that there are a large number of dusty AGNs in the local Universe. The result mentioned above is consistent with previous work using X-ray observations (Maiolino *et al.* 1998; Risaliti *et al.* 1999; Malizia *et al.* 2009).

4.2. *Luminosity dependence of the fraction of Sy2s*

In Figure 2 we show the fraction of Sy2s as function of 18 μm luminosity. The fraction of Sy2s seems to be anti-correlated with the 18 μm luminosity. This anti-correlation probably means that the emission from the central engine has influenced the structure of the torus. Therefore, this relation indicates that the torus structure may be different for AGNs with different 18 μm luminosities.

References

Abazajian, *et al.* 2009, *ApJS*, 182, 543A

Huchra, J. & Burg, R. 1992, *ApJ*, 393, 90

Huchra, J. P., Geller, M. J., & Corwin, H. G. 1995, *ApJS*, 99, 391

Ishihara, D., *et al.* 2010, *A&A*, 514, A1

Kataza, H., Alfageme C., Cassatella, A., Cox. N., Fujiwara, H., Ishihara, D., Oyabu, S., Salama, A., Takita, S., & Yamamura I., AKARI/IRC All-Sky Survey Point Source Catalogue Version 1.0 -Release Note (Rev.1)-, 2010

Maiolino R., Salvati M., Bassani L., Dadina M., della Ceca R., Matt G., Risaliti G., & Zamorani G. 1998, *A&A*, 338, 781

Malizia, A., Stephen, J. B., Bassani, L., Bird, A. J., Panessa, F., & Ubertini, P. 2009, *MNRAS*, 399, 944

Murakami, H., *et al.* 2007, *PASJ*, 59, 369

Risaliti G., Maiolino R., & Salvati M. 1999, *ApJ*, 522, 157

Rush, B., Malkan, M. A., & Spinoglio, L. 1993, *ApJS*, 89, 1

Schmidt M. 1968, *ApJ*, 151, 393

The Spectral Energy Distribution of Galaxies
Proceedings IAU Symposium No. 284, 2011
R.J. Tuffs & C.C. Popescu, eds.

doi:10.1017/S174392131200912X

Optically faint radio sources: reborn AGN?

Mercedes E. Filho[1], **Jarle Brinchmann**[1,2], **Catarina Lobo**[1,3] **and Sonia Antón**[4,5]

[1]CAUP, Rua das Estrelas, 4150–762 Porto, Portugal [2]Leiden Observatory, University of Leiden, PO Box 9513, NL–2300 RA Leiden, The Netherlands [3]Departamento de Física e Astronomia, FCUP, Rua do Campo Alegre, 687, 4169–007, Porto, Portugal [4]CICGE, FCUP, Porto, Porto, Portugal [5]SIM, FCUL, Lisboa, Portugal

email: mfilho@astro.up.pt

Abstract. We have discovered eight relatively strong radio sources that have no optical counterparts. A NIR follow-up has detected faint (17–20 mag) host galaxies in all targets. In general, the radio properties are similar to those observed in 3CRR sources but the optical-radio slopes are consistent with moderate to high redshift (z <4) GHz-peaked spectrum sources. Our results suggest that these are galaxies whose black hole has been recently re-ignited into activity but that retain large-scale radio structures, signatures of previous AGN activity.

Keywords. Galaxies: active

1. The Sources

The SDSS maxBCG cluster catalog (Koester *et al.* 2007) was used as the seed catalog for our study. We first cross-correlated the cluster sample with the FIRST radio catalog (White *et al.* 1997). We retained 291 cluster fields that contained at least one FIRST radio source within 1 Mpc in projection from the BCG.

During this process, we identified eight radio sources with no optical SDSS counterpart, indicating that $r_{\rm AB}$ >22 mag. Three of the sources, however, have been identified on Stripe 82 (with $r_{\rm AB}$-band magnitudes in the range 23–25 mag. The radio sources are further characterized by their radio-loudness (large radio-to-optical ratio), arcsec-scale FR II radio morphology, and relatively strong FIRST flux densities (1 mJy<$F_{1.4\,\rm GHz}$ < 80 mJy).

The unidentified radio sources were followed-up with NIR imaging using the wide-field imager HAWK-I. The HAWK-I observations were successful, as NIR emission, coincident with the centers of the radio structures and that we interpret as host-related emission, was detected in all the targets. The NIR magnitudes of these sources are in the range 17–20 mag (Vega system) and their NIR sizes are typically $\sim$1.5 arcsec (Fig. 1).

2. Results

Overall, if we focus on the radio properties alone, we may conclude that our optically faint radio sources appear similar to the radio sources found in the 3CRR catalog†, grazing the upper envelope of the 3CRR distribution in terms of radio morphology (Fig. 1), LLS, and (total) radio power (Fig. 2).

We have also explored the radio-to-optical SED, adopting a similar approach to Huynh *et al.* (2010). We have compared our sample data with a set of SED templates from the 3CR catalog (Spinrad 1985), a GPS sample (Labiano *et al.* 2007) and galaxies that span

† http://3crr.extragalactic.info/cgi/database

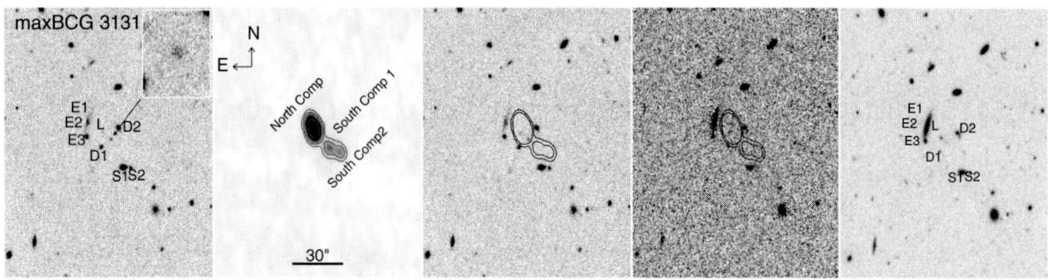

Figure 1. The radio source in the maxBCG cluster field 3131. From left o right: the NIR image, the FIRST image with radio contours, radio contours superimposed on the NIR, SDSS r_{AB}-band and Stripe 82 r_{AB}-band image. The arrow and inlay shows the position of the NIR host galaxy. Other designations refer to the radio components and secondary (unrelated) NIR/optical sources in the field.

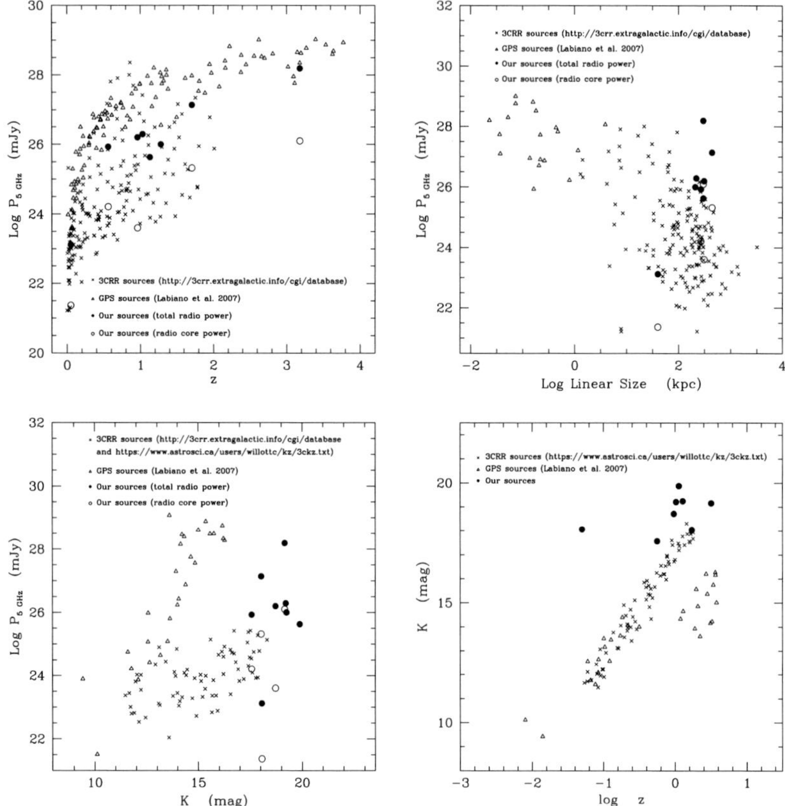

Figure 2. The redshift, 5 GHz radio power, the largest linear size (LLS), and K-band magnitude plots for the 3CRR (crosses), GPS (triangles), and our sample total radio (closed circles) and radio core (open circles) flux density.

a range of MIR galaxy classifications (Spoon *et al.* 2007): M82 a star-forming FR I, Arp 220 a Sy-type ULIRG, Mrk 231 a dusty AGN-dominated Sy 1 ULIRG, Mrk 1501 a Sy 1.2 flat-spectrum radio source, 3C 305 a Sy 2 FR I, Mrk 668 a Sy 1.5 GPS source, 3C 273 a radio-loud quasar, and 3C 295 a narrow-line FR II (Fig. 3).

Fig. 3 shows that viewed as a class, the radio-to-optical slope of our sample of sources are more consistent with those of "young" (GPS) radio sources at $z < 4$. Using Mrk 668 as a GPS template, we can obtain redshift estimates for our sample sources. We can

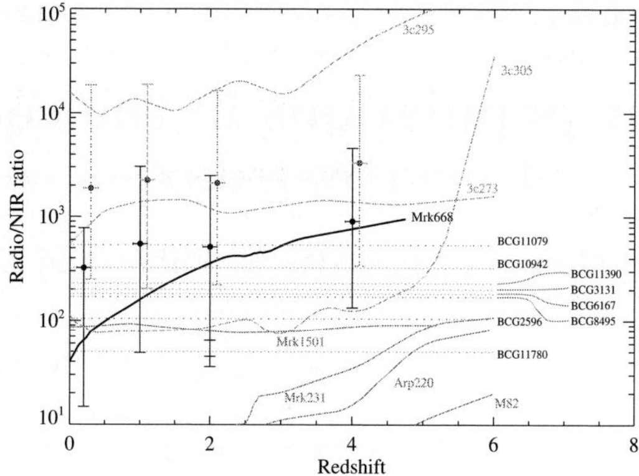

Figure 3. The radio-NIR ratio as a function of redshift for different galaxy templates (solid and dashed lines). The dotted lines are the radio-NIR ratios of our sample. The large dots are the median redshift values for the 3CR (dashed; Spinrad 1985) and GPS (solid; Labiano *et al.* 2007) sample sources, with a one σ errorbar.

Table 1. Redshift estimates.

maxBCG	$z_{\rm Mrk668}$	$z_{\rm Le\,Phare}$
2596	0.56	...
3131	1.03	$1.57^{+1.14}_{-0.43}$
6167	1.13	$1.00^{+3.81}_{-0.40}$
8495	0.96	$0.88^{+0.09}_{-0.08}$
10942	1.71	...
11079	3.18	...
11390	1.28	...
11780	0.05	...

also obtain an independent check by exploiting that three of the NIR sources have faint counterparts in the Stripe 82 images. By combining the Stripe 82 and NIR photometry, we have estimated photometric redshifts using the Le photometric redshift code (Arnouts *et al.* 1999; Ilbert *et al.* 2006; http://www.cfht.hawaii.edu). The resulting redshift estimates are in very good agreement (Table 1).

The way to reconcile the "young" radio source (GPS-type radio-optical slope) with the large-scale radio structure is to assume a "double-double" radio galaxy scenario (Schoenmakers *et al.* 2000a) of intermittent jets (Saikia & Jamrozy 2009), where our sample sources are recently re-ignited AGN, and the large-scale structure, a relic of a previous cycle of black hole activity.

References

Arnouts, S. *et al.* 1999, *MNRAS*, 310, 540

Huynh, M. T., Norris, R. P., Siana, B., & Middelberg, E. 2010, *ApJ*, 710, 698

Ilbert, O. *et al.* 2006, *A&A*, 457, 841

Koester, B. P. *et al.* 2007, *ApJ*, 660, 239

Labiano, A. *et al.* 2007, *A&A*, 463, 97

Saikia, D. J. & Jamrozy M. 2009, *BASI*, 37, 63

Schoenmakers, A. P. *et al.* 2000c, *MNRAS*, 315, 381

Spinrad, H., Marr, J., Aguilar, L., & Djorgovski, S. 1985, *PASP*, 97, 932

Spoon, H. W. W. *et al.* 2007, *ApJ*, 654, 77

White, R. L., Becker, R. H., Helfand, D. J., & Gregg, M. D. 1997, *ApJ*, 475, 479

The Spectral Energy Distribution of Galaxies
Proceedings IAU Symposium No. 284, 2011
R.J. Tuffs & C.C. Popescu, eds.

doi:10.1017/S1743921312009131

Stellar populations in the centers of nearby galaxies

Jean Michel Gomes[1], Mercedes E. Filho[1] and Luis C. Ho[2]

[1]Centro de Astrofísica, Universidade do Porto, Rua das Estrelas, 4150-762 Porto, Portugal

[2]The Observatories of the Carnegie Institution for Science, 813 Santa Barbara Street, Pasadena, CA 91101, USA

email: `jean@astro.up.pt`

Abstract. The great amount of data observed in recent years coupled with modelling using evolutionary synthesis codes (BPASS, COELHO, GALAXEV, GALEV, MILES, PÉGASE, etc...) to compute Single Stellar Populations (SSPs) and the availability of fast and ingenious spectral synthesis codes such as STARLIGHT, ULySS and VESPA, have significantly shed light on our knowledge about the formation and evolution of galaxies. However, there are still open issues concerning the stellar populations in nearby galaxies, particularly those harbouring Active Galactic Nuclei (AGN): can stellar populations mimic nuclear activity, leading to a misclassification based on optical emission line ratios (Stasińska *et al.* 2008)? We have applied the STARLIGHT code (Cid Fernandes *et al.* 2005) to a well studied sample of nearby galaxies' nuclear spectra (r $<\sim$ 200 pc), observed with the Hale 5 m telescope at Palomar Observatory in two different regions: $\sim$ 4230 – 5110 Å and $\sim$ 6210 – 6860 Å (Ho *et al.* 1995), with spectral resolutions of approximately 4 Å and 2.5 Å. The aim is to properly derive the star-formation history (SFH), mean stellar age and metallicity and total stellar mass. Our results show that the star-formation history of Seyfert galaxies are very heterogeneous, i.e. these are composed of young, intermediate and old stellar populations, while the SFH of Low-Ionization Nuclear Emission-Line Regions (LINERs) are basically composed of old stellar populations. The absence of young stars in LINERs indicates that these are not responsible for the observed low-ionization emission lines. Furthermore, although a significant fraction of AGN spectra require a featureless continuum in their Spectral Energy Distribution (SED) modelling, this is not an indicative of the presence of an AGN, instead the continuum may simulate the presence of young stellar populations. The main objective of this research is to complement the study of spectroscopic parameters from 486 galaxies analyzed by Ho *et al.* (1995) that are public available in the VizieR catalog (Ho *et al.* 1997, 2009) and provide information about their stellar population content by means of the STARLIGHT. The base of Simple Stellar Populations used here was taken from Bruzual & Charlot (2003) and spans 25 ages (from 1 Myr to 18 Gyr) and 6 metallicities (Z = 0.005, 0.02, 0.2, 0.4, 1 & 2.5 $Z_{\odot}$).

Keywords. galaxies: evolution, galaxies: formation, galaxies: nuclei, galaxies: active, galaxies: stellar content, galaxies: Seyfert

1. Introduction

We have applied STARLIGHT to the entire sample of galaxies analyzed by Ho *et al.* (1995), which comprises 486 spectra observed with the Hale 5 m telescope at Palomar Observatory in two different regions: $\sim$ 4230 – 5110 Å (BLUE) and $\sim$ 6210 – 6860 Å (RED). After a series of tests, we decided to only model the BLUE part of the spectrum to derive the star-formation histories, mean stellar ages and metallicities, because the spectral information concerning young stellar populations is located mostly in the BLUE range. However, emission lines from both parts of the spectrum have been used in order to

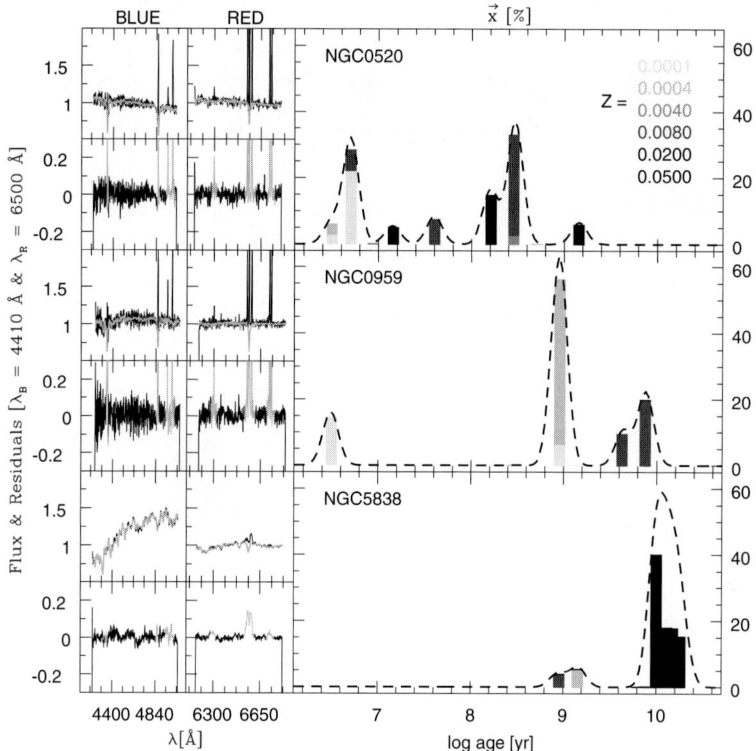

Figure 1. STARLIGHT spectral analysis of NGC 520 (Star-Forming), NGC 959 (Composite), NGC 5838 (AGN/LINER) from top to bottom. Left panels: we show the blue and red parts of spectrum in black and the best fit in grey colour. For clarity, residuals (observed – modeled) are shown bellow each spectrum window with emission lines in grey colour. The Star Formation History (SFH), characterized by the population vector $\vec{x}$ (i.e. the fractional contribution of each SSP), is shown as a function of log age in the right panels. The greyscale stands for distinct metallicities (see label). The black dashed lines are the corresponding smoothed SFHs (re-scaled) with a Gaussian FWHM of 0.2 dex to better show the trends for the different galaxies. Star-Forming galaxies tend to have younger stellar populations, Composites seem to have intermediate ones and AGN galaxies harbour older stellar populations.

distinguish between star-forming galaxies and AGN, based on the diagnostic emission-line ratios: [OIII]λ 5007/Hβ and [NII]λ 6583/Hα in BPT diagrams (Baldwin, Phillips & Terlevich 1981). In Fig. 1, we show three examples of spectral fitting, the respective residuals and SFHs for a Star-Forming, Composite and AGN/LINER.

2. Main result: Age and Metallicity distributions

The principal result retrieved by the STARLIGHT is depicted in Fig. 2. The histograms correspond to the fractional contribution of Young ($\vec{x}_{\rm Young}$, $< 9 \times 10^7$ years), Intermediate ($\vec{x}_{\rm Intermediate}$, between 9×10^7 and 10^9 years) and Old ($\vec{x}_{\rm Old}$, $> 10^9$ years) stellar populations. The histograms stand for Star-Forming (shaded histogram; light grey), Composites (solid line; medium grey) and AGN (dashed line; black) galaxies, classified according to the classical BPT diagram. For clarity, the mean stellar age and metallicity are also shown for the whole sample. We can see that Star-Forming galaxies tend to harbour younger stellar populations, Composites seem to have intermediate ones and AGN galaxies have older stellar populations in the centers of galaxies.

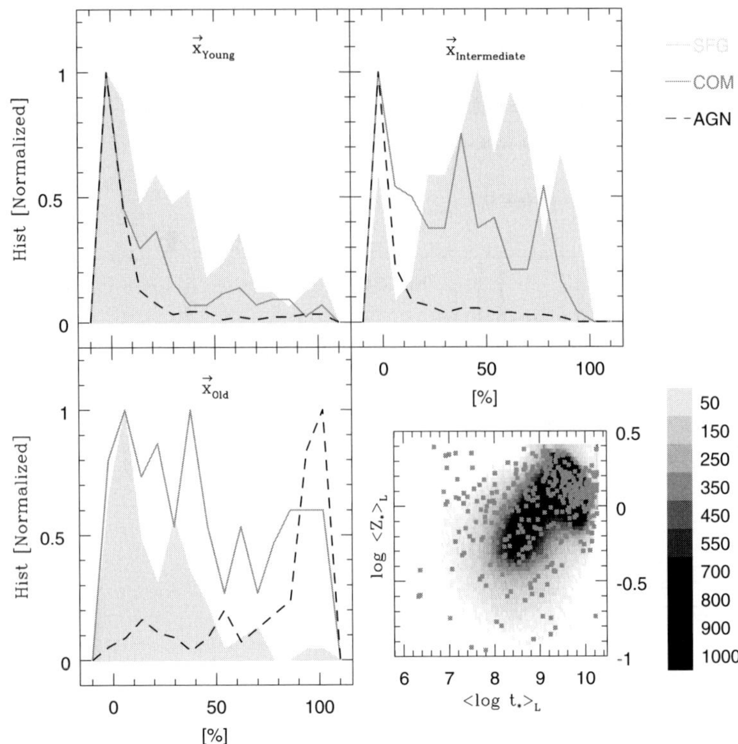

Figure 2. The histograms are the fractional contribution of **Young**, **Intermediate** and **Old** stellar populations, as explained in the text for three distinct galaxies of the sample: Star-Forming (shaded histogram; light grey), Composites (solid line; medium grey) and AGN (dashed line; black) galaxies, classified according to the classical BPT diagram. We also show the mean stellar age and metallicity distributions for the Palomar sample using grey squares. For comparison, the greyscale depicts the number of SDSS galaxies in each bin. We can reach the same conclusions as Fig. 1, where statistically star-forming galaxies tend to have younger stellar populations, while Composites seem to have intermediate ones and AGN galaxies harbour older stellar populations.

Acknowledgements

J. M. Gomes is supported by a Post-Doctoral grant, funded by FCT/MCTES (Portugal) and POPH/FSE (EC).

References

Baldwin, J. A., Phillips, M. M. & Terlevich, R. 1981, *PASP*, 93, 5
Bruzual, G. & Charlot, S., *MNRAS*, 344, 1000
Cid Fernandes, R. *et al.* 2005, *MNRAS*, 358, 363
Ho, L. C. *et al.* 1995, *ApJ*, 98, 477
Kauffmann, G. *et al.* 2003, *MNRAS*, 346, 1055
Kewley, L. J. *et al.* 2001, *ApJ*, 556, 121
Stasińska, G. *et al.* 2008, *MNRAS*, 391L, 29S

The Spectral Energy Distribution of Galaxies
Proceedings IAU Symposium No. 284, 2011
R.J. Tuffs & C.C. Popescu, eds.

doi:10.1017/S1743921312009143

Multiwavelength data for bright active galaxies

Areg M. Mickaelian[1], Hayk V. Abrahamyan[1], Gurgen M. Paronyan[1] and Gohar S. Harutyunyan[2]

[1]Byurakan Astrophysical Observatory (BAO), Byurakan 0213, Aragatzotn province, Armenia
email: aregmick@aras.am, abrahamyanhayk@gmail.com, pgurgen@bao.sci.am

[2]Yerevan State University (YSU), A. Manoogian 1, Yerevan 0025, Armenia
email: goharutyunyan@gmail.com

Abstract. The spectral energy distribution (SED) gives a complete picture of the radiation of space objects and may result in correct classifications compared to those based only on optical (or other local) spectra. This is especially crucial for active galaxies, both AGN and Starbursts (SB). For this, multiwavelength (MW) data are needed taken from available surveys and catalogs. We have cross-correlated the Catalogue of quasars and active galaxies with all-sky or large-area MW catalogues, such as X-ray ROSAT (BSC and FSC), UV GALEX (MIS and AIS), optical APM, MAPS, USNO-B1.0, GSC 2.3.2, and SDSS DR8, NIR 2MASS, MIR/FIR WISE, IRAS (PSC and FSC) and AKARI (IRC and FIS), radio GB6, NVSS, FIRST, and WENSS. We have established accurate positions and photometry for a few thousands of objects that appeared in the catalog with poor data, as well as achieved the best astrometric and photometric data for all objects. This allowed correct cross-correlations and establishing correct MW data for these objects. As a result, we obtained 34 photometric points from X-rays to radio and using VO tools built SEDs for some 10,000 bright objects. Some data from other surveys were also used, such as Chandra, XMM, Spitzer, etc. All objects were grouped into several forms of SED and were compared to the known optical classes given in the catalog (QSO, BLL, Sy1, Sy1.2–1.9, Sy2, LINER, SB, and HII). This allowed reveal obscured AGN, as well as find previously misclassified objects. A homogeneous classification for these objects was established. The first part of this project is presented; establishment of accurate positions and photometry and cross-correlations with MW catalogs.

Keywords. catalogs, surveys, galaxies: active, galaxies: starburst

1. The Catalogue of active galaxies and its statistical use

The Catalogue of quasars and active nuclei (13th edition; Veron-Cetty & Veron 2010; hereafter VCV-13) is the fullest collection of AGN (and other active galaxies). It contains 168,941 objects, including 133,336 QSO, 1,374 BL Lac objects, and 34,231 active galaxies (including 16,517 Sy1). Most of objects come from SDSS DR7 (Abazajian *et al.* 2009) and 2QZ/6QZ (Croom *et al.* 2004) surveys. The latest SDSS-based (SDSS-I and SDSS-II) QSO catalog was published by Schneider *et al.* (2010) and contains 105,783 spectroscopically confirmed QSOs over 9380 deg^2 surface. A Large quasar astrometric catalogue (LQAC) was recently built by Souchay *et al.* (2009) based on 12 optical and radio QSO catalogues. It contains 113,666 objects. In addition, Massaro *et al.* (2009) have published a multi-frequency catalogue of blazars with the fullest collection (3061 objects) of BL Lac and flat radio spectrum QSOs (last update: 1 July 2011). All these lists partially cover each other and for the users it is not easy to have a full set of data or even carry out cross-correlations between these catalogs (mainly, due to positional uncertainties).

Table 1. All-sky / large area catalogues and their data.

Catalogs	Range	Area, surface	Accuracy	Sources	Release	Reference
ROSAT BSC	X-ray	All-sky	60 as	18806	1999-11-05	Voges *et al.* 1999
ROSAT FSC	X-ray	All-sky	60 as	105924	2000-05-26	Voges *et al.* 2000
GALEX MIS	UV	All-sky	2 as	12600000	2011-03-14	Bianchi *et al.* 2011
GALEX AIS	UV	All-sky	2 as	65300000	2011-03-14	Bianchi *et al.* 2011
APM	opt	δ >-33°, $\|b\|$ >20°	619 mas	166466987	2000-12-05	McMahon *et al.* 2000
MAPS	opt	δ >-33°, $\|b\|$ >20°	519 mas	89234404	2003-07-00	Cabanela *et al.* 2003
USNO-B1.0	opt	All-sky	387 mas	1045913669	2005-11-17	Monet *et al.* 2003
GSC 2.3.2	opt	All-sky	258 mas	945592683	2007-04-19	Lasker *et al.* 2008
2QZ/6QZ	opt	SGP/NGP,	338 mas	49425	2006-06-01	Croom *et al.* 2004
SDSS-II	opt	δ >0°, $\|b\|$ >30°	67 mas	357175411	2009-06-03	Abazajian *et al.* 2009
2MASS	NIR	All-sky	479 mas	470992970	2003-06-10	Cutri *et al.* 2003
WISE	NIR/MIR	All-sky	500 mas	257310278	2011-04-14	Wright *et al.* 2010
IRAS PSC	MIR/FIR	All-sky	28 as	245889	1994-01-29	IRAS 1986
IRAS FSC	MIR/FIR	All-sky	13 as	173044	1993-02-24	Moshir *et al.* 1989
AKARI IRC	MIR	All-sky	300 mas	870973	2010-04-21	Ishihara *et al.* 2010
AKARI FIS	FIR	All-sky	800 mas	427071	2010-04-21	Yamamura *et al.* 2010
GB6	radio	0° < δ <+75°	210 as	75162	1997-01-07	Gregory *et al.*1996
NVSS	radio	δ >-40°,	45 as	1773484	2002-09-27	Condon *et al.* 1998
FIRST	radio	δ >0°, $\|b\|$ >30°	5 as	811117	2011-04-03	Becker *et al.* 2003
WENSS	radio	+28° < δ <+76°	54 as	229420	2004-05-14	de Bruyn *et al.* 1998

VCV-13 contains quasars with measured redshift known prior to July 1, 2009 (including SDSS DR7). As in the preceding editions, it does not give any information about absorption lines or X-ray/UV/IR properties. But absolute magnitudes and, when available, the 6 and 21 cm flux densities are given. The authors warn that the catalogue should not be used for any statistical analysis as it is not complete in any sense. One of the main problems is the absence of homogeneous accurate positional and photometric data: they are coming from the original papers and have extremely different accuracy and (in case of the magnitudes) bands (B, V, R, and other, making the magnitudes not useful for statistics). SDSS and 2QZ/6QZ provide reliable photometry, and in addition, homogeneous optical photometry is given in APM/MAPS/USNO/GSC catalogs measured from POSS1/POSS2. Absence of reliable photometry leads to misclassification of objects into QSO/Sy1 (as it is connected with the M_B=-22.25 limit of the absolute magnitude), as well as impossibility of studies of luminosity functions, luminosity evolution, etc. Moreover, MW data for active galaxies would provide possibility to compare there properties, build MW SEDs, make up better classifications, and better understand their physics and evolution. Thus there is a need for cross-correlations of the AGN Catalogue with optical and MW data.

2. Cross-correlations of AGN catalogue with MW data

In general, cross-correlations lead to numerous misidentifications and doubtful associations. The problem arises due to selection of the search radius and final acceptance of found objects. Typically, a standard search radius is being given (like in Vizier and elsewhere) based on typical positional errors of the given catalog. We have developed a new method of calculation of the search radius based on 3σ rms of individual objects/sources from each catalogue (Knyazyan *et al.* 2011). Such errors differ up to 10-15 times, so the traditional method leads to finding many associations out of 3σ or losing many associations for objects with larger errors. Moreover, we use accurate rms values for optical catalogs to achieve the best matches (Mickaelian & Sinamyan 2010). Accurate coordinates and magnitudes were derived using our methods developed for APM/MAPS/USNO/GSC data (Mickaelian & Sinamyan 2010, Mickaelian *et al.* 2011).

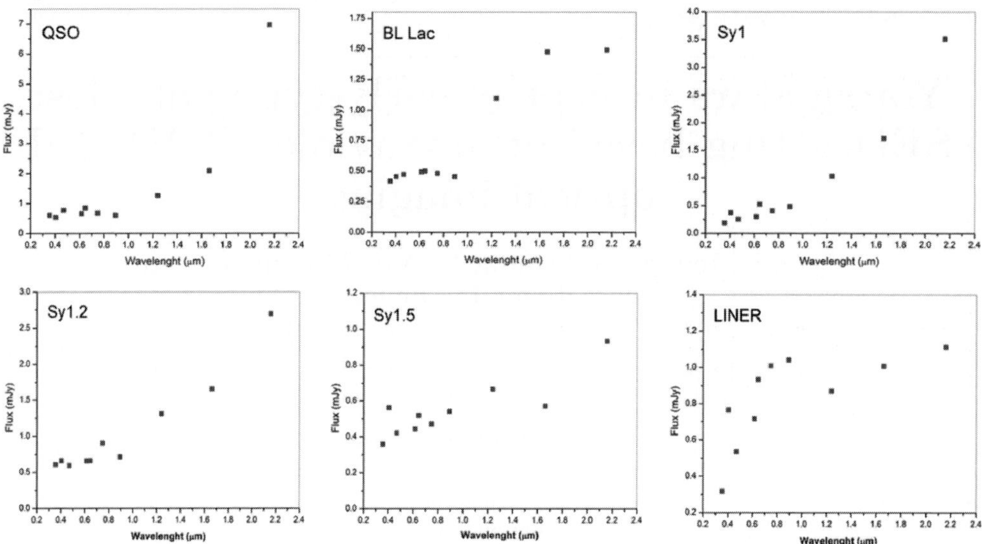

Figure 1. SEDs of active galaxies in the optical and NIR ranges by means of SDSS, POSS, and 2MASS photometric data.

Cross-correlations of the catalog of Quasars and AGN (VCV-13) have been carried out with the following surveys and catalogues given in Table 1.

3. Results: MW data and SEDs for active galaxies

MW SEDs have been built as shown in our previous papers (Sargsyan 2009; Hovhannisyan *et al.* 2011; Sargsyan *et al.* 2011). For the beginning, we have correlated POSS1/POSS2 B and R, SDSS ugriz and 2MASS JHK magnitudes. Different activity type objects have been selected: QSO, BLL, Sy1, Sy1.2, Sy1.5, and LINER. The results are shown in Figure 1. Slight differences are present in SEDs, however, average SEDs will reveal significant changes in different activity types.

References

Hovhannisyan A., Sargsyan, L. A., Mickaelian, A. M., & Weedman, D. W. 2011, *Ap*, 54, 147

Knyazyan, A., Mickaelian, A., & Astsatryan, H. 2011, *Proc. CSIT-2011*, in press

Massaro, E., Mickaelian, A. M., Nesci, R., & Weedman, D. 2008, *The Digitized First Byurakan Survey*, ARACNE Editrice, Rome, Italy, 78 p.

Massaro, E., Giommi, P., Leto, C., *et al.* 2009, *A&A*, 495, 691

Mickaelian, A. M., Nesci, R., Rossi, C., *et al.* 2007, *A&A*, 464, 1177

Mickaelian, A. M. & Sinamyan, P. K. 2010, *MNRAS*, 407, 681

Mickaelian, A. M., Mikayelyan, G. A., & Sinamyan, P. K. 2011, *MNRAS*, 415, 1061

Sargsyan, L. A. 2009, *Ap*, 52, 377

Sargsyan, L., Weedman, D., Lebouteiller, V., *et al.* 2011, *ApJ*, 730, 19

Schneider, D. P., Richards, G. T., Hall, P. B., *et al.* 2010, *AJ*, 139, 2360

Souchay, J., Andrei, A. H., Barache, C., *et al.* 2009, *A&A*, 494, 799

Veron-Cetty, M.-P. & Veron, P. 2010, *A&A*, 518, 10

The Spectral Energy Distribution of Galaxies
Proceedings IAU Symposium No. 284, 2011
R.J. Tuffs & C.C. Popescu, eds.

doi:10.1017/S1743921312009155

Young stars in nearby early-type galaxies: SED fitting based on ultraviolet (UV) and optical imaging

Hyunjin Jeong[1], **Sukyoung K. Yi**[2], **Martin Bureau**[3] **and Roger L. Davies**[3]

[1]Korea Astronomy and Space Science Institute, Daejeon 305-348, Korea
[2]Department of Astronomy, Yonsei University, Seoul 120-749, Korea
[3]Sub-Department of Astrophysics, University of Oxford, Oxford OX1 3RH, UK

email: hyunjin@kasi.re.kr

Abstract. Recent studies from the *Galaxy Evolution Explore* (*GALEX*) ultraviolet (UV) data have demonstrated that the recent star formation is more common in early-type galaxies (ETGs) than we used to believe. The UV is one order of magnitude more sensitive than the optical to the presence of young stellar populations. The near-ultraviolet (NUV) lights of ETGs, especially, are used to reveal their residual star formation history. Here we used the *GALEX* UV data of 34 nearby early-type galaxies from the SAURON sample, all of which have optical data from MDM Observatory. At least 15% of the galaxies in this sample show blue UV−optical colours suggesting recent star formation (Jeong *et al.* 2009). These NUV blue galaxies are generally low velocity dispersion systems and change the slopes of scaling relations (colour-magnitude relations and fundamental planes) and increase the scatters. To quantify the amount of recent star formation in our sample, we assume two bursts of star formation, allowing us to constrain the age and mass fraction of the young component pixel by pixel (Jeong *et al.* 2007). The pixel-by-pixel SED fitting based on UV and optical imaging reveals that the mass fraction of young ($<$ 1 Gyr old) stars in ETGs varies between 1 and 3% in the nearby universe (Jeong *et al.* in prep.). We will compare our results with the prediction from the hierarchical merger paradigm to understand the mechanism of low-level recent star formation observed in early-type galaxies.

Keywords. galaxies: elliptical and lenticular, cD – galaxies: evolution – galaxies: photometry – galaxies: structure – ultraviolet: galaxies.

1. Introduction

One of the greatest goals of modern astronomy is to understand how galaxies form and evolve. While a wealth of information is now available at medium and high redshifts, those studies rely on a detailed understanding of the local galaxy population, and much insight can be gained from studying the fossil record in nearby objects. To probe the mass assembly and star formation histories of galaxies in a hierarchical context, simultaneously constraining their dynamics and stellar population is the key. In particular, a thorough understanding of the star formation history of early-type galaxy promises key insights into the evolution of galaxies as a whole, and it strongly constrains galaxy formation models.

The exact formation mechanism of early-type galaxies remains one of the long-standing debates of modern astrophysics. Optical colour-magnitude relations (CMRs) of early-type galaxies consistently reveal a small scatter around the mean relation for the past several billion years (e.g. Bower, Lucey & Ellis 1992), apparently supporting the classical collapse model. Similarly, it is well known that early-type galaxies occupy a two-dimensional plane

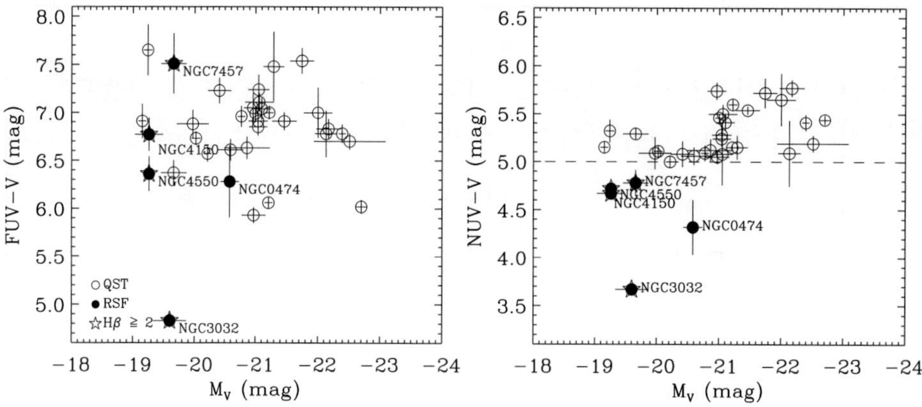

Figure 1. UV CMRs of our 34 SAURON early-type galaxies. The dashed line indicates the NUV−V = 5.0 cut-off for recent star formation. Using this standard, we divided the sample into quiescent (QST; open symbols) early-type galaxies and recent star formation (RSF, filled symbols) galaxies.

in the three-dimensional manifold of their global structural parameters (effective radius, velocity dispersion and effective surface brightness) which is called the "fundamental plane (FP)" (e.g. Djorgovski & Davis 1987, Dressler *et al.* 1987). The scatter in the fundamental plane was first thought to be small enough to represent further evidence for the monolithic scenario. However, the observed discrepancy or tilt of the fundamental plane with respect to the virial expectation has prompted many debates to explain its origin (see e.g. D'Onofrio *et al.* 2006 and references therein).

Recent advances in observational techniques are making the picture increasingly clear. The first is the advent of deep imaging facilities. Deep imaging surveys have shown that many of the "simple systems" do possess shells, tidal features and signatures of ongoing or recent star formation. The second advance is integral-field spectroscopy that makes it possible to obtain spatially resolved maps of various properties of galaxies. For example, the SAURON team (see e.g. Bacon *et al.* 2001, de Zeeuw *et al.* 2002) surveyed the two-dimensional stellar and ionised-gas kinematics and stellar populations of 72 nearby early-type galaxies and found a rich diversity in their dynamics, discovering numerous central disks and kinematically-decoupled cores. Such properties may be taken as evidence against the classical monolithic collapse model. Furthermore, the recent availability of survey data in the ultraviolet (e.g. *GALEX*) has reinvigorated the study of early-type galaxies and forced us to revisit assumptions about their evolution (see e.g. Yi *et al.* 2005).

2. Observations and data reduction

We observed 34 early-type galaxies from the SAURON project with the *GALEX* and the MDM Observatory 1.3-m McGraw-Hill Telescope. The *GALEX* UV images are delivered pre-processed but we undertook our own estimate of the sky values. And then, we carried out surface photometry by measuring the surface brightness along elliptical annuli after masking-out Sextractor-detected sources. From the radial surface brightness profiles, we derived total apparent magnitudes by extrapolating the growth curves to infinity.

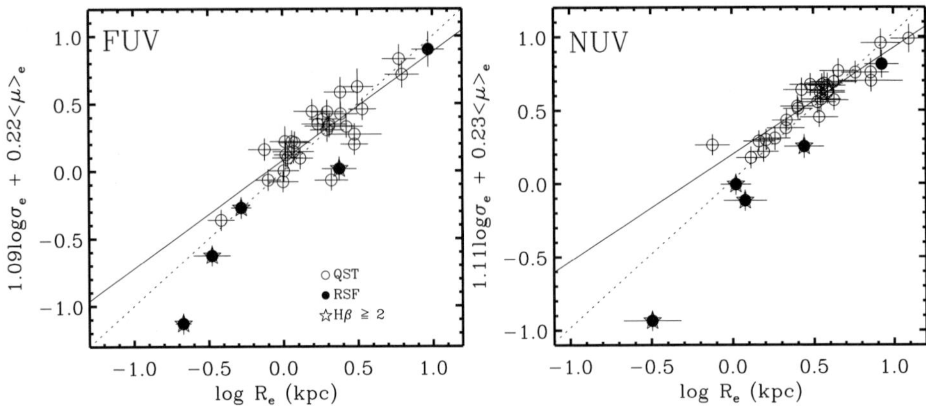

Figure 2. Fundamental Planes in the FUV and NUV bands. Symbols are the same as in Figure 1 and linear fits to the two subsamples (whole sample, quiescent galaxies only) are shown as dotted and solid lines, respectively.

3. Results

Numerous investigations can be performed with such a database. But, in this paper, we focus on the effects of star formation on the scaling relations. CMRs of early-type galaxies have been used as a tool for tracing their star formation histories. The NUV band is particularly sensitive to the presence of young stellar populations. The NUV CMR, therefore, is a good tool for tracking recent star formation. Figure 1 shows the UV−optical CMRs. The empirical demarcation of NUV−V = 5.0 for identifying recent star formation galaxies is shown as a horizontal line in the NUV−V CMR. Using this empirical criterion, we have identified the galaxies that are likely to have experienced a recent star formation as filled symbols. The overall fraction of galaxies with recent star formation is 15 % (5/34). We label galaxies with NUV−V $\geqslant$5.0 as quiescent (QST) galaxies (open symbols). Furthermore, stars indicate Balmer absorption line strength H$\beta \gtrsim$ 2 Å. Supporting our empirical threshold, four out of five recent star formation candidates also show an enhanced Hβ line strength. Our star formation interpretation is also consistent with the results of the molecular gas emission surveys. The CO detection rate decreases monotonically from recent star formation galaxies to quiescent galaxies. However, there is no clear dependence of FUV−V on M_V in our sample.

The fundamental plane is another key scaling relation of early-type galaxies. It has long been recognised that the effective radius (R_e), the *mean* effective surface brightness ($\langle\mu\rangle_e$) and the velocity dispersion (σ_e) are mutually correlated and unified in a two-dimensional manifold. Under the assumption of structural homology (i.e. all early-type galaxies have the same mass distribution and kinematics), the virial theorem predicts that the fundamental plane parameters should scale in a specific manner, but observations reveal a tilt away from the virial prediction. In Figure 2, we present the fundamental planes in the FUV and NUV bands, using the same symbols as in Figure 1. Recent star formation galaxies systematically deviate from the best-fit planes (dotted lines) so as to create shallower slopes, and they significantly increase the scatter. Considering our NUV fundamental plane fit for the entire sample, our slope is significantly different from the virial expectation. However, if we exclude recent star formation galaxies we recover almost all of the virial prediction. We thus conclude that a significant fraction of the fundamental plane tilt and scatter is due to low-mass early-type galaxies with young stellar populations.

In order to constrain the age and mass fraction of the young component pixel by pixel, we fit the observed colours (FUV−V and NUV−V) to simple two-population (young

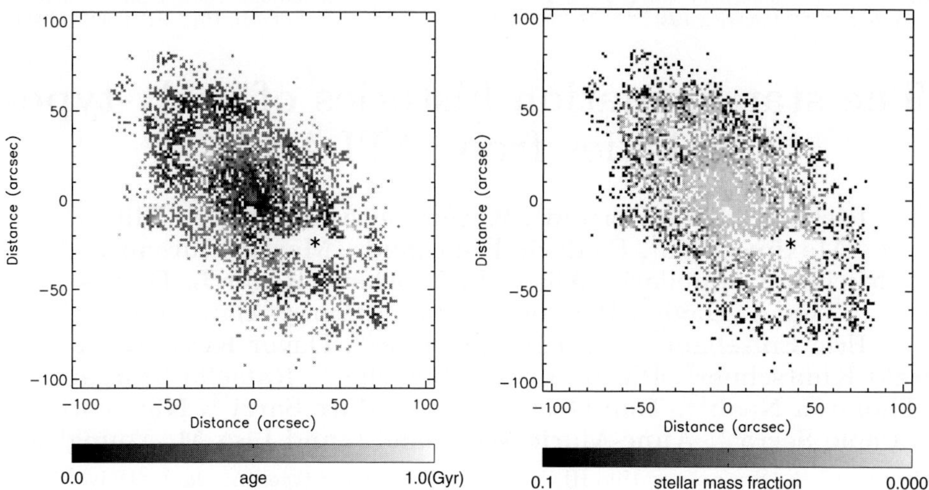

Figure 3. Two-component fit maps. A star masks the area contaminated by the foreground star.

and old) models from Yi (2003) and obtain a probability distribution of the age and mass fraction of the young stellar population. The UV data set is able to provide good constraints on the age and mass estimates of the young stellar populations in early-type galaxies. The inclusion of FUV data particulary yields a better fit to the model. The best estimates of one sample galaxy (NGC 2974) are shown in Figure 3. The age map shows very young stars (< 500 Myr) in both the central regions and in and around the UV rings. On the other hand, the mass fraction map shows that the fractional contribution of the young component increases outward. We calculate the total mass fraction of young stars in this galaxy to be around 1 %.

4. Future works

Star formation seems to have been more common in early-type galaxies than we used to believe. Early-type galaxies in the nearby universe have about 1–3% of their stellar mass in stars younger than 1 Gyr old. However, it is still unclear what is the main driver of star formation in early-type galaxies. We will compare our results with the prediction from the hierarchical merger paradigm and attempt to understand the overall assembly history of early-type galaxies.

References

Bacon, R. *et al.* 2001, *MNRAS*, 326, 23

Bower R. G., Lucey J. R., & Ellis R. 1992, *MNRAS*, 254, 589

de Zeeuw, P. T. *et al.* 2002, *MNRAS*, 329, 513

Djorgovski, S. & Davis, M. 1987, *ApJ*, 313, 59

D'Onofrio, M., Valentinuzzi, T., Secco, L., Caimmi, R., & Bindoni, D. 2006, *NewA Rev.*, 50, 447

Dressler, A., Lynden-Bell, D., Burstein, D., Davies, R. L., Faber, S. M., Terlevich, R. J., & Vegner, G. 1987, *ApJ*, 313, 42

Jeong, H., Bureau, M., Yi, S. K., Krajnovi'c D., & Davies, R. L. 2007, *MNRAS*, 376, 1021

Jeong, H., *et al.* 2009, *MNRAS*, 398, 2028

Yi, S. K. 2003, *ApJ*, 582, 202

Yi, S. K. *et al.* 2005, *ApJ*, 619, L111

The Spectral Energy Distribution of Galaxies
Proceedings IAU Symposium No. 284, 2011
R.J. Tuffs & C.C. Popescu, eds.

doi:10.1017/S1743921312009167

The star-formation histories of early-type galaxies from ATLAS^{3D}

Richard M. McDermid[1], Katherine Alatalo[2], Leo Blitz[2], Maxime Bois[3], Frédéric Bournaud[4], Martin Bureau[5], Michele Cappellari[5], Alison F. Crocker[6], Roger L. Davies[5], Tim A. Davis[7], P.T. de Zeeuw[7,8], Pierre-Alain Duc[4], Eric Emsellem[7,9], Sadegh Khochfar[10], Davor Krajnović[7], Harald Kuntschner[7], Pierre-Yves Lablanche[7,9], Rafaella Morganti[11,12], Thorsten Naab[13], Tom Oosterloo[11,12], Marc Sarzi[14], Nic Scott[15], Paolo Serra[11], Anne-Marie Weijmans[16], and Lisa M. Young[17]

[1]Gemini North Observatory, Hilo HI, USA (email: rmcdermid@gemini.edu), [2]University of California, Berkeley, USA, [3]Observatoire de Paris, Paris, France, [4]Laboratoire AIM Paris-Saclay, Gif-sur-Yvette Cedex, France, [5]University of Oxford, UK, [6]University of Massachusetts, Amherst, USA, [7]European Southern Observatory, Garching, Germany, [8]Sterrewacht Leiden, the Netherlands, [9]Observatoire de Lyon, Saint-Genis Laval, France, [10]Max Planck Institut für extraterrestrische Physik, Garching, Germany, [11]Netherlands Institute for Radio Astronomy (ASTRON), Dwingeloo, The Netherlands, [12]University of Groningen, The Netherlands, [13]Max-Planck-Institut für Astrophysik, Garching, Germany [14]University of Hertfordshire, UK [15]Centre for Astrophysics and Supercomputing, Swinburne University of Technology, Australia [16]Dunlap Institute for Astronomy & Astrophysics, Toronto, Canada, [17]New Mexico Institute of Mining and Technology, Socorro, USA

Abstract. We present an exploration of the integrated stellar populations of early-type galaxies (ETGs) from the ATLAS^{3D} survey. We use two approaches: firstly the application of line-indices interpreted through single stellar population (SSP) models, which provide a single value of age, metallicity and abundance ratio. And secondly, by fitting a linear combination of SSP *spectra* to our data, smoothly weighted in the free parameters of age and metallicity, thereby inferring a star-formation history of these galaxies. Despite the significant differences in these approaches, we obtain generally consistent results, such that galaxies that are more massive appear older with enhanced abundance ratios using line indices, and have shorter star-formation histories weighted to early times. We highlight two limitations of the index-SSP approach. Firstly the SSP-equivalent ages belie the fact that ETGs are overwhelmingly composed of ancient stars. Secondly, the young stellar contributions implied in our star formation histories are *required* to obtain realistic UV-optical colours. We remark that, even fitting solar-abundance models, we can recover a star-formation duration that correlates with the measured alpha-enhancement, in agreement with other recent work.

Keywords. galaxies: elliptical and lenticular, cD; evolution; formation; stellar content

---

1. Introduction

The measurement of absorption features in the integrated light of galaxies has been a key tool to understanding the chemistry and star-formation history of these complex systems. Early-type galaxies (ETGs: ellipticals and lenticulars) have been particularly well studied with this technique, thanks to their relative lack of dust extinction and nebular emission which obfuscate the stellar continuum. These galaxies contain the so-called 'fossil record' of a Hubble Time of evolution, being among the most massive and ancient systems in the Universe. Untangling 13 Gyr of star formation and evolution is

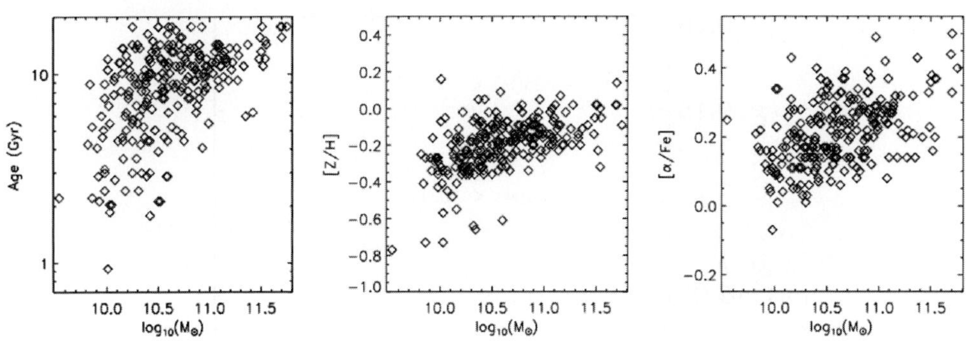

Figure 1. *From left to right:* SSP-equivalent values of stellar age, metallicity and abundance ratio for the ATLAS^{3D} sample as inferred from the measurement of absorption line indices, and finding the best-fitting SSP model. All population parameters are plotted as a function of dynamical mass. Indices were measured for each galaxy using a single spectrum integrated within an aperture of one effective radius.

complicated by the strong degeneracies of the key free parameters, such as age, metallicity and abundance ratios, not to mention that we observe these properties integrated in space and time. The degeneracies can be broken by using specific absorption lines to track the stellar chemistry, and relate these combined properties to predictions from stellar population models. These models mimic the time-varying properties of ensembles of stars consistent with an assumed initial mass function (IMF) and a single shared chemistry. In this way, a set of measured absorption line strengths, or *indices*, can be compared to these single stellar population (SSP) models, and the free parameters of age, metallicity and abundance ratios inferred.

In the past decade, significant effort has gone into producing predictions for stellar population *spectra*, instead of single line indices. These model spectra can be directly fitted to the observational data, avoiding many of the limitations of measuring indices. This is a powerful approach, but is more directly limited to the properties of the models, who's deficiencies may be hidden by the wish to numerically fit all features of the data set - even those containing no interesting information. In particular, the current generation of population model spectra are largely confined to solar element abundance ratios. These limitations notwithstanding, SSP spectra allow a key development, which is to solve the inverse problem of a galaxy's star-formation history (SFH), being naturally amenable to computing combinations of populations to reproduce the observed galaxy properties.

In this contribution, we explore the application of both techniques to spectroscopy from the ATLAS^{3D} complete survey of 260 nearby ETGs obtained using the SAURON integral-field spectrograph. To measure the SFH, we use the penalized pixel-fitting code pPXF (Cappellari & Emsellem 2004) with the addition of linear regularization constraints (Press *et al.* 1992) to fit linear combinations of SSP spectra from Vazdekis *et al.* (2010), such that variations in the contributions from neighbouring time and metallicity intervals are minimized. The degree of regularization is increased, giving a smoother SFH, until the fit to the spectra is no longer acceptable, as measured via chi-squared. In this way we obtain a 'maximally-smooth' SFH that is consistent with the data.

2. Results

Figure 1 shows the distribution of SSP-equivalent ages, metallicities and abundance ratios of the ATLAS^{3D} galaxies as derived from the line indices measured from an integrated aperture spectrum for each galaxy, as a function of mass. The known trends of increasing age, metallicity and abundance ratios with mass are clearly recovered Thomas

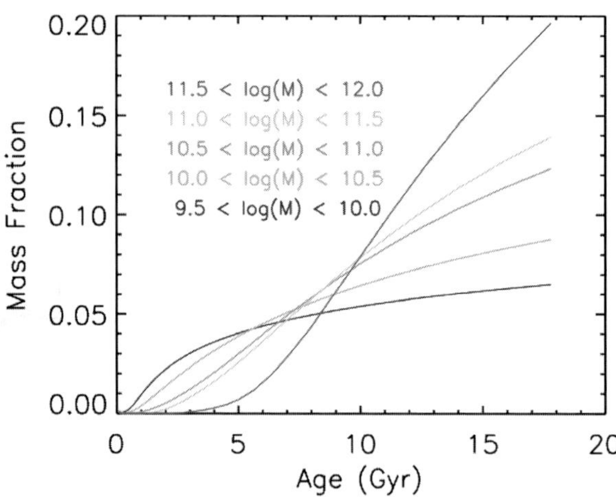

Figure 2. Average star formation histories of ATLAS^{3D} galaxies for five bins of galaxy mass.

et al. (2005). We note the large scatter in age values, spanning 1 to 13 Gyr at some masses. These are SSP-equivalent values of age which do not necessarily reflect the mass-contributions from different populations, making it hard to interpret the distributions in terms of the history of star formation. There is, however, a clear trend that less massive galaxies are younger and have lower abundance ratios, implying a more protracted SFH.

Figure 2 shows the average measured SFH derived from our spectral fitting technique in five bins of mass. The SFHs are normalized within each mass bin to have an integral of unity. This shows what we expect from our SSP analysis: as we decrease in mass, the star formation histories become broader, extending to younger ages. Note, however, that contrary to the classic picture of Thomas *et al.* (2005), the largest stellar age contributions by mass are in the oldest bins, on average. This seems consistent with recent simulations finding galaxy formation is characterized by two-phases: a rapid early phase of gas infall at $z > 2$, followed by extended accretion at later times (Oser *et al.* 2010).

It is remarkable that, despite fitting solar-abundance models, we recover essentially the same star-formation timescale behavior as implied by the degree of alpha-enhancement measured from the indices and non-solar abundance models. The spectral fits are quite accurate around the Mg *b* feature, though clearly they cannot fit accurately both Fe and Mg *b* in the most over-abundant cases. We note that recently this has even been proposed as a *calibration* of the $[\alpha/\mathrm{Fe}]$ parameter (de la Rosa *et al.* 2011). The success in recovering a sensible star-formation timescale with solar models is likely due to the coupling of the age and alpha-enhancement parameters, as evident in Figure 1.

Our SAURON spectra cover a very short wavelength range (≈4800-5400Å). It is therefore interesting to know how well our derived SFHs predict the galaxy SED at other wavelengths. Figure 2 compares the central spectrum of NGC3379 from the SDSS with the SSP spectrum implied from our SSP analysis (green) and the SFH derived from fitting the independent SAURON spectrum (magenta), using the models of Bruzual & Charlot (2003). The fit region is indicated by vertical dashed lines. The SFH is shown in the insert, where the green vertical line shows the SSP-equivalent age. This shows some typical features seen in other cases, namely that the SED inferred from fitting the short SAURON spectrum does a remarkably good job at predicting the SED at other wavelengths. More importantly, the SFH and SSP-based predictions differ most significantly in the UV, due to the small (by mass) contribution of young populations afforded by the SFH approach.

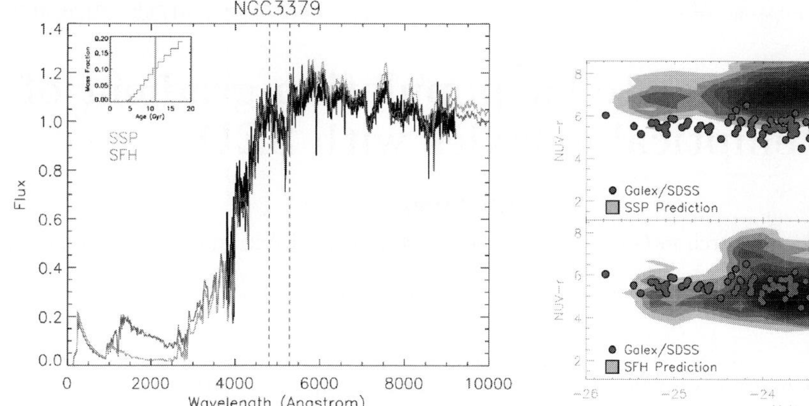

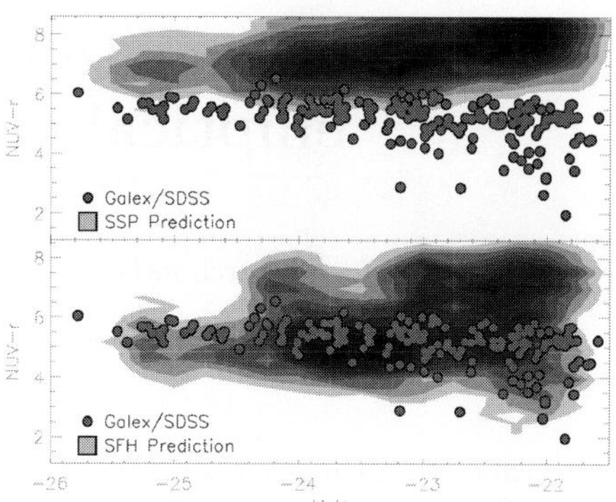

Figure 3. *Left*: Comparison of SDSS spectrum of NGC3379 with stellar population models derived from our **SAURON** spectra, assuming a single SSP and composite SFH. *Right*: Comparison of observed total UV-optical colours from GALEX and SDSS versus K-band absolute magnitude from 2MASS for the ATLAS^{3D} sample (circles) and predictions (regions) from the SSP-equivalent spectrum (top) and the fitted SFH (bottom).

The right-hand panel of Figure 2 shows two (total) UV-r colour-magnitude diagrams using archival GALEX, SDSS and 2MASS data for our sample. Shaded regions indicate the predicted colour distributions based on the SSP-equivalent spectra (*top*) and the SFH spectra inferred from our SAURON data (bottom). The SFH does a significantly better job in predicting the UV-r colours. For some galaxies, this could be a feature of the stellar population models used, as the UV range has uncertain contributions from old metal-poor stars and is still a matter for debate. However, the improvement occurs for all objects, suggesting that the need for some small fraction of recently-formed stars is real. The photometry apertures are not matched to the spectroscopy in this analysis, but most ETGs are centrally blue in the UV, thus we underestimate the discrepancy here.

3. Conclusions

In this short contribution, we compare the SSP and SFH properties of the ATLAS^{3D} sample of ETGs, and conclude that the measured SFHs are consistent with the picture inferred from SSPs, namely that galaxies of increasing mass have a shorter star-formation duration, completed at early times. In addition, we directly show that, on average, the majority of stellar mass in all masses of ETGs in our sample are dominated by old stars. We also show that the spread in stellar ages afforded by considering the SFH generally gives a much better prediction of the UV-optical colours of ETGs as compared to SSPs.

References

Bruzual, G. & Charlot, S. 2003, *MNRAS*, 344, 1000
Cappellari, M. & Emsellem, E. 2004, *PASP*, 116, 138
de La Rosa, I. G., La Barbera, F., Ferreras, I., & de Carvalho, R. R. 2011, *MNRAS*, 418, L74
Oser, L., Ostriker, J. P., Naab, T., Johansson, P. H., & Burkert, A. 2010, *ApJ*, 725, 2312
Press, W. H., Teukolsky, S. A., Vetterling, W. T., & Flannery, B. P. 1992, Cambridge: University Press, 1992, 2nd ed.
Thomas, D., Maraston, C., Bender, R., & Mendes de Oliveira, C. 2005, *ApJ*, 621, 673
Vazdekis, A., Sánchez-Blázquez, P., Falcón-Barroso, J., *et al.* 2010, *MNRAS*, 404, 1639

The Spectral Energy Distribution of Galaxies
Proceedings IAU Symposium No. 284, 2011
R.J. Tuffs & C.C. Popescu, eds.

doi:10.1017/S1743921312009179

Constrain the stellar population gradient of elliptical galaxies with SED

Zhengyi Shao

Key Laboratory for Research in Galaxies and Cosmology, Shanghai Astronomical Observatory, CAS, Shanghai 200030, China
Key Lab for Astrophysics, Shanghai 200234,China
email: zyshao@shao.ac.cn

Abstract. We use multiple wavelength photometric data, from Optical to NIR, to determine the radial gradient of stellar populations of nearby elliptical galaxies. Empirical model, such as the Sersic profile, is assumed to describe the radial stellar distribution, and the metallicity gradient is parameterlized in this model. The PSF effects of different wavebands are also involved to predict the photometric profiles which can be compared with observational surface brightness profile directly. For most of elliptical galaxies, they have negative gradients, which means their central parts is metal rich than outer parts. We also found the gradient is significantly correlated to the stellar mass of the whole galaxy. More massive elliptical galaxies have shallower gradients. But for our sample galaxies, their is no correlations between Sersic n and mass, and between Sersic n and gradient can be found.

Keywords. galaxies: elliptical galaxies: photometry galaxies: abundances

---

1. Introduction

It is well-known that there exist the radial stellar population gradients in the early type galaxies (ETGs). This can be seen in the spectral absorption line indices and the shallow broad band color gradients. Although the spectral absorption line indices are the best tracers of the stellar population, the difficulties of the line observation confine the measurement within a range of one effective radius of galaxies. However, the measurement of the color gradient can extend to several effective radius or even farther. The moderate consistent of the dust free spectral absorption line indices gradient and the broadband color gradient indicate that photometric measurement is still one of the most important method to study the stellar population gradient in ETGs. It will give restrictions on the model of galaxy formation and evolution.

The Sloan Digital Sky Survey (SDSS) and the Two Micron ALL Sky Survey (2MASS) produce large and homogeneous databases and high quality images in five optical and three near infrared bands. This provides an opportunity to study the stellar population gradients with a large sample of nearby galaxies. The spectral energy distribution (SED) from the optical to NIR, at different radius will give us much more information about the stellar population distribution than before. Thus we can quantitatively constrain stellar population gradients of individual galaxies.

2. Sample galaxies and photometric data

Nair & Abraham (2010) classified SDSS(DR4) galaxies with $g < 16.0$ and $0.01 < z < 0.1$ by naked eye. We further select brighter ($g < 15.0$) ETGs ($T < 0$) from their sample, and matched them with 2MASS extend sources. Finally, we got 2123 ETGs with photometric data of eight broad wavebands from optical to NIR ($ugrizJHK_s$). Photometric profiles are used in this work. For SDSS, we use *profmean* values, and for 2MASS we use round aperture magnitudes. Typically, the outer most SDSS annulus and 2MASS aperture can reach 30 and 50 arcsec respectively. It is almost $6 \sim 10$ times as large as the effective radius of our sample galaxies.

3. Method

In this work, we introduce a new approach to modelling the galaxy profile. Instead of fitting galaxies in different bands separately, we tried to combine observational data of all given wave bands.

We assume the stellar column density profile (not the surface brightness of a single band as commonly used before) of an ETG follows the Sersic profile. Giving a set of parameters, such as stellar mass (M_{star}), Sersic index (n), effective radius (R_e), metallicity (Z_{met}) and age (t), and use single stellar population (SSP) model of synthesized spectra (BC03), we can predict the observational profiles in each band, such as *profmens* of SDSS and aperture magnitudes of 2MASS. Using this method, we can fit these parameters simulatively. Also, an additional parameter should be involved to describe the gradient of stellar population. For simplicity, we only use the radial metallicity gradient ($g_Z \equiv \Delta \log Z_{met}/\Delta \log R$) to represent all gradient effects. Because as Wu et. al. (2005) concluded that metallicity dominates the radial gradients of stellar population of ETGs.

During the fitting, PSFs of different bands are all considered in calculating their surface brightness profiles separately. The axis ratio of isophote at g band is also involved to correct the elliptical shape effect of the galaxy.

4. Results

Fig. 1 shows our fitting results of ETG's shape of stellar column density (n) and metallicity gradient. Sersic index n are mainly distributed from 3.1 To 4.7, with median value of 3.8, which pretty well appears as de Vaucouleurs shape. Radial metallicity gradients are distributed from -0.49 to -0.03, with median value of -0.24. This is in good agreement with Wu et. al. (2005). There are 12% sample galaxies have positive gradients, corresponding to the bluer core ETGs. This could be due to a recent star forming in the central region, though we only assume a metallicity gradients.

In Fig. 2, we can find a significant correlation between stellar mass and radial gradient. It shows more massive ETGs have shallower gradients. Also, form this figure, we can find

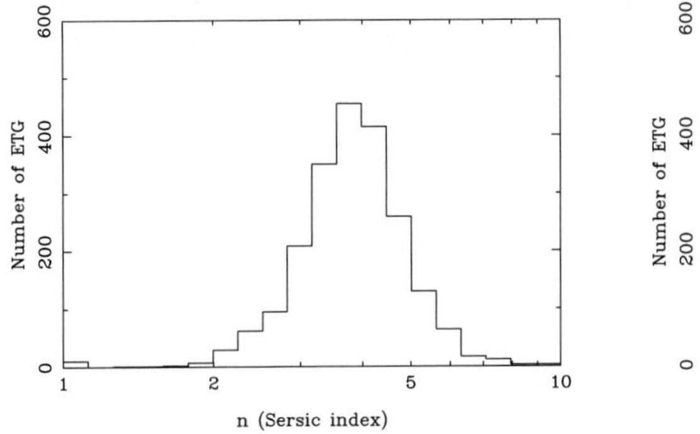

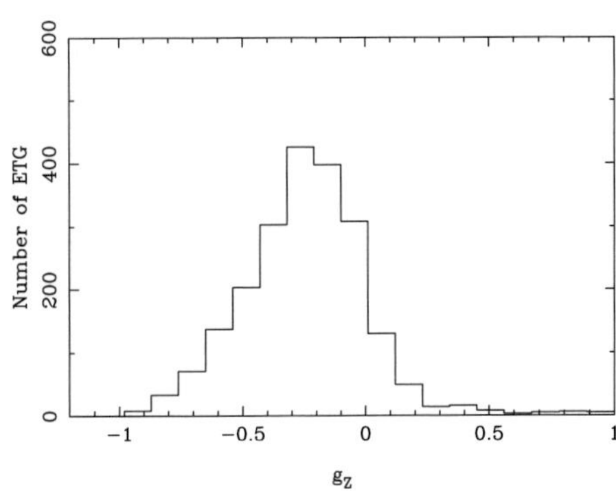

Figure 1. Distribution of fitted Sersic index n and metallicity gradient of ETGs

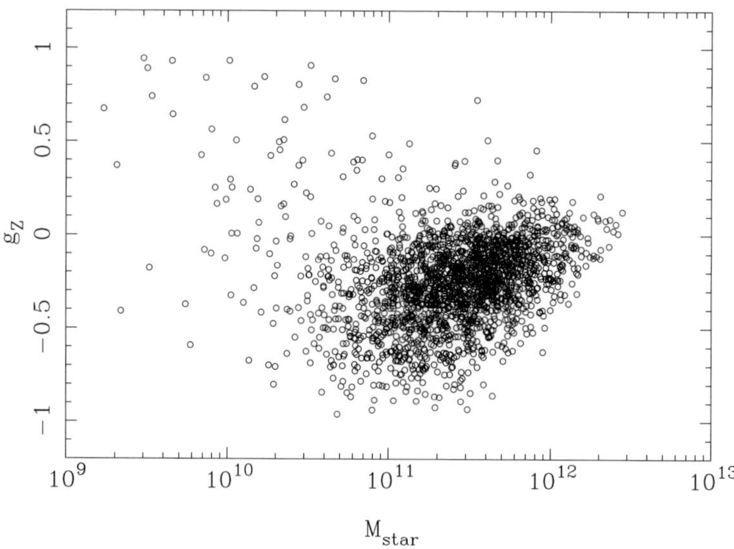

Figure 2. Correlation between stellar mass and metallicity gradient

most of the galaxies with positive value of gradients are smaller ones, and they do not follow this relationship of bigger galaxies.

5. Discussions

Our results confirm that only assume the stellar column density profile of ETGs, their observational surface brightness profile can be well predicted. The radial distribution of stellar mass can also be well modelled by Sersic profile, and their radial gradient of stellar population can be constrained.

The scatter of g_Z, $\sigma_{g_z} \geqslant 0.2$ is relatively large that can not be only explained by the uncertainties of photometric measurement or fitting approach. So, the scatter must be intrinsic. The formation process such as major and minor merger may all play as important roles, and the environment of these galaxies are also need to be taken into account.

An significant correction between stellar mass and radial gradient is found for most of our sample galaxies, except for those smallest ETGs with positive gradient. There is no correction between stellar mass and Seraic n, and no correction between gradient and n. These all imply that the mass is the most dominated factor during the ETG formation and evolution process.

References

Bruzual G. & Charlot S. 2003, *MNRAS*, 344, 1000

Nair, P. B. & Abraham, R. G. 2010, *ApJS*, 186, 427

Wu H., Shao Z. *et al.* 2005, *ApJ*, 622, 244

The Spectral Energy Distribution of Galaxies
Proceedings IAU Symposium No. 284, 2011
R.J. Tuffs & C.C. Popescu, eds.

doi:10.1017/S1743921312009180

Outer stellar disks of lenticular galaxies

Olga K. Sil'chenko

Sternberg Astronomical Institute of the Lomonosov Moscow State University,
University av. 13, 119991, Moscow, Russia
email: olga@sai.msu.su

Abstract. By studying the radial distribution of the properties of stellar populations in 15 nearby S0 galaxies, I have found that the outer stellar disks are mostly old, with SSP-equivalent ages of 8–15 Gyr, often being older than the bulges. This fact throws doubt on the currently accepted paradigm that S0 galaxies were formed at $z = 0.4$ through the quenching of star formation in spiral galaxies.

Keywords. galaxies: elliptical and lenticular, cD, galaxies: evolution, galaxies: formation

1. Introduction

Lenticular galaxies were introduced by Edwin Hubble (1936) as an intermediate type between ellipticals and spirals: they have large-scale stellar disks, as spirals do, but lack blue spiral arms and HII-regions, and they appear smooth and red, as ellipticals do. Nowadays, lenticulars are thought to have been (trans-)formed from spirals through the removal of gas from their disks, and the quenching of star formation. Also, dynamical heating is required to make the S0 disks stable against spiral wave perturbations. There is some evidence that this transformation might have taken place at a redshift of $z \approx 0.4$, within dense environments, groups or clusters, through a sudden shift in the dominance of blue (spiral?) galaxies to a dominance of S0 galaxies (e.g., Fasano *et al.* 2000, Wilman *et al.* 2009). But if this scenario is valid, the star formation in the outer disks of nearby S0s should have been going on up to only 4 Gyr ago, and we ought to be able to see intermediate-age stellar populations there now. Using deep long-slit spectroscopy, I have studied the stellar population properties of several nearby S0 galaxies as a function of radial position over several scalelengths of their large-scale disks to check the validity of this paradigm.

2. Sample

The sample consists of nearby lenticular galaxies for which deep long-slit spectra were obtained at the Russian 6m telescope over the last five years in the context of several observational programs. The main part of the sample are edge-on lenticular galaxies selected for kinematical study by Natalia Sotnikova which have been observed in the context of her observing proposal; I use here these data to derive Lick indices. Four moderately inclined S0 galaxies represent a part of our sample of nearby early-type disk galaxies – group members whose central parts have been studied earlier with the Multi-Pupil Fiber Spectrograph of the 6m telescope (Sil'chenko 2006). The galaxies are homogeneously distributed over luminosity - their blue absolute magnitudes being spread from –19 to –21 - and density of environment. We have one galaxy (NGC 4570) in the Virgo cluster where the influence of the intracluster medium is unavoidable, and one galaxy (NGC 4111) in the Ursa Major Cluster, where X-ray gas is not detected. Among group galaxies, NGC 524 and NGC 5353 are central group galaxies embedded in X-ray haloes; NGC 5308 is a member galaxy of an X-ray bright group, and NGC 502 and IC 1541, though members of rich groups, lie outside their X-ray halo (Osmond & Ponman 2004, according to our checks of archival ASCA images. NGC 3414 is the central galaxy of a rich group undetected in X-ray, while NGC 2732 is a host of a few faint satellites. NGC 1029, NGC 2549, and NGC 7332 are in triplets. By using the NED environment searcher, we found no galaxies within 300 kpc of NGC 1032 and NGC 1184, so take them to be isolated field galaxies.

3. Observations

The long-slit spectral observations were made with the focal reducer SCORPIO (Afanasiev & Moiseev 2005) installed at the prime focus of the Russian 6m telescope (at the Special Astrophysical Observatory of the Russian Academy of Sciences). We exposed a rather narrow spectral range rich in absorpion lines, 4800–5500 Å, which is quite suitable for the study of stellar kinematics and stellar population properties. The slit width was one arcsecond and the spectral resolution about 2 Å. The 2k×2k and 2k×4k CCDs were used as detectors, and the scale along the slit was 0.36 arcsec per pixel. The slit length was about 6 arcminutes so at the edges of the slit we could measure the sky background to subtract from the galaxy spectra. Inhomogeneties in the transparency of the optics and in spectral resolution along the slit were checked with the twilight exposures. The Lick index system was calibrated by observing the standard Lick stars (Worthey *et al.* 1994) with the same instrumental setup as used for the observations of the galaxies. We calculated the Lick indices H-beta, Mgb, Fe5270, and Fe5335 along the slit up to several exponential scalelengths of the disks. To estimate the radial variations of the SSP-equivalent ages, metallicities and magnesium-to-iron ratios of the stellar populations, I compared the measured Lick indices to the models of old stellar populations by Thomas *et al.* (2003).

4. Results and Discussion

I have measured the Lick indices in 15 S0s along the radius beyond 2–4 exponential scalelengths of their disks, and have estimated the ages and abundances of the stellar populations for the bulges and for the disks. The bulges have solar metallicities or higher, and the disks have solar metallicities or lower. The disk stellar populations demonstrate very high magnesium-to-iron ratios so cannot be descended from prolonged star-formation in spiral galaxies. The SSP-equivalent ages of the disks are mostly old – 60% of the sample demonstrate ages older than 10 Gyr, – and are almost always (except in one case) larger than the SSP-equivalent ages of their corresponding bulges (Fig. 1).

The parameters of the stellar populations of the disks do not show any correlation with the luminosities or masses of the disks. The only correlation found was with the photometric disk scaleheight taken from the decomposition results by Mosenkov *et al.* (2010). The whole scaleheight range is from 0.3 kpc (found in the disks with the ages of 2–3 Gyr and [Mg/Fe]$< +0.2$) to 0.6–0.9 kpc (found mostly in the disks with the ages > 10 Gyr and [Mg/Fe]$> +0.3$). All the old disks of our sample galaxies belong to the latter thick disk catagory, while the only two young outer disks, those of NGC 4111 and NGC 7332, are certainly thin disks. The disk of NGC 2732, with its SSP-equivalent age of 8 Gyr and scaleheight of 0.5 kpc, is halfway between the thin and thick disks.

Although, many arguments have provided evidence that S0s and spiral galaxies are related, these present results indicate that S0s are progenitors of spirals, opposite to what was thought before. Indeed, if we compare stars of the thick disk of our own Galaxy with

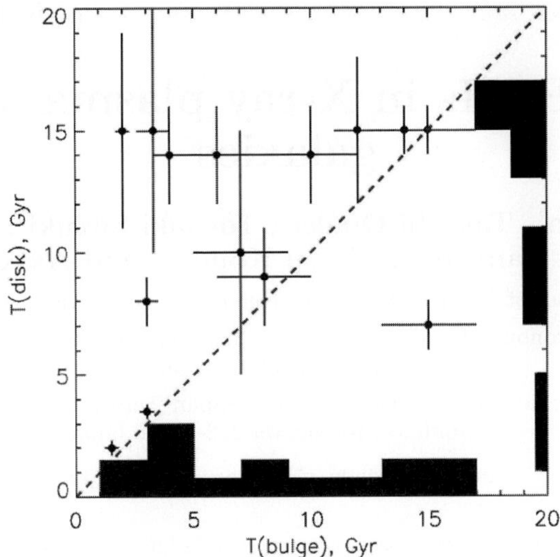

Figure 1. Comparison between the SSP-equivalent ages of the bulges, at $0.5r_e$, and of the large-scale disks for the sample lenticulars. The dashed line is the line of equality; the filled histograms attached to the axes characterize the distributions of the ages of the bulges and disks.

those in the thick outer stellar disks of the S0s studied here we see a full resemblance, with ages of > 10 Gyr, $[\mathrm{Mg/Fe}] > +0.2$, and a total metallicity $[\mathrm{Z/H}]$ between 0.0 and -0.7 (such parameters of the thick disk of our Galaxy are found by e.g. Bernkopf & Fuhrmann 2006 or by Schuster *et al.* 2006). So, if one provides fresh cold gas accretion into the disks of these S0s, after several Gyrs of star formation we would get a typical spiral galaxy, with the thick old stellar disk and thin younger stellar disk. I propose now the following evolutionary sequence: All disk galaxies were S0s immediately after their birth; later, at $z < 1$, some of them acquired cold gas accretion sources – these became spirals, – and some of them failed to find such sources – those remained lenticulars. In large cluster-size and group-size dark haloes, there is little chance to find external sources of cold gas accretion, due to the hot gas effect, – so in nearby clusters the dominant disk-galaxy population are S0s. It remains an open question as to the nature of these sources of cold-gas prolonged accretion. They may be filaments formed in the cosmological development of structure (Dekel & Birnboim 2006) or rich systems of irregular-type dwarf satellites which have been merging with the host one after another.

References

Afanasiev, V. L. & Moiseev A. V. 2005, *Astron. Lett.*, 31, 194
Bernkopf, J. & Fuhrmann K. 2006, *MNRAS*, 369, 673
Dekel, A. & Birnboim, Yu. 2006, *MNRAS*, 368, 2
Fasano, G., Poggianti, B. M., Couch, W. J., *et al.* 2000, *ApJ*, 542, 673
Hubble, E. 1936, *The Realm of the Nebulae* (New Haven: Yale Univ. Press)
Mosenkov, A. V., Sotnikova, N. Ya., & Reshetnikov, V. P. 2010, *MNRAS*, 401, 559
Osmond, J. P. F. & Ponman T. J. 2004, *MNRAS*, 350, 1511
Schuster, W. J., Moitinho, A., Marquez, A., Parrao, L., & Covarrubias, E. 2006, *A&A*, 445, 939
Sil'chenko, O. K. 2006, *ApJ*, 641, 229
Thomas, D., Maraston, C., & Bender, R. 2003, *MNRAS*, 339, 897
Wilman, D. J., Oemler, A., Mulchaey, J. S., *et al.* 2009, *ApJ*, 692, 298
Worthey, G., Faber, S. M., González, J. J., & Burstein, D. 1994, *ApJS*, 94, 687

The Spectral Energy Distribution of Galaxies
Proceedings IAU Symposium No. 284, 2011
R.J. Tuffs & C.C. Popescu, eds.

doi:10.1017/S1743921312009192

Dust and PAHs in X-ray plasma of elliptical galaxies

Hidehiro Kaneda[1], Takashi Onaka[2], Toyoaki Suzuki[3], Tatsuya Mori[1], Mitsuyoshi Yamagishi[1], Toru Kondo[1], and Akiko Yasuda[1]

[1]Graduate School of Science, Nagoya University, Nagoya, Aichi 464-8602, Japan

[2]Department of Astronomy, Graduate School of Science, University of Tokyo, Bunkyo-ku, Tokyo 113-0003, Japan

[3]Institute of Space and Astronautical Science, Japan Aerospace Exploration Agency, Sagamihara, Kanagawa 252-5210, Japan

email: `kaneda@u.phys.nagoya-u.ac.jp`

Abstract. Many elliptical galaxies possess an appreciable amount of X-ray-emitting hot plasma, providing a harsh interstellar environment for the survival of dust grains and polycyclic aromatic hydrocarbons (PAHs). Despite such a hostile environment, it has been found that a significant fraction of X-ray elliptical galaxies contain a considerable amount of dust, which cannot be explained solely from replenishment by old stars. Some of them even show the presence of PAHs. We present the results of AKARI and Spitzer observations of dust and PAHs in X-ray elliptical galaxies. We investigate their possible origins and discuss the implications of their presence for the evolution of elliptical galaxies.

Keywords. ISM: dust, extinction — infrared: galaxies — galaxies: elliptical and lenticular, cD — galaxies: ISM

1. Introduction

The interstellar environment of elliptical galaxies is characterized by the dominance of X-ray-emitting hot plasma and old stellar radiation fields with little UV. Submicron dust in X-ray plasma is easily destroyed by sputtering on timescales of $10^6 \sim 10^7$ yr (Draine & Salpeter 1979), while old stars cannot replenish a large amount of dust into the interstellar space (Knapp *et al.* 1992). Despite such hostile conditions, a significant fraction of elliptical galaxies contain a considerable amount of dust (e.g., Knapp *et al.* 1989; Goudfrooij & de Jong 1995; Temi *et al.* 2007a). Spitzer revealed the presence of polycyclic aromatic hydrocarbons (PAHs; Kaneda *et al.* 2008), which was even more surprising because PAHs are much more easily destroyed on timescales of $\sim 10^2$ yr (Micelotta *et al.* 2010). In this paper, we present the results of AKARI and Spitzer observations of dust and PAHs in elliptical galaxies to investigate their possible origins.

2. Observations

We observed 18 nearby IRAS-detected elliptical galaxies with AKARI and Spitzer. For each target, we obtained near- to far-infrared (IR) 10-band images from the AKARI nearby galaxy program (Kaneda *et al.* 2009), and mid-IR ($5 - 36\ \mu$m) spectra from the 3 Guest Observers programs of Spitzer (PI: HK; PIDs: 3619, 30483, 50369). In addition, AKARI surveyed all the sky in the mid- and far-IR 6 bands, from which we obtained a complete sample of early-type galaxies with the far-IR flux limit of ~1 Jy. Figure 1 shows typical spectral energy distributions (SEDs) of elliptical galaxies in the IR to

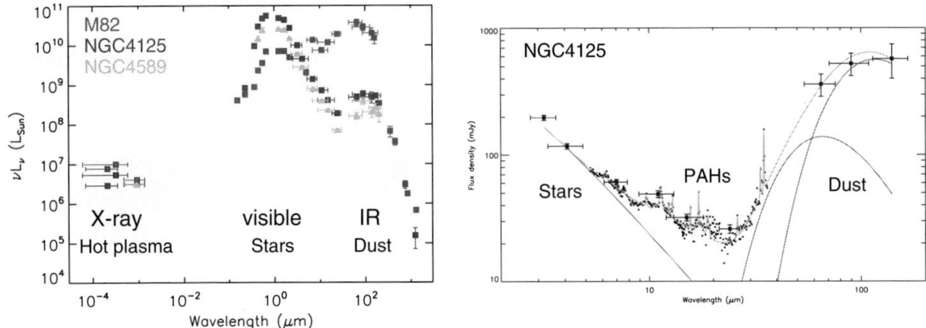

Figure 1. (Left) SEDs of the elliptical galaxies, NGC4125 and 4589, compared with that of the starburst galaxy M82. (Right) IR (2–200 μm) SED of the central 30″ region of NGC4125, composed of the AKARI photometric fluxes and the Spitzer/IRS spectrum (Kaneda *et al.* 2011).

X-ray, where the AKARI 10-band fluxes are plotted in the IR. As compared with the SED of the starburst galaxy M82, the elliptical galaxies are indeed very bright in the stellar emission in the visible to near-IR relative to dust emission in the mid- to far-IR, while their total X-ray luminosities are comparable with each other. The right panel in Fig. 1 shows the detailed IR (2–200 μm) SED of NGC4125, where we can recognize the presence of the PAH features. It should be noted that the spectrum shows an unusually faint PAH 7.7 μm feature. Most of our sample elliptical galaxies show similarly faint 7.7 μm features relative to prominent 11.3 μm and 17 μm features, probably reflecting the dominance of neutral PAHs due to soft radiation fields.

3. Results

The left panel in Fig. 2 shows the dust masses derived from the AKARI all-sky survey 90 and 140 μm fluxes for the sample of early type galaxies with the ~1 Jy far-IR flux limit, plotted against the B-band luminosity. The asterisks correspond to the results of the above 18 galaxies, for which we performed deep pointed observations with AKARI and Spitzer. The lines in the figure represent the dust masses expected by the balance between supply from old stars and sputtering in X-ray plasma with the denoted destruction timescales (Knapp *et al.* 1992). Thus the presence of the excess dust is beyond doubt. Besides the figure suggests a notable trend that the dust mass increases with the stellar luminosity, implying some connection between dust supplying sources and galaxy evolution. Most of the excess dust may originate in past galaxy mergers, because many elliptical galaxies are believed to have been evolved through mergers. However, considering the current smooth stellar distributions, look-back times for the mergers are typically $\gtrsim 10^9$ yr (e.g. Bournaud *et al.* 2008), much longer than the lifetime of dust. Thus the presence of dust suggests that they were likely supplied very recently.

It is known that the intensities of the PAH features are correlated well with those of the dust emission from galaxy to galaxy (Kaneda *et al.* 2008). The presence of PAHs even suggests that there are some ways for them to avoid interaction with the X-ray plasma. Therefore spatial information on the dust and PAHs is crucial to understand their origins. The right panels in Fig. 2 show examples of 90 μm contour maps of our sample galaxies, together with their 15 μm (PAHs + stars) contours and 3 μm (stars) images. Our results suggest that some galaxies reveal extended far-IR distributions of dust. Figure 3 shows the results of our Spitzer/IRS spectral mapping. With the spectroscopy, we unambiguously obtain the spatial distribution of PAHs in elliptical galaxies for the first time, which

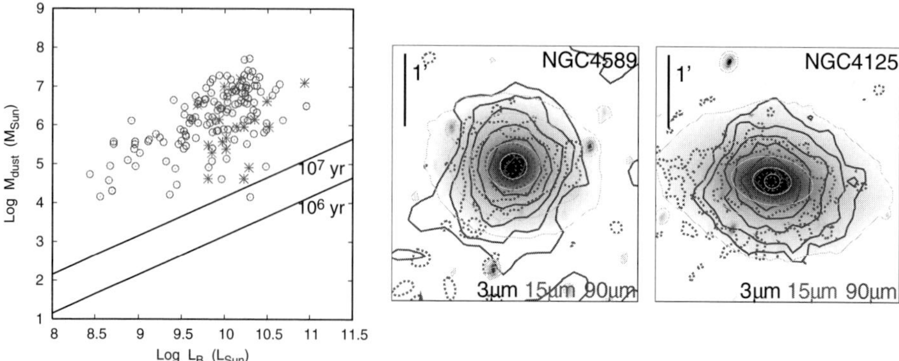

Figure 2. (Left) dust masses of early-type galaxies derived from the AKARI all-sky survey, plotted against their B-band total luminosity. (Right 2 panels) AKARI images of 2 elliptical galaxies in our sample; the 90 μm (solid lines) and 15 μm (dashed lines) contours are overlaid on the 3 μm band grey image, with logarithmic spacings from 80 to 10 % (90 μm) and 2 % (15 μm) of the peaks.

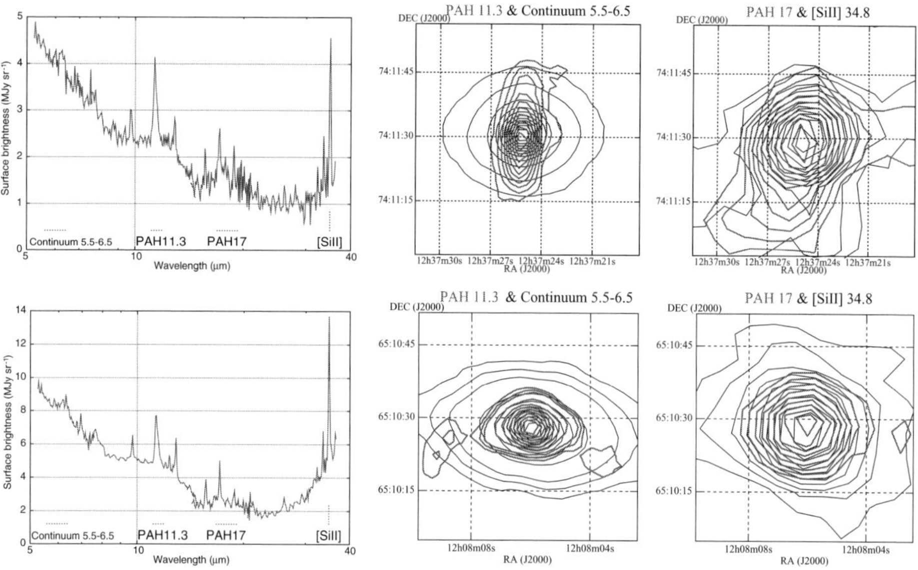

Figure 3. PAH spectral maps obtained with the Spitzer/IRS for NGC4589 in the upper and NCG4125 in the lower panels (Kaneda *et al.* 2010; 2011). (Left) the spectra of the central 15″ regions. (Middle) the contour maps of the PAH 11.3 μm feature (thick lines) on those of the 5.5–6.5 μm stellar continuum emission (thin lines). (Right) the contour maps of the PAH 17 μm feature (thick lines) on those of the [SiII] 34.8 μm line emission (thin lines). In each map, the contours are drawn on a linear scale from 10% to 90% of the peak.

reveals that the PAHs are compactly distributed near the centers of the galaxies. Large PAHs responsible for the 17 μm feature (Peeters *et al.* 2004) seem to be more extended than PAHs of typical sizes emitting the 11.3 μm feature, and far-IR dust grains are far more extended than PAHs because their emission is significantly extended even with poor spatial resolution in the far-IR. Furthermore the distributions of the dust and PAHs are entirely different from the stellar distribution (Figs. 2 and 3).

4. Discussion

One possibility to explain the extended excess dust is very minor mergers, i.e. accretion of dust-rich dense cold gas, which do not appreciably disturb the stellar distribution. Another possibility is outflows assisted by a low-luminosity AGN (LLAGN); we expect the presence of dust reservoirs near the center by analogy with the dusy tori of AGNs or from optical dust lanes. LLAGNs may produce buoyant outflows in response to infalling gas, distributing the dust from the central reservoirs to outer regions (Temi *et al.* 2007b). Alternatively X-ray plasma may not dominate the interstellar space, possibly localized by magnetic confinement, whereby dust survives for a long time. Our observational results suggest that larger grains are extended farther from the center, supporting the second possibility; for the outflow velocity of $\sim$400 km^{-1}, dust of 0.1 μm size reaches $\sim$ 5 kpc in a sputtering lifetime of $\sim 10^7$ yr (Temi *et al.* 2007b), while PAHs cannot be distributed because their destruction timescales are too short ($\sim 10^2$ yr; Micelotta *et al.* 2010). Near the center, the PAHs and dust are likely to have been protected from the interaction with the X-ray plasma inside the dense molecular gas (i.e. dust reservoirs). The trend seen in Fig. 1 would imply that the dust may have been recycled from the ISM brought in by mergers and being accumulated near the center. For NGC4125, softer X-ray emission is extended toward directions similar to the dust emission, indicating that the dust is interacting with the X-ray plasma, thus cooling the plasma (Kaneda *et al.* 2011). In fact, Spitzer detected very strong [SiII] 34.8 μm emission from many elliptical galaxies (see Fig. 3), suggesting abundant gas-phase Si through sputtering of silicate dust.

5. Summary

With AKARI and Spitzer, we have shown that considerable amounts of dust and PAHs exist in X-ray elliptical galaxies, which cannot be accounted for by stellar mass loss alone. For NGC4125 and 4589, we have spatially resolved PAH and dust emissions; PAHs are centrally concentrated, while dust is more distributed. We have obtained signature of interaction between the dust and X-ray plasma. Reservoirs of dust and PAHs seem to exist near the centers of the galaxies. The dust may be distributed by LLAGN-assisted outflow from the central reservoir to outer interstellar space, where the X-ray plasma is cooled. The PAHs cannot be distributed due to their short destruction timescale. To further verify this interpretation, we require a larger sample of elliptical galaxies in which both dust and PAHs are spatially resolved.

Acknowledgements

AKARI is a JAXA project with the participation of ESA.

References

Bournaud, F. Duc, P.-A., & Emsellem, E. 2008, *MNRAS*, 389, L8
Draine, B. T. & Salpeter, E. 1979, *ApJ*, 231, 77
Kaneda, H., Onaka, T., Sakon, I., *et al.* 2008, *ApJ*, 684, 270
Kaneda, H., Koo, B. C., Onaka, T., & Takahashi, H. 2009, *AdSpR*, 44, 1038
Kaneda, H., Onaka, T., Sakon, I., *et al.* 2010, *ApJ*, 716, L161
Kaneda, H., Ishihara, D., Onaka, T., *et al.* 2011, *PASJ*, 63, 601
Knapp, G. R., Guhathakurta, P., Kim, D.-W., & Jura, M. 1989, *ApJS*, 70, 329
Knapp, G. R., Gunn, J. E., & Wynn-Williams, C. G. 1992, *ApJ*, 399, 76
Goudfrooij, P. & de Jong, T. 1995, *A&A*, 298, 784
Micelotta, E. R., Jones, A. P., & Tielens, A. G. G. M. 2010, *A&A*, 510, A37

Peeters, E., Mattioda, A. L., Hudgins, D. M., & Allamandola, L. J. 2004, *ApJ*, 617, L65
Temi, P., Brighenti, F., & Mathews, W. G. 2007a, *ApJ*, 660, 1215
Temi, P., Brighenti, F., & Mathews, W. G. 2007b, *ApJ*, 666, 222

Discussion

WANG: In one of your 70 μm images, the dust emission distribution seems to be much narrower than that of the X-ray morphology. Do you have an explanation?

KANEDA: The dust is likely to be outflowing from the centre, being destroyed by spultering in X-ray plasma. The 70 μm image traces smaller grains, and their destruction timescales are shorter. That's why the 70 μm image shows a narrower distribution.

GALLAGHER: There seems to be a range in the optically detected dust in the Ellipticals - some cases with only compact central structures, and others with spatially extended dust lanes. Do you see any differences in the PAH properties associated with the scale of dust distributions in ellipticals?

KANEDA: We don't see significant differences in the PAH properties.

FERRERAS: In the M_{dust} vs. L_B plot, all galaxies fall above the expected value for mass loss. How do you reconcile this result with the fact that 2/3 of the sample from Martin Bureau have no misalignement of the CO kinematics?

KANEDA: It is probably due to the relatively high detection limit of the AKARI all-sky survey. The galaxies must be relatively bright in the FIR.

The Spectral Energy Distribution of Galaxies
Proceedings IAU Symposium No. 284, 2011
R.J. Tuffs & C.C. Popescu, eds.

doi:10.1017/S1743921312009209

Dusty early-type galaxies and passive spirals

Kate Rowlands, Loretta Dunne, Steve Maddox & the Herschel-ATLAS and GAMA collaborations

School of Physics & Astronomy, The University of Nottingham, University Park Campus, Nottingham, NG7 2RD, UK

email: ppxkr@nottingham.ac.uk

Abstract. Early-type galaxies (ETGs) are thought to be devoid of dust and star-formation, having formed most of their stars at early epochs. We present the detection of the dustiest ETGs in a large area blind submillimetre survey with Herschel (H-ATLAS, Eales *et al.* 2010), where the lack of pre-selection in other bands makes it the first unbiased survey for cold dust in ETGs. We compare to a control sample of optically selected ETGs to investigate how the two populations are different. We also highlight the properties of an interesting population of passive spirals detected by Herschel.

Keywords. galaxies: evolution, galaxies: elliptical and lenticular, cD, (ISM:) dust, extinction, infrared: galaxies, submillimeter

---

1. Sample selection

The parent sample of H-ATLAS galaxies in this study have a 5σ detection at 250μm, a reliable optical counterpart to the submillimetre source (see Smith *et al.* 2011a) and a spectroscopic redshift, many of which are taken from the GAMA survey (Driver *et al.* 2011). This gave rise to a total of 1087 galaxies, which were visually classified by eye using SDSS images into four categories: early-type, late-type, merger and unknown. ETGs were identified by looking for a dominant bulge and a complete lack of spiral arms, and late-types were identified by the presence of spiral arms. Due to the shallow depth of the SDSS images, we do not discriminate between E and S0 types. In addition, a control sample of 1052 galaxies which were not detected at 250μm was constructed; these galaxies were chosen to have the same redshift and r-band magnitude distribution (r, z) as the H-ATLAS galaxies to eliminate selection effects between the samples. Further details can be found in Rowlands *et al.* (2011).

2. SED fitting

Physical parameters such as stellar mass, stellar population age and dust mass are derived for each galaxy using the multiwavelength SED fitting code of da Cunha, Charlot and Elbaz (2008). An energy balance argument is used to account for the UV-optical radiation from stars which is attenuated by dust and re-radiated in the far-infrared (FIR) and submillimetre. A Bayesian approach is used to derive statistical constraints on each physical parameter, as each observed galaxy SED is compared to a large library of models which encompass all plausible parameter combinations. For each galaxy, we build the marginalised likelihood distribution of any physical parameter by evaluating how well each model in the library can account for the observed properties of the galaxy (by computing the χ^2 goodness of fit). This method ensures that possible degeneracies between model parameters are included in the final probability density function (PDF) of each parameter. An example SED with UV-submillimetre photometry is shown in Fig. 1.

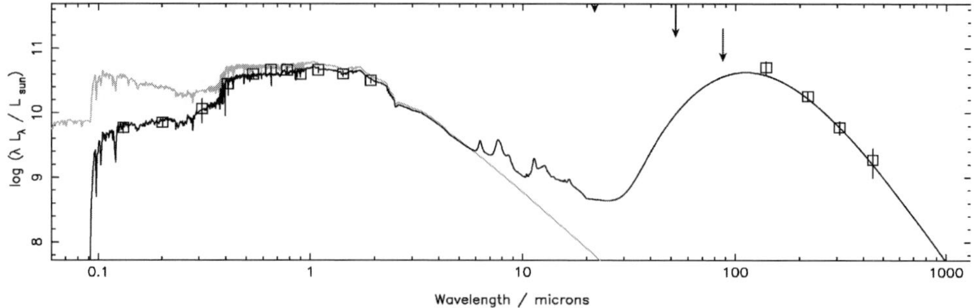

Figure 1. Example SED of an ETG, with observed photometry from the UV–submillimetre (squares), 5σ upper limits are shown as arrows. Errors on the photometry are described in Smith *et al.* (2011b). The black line is the best fit model, and the grey line is the attenuated optical model.

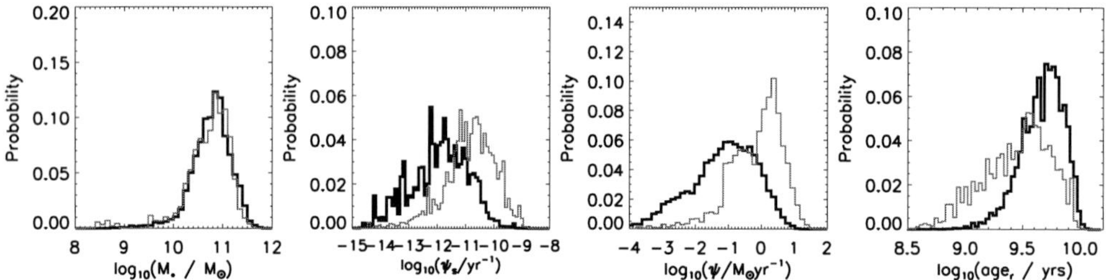

Figure 2. Average PDFs of the SED parameters of 42 H-ATLAS ETGs (thin line) compared to 222 control ETGs (thick line). The parameters are (from left to right): M_*, stellar mass; ψ_S/yr, SSFR; $\psi/M_\odot yr^{-1}$, SFR; and age_r, the r-band light-weighted age of the stellar population.

3. How are H-ATLAS ETGs different to control ETGs?

A comparison of the average PDFs of the H-ATLAS and control ETGs in Fig. 2 shows that the stellar mass distributions are not significantly different, since the control sample is selected to have the same r, z distribution. Conversely, on average the H-ATLAS ETGs are more than an order of magnitude dustier than the control ETGs. The dust mass of the 233 control ETGs inferred from stacking at optical positions on the 250μm map (see Bourne *et al.* 2011) is $(0.8-4.0)\times10^6 M_\odot$ for 25-15 K dust, whereas the mean dust mass of the 42 H-ATLAS ETGs derived from the SED fitting is $5.5\times10^7 M_\odot$. The range of M_d/M_* for the control ETGs is $(1.4-6.8)\times10^{-5}$ for 25-15 K dust, and on average, the mean SSFR of the control ETGs is 1.1 dex lower than that of H-ATLAS ETGs. A similar trend is found when comparing the mean SFR of the ETGs. For our control ETGs the mean r-band light-weighted age of the stellar population is 1.8 Gyr older than the H-ATLAS ETGs. The rest-frame $NUV-r$ colours of the control ETGs are 1.0 magnitudes redder than the H-ATLAS ETGs, and as they contain less dust their red colour must be due to an old stellar population. Control ETGs seem to inhabit higher density environments on average than the H-ATLAS ETGs, however the environments of the two populations are only different at the 1.8σ level. We do not have strong evidence that environment has an affect on whether an ETG is dusty at low-intermediate densities, as a larger sample is needed to test this.

4. Origin of dust in ETGs

Dust in ETGs has two possible origins: internal sources from asymptotic giant branch (AGB) stars, or an external supply from mergers. Using the chemical evolution code

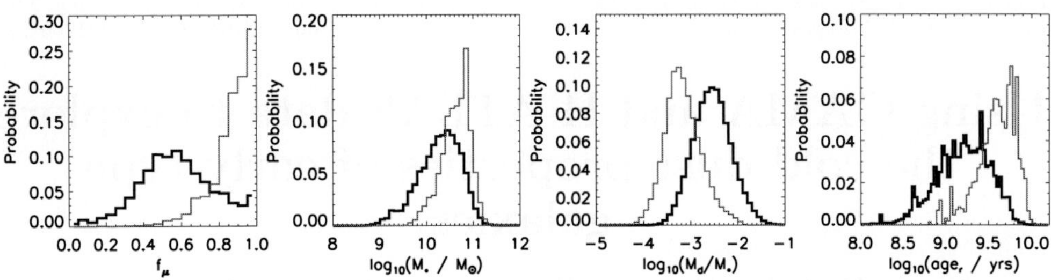

Figure 3. Average PDFs of the SED parameters of 19 passive spirals (SSFR$< 10^{-11}$/yr, thin line) compared to 431 normal spirals with SSFR$\geqslant 10^{-11}$/yr (thick line). The parameters are (from left to right): f_μ, the fraction of total dust luminosity contributed by the diffuse ISM; M_*, stellar mass; M_d/M_*, dust to stellar mass ratio; and age$_r$, the r-band light-weighted age of the stellar population.

of Gomez *et al.* (in prep), it is found that AGB stars cannot produce enough dust to account for that observed in the H-ATLAS ETGs. Even if dust production in Type II SNe is a significant dust source, the required dust injection rate for an average H-ATLAS ETG cannot be achieved. This indicates that an external source of dust from mergers is required, a substantial amount of dust grain growth must occur in the ISM or dust destruction is less efficient than predicted.

5. Properties of passive spirals

We also highlight the properties of a population of 19 passive spirals detected in H-ATLAS. There is discussion in the literature whether some spirals are red due to dust extinction or an old stellar population (Wolf *et al.* 2009). By selecting a sample of H-ATLAS spirals with SSFR$< 10^{-11}$/yr and comparing them to 'normal' spirals with SSFR$\geqslant 10^{-11}$/yr, Fig. 3 shows that the majority of the dust luminosity is produced in the ISM (f_μ), which is heated by old stars. The passive spirals have higher stellar masses and less dust per stellar mass, meaning that the passive spirals are dust deficient compared to normal spirals. This could be a selection bias in that we can only detect dust in the most massive passive spirals as their M_d/M_* ratio is much lower than the normal spiral population. Alternatively, it could be that passive galaxies tend to be massive (Masters *et al.* 2010). We find passive spirals have green-red rest-frame $NUV - r$ colours that are due to an old stellar population and not due to increased dust reddening.

References

Bourne, N. *et al. MNRAS*, submitted
da Cunha, E., Charlot S., & Elbaz, D. 2008, *MNRAS*, 388, 1595
Driver, S. *et al.* 2011, *MNRAS*, 413, 971
Eales, S. *et al.* 2010, *PASP*, 122, 499
Masters, K. *et al.* 2010, *MNRAS*, 405, 783
Rowlands, K. *et al.* 2011, *MNRAS*, in press
Smith, D. J. B. *et al.* 2011, *MNRAS*, 416, 857
Smith, D. J. B. *et al.* 2011, *MNRAS*, submitted
Wolf, C. *et al.* 2009, *MNRAS*, 393, 1302

The Spectral Energy Distribution of Galaxies
Proceedings IAU Symposium No. 284, 2011
R.J. Tuffs & C.C. Popescu, eds.

doi:10.1017/S1743921312009210

Using GAMA and H-ATLAS data to explore the cold dust properties of early-type galaxies

Nicola K. Agius, Anne E. Sansom, and Cristina C. Popescu

Jeremiah Horrocks Institute, University of Central Lancashire,
Preston, PR1 2HE, Lancashire, United Kingdom
email: NKAgius@uclan.ac.uk

Abstract. Hierarchical galaxy formation models predict the development of elliptical galaxies through a combination of the mergers and interactions of smaller galaxies. We are carrying out a study of Early-Type Galaxies (ETGs) using GAMA multi-wavelength and Herschel-ATLAS sub-mm data to understand their intrinsic dust properties. The dust in some ETGs may be a relic of past interactions and mergers of galaxies, or may be produced within the galaxies themselves. With this large dataset we will probe the properties of the dust and its relation to host galaxy properties. This paper presents our criteria for selecting ETGs and explores the usefulness of proxies for their morphology, including optical colour, Sérsic index and Concentration index. We find that a combination of criteria including r band Concentration index, ellipticity and apparent sizes is needed to select a robust sample. Optical and sub-mm parameter diagnostics are examined for the selected ETG sample, and the sub-mm data are fitted with modified Planck functions giving initial estimates for the cold dust temperatures and masses.

Keywords. Statistical, techniques: spectroscopic, Surveys, ISM: dust, Galaxy: fundamental parameters

1. Introduction

Early-Type Galaxies (ETGs) are thought to be a homogeneous class of E and S0 galaxies (Baldry *et al.* 2004). They are predominantly old and inert, have a high central surface brightness, and cover a wide range of luminosities (Driver *et al.* 2006). ETGs are classed as smooth, highly concentrated, and spheroidal systems with a lack of spiral arms. Their luminosity profiles tend to follow the de Vaucouleurs $I(r) \sim r^{1/4}$ law or a more general Sérsic law distribution (D'Onofrio *et al.* 2011).

ETGs are thought to be quiescent at zero redshift, leading to the assumption that they are devoid of both cold gas and dust (Bregman, Hogg & Roberts 1992). Recent detections of ETGs in infrared and sub-mm regimes have revealed largely unexpected amount of cold gas and dust (e.g. Leeuw *et al.* 2004). Our interests lie in exploring this dusty presence, with the aims of gaining further understanding about the processes which form ETGs.

2. Selection Techniques

With the aim of creating a highly robust sample of sub-mm selected ETGs, we experimented with multiple proxies for morphology. Proxies tested were optical colour (e.g. Kaviraj *et al.* 2010), r band Sérsic index (e.g. Leeuw *et al.* 2008) and r band Concentration index (e.g. Blanton *et al.* 2003). We used Sérsic index values from the GAMA single Sérsic fits to the objects, using GALFIT3 (Kelvin *et al.* submitted). The Concentration

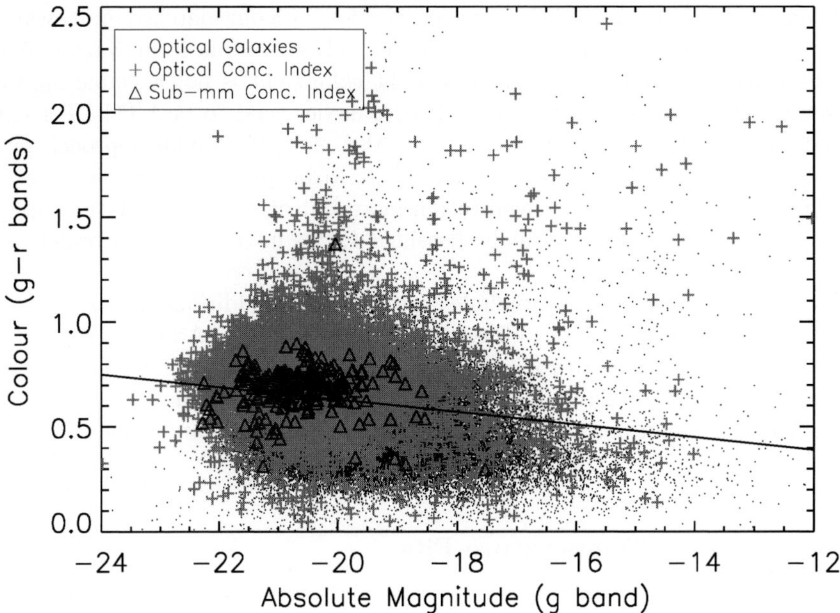

Figure 1. Optical Colour-Magnitude Diagram for our samples. The highly concentrated (C⩾2.86) optical GAMA sources are highlighted by plus signs above a background of all the GAMA sources (dots). On top of this are the sub-mm selected sources from the combined GAMA/H-ATLAS catalogues, shown as open triangles. The Red Sequence (RS) line (Bernardi *et al.* 2010) is shown to indicate the regime above which we expect RS galaxies to sit. This plot shows that our sample of ETGs could not have been selected with a direct colour cut, as both blue and red ETGs are showing up. This is confirmed through visual classifications of the samples.

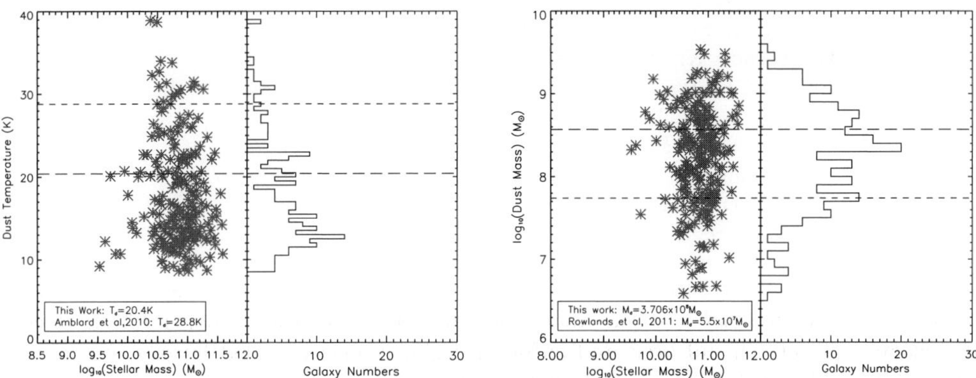

Figure 2. [Left] The distribution of dust temperatures for the ETG sample with concentration index ⩾ 2.86. Our mean temperature is the long dashed line and the short dashed line is the mean temperature from Amblard *et al.* (2010) (for the H-ATLAS Science Demonstration Phase (SDP) sample). [Right] The distribution of dust masses fit by modified Planck functions, emphasising our mean dust mass (long dashed line) and that of the SDP ETG sample of Rowlands *et al.* (2011) (short dashed line).

values are the ratio of SDSS Petrosian 90 radius to Petrosian 50 radius for each galaxy. We combined the GAMA I multi-wavelength survey with Herschel-ATLAS Phase 1 far infrared and sub-mm data to create a catalogue of sub-mm detected ETGs.

We examined the behaviour of the three proxies for both our matched multi-wavelength sample and an optical morphologically classified SDSS sample from Nair & Abraham (2010). From this we concluded that a straight colour cut is not a viable approach to take, as ETGs tend to be red, but not all red galaxies tend to be ETGs. For example, many red spirals were present in such a cut. Also, as our sub-mm approach picks out dusty galaxies, we may expect some blue ETGs to be present. A Sérsic index cut at $\geqslant$3.5 seemed reasonable, but a concentration index cut of $\geqslant$2.86 gave us a sample which matched best with the optical visually classified ETGs in the Nair & Abraham control sample.

We have applied our concentration index criterion, with additional criteria for redshift, ellipticity and apparent effective radius, to our GAMA/H-ATLAS data. With the addition of visual classifications to root out any remaining late-type galaxies, this resulted in a condensed sample of 239 lenticular and elliptical galaxies, shown as open triangles in Fig. 1. These sub-mm detected ETGs mostly have intermediate absolute magnitudes (-19.5$\sim$$M_g$ $\sim$-21.5).

3. Sub-mm Single Temperature Fits

Modified Planck functions were fitted to the sub-mm regime of our sample's observed data. The resultant cold dust temperatures and masses are displayed in Fig. 2. We find a mean dust temperature of 20.4^{+6}_{-9}K (for β =2.0) and a mean dust mass of $3.7\times10^{8}M_{\odot}$. Fig. 2 shows that the sub-mm sample contains some ETGs with high dust masses and mainly low dust temperatures compared to these references. Unlike the H-ATLAS galaxies which have been studied up to this point, our sample exhibits only intermediate stellar masses, not high stellar masses.

References

Amblard, A., Cooray, A., Serra, P. *et al.* 2010, *A&A*, 518, L9

Baldry, I. K., Glazebrook, K., Brinkmann, J. *et al.* 2004, *ApJ*, 600, 681

Bernardi, M., Shankar, F., Hyde, J. B., Mei *et al.* 2010, *MNRAS*, 2087-2122, 404

Blanton, M. R., Hogg, D. W., Bahcall, N. A. *et al.* 2003, *ApJ*, 594, 186

Bregman, J. N., Hogg, D. E., & Roberts, M. S. 1992, *ApJ*, 387, 484

D'Onofrio, M., Valentinuzzi, T., Fasano, G. *et al.* 2011, *ApJ* (Letters), 727, L6+

Driver, S. P., Allen, P. D., Graham, A. W. *et al.* 2006, *MNRAS*, 368, 414

2010)]kaviraj2010 Kaviraj, S., Schawinski, K., Silk, J., & Shabala, S. 2010, *ArXiv e-prints*, 1008.1583

Kelvin, L. S., Driver, S. P., Robotham, A. S. G. *et al.* in prep,

Leeuw, L. L., Sansom, A. E., Robson, E. I. *et al.* 2004, *ApJ*, 612, 837

Leeuw, L. L., Davidson, J., Dowell, C. D., & Matthews, H. E. 2008, *ApJ*, 677, 249

Nair, P. B. & Abraham, R. G. 2010, *VizieR Online Data Catalog*

Rowlands, K. Dunne, L., Maddox *et al.* 2011, *ArXiv e-prints*, 1109.6274

The Spectral Energy Distribution of Galaxies
Proceedings IAU Symposium No. 284, 2011
R.J. Tuffs & C.C. Popescu, eds.

doi:10.1017/S1743921312009222

Age-dating Stellar Populations of Luminous Red Galaxies

Ando Ratsimbazafy[1], Catherine Cress[1], Steve Crawford[2], Claudia Maraston[3], Robert Nichol[3], and Daniel Thomas[3]

[1]Physics Department, University of the Western Cape, Private bag X17, Cape Town 7535, South Africa

[2]South African Astronomical Observatory, PO Box 9, Observatory 7935, Cape Town, South Africa

[3]Institute of Cosmology and Gravitation, University of Portsmouth, Dennis Sciama Building, Burnaby road, Portsmouth PO1 3FX, The United Kingdom

email: `raljha.a@gmail.com`

Abstract. We investigate the possibility of using Luminous Red Galaxies (LRGs) as "Cosmic chronometers" to measure the expansion rate of the universe to 3% over a redshift range $0.1 < z < 1.0$. In this method, H(z) is directly measured by using the ages of passively evolving galaxies to determine dz/dt. We first present results from our study of LRGs in simulations Crawford *et al.* (2010) where we explore the impact of extended star formation histories on the measurements of the Hubble parameter. We then extract a carefully selected sample of LRGs taken from Sloan Digital Sky Survey (SDSS) Data Released Seven (DR7), stack spectra in redshift bins to increase the signal-to-noise, and use the Lick index modelling presented in Thomas *et al.* (2011) to age-date the sample. We discuss the implications for expansion rate measurements and outline a proposal to observe LRGs with the Southern African Large Telescope (SALT).

Keywords. galaxies: evolution , cosmology: cosmological parameters , cosmology: observations

1. Introduction

The expansion rate of the Universe can potentially be measured by age-dating LRGs and using them as "Cosmic Chronometers" (CC). Most cosmological probes only measure the expansion rate integrated along a line-of-sight, however, the CC method allows a measurement at a specific redshift and this can provide tighter constraints on cosmological parameters (Jimenez & Loeb 2002). In broad terms the basic premise is that stars in LRGs are formed at a similar time in the past, and by measuring the age difference between two ensembles of passively-evolving LRGs at different redshifts one can calculate the time interval, Δt, associated with a given redshift interval, Δz. The Hubble parameter within the redshift interval is then given by

$$H(z) = -\frac{1}{1+z}\frac{\Delta z}{\Delta t} \tag{1.1}$$

A number of scientists have attempted to use this method to track the evolution of H(z) as a function of redshift and place constraints on cosmological parameters (e.g. Jimenez *et al.* 2003, Simon *et al.* 2005, Carson & Nichol 2010). For example in Jimenez *et al.* 2003, using SDSS data, they measured the Hubble constant of $H_0 = 69 \pm 12\ km\ s^{-1}\ Mpc^{-1}$.

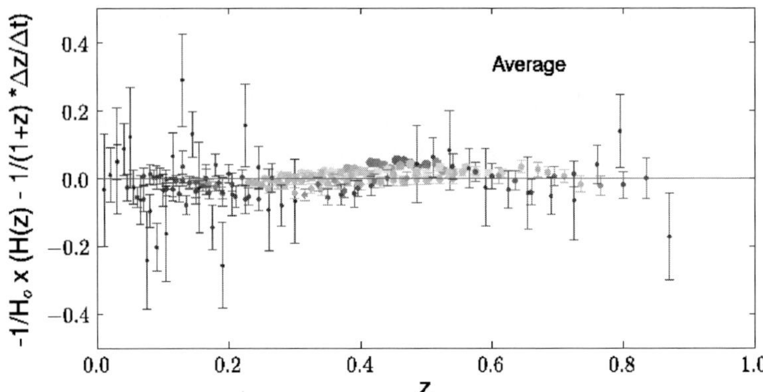

Figure 1. The calculated values of $\Delta z/\Delta t$ (data points) and the expected value for H(z) (solid line). $\Delta z/\Delta t$ were calculated from the age of simulated LRGs in MS. Each point represents the distance between the redshift gaps. The error in H(z) is associated with the inverse of difference in age hence the closer the redshift bins, the larger the error.

2. LRGs in simulations

In Crawford *et al.* (2010), we explored the validity of the assumptions implicit in the CC method using LRGs identified in the Millennium Simulation (MS). We used stellar population modelling and spectral synthesis to estimate the errors on ages that can be expected. Based on these we discussed optimization of such an experiment using SALT. Using our simulated set of galaxies, we found that H(z) can be recovered with a high accuracy of $< 3\%$ at $z \sim 0.4$ (Figure 1).

3. High signal-to-noise LRG spectra

We selected a homogeneous sample of LRGs from the SDSS catalogue in the redshift range of $0.31 < z < 0.55$. We then extracted galaxies which do not show either star formation or AGN-like activity. In order to achieve this, we measured the different emission lines in these galaxies and stellar kinematics using GANDALF (Gas AND Absorption Line Fitting) routine. We considered objects without $H\beta$, $H\alpha$ and [OIII]$\lambda\lambda$5007 emission lines. In the fitting process to recover the stellar velocity (V) and the velocity dispersion (σ), we used the Bruzual & Charlot (2003) stellar population model used by Tremonti *et al.* (2004) combined with stellar templates based on MILES stellar library (Sánchez-Blázquez *et al.* 2006). We imposed all forbidden lines to have the kinematics of [NII]$\lambda\lambda$6583 and all recombination lines to have the kinematics of $H\alpha$. In addition, the NaD and skyline regions were excluded during the procedure.

To have accurate and precise derived SSP parameters, we combined the individual spectra into high signal-to-noise spectrum at each redshift bin. The effect of co-adding spectra before the line-strength measurement is that skyline residuals and the telluric absorption features have less impact. The line-strengths of the Lick indices were measured using the INDEXF program on co-added spectra. We included the contribution of the measured line-strength error from the uncertainty in the radial velocity. This is added in quadrature to the random error in the index line-strength arising from photon statistics and read-out noise.

The Lick index measurements were also corrected for kinematic broadening caused by the line-of-sight velocity distribution. We compared our measured Lick indices: $H\beta$, Mgb, Fe5270, Fe5335 to the predicted indices of Thomas *et al.* (2011). This calibrated

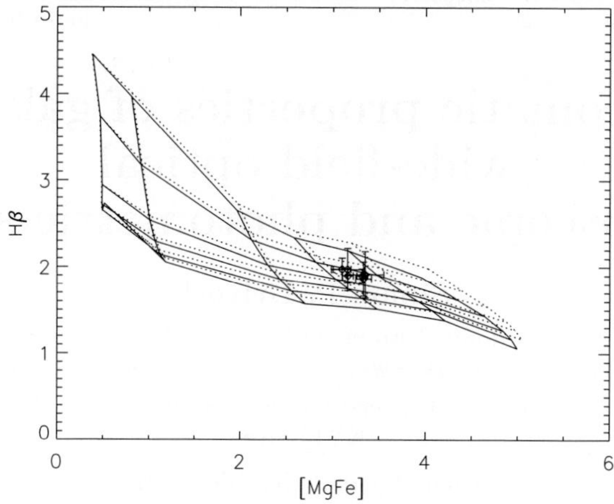

Figure 2. Index -index plots. Measured line strengths plotted over Thomas *et al.* (2011) model grids with [E/Fe]=0 (solid lines) and [E/Fe]=0.3 (dashed lines). From the bottom, age lines are 15, 12, 10, 8, 5, 3 Gyr. From right to left, [Z/H]=0.67, 0.35, 0.00, -1.35, -2.25. The composite index [MgFe] is calculated as defined in Thomas *et al.* (2011).

model makes use of the MILES stellar library, includes alpha element enhancements and is found to be comparable to the SDSS spectral resolution.

4. Summary and future work

We have stacked LRG spectra within redshift bins of $\Delta z = 0.10$ in order to have high SNR spectra. We have measured Lick indices and compared them with the model of Thomas *et al.* (2011) to extract all stellar parameters. Our preliminary results show that ages vary from $8 - 11$ *Gyr* for LRGs in the range of $0.31 < z < 0.37$ (Figure 2). Age measurements up to $z = 0.55$ are ongoing which will be combined with these to estimate the Hubble parameter.

An observation of LRGs at $z = 0.40$ and $z = 0.55$ with SALT as a pilot survey is underway. A minimum of SNR=10 per resolution element with a rest frame resolution of $3\mathring{A}$ will be obtained to age date the LRGs to 10% by means of matching the resolution of model SEDs. The use of BOSS data in 2012 is also a possibility.

References

Bruzual, G. & Charlot, S. 2003, *MNRAS*, 344, 1000
Carson, D. & Nichol, R. 2010, *MNRAS*, 408, 213
Crawford, S. M., Ratsimbazafy, A. L., Cress, C. M. *et al.* 2010, *MNRAS*, 406, 2569
Jimenez, R., Verde, L., Treu, T., & Stern, D. 2003, *ApJ*, 593, 622
Jimenez, R. & Loeb, A. 2002, *ApJ*, 573, 37
Sánchez-Blázquez P., Peletier, R. F., Jiménez-Vicente, J. *et al.* 2006, *MNRAS*, 371, 703
Simon, J., Verde, L., & Jimenez, R. 2005, *PhRvD*, 71, 123001
Thomas, D., Maraston, C., & Johansson, J. 2011, *MNRAS*, 412, 2183
Tremonti C. A., Heckman, T. M., Kauffmann, G. *et al.* 2004, *ApJ*, 613, 898

The Spectral Energy Distribution of Galaxies
Proceedings IAU Symposium No. 284, 2011
R.J. Tuffs & C.C. Popescu, eds.

doi:10.1017/S1743921312009234

Panchromatic properties of galaxies in wide-field optical spectroscopic and photometric surveys

Simon P. Driver[1,2]

[1]International Centre for Radio Astronomy Research (ICRAR), University of Western Australia, 35 Stirling Highway, Crawley, Perth, Australia, WA 6009

[2]School of Physics and Astronomy, University of St Andrews, North Haugh, St Andrews, UK, KY16 8RS

email: Simon.Driver@icrar.org

Abstract. The past 15 years have seen an explosion in the number of redshifts recovered via wide area spectroscopic surveys. At the current time there are approximately 2 million spectroscopic galaxy redshifts known (and rising) which represents an extraordinary growth since the pioneering work of Marc Davis and John Huchra. Similarly there has been a parallel explosion in wavelength coverage with imaging surveys progressing from single band, to multi-band, to truly multiwavelength or pan-chromatic involving the coordination of multiple facilities. With these empirically motivated studies has come a wealth of new discoveries impacting almost all areas of astrophysics. Today individual surveys, as best demonstrated by the Sloan Digital Sky Survey, now rank shoulder-to-shoulder alongside major facilities. In the coming years this trend is set to continue as we begin the process of designing and conducting the next generation of spectroscopic surveys supported by multi-facility wavelength coverage.

Keywords. galaxies: distances and redshifts galaxies: evolution galaxies: formation galaxies: fundamental parameters (classification, colors, luminosities, masses, radii, etc.)

---

1. Introduction

This article briefly summarises the development of wide field spectroscopic survey programs in extragalactic astronomy (see Fig. 1, Table. 1), and their impact as measured in publications and citations (see Table. 2). The statistics shown in Table. 2 are necessarily crude but indicative of the major revolution which is overtaking our subject — the rise of surveys over facilities. Whereas once the key question senior astronomers might ask of each other seemed to be "so what are you building?", in this age of internationally funded mega-facilities, the more pertinent question might be "so what survey are you designing?". The likely answer will be one which combines both spectroscopic information with panchromatic imaging. Traditionally, imaging surveys, as pioneered by the various Schmidt surveys, were single bandpass (b_J) and then later double bandpass (b_J, r_F). These were followed by a move to multi-band programs (e.g., $BVRI$, $ugriz$, $YJHK$) which used multiple filters to fill in the wavelength gaps but still remained single facility imaging surveys. Truly multiwavelength surveys span more than one imaging facility, and panchromatic surveys span multiple facilities covering a significant fraction of the em-spectrum (e.g., GOODS; Giavalisco *et al.* 2004; GAMA; Driver *et al.* 2011). This trend towards broader spectral coverage is driven not only by technological advancement but also by an appreciation that galaxies in particular emit significant levels of radiation at almost all wavelengths and to truly understand their nature will require robust distances combined with total energy measurements.

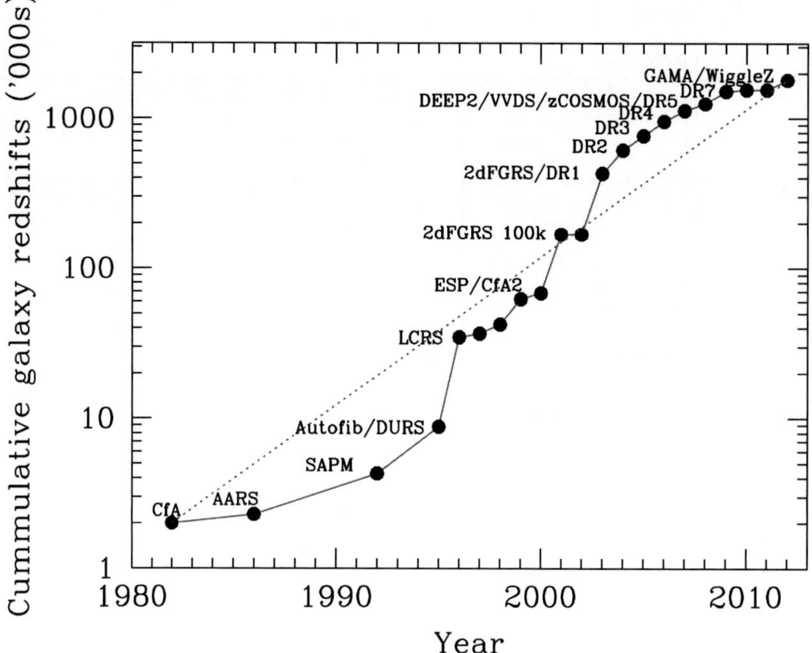

Figure 1. The rise in galaxy redshifts over the past 30 years.

2. Do we still need spectroscopic surveys?

Figure 1 shows the increase in the cumulative number of known galaxy redshifts (AGN are not included here) and the prominent surveys are listed in Table 1. The most notable are of course the Sloan Digital Sky Survey (SDSS; York *et al.* 2000), and the two-degree field galaxy redshift survey (2dFGRS; Colless *et al.* 2001). Between them these surveys account for over half the known redshifts and, even more importantly, both surveys have very clean selection criteria ($r_{AB} < 17.77$ and $b_J < 19.6$ respectively).

However this rise in spectroscopic surveys has been matched by an expansion in wavelength coverage which has led to the realisation of photometric redshifts (e.g., ANNz; Collister & Lahav 2004, see Fig. 2 (left)). Photometric redshifts allow one to bypass the laborious process of conducting a spectroscopic program by using the continuum shape (and in particular the 4000Å break and the Lyman-limit) to estimate distance. Comparisons between photometric and spectroscopic surveys are impressive with typical accuracies running at $\Delta z/(1+z) \pm 0.03$ (Collister & Lahav 2004). Purpose built photo-z surveys such as COMBO-17 (Wolf *et al.*2003), which uses a 17 medium band filter set — essentially producing a very low dispersion spectrum — can improve upon this accuracy even further ($\Delta z/(1+z) \sim 0.01$; Wolfe *et al.* 2008). This rise in photometric redshifts begs the question as to whether we still actually need large scale spectroscopic programs rather than simply conducting spot test checks of photo-z cats. For large scale statistical studies this is probably true but the following three points argue for some caution:

1) At low redshift: At low redshift the typically photo-z error $\Delta z/(1+z) \sim \pm 0.03$ becomes significant. This is easily demonstrated by constructing a luminosity function using entirely photometric or spectroscopic redshifts via a simple $1/V_{\rm max}$ estimator. Fig. 2 (left) shows the photo-z v spec-z comparison for 53k galaxies from the GAMA survey which generally appears well behaved. However Fig. 2 (right) shows the corresponding

Table 1. Major spectroscopic surveys motivated from UV/optical or near-IR imaging data

Survey	Reference	Facility	Selection	Redshifts
CfA	Davis *et al.* (1982)	Mt Hopkins 1.5m/Z-machine	$B < 14.5$	2k
AARS	Peterson *et al.* (1986)	AAT/RGO	$B_J < 17.0$	0.3k
SARS	Loveday *et al.* (1992)	MSSSO 2.3m/DBS	$b_J < 17.15$	2k
Autofib	Ellis *et al.* (1996)	AAT/Autofib	$b_J < 22.0$	1k
DURS	Ratcliffe *et al.* (1996)	UKST/FLAIR	$b_J < 17.0$	2.5k
LCRS	Schectman *et al.* (1996)	DuPont/MOS	$R < 17.5$	26k
CFRS	Lilly *et al.* (1996)	CFHT/MOS	$I_{AB} < 22.5$	1k
CS	Geller *et al.* (1997)	Various	$r < 16.13$	2k
ESP	Vettolani *et al.* (1997)	ESO 3.6m/OPTOPUS	$b_J < 19.4$	4k
SSRS2	da Costa *et al.* (1998)	Various	$B < 15.5$	5.5k
CfA2	Falco *et al.* (1999)	Various	$B < 15.5$	20k
CNOC2	Yee *et al.* (2000)	CFHT/MOS	$R < 21.5$	6k
2dFGRS	Colless *et al.* (2001)	AAT/2dF	$b_J < 19.6$	227k
SDSS Main	Strauss *et al.* (2002)	SDSS 2.5m/Spectrographs	$r < 17.77$	930k
SDSS LRG	Eisenstein *et al.* (2002)	SDSS 2.5m/Spectrographs	$r < 19.5$	120k
DEEP1&2	Davis *et al.* (2003)	Keck/Deimos	$R_{AB} < 24.1$	51k
H-AAO	Huang *et al.* (2003)	AAT/2dF	$K < 15.0$	1k
MGC	Driver *et al.* (2005)	AAT/2dF	$B < 20.0$	10k
SDSS Stripe82	Baldry *et al.* (2005)	SDSS/Spectrographs	$u < 20.5$	70k
2SLAQ-LRG	Cannon *et al.* (2006)	AAT/2dF	$i < 19.8$	13k
6dfGRS	Jones *et al.* (2009)	UKST/6dF	$K < 12.75+$	110k
VVDS-wide	Garilli *et al.* (2008)	VLT/VIMOS	$I_{AB} < 22.5$	35k
VVDS-deep	Le Fevre *et al.* (2005)	VLT/VIMOS	$I_{AB} < 24.0$	12k(150k)
VVDS-ultradeep	Le Fevre *et al.* (2005)	VLT/VIMOS	$I_{AB} < 24.75$	0k(1k)
zCOSMOS-bright	Lilly *et al.* (2007)	VLT/VIMOS	$I_{AB} < 22.5$	10k(20k)
zCOSMOS-deep	Lilly *et al.* (2007)	VLT/VIMOS	$I_{AB} < 24.0$	1k(10k)
WiggleZ	Drinkwater *et al.* (2010)	AAT/AAOmega	$r < 22.5+$	250k
AGES	Kochaneck *et al.* (2011)	MMT/Hectospec	$R < 20+$	19k
GAMA	Driver *et al.* (2011)	AAT/AAOmega	$r < 19.8$	180k(350k)
Vipers	N/A	VLT/VIMOS	$I_{AB} < 22.5$	20k(100k)
BOSS	N/A	SDSS 2.5m/Spectrographs	$i < 20$	500k(1500k)

The majority of the information shown in this table was kindly provided by Ivan Baldry (see http://www.astro.ljmu.ac.uk/~ikb/research/galaxy-redshift-surveys.html) which was used to produce Fig. 1 in Baldry *et al.* (2010).

derived luminosity functions with somewhat disastrous implications for the faint-end slope. This is because photo-z's are generally calibrated for the most numerous galaxy type within ones' training set. Galaxies with divergent properties can have significant systematic errors leading to the kind of severe bias shown in Fig. 2 (right).

2) When fidelity is required: Fig. 3 shows a typical cone diagram for the GAMA 12hr region using photometric (upper) or spectroscopic (lower) redshifts. While statistically one can still recover some sense of the generic clustering properties (suitable for cosmology) it is clearly impractical to use photo-z's to identify individual filaments, clusters, or groups and their associated masses. High fidelity studies are likely to be a key focus area in the coming years as we look to distinguish between the influence of environment and halo mass (Haas, Schaye & Jeeson-Daniel 2011).

3) When the test galaxies extend beyond the calibration data. Photometric redshifts make use of the 4000Å break and are typically calibrated to bright and local spectroscopic redshift samples which exhibit strong 4000Å breaks. However the break is less apparent in star-forming systems. As one progresses either to lower mass systems in the nearby Universe or to high redshift massive systems one moves to more star-forming populations which exhibit flatter spectra and the likelihood of confusion increases.

3. High impact photo-z and spectro-z surveys

Despite these concerns photo-z's have proved invaluable in many areas and in particular for the interpretation of data seen in the very deep fields obtained by the Hubble

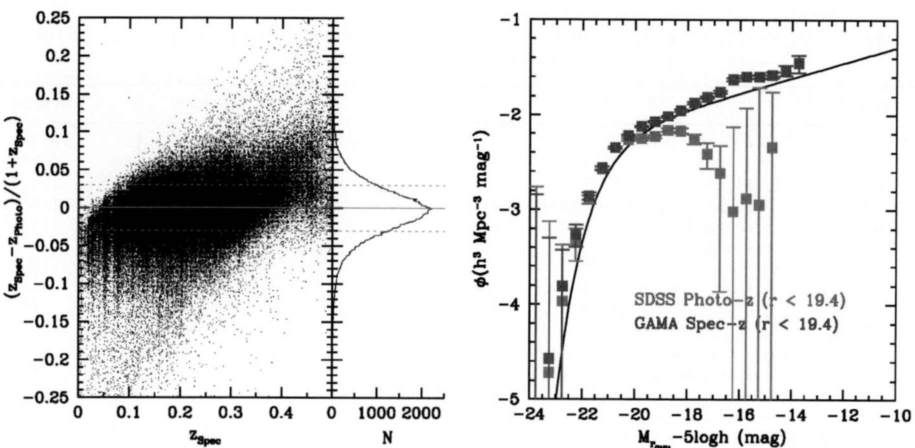

Figure 2. (left) Photometric versus Spectroscopic redshifts for 170,000 galaxies drawn from the GAMA database. The accuracy is ± 0.03. (right) the resulting luminosity functions derived for the two samples indicating good agreement at L^* but dramatically different faint-ends.

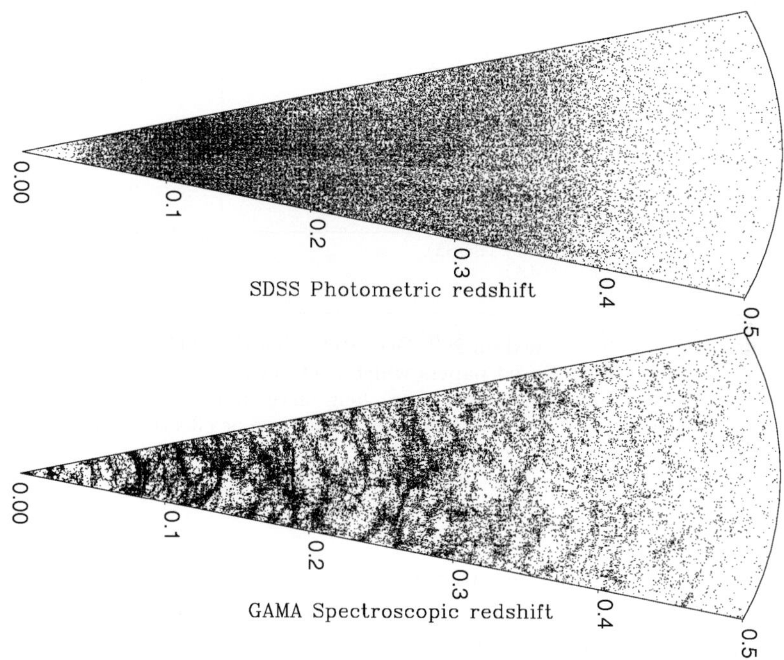

Figure 3. A comparison of photometric redshifts (upper) and spectroscopic redshifts (lower) derived from the SDSS and GAMA surveys.

Space Telescope. Table 2 attempts to provide some indication of the impact of many of the spectroscopic and photometric redshift surveys over the past few decades. The landscape is generally dominated by two nearby spectroscopic programmes (SDSS, 2dF-GRS), coupled with deep HST studies (HDF, GOODS, COSMOS), and forays into new wavelengths (2MASS, GOODS, GALEX) all of which build upon the pioneering work on the CfA surveys.

Table 2. Major extragalactic optical/near-IR surveys and their impact in terms of refereed papers and citations

Survey	Papers	Citations
Sloan Digital Sky Survey (SDSS)	3254	135330
Hubble Deep Field (HDF)	590	46360
Two Micron All Sky Survey (2MASS)	968	27253
Two degree Field Galaxy Redshift Survey (2dFGRS)	197	23219
Great Observatories Origins Deep Survey (GOODS)	498	24578
Centre for Astrophysics (CfA)	263	11989
Galaxy Evolution Explorer GALEX)	465	11560
Canada France Redshift Survey (CFRS)	90	8493
Cosmic Evolution Survey (COSMOS)	241	7354
Ultra Deep Field (UDF)	188	7237
Classifying Objects by Medium-Band Observations (COMBO-17)	86	6088
Deep Evolutionary Exploratory Probe (DEEP)	113	5118
Las Campanas Redshift Survey (LCRS)	96	4063
APM Galaxy Survey (APMGS)	57	4058
VIMOS VLT Deep Survey (VVDS)	75	3689
UKIRT Infrared Deep Sky Survey (UKIDSS)	105	3444
Southern Sky Redshift Survey 2 (SSRS2)	72	3421
Point Source Catalogue Survey (PSCz)	94	3270
2dF QSO redshift Survey (2QZ)	53	2930
Galaxy Evolution from Morphologies and SEDs (GEMs)	62	2696
NOAO Deep Wide Field Survey (NDFWS)	77	2555
CFHT Legacy Survey (CFHTLS)	72	2442
Millennium Galaxy Catalogue (MGC)	32	1590
zCOSMOS	47	1567
Stromlo-APM Redshift Survey (SARS)	19	1389
AGN and Galaxy Evolution Survey (AGES)	31	1337
Gemini Deep Deep Survey (GDDS)	20	1304
The 2dF-SDSS LRG And QSO Survey (2SLAQ)	33	1280
Canadian Network for Observational Cosmology Field Galaxy Redshift Survey (CNOC2)	37	1249
ESO Slice Project (ESP)	24	939
The 6dF Galaxy Survey (6dFGS)	22	535
Baryonic Acoustic Oscillations Survey (BOSS)	20	370
Galaxy And Mass Assembly (GAMA)	19	213
WiggleZ	14	194

Note these numbers were determined on 20th Dec 2011 using the SOA/NASA ADS Astronomy Query Form by searching for refereed papers which contained abstract keywords based on the following boolean logic: (galaxy or galaxies) and (<long survey name> or <short survey name>). The table is purely indicative and not weighted by survey age or effective cost. My apologies in advance for the many surveys not included.

4. Selected Highlights

It is obviously an impossible task to try and summarise all the papers covered by the surveys shown in Table 2 but below I provide some personal reflections on the areas which have most interested me (apologies in advance for the obvious bias):

4.1. *Luminosity Functions*

An original motivation for many surveys has been the measurement of the galaxy luminosity function which describes the space density of galaxies. Over the past 10 years this has now been measured in UV/optical and near-IR bands and from redshift zero to relatively high-redshifts (see for example: Cole *et al.* 2001; Bell *et al.* 2003; Blanton *et al.* 2003, 2005; Bouwens *et al.* 2006, 2007; Faber *et al.*, 2007; Hill *et al.* 2010; Robotham & Driver 2011). Although the bright-end is generally well defined and well behaved the faint-end and the implied space-density of dwarf systems remains elusive. This is because of both the Eddington bias as well as the intrinsic low surface brightness nature of these systems which places their peak central surface brightnesses below the detection thresholds of the imaging surveys. However it is also becoming clear that while the Universe is not filled with giant low surface brightness galaxies a key issue is our ability to accurately recover the fluxes of even the most luminous systems. Two key questions are: how much light are we missing around luminous systems, and how many dwarfs lie below our thresholds. The upcoming deep optical surveys (VST, PanSTARRs, DES, LSST) should be able to clarify both issues.

4.2. *Stellar Mass Measurements and the Galaxy Stellar Mass Function*

The past ten years has also seen a movement away from luminosity functions to the more fundamental stellar mass functions. Motivated by credible stellar mass estimates (e.g., Bell & de Jong 2001; Kauffmann *et al.* 2003a; Bundy *et al.* 2006; Taylor *et al.* 2011). Typically alternative mass estimates agree to within a factor of 2 which has allowed the construction of relatively consistent, and therefore presumably robust, stellar mass functions (e.g., Bell *et al.* 2003; Baldry *et al.* 2008, 2011; Fontana *et al.* 2004; Ilbert *et al.*, 2010). Our Universe appears to have a stellar mass density of order $\Omega_* = 0.0017$ (4% of the baryonic mass density, Baldry *et al.*, 2011) with the evolution of the mass functions implying a relatively linear build-up of log stellar mass with redshift (see Sawicki 2011). One key issue is that the current estimates of stellar mass appear to be a factor of two higher than what one expects from the cosmic star-formation history using a standard IMF (Wilkins, Trentham & Hopkins 2008). This issue had also been explored earlier by Baldry & Glazebrook (2003) who proposed a modified-IMF to resolve this issue which opens Pandora's Box on the question of an evolving IMF (see for example Dave 2008).

4.3. *Galaxy Bimodality - Bimodality or Duality?*

Perhaps one of the strongest themes has been the rediscovery (c.f. Baum 1959) of galaxy bimodality (e.g., Strateva *et al.* 2001; Baldry *et al.* 2004, 2006; Driver *et al.* 2006). Samples are now routinely divided into red and blue subsets. Like Christmas LEDs which spontaneously change colour as star-formation is quenched, unquenched and quenched again. Looking at the images in Fig. 4 I'm tempted to think the key point has been missed. On the whole the red and blue samples show not only a marked difference in colour but also in morphology. While quenching and unquenching may be a plausible explanation for the colour change it is significantly harder to identify a mechanism which readily modifies, compresses, and stretches the morphologies (defined by the energy in the stellar orbits). In my mind it seems the more elegant way forward is to recognise not the *bimodality* of galaxies but the *duality*, or dual nature, with galaxies typically comprising of a bulge and/or disc component. In looking at Fig. 4 it is also clear that the red sample is a mixed bag containing: spheroids, anemic spirals, and reddened spirals. This becomes more apparent when one divides the sample by Sérsic index rather than colour and finds an equally distinct bimodality but with less overlap with the colour-split sample than one would like.

Unfortunately considering the dual nature of galaxies is far harder than a simple global bimodality split as it requires bulge-disc decomposition which in turn requires high spatially resolved high signal-to-noise data. Numerous groups are now embarked on this endeavour both at low and high redshift (Kelvin *et al.* 2011; Simard *et al.* 2011). One surprise which did stem from early work in this direction (Allen *et al.* 2006; Driver *et al.* 2007a) is that while the red peak might well contain 60% of the stellar mass only half of this is in spheroid structures such that only 40% of the stellar mass in total resides in spheroids and the remaining 60% in discs. This is perhaps at odds with a CDM hierarchical merger picture but, confirmed by Gadotti (2009) and Tasca & White (2011),

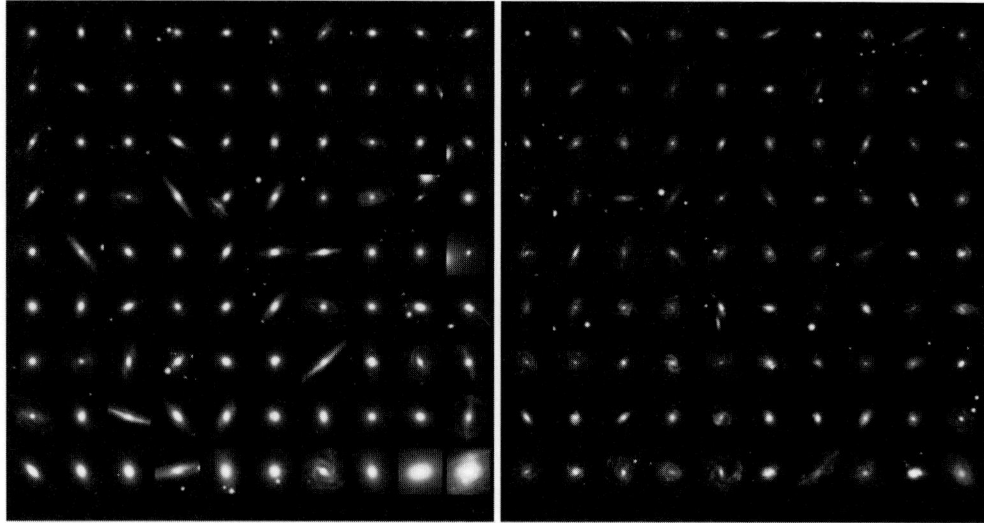

Figure 4. A sample of red (left) and blue (right) galaxies

appears to be a result that is here to stay. So how then in a backdrop of hierarchical merging, in which discs are easily disrupted, can the majority of the stellar mass reside in dynamically thin fragile discs? It is also worth noting that this must be a lower limit as numerous discs must form which ultimately merge into spheroids such that some component of the spheroids stellar mass also formed via the disc formation process. In fact it begs the questions to whether merger-induced star-formation is an almost negligible phenomena and that the majority of stars are actually formed via direct gas infall – as now argued by some simulations (L'Huillier, Combes & Semelin 2011) and directly observed (e.g., Sancisi *et al.* 2008).

4.4. *Dust Attenuation*

As highlighted by the obvious very red edge-on spiral interlopers in Fig. 4 we see than dust attenuation is a significant factor typically contaminating red samples at the 5-10% level (simply from counting the obvious edge-on systems in Fig. 4). While we now have a good understanding of the dust attenuation law (e.g., Calzetti *et al.* 2000), and viable radiative transfer models (Popescu *et al.* 2011), what is not clear is how the dust properties vary with galaxy type, environment, or redshift. Although somewhat low profile work Choi, Park & Vogeley (2007), Shao *et al.* (2007), Driver *et al.* (2007b; 2008) and Masters *et al.* (2010) have all highlighted the severe impact that dust attenuation can have with implied optical bandpass corrections measured in magnitudes. In Driver *et al.* (2008) we showed that in the B band discs may be attenuated by as much as 1 magnitude while bulges by up to 2-3 magnitudes — depending on inclination. This is a result which is shocking to the optical astronomers who routinely ignore intrinsic dust attenuation, but well established to the far-IR community. Surveys such as GAMA (Driver *et al.*, 2011) which aim to combine UV, optical, near-IR, mid-IR and far-IR with a high fidelity spectroscopic data will be vital to untangle this mess.

4.5. *The mass-metalicity relation*

A particularly elegant result of the past ten years has been the beautiful mass-metalicity relation identified in SDSS and other datasets (see in particular: Tremonti *et al.* 2004;

Savaglio *et al.* 2005; Erb *et al.* 2006). Baldry, Glazebrook & Driver (2008) argue that this relation is an inevitable consequence of the variation in star-formation efficiency with stellar mass. It could also be a combination of star-formation efficiency (i.e., fraction of gas used, inflows/outflows and IMF variation). A possible complicating factor comes from consideration of the IGIMF (Wiedner & Kroupa 2006) in which more massive galaxies preferentially host more massive star-formation regions leading to more enriched gas because the upper end of the IMF is more likely to be populated

4.6. *The growth of galaxy sizes*

Slightly lower profile but entirely orthogonal is the work on mass-size relations (e.g., Shen *et al.* 2003; Blanton *et al.* 2005; Driver *et al.* 2005). This has the potential to connect observable galaxy sizes with the underlying halo spin parameter (i.e., the old Fall & Efstathiou 1988 connection, see also Dalcanton, Spergel & Summers 1998 and Mo, Mao and White 1998). More effort needs to be invested in the numerical and simulation side however very nice clean cut empirical results seem to be emerging the work of Trujillo *et al.* (2006; 2007) which finds significant evolution in galaxy sizes with redshift. The obvious implication is that galaxies are visibly growing from the inside out. In my world of duality this fits rather nicely with the idea that we're witnessing the late formation and growth of discs around the old pre-formed bulges. However other ideas which involve a dynamical relaxation in which the galaxy core moves inward while the outer regions move outwards have also been proposed.

4.7. *The cosmic star-formation rate*

Finally one cannot not fail to mention the great advances in measuring star-formation rates starting from the seminal work of the CFRS and HDF studies (Lilly *et al.* 1996; Madau *et al.* 1996), coupled with the more local and exhaustive studies by Kauffmann *et al.* (2003b); Brinchmann *et al.* (2004); Juneau *et al.* (2005) Salim *et al.* (2007). The compilation of Hopkins & Beacom (2006) provides what appears to be a very firm and mature insight into the overall cosmic star-formation history history with subsequent studies now finding significant distinction by mass (e.g., Pozzetti *et al.* 2007) — i.e., cosmic downsizing.

5. Summary

The above is a very brief précis which tries in some way to acknowledge the efforts of those who have designed and led these surveys, to temper our appetite for photometric redshifts with some words of caution, and to mention in passing a small fraction of the high-note papers and results in this subject over the past decade. The next decade will see survey expansion continue as we push deeper at optical wavelengths (VST, VISTA, PanSTARRs, DES, LSST, Euclid), but perhaps more significant the expansion in wavelength with public releases imminent of the WISE mid-IR and Herschel Far-IR data. These in turn will be followed by major X-ray (XMM-XLL) and Radio programs (ASKAP, MeerKAT) which will enable the start of truly panchromatic surveys which will allow for the comprehensive and simultaneous study of the AGN, gas, dust, and stellar components as a function of environment, and distance. These mega-surveys will then feed Integral Field Unit (IFU) and multi-IFU follow-up which will provide spatial studies of the internal dynamics and chemistry of well defined sub-samples. Exciting times.

I'd like to thank the organisers for a very enjoyable meeting and conclude by dedicating this article to the memory of John Huchra who really started something.

References

Allen. P. D., Driver, S. P., Graham, A. W., Cameron, E., Liske, J., & De Propris, R. 2006, *MNRAS*, 371, 2

Baldry, I. K. & Glazebrook, K. 2003, *ApJ*, 593, 258

Baldry, I. K., Glazebrook, K., Brinkmann, J., Ivezic, Z., Lupton, R. H., Nichol, R. C., & Szalay, A. S. 2004, *ApJ*, 600, 681

Baldry, I. K. *et al.* 2005, *MNRAS*, 358, 441

Baldry, I. K., Balogh, M. L., Bower, R. G., Glazebrook, K., Nichol, R. C., Bamford, S. P., & Budavari, T. 2006, *MNRAS*, 373, 469

Baldry, I. K., Glazebrook, K., & Driver, S. P. 2008, *MNRAS*, 388, 945

Baldry, I. K. *et al.* 2010, *MNRAS*, 404, 86

Baldry, I. K. *et al.* 2011, *MNRAS*, in press (arXiv:1111.5707)

Baum, W. A. 1959, *PASP*, 71, 106

Bell, E. F. & de Jong, R. S. 2001, *ApJ*, 550, 212

Bell, E. F., McIntosh, D. H., Katz, N., & Weinberg, M. D. 2003, *ApJS*, 149, 289

Blanton, M. R. *et al.* 2003, *ApJ*, 592, 819

Blanton, M. R. *et al.* 2005, *AJ*, 129, 2562

Bouwens, R. J., Illingworth, G. D., Blakeslee, J. P., & Franx, M. 2006, *ApJ*, 653, 53

Bouwens, R. J., Illingworth, G. D., Franx, M., & Ford, H. 2007, *ApJ*, 670, 928

Brinchmann, J., Charlot, S., White, S. D. M., Tremonti, C., Kauffmann, G., Heckman, T., & Brinkmann, J. 2004, *MNRAS*, 351, 1151

Bundy, K. *et al.* 2006, *ApJ*, 651, 120

Calzetti, D., Armus, L., Bohlin, R. C., Kinney, A. L., Koornneef, J., & Storchi-Bergmann, T. 2000, *ApJ*, 533, 682

Choi, Y-Y., Park, C., & Vogeley, M. S. 2007, *ApJ*, 658, 884

Cole, S. *et al.* 2001, *MNRAS*, 326, 255

Collister, A. & Lahav, O. 2004, *PASP*, 116, 345

Canon, R. *et al.* 2006, *MNRAS*, 372, 425

Colless, M. M. *et al.* 2001, *MNRAS*, 328, 1039

da Costa, L. *et al.* 1998, *AJ*, 116, 1

Dave, R. 2008, *MNRAS*, 385, 147

Davis. M., Huchra, J., Latham, D. W., & Tonry, J. 1982, *ApJ*, 253, 423

Davis, M. *et al.* 2003, *SPIE*, 4843, 161

Dalcanton, J., Spergel, D. N., & Summers, F. J. 1997, *ApJ*, 482, 659

Drinkwater, M. J. *et al.* 2010, *MNRAS*, 401, 1429

Driver, S. P., Liske, J., Cross, N. J. G., De Propris, R., & Allen, P. D. 2005, *MNRAS*, 360, 81

Driver, S. P., Allen, P. D., Graham, A. W., Cameron, E., Liske, J., Ellis, S. C., Cross, N. J. G., De Propris, R., Phillipps, S., & Couch, W. J. 2006, *MNRAS*, 368, 414

Driver, S. P., Allen, P. D., Liske, J., & Graham, A. W. 2007, *ApJ*, 657, 85

Driver, S. P., Popescu, C. C., Tuffs, R. J., Liske, J., Graham, A. W., Allen, P. D., & De Propris, R. 2007, *MNRAS*, 379, 1022

Driver, S. P., Popescu, C. C., Tuffs, R. J., Graham, A. W., Liske, J., & Baldry, I. K. 2008, *ApJ*, 678, 101

Driver, S. P. *et al.* 2011, *MNRAS*, 413, 971

Eisenstein, D. J. *et al.* 2001, *AJ*, 122, 2267

Ellis, R. S., Colless, M., Broadhurst, T., Heyl, J., & Glazebrook, K. 1996, *MNRAS*, 280, 235

Erb, D. S., Shapley, A. E., Pettini, M., Steidel, C. C., Reddy, N. A., & Adelberger, K. L. 2006, *ApJ*, 644, 813

Faber, S. *et al.* 2007, *ApJ*, 665, 265

Falco, E. E. *et al.* 1999, *PASP*, 758, 438

Fall, S. M. & Efstathiou, G. 1980, *MNRAS*, 193, 189

Fontana, A. *et al.* 2004, *A&A*, 424, 23

Gadotti, D. 2009, *MNRAS*, 393, 1531

Garilli, B. *et al.* 2008, *A&A*, 486, 683

Geller, M. *et al.* 1997, *AJ*, 114, 2205
Giavalisco, M. *et al.* 2004, *ApJ*, 600, L93
Gunawardhana, M. L. P. *et al.* 2011, *MNRAS*, 415, 1647
Haas, M. R., Schaye, J., & Jeeson-Daniel, A. 2011, *MNRAS*, in press (arXiv: 1103.0547)
Hopkins, A. M. & Beacom, J. F. 2006, *ApJ*, 651, 142
Hill, D. T., Driver, S. P., Cameron, E., Cross, N. J. G., Liske, J., & Robotham, A. 2010, *MNRAS*, 404, 1215
Huang, J.-S., Glazebrook, K., Cowie, L. L., & Tinney, C. 2003, *ApJ*, 584, 203
Ilbert, O. *et al.* 2010, *ApJ*, 709, 644
Jones, H. *et al.* 2009, *MNRAS*, 399, 683
Juneau, S. *et al.* 2005, *ApJ*, 619, 135
Kauffmann, G. *et al.* 2003a, *MNRAS*, 341, 33
Kauffmann, G. *et al.* 2003b, *MNRAS*, 341, 54
Kelvin, L. S. *et al.* 2012, *MNRAS*, in press (arXiv:1112.1956)
Kochaneck, C. S. *et al.* 2011, *ApJS*, in press (arXiv:1110.4371)
L'Huillier, B., Combes, F., & Semelin, B. 2011, *A&A*, in press (arXiv:1108.4247)
Le Fèvre, O. *et al.* 2005, *A&A*, 439, 845
Lilly, S. J., Fevre, O., Crampton, D., Hammer, F., & Tresse, L. 1995, *ApJ*, 455, 50
Lilly, S. J., Le Fevre, O., Hammer, F., & Crampton, D. 1996, *ApJ*, 460, 1
Lilly, S. J. *et al.* 2007, *ApJS*, 172, 70
Loveday, J., Peterson, B. A., Efstathiou, G., & Maddox, S. J. 1992, *ApJ*, 390, 338
Madau, P., Ferguson, H. C., Dickinson, M. E., Giavalisco, M., Steidel, C. C., & Fruchter, A. 1996, *MNRAS*, 283, 1388
Masters, K. *et al.* 2010, *MNRAS*, 404, 792
Mo, H. J., Mao, S., & White, S. D. M. 1998, *MNRAS*, 295, 319
Peterson, B. A., Ellis, R. S., Efstathiou, G., Shanks, T., Bean, A. J., Fong, R., & Zen-Long, Z. 1986, *MNRAS*, 221, 233
Popescu, C. C., Tuffs, R. J., Dopita, M. A., Fischera, J., Kylafis, N. D., & Madore, B. F. 2011, *A&A*, 527, 109
Pozzetti, L. *et al.* 2007, *A&A*, 474, 443
Ratcliffe, A., Shanks, T., Broadbent, A., Parker, Q. A., Watson, F. G., Oates, A. P., Fong, R., & Collins, C. A. 1996, *MNRAS*, 281, 47
Robotham, A. S. G. & Driver, S. P. 2011, *MNRAS*, 413, 2570
Salim, S. *et al.* 2007, *ApJS*, 173, 267
Sancisi, R., Fraternalli, F., Oosterloo, T., & van der Hulst, J. M. 2008, *A&A Rv*, 15, 189
Sawicki, M. 2011, *MNRAS*, in press (arXiv:1108.5186)
Savaglio, S. *et al.* 2005, *ApJ*, 635, 260
Schectman, S. A., Landy, S. D., Oemler, A., Tucker, D. L., Lin, H., Kirschner, R. P., & Schechter, P. L. 1996, *ApJ*, 470, 172
Shao, Z., Xiao, Q., Shen, S., Mo, H. J., Xia, X., & Deng, Z. 2007, *ApJ*, 659, 1159
Shen, S. *et al.* 2003, *MNRAS*, 343, 978
Simard, L., Mendel, J. T., Patton, D. R., Ellison, S. L., & McConnachie, A. W. 2011, *ApJS*, 196, 11
Strateva, I. *et al.* 2001, *AJ*, 112, 1861
Strauss, M. A. *et al.* 2002, *AJ*, 124, 1810
Tasca, L. A. M. & White, S. D. M. 2011, *A&A*, 530, 106
Taylor, E. *et al.* 2011, *MNRAS*, 418, 1587
Tremonti, C. A. *et al.* 2004, *ApJ*, 613, 898
Trujillo, I. *et al.* 2006, *ApJ*, 650, 18
Trujillo, I. *et al.* 2007, *MNRAS*, 382, 109
Vettolani, G. *et al.* 1997, 325, 954
Wilkins, S. M., Trentham, N., & Hopkins, A. 2008, *MNRAS*, 385, 687
Wolf, C., Meisenheimer, K., Rix, H.-W., Borch, A., Dye, S., & Kleinheinrich, M. 2003, *A&A*, 401, 73
Wolf, C., Hildebrandt, H., Taylor, E. N., & Meisenheimer, K. 2008, *A&A*, 492, 933

Yee, H. K. C. *et al.* 2000, *ApJS*, 129, 475
York, D. *et al.* 2000, *AJ*, 120, 1579

Discussion

CHILINGARIAN: I have two remarks:
1. In our recently accepted paper (Chilingarian & Zolotukhin arXiv: 1102.1159) we clearly demonstrated WHY the photo-z works at low redshifts - this is a mathematical consequence of the existence of a tight colour-colour-magnitude relation for all non-active galaxies, both blue & red (M_r, $g-r$, $NUV-r$).
2. In the same paper we also show that if one adds the restframe UV (GALEX NUV) as the 2nd colour, the contamination of the red sequence by dusty late type galaxies decreases from 25% to $2-3$%. This is an important factor in colour-based galaxy selection, e.g. colour cuts.

MATTILA: This is a question about the influence of dust on the SEDs. Part of the removed (extincted) UV-Opt-NIR starlight is reshuffled into the (far) IR and is well known. However, about an equal amount of starlight is scattered by dust and remains at the original UV-Opt-NIR wavelengths. How do you account for that? Does it have an effect on your calculations?

DRIVER: I'll defer this to Cristina...

POPESCU: The radiation transfer calculations of Tuffs *et al.* (2004; A&A 419, 821) used to calculate, and correct for, the attenuation of starlight in spiral galaxies measured in the GAMA survey incorporate a calculation of anisotropic scattered light in the UV/optical/NIR range. The dust and star geometry used for these calulations is constrained empirically, using both resolved optical imaging (for the translucent components) and the amplitude and colours of the re-radiated dust emission (for the optically thick components), according to the models for the panchromatic UV/optical-FIR/submm SED by Popescu *et al.* (2011; A&A 527, 109). The scattered light makes a very substantial difference to the attenuation and its dependence on inclination because much of the dust is found to reside in extended, translucent structures, leading to a high percentage of UV/optical photons suffering either an absorption or a single scattering before escaping.

ELMEGREEN: You discussed the clumpy irregular galaxies in terms of spheroid formation but these galaxies also made disks which are still with us today, the thick disks. So mergers have not destroyed those disks, probably because subsequent galaxy growth is by cold flows, which is a gentler process. So, yes these galaxies make spheroids, but not only spheroids.

DRIVER: Agreed. The model presented is a generalisation. We see an old thick disk in the Milky Way. So the two processes (disk & spheroid formation) do overlap. The key point here is that $z > 1.5$ spheroid formation dominates and at $z < 1.5$ disk formation dominates, but both are occurring all epochs.

The Spectral Energy Distribution of Galaxies
Proceedings IAU Symposium No. 284, 2011
R.J. Tuffs & C.C. Popescu, eds.

doi:10.1017/S1743921312009246

Far-IR to Submm SEDs for local galaxies: Herschel, Planck and the HRS

David L Clements[1] and
The Herschel Reference Sample Team

[1]Astrophysics Group, Imperial College, Blackett Laboratory,
Prince Consort Road, London, SW7 2AZ,UK.
email: d.clements@imperial.ac.uk

Abstract. The Herschel Space Observatory and the Planck satellite are providing radical improvements to our knowledge of the spectral energy distributions of galaxies in the far-IR and submm. We here present the results of the first combination of Herschel and Planck fluxes of local galaxies from the Herschel Reference Sample (HRS) survey, covering galaxies at distances between 15 and 25 Mpc. This combination provides information on SEDs in eight bands from 60μm, using IRAS, to 1.4mm using Planck. We apply a similar fitting procedure to this data as applied to the Planck ERCSC-detected nearby galaxies and confirm the result that dust significantly colder than 20K is common in local galaxies. It is early days for this kind of study, but it is clear that the new generation of satellites are already adding considerably to our knowledge of the far-IR/submm properties of galaxies.

Keywords. galaxies:dust, Herschel Space Observatory, Planck, galaxies:infrared, galaxies:submm

---

1. Introduction

The last 15 years have seen a huge increase in our capability for studying the long wavelength far-IR ($\lambda > 100\mu$m) and submm spectral energy distributions (SEDs) of galaxies. Where once we would have to rely on IRAS data that cuts off at 100μm and perhaps one or two points at longer wavelengths with large error bars from single element bolometer detectors, such as UKT-14 on JCMT (eg. Clements *et al.* 1993), we can now extend our coverage to longer wavelength far-IR points by using ISO and Spitzer, while our submm capabilities have been immensely improved by bolometer arrays such as SCUBA and MAMBO. Our capabilities are now once more being boosted, and this time the important wavelength region between 200 and 160μm, respectively the longest wavelength ISO and Spitzer bands, and 450μm, the longest wavelength band for which significant ground based data is available, is being filled. The Herschel Space Observatory (Pilbratt *et al.* 2010) and the Planck satellite (Planck Collaboration, 2011a) are each playing their role. Herschel provides detailed sensitive pointed observations from 70 to 500μm, with the SPIRE instrument (Griffin *et al.* 2010) covering 3 bands from 250 to 500μm. Planck, meanwhile, is providing all sky coverage in nine bands from 350μm (857GHz in Planck parlance) to 30 GHz. This is the first all sky survey for many of these bands, providing especially important constraints on both the cool dust that can be seen in the submm (eg. Planck Collaboration, 2011b) and on free-free and anomalous microwave emission at wavelengths longer than about 1mm (Planck Collaboration, 2011c; Peel *et al.* 2011).

The full impact of the new capabilities that Herschel and Planck provide has still to arrive, but first results can already be seen in this volume and elsewhere. In this contribution we will be looking at the results of combining Herschel and Planck data for nearby galaxies that have been observed as part of the Herschel Reference Survey (HRS; Boselli *et al.* 2010).

2. The HRS and the Planck ERCSC

The HRS (Boselli *et al.* 2010) is a survey of a statistically complete sample of local galaxies with the Herschel SPIRE instrument. Images of 323 galaxies at 250, 350 and 500μm have been obtained and additional observations to 100 and 160μm have been proposed. The sample is volume limited with distances between 15 and 25 Mpc and covers the whole range of morphological types, from ellipticals to spirals, and environments, from the field to the centre of the Virgo cluster.

The Planck Early Release Compact Source Catalog (ERCSC; Planck Collaboration, 2011d) is one of the first public data releases from the Planck mission. It includes fluxes for all compact sources detected in Planck bands with high reliability (90% or greater) during the first Planck all sky survey. Of specific interest to the present work is the 857GHz catalog in the ERCSC since this includes additional band merged fluxes at 545, 353 and 217 GHz. It is thus a 350μm (857GHz) selected sample with additional fluxes at 550, 850 and 1400μm. This makes it an ideal resource for studying the far-IR to submm SEDs of galaxies (see e.g. Planck Collaboration, 2011b).

We combine the ERCSC and HRS catalogs by searching for all sources in the HRS that appear in the ERCSC 857GHz catalog. We find 152 matches. Of these, there is a $> 1\sigma$ difference between the Planck 350μm fluxes and the currently available HRS fluxes. This is because the current HRS flux extraction is not determining the fluxes within the large ($\sim 5'$) Planck beams. Instead it is generated using a much smaller beam, so that extended emission and companion sources are not properly accounted for. The production of HRS fluxes optimised for matching to Planck data is currently underway, but is not yet complete.

3. HRS & Planck SEDs

We combine our HRS and Planck fluxes with the IRAS fluxes for these sources, and use the same Bayesian SED fitting approach that was used in Planck Collaboration (2011b) to fit the SEDs of Planck sources. The model SEDs that we fit are based on the usual empirical $F_\nu \propto \nu^\beta B_\nu(T)$ SED, where F_ν is the flux of the source at frequency ν, β is the emissivity, usually between 1 and 2, B_ν is a Planckian function and T is the temperature of the dust. We try fits using a single dust component with T and β as free parameters, two dust components at different temperatures, T_1 and T_2, and β fixed to 2 (eg. Dunne & Eales 2001), and two components with different temperatures and βs. Our Bayesian fitting approach not only allows the best fit and accurate errors on fitted parameters to be calculated, even in the presence of correlated errors in the parameters, but it also allows the Bayesian Evidence (Jaffe 1996; Jaynes 2003) for specific fits and models to be calculated. In this way we can use the Bayes factor to determine which of the models is the better fit for a given object, taking into account the increasing numbers of free parameters where there are more components. There is also the prospect of combining the Bayesian Evidence for all objects for a given fit to determine which model is best at describing the whole sample.

In Fig.1 we show the results of the single component and two component $\beta = 2$ fits for the combined HRS and Planck fluxes. We also show the results of similar fits using IRAS and Planck data alone, and the results of SED fits to IRAS and SCUBA

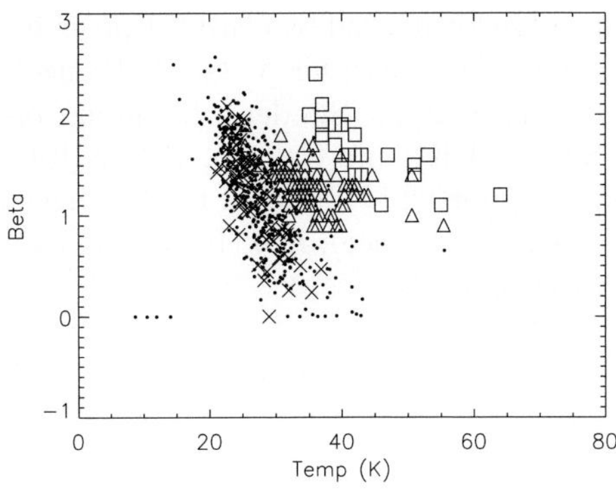

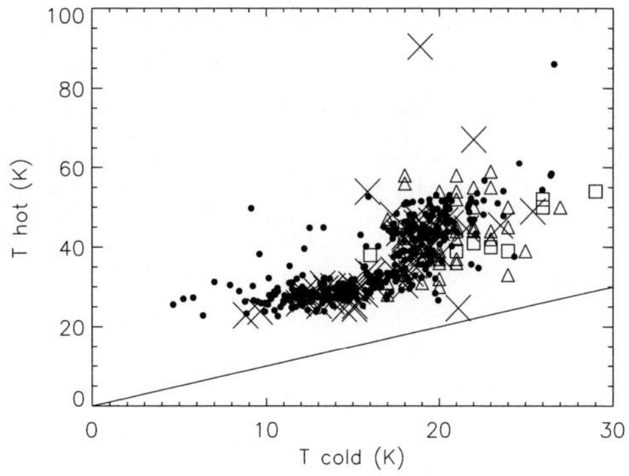

Figure 1. Comparison of the results of parametric fits to the IRAS, HRS and Planck fluxes of the HRS sources (crosses) to similar fits to Planck ERCSC sources (black dots) and to sources observed as part of the SLUGS survey (triangles; Dunne *et al.* 2000; Dunne & Eales 2001) and local ULIRGs (Clements *et al.* 2010).

submm data. As can be seen, the results of our parametric fits to those galaxies where Herschel data is available are broadly similar to the fits obtained using just Planck and IRAS data. We thus confirm the key result on nearby galaxies from the Planck ERCSC (Planck Collaboration, 2011b) that cold dust, with temperatures <<20K, is a significant constituent of the interstellar medium of galaxies.

4. Discussion and Conclusions

We have confirmed the earlier results on these galaxies obtained from just Planck and IRAS data that cold dust with T<<20K is an important constituent of the interstellar medium of nearby galaxies, but there is still much to be done. With the resolved Herschel maps of these sources it will be possible to determine the spatial distribution of different dust components, warm and cold. Meanwhile, once the beam-matched photometry for HRS galaxies is complete, the larger sample size will allow us to cross correlate dust properties to other galaxy properties such as environment, morphology and star formation history. More broadly, the overall methodology used here establishes a framework in which a wide range of SED models, physical as well as parametric, may be tested against the data. Given a large enough sample of galaxies we will be able to draw conclusions about the applicability of these different models using Bayesian Evidence and Bayesian model selection. In this way we hope that the current large range of different models for dust SEDs can be quantitatively compared, and an optimum set of models will emerge.

Acknowledgements

We would like to thank Laure Cisela for her work on the HRS photometry used in this paper and Andrew Jaffe for the provision of the Bayesian SED fitting code. This work is funded in part by STFC. This work is based on observations obtained with Planck, an ESA science mission with instruments and contributions directly funded by ESA Member States, NASA, and Canada. The development of Planck has been supported by: ESA; CNES and CNRS/INSU-IN2P3-INP (France); ASI, CNR, and INAF (Italy); NASA and DoE (USA); STFC and UKSA (UK); CSIC, MICINN and JA (Spain); Tekes, AoF and CSC (Finland); DLR and MPG (Germany); CSA (Canada); DTU Space (Denmark); SER/SSO (Switzerland); RCN (Norway); SFI (Ireland); FCT/MCTES (Portugal); and DEISA (EU). Herschel is an ESA space observatory with science instruments provided by European-led Principal Investigator consortia and with important participation from

NASA. SPIRE has been developed by a consortium of institutes led by Cardiff University (UK) and including Univ. Lethbridge (Canada); NAOC (China); CEA, LAM (France); IFSI, Univ. Padua (Italy); IAC (Spain); Stockholm Observatory (Sweden); Imperial College London, RAL, UCL-MSSL, UKATC, Univ. Sussex (UK); and Caltech, JPL, NHSC, Univ. Colorado (USA). This development has been supported by national funding agencies: CSA (Canada); NAOC (China); CEA, CNES, CNRS (France); ASI (Italy); MCINN (Spain); SNSB (Sweden); STFC (UK); and NASA (USA).

References

Boselli, A. *et al.* 2010, *PASP*, 122, 261
Clements, D. L., Andreani, P., & Chase, S. T. 1993, *MNRAS*, 261, 299
Clements, D. L., Dunne, L., & Eales, S. 2010, *MNRAS*, 403, 274
Dunne, L. & Eales, S. A. 2001, *MNRAS*, 327, 697
Dunne, L. *et al.* 2000, *MNRAS*, 315, 115
Griffin, M. J. *et al.* 2010, *A&A*, 518, L3
Jaffe, A. 1996, *ApJ*, 471, 24
Jaynes, E. 2003, *Probability Theory: the Logic of Science* ed. L. Bretthorst (Cambridge University Press)
Peel, M. W. *et al.* 2011, *MNRAS*, 416, L99
Pilbratt, G. L. *et al.* 2010, *A&A*, 518, L1
Planck Collaboration 2011a, *A&A* 536, A1
Planck Collaboration 2011b, *A&A*, 536, A16
Planck Collaboration 2011c, *A&A*, 536, A17
Planck Collaboration 2011d, *A&A*, 536, A7

Discussion

MADDEN: The data you are using, especially the Planck data, contains dust + CO + free- free/synchrotron. When you perform your Bayesian fits, can you add a radio continuum (free-free, synchrotron) component, so that your resultant B-T solutions are really applicable to the dust? Otherwise your results for B-T cannot be related to dust temperature or dust emissivity.

CLEMENTS: We don't add extra components like CO, free- free, synchrotron at this stage but it would certainly be possible. At the moment we do not push to the long wavelengths where these terms are large. But for some bright sources this will be possible and necessary in the end.

The Spectral Energy Distribution of Galaxies
Proceedings IAU Symposium No. 284, 2011
R.J. Tuffs & C.C. Popescu, eds.

doi:10.1017/S1743921312009258

The spectral energy distributions of the entire *Herschel* Reference Survey

Laure Ciesla[1] and the Herschel-SPIRE Local Galaxies Guaranteed Time Programs

[1]Laboratoire d'Astrophysique de Marseille, UMR6110 CNRS,
38 rue F. Joliot-Curie, F-13388 Marseille France
email: laure.ciesla@oamp.fr

Abstract. We present the spectral energy distributions (SED) of the 323 galaxies of the *Herschel* Reference Survey. In order to provide templates for nearby galaxies calibrated on physical parameters, we computed mean SEDs per bin of morphological types and stellar masses. They will be very useful to study more distant galaxies and their evolution with redshift. This preliminary work aims to study how the most commonly used libraries (Chary & Elbaz 2001, Dale & Helou 2002 and Draine & Li 2007) reproduce the far-infrared emission of galaxies. First results show that they reproduce well the far-infrared part of mean SEDs. For single galaxies the Draine & Li (2007) models seem to reproduce very well the far-infrared emission, as does the Dale & Helou (2002).

Keywords. infrared: galaxies, galaxies: statistics, galaxies: elliptical and lenticular, galaxies: spiral

1. Introduction

The spectral energy distributions (SEDs) of galaxies are our primary source of information about their physical properties. All the different physical processes occurring in galaxies leave their imprint on the global and detailed shape of the SED, each dominating at different wavelengths. Dust is produced by the aggregation of metals injected into the interstellar medium by massive stars (through stellar winds). It absorbs the stellar light and re-emits it in the infrared, from 5 μm to 1mm. *IRAS* (Neugebauer *et al.* 1984), *ISO* (Kessler *et al.* 1996) and *Spitzer* (Werner *et al.* 2004) provided us with data up to 200 μm allowing us to study the emission of the warm dust. The *Herschel* space observatory (Pilbratt *et al.* 2010), launched in May 2009, opened a new window on the far infrared (55 to 672 μm). This new domain of the SED is fundamental to study the emission from cold dust, which dominates by mass. To study the properties of the cold dust, we built up a statistically complete, K-band-selected and volume-limited sample: the guaranteed time key project *Herschel* Reference Survey (HRS; PI: Eales. S.; Boselli *et al.* 2010a) which offers an unique view from 250 μm to 500 μm of nearby galaxies.

2. SEDs of the HRS galaxies: data and mean SEDs

Since HRS galaxies are well known nearby galaxies, HRS benefits from a large amount of ancillary data collected at other wavelengths (ultraviolet: GALEX; optical: SDSS; near-infrared: 2MASS, Spitzer/IRAC; radio) indispensable to extend the study of this proceeding to the whole spectrum in a future work. We focus only on the infrared part of the SED from 24 to 500 μm. We use IRAS 60 and 100 μm data collected from different sources (Boselli *et al.* 2010a) and from NED. Spitzer/MIPS data come from Bendo *et al.*

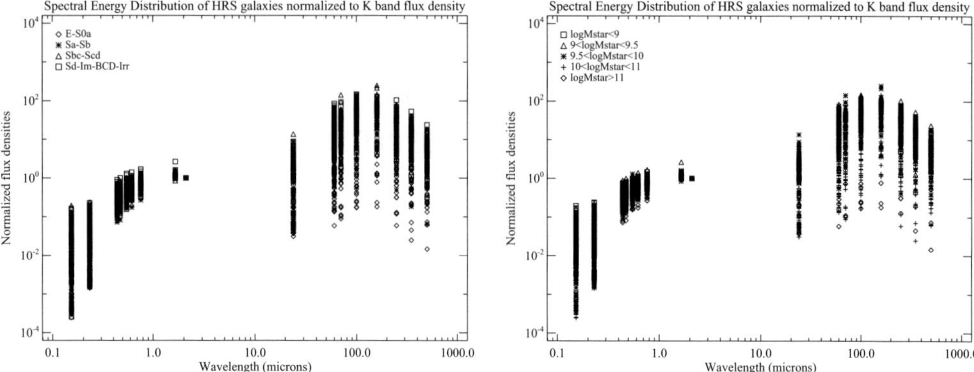

Figure 1. Spectral Energy Distributions of all the HRS galaxies normalized to the K band flux densities. Left panel: Galaxies are coded according to morphological type. Right panel: Galaxies are coded according to stellar mass.

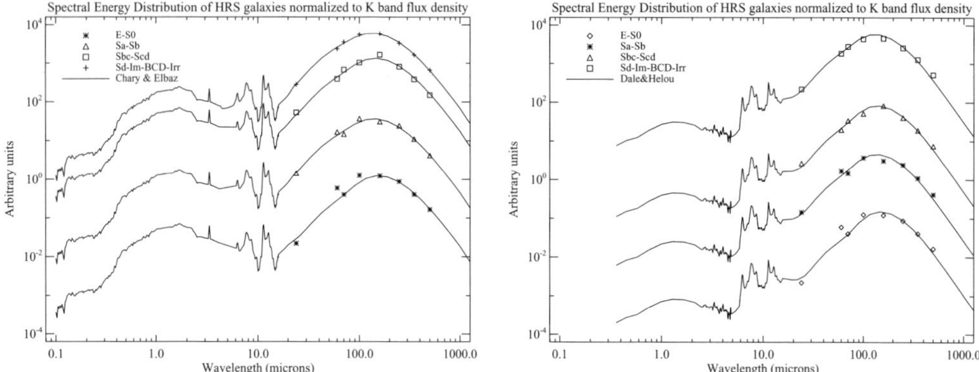

Figure 2. SED fitting of the galaxies binned by morphological type using Chary & Elbaz (2001) (left panel) and Dale & Helou (2002) (right panel).

(2011), who collected and performed the photometry of 168 HRS galaxies at 24, 70 and 160 μm. We have *Herschel*/SPIRE data for all 323 galaxies at 250, 350 and 500 μm. The photometry is fully described in Ciesla *et al.* (2012).

Because these SEDs come from a complete statistically sample, they are ideal to construct templates as a function of different parameters (morphological type, stellar mass, dust mass, specific star formation rate (sSFR), interstellar radiation field intensity, environment). In this very preliminary work, we focus on the morphological type and the stellar mass. Figure 1 shows the SEDs of all the HRS galaxies normalized to the K-band flux densities, per bins of morphological type (left panel) and stellar mass (right panel). For a first approach, we treat upper limits as real measurements of flux densities, and calculate the mean SED for each bin of morphological type and stellar mass.

3. Comparison with literature

We compare our observed SEDs with the popular infrared templates of Dale & Helou (2002) and Chary & Elbaz (2001). As shown on Figure 2, both libraries reproduce well the mean SEDs. Indeed, the particularities that single galaxies have is smoothed by computing the mean flux density in the bins. Elbaz *et al.* (2011) proposed two templates: one for starburst and one for main sequence (normal) galaxies. We find that the main

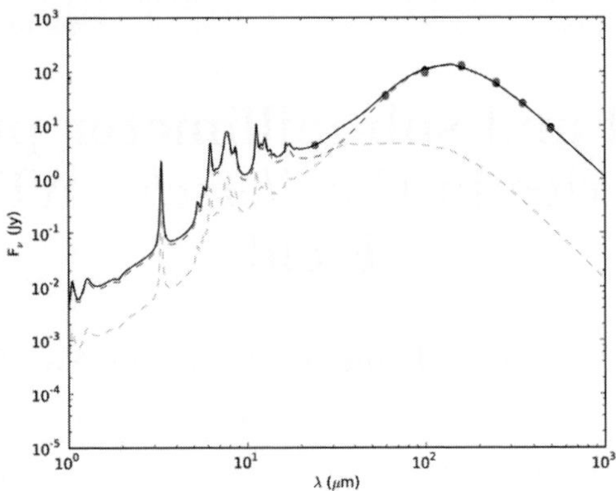

Figure 3. SED fitting of M99 (HRS102) using Draine & Li (2007) models. The parameters of the model are the dust composition (LMC is used in this example), the minimum intensity of the radiation field, the fraction of the total dust mass that is heated by the distribution of starlight intensities, and the dust mass.

sequence template reproduces well the far-infrared part of the mean SED of late-type galaxies, unlike the starburst template. This is to be expected as the HRS sample doesn't contain strong starburst galaxies. However, fitting the two libraries on every single SED, we find that the Dale & Helou (2002) library reproduces better the far-infrared emission of late-type galaxies than the Chary & Elbaz (2001) one. We are now starting to fit the Draine & Li (2007) models to the single SEDs (example with M99 in Figure 3). First preliminary results show that Draine & Li (2007) models are the most appropriate models to reproduce the far-infrared emission of our galaxies.

4. Conclusion

HRS is a complete sample, small enough to study the galaxies one by one and large enough to permit statistical studies. First results on SED fitting show that the Dale & Helou (2002) library seems to reproduce well the FIR SEDs of late types. While the Chary & Elbaz (2001) library also seems to reproduce well the mean SEDs of late types, a further investigation has to be made with single galaxies. The Elbaz *et al.* (2011) main sequence template reproduces well the mean SEDs of late types. First preliminary results show that Draine & Li (2007) seems to better reproduce HRS SEDs of late types.

References

Boselli, A., Eales, S., Cortese, L. *et al.* 2010a, *PASP*, 122, 261
Bendo G., Galliano F. & Madden, S. 2011, *MNRAS* (in preparation)
Chary, R. & Elbaz, D. 2001, *ApJ*, 556, 562
Ciesla, L., Boselli, A., Bendo, G. *et al.* 2012, *A&A* (in preparation)
Dale, D. & Helou, G. 2002, *A&A*, 576, 159
Draine, B. T. & Li, A. 2007, *ApJ*, 657, 810
Elbaz, D. *et al.* 2011, *A&A*, 533, A119
Kessler, M. F. *et al.* 1996, *A&A*, 315, L27
Neugebauer, G. *et al.* 1984, *ApJ*, 278, L1
Pilbratt, G. *et al.* 2010, *A&A*, 518, 3
Werner, M. *et al.* 2004, *ApJS*, 154, 1

The Spectral Energy Distribution of Galaxies
Proceedings IAU Symposium No. 284, 2011
R.J. Tuffs & C.C. Popescu, eds.

doi:10.1017/S174392131200926X

Far-infrared and sub-millimeter properties of SDSS galaxies in the *Herschel* ATLAS SDP Field

Man I. Lam[1,2], Hong Wu[1] and Yi-Nan Zhu[1]

[1]National Astronomical Observatories, CAS Beijing, 100012, P. R. China
[2]Graduate University of the Chinese Academy of Sciences, Beijing, 100049, P. R. China

email: hwu@bao.ac.cn

Abstract. Using data from the new infrared facility the *Herschel* Space Observatory, we have analyzed correlations between morphological type, far-infrared (FIR) luminosity, and Hα luminosity for star-forming galaxies, composite galaxies, and AGNs. We found a trend in scatter from 100μm to 500μm, which indicates that the submillimeter bands are not a good star formation tracer in these galaxies, being contaminated either by the old stellar population or by the interstellar medium (ISM). AGNs have no significant effect on our fitting results since the far-infrared to submillimeter emission is from cold dust/large dust grains.

Keywords. galaxies: starburst, galaxies: ISM, galaxies:classification, infrared: galaxies

1. Introduction

Infrared astronomy developed spectacularly since *IRAS* was launched in 1983, and discovered a type of peculiar object called ultra luminous infrared galaxies (ULIRGs) (Sanders & Mirabel 1996). More space-based infrared telescopes have helped us to study the MIR to FIR emission properties of galaxies from that time, such as *ISO*, and *Spitzer* (Lonsdale *et al.* 2006, Soifer *et al.* 2008). *Herschel* is the first FIR space telescope which can extend the observation from FIR to the submm range to reveal the cold dust properties in our universe.

The MIR and FIR emission of galaxies are usually an indirect detection of extragalactic star formation activities, since infrared emission originates from the dust re-radiation of UV photons emitted by young stellar population. Most of the stellar light is emitted in the UV to near-IR domain, with the short-lived, massive stars dominating the UV and the more numerous older stars the near-IR. Dust, produced by the aggregations of metals injected into the ISM by massive stars through stellar winds and supernovae, absorbs the stellar light and reemits it in the IR and submm domains. At shorter FIR wavelengths (e.g. less than 200μm) luminosity has been proved to be a good star formation tracer of galaxies by *IRAS* observations. Also, previous studies in the MIR using *Spitzer* observations indicate that the MIR emission is well correlated with the 1.4GHz and Hα luminosities. The longer wavelength properties of galaxies are however still unclear, especially in the bands which are dominated by the continuum from big grains. Some clues offered by nearby galaxies show that the older stellar population can also heat dust grains, and dominate the far-infrared luminosity.

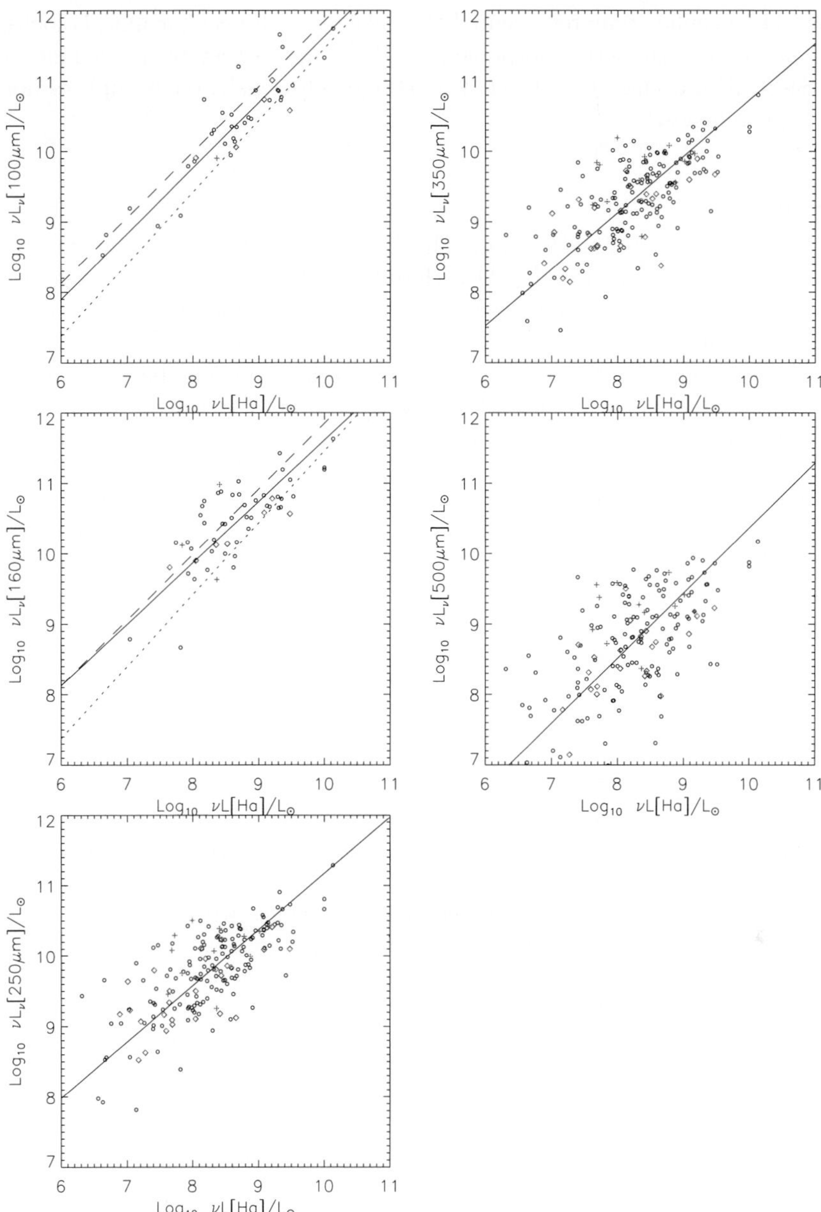

Figure 1. The L(FIR)-L(Hα) correlation. Circles: Spiral galaxies. Crosses: E/S0 galaxies (which show an obvious bias with respect to the fitted line. Diamonds: AGN-dominated galaxies.

2. Sample

We used the first release of data from the *Herschel* ATLAS survey. In order to compare the IR/submillimeter properties of galaxies with different galamorphological types, only sample galaxies with SDSS optical spectra and redshift $z < 0.3$ were selected in our final sample. This sample included 297 objects without obvious lensing.

We divided our sample galaxies into six different morpholocial groups based on de Vaucouleurs *et al.* (1991), which included E/S0, early than Sb, later than Sb, Peculiar, and

Compact. The peculiar type represented the galaxies with peculiar morphologies, either irregular galxies or asymmetric morphologies. The sample is thereby divided into fractions E/S0:S(Sa-Sd):Pec:Compact = 0.08:0.38:0.34:0.08. Obviously, our sample is dominated by later type galaxies.

3. Results

L/M Ratio in different Hubble Types The L/M ratio is influenced by the fraction of gas to stars. In our sample, L(FIR)/M with different Hubble types show a shift between different FIR/Submm bands. The average L/M ratios become higher with later Hubble types. The peculiar galaxies have markedly higher L(FIR)/M ratio than the other five types, because these galaxies have larger gas reservoirs. Also, the peculiar galaxies have higher L(Hα)/M ratios, since these galaxies have a larger fraction of young stars.

L(FIR)-L(Hα) relations Far-infrared and Hα emissions are usually used as star-formation tracers. We present in Fig. 1 the correlations between 'total' FIR and 'total' Hα luminosity after K-correction, dust extinction correction and aperture correction. The L(FIR)-L(Hα) relation shows a trend with Spearman rank coefficient from 0.89 on 100μm to 0.56 on 500μm. The equivalent width has an obvious influence on the fitting. Galaxies with lower equivalent widths deviate from the the best fitting line in our sample. AGNs do not significantly affect the FIR/submm luminosites, since the SEDs of AGNs are power laws from the UV to the infrared, so that the far-infrared amplitude is small.

4. Discussion

L(FIR)-L(Hα) Scatter The trend of L(FIR)-L(Hα) correlation seems to show that the flux in FIR are not just from star formation. Therefore, FIR/Submm emission may be not a good star formation tracer. The FIR emission is more complicated than our previous assumption.

FIR from Early Type Galaxies Some early type galaxies with a very red colour were found in our sample. They are lacking in Hα emission, instead exhibiting stronger Hβ and sodium absorption lines. These galaxies indicate that we cannot ignore the contribution from old stellar populations in the far-infrared. Besides, we also found some sources with weak Hα line emission and a strong continuum from older stellar populations. These sources may be either post-starburst galaxies or undergoing a minor merger event. Although still involved in a starburst, a smaller fraction of the far-infrared luminosity is contributed by young stars in these galaxies.

5. Acknowledgment

We gratefully thank *Hershel* ATLAS group, SDSS, and the SED2011 conference organizer.

References

Sanders, D. B. & Mirabel, I. F. 1996, *ARAA*, 34, 749S
Lonsdale, C. J., Farrah, D., & Smith, H. E. 2006, *asup.book*, 285L
Soifer, B. T., Helou, G., & Werner, M 2008, *ARAA*, 46, 201S
de Vaucouleurs, G., de Vaucouleurs, A., Corwin, H. G., Buta, R. J., Paturel, G., & Fouqué, P., *Third Reference Catalogue of Bright Galaxies*, 1991, Springer

The Spectral Energy Distribution of Galaxies
Proceedings IAU Symposium No. 284, 2011
R.J. Tuffs & C.C. Popescu, eds.

doi:10.1017/S1743921312009271

Far Infrared Luminosity Function of Local Galaxies in the AKARI Deep Field South

Chris Sedgwick[1], Stephen Serjeant[1], Chris Pearson[4,2,1], Shuji Matsuura[3], Mai Shirahata[3], Shinki Oyabu[8], Tomotsugu Goto[5,6], Hideo Matsuhara[3], D.L. Clements[7], Mattia Negrello[1], Toshinobu Takagi[3], and Glenn J. White[1,2]

[1]Department of Physical Sciences, The Open University, Milton Keynes MK7 6AA
[2]Rutherford Appleton Laboratory, Chilton, Didcot, Oxfordshire OX11 0QX
[3]Institute of Space and Astronautical Science, JAXA, Sagamihara, Kanagawa, 252 5210, Japan
[4]Institute for Space Imaging Science, U. of Lethbridge, Lethbridge, Alberta, T1K 3M4, Canada
[5]Institute for Astronomy, U. of Hawaii, 2680 Woodlawn Drive, Honolulu, HI 96822, USA
[6]Subaru Telescope, 650 North A'ohoku Place, Hilo, HI 96720, USA
[7]Astrophysics Group, Imperial College, Blackett Lab., Prince Consort Rd, London SW7 2AZ
[8]Graduate School of Science, Nagoya U., Furo-cho, Chikusa-ku, Nagoya, Aichi 464-8602, Japan

Abstract. We present the first far-infrared luminosity function in the AKARI Deep Field South, a premier deep field of the AKARI Space Telescope, using spectroscopic redshifts obtained with AAOmega. To date, we have found spectroscopic redshifts for 389 galaxies in this field and have measured the local ($z < 0.25$) 90 μm luminosity function using about one-third of these redshifts. The results are in reasonable agreement with recent theoretical predictions.

Keywords. galaxies: evolution, galaxies: luminosity function, infrared: galaxies

---

1. AKARI Deep Field South

The AKARI Deep Field South (ADF-S) is centered on RA 4h 44m 00s, Dec -53° 20' 00" (J2000), and is a low-cirrus ($I_{100\mu m} < 0.5$ MJy sr^{-1}) region of about 12 square degrees near the South Ecliptic Pole (Matsuhara *et al.* 2006). It has been extensively studied over the full area by the AKARI-FIS (at 60, 90, 140, and 160 μm, Shirahata *et al.* 2008), Spitzer (24 and 70 μm, Clements *et al.* 2011), BLAST (250, 350, 500 μm, Valiante *et al.* 2010), and over part of the area by CTIO (UVBRI), AKARI-IRC (7 mid-infrared bands, Pearson *et al.* in preparation), ATCA (deep radio survey at 1.4 GHz, White *et al.* 2011), ASTE/AzTEC (1.1 mm, Hatsukade *et al.* 2011) and APEX/LABOCA (870 μm, Khan *et al.* in preparation). Other work includes CIB detection (Matsuura *et al.* 2011) and cross-identification with public data and SEDs (Malek *et al.* 2010). A large part of the field is being mapped by Herschel as part of the HerMES project.

The 90 μm catalogue from the AKARI Far Infrared Surveyor (FIS) deep survey lists 2,282 sources at a signal-to-noise ratio SNR > 5 (giving a minimum flux of 12.81 mJy). Since this wavelength is within the range at which the re-processed radiation received from dust in Star Forming Galaxies (SFGs) is expected to peak, the evolution of the 90 μm luminosity function provides an excellent proxy for the study of star formation.

This poster (which is based on Sedgwick *et al.* 2011) describes the first luminosity function from this data, for $z < 0.25$ galaxies, and combines this data with earlier ISO data from the ELAIS survey to provide the highest signal-to-noise deep luminosity function at this wavelength to date. We also compare our results with predictions from a recent backward galaxy evolution model, finding a reasonable agreement. Our aim is to provide

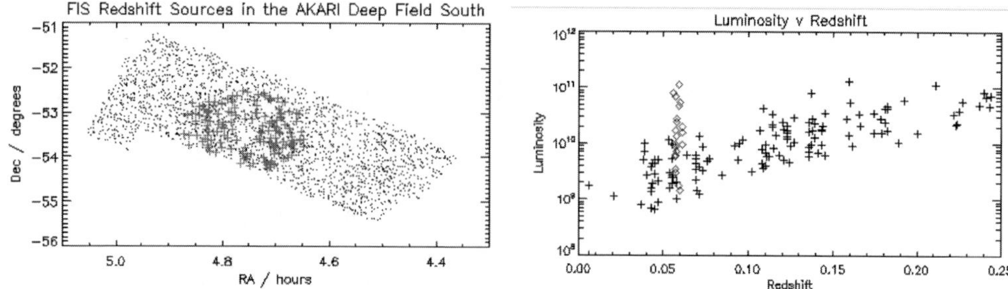

Figure 1. Left: FIS sources with z<0.25 (plus symbols) on all FIS sources (dots) in the AKARI Deep Field South. Right: the redshift - luminosity plane for all sources used, with non-cluster sources shown as plus symbols and the cluster sources shown as diamonds. Luminosity is $\nu L_\nu / L_\odot$ as calculated for the luminosity functions in this paper.

a local benchmark against which to compare luminosity functions of more distant SFGs based on more detailed, deeper data which will become available in the near future.

2. AAOmega Spectroscopy

We have used AAOmega to obtain spectra of selected infrared and sub-millimetre sources with optical counterparts in 3.14 square degrees in the centre of the ADF-S. Data was obtained between October 2007 and November 2008 and redshifts identified for 389 sources, of which we have used 130 for our luminosity function.

Redshifts were measured using the [NII]/Hα/[NII], and Hβ/ [OIII]/[OIII] emission lines. Redshifts were only taken where one or other (or both) of these two triplets of emission lines were identified. Unless out of range, the [SII] doublet was also seen. The [OII] emission line resulting from the $\lambda\lambda$3726, 3729 was often found at the same redshift to support the identifications. AGNs and SFGs were segregated using a BPT diagram. Figure 1 (left) shows the redshifts $z < 0.25$ which were used for this paper.

As shown in the luminosity/redshift plot (Figure 1 right), we have identified a new cluster in this field at $z \sim 0.06$, centred on RA 4h 42m 12.5s, Dec -53° 30' 46" which is discussed in Sedgwick *et al.* (2011) and excluded from the luminosity function.

3. Far-infrared Local Luminosity Function

In the absence of completeness as a function of flux based directly on AKARI data (under construction, Shirahata *et al.*), we have used the ultra-deep Spitzer 70 μm observations of the GOODS-N field which reached down to a flux of 1.2 mJy. Frayer *et al.* (2006) showed that this is well fitted by the model of Lagache *et al.* (2004), which we used to generate number counts against which to measure AKARI completeness. We have adjusted for our 90 μm case using the model described in Pearson (2001), scaling the fluxes by a factor of 0.673 on the basis of its prediction that N($S_{70\mu m}$ > 0.0673 Jy) = N($S_{90\mu m}$ > 0.1 Jy). In addition, we have estimated spectroscopic completeness by comparing the magnitude distribution of AAOmega redshift sources with that of APM B-magnitude sources which have counterparts in the total ADFS-FIS catalogue. The area of the sample was re-normalised to account for large scale cosmic variance, using the number density of objects > 0.1 Jy in the 50 deg^2 SWIRE survey (following the procedure in ELAIS, Serjeant *et al.* 2001). We calculated the 1/V_{max} luminosity function (Schmidt 1968) following the methodology in Serjeant *et al.* (2004). K-corrections were made assuming the M82 star-forming spectral energy distribution. We have assumed cosmological parameter values of H_0=72 kms^{-1}Mpc^{-1}, Ω_M=0.3 and Ω_Λ=0.7.

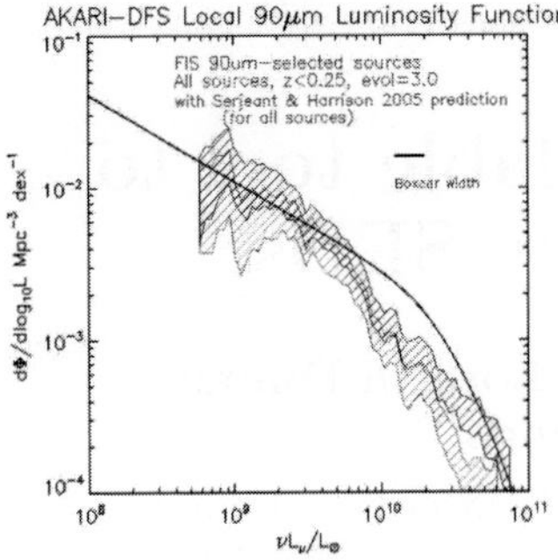

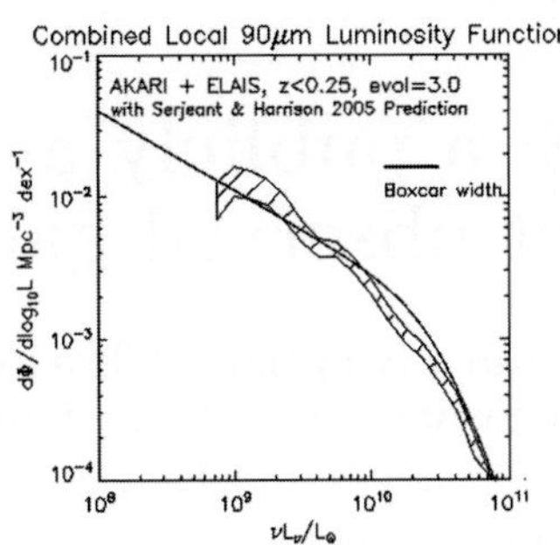

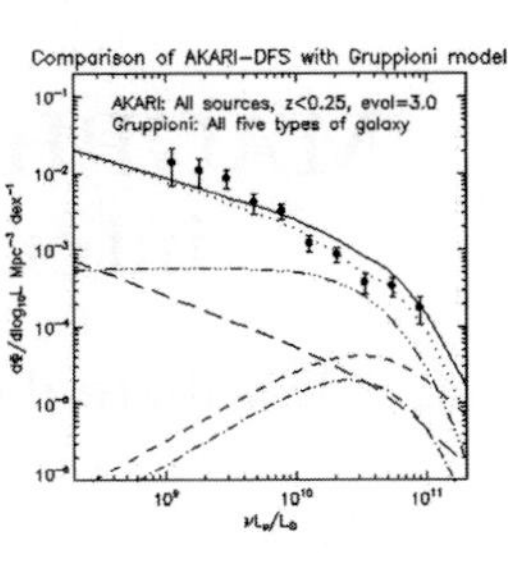

Figure 2. Left: Our new ADF-S luminosity function (Poisson error bands hatched; SFGs-only in shadow). The line is the double power-law prediction by Serjeant & Harrison 2005. Centre: Combined AKARI and ELAIS luminosity function. Right: Comparison of our (unsmoothed) data with predictions from the Gruppioni *et al.* (2011) model. Lines (from top at 10^{10}): total (solid), spirals (dotted), Seyfert2s (dashed-three dots), AGN1s (long-dashed), starbursts (short-dashed), obscured AGNs (dashed-dotted).

In Figure 2, the left panel shows our luminosity function assuming pure luminosity evolution of $(1+z)^3$. We have combined our results with the 90 μm luminosity function of the ELAIS-N fields presented in Serjeant *et al.* (2004) in the centre panel (as described in detail in Sedgwick *et al.* 2011), and in the right panel we show that our results are in reasonable agreement with the theoretical prediction for the 90 μm luminosity function for $0 < z < 0.25$ galaxies from a new backward evolution model of Gruppioni *et al.* (2011) which includes separate evolution for five types of galaxy. The figure also shows that our result is close to the Serjeant & Harrison (2005) double power-law prediction, with a slight faint-end excess and a deficit just below L*, attributed to large-scale structure.

Acknowledgements. This research is based on observations with AKARI, a JAXA project with the participation of ESA. This work was funded in part by STFC (grant PP/D002400/1), the Royal Society (2006/R4-IJP), the Sasakawa Foundation (3108) and KAKENHI (19540250 and 21111004). We extend our thanks to Carlotta Gruppioni.

References

Clements, D. L., Bendo, G., Pearson, C. *et al.* 2011, *MNRAS*, 411, 373
Frayer, D. T., Huynh, M. T., Chary, R. *et al.* 2006, *ApJ*, 647, L9
Gruppioni, C., Pozzi, F., Zamorani, G., & Vignali, C. 2011, *MNRAS*, 416, 70
Hatsukade, B., Kohno, K., Aretxaga, I. *et al.* 2011, *MNRAS*, 411, 102
Lagache, G., Dole, H., Puget, J.-L., *et al.* 2004, *ApJS*, 154, 112
Malek, K., Pollo, A., Takeuchi, T. T. *et al.*. 2010, *A&A*, 514, 11
Matsuhara, H., Wada, T., Matsuura, S. *et al.* 2006, *PASJ*, 58, 673
Matsuura, S., Shirahata, M., Kawada, M. *et al.* 2011, *ApJ*, 737, 2
Pearson, C. 2001, *MNRAS*, 325, 1511
Schmidt, M. 1968, *ApJ*, 151, 393
Sedgwick, C., Serjeant, S., Pearson, C. *et al.* 2011, *MNRAS*, 416, 1862
Serjeant, S., Carraminana, A., Gonzales-Solares, E. *et al.* 2004, *MNRAS*, 355, 813
Serjeant, S., Efstathiou, A., Oliver, S. *et al.* 2001, *MNRAS* 322, 262
Serjeant, S. & Harrison, D. 2005, *MNRAS*, 356,192
Shirahata, M., Matsuura, S., Takagi, T. *et al.* 2008, *ASPC*, 399, 290
Valiante, E., Ade, P. A. R., Bock, J. J. *et al.* 2010, *ApJS*, 191, 222
White, G. J., Hatsukade, B., Pearson, C. *et al.* 2011, *MNRAS, submitted*

The Spectral Energy Distribution of Galaxies
Proceedings IAU Symposium No. 284, 2011
R.J. Tuffs & C.C. Popescu, eds.

doi:10.1017/S1743921312009283

MAGPHYS: a publicly available tool to interpret observed galaxy SEDs

Elisabete da Cunha[1], Stéphane Charlot[2], Loretta Dunne[3], Dan Smith[4], and Kate Rowlands[3]

[1]Max Planck Institute for Astronomy, Königstuhl 17, 69117 Heidelberg, Germany

[2]UPMC Univ. Paris 6/CNRS, UMR 7095, Institut d'Astrophysique de Paris, France

[3]School of Physics & Astronomy, Nottingham University, University Park Campus, Nottingham NG7 2RD, UK

[4]Centre for Astrophysics, Science & Technology Research Institute, University of Hertfordshire, Hatfield, Herts, AL10 9AB, UK

email: cunha@mpia.de

Abstract. We present a simple, physically-motivated model to interpret consistently the emission from galaxies at ultraviolet, optical and infrared wavelengths. We combine this model with a Bayesian method to obtain robust statistical constraints on key parameters describing the stellar content, star formation activity and dust content of galaxies. Our model is now publicly available via a user-friendly code package, MAGPHYS at www.iap.fr/magphys. We present an application of this model to interpret a sample of ~ 1400 local ($z < 0.5$) galaxies from the H-ATLAS survey. We find that, for these galaxies, the diffuse interstellar medium, powered mainly by stars older than 10 Myr, accounts for about half the total infrared luminosity. We discuss the implications of this result to the use of star formation rate indicators based on total infrared luminosity.

Keywords. dust, extinction galaxies: ISM galaxies: stellar content galaxies: statistics.

---

1. A simple model to interpret galaxy SEDs

Multi-wavelength surveys of large samples of galaxies both in the local and high-redshift Universe have become widely available in recent years. To understand these observations in the framework of galaxy formation and evolution, we must be able to extract key physical parameters from their observed spectral energy distributions (SEDs). In da Cunha, Charlot & Elbaz (2008), we have developed a simple model to interpret consistently the ultraviolet, optical and infrared emission from galaxies in terms of their star formation histories and dust content.

1.1. *Description of the model*

We compute the spectral evolution of stellar populations in galaxies using the state-of-the-art population synthesis model of Bruzual & Charlot (2003). This model is based on the property that stellar populations with any star formation history can be expanded in a series of instantaneous bursts, simple stellar populations (SSPs). The spectral energy distribution of a galaxy is then computed by adding the individual spectra of all SSPs weighted by the star formation rate over time since the galaxy was formed.

The spectra of galaxies also contain valuable information about the interaction of starlight with interstellar gas and dust, and the physical properties of the interstellar medium (ISM), such as dust content. Following Charlot & Fall (2000), we describe the

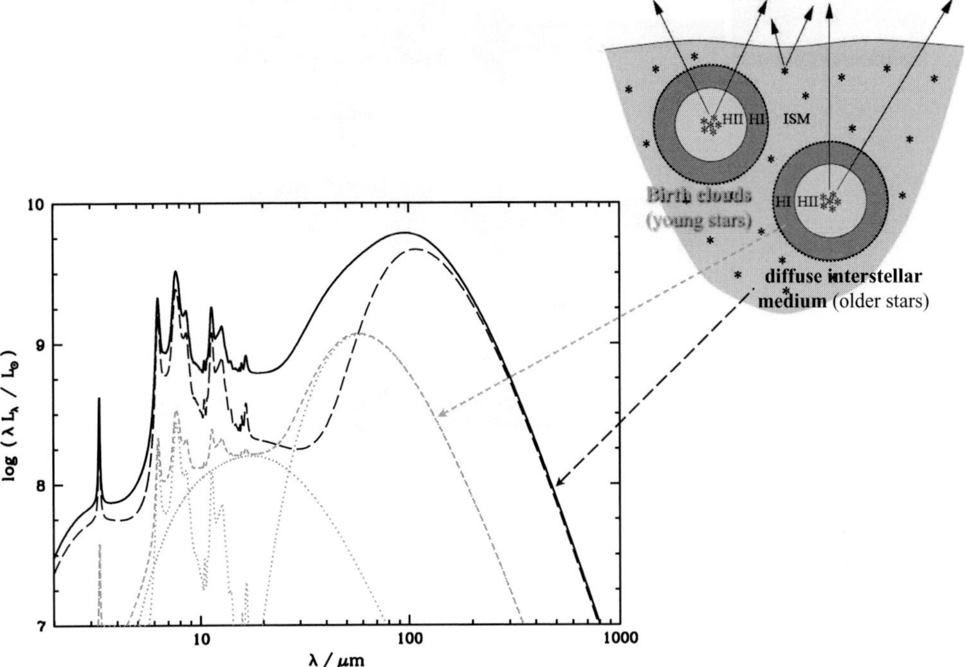

Figure 1. Schematic view of the two-component ISM by Charlot & Fall (2000) (top-right): stars are born in dense molecular clouds – birth clouds – and later (after 10^7 yr) migrate to the ambient diffuse ISM. The left-bottom plot shows an example total dust emission SED (solid line) in the infrared range, constructed using the emission components described in da Cunha, Charlot & Elbaz (2008). The contributions by the birth clouds and the ambient ISM are plotted with short-dashed and long-dashed lines, respectively.

ISM of galaxies in our model using two main components: the ambient (diffuse) interstellar medium and the star-forming regions (birth clouds). Stars are born in dense molecular clouds which dissipate typically on a time-scale of 10^7 years. As a result, the non-ionizing continuum emission from young OB stars and line emission from their surrounding HII regions may be absorbed by dust in these birth clouds and then in the ambient ISM, while the light emitted by stars older than 10^7 yr propagates only through the diffuse ISM. This simple model successfully accounts for the different attenuation of line and continuum emission in star-forming galaxies. We use this prescription to compute the total energy absorbed by dust in the birth clouds and in the ambient ISM. We define the total dust luminosity re-radiated by dust in the birth clouds and in the ambient ISM as $L_{\rm d}^{\rm BC}$ and $L_{\rm d}^{\rm ISM}$, respectively. The total luminosity emitted by dust in the galaxy is $L_{\rm d}^{\rm tot} = L_{\rm d}^{\rm BC} + L_{\rm d}^{\rm ISM}$. We distribute $L_{\rm d}^{\rm BC}$ and $L_{\rm d}^{\rm ISM}$ in wavelength over the range from 3 to 1000 μm using four main dust components: *(i)* the emission from polycyclic aromatic hydrocarbons (PAHs); *(ii)* the mid-IR continuum from hot dust; *(iii)* the emission from warm dust (30–60 K) in thermal equilibrium; *(iv)* the emission from cold dust (15–25 K) in thermal equilibrium. In stellar birth clouds, the relative contributions to $L_{\rm d}^{\rm BC}$ by PAHs, the hot mid-infrared continuum and warm dust are kept as adjustable parameters. These clouds are assumed not to contain any cold dust. In the ambient ISM, the contribution to $L_{\rm d}^{\rm ISM}$ by cold dust is kept as an adjustable parameter. The relative proportions of the other 3 components are fixed to the values reproducing the mid-infrared cirrus emission of the Milky Way. We find that this minimum number of components is required to

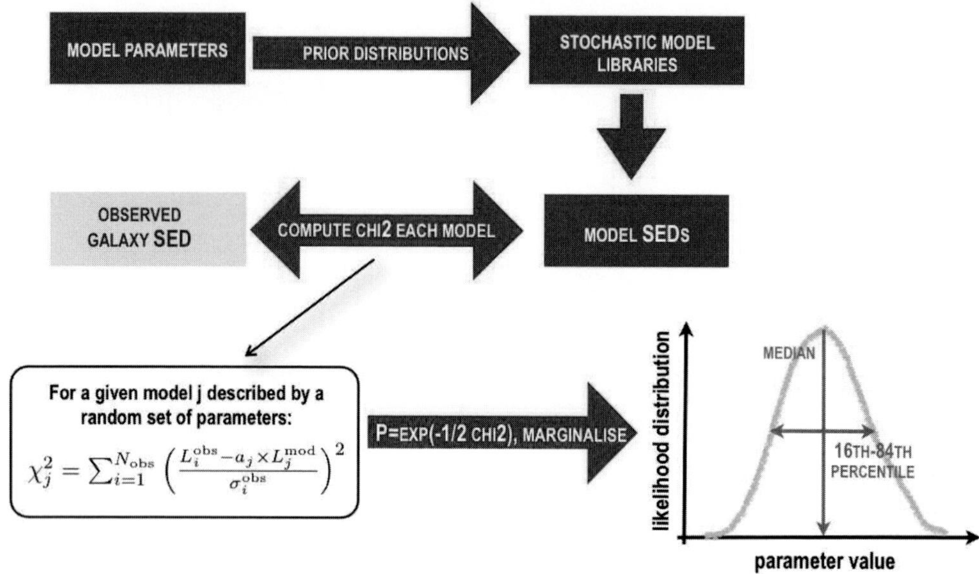

Figure 2. Summary of the Bayesian approach we use to compare observed galaxy SEDs with our model and derive statistical constraints on the physical parameters of observed galaxies (see da Cunha *et al.* 2008 for details).

account for the infrared SEDs of galaxies in a wide range of star formation histories (see da Cunha *et al.* 2008 for details). And example model SED is shownin in Fig. 1.

1.2. *Statistical constraints of physical parameters*

Our model is optimized to derive statistical constraints of fundamental parameters related to star formation activity and dust content (e.g. star formation rate, stellar mass, dust attenuation, dust temperatures) of large samples of galaxies using a wide range of multi-wavelength observations (e.g. da Cunha *et al.* 2008, 2010). We use a Bayesian approach, summarized in Fig. 2, to interpret the SEDs all the way from the ultraviolet/optical to the far-infrared. A similar approach has been previously used mostly to interpret optical galaxy spectra from the ultraviolet to the near-infrared, i.e. not including the dust emission. This approach allows us to understand in detail our parameter constraints by identifying degeneracies and exploring what observations are required to constrain each parameter.

2. The MAGPHYS package

The da Cunha, Charlot & Elbaz (2008) model is **publicly available as the MAGPHYS package at** www.iap.fr/magphys. MAGPHYS - *Multi-wavelength Analysis of Galaxy Physical Properties* - is a self-contained, user-friendly model package to interpret observed spectral energy distributions of galaxies in terms of galaxy-wide physical parameters pertaining to the stars and the interstellar medium. The analysis of the spectral energy distribution of an observed galaxy with MAGPHYS is done in two steps: (i) The assembly of a comprehensive library of model SEDs at the same redshift and in the same photometric bands as the observed galaxy, for wide ranges of plausible physical parameters pertaining to the stars and ISM. (ii) The build-up of the marginalised likelihood distribution of each physical parameter of the observed galaxy, through the comparison of the observed SED with all the models in the library (Fig. 2).

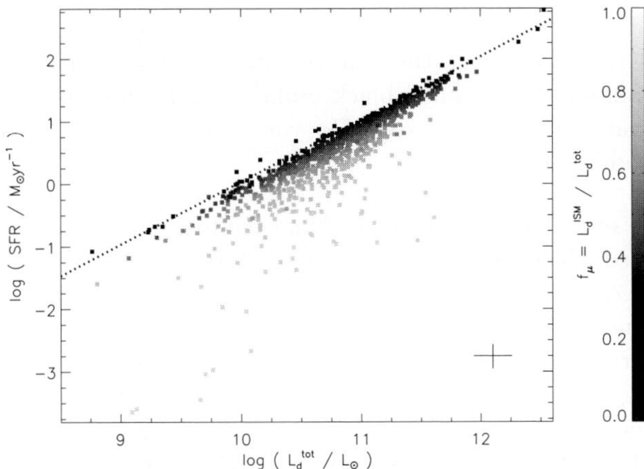

Figure 3. Star formation rate, SFR, plotted against total dust luminosity, $L_{\rm d}^{\rm tot}$, for 1404 galaxies from the H-ATLAS survey (da Cunha *et al.*, in prep.). Each galaxy is colour-coded (on a grey-scale) according to the fraction of total infrared luminosity contributed by the diffuse ISM, f_μ. The dotted line shows, for reference, the conversion between total IR luminosity and SFR from Kennicutt (1998).

3. Dust heating by stars older than 10 Myr in H-ATLAS galaxies

The H-ATLAS survey is detecting thousands of galaxies in the Herschel/SPIRE bands over a large area of the sky. Thanks to the overlap with other multi-wavelength surveys, complete UV-to-IR SEDs are available for large samples of galaxies (Eales *et al.* 2010). We have used MAGPHYS to extract statistical constraints on star formation rates (SFRs), stellar masses, dust luminosities and dust masses of a sample of 250-μm detected galaxies at $z < 0.5$ from their observed SEDs (Smith *et al.*, submitted). We are investigating the use of the IR luminosity as a SFR indicator in these galaxies. The total IR is often used as a SFR tracer but observational evidence shows that dust in galaxies is not exclusively heated by newly-formed stars (e.g. Bendo *et al.* 2010). We find that the SFR derived from fits to the total SED with our model, which includes heating by stars older than 10^7 yr in the diffuse ISM, is not exactly traced by the dust luminosity (as, for example, when using the Kennicutt 1998 IR to SFR conversion; see Fig. 3). A significant fraction (typically 50%) of $L_{\rm d}^{\rm tot}$ in H-ATLAS galaxies comes from the diffuse ISM, mainly powered by stars older than 10 Myr. This implies that using the total IR as a SFR tracer may lead to overestimating the SFR unless the contribution by the diffuse ISM to the total IR is properly taken into account (see also e.g. Kennicutt 2009).

Acknowledgements

We thank the H-ATLAS team for providing the photometry for the work described in Section 3; the H-ATLAS website is www.h-atlas.org.

References

Bendo G. J. *et al.* 2010, *A&A*, 518, L65
Bruzual G. & Charlot S. 2003, *MNRAS*, 344, 1000
Charlot S. & Fall S. M. 2000, *ApJ*, 539, 718
da Cunha E., Charlot S., & Elbaz D. 2008, *MNRAS*, 388, 1595
da Cunha E., Eminian C., Charlot S., & Blaizot J. 2010, *MNRAS*, 403, 1894
Eales S. *et al.* 2010, *PASP*, 122, 499
Kennicutt R. C., Jr. 1998, *ARA&A*, 36, 189
Kennicutt R. C., Jr. *et al.* 2009, *ApJ*, 703, 1672

Discussion

MADDEN: Could you describe further the parameters of the dust components in your model? For example for the modified black bodies you use for the FIR-submm, are the beta parameter and temperature fixed? Do you solve for PAHs, or do you use fixed templates from models/obervations?

DA CUNHA: The emission by dust in thermal equilibrium in our model is described by a set of modified black bodies. In the stellar birth clouds, we adopt an emissivity index of $\beta = 1.5$ to describe the emission by warm dust; the temperature of this dust can vary between 30 and 60 K. Cold dust in the diffuse ISM is described by $\beta = 2$ and the equilibrium temperature can vary between 15 and 25 K. The emissivity index is fixed for simplicity, to keep the number of adjustable parameters of the model minimal. The PAH emission is described by an empirical template for simplicity: we use the mid-IR spectrum of the M17 PDR. The contribution of PAHs to the total IR of the diffuse ISM is fixed according to the observed SED of the Milky Way cirrus emission. In the birth clouds the contribution of PAHs to the total IR is allowed to vary (more details are given in da Cunha, Charlot & Elbaz 2008).

BENDO: Did you test your model fitting code on model spectra without some SED data points (similar to the tests performed with observational data)?

DA CUNHA: Yes, we did such tests in da Cunha, Charlot & Elbaz (2008), to test how well we can recover known model parameters from a set of photometric observations. We have not done this in the context of higher redshift galaxies ($z \sim 2$) and two-dimensional pdfs as I showed in this talk, but this is definitely a test we plan to do in the near future.

SCHAERER: I have a comment regarding the contribution from "old" stars to the IR emission. One has to be careful: I suppose that by "old" you mean younger than $\sim$ 10 Myr? Remember that stars older than this still contribute to the UV. That may explain the very large contribution of "old" stars you find.

DA CUNHA: Indeed that is a good point. By "old stars" I mean stars older than 10 Myr, which have migrated from the stellar birth clouds into the diffuse ISM component. These stars still radiate a large amount of non-ionizing UV radiation which heats the dust. But even stars with little UV emission (typically older than 100 Myr) still contribute to the dust heating in a significant way. But I agree that if we considered a large time-scale, the fraction contributed to the IR luminosity might be smaller. The main point of the plot I showed was to emphasize that a non-negligible fraction of the total IR is not powered by recent star formation, and this should be taken into account when deriving SFR from the observed IR luminosity.

The Spectral Energy Distribution of Galaxies
Proceedings IAU Symposium No. 284, 2011
R.J. Tuffs & C.C. Popescu, eds.

doi:10.1017/S1743921312009295

Fitting the full SED of galaxies to put constraints on dust attenuation and star formation determinations

Veronique Buat, Elodie Giovannoli, Mederic Boquien and Sébastien Heinis

Laboratoire d'Astrophysique de Marseille
Aix Marseille University and CNRS 38 rue Joliot Curie, 13013 Marseille, France
email: Veronique.Buat@oamp.fr

Abstract. The combination of far-IR and UV-optical rest-frame data has proved to be very efficient to extract physical parameters from the SEDs of galaxies. Using *Herschel* and ancillary data from the *Herschel* Reference Survey and GOODS-*Herschel* Key Projects, we show how dust attenuation properties can be estimated inside local galaxies as well as in the distant Universe.

Keywords. infrared: galaxies, dust: extinction, ultraviolet: galaxies

---

1. Introduction

The broad-band spectral energy distributions (SED) of galaxies are the combination of the emission from stars of all ages and interstellar dust which interact in a complex way through absorption and scattering. Stars emit from the UV to the near-IR whereas the mid- and far-IR emissions come from interstellar dust heated by the stellar emission. With the availability of mid- and far-IR data for large samples of galaxies, codes combining stellar and dust emissions to analyze SEDs are particularly useful to constrain dust attenuation and provide robust star formation rates. *Herschel* observations coupled to ancillary data provide us with UV-to-far-IR SEDs for galaxies from the local to the high redshift Universe. Here we use data from two key projects: the *Herschel* Reference Survey (Boselli *et al.* 2010) and the GOODS-*Herschel* project (Elbaz *et al.* 2011).

2. CIGALE: a physically-motivated code

We use the code CIGALE (Code Investigating GALaxy Emission) (http://cigale.oamp.fr) which provides physical informations about galaxies by fitting their UV-to-far-IR SED (Noll *et al.* 2009, Giovannoli *et al.* 2011). CIGALE combines a UV-optical stellar SED and an IR-emitting dust component. First, models are built, then each model is quantitatively compared to the observed SED, accounting for uncertainties in the observed fluxes. The probability function of each parameter is calculated and the estimated value of the parameter and its error correspond to the mean and standard deviation of this distribution. The robustness of the parameter estimations is checked with artificial catalogues constructed for each set of input data. Models are generated with a stellar population synthesis code, assuming a star formation history and a dust attenuation scenario as inputs. The star formation history implemented in CIGALE is the combination of two stellar components which mimic an old and a young stellar population. To model the

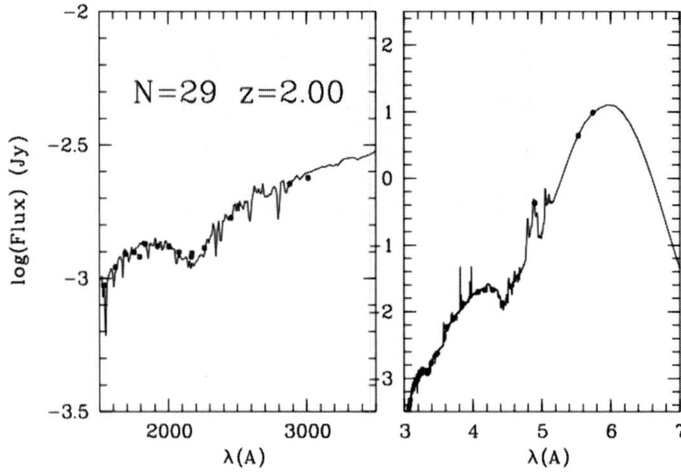

Figure 1. Example of SED fitting with CIGALE in the CDFS field for a galaxy at z=2. The whole SED is plotted in the right panel. The left panel is a zoom of the UV-optical range

attenuation by dust, the code uses the attenuation law of Calzetti *et al.* (2000) as a baseline, and offers the possibility of varying the steepness of this law and adding a bump centred at 2175 Å. We refer to Noll *et al.* (2009) for a complete description of the dust attenuation prescription. The energetic balance between dust-enshrouded stellar emission and re-emission in the IR is carefully conserved by combining the UV-optical and IR SEDs. The IR SEDs are built from the Dale & Helou (2002) templates. CIGALE allows for additional dust emission from a dust-enshrouded AGN but does not include the unobscured emission of an active nucleus.

3. Spatially resolved galaxies from the *Herschel* Reference Survey

We exploit the spatial resolution of SPIRE maps to perform a local study of dust attenuation in a sub-sample of 7 galaxies extracted from the Herschel Reference Survey (Boselli *et al.* 2010). 11 photometric bands are used, FUV, NUV (GALEX), u, g, r, i, z (SDSS), h (2MASS), 70 μm (SPITZER), 250 and 350 μm (Herschel/SPIRE) are used. The method is illustrated in Figure 2 with HRS102 (Messier 99): the data are convolved and regridded to the same reference frame (at 350 microns) and a pixel-to-pixel analysis of the emission is performed by fitting the SED of each pixel with CIGALE. Maps are built for each output parameter of the code. Results of the SED fitting are used to study local variations of dust attenuation and their link to the UV emission (Boquien *et al.*, in preparation).

4. Dust attenuation curves in high redshift galaxies from the GOODS-*Herschel* survey

Dust attenuation curves in external galaxies are useful for studying their dust properties as well as interpreting their intrinsic SEDs. These functions are not very well known in the UV range whether at low or high redshift. In particular the presence or absence of a UV bump at 2175 Å remains an open issue that has consequences on the interpretation of broad band colours of galaxies involving the UV range. As part of the GOODS-*Herschel* Key Programme (Elbaz *et al.* 2011), the *Herschel* Space Observatory (Pilbratt *et al.* 2010) surveyed a small area of the Great Observatories Origins Deep Survey Southern

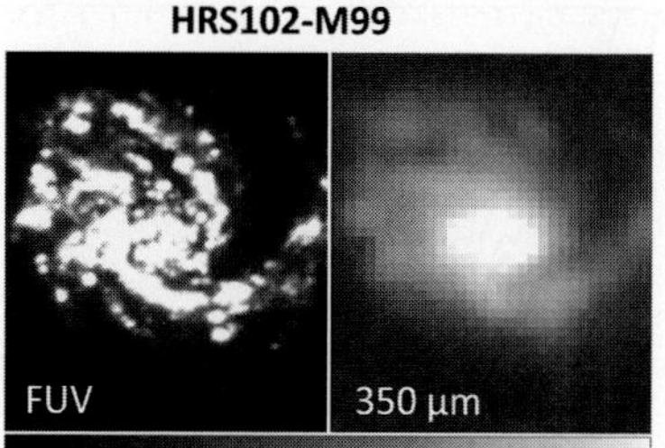

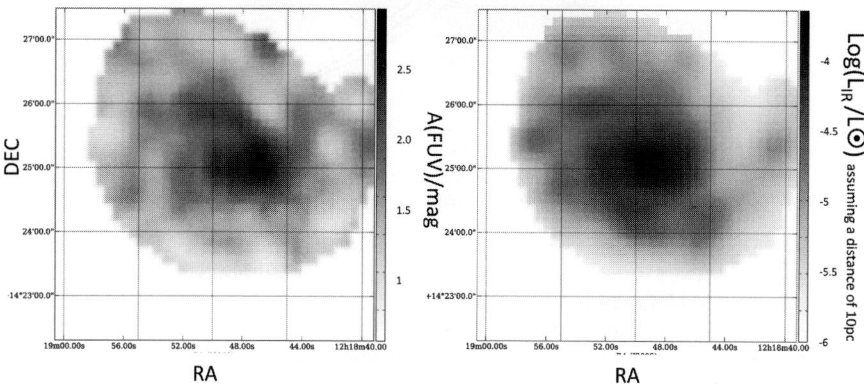

Figure 2. Application of CIGALE to HRS102

field (GOODS-S): $10' \times 10'$ centred on the CDFS were observed at 100 and 160 μm over 264 hours by the PACS instrument (Poglitsch *et al.* 2010). Source extraction on the PACS images was performed at the prior position of *Spitzer* 24 μm sources as described in Elbaz *et al.* (2011).

We use deep photometric data of the Chandra Deep Field South obtained with intermediate and broad band filters by the MUSYC project (Cardamone *et al.* (2010)) to sample the UV rest frame of galaxies with $1 < z < 2$. *Herschel*/PACS and *Spitzer*/MIPS data are used to measure the dust emission. We select 30 galaxies with high S/N in all bands. Their SEDs from the UV to the far-IR are fitted using the CIGALE code and the characteristics of the dust attenuation curves obtained as outputs of the SED fitting process.

The average dust attenuation curve, shown in Fig. 3, is described well by a modified Calzetti *et al.* law slightly steeper than the original one and with a UV bump at 2175 Å whose amplitude is $\sim$ 35% of the MW one. We propose an analytical expression of the average attenuation curve, which can be used to correct the SEDs of galaxies for dust attenuation.

$$\frac{A(\lambda)}{A_{\rm V}} = \frac{k'(\lambda) + D_{\lambda_0,\gamma,E_{\rm b}}(\lambda)}{4.05}\left(\frac{\lambda}{\lambda_V}\right)^{-0.13}, \tag{4.1}$$

where $\lambda_V = 5500$ Å, $k'(\lambda)$ is given in Calzetti *et al.* (2000) (Eq.4), and

$$D_{\lambda_0,\gamma,E_{\rm b}}(\lambda) = \frac{1.26 \times 356^2 \lambda^2}{(\lambda^2 - 2175^2)^2 + \lambda^2 \times 356^2}. \tag{4.2}$$

The moderate amplitude of the bump, together with the substantial slope of our average attenuation curve in the UV, argues for a deficit of UV bump carriers in our sample galaxies as compared to the MW. For more details we refer to Buat *et al.* (2011).

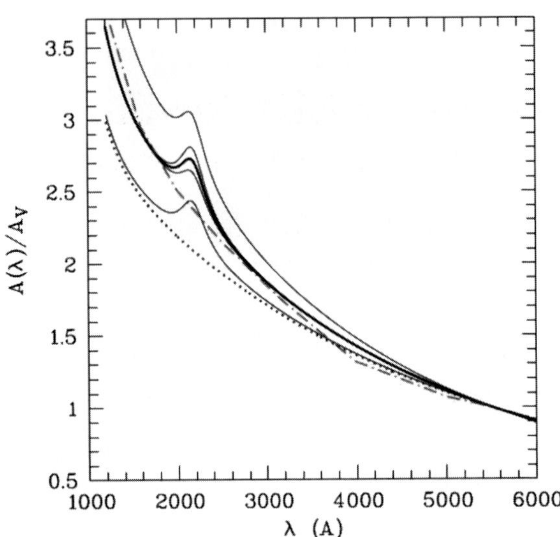

Figure 3. Average dust attenuation curve: thick line. The thin lines correspond to variations of $\pm 1\sigma$ of the amplitude of the bump, and the slope of the continuum. The Calzetti *et al.* (2000) attenuation curve (dotted line) and the effective absorption curve of Charlot & Fall (2000) for a starburst age of 3 10^8 years (dot-dashed line) are shown in the figure for comparison.

References

Boselli, A. *et al.* 2010, *PASP*, 122, 261
Buat, V. *et al.* 2011, *A&A*, 533, 93
Calzetti, D., Armus, L., Bohlin, R. C. *et al.* 2000, *ApJ*, 533, 682
Cardamone, C. N., van Dokkum, P. G., Urry, C. M. *et al.* 2010, *ApJSS*, 189, 270
Charlot, S. & Fall, S. M. 2000, *ApJ*, 539, 718
Dale, D. A. & Helou, G. 2002, *ApJ*, 576, 159
Elbaz, D. *et al.* 2011, *A&A*, 533, 119
Giovannoli, E., Buat, V., Noll, S., Burgarella, D., & Magnelli, B. 2011, *A&A*, 525, 150
Noll S., Burgarella D., Giovannoli E. *et al.* 2009b, *A&A*, 507, 1793
Pilbratt, G., Riedinger, J. R., Passvogel, T., *et al.* 2010, *A&A*, 518, L1
Poglitsch, A., Waelkens, C., Geis, N. *et al.* 2010, *A&A*, 518, L2

Discussion

CLEMENTS: Naively, how many free parameters go into the CIGALE fits?

BUAT: The number of free parameters has to be fixed by the user. It clearly depends on the number of data available for each galaxy, for example one can fix the attenuation curve or use only one stellar component. Mock catalogues are very useful to decide which parameter can be free or have to be fixed.

GALLAGHER: Does CIGALE make linear combinations of templates, and if so are there issues in selecting templates to combine?

BUAT: In fact templates are based on a set of input parameters and built for each set of parameters. Templates are only pre-defined for the IR component. The set of parameters is chosen in order that the SED templates used for the fits reproduce the colours of the galaxies to be fitted. Therefore, CIGALE does not make a linear combination of templates.

The Spectral Energy Distribution of Galaxies
Proceedings IAU Symposium No. 284, 2011
R.J. Tuffs & C.C. Popescu, eds.

doi:10.1017/S1743921312009301

Measuring SEDs for individual galaxy components

Steven P. Bamford[1], Boris Häußler[1], Alex Rojas[2], Marina Vika[2] and Jim Cresswell[2]

[1]School of Physics and Astronomy, University of Nottingham, University Park, Nottingham, NG7 2RD, UK
email: `steven.bamford@nottingham.ac.uk`

[2]Carnegie Mellon University in Qatar, PO Box 24866, Doha, Qatar

Abstract. Our project, 'MegaMorph', is developing a next-generation tool for decomposing galaxies, in terms of both their structures and stellar populations. By combining data from UV to NIR wavelengths, accounting for morphological peculiarities using non-parametric components, and utilising efficient likelihood sampling methods, we are working to significantly improve the robustness and accuracy of galaxy decomposition. Applying these new techniques to modern large surveys will provide us with a deeper understanding of galaxies.

Keywords. galaxies: fundamental parameters, galaxies: structure, galaxies: photometry, methods: data analysis, techniques: image processing

---

1. Galaxy components

Galaxies comprise a complex mix of stars, gas, dust and dark matter, with their proportions, distributions and kinematics resulting from a wide variety of physical processes. Even considering just the stars, galaxies typically possess distinct stellar populations, which formed at different times, under varied conditions. To gain more than a superficial understanding of the galaxy population, and the processes that shape it, we need to disentangle these physical components within individual galaxies.

This conference primarily focused on separating and characterising galaxy components through their spectral energy distributions. These SEDs are usually, with a few notable exceptions, integrated over the spatial extent of each galaxy. For separating components which dominate at different wavelength ranges, e.g., stars and dust, this approach works well. However, when dividing components with only subtle differences in their SEDs, e.g., multiple stellar populations, any additional discriminatory information is valuable.

The various stellar populations within a galaxy often have contrasting spatial distributions. In our simplest models, a galaxy grows through two principal mechanisms: the gradual accretion of gas, which cools and forms a thin disk of stars; and the merging of existing stellar systems, which produces a spheroid. These two structural components thus form through disparate mechanisms, contain different stellar populations, and represent disjoint periods in a galaxy's history (e.g., Cole *et al.* 2000; Cook *et al.* 2009; Benson 2010). Using information regarding both the spatial distributions and SEDs of these components will clearly enable a more robust separation of their properties, when compared with trying to use just one of these pieces of information alone. The dichotomy between disk and spheroid stellar components (e.g., Allen *et al.* 2006; Benson & Devereux 2010) is at the heart of our understanding of galaxy development: measuring their individual properties will provide a deeper understanding of the galaxy population and more effective comparison with theoretical models.

2. Present structural decomposition solutions

A number of solutions exist for decomposing galaxy structures (e.g., GALFIT, Peng *et al.* 2002, 2010; GIM2D, Simard *et al.* 2002, etc.), which generally fit smooth, parametric models to a single image of each galaxy. These have been applied to large samples (e.g., Allen *et al.* 2006; Simard *et al.* 2011), confirming the value of the approach. However, there remain significant problems regarding the robustness and physical interpretation of the fitted model parameters, e.g., Allen *et al.* find that ~15 per cent of their attempted bulge+disk fits result in unphysical parameters. These difficulties are due to degeneracies between the parameters of the two galaxy components in a single wavelength-band model, exacerbated by low signal-to-noise imaging, and the presence of features in the data (e.g., star-formation regions, spiral arms, bars, etc.) which are not included in the model.

A further issue is the measurement of SEDs for the separate components. Surveys now typically produce imaging with comparable quality across many photometric bands. For example, the SDSS (Abazajian *et al.* 2009) and UKIDSS LAS (Lawrence *et al.* 2007) surveys together provide imaging spanning nine optical and near-infrared bands. One may treat each band independently, but the recovered components will differ in size and shape and thus not correspond to the same population of stars at each wavelength, compromising the physical meaning of their colours. More usefully, one can perform a full fit in a single band, and then rescale the flux of the resulting model to fit each additional band. This produces meaningful colours for the physically distinct components, but their quality depends upon the reliability of the initial single-band decomposition. As mentioned above, a significant fraction of these will be physically unrealistic.

3. Next-generation galaxy decomposition

We have embarked on a project, named MegaMorph, which aims to improve our ability to separate galaxy components, particularly in large surveys. Our approach is to build upon an existing, tried-and-tested system which takes care of the numerous details that are essential for reliable galaxy fitting, but which do not require significant alteration. We therefore base our developments on GALAPAGOS (Häussler *et al.* 2007), which performs detection, deblending and preliminary object measurements (using SExtractor; Bertin & Arnouts 1996), measures the sky level, extracts galaxy images, creates masks, fits each galaxy using GALFIT (Peng *et al.* 2002), and finally collates the results.

Below we describe the developments we are considering, in order to address the issues discussed in the previous section. This work is ongoing, but we have already made substantial progress implementing these methods, and are now focusing on testing and refining our techniques, while starting work on key science examples to demonstrate their practical benefits. We assess our modifications to GALAPAGOS and GALFIT using a variety of data, including: (1) nearby galaxies with SDSS imaging, artificially redshifted using FERENGI (Barden *et al.* 2008), (2) the Galaxy And Mass Assembly survey (Driver *et al.* 2009, 2010) which provides homogenised imaging over nine optical to near-infrared bands, redshifts, and a great deal of other data, together with visual morphologies from Galaxy Zoo (Lintott *et al.* 2010), and (3) simulations of the GAMA imaging created by extending the method of Häussler *et al.* (2007).

3.1. *Multiband fitting*

We propose that the reliability of the decomposition process can be greatly increased, while ensuring physically meaningful component colours, by constraining a single wavelength-dependent model using all of the available multiband imaging simultaneously. Firstly, this increases the signal-to-noise of the data used to constrain the fit, for a

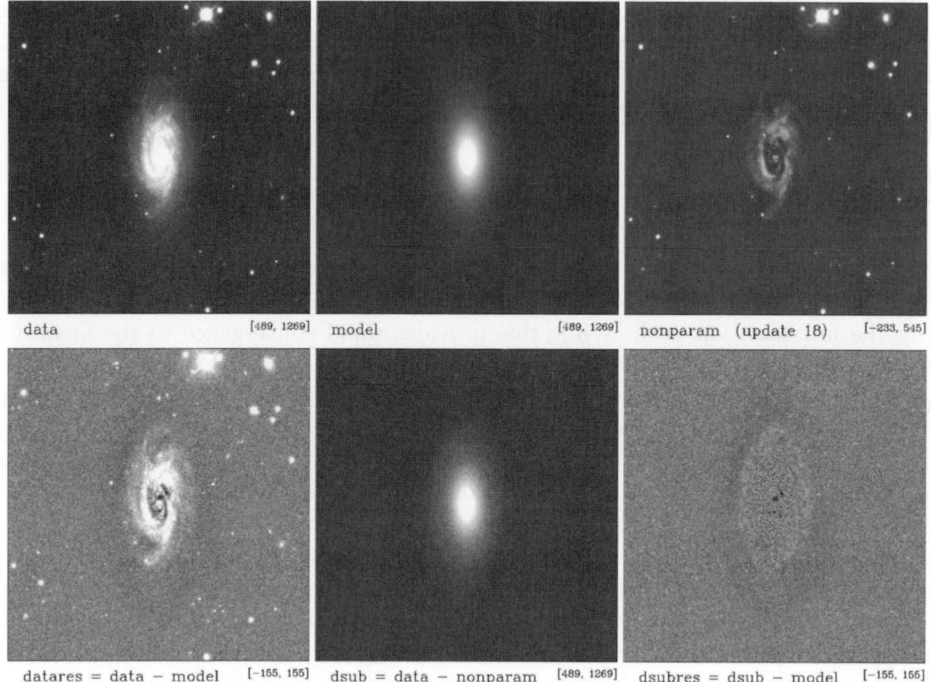

Figure 1. An example of fitting a spiral galaxy with a non-parametric component. The panels show (data) the original image; (model) the final parametric model, comprising a Sérsic bulge and a exponential disk; (nonparam) the non-parametric component; (datares) the residual of the original image minus the parametric model; (dsub) the original image minus the non-parametric component, note that the smooth profile, pixel noise and some low contrast features remain in this image; and (dsubres) the residual of the original image with the non-parametric component and parametric model subtracted, an unmodelled outer disk truncation is just visible. The numbers to the bottom right of each panel give the limits of the linear intensity scale.

comparatively small increase in the number of fitted parameters. Secondly, and more importantly, multi-wavelength imaging provides information that is not available to techniques which operate on only a single band at a time: the wavelength dependence of the flux of each component. We expect this to be critically important for overcoming the degeneracies inherent in multicomponent fitting, particularly in the common case of an exponential disk and Sérsic bulge. Finally, this approach enables the use of SED priors for each component, reducing the effective number of free parameters.

3.2. *Non-parametric components*

The observed surface brightness distribution of galaxies, particularly at blue wavelengths, is strongly modulated by star formation, dust and resonance structures such as spiral arms, bars and rings. However, in general, most of the stellar mass may be identified with a small number of relatively smooth, approximately azimuthally-symmetric components, and it is therefore these components which we primarily wish to characterise. The conventional approach is to fit idealised, smooth models to the images of real, messy, galaxies. However, it is generally unwise to fit data with a model which is unable to adequately represent the behaviour of that data, and then attempt to interpret the resulting parameters in a meaningful way (e.g., see Hogg *et al.* 2010).

To solve this dilemma, we are developing a technique which adds non-parametric components to the galaxy model, to account for features that are not represented by the usual parametric components. An exciting aspect of this work is the combination with the multiband approach, which enables one to fit for multiple non-parametric components with different colours. We have produced encouraging results so far, see Figure 1, and we hope that this approach will substantially increase the reliability of the recovered parameters (and their uncertainties) for the spheroid and disk components of galaxies.

3.3. *Parameter uncertainties and model selection*

GALFIT uses the Levenberg-Marquardt algorithm to find the optimal model fit to the data. This is a downhill method, and therefore only samples a small number of points along the minimisation path. While this provides rapid convergence to the nearest local minimum, it does mean that the result may depend upon the starting parameters, that no knowledge of the overall likelihood landscape is acquired, and that the resulting parameter uncertainties are highly approximate.

In certain cases it is worth sacrificing speed for a better understanding of parameter degeneracies, accurate uncertainties, and an estimate of the Bayesian evidence. The latter helps select the most suitable model from a range of possibilities, e.g. single Sérsic, bulge+disk, etc. Obtaining all this extra information requires the parameter space to be more generally sampled, which, to be completed in an reasonable time, requires the use of efficient algorithms. One excellent example, well used in cosmology, is provided by the MultiNest (Feroz *et al.* 2009) library. We have incorporated this sampler as an option in GALFIT, thereby providing the additional details described above. The cost is a running time 100 – 1000 longer than usual, but this computational increase is surmountable, for reasonably large samples, through the use of supercomputers.

Acknowledgements

This work was made possible by a NPRP grant from the Qatar National Research Fund (a member of the Qatar Foundation). The statements made herein are solely the responsibility of the authors. SPB holds an STFC Advanced Fellowship.

References

Abazajian, K. N., *et al.* 2009, *ApJS*, 182, 543
Allen, P. D., *et al.* 2006, *MNRAS*, 371, 2
Barden, M., Jahnke, K., & Häußler, B. 2008, *ApJS*, 175, 105
Benson, A. J. 2010, *Phys. Rep.*, 495, 33
Benson, A. J. & Devereux, N. 2010, *MNRAS*, 402, 2321
Bertin, E. & Arnouts, S. 1996, *A&AS*, 117, 393
Cole, S., *et al.* 2000, *MNRAS*, 319, 168
Cook, M., Lapi, A., & Granato, G. L. 2009, *MNRAS*, 397, 534
Driver, S. P., *et al.* 2009, *A&G*, 50, 5.12
Driver, S. P., *et al.* 2011, *MNRAS*, 413, 971
Feroz, F., Hobson, M. P., & Bridges, M. 2009, *MNRAS*, 398, 1601
Häussler, B., *et al.* 2007, *ApJS*, 172, 615
Hogg, D. W., Bovy, J., & Lang, D. 2010, *ArXiv e-prints*, arXiv:1008.4686
Lawrence, A., *et al.* 2007, *MNRAS*, 379, 1599
Lintott, C., *et al.* 2011, MNRAS, 410, 166
Peng, C. Y., *et al.* 2002, *AJ*, 124, 266
Peng, C. Y., *et al.* 2010, *AJ*, 139, 2097
Simard, L., *et al.* 2002, *ApJS*, 142, 1
Simard L., *et al.* 2011, *ApJS*, 196, 11

Discussion

BUREAU: If I may make a comment, decomposing galaxies into independent components is fine as a practical/numerical tool, but one should be careful about assigning a physical meaning and interpretation. Galaxies are not static objects with stars as building blocks but dynamic objects with orbits as building blocks. So separating a galaxy into e.g. a disk and a bulge creates two entities that could not exist separately. Especially, bulges are rarely "mini-ellipticals" but mostly disk phenomena: thickened bars ("peanut-shaped" bulges) and disk-like ("pseudo") bulges. So associating steep light profiles (from bulge-disk photometric decompositions) with spheroids is fraught with danger and can lead to flawed interpretations. Use with caution!

The Spectral Energy Distribution of Galaxies
Proceedings IAU Symposium No. 284, 2011
R.J. Tuffs & C.C. Popescu, eds.

doi:10.1017/S1743921312009313

Dust effects on the derived Sérsic indexes of disks and bulges in spiral galaxies

Bogdan A. Pastrav[1], Cristina C. Popescu[1], Richard J. Tuffs[2] and Anne E. Sansom[1]

[1]Jeremiah Horrocks Institute, University of Central Lancashire, Preston PR1 2HE, UK
[2]Max Planck Institut für Kernphysik, Saupfercheckweg 1, D-69117 Heidelberg, Germany

email: bapastrav@uclan.ac.uk

Abstract. We present a theoretical study that quantifies the effect of dust on the derived Sérsic indexes of disks and bulges. The changes in the derived parameters from their intrinsic values (as seen in the absence of dust) were obtained by fitting Sérsic distributions on simulated images of disks and bulges produced using radiative transfer calculations and the model of Popescu *et al.* (2011). We found that dust has the effect of lowering the measured Sérsic index in most cases, with stronger effects for disks and bulges seen through more optically thick lines of sight.

Keywords. Galaxy: disk, galaxies: bulges, Galaxy: fundamental parameters, galaxies: spiral, galaxies: ISM, galaxies: structure, (ISM:) dust, extinction, radiative transfer

---

1. Introduction

In recent years deep wide field spectroscopic and photometric surveys of galaxies (e.g. GAMA, Driver *et al.* 2011) are providing us with large statistical samples of galaxies with good quality imaging up to z=0.1. In parallel, automatic routines like GALFIT (Peng *et al.* 2002, Peng *et al.* 2010) or GIM2D (Simard *et al.* 2002) have been developed to address the need of fitting large number of images of galaxies with 2D analytic functions to characterise the surface brightness distribution of their stellar components. In particular Sérsic functions are the most common distributions that are used to describe the profiles of galaxies and their constituent morphological components. The derived Sérsic indexes are then used to classify galaxies as disk or bulge dominated ones (e.g. Kelvin *et al.* 2012) or in terms of a bulge-to-disk ratio when bulge/disk decomposition is performed (Simard *et al.* 2011).

One potential problem with the interpretation of the results of Sérsic fits is that the measured Sérsic parameters differ from the intrinsic ones (as would be derived in the absence of dust). This is because real galaxies, in particular spiral galaxies contain large amount of dust, and this dust changes their appearance from what would be predicted to be seen in projection based on only their intrinsic stellar distributions. Determining the changes due to dust is thus essential when characterising and classifying galaxies based on their fitted Sérsic indexes.

Here we present results of a theoretical study to quantify the effects of dust on the measured fitted Sérsic indexes, as a function of the relevant parameters: dust opacity, disk inclination and wavelength. The results are presented for pure disks and bulges, as seen through a common distribution of dust (in the disk of the parent galaxy). These results are part of a larger study that aims to quantify the effects of dust on all photometric parameters of young stellar disks, old stellar disks and bulges (Pastrav *et al.*, in prep) and builds on our previous work of Möllenhoff *et al.* (2006).

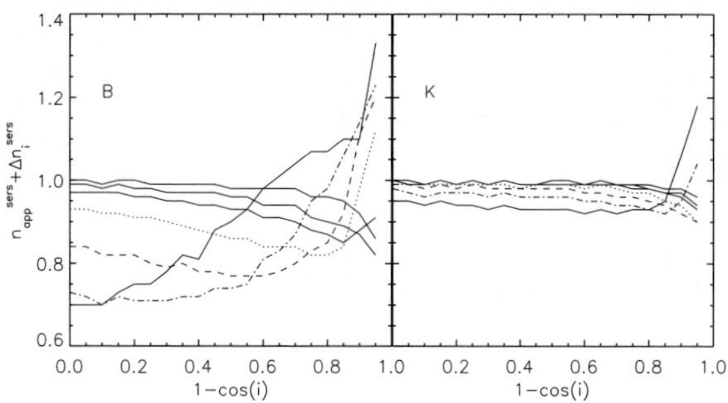

Figure 1. The inclination dependence of the derived Sérsic index, n_{app}^{sers}, corrected for projection effects, Δn_i^{sers}, for disk images in the B and K bands. From top to bottom, the curves are plotted for central face-on opacity in the B band $\tau_B^f = 0.1, 0.3, 0.5$ (solid lines), 1.0 (dotted), 2.0 (dashed), 4.0 (dashed-dotted), 8.0 (dashed-3 dotted).

2. The method

We based our study on simulated images of dust-attenuated disks and bulges for which their intrinsic parameters are known, being input for the simulations. We then measured the perceived (apparent) photometric parameters using widely used 2D fitting routines, in an attempt to exactly mimic what an observer would measure for a real galaxy. The difference between the measured parameters (affected by a combination of projection and dust effects) and the intrinsic ones (as used to construct the 3D distributions of stellar emissivity) provides knowledge on the effects of dust (and also on projection effects). The aim of the study is thus to provide observers with corrections for the "simplicity" of the templates commonly used to analyse images of galaxies, "simplicity" which is nevertheless necessary for practical purposes when dealing with large numbers of objects.

The simulated images were produced using radiative transfer calculations and the model of Popescu *et al.* (2011) (see also Popescu *et al.* 2000 and Tuffs *et al.* 2004). In brief the disks were built using a distribution of stellar emissivity described by a double exponential (in radial and vertical direction) while bulges were built using the deprojection of general Sérsic distributions, in particular of a distribution with a Sérsic index $n_0^{sers} = 4$ (de Vaucouleurs) and a Sérsic index $n_0^{sers} = 1$ (exponential). Both disks and bulges were attenuated by a common distribution of dust (present only in the disk). More details about the geometry of the model can be found in Popescu *et al.* (2011).

To fit the simulated images, we applied the commonly used GALFIT data analysis algorithm (Peng *et al.* 2002, Peng *et al.* 2010). Both disks and bulges were fitted by the commonly used 2D Sérsic distributions, corresponding to an infinitely thin disk. In both cases the Sérsic index of the fitted function was left as a free parameter of the fit.

3. Results

The results on the effects of dust on the derived Sérsic indexes are summarised in Fig. 1 (for the disk) and Fig. 2 (for the bulge). In each figure we plot the results on the measured Sérsic index corrected for projection effects. By projection effects we mean the difference between the projection of a distribution of stellar emissivity that has a vertical distribution in addition to a radial distribution, and the projection of an infinitely thin disk, which is assumed when fitting the images with analytical 2D Sérsic functions. This results in an alteration of the measured Sérsic index, even in the absence of dust. The corrections for projection effects are discussed and listed in Pastrav *et al.* (in prep). Here

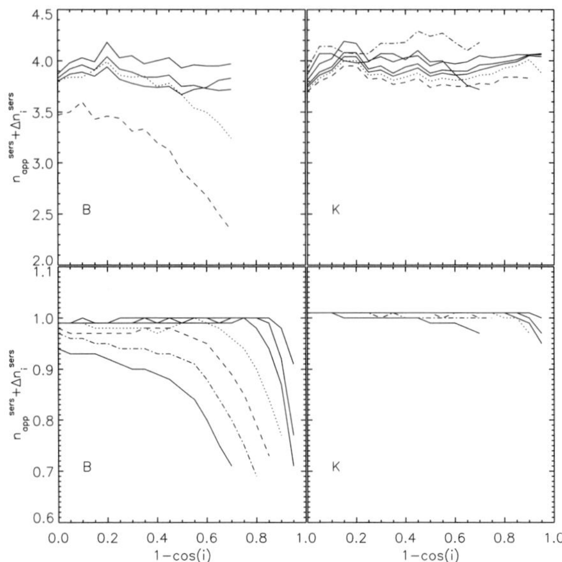

Figure 2. Upper row: The inclination dependence of the derived Sérsic index, n_{app}^{sers}, corrected for projection effects, Δn_i^{sers}, for de Vaucouleurs bulges ($n_0^{sers} = 4$), in the B and K optical bands. Lower row: The same, but for exponential bulges ($n_0^{sers} = 1$). Line styles are as in Fig. 1.

we only show the results on the measured Sérsic index after correction for projection effects.

In each figure we show the results as a function of inclination and for different values of central face-on opacity in the B band, τ_B^f, which are the two parameters of the radiative transfer model. In addition we show results at two different wavebands (B and K).

One can see that at longer wavelengths, in the K band, the effects of dust are negligible. Essentially one can measure the intrinsic Sérsic indexes for both disks and bulges, once corrected for projection effects. For example for intrinsically exponential disks the derived Sérsic index is ~ 1. Similarly, for intrinsically de Vaucouleurs and exponential bulges the derived Sérsic indexes are ~ 4 and ~ 1, respectively. At shorter wavelength, in the B band, the dust starts to affect the derived Sérsic indexes, with stronger effects for higher τ_B^f and larger inclinations. In most cases the effect of dust is *to lower the Sérsic index* from its intrinsic value. This is because of the large scale distribution of dust, which decreases exponentially with increasing radial distance, making the profiles flatter in the central parts.

Another result is that in most cases the *Sérsic index decreases with increasing inclination and* τ_B^f, except for very optically thick disks, which show a reverse trend.

References

Driver, S. P., Hill, D. K., Kelvin, L. S. *et al.* 2011, *MNRAS*, 413, 971
Kelvin, L. S., Driver, S. P., Robotham, A. S. G. *et al.* 2012, *MNRAS*, in press (arXiv:1112.1956)
Möllenhoff, C., Popescu, C. C., & Tuffs, R. J. 2006, *A&A* 456, 941
Peng, C. Y., Ho, L. C., Impey, C. D., & Rix, H.-W. 2002, *ApJ* 124, 266
Peng, C. Y., Ho, L. C., Impey, C. D., & Rix, H.-W. 2010, *ApJ* 139, 2097
Popescu, C. C., Misiriotis, A., Kylafis, N., Tuffs, R. J., & Fishera, J. 2000, *A&A*, 362, 138
Popescu, C. C., Tuffs, R. J., & Dopita, M. *et al.* 2011, *A&A*, 527, A109
Simard, L., Willmer, C. N. A., Vogt, N. P. *et al.* 2002, *ApJS*, 142, 1
Simard, L., Mendel, J. T., Patton, D. R. *et al.* 2011, *ApJS*, 196, 11
Tuffs, R. J., Popescu, C. C., Völk, H. J., Kylafis, N. D., & Dopita, M. A. 2004, *A&A*, 419, 835

The Spectral Energy Distribution of Galaxies
Proceedings IAU Symposium No. 284, 2011
R.J. Tuffs & C.C. Popescu, eds.

doi:10.1017/S1743921312009325

Publicly available database for spectral line measurements of SDSS DR7 galaxies

Kyuseok Oh[1], Marc Sarzi[2], Kevin Schawinski[3,4] and Sukyoung K. Yi[1]

[1]Department of Astronomy, Yonsei University, Seoul 120-749, Korea
[2]Centre for Astrophysics Research, University of Hertfordshire, Hatfield AL10 9AB, UK
[3]Yale Center for Astronomy and Astrophysics, Yale University, P.O. Box 208121, New Haven, CT 06511, U.S.A. [4]Einstein Fellow

email: ksoh@galaxy.yonsei.ac.kr

Abstract. We present a new database of absorption and emission-line measurements based on the Sloan Digital Sky Survey (SDSS) 7th data release of galaxies within a redshift of 0.2. Using the publicly available penalized pixel-fitting (pPXF) and gas and absorption line fitting (gandalf) codes, our work improve the existing measurements for stellar kinematics, the strength of various absorption line features, and the flux and width of the emissions from different species of ionised gas. Most notable of our work is that, we provide quality of the fit to assess reliability of the measurements. The quality assessment can be highly effective for finding new classes of objects. For example, based on the quality assessment around the Ha and [NII] nebular lines, we found approximately 1% of the SDSS spectra which classified as galaxies by the SDSS pipeline are in fact type I Seyfert AGN. This paper presents a summary of the recent paper, Oh *et al.*(2011). The database is publicly available at http://gem.yonsei.ac.kr/ossy/.

Keywords. catalogs, galaxies:abundances, galaxies:stellar content, galaxies:statistics

1. Introduction

Spectral lines that reveal various properties of galaxies have been widely used to understand the physical processes of the evolution of galaxies. Absorption features reveal vital information on the galaxy formation and evolution, such as luminosity-weighted ages, metallicities and abundance ratios of early-type galaxies (Davies, Sadler & Peletier 1993; Fisher, Franx & Illingworth 1995; Fisher, Franx & Illingworth 1996; Jorgensen (1999); Kuntschner 2000; Kuntschner *et al.* 2006). Nebular emission lines, on the other hand, indicate the status of nuclear (Baldwin, Phillips & Terlevich 1981; Veilleux & Osterbrock 1987; Kewley *et al.* 2001; Kauffmann *et al.* 2003; Schawinski *et al.* 2007) and star formation activities (Osterbrock & Pogge 1985).

Thanks to the Sloan Digital Sky Survey (SDSS) that established one of the largest and homogeneous database of photometric and spectroscopic data, it is possible to systematically study various types of galaxies with little bias. However, the SDSS pipeline measurements on the absorption line strengths are not corrected for the contamination of nebular emissions. Also, internal extinction is not considered, which leads to inaccurate measurements of nebular emission fluxes. In this work, we present improved absorption and emission line measurements of the seventh data release of SDSS (Abazajian *et al.* 2009) galaxies, as well as providing quality assessments of our fits.

2. Method

To properly measure the strength of stellar kinematics, absorption and nebular emission lines, this work adopted a very similar strategy to that used for the SAURON integral field spectroscopic survey (Emsellem *et al.* 2004; Sarzi *et al.* 2006; Kuntschner *et al.* 2006). First, the stellar kinematics were extracted from direct matching of spectra with a set of stellar templates. In this process, spectral regions that are potentially affected by nebular emission are excluded. Second, the strengths of gaseous kinematics and each emission line were measured by fitting the entire spectrum. Finally, our best model for the nebular emission was subtracted from the SDSS spectrum and the strengths of the stellar absorption lines were measured following the standard line-strength index definitions. We used both theoretical stellar population synthesis models (Bruzual & Charlot 2003) and MILES stellar templates (Sánchez-Blázquez *et al.* 2006). Foreground dust Galactic extinction was treated following (Schlegel, Finkbeiner & Davis 1998). Using Balmer decrement, we considered internal reddening that affects only the nebular spectrum.

3. Fitting Quality Assessment

A novel feature is the provision of a fitting quality assessment. This is key to our measurements because a good quality of an observed spectrum (signal-to-noise ratio, S/N) itself does not guarantee a good fit. The most frequent failures of our fits occur in Broad Line Region (BLR) galaxies. Some of these show extreme broad lines similar to those found in QSOs. In addition, template mismatch and telluric atmospheric features induce low quality fits even when the object has a high S/N . The level of fluctuations in the fit residuals (rN, residual noise) to the expected statistical fluctuations (sN, statistical noise), rN/sN ratio well represents the quality of our fit. In order to avoid various stellar absorption and nebular emission line contaminations, we adopted the 4500-4700Å, 5400-5600Å, 6000-6200 Å, 6800-7000Å wavelength intervals and combined the rN/sN ratio of those continuum regions into a single average, excluding the largest value to avoid a spurious contribution from the partially broken SDSS spectra. The distance of rN/sN from the median distribution expresses the goodness of fits in units of sigma, N_σ.

4. Broad Line Regions

The presence of broadened Hα and Hβ lines frequently causes a mismatch in both the emission lines and the stellar continuum. As a result, highly broadened galaxies appear as an unphysical clump beside star-forming sequence in standard BPT diagrams due to poor fit and corresponding line measurements. After identifying such cases, we properly dealt with the broadened component by giving a broader mask around Balmer lines during the `pPXF` fit (Cappellari & Emsellem 2004) and measured the nebular emission line strengths. The objects turned out to be Seyfert I nuclei. Roughly 0.6% of SDSS galaxies are classified as BLRs. Their Hα line width ranges from ~1,000 to ~5,000Å.

5. [NI] $\lambda\lambda$ 5198,5200 doublet correction for elemental abundances

Mgb and Fe absorption strengths are often used to investigate the formation timescale of galaxies. Mgb index strength, however, can be significantly affected by the presence of [NI] $\lambda\lambda$ 5198,5200 doublet. We measured Mgb absorption line strength considering [NI] $\lambda\lambda$ 5198,5200 doublet. After making sure of morphology by applying FracDeV and concentration index parameters, we explored how [NI] $\lambda\lambda$ 5198,5200 doublet enlarges

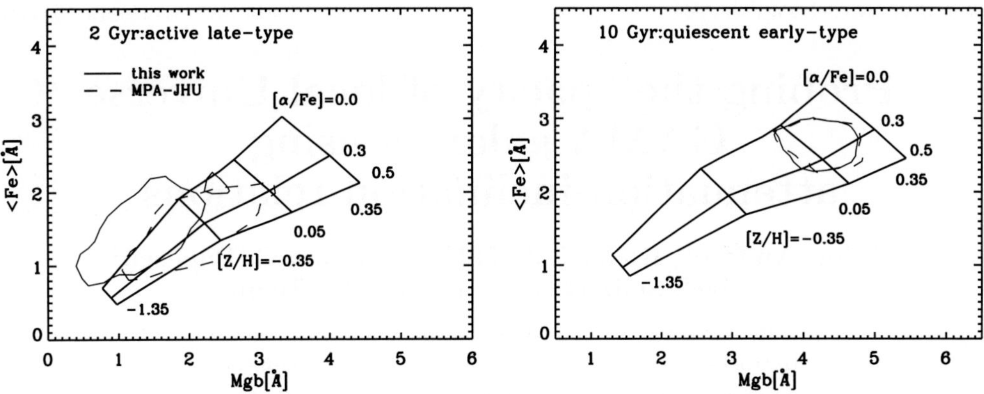

Figure 1. Effect of [NI] $\lambda\lambda$5198,5200 emission correction on Mgb index. The stellar population models(Thomas *et al.* 2003, grids) on $[\alpha/\mathrm{Fe}]$ and metallicity spaces for active late-(left) and quiescent early-type galaxies(right) assuming 2 Gyr and 10 Gyr of age, respectively. Contours(solid and dashed) represent 1σ distribution.

Mgb index strength(Goudfrooij & Emsellem 1996). Fig. 1 confirms the importance of the correction of [NI]$\lambda\lambda$5198,5200 doublet on Mgb index passband. Unlike the MPA-JHU catalog, our measurements conclude that only early-type galaxies are α-enhanced.

6. Database

The database (http://gem.yonsei.ac.kr/ossy/) is mainly organized in three parts ; stellar absorption, nebular emission line strengths and fitting quality assessment. All the values are provided in both plain ascii and binary fits formats.

References

Abazajian, K. N., *et al.* 2009, *ApJS*, 182, 543
Baldwin, J. A., Phillips, M. M., & Terlevich, R. 1981, *PASP*, 93, 5
Bruzual, G. & Charlot, S. 2003, *MNRAS*, 344, 1000
Cappellari, M. & Emsellem, E. 2004, *PASP*, 116, 138
Davies, R. L., Sadler, E. M., & Peletier, R. F. 1993, *MNRAS*, 262, 650
Emsellem, E., *et al.* 2004, *MNRAS*, 352, 721
Fisher, D., Franx, M., & Illingworth, G. 1995, *ApJ*, 448, 119
Fisher, D., Franx, M., & Illingworth, G. 1996, *ApJ*, 459, 110
Goudfrooij, P. & Emsellem, E. 1996, *A&A*, 306, L45
Jorgensen, I. 1999, *MNRAS*, 306, 607
Kauffmann, G., *et al.* 2003, *MNRAS*, 346, 1055
Kewley, L. J., Dopita, M. A., Sutherland, R. S., Heisler, C. A., & Trevena, J. 2001, *ApJ*, 556, 121
Kuntschner, H. 2000, *MNRAS*, 315, 184
Kuntschner, H., *et al.* 2006, *MNRAS*, 369, 497
Oh, K., Sarzi, M., Schawinski, K., & Yi, S. K. 2011, *ApJS*, 195, 13
Osterbrock, D. E. & Pogge, R. W. 1985, *ApJ*, 297, 166
Sánchez-Blázquez, P., *et al.* 2006, *MNRAS*, 371, 703
Sarzi, M., *et al.* 2006, *MNRAS*, 366, 1151
Schawinski, K., *et al.* 2007, *MNRAS*, 382, 1415
Schlegel, D. J., Finkbeiner, D. P., & Davis M. 1998, *ApJ*, 500, 525
Thomas, D., Maraston, C., & Bender, R. 2003, *MNRAS*, 339, 897
Veilleux S. & Osterbrock D. E. 1987, *ApJS*, 63, 295

The Spectral Energy Distribution of Galaxies
Proceedings IAU Symposium No. 284, 2011
R.J. Tuffs & C.C. Popescu, eds.

doi:10.1017/S1743921312009337

Probing the opacity of local Universe GAMA galaxies using attenuation-inclination relations

Ellen Andrae[1], Richard J. Tuffs[1], Cristina C. Popescu[2], Mark Seibert[3] and the GAMA Team

[1]Max-Planck-Institut für Kernphysik, Heidelberg, Germany

[2]Jeremiah Horrocks Institute for Astrophysics and Supercomputing, Preston, UK

[3]Observatories of the Carnegie Institution of Washington, Pasadena, USA

email: Ellen.Andrae@mpi-hd.mpg.de

Abstract. Even though attenuation by dust strongly attenuates and reddens the UV/optical spectral energy distributions of spiral galaxies, quantifying these effects is challenging, particularly when, as is most often the case, infrared measurements of the absorbed energy are not available. We have initiated a study to model the dependence on inclination of the attenuation of light from galaxies measured in the Galaxy and Mass Assembly (GAMA) survey. We present preliminary results using SDSS galaxies in the phase-I GAMA catalogue, comparing observed attenuation-inclination relations with the predictions of the radiation transfer model of Popescu *et al.* (2011) to derive a statistical measurement of the mean face-on B-band optical depth of disks τ_B^f in a sample of spiral galaxies complete to the SDSS spectroscopic r-band limit of 17.8 mag.

Keywords. ISM: dust, extinction, galaxies: fundamental parameters, galaxies: ISM, galaxies: luminosity function, galaxies: statistics

---

1. Introduction

The presence of dust changes the spectral energy distributions (SEDs) of galaxies, as grains absorb optical / UV light, re-emitting in the infrared (primarily in the 50 - 300 μm far-IR range). This markedly changes the perceived SEDs in the UV-optical range, even for local Universe galaxies (e.g. Popescu & Tuffs 2002; Driver *et al.* 2007). There are basically three methods for quantifying the attenuation by dust of the spatially integrated stellar emission from galaxies, as routinely measured in broad-band UV/optical surveys. The most direct method is to measure the absorbed UV/optical light in the far-IR, and use a radiation transfer (RT) analysis of the combined UV/optical-FIR/submm SED to infer the attenuation of starlight as a function of wavelength in the UV/optical range. However, the limited sky coverage and depth of contemporary FIR/submm surveys effectively limit the application of this technique to incomplete subsets of the galaxy population. The second method is to use measurements of nebular emission lines with a known intrinsic luminosity ratio, such as hydrogen recombination lines, to probe the dust attenuating the broad-band emission. This is potentially a very powerful technique, but requires a careful treatment of the strong local attenuation by dust in star-formation regions and a combined spatial and spectral coverage of the whole galaxy, which is difficult to realise for large statistical samples. The third method is to statistically measure the dimming in integrated luminosity of starlight from dusty galaxies as the dust-bearing

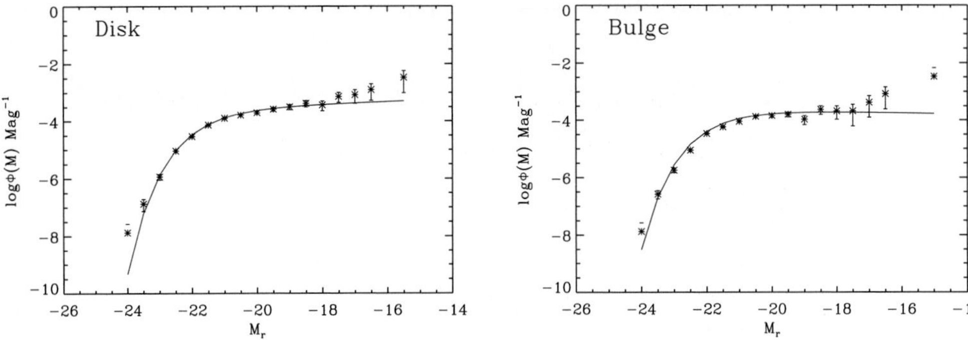

Figure 1. r-band luminosity distributions of the disks (left) and bulges (right) for inclinations $0.2 < 1 - \cos(i) < 0.3$. The curves indicate the fitted Schechter functions with the parameter α constrained from fits to disks and bulges for all spiral galaxies with $1 - \cos(i) < 0.3$ (see text).

disks are progressively observed from more edge-on viewpoints. We have embarked upon a project to model this so-called "attenuation-inclination" relation using the full wavelength range (UV to Near-IR) of the phase-I data on disk-dominated galaxies from the GAlaxy and Mass Assembly spectroscopic survey (GAMA; Driver *et al.* 2011). Here, we present preliminary results for bright (r < 17.77) spiral galaxies in the SDSS.

2. Attenuation-Inclination relation

The basis for this work are galaxies with available bulge-disk decompositions and inclinations from Simard *et al.* (2011), with r-band magnitudes and redshifts taken from the GAMA catalogue. This results in a sample of ~14k galaxies with $m_r < 17.77$ mag. In deriving the attenuation-inclination relation, it is critical to take into account the different behaviour of bulges and disks. As the (dustless) bulge is more centrally placed in the disk, it is more strongly attenuated (e.g. Tuffs *et al.* 2004). We employ the approach by Driver *et al.* (2007) in which luminosity functions are fitted to subsamples of galaxies devided into inclination bins to find the turn-over magnitude, M_*, in a Schechter function as a function of inclination i [i.e. M_* versus $1 - \cos(i)$]. This is an iterative process in which the mean face-on B-band optical depth of the disks τ_B^f is derived by fitting the observed attenuation-inclination relations with the RT model of Tuffs *et al.* (2004). This is done separately for disks and bulges, following the procedure outlined below. Iteration is required because the bulge-to-total luminosity ratios (B/T) should be based on intrinsic rather than apparent luminosity ratios.

(*a*) Apply the current best attenuation–inclination correction estimates of the disks and bulges (no correction applied for the first iteration).

(*b*) Recompute B/T. Apply B/T < 0.8 cut (i.e. excluding elliptical galaxies).

(*c*) Derive the luminosity distribution (LD) using the 1/Vmax method for all disks and bulges with low inclination $[1 - \cos(i) < 0.3]$. Fit Schechter functions to the LDs in order to obtain the faint-end slope, α, for low inclination.

(*d*) Derive the disk and bulge LDs for the full sample in equidistant $\cos(i)$ intervals. Fit Schechter functions with α fixed to the value from step (3). The final iteration example is shown in Fig. 1.

(*e*) Plot the recovered turnover magnitude M_* versus $1 - \cos(i)$. Determine the improved attenuation-inclination correction estimates by fitting the RT model (as a function of τ_B^f) to the bulge and disk data (leaving out the incomplete interval $[0.9 < 1 - \cos(i) < 1]$). The final iteration example is shown in Fig. 2.

(*f*) Repeat *a*–*e* until convergence.

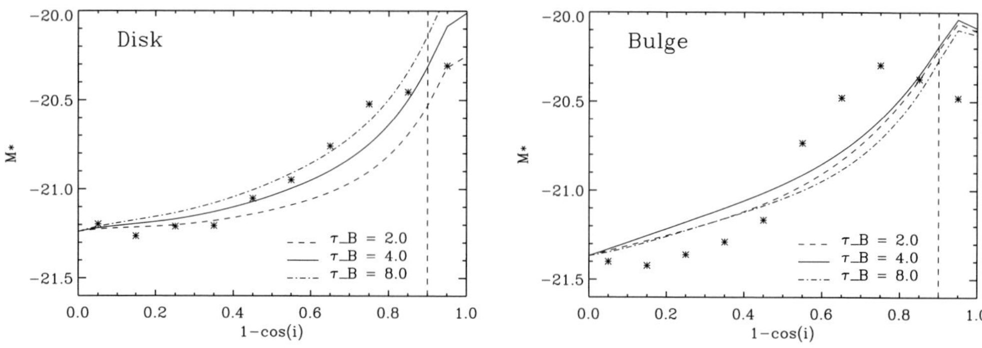

Figure 2. The r-band attenuation-inclination relation for disks (left) and bulges (right). The lines represent predicted relations derived using the model of Tuffs *et al.* (2004) for values of $\tau_B^f = 2$, 4, and 8. The best fitting values are $\tau_{B,disk} = 4.1$ and $\tau_{B,bulge} = 4.0$.

The best fitting values for τ_B^f derived from the disk and bulge attenuation–inclination relations are $\tau_B^f = 4.1$ and $\tau_B^f = 4.0$, respectively. These results are consistent with the value $\tau_B^f = 3.8 \pm 0.7$ found by Driver *et al.* (2007) using the B-band Millenium sample. As the inclination $1 - \cos(i)$ is biased low (i.e. galaxies appear rounder than they are, due to resolution effects and the intrinsic thickness of the disk), high inclination galaxies with high attenuations appear at too low inclinations. Consequently, the derived values of τ_B^f may be considered as upper limits. While the best fitting RT model for the disk is nevertheless a good representation of the data, this bias becomes obvious for the bulge. Future higher resolution data will mitigate this effect as well as improve the bulge-disk decomposition.

3. Outlook

The attenuation-inclination relation will be derived for the full GAMA dataset from the FUV to the near-IR with its excellent statistics and spectroscopic depth (340k galaxies complete to r 19.8 over ca. 320 sq deg out to z∼0.4). In addition, for a subset of galaxies measured by WISE and Herschel in the MIR/FIR/submm and with quantitative optical spectroscopy from GAMA it will be possible to test the self-consistency of the values for τ_B^f gained from the attenuation-inclination relation with those found from RT modelling of nebular emission lines and the broad-band UV-submm SED. If consistency can be achieved for a fixed geometry of stars and dust, the attenuation-inclination relation could be used to predict the face-on attenuation. The multicomponent morphological decompositions and quantitative spectroscopy of the GAMA dataset will ultimately allow separate analyses of different categories of rotationally-supported galaxies separated according to size, stellar mass, star-formation rate, star-formation history, morphology and environment.

References

Driver, S. P., Hill, D. T., Kelvin, L. S., *et al.* 2011, *MNRAS*, 413, 971
Driver, S. P., Popescu, C. C., Tuffs, R. J., *et al.* 2007, *MNRAS*, 379, 1022
Popescu, C. C. & Tuffs, R. J. 2002, *Reviews in Modern Astronomy*, 15, 239
Popescu, C. C., Tuffs, R. J., Dopita, M. A., *et al.* 2011, *A&A*, 527, A109
Simard, L., Mendel, J. T., Patton, D. R., *et al.* 2011, *ApJS*, 196, 11
Tuffs, R. J., Popescu, C. C., Völk, H. J., *et al.* 2004, *A&A*, 419, 821

The Spectral Energy Distribution of Galaxies
Proceedings IAU Symposium No. 284, 2011
R.J. Tuffs & C.C. Popescu, eds.

doi:10.1017/S1743921312009349

The unusual multi-wavelength SED of two optical dropout galaxies

Daniel Schaerer[1,2], Frederic Boone[2] and Nicolas Laporte[2]

[1]Observatoire de Genève, Université de Genève, 51 Ch. des Maillettes, 1290 Versoix, Switzerland

[2]CNRS, IRAP, 14 Avenue E. Belin, 31400 Toulouse, France

email: daniel.schaerer@unige.ch

Abstract. We have used deep optical, near-IR, and IR observations from VLT, Spitzer, Herschel, and LABOCA in strong lensing clusters to study distant galaxies. In searches for optical-dropout galaxies (i.e. for $z \gtrsim 7$ candidates) we have found several galaxies with very unusual SEDs characterised by a strong spectral break, presumably indicative of high-z, although the objects are detected even in the Herschel bands between 160 and 500 μm and at 870 μm. The latter indicates, from simple estimates of the bolometric luminosity and from the IR SED, that these ob jects are most likely at $z \sim$ 2–2.5.

The resulting SEDs imply very high IR/UV ratios, indicative of very large attenuation. Despite this, the large spectral break observed between the optical and near-IR data is difficult to understand with currently know spectral templates from galaxies, EROs, SMGs, and others, both empirical and theoretical ones.

Keywords. galaxies: starburst, galaxies: high-redshift, galaxies: evolution, galaxies: formation

1. Introduction

Recently we have identified ten $z > 7$ candidates in the field of the cluster Abell 2667 using photometric dropout criteria based on deep observations with HAWK-I on the ESO VLT (Laporte *et al.* 2011). Two galaxies of this sample were later found to be clearly detected in the far-IR with Herschel and one of them with LABOCA (at 870 μm), as described in Boone *et al.* (2011). Most likely, this indicates that these galaxies are "low"redshift ($z \sim$ 2-2.5) interlopers. However, their SED including a well-defined spectral break appears quite unusual, and extreme. The origin of the spectral break and the nature of these objects remains puzzling, as we briefly describe here. For more details see Boone *et al.* (2011).

2. Observed SEDs of z- and Y- dropout galaxies

Targetting the strong lensing cluster Abell 2667 we have obtained deep optical imaging (I, z with FORS2/VLT), deep near-IR data in the Y, J, H, and Ks bands with HAWK-I/VLT. For the cluster, which is part of our ongoing Herschel Open Time Key project, the "Herschel Lensing Survey" (Egami *et al.* 2010), we also have data from Spitzer (from 3.6 to 8 μm), MIPS 24 m, LABOCA 870 μm, and the VLA (1.4 GHz) (see Boone *et al.* 2011). Our galaxies are selected from H+K images after applying z- and Y-band dropout criteria (i.e. Lyman break selection + visual inspection). All 10 optical dropout candidates found in this field are discussed in Laporte *et al.* (2011). Two of them, denoted z1 and Y5, are detected at 24 μm, and/or in several Herschel bands. The SED of z1 is shown in Fig. 1. The near-IR magnitudes of the two ob jects is 23.5-24 (H to K band

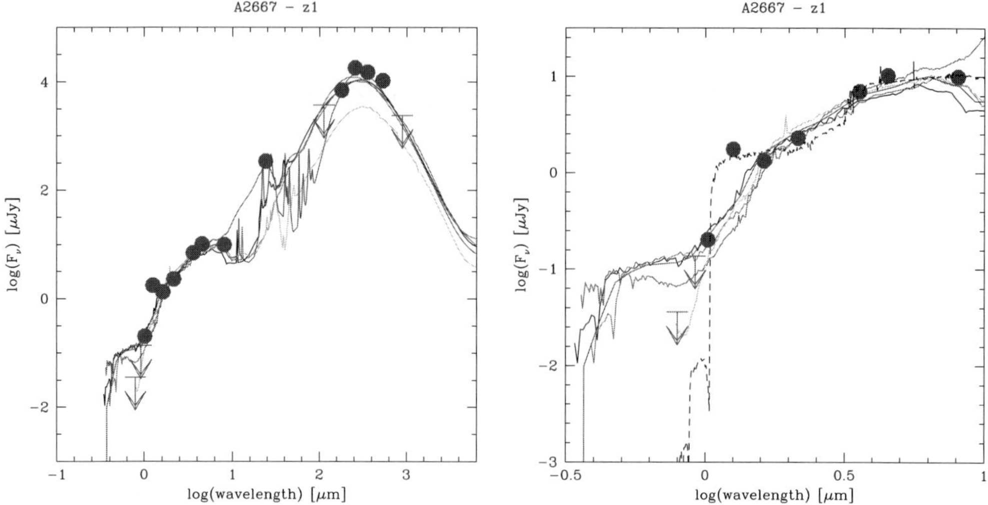

Figure 1. Best-fit SEDs for z1 using different templates: Chary & Elbaz (2001), Polletta *et al.* (2007), and Michalowski *et al.* (2010). Left: global SED, Right: zoom on region with the spectral break. Figure from Boone *et al.* (2011)

in AB). This would imply very luminous objects, if at $z > 7$! Still high-z is formally favoured for z1 from photometry redshift including luminosity prior (cf. Laporte *et al.*) The IR SED would imply a very high $L_{\rm IR} \sim (0.5 - 1.5) \times 10^{14}$ L$_\odot$, and would requite a very high dust temperature (> 100-130 K) if at high z , which are both very unlikely. The overall multi-wavelength SED indicates best-fit redshifts at $z \sim$ 2–2.5. Then one has $L_{\rm IR} \sim (2 - 3) \times 10^{12}$ L$_\odot$, corresponding to SFR $\sim$ 400–500 M$_\odot$ yr^{-1}.

The galaxies exhibit a very large flux ratio between IR and rest-UV (comparable to extreme sub-mm galaxies). The observed spectral break (used as selection criterium) is stronger than predicted by current empirical and theoretical spectral templates, as shown in Fig. 1. The origin of such a strong spectral break is unknown. Extinction, emission lines, and multiple populations appear difficult to explain it (see Boone *et al.* 2011). Our two galaxies may be comparable to other extreme objects, such as the sub-mm galaxy GN10=GOODS 850-5 or others (cf. Wang *et al.* 2009, Daddi *et al.* 2009, Huang *et al.* 2011). How frequent such possible contaminant are in high-z searches remains to be seen.

This work received support from the Agence Nationale de la Recherche bearing the reference ANR-09-BLAN-0234.

References

Boone, F., Schaerer, D., *et al.* 2011, *A&A*, 534, A124
Chary, R. & Elbaz, D. 2001, *ApJ*, 556, 562
Daddi, E., Dannerbauer, H., Krips, M., *et al.* 2009, *ApJ*,, 695, L176
Egami, E., *et al.* 2010, *A&A*, 518, L12
Huang, J.-S., *et al.* 2011, *ApJ*, 742, L13
Laporte, N., Pelló, R., Schaerer, D., *et al.* 2011, *A&A*, 531, A74
Michałowski, M. J., Watson, D., & Hjorth, J. 2010, *ApJ*, 712, 942
Polletta, M., Tajer, M., Maraschi, L., *et al.* 2007, *ApJ*, 663, 81
Wang, W., Barger, A. J., & Cowie, L. L. 2009, *ApJ*,, 690, 319

The Spectral Energy Distribution of Galaxies
Proceedings IAU Symposium No. 284, 2011
R.J. Tuffs & C.C. Popescu, eds.

doi:10.1017/S1743921312009350

What triggers star formation in galaxies?

Bruce G. Elmegreen

IBM Research Division, T.J. Watson Research Center
1101 Kitchawan Road, Yorktown Heights, NY 10598 USA

email: bge@us.ibm.com

Abstract. Processes that promote the formation of dense cold clouds in the interstellar media of galaxies are reviewed. Those that involve background stellar mass include two-fluid instabilities, spiral density wave shocking, and bar accretion. Young stellar pressures trigger gas accumulation on the periphery of cleared cavities, which often take the form of rings by the time new stars form. Stellar pressures also trigger star formation in bright-rim structures, directly squeezing the pre-existing clumps in nearby clouds and clearing out the lower density gas between them. Observations of these processes are common. How they fit into the empirical star formation laws, which relate the star formation rate primarily to the gas density, is unclear. Most likely, star formation follows directly from the formation of cold dense gas, whatever the origin of that gas. If the average pressure from the weight of the gas layer is large enough to produce a high molecular fraction in the ambient medium, then star formation should follow from a variety of processes that combine and lose their distinctive origins. Pressurized triggering might have more influence on the star formation rate in regions with low average molecular fraction. This implies, for example, that the arm/interarm ratio of star formation efficiency should be higher in the outer regions of galaxies than in the main disks.

Keywords. stars: formation, ISM: bubbles, galaxies: spiral

1. The Galactic Scale

When we observe a star-forming region we sometimes wonder how it got there, or why stars formed there, or anywhere for that matter. The average gas density in most galaxies is very low, close to the tidal limit of $\rho_{\rm tidal} = -1.5\Omega R(d\Omega/dR)/(\pi G) \sim 1$ cm^{-3} for galactic rotation rate Ω and radius R. This is also about the gas surface density limit where Toomre $Q = \sigma_{\rm gas}\kappa/(\pi G\Sigma_{\rm gas}) \sim 1$ for velocity dispersion $\sigma_{\rm gas}$, epicyclic frequency κ, and mass column density $\Sigma_{\rm gas}$. Stellar explosions produce denser gas in supernova remnants, but this gas is usually too tenuous to be strongly self-gravitating, as in the Veil Nebula, and it is also too short-lived in that state for self-gravity to operate (Desai *et al.* 2010). The two-phase instability makes cool gas, but this operates only between temperatures of $\sim 10,000$K and ~ 100K, which is not cold enough to make stars. Star formation requires a thermal Jeans mass close to a solar mass. The thermal Jeans mass is $M_{\rm J,th} = \rho k_{\rm J}^{-3} \sim 7(T/10\ {\rm K})^{1.5}n^{-0.5}M_\odot$ for thermal Jeans wavenumber $k_{\rm J} = (4\pi G\rho)^{1/2}/c$ and isothermal sound speed $c = (kT/\mu)^{1/2}$ with mean weight μ of atomic gas, and density $n = \rho/\mu$. For the range of temperatures expected from the thermal instability and for the average density, $M_{\rm J,th} \sim 200M_\odot$, which is too large for a star. Star formation requires gas that is both cold and dense to bring $M_{\rm J,th}$ down. Processes that do this may be thought of as star-formation triggers.

A fair question is whether the ISM can make cold and dense gas on its own, without non-ISM processes, such as stellar density waves, galactic shear, superbubbles, etc.. If there were no stars to compress the gas or supplement its gravitational self-attraction with additional mass, would new stars form? The answer is probably yes because galaxies once had no stars, and the gas formed stars anyway. Still, galaxy collisions in the young universe could have triggered the star formation by forcing the gas to be dense in the shocked overlap region and in a central concentration that formed after angular momentum loss.

During the last few years, observations of very young galaxies seem to suggest that they can form giant clumps of star formation on their own, even with no observable underlying disk of older stars and no evidence for an interaction with another galaxy (Elmegreen *et al.* 2009ab, Genzel *et al.* 2011). The only possible process for this seems to be a gravitational instability in the whole gas disk, with a resulting clump mass comparable to the turbulent Jeans mass, $M_{\rm J,turb} = \Sigma_{\rm gas} k_{\rm J,2D}^{-2}$ for $k_{\rm J,2D} = \pi G \Sigma_{\rm gas}/\sigma_{\rm gas}^2$. The clumps in young galaxies are large relative to the galaxy radii because the turbulent speeds are large relative to the rotation speeds $v_{\rm rot}$ (Davies *et al.* 2011). This follows from $Rk_{\rm J,2D} \sim GM/(R\sigma_{\rm gas}^2) \sim (v_{\rm rot}/\sigma_{\rm gas})^2$.

This process of gravitational collapse of the ISM is the most fundamental of triggering mechanisms. To carry the initial collapse all the way to stars in a young ISM may have also relied on the pervasive presence of CO molecules (Daddi *et al.* 2010, Tacconi *et al.* 2010), which allows for cooling to $T < 100$K, and on the extremely high $\Sigma_{\rm gas}$, which lowers $M_{\rm J,th}$ by increasing the average midplane density. If we write the midplane pressure in a pure-gas disk as $P = 0.5\pi G \Sigma_{\rm gas}^2$ and the thermal Jeans mass as $M_{\rm J,th} = c^4 (4\pi G)^{-1.5} P^{-0.5}$, then

$$M_{\rm J,th} = 3.8 (T/100 \text{ K})^2 (\Sigma_{\rm gas}/100 M_\odot \text{ pc}^{-2})^{-1} M_\odot \tag{1.1}$$

from which it can be seen that the observed ISM column densities in the clumps of young galaxies, $\Sigma_{\rm gas} > 100 M_\odot \text{pc}^{-2}$ (Tacconi *et al.* 2010), raise the pressure so much that even temperatures of 100 K may be low enough for star formation. Collapse inside these clouds raises the pressure more because of the higher local column density in the resulting core, and it lowers the temperature more because of opacity.

When stars are present, gravitational instability in the ISM is augmented by stellar gravity. This happens most commonly in three ways: by a two-fluid swing-amplified instability (Toomre 1981), in the shocks of global spiral density waves (Roberts 1969), and in the dense central regions formed by gas accretion in a bar potential (Matsuda & Nelson 1977).

The first of these involves the simultaneous collapse of gas and stars, which produces a moderately dense clump of both components on a timescale $(G\rho_{\rm total})^{-1/2} \sim \kappa^{-1}$ for $Q \sim 1$. The stellar part of this collapse bounces as the stars move away with enhanced energy in epicycles. Nearby stars form a spiral wake (Julian & Toomre 1966). The dissipative gas remains as a dense clump at the center of gravity of the instability, with a gas filament also forming in the wake. This is the two-fluid swing-amplified instability (Jog & Solomon 1984, Rafikov 2001, Romeo & Wiegert 2011, Elmegreen 2011). It appears to be common in galaxies, taking the form of multiple spiral arms that are long and irregular in the outer parts. Usually these arms are more regular in the inner regions, where multiple-arms often merge into symmetric two-arm structures midway in the disk (Elmegreen & Elmegreen 1995).

The second way in which stellar gravity augments the condensation of gas into star-forming clumps is through stellar density waves, which arise in the gas+star mixture as a result of perturbations from the outer disk, companion galaxies, or bars, and which may last for several rotations (Lin & Shu 1964) before moving to the center (Toomre 1969). These waves move through the gas and stars (unlike the swing amplifier instability which follows the gas+star mixture for a time κ^{-1}) and they shock the gas as it passes

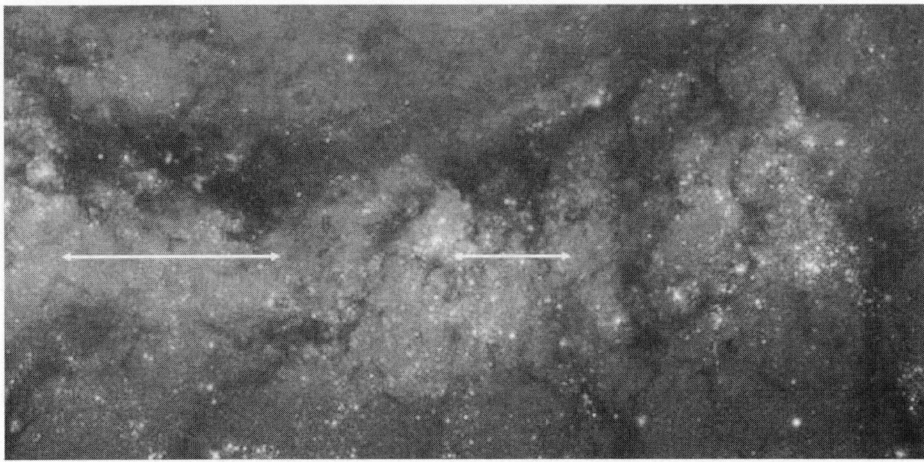

Figure 1. HST image of the southern arm in M51, showing large dark clouds, star formation in the clouds, and young stars downstream with shells and rings around them. The gas flows from the top left to the bottom right in the figure. The large dark clouds contain around $10^7 M_\odot$ of gas. The shells and rings appear to contain new star formation along the periphery.

through. The shock appears as a thin dust lane. If the shocked gas is dense enough, it can become gravitationally unstable and form giant clumps and star complexes. It can also form stars by cooling interarm gas and squeezing interarm clouds. Most star formation in the Milky Way is in giant molecular clouds (GMCs) that are parts of $10^7 M_\odot$ HI+CO complexes in spiral arms (Grabelsky *et al.* 1987). Because shear is low in spiral density wave arms, the process of gas collapse can be augmented by magnetic tensional forces, which remove angular momentum (Elmegreen 1987, Kim *et al.* 2002).

Some galaxies have a stellar component that is either too warm or too rapidly rotating for its column density to form strong spiral arms, i.e., $Q_{\rm stars} >> 1$; then the gas+star mixture is not particularly unstable. The gas is dissipative, however, and can still collapse on its own, or with only a small contribution from underlying stars. The result is a network of thin and short gaseous arms that form stars, giving a flocculent appearance. These arms should be co-moving with the material close to the site of the initial collapse, although they can be wave-like far from this site, in the spiral wake.

Figure 1 shows a piece of the galaxy M51 from an image taken with the Hubble Space Telescope. There is a spiral density wave with two large concentrations of gas in the dust lane, $10^7 M_\odot$ each, and star formation in each concentration. When there are several of these concentrations along part of an arm, they look like "beads on a string". Efremov (2010) studied the relation between such regular patterns and the magnetic field. For several magnitudes of extinction, and for a thickness through the plane of $\sim$ 100 pc, the average density in the dustlane is $\sim$ 10 cm^{-3} (D. Elmegreen 1980) and so the average dynamical time is $\sim (G\rho)^{-1/2} \sim$ 30 Myr. This average density is about what is expected for a density wave shock where the average incident density is the intercloud value ($\sim$ 0.3 cm^{-3}) and the ratio of the shock velocity to the internal turbulent speed is $\sim$ 5. The average dynamical time is too long for gravitational collapse to develop much while the gas is in the dust lane, but the interarm gas is clumpy and these clumps feel a pressure increase when they enter the arm, causing them to collapse. Kim & Ostriker (2002) showed the process of dustlane collapse starting with a smooth ISM. It took several rotations to build up enough density structure for a self-gravitating cloud to form in a spiral arm.

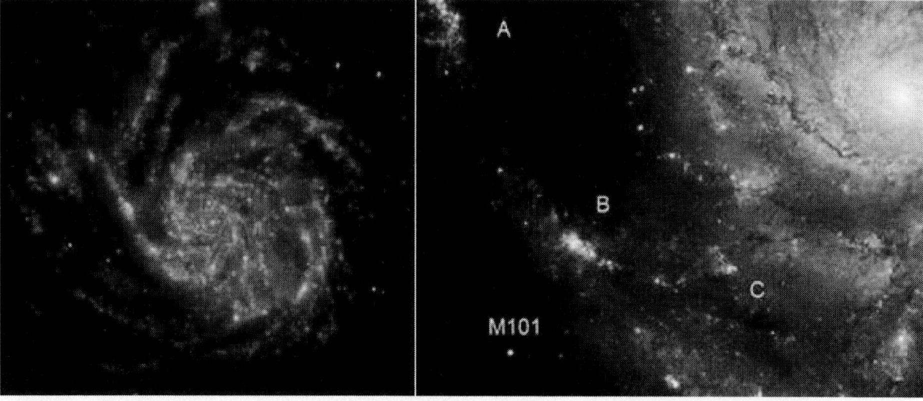

Figure 2. (Left) GALEX image in the uv of M101 from Gil de Paz *et al.* (2007). (Right) HST image of the western part of M101, showing two giant star complexes (A and B) and a dark cloud without much star formation yet (C). Young stars are centered in the spiral arms of M101, suggesting these arms are material patterns made from local gravitational collapse.

The structure of gas in the interarms is a remnant of the structure it had leaving the previous arm, which can also be seen in Figure 1. As the ISM moves through the arm, star formation pressures from HII regions, winds, and multiple supernovae make superbubbles and super-rings. The figure shows many rings of dust. The small rings can be three-dimensional shells seen with enhanced absorption along the lines of sight through the edges, but the large rings are much bigger than the ISM scale height and have to be two-dimensional. All of these structures seem to contain active star formation in the peripheral dense gas, in addition to OB associations and star complexes inside the cleared regions. Some of the active star formation is *lingering* in the dense clouds, which means that it follows the dense gas as it moves. Other active star formation on the periphery is probably *triggered* by the high pressures that made the cavities. This is a second type of triggering for star formation: sequential triggering by previous generations of stars. We review this in the next section.

Figures 2 and 3 show a multiple-arm galaxy, M101, and a flocculent galaxy, NGC 5055. The spiral structure in M101 is more irregular than in the grand design galaxy M51, but there are still long spiral arms and each arm has several regularly-spaced giant star complexes in it, in various stages of collapse. The right-hand side of Figure 2 highlights the western arm in M101, suggesting that the giant dust cloud toward the bottom is in a younger stage than the two other star complexes toward the west. Unlike in M51, the star formation in M101 seems to be centered on the arms with no rings offset to one side. This is expected when the whole arm is the result of a gravitational instability, because it all twists around with local shear and has little relative motion between the pattern speed and the gas. Figure 3 shows numerous long and thin spiral arms with more of the "beads on a string" pattern. These arms are so thin and irregular that they should be mostly gas. They shear around, stretch out, and disperse over time, most likely forming stars by gravitational collapse along their length. Rings and shells are not visible in this image. It would be interesting to observe further whether flocculent galaxies produce the same type of ring pattern as observed down stream from the spiral arms in M51.

The third common way in which stars in a galaxy promote gravitational collapse in the gas is through bar-driven inflows. Bars exert strong negative torques inside their corotation radii, which are typically at 1.2 to 1.4 bar radii. Inside this, the gas hits the bar from behind and shocks, forming a dustlane (Athanassoula 1992). The shock strips

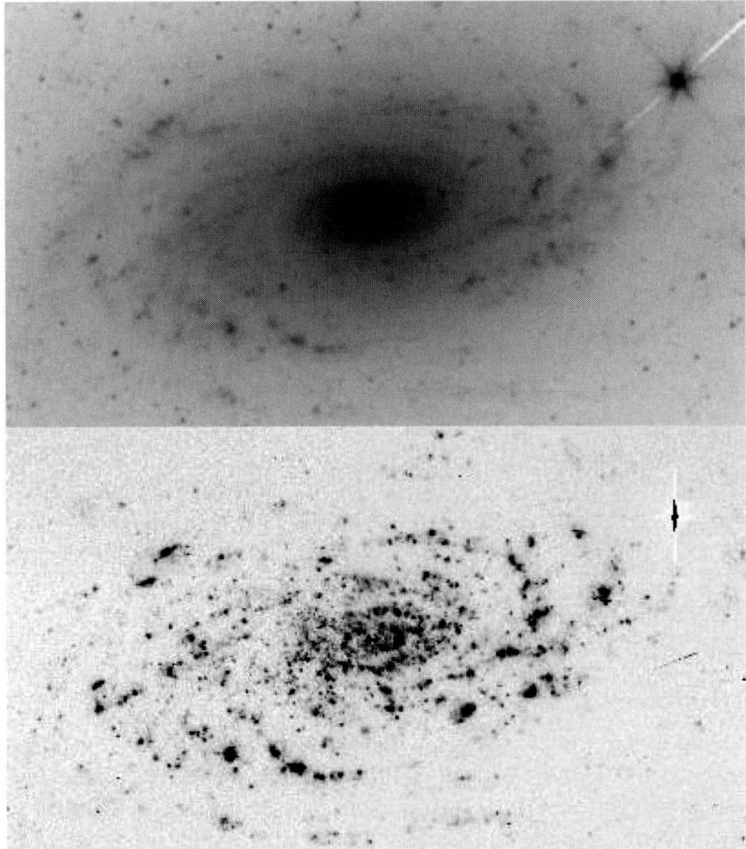

Figure 3. (top) Spitzer 3.6μ image from Elmegreen *et al.* (2011) showing weak and irregular underlying stellar waves. (bottom) Hα image from Dale *et al.* (2009) showing numerous beads on a string of star formation.

angular momentum and orbital energy from the gas, which then falls along the bar to the region of the inner Lindblad resonance (ILR). After sufficient accumulation, the gas becomes unstable and form stars. Often it does this in a ring or tight two-arm spiral. There may be a second, smaller bar inside the ILR which affects the orbits there and brings the gas in further.

Figure 4 shows how the gas flows to the center of NGC 1672. Dust streamers plunge from near the end of the bar into the bar itself, and then directly to the center. This direct path means that gas inflow is rapid. A comparison of the inflow rate to the gas reservoir gives the timescale for accretion. In a study of NGC 1365 (Elmegreen *et al.* 2009c), which looks like NGC 1672, the timescale for gas exhaustion is only ~ 0.5 Gyr. If the initial reservoir of gas was comparable to what is there today, then this short time is also comparable to the age of the bar. That is, bars like these have to be fairly young, ~ 1 Gyr, or the gas would have been cleared out by now.

NGC 1365 and many other barred galaxies have massive clusters in their ILR regions – much more massive than typical disk clusters or clusters forming in spiral density wave arms. Perhaps this difference is only a size-of-sample effect if the ILR regions of bars have higher overall star formation rates. Or, star formation in bar ILRs may differ from that in main disks.

Figure 4. Barred galaxy NGC 1672 from the Hubble Heritage Team. Dust lanes suggest the gas flow pattern toward the center (arrows), where star formation is very active.

2. Super-Bubbles, Super-Rings, and Colliding Flows

Figure 1 shows a spiral arm segment with large rings of gas and star formation downstream from the arms. These are giant cavities are made by star formation, as discussed above. The solar neighborhood is downstream from the Carina spiral arm and also has a large cavity apparently made by star formation. This is associated with the expanding HI ring discovered by Lindblad *et al.* (1973) with a hole in the diffuse dust (Lallement *et al.* 2003). The size of the dust hole is 100 pc by 300 pc, and it extends all the way through the disk. The cavities in Figure 1 are about the same size, and larger than the ISM thickness. Thus they are rings rather than shells. The smaller regions could be three-dimensional shells.

Expanding rings are more unstable than expanding shells because the self-gravitational force vectors that drive gas collection in a one-dimensional section of a ring are all directed toward the growing condensation. According to Elmegreen (1994), the time for significant collapse in an expanding shell is $\sim 100(n\mathcal{M})^{-0.5}$ Myr for ambient density n in cm^{-3} and Mach number $\mathcal{M}$, and the time in an expanding ring is $124(n\mathcal{M}^2)^{-0.5}$ Myr. The ring expression contains a stronger dependence on the Mach number, and this give the ring a shorter collapse time.

Generally, the expansion scale equals the shock speed multiplied by the time, and the shock speed is about $(P/\rho_0)^{0.5}$ for driving pressure P and preshock density ρ_0. If the relevant time is the collapse time, $(G\rho_{\mathrm{comp}})^{-0.5}$ for compressed density ρ_{comp}, then the expansion scale is, after rearrangement, $(P/\rho_{\mathrm{comp}})^{0.5}(G\rho_0)^{-0.5}$, which is the velocity dispersion in the compressed region multiplied by the dynamical time in the ambient gas. Details about the compression source drop out.

If the expansion scale is less than the cloud scale, then pillars and bright rims form by the push-back of interclump gas (Elmegreen, Kimura & Tosa 1995, Gritschneder *et al.* 2009). Star formation is a fast process (squeezing pre-existing clumps), the velocity of triggered stars is small, and causality is difficult to determine, i.e., stars could have formed in the clumps anyway. If the expansion scale is larger than the cloud scale, then shells and rings form by the push-back of all ISM gas (e.g. Dale *et al.* 2011), triggering is a slow process because new clumps have to form on a timescale of $(G\rho_{\rm shell})^{-0.5}$, and the velocity of triggered stars can be large, on the order of the shock speed, $(P/\rho_0)^{1/2}$. There is also a clear causality condition because two stellar generations have to be separated by a distance equal to the age times the mean velocity.

A popular cloud formation scenario considers the compression of gas between two "colliding flows" (e.g., Heitsch *et al.* 2008, Audit & Hennebelle 2010). Shell or ring accumulation forms clouds too, but is different in several ways. Shells and rings have a lateral expansion as they expand radially, and this lateral expansion resists gravitational collapse. Shells and rings also decelerate so that newly formed condensations protrude out the front and have the potential to erode. Shells and rings have a shock on only one side. A shock boundary condition causes a diverging flow at each clump, and this divergence resists collapse. The pressure boundary condition on the other side of the shell or ring squeezes the perturbations and aids collapse. For colliding flows with constant velocity streams, there is a shock on each side and no acceleration of the condensation between them when it is in equilibrium. We observe shells and rings commonly, as shown in the figures and in surveys (e.g., Könyves *et al.* 2007, Ehlerová & Palouš 2005), but there is no clear evidence yet for cloud formation on GMC-scales by colliding flows (on much larger scales, the collision between two galaxies can have a colliding flow; Herrara *et al.* 2011). Still, colliding flows are a good model to study molecule formation and fragmentation in a dynamic environment.

Triggering in the ring RCW 79 was studied in detail by Zavagno *et al.* (2006), who suggested there was a collapsed neutral region containing young stars, 0.1 Myr old, along the periphery of a swept-up region that was 1.7 Myr old. There are several neutral condensations in this shell, somewhat equally spaced around part of the projected edge. Deharveng *et al.* (2010) studied 102 Milky Way bubbles and suggested that 18 of them have either ultracompact HII regions or methanol masers along their edge, suggesting triggering of massive stars in swept-up gas.

3. Bright Rims and Pillars

Most regions of massive star formation have bright ionized rims and pillars of neutral gas pointing to the sources of radiation. A well-studied example is IC 1396, which is a circular HII region 12 pc in radius, with an expansion speed of 5 km s^{-1} and an age of 2.5 Myr (Patel *et al.* 1995). On the western edge is a neutral pillar several pc long (Fig. 5) with very young (class I – diamonds) stars in head-like protrusions and other young stars (class II – circles) all around (Reach *et al.* 2009). This is a typical case.

There is often an age gradient in pillars with the oldest stars closer to the center of the HII region (Sugitani *et al.* 1995). The timescale for triggering can be very fast, 10^4 yrs (Sugitani *et al.* 1989), but the age gradient can span a time of 10^6 years or longer during which the compression moves down the pillar (Smith 2010).

This mechanism of triggering was originally illustrated in models by Klein *et al.* (1983). Recent simulations by Bisbas *et al.* (2011) compare radiative implosion in a variety of conditions. At low incident flux, the implosion is slow, the bright rim is wide, and star formation is far from the tip in the center of a converging compression front. At high incident flux, the implosion is fast, the pillar is narrow, and star formation is close to the tip where the initial compression was concentrated. At late times, the pillar can be long and thin, with bare stars near the head and other stars embedded throughout. If there are many small clumps in the original cloud, then there can be many small pillars, one for each clump, with star formation in each.

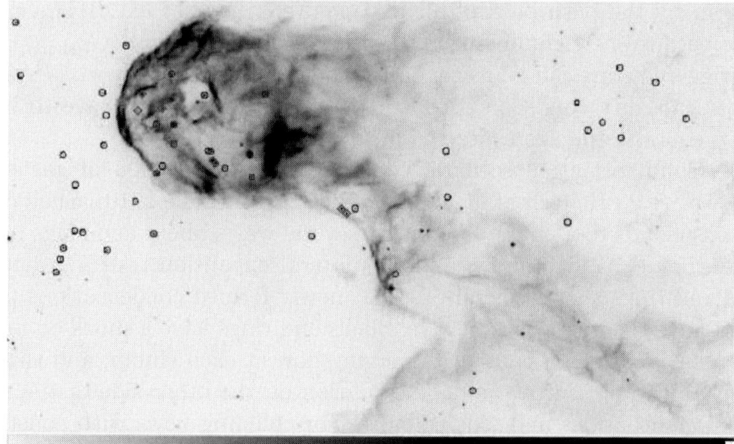

Figure 5. Bright rim at the edge of the HII region IC 1396, from Reach *et al.* (2009). The youngest stars are denoted by diamonds. They were probably triggered to form in compressed cloud clumps that were exposed by ionization and movement of the lowest density, surrounding regions.

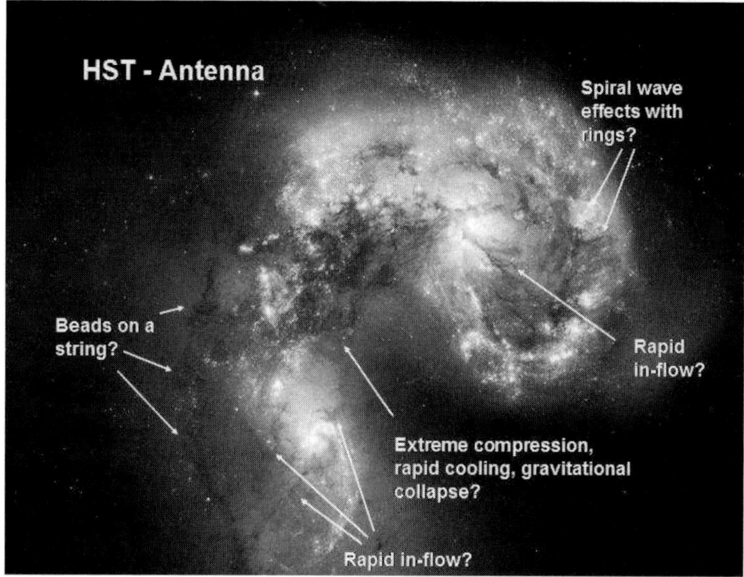

Figure 6. HST image of the Antenna galaxy showing several types of triggering discussed in the text.

4. The Antenna Galaxy: a Mixture of Triggering Processes

Several of the triggering mechanisms discussed above are illustrated in the Antenna galaxy, which is a merger. Figure 6 shows the Antenna labeled with suggested analogies: spiral arm shock triggering with downstream rings, beads on a string from gravitational collapse in a filament, radial inflow in elongated, bar-like regions, and a converging flow that produces extreme compressions between the two gas disks.

5. Triggering and Empirical Star Formation Laws

Star formation triggering is about when, where, and how star formation begins. The global laws of star formation, such as the Kennicutt (1998) or Bigiel *et al.* (2008) relations, show a correlation between the star formation rate and the supply of gas of various types. The Kennicutt (1998) study shows a star formation rate per unit area that scales with the 1.4 power of the total gas column density in the main disks and central regions of spiral galaxies, and the Bigiel *et al.* (2008) and Leroy *et al.* (2008) studies show a linear relation between the star formation rate per unit area and the CO emission per unit area. A linear relation for dense gas traced by HCN was first shown by Gao & Solomon (2004) and Wu *et al.* (2005).

In all cases, star formation is assumed to occur only in cold dark gas, which is traced by CO at low density and HCN and other molecules at high density. The empirical laws suggest that the star formation rate is independent of how the gas is assembled, and therefore independent of the triggering mechanisms. This is particularly true if the average pressure and radiation field determine the general molecular state of the gas, and the pressures which make shells and rings do not change this state much, they just push the gas around. Triggering will also not influence the empirical laws much if the various types of triggering are always present to the same degree. For example, if every initial incidence of star formation is followed by a certain proportion of additional stars that form by radiative implosion or in shell and ring collapse, then the conversion efficiency from ambient gas to young stars will contain all of these effects combined. Each triggering process will not stand out separately in the total star formation rate.

According to Gnedin & Kravtsov (2011) and others, CO and star formation are predictable from the average pressure, self-shielding, radiation field, and other quantities. The pressure that determines molecular self-shielding does not vary much on the scale of shells and rings once they are old enough for star formation to begin. Regions with radiative implosion have high pressure, but this may be a relatively minor star formation event. The pressure varies in spiral arms, but this should only put scatter in the empirical relations. Thus the signatures of localized triggering may be imperceptible in the empirical star formation laws.

6. All you need is cold gas: the legacy of K.E. Edgeworth

The universality of the empirical laws combined with the evidence for star formation in shells, spiral arms, pillars, and other pressurized structures suggests that the primary ingredient for star formation is cold gas and not a particular geometry. Cold gas means colder than thermal equilibrium commonly gives for atomic gas in the background galactic radiation field. Cold gas for star formation implies shielding from starlight and the presence of molecules for rotational cooling, which can operate down to a few degrees above the microwave background. Once the gas is shielded from starlight and turns molecular, background pressure and turbulent motion will compress it to a high density in small regions. Then gravity becomes important and the thermal Jeans mass drops to be comparable to the region mass. Star formation usually follows.

Shell, ring, or pillar-like triggering and spiral waves might be more important as a net positive contributor to star formation in regions where the molecular fraction is moderately low on average, i.e., at the boundary between highly molecular inner disks of galaxies and highly atomic outer disks. In the inner disks, the gas pretty much stays molecular everywhere because of the high pressure from its weight in the disk, and because of the high opacity. Starlight ionizes and photodissociates molecules at the cloud edges (Heiner *et al.* 2011) and between the clouds. Cold gas is present in most clouds, so additional pressures from young stars or spiral waves will not make the gas much colder or more molecular. In the far-outer regions, however, it is rather difficult to turn atomic gas into molecular gas because the average pressure and opacity are very low. Then superbubbles might not trigger new star formation, but only make diffuse atomic shells which have little cold gas. Between these two zones, or perhaps in dwarf irregular galaxies where the same intermediate conditions apply, triggering by the pressures of young stars might make more of a difference in the total star formation rate.

An observational test of this prediction would be to measure the radial variation of the ratio of the normalized star formation efficiency in compressed and non-compressed regions. The normalized efficiency is the star formation rate per unit gas mass per unit dynamical rate. An example would be the ratio of the efficiency in the spiral arms to the interarm regions. If this ratio increases with radius, then spiral arm triggering is doing more to enhance star formation in the outer parts of galaxies, where the gas is mostly atomic, than in the inner parts, where the gas is mostly molecular. Alternatively, one could plot this ratio versus the average molecular fraction instead of the radius.

The overriding importance of cold gas for star formation was originally recognized by Kenneth Essex Edgeworth in 1946 – long before the modern star formation era. (For a biography of Edgeworth, see Hollis 1996.) At that time, most stars, like the Earth and the whole universe, were thought to be about 3 Gyr old, star formation was not considered to happen in the present day, except possibly for the most massive stars (which were known to burn their fuel quickly), and the stability of galactic disks was not yet understood. Edgeworth and others were thinking about star formation mostly in the context of the beginning of the universe. He noted from a thermal Jeans analysis that star formation by gravitational instability in a galaxy disk, using an equation like eq. 1.1, requires a temperature of $\sim$ 2.8K, which he considered "very improbable." He also said that angular momentum from galactic rotation was too large for the solar system to form, and suggested that stars might form instead by successive condensations to overcome the angular momentum problem. This led him to "expect to find that the majority of the stars were members of star clusters" which is "not in agreement with observation." Finally, he suggested that the rotating gas disk of a galaxy, at $\sim$ 1000 K, breaks up into azimuthal filaments, which then break up into stars following the removal of heat. He went on to suggest that the residual material around each star makes planets.

These ideas are all essentially correct by modern standards – even though Edgeworth did not believe or observe them at the time. Edgeworth understood that star formation requires very cold gas, it most likely occurs in massive aggregates, and it should be patterned as beads on a string for the galactic scale. These ideas were too far ahead of their time to have much influence. Star formation as we know it was discovered several years later when Ambartsumian (1949) showed that local OB associations are expanding away from a common center. This limited their age to 10 Myr.

7. Summary

A variety of processes cause interstellar gas to become cold enough and dense enough to form stars. On galactic scales, stellar instabilities, spiral waves, and global perturbations like bars can move the gas around supersonically and cause shocks to form that are larger than the characteristic size of a gravitational instability in the gas. Then giant cloud complexes form from the ambient gas. As these complexes dissipate their internal turbulent energy, they contract gravitationally and fragment because of converging and diverging turbulent motions until dense, thermally-dominated cold cloud cores form.

Stellar pressures also compress the gas supersonically. On sufficiently large scales, these compressions lead to collapse in shells and rings. On small scales, stellar pressures can turn in pre-existing clumps unstable to collapse, especially along the edges of HII regions and super-bubbles.

The empirical laws of star formation have no obvious connection to the details of these triggering mechanisms. The empirical laws state mostly that star formation requires cold and dense gas. In the case of the Kennicutt (1998) law, with its non-linear dependence of star formation rate on total gas density, empirical evidence suggests also that general dynamical processes in the ISM are involved in determining the time scale. Since triggering on the length scale of these laws has the same dynamical time scale, all of the various triggering processes can mix together without much distinction.

The distinct contribution that triggering makes to the star formation rate might be most evident in large-scale regions where the average molecular fraction is neither very high nor very low. There the dynamical processes related to cloud formation could have a significant influence on the abundance of cold gas in clouds. There might still be a linear relation between cold gas mass and star formation rate at these places, because star formation follows cold gas no matter what forms the cold gas, but the rate of both cold gas formation and star formation could be modulated by dynamical processes more there than elsewhere.

The author is grateful to the National Science Foundation for support from grant AST-0707426, and to the conference organizers, particularly Professors Cristina Popescu and Richard Tuffs, for their support and hospitality.

References

Ambartsumian, V. A. 1949, *Soviet AJ*, 26, 3
Athanassoula, E. 1992, *MNRAS*, 259, 345
Audit, E. & Hennebelle, P. 2010, *A&A*, 511, 76
Bigiel, F., Leroy, A., Walter, F., *et al.* 2008, *AJ*, 136, 2846
Bisbas, T. G., Wünsch, R., Whitworth, A. P., Hubber, D. A., & Walch, S. 2011, *ApJ*, 736, 142
Daddi, E. *et al.* 2010a, *ApJ*, 713, 686
Dale, D. A. *et al.* 2009, *ApJ*, 703, 517
Dale, J. E., Wünsch, R., Smith, R. J., Whitworth, A., & Palouš, J. 2011, *MNRAS*, 411, 2230
Davies, R., Förster Schreiber, N. M., Cresci, G., *et al.* 2011, *ApJ*, 741, 69
Deharveng, L., Schuller, F., Anderson, L. D., *et al.* 2010, *A&A*, 523, 6
Desai, K. M., Chu, Y.-H., *et al.* 2010, *AJ*, 140, 584
Edgeworth, K. E. 1946, *MNRAS*, 106, 470
Efremov, Yu. N. 2010, *MNRAS*, 405, 1531
Ehlerová, S. & Palouš, J. 2005, *A&A*, 437, 101
Elmegreen, B. G. 1987, *ApJ*, 312, 626
Elmegreen, B. G. 1994, *ApJ*, 427, 384
Elmegreen, B. C. 2011, *ApJ*, 737, 10
Elmegreen, B. G., Kimura, T., & Tosa, M. 1995, *ApJ*, 451, 675
Elmegreen, B. G., Elmegreen, D. M., Fernandez, M. X., & Lemonias, J. J. 2009a, *ApJ*, 692, 12
Elmegreen, B. G., Galliano, E., & Alloin, D. 2009, *ApJ*, 703, 1297
Elmegreen, D. M. 1980, *ApJS*, 43, 37
Elmegreen, D. M. & Elmegreen, B. G. 1995, *ApJ*, 445, 591
Elmegreen, D. M., Elmegreen, B. G., Marcus, M., *et al.* 2009b, *ApJ*, 701, 306
Elmegreen, D. M., Elmegreen, B. G., Yau, A. *et al.* 2011, ApJ, 737, 32
Gao, Y. & Solomon, P. M. 2004, *ApJ*, 609, 271
Genzel, R., Newman, S., Jones, T., *et al.* 2011, *ApJ* , 733, 101

Gil de Paz, A. *et al.* 2007, *ApJS*, 173, 185
Gnedin, N. Y. & Kravtsov, A. V. 2011, *ApJ*, 728, 88
Grabelsky, D. A., Cohen, R. S., Bronfman, L., Thaddeus, P., & May, J. 1987, *ApJ*, 315, 122
Gritschneder, M., Naab, T., Walch, S., Burkert, A., & Heitsch, F. 2009, *ApJ*, 694, L26
Heiner, J. S., Allen, R. J., & van der Kruit, P. C. 2011, *MNRAS*, 416, 2
Heitsch, F., Hartmann, L. W., & Burkert, A. 2008, *ApJ*, 683, 786
Herrera, C. N., Boulanger, F., & Nesvadba, N. P. H. 2011, *A&A*, 534, 138
Hollis, A. J. 1996, *J. British Astron. Assoc.*, 106, 354
Jog, C. J. & Solomon, P. M. 1984, *ApJ*, 276, 114
Julian, W. H. & Toomre, A. 1966, *ApJ*, 146, 810
Kennicutt, R. C., Jr. 1998, *ApJ*, 498, 541
Kim, W.-T. & Ostriker, E. C. 2002, *ApJ*, 570, 132
Kim, W.-T., Ostriker, E. C., & Stone, J. M. 2002, *ApJ*, 581, 1080
Klein, R. I., Sandford, M. T., & II, Whitaker, R. W. 1983, *ApJL*, 271, 69
Könyves, V., Kiss, Cs., Moór, A., Kiss, Z. T., & Tóth, L. V. 2007, *A&A*, 463, 1227
Lallement, R., Welsh, B. Y., Vergely, J. L., Crifo, F., & Sfeir, D. 2003, *A&A*, 411, 447
Leroy, A. K., Walter, F., Brinks, E. *et al.* 2008, *AJ*, 136, 2782
Lin, C. C. & Shu, F. H. 1964, ApJ, 140, 646
Lindblad, P. O., Grape, K., Sandqvist, A., & Schober, J. 1973, *A&A*, 24, 309
Matsuda, T. & Nelson, A. H. 1977, *Nature*, 266, 607
Patel, N. A., Goldsmith, P. F., Snell, R. L., Hezel, T., & Xie, T. 1995, *ApJ*, 447, 721
Rafikov, R. R. 2001, *MNRAS*, 323, 445
Reach, W. T., *et al.* 2009, *ApJ*, 690, 683
Roberts, W. W. 1969, *ApJ*, 158, 123
Romeo, A. B. & Wiegert, J. 2011, *MNRAS*, 416, 1191
Smith, N. 2010, *MNRAS*, 406, 952
Sugitani, K., Fukui, Y., Mizuni, A., & Ohashi, N. 1989, *ApJL*, 342, 87
Sugitani, K., Tamura, M., & Ogura, K. 1995, *ApJL*, 455, 39
Tacconi, L., *et al.* 2010, *Nature*, 463, 781
Toomre, A. 1969, ApJ, 158, 899
Toomre, A. 1981, in: S.M. Fall (ed.), *The structure and evolution of normal galaxies*, (Cambridge: Cambridge University Press), p. 111.
Wu, J., Evans, N. J., I. I., & Gao, Y., *et al.* 2005, *ApJ*, 635, L173
Zavagno, A., Deharveng, L., Comerón, F., *et al.* 2006, *A&A*, 446, 171

Discussion

GAO: You mentioned that supernovae are not important in triggering star formation on galactic scales. Could they be important in extreme starbursts where essentially most of the molecular gas concentrated within a few hundred parsec, experiencing constant impacts of supernovae due to the several orders of magnitude higher star formation ratee?

ELMEGREEN: There is no observational answer to this question now, but a guess based on theory is that supernova remnants in high density environments are very short-lived because the radiation rate is high. If their lifetime is less than the dynamical time of the surrounding medium, then triggering gets different. The prior HII region of that supernova star might have had a bigger triggering effect, if it was a massive star.

TUFFS: We know that more massive galaxies form stars more efficiently (out of available gas) than low mass galaxies. Can you comment on how the small scale processes you have described relate to global galaxian properties? For example, does the expected differing vertical scale of gas in different gravitational potentials affect the confinement of the ring structures you described?

ELMEGREEN: The ISM Jeans length determines both the thickness of a disk and the size of its largest star forming regions. So relatively thick gas disks, as in late type galaxies, have relatively bigger star complexes. Their blow out from shells and rings should be about the same as in relatively thin gas disks. The reason why massive galaxies form stars with a higher specific star formation rate during their most active period is probably related to the average density and average intergalactic accretion rate during this period in their evolution.

The Spectral Energy Distribution of Galaxies
Proceedings IAU Symposium No. 284, 2011
R.J. Tuffs & C.C. Popescu, eds.

doi:10.1017/S1743921312009362

The energetics of turbulent molecular gas and star formation

François Boulanger

Institut d'Astrophysique Spatiale (IAS), CNRS & Université Paris-Sud, France

email: Francois.Boulanger@ias.u-psud.fr

Abstract. The role interstellar turbulence plays in regulating star formation is a much debated research topic. In this paper, I take an observational view point in presenting observations of H_2 line emission from extragalactic sources. I highlight key results from these observations. (1) H_2 line emission is a main cooling channel of molecular gas. It is a tracer of mechanical energy dissipation complementing mass tracers in describing the dynamical state of molecular gas in galaxies. (2) Spectroscopy of warm H_2 observations with the Spitzer Space Telescope and the SINFONI spectro-imager at ESO provide evidence of shock excited H_2 line emission in galaxies that exemplify the main agents of galaxy evolution. (3) The dissipation of mechanical energy involves a turbulent energy cascade and the cycling of interstellar matter across ISM phases, including the formation of H_2 gas from warm atomic gas. (4) In Stephan's Quintet and the radio galaxy 3C326, two sources with a high H_2 luminosity to mass ratio (i.e. a high dissipation rate per unit mass), turbulence is observed to quench star formation. In the Antennae merger, star formation is observed to proceed where the turbulent kinetic energy is being dissipated.

Keywords. turbulence, ISM: molecules, ISM: evolution, ISM: structure, stars: formation, galaxies: evolution, infrared: galaxies

---

1. Introduction

What regulates the star formation rate in galaxies? This is a long standing question within the field of galaxy evolution. The low efficiency of star formation ($\sim 1\%$ per free-fall time, Krumholz *et al.* 2012) is most often associated with interstellar turbulence. Turbulence has a dual effect on star formation. On the one hand, it opposes gravity by continuously shearing apart gas, and on the other hand, where the gas kinetic energy is dissipated (e.g. in shocks), it creates structures that can become gravitationally bound and form stars (Mac Low & Klessen 2004). This paradigm establishes a direct connection between star formation and the energetics of molecular clouds, i.e. the energy injection that feeds turbulence and dissipation. To get a full picture of star formation we must combine observations which trace both the gas and the dissipation rate of its turbulent kinetic energy.

To characterize molecular gas, we primarily rely on rotational line emission of trace molecular species. The rotational lines of CO, are used to trace the spatial distribution and the kinematics of the bulk of the molecular gas. Lines from molecules with higher dipole moment, such as HCN, serve to trace the dense H_2 gas that is directly associated with star formation. It is far less straightforward to quantify the energy radiated away by molecular gas, and, thereby, the energetics of molecular clouds. Spitzer observations of H_2 rotational lines have contributed to open this missing perspective on molecular gas in galaxies.

H_2 emission from the interstellar medium is most often thought to be associated with localized heating of H_2 in photo-dissociation regions and shocks within star forming regions, but observations from the Infrared Space Observatory and more recently the Spitzer Space Telescope provide evidence for diffuse H_2 emission tracing the dissipation of turbulence kinetic energy. Bright H_2 line emission powered by interstellar turbulence, rather than by star formation, has been identified in the Milky Way diffuse interstellar medium (Falgarone *et al.* 2005, Ingalls *et al.* 2011), as well as in a number of extragalactic systems (Egami *et al.* 2006, Appleton *et al.* 2006, Ogle *et al.* 2010, Zakamska 2010, Herrera *et al.* 2011). These observations are the main focus of the paper.

2. H_2 Luminous Gas in Galaxies

Spitzer observations have provided fluxes of H_2 rotational lines for a number of extragalactic sources including sources with bright H_2 line emission with no, or relatively weak, spectroscopic signatures (dust and ionized gas lines) of star formation. Molecular gas is found to act as a cooling agent in a diverse set of objects which exemplify violent phases in the evolution of galaxies. In Fig.1, the H_2 emission in the four first rotational lines S(0) to S(3) is compared to the flux in the 7.7 μm PAH band. This ratio indicates what is powering the H_2 line emission. A tight correlation between the H_2 and PAH fluxes is observed in star forming galaxies, e.g. the SINGS sample (Roussel *et al.* 2007). These galaxies are distributed along the bottom horizontal dashed line in the plot. A number of galaxies lie significantly above this line. For these galaxies, the *excess* H_2 emission cannot be powered by UV heating of molecular gas. Sources with *excess* H_2 emission include galaxies with active galactic nuclei, cooling flows, star-burst winds, interacting galaxies and mergers. A common characteristic of these sources is that energy is injected in the interstellar medium (ISM) on galaxy-wide scales by a violent process.

In several of the H_2 galaxies the star formation rate per unit surface is much below the standard Schmidt-Kennicutt relation. For example, in the radio galaxy 3C326 (measured gas surface density 250 $M_\odot$ pc^{-2}), the Schmidt-Kennicutt law implies a star formation rate of 4.5 $M_\odot$ yr^{-1}, a value $\sim$ 60 times greater than the observed upper limit from Spitzer 70μm imaging (Nesvadba *et al.* 2010, Nesvadba *et al.* 2011). A similar difference is observed for the molecular gas in the Stephan's Quintet (SQ) shock (Guillard *et al.* 2012). The fact that there is little star formation in these sources suggests that the H_2 gas is too turbulent to permit the formation of gravitationally unstable condensations. In other words it is sheared apart and shredded by turbulence on timescales shorter than the local free-fall time.

3. Turbulent energy cascade and H_2 cooling

A range of H_2 temperatures is required to account for the H_2 excitation diagram. The bulk of the warm H_2 gas mass is at the lowest temperature $\sim$ 150 K. The mass of warm H_2 is estimated to be 5×10^8, 10^9, 2×10^9, and $10^{10}\,M_\odot$ in SQ, the radio galaxy 3C326, the merger NGC 6240 and the brightest cluster galaxy ZW3146, respectively. For these four galaxies that are archetype sources, the warm H_2 gas is observed to account for a significant fraction (> 10%) of the cold H_2 mass inferred from the CO luminosity. This is a significant result because the cooling time of H_2 gas at a temperature of 150 K is $\sim 10^4$ yr, a value much smaller than the dynamical timescales associated with energy injection. To account for the high fraction of warm H_2, the gas must be repeatedly heated. This implies that the dissipation of mechanical energy does not occur in shocks directly driven by bulk motions, but involves an energy cascade which feeds turbulence within molecular clouds. The galaxy-wide shock in the SQ compact group is a template source which highlights this key idea.

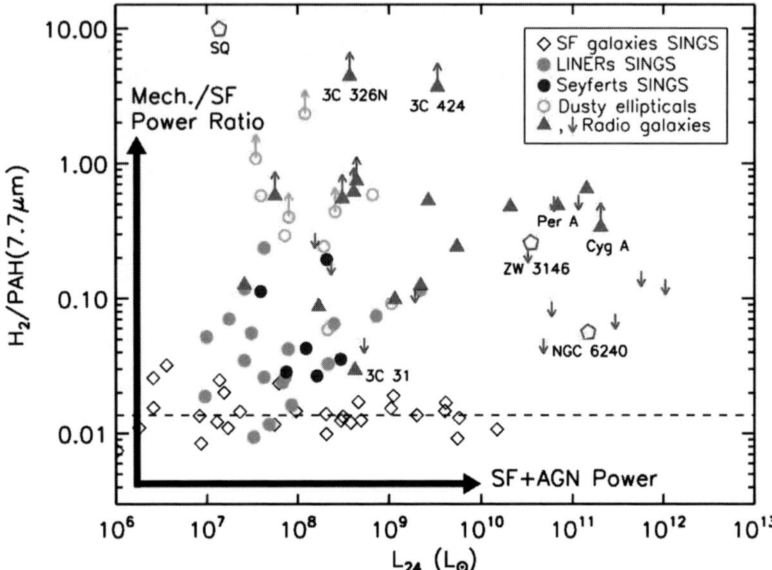

Figure 1. The ratio between the H_2 fluxes summed over the 0-0 S(0)-S(3) lines to the 7.7 μm PAH flux is plotted versus the 24 μ rest frame luminosity. The gas to dust, H_2 to PAH, emission ratio indicates the relative importance of mechanical heating and UV (star formation) as energy sources powering the H_2 emission. This figure is borrowed from Ogle *et al.* (2010).

In SQ, the kinetic energy from a high speed ($\sim$ 1000 km s^{-1}) collision, between an intruding galaxy and a tidal arm within the galaxy group, is observed to be powering X-ray emission with an emission temperature of 5×10^6 K corresponding to a shock velocity of $\sim$ 600 km s^{-1}. This shock velocity is consistent with the gas velocity in the center of mass frame; it is about half of the observed relative velocity, as expected if the gas mass on the intruder and SQ tail sides are similar. The H_2 line emission has the same spatial extent as the plasma, and is a few times more luminous than the X-ray luminosity (Appleton *et al.* 2006, Cluver *et al.* 2010). CO observations show that the turbulent energy of the post-shock molecular gas is larger than the thermal energy of the shock-heated X-ray emitting plasma (Guillard *et al.* 2012). Most of the kinetic energy of the collision has not been thermalized. It is observed to be in bulk and turbulent motions of the molecular gas. It is the progressive dissipation of this energy that powers the H_2 emission.

The presence of large amounts of warm H_2 gas in the post-shock gas implies an energy cascade transferring a significant fraction of the kinetic energy of the collision to turbulent motions of much smaller amplitude within molecular clouds. The amplitude of turbulence may be estimated from the luminosity to mass ratio of the H_2 gas which is observed to be $4\times10^{-2}\, L_\odot/M_\odot$ (Guillard *et al.* 2012). The H_2 luminosity is summed over the S(0) to S(5) rotational lines and the H_2 mass inferred from the CO(1-0) luminosity assuming that the Milky Way conversion factor applies to SQ. The L_{H_2}/M_{H_2} ratio may be used to estimate the amplitude of turbulent motions within SQ molecular clouds. We consider that the H_2 luminosity is powered by the dissipation of the gas turbulent kinetic energy over a timescale $t_{diss} = R/\sigma_v$, where R is the cloud radius and σ_v the 1D velocity dispersion. Since the dissipation of the gas kinetic energy does not occur only through the S(0) to S(5) lines used to estimate the H_2 luminosity, we introduce the factor $f_{H_2} = L_{H_2}/L_{diss}$ where L_{diss} is the bolometric luminosity.

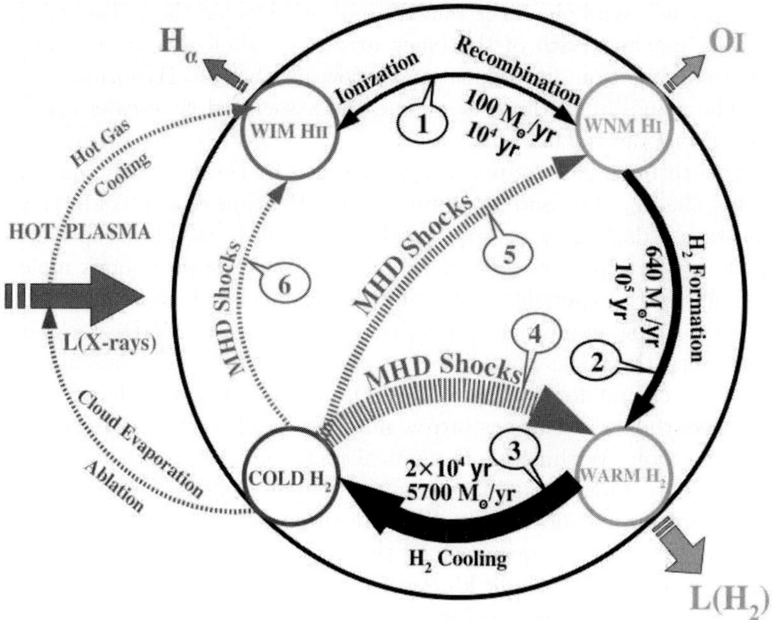

Figure 2. Schematic view at the gas evolutionary cycle proposed in our interpretation of Stephan's Quintet optical and H_2 observations. Arrows represent the mass flows between the HII, warm HI, warm and cold H_2 gas components. They are numbered for clarity. The dynamical interaction between gas phases drives the cycle. The mass flow values and associated timescales are derived from the H_α, [OI], and H_2 luminosities and model calculations. Heating of the cold H_2 gas (dark grey arrows) is necessary to account for the increasing mass flow from the ionized gas to cold H_2 phases. This figure is borrowed from Guillard *et al.* (2009).

$$\begin{aligned} L_{H_2}/M_{H_2} &= f_{H_2}\, L_{diss}/M_{H_2} = 3/2\, f_{H_2}\, \sigma_v^3/R \\ &= 4\times10^{-2}\, f_{H_2}\, (\sigma_v/20\,\mathrm{km\,s^{-1}})^3\, (R/50\,\mathrm{pc})^{-1} \end{aligned} \quad (3.1)$$

Since $f_{H_2} < 1$, a turbulence amplitude $> 20\,\mathrm{km\,s^{-1}}$, for a cloud radius $R = 50\,\mathrm{pc}$, is required to account for the H_2 emission. The lower limit on σ_v is a factor 6 larger than that measured for giant molecular clouds in the Milky Way (Heyer *et al.* 2009). The modeling of the H_2 excitation diagram with MHD shocks yields comparable velocities (Guillard *et al.* 2009).

4. The dynamical interaction between gas phases

For a few of the H_2 sources we have imaging data displaying the spatial distribution of the H_2 emitting gas. In the Perseus cluster and the M82 galactic wind, the data show that the warm H_2 extends to the halo where it is mixed with hot and warm gas traced by X-ray and optical and infrared emission lines. This is also true for SQ where the H_2 and X-ray emission have similar spatial extent (Cluver *et al.* 2010). These imaging data show a close association between the cold and hot ISM. Based on multi-wavelength observations of SQ, Guillard *et al.* (2009) showed that the energy dissipation must involve gas cycling across ISM phases, including formation of H_2 out of warm atomic gas.

A schematic cartoon of the dynamical evolution of the post-shock gas in SQ is presented in Fig. 2. Black and dark grey arrows represent the mass flows between the HII, warm HI, warm and cold H_2 gas components of the post-shock gas. The large inward-pointing grey arrow to the left symbolizes the relative motion between the warm and cold gas and the surrounding plasma. Each of the black arrows is labelled with its main associated process: gas recombination and ionization (arrow number 1), H_2 formation (2) and H_2 cooling (3). The values of the mass flows and the associated timescales are derived from observations and model calculations.

A single cycle through gas components is excluded by the increasing mass flow needed to account for the H_α, OI, and H_2 luminosities. Heating of the cold H_2 gas towards warmer gas states (dark grey arrows) needs to occur. It is the dissipation of the gas mechanical energy that powers these dark grey arrows. The post-shock molecular cloud fragments are likely to experience a distribution of shock velocities, depending on their size and density. Arrow number 4 represents the low velocity MHD-shocks excitation of H_2 gas. More energetic shocks may dissociate the molecular gas (arrows number 5). They are necessary to account for the low H_α to O I luminosity ratio. Even more energetic shocks may ionize the molecular gas (arrow number 6). This would bring cold H_2 directly into the H II reservoir. Within this dynamical picture of the post-shock gas, molecular gas is not necessarily a mass sink. The mass accumulated in H_2 gas depends on the ratio between the excitation by shocks trough one of the dark grey arrows 4, 5 and 6 and the H_2 formation and cooling timescales along the black arrows 2 and 3.

An efficient transfer of the bulk kinetic energy to turbulence in molecular clouds is required to make H_2 a dominant coolant of the post-shock gas. Dissipation within either the hot plasma or warm atomic gas, leading to radiation at X-rays or optical wavelengths that is not efficiently absorbed by the molecular gas, has to be minor.

We are far from being able to describe how the energy transfer occurs. We just make qualitative statements that would need to be quantified with numerical simulations in future studies. The fragmented, multiphase structure of the clouds is likely to be a main key of the energy transfer. The dynamical interaction with the background plasma flow generates relative velocities between cloud fragments, because their acceleration depends on their mass and density. The relative motions between cloud fragments can lead to collisions, which will result in the dissipation of the kinetic energy. This dissipation will occur preferentially in the molecular gas, because it is the coldest component with the lowest sound speed. The magnetic field is likely to be important, because it weaves the cloud H_2 fragments to the lower density cloud gas which is more easily entrained by the background plasma flow. The mass cycle between cloud gas states contributes to the energy transfer. The gas that cools carries its turbulent kinetic energy to the colder gas phase it cools to. In other words, gas cooling transfers the turbulent velocities of the warm H II and H I to the colder and denser H_2 gas that is formed.

5. From the dissipation of gas kinetic energy to star formation

Herrera *et al.* (2011 & 2012) have presented CO and H_2 observations of the overlap region in the Antennae, obtained with the near-infrared imaging spectrograph SINFONI on the ESO Very Large Telescope and ALMA during science verification. These observations give a foretaste of the power of combining mass and energy tracers to study the dynamical state of molecular gas in galaxy mergers and the early stages of star formation.

Most of the star formation in the overlap region occurs in massive super-star clusters (SSCs) with masses up to a few 10^6 $M_\odot$ (Whitmore *et al.* 2010). The large surface density of molecular gas and its fragmentation in super giant molecular complexes (SGMCs) with masses of several 10^8 $M_\odot$, two orders of magnitude larger than masses of giant molecular clouds in spiral galaxies, are evidence for cooling and gravitational fragmentation of the diffuse gas compressed in the galaxies collision (Teyssier *et al.* 2010). The data show that

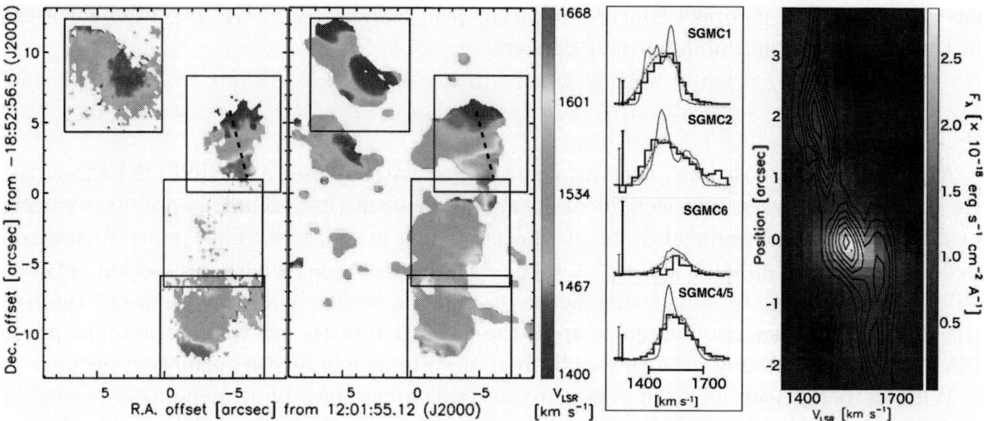

Figure 3. *Left panel:* Velocity maps derived from the H_2 (left) and CO (right) line emission. Dotted lines mark the position-velocity cut shown in the right panel. *Mid panel:* Integrated line profiles of CO (thin solid line) and H_2 (thick black solid line with steps) for each SGMC. Dashed spectra show the CO lines convolved to the spectral resolution of SINFONI. Black bars correspond to 5×10^{-16} erg s^{-1} cm^{-2} $\AA^{-1}$ for the 1-0 S(1) H_2 line, and grey bars to 1.5 Jy for CO, for all of the SGMCs. Offsets between the black and dashed spectra indicate variations in the CO-to-H_2 ratio between velocity components. *Right panel:* H_2 position-velocity diagram, across the bright compact H_2 source discussed in the text, with CO emission shown as contours in steps of 0.02 Jy beam^{-1} starting at 0.03 Jy beam^{-1}. This figure is borrowed from Herrera *et al.* (2012).

all SGMCs have H_2 1−0 S(1) line emission and the H_2 1−0 S(1) kinematics match those of CO(3−2) well (Fig. 3).

Herrera *et al.* (2011) present an interpretation of the H_2 emission, which relates the emission to the energy dissipation required to form the SGMCs and clouds within the SGMCs from gas driven by the galaxy interaction. This interpretation introduces a dynamical view of the present state of the complexes. They are physically associated with gas which is too turbulent to be bound. This unbound gas is dynamically fed by the gas dynamics on larger scales. This idea is supported by the gas kinematics. The CO data show that all complexes have two spatially separated velocity components. The velocity difference between components within an individual SGMC is up to 150 km s^{-1} (Fig. 3). Given the size and mass of the SGMCs, this is too large to be accounted for by the gas self-gravity. The gas kinematics is most likely driven by the galaxy interaction. In single-dish observations similar velocity gradients are found in the extended emission around the SGMCs, which further supports this idea.

Herrera *et al.* (2011 & 2012) also discuss the nature of the bright, compact source of near-IR H_2 line emission discovered with their SINFONI data. This source shows no Brγ emission. It has the highest H_2/CO line emission ratio, and coincides with the steepest CO velocity gradient of the entire overlap region. With a size of 50 pc and a virial mass of $5 \times 10^7\ M_\odot$, it is perhaps a pre-cluster cloud that has not yet formed significant numbers of massive stars. The bolometric luminosity of this source is $\sim 10^7\ L_\odot$. It may be accounted for by the dissipation of the cloud kinetic energy for a cloud mass of a few $10^7\ M_\odot$ – a value comparable to the virial mass – and a dissipation timescale of 1 Myr. This timescale is comparable to the cloud crossing time, and also the dynamical time scale associated with the velocity gradient of 1 km s^{-1} pc^{-1} at the position of the cloud. The similarity of both timescales indicates that the cloud may still be forming by accreting gas, and thus that a significant part of the cloud luminosity may be powered by gas accretion. The compact source could represent a short ($\sim 1\,$Myr) evolutionary stage in the early formation of super-star clusters.

6. Conclusion

Analysis of spectroscopic observations obtained with Spitzer and SINFONI show that H_2 emission is a tracer of mechanical energy dissipation that complements mass tracers in describing the dynamical state of molecular gas in galaxies. This paper focuses on two objects, the Stephan's Quintet and the Antennae merger, where interaction between galaxies is feeding interstellar turbulence on galactic scales. The data highlight the role the H_2 gas plays as a cooling agent, from the formation of H_2 gas within the multi-phase ISM to the formation of gravitationally bound systems where star formation occurs.

What is being learned from these archetype sources may be of general relevance to violent phases of galaxy evolution, when merging, gas accretion, AGNs or star bursts release large amounts of mechanical energy on galactic scales. Observations of these and other H_2 luminous sources with Herschel and ALMA will soon complement the results I have presented. Data to be obtained with Herschel – spectroscopy of far-IR cooling lines and dust imaging – will provide key information on the energy dissipation process and the physical state of the H_2 luminous gas. ALMA offers the required combination of spectral and angular resolution to relate the gas energetics to the mass distribution and gas dynamics. What we will learn from these observations is likely to be of general relevance to understanding the energetics of the turbulent molecular gas in galaxies and what regulates its efficiency of forming stars.

Acknowledgements

I thank the collaborators who have contributed to the work I have presented in this paper: Phil Appleton, Edith Falgarone, Pierre Guillard, Cinthya Herrera, Nicole Nesvadba, Patrick Ogle, and Guillaume Pineau des Fôrets.

References

Appleton, P. N., Xu, K. C., Reach, W. T. *et al.* 2006, *ApJ*, 639, L51
Cluver, M. E., Appleton, P. N., Boulanger, F. *et al.* 2010, *ApJ*, 710, 248
Falgarone, E., Verstraete, L., & Pineau des Forêts, Hily-Blant, P. 2005, *A&A*, 433, 997
Egami, E., Rieke, G. H., Fadda, D., & Hines, D. C. 2006 *ApJ*, 652, L21
Guillard, P., Boulanger, F., Pineau des Forêts, G., & Appleton, P. N. 2009, *A&A*, 502, 515
Guillard, P., Boulanger, F., Gusdorf, A. *et al.* 2011, *A&A*, in press
Heyer, M., Krawczyk, C., Duval, J., & Jackson, J. M. 2009, *ApJ*, 699, 1092
Herrera, C., Boulanger, F., & Nesvadba, N. P. H. 2011, *A&A*, 534, 138
Herrera, C., Boulanger, F., Nesvadba, N. P. H., & Falgarone, E. 2012, *A&A*, in press
Ingalls, J. G., Bania T. M., Boulanger, F. *et al.* 2011, *ApJ*, 743, 174
Krumholz, M. R., Dekel, A., & McKee, C. F. 2012, *ApJ*, 745, 69
Mac Low, M.-M. & Klessen, R. S. 2004, *Reviews of Modern Physics*, 76, 125
Nesvadba, N. P. H., Boulanger, F., Salomé, P., *et al.* 2010, *A&A*, 521, A65
Nesvadba, N. P. H., Boulanger, F., Lehnert, M., *et al.* 2011, *A&A*, 536, L5
Ogle, P., Boulanger, F., Guillard, P., *et al.* 2010, *ApJ*, 724, 1193
Roussel, H., Helou, G., Hollenbach, D. J. *et al.* 2007, *ApJ*, 669, 959
Teyssier, R., Chapon, D., & Bournaud, F. 2010, *ApJ*, 720, L149
Whitmore, B. C., Chandar, R., Schweizer, F., *et al.* 2010, *AJ*, 140, 75
Zakamska, N. L. 2010, *Nature*, 465, 60

The Spectral Energy Distribution of Galaxies
Proceedings IAU Symposium No. 284, 2011
R.J. Tuffs & C.C. Popescu, eds.

doi:10.1017/S1743921312009374

The dust emission SED of X-ray emitting regions in Stephan's Quintet

G. Natale[1], R. J. Tuffs[2], C. K. Xu[3], C. C. Popescu[1], J. Fischera[4], U. Lisenfeld[5], N. Lu[3], P. Appleton[6], M. Dopita[7], P.-A. Duc[8], Y. Gao[9], W. Reach[10], J. Sulentic[11], and M. Yun[12]

[1]University of Central Lancashire, Preston, PR1 2HE, UK; [2]Max Planck Institute für Kernphysik, Heidelberg, Germany; [3]Infrared Processing and Analysis Center, Caltech, Pasadena, USA; [4]Canadian Institute for Theoretical Astrophysics, University of Toronto, Toronto, Canada; [5]Department de Física Teórica y del Cosmos, Universidad de Granada, Granada, Spain; [6]NASA Herschel Science Center, IPAC, Caltech, Pasadena, USA; [7]Research School of Astronomy & Astrophysics, The Australian National University, Weston Creek, Australia; [8]Laboratoire AIM, CEA/DSM-CNRS-Université Paris Diderot, Dapnia/Service d'Astrophysique, CEA-Saclay, France; [9]Purple Mountain Observatory, Chinese Academy of Sciences, Nanjing, China; [10]Spitzer Science Center, IPAC, Caltech, Pasadena, USA; [11]Instituto de Astrofísica de Andalucía, Granada, Spain; [12]Department of Astronomy, University of Massachusetts, Amherst, USA

email: `gnatale@uclan.ac.uk`

Abstract. We analysed the Spitzer maps of Stephan's Quintet in order to investigate the nature of the dust emission associated with the X-ray emitting regions of the large scale intergalactic shock and of the group halo. This emission can in principle be powered by dust-gas particle collisions, thus providing efficient cooling of the hot gas. However the results of our analysis suggest that the dust emission from those regions is mostly powered by photons. Nonetheless dust collisional heating could be important in determining the cooling of the IGM gas and the large scale star formation morphology observed in SQ.

Keywords. galaxies: interactions, galaxies: intergalactic medium, infrared: galaxies: groups: individual(HCG92)

---

1. Introduction

Most of the MIR and FIR dust emission from galaxies is powered by dust absorption of UV/optical photons coming from different sources, predominantly young stars in star formation regions, more widely distributed older stellar populations and active galactic nuclei. However dust-gas particle collisions can also be important for the dust heating in some cases. In particular, for gas densities $\approx 10^{-3} - 10^{2}\ cm^{-2}$ and gas temperatures $\gtrsim 10^{6}\ K$, a range covering both the shocked interstellar medium (ISM) and the group/cluster intergalactic medium (IGM), collisionally heated dust temperatures are of order of $\approx 10 - 10^{2}\ K$. Therefore, emission from collisionally heated dust is predicted to peak in the MIR/FIR range, as in the case of dust heated by galaxy interstellar radiation fields Popescu *et al.* (2000).

Dust-gas particle collisions are potentially very important for the cooling of hot ISM/IGM gas. In fact, for a dust-gas mass ratio higher than $\approx 10^{-4}$, the "dust cooling" of the hot gas predominates over standard bremmstrahlung cooling (see e.g. Montier & Giard (2004)). However its efficiency is highly reduced by the rapid removal of dust through dust sputtering Draine & Salpeter (1979).

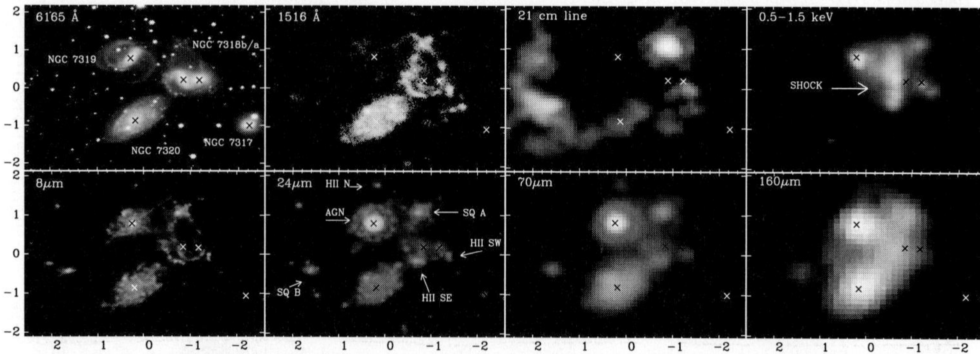

Figure 1. A multiwavelength view of Stephan's Quintet. Top-panels: SDSS r-band, GALEX FUV, VLA 21 cm line, XMM-NEWTON soft X-ray. Bottom-panels: Spitzer IRAC 8 μm and MIPS 24, 70 and 160μm.

The presence of dust in the hot IGM of groups and clusters has long been searched for, both in direct FIR emission (e.g. Stickel *et al.* 1998, Giard *et al.* 2008, Kitayama *et al.* 2009) and through the attenuation of optical light from background galaxies and QSOs (e.g. Chelouche07). Although good evidence for grains in the IGM has been found through the extinction technique, it has proven challenging to detect IGM dust in the FIR. This is largely due to the need for extreme detector stability to measure faint extended structure in well resolved clusters, or, conversely, due to confusion with foreground cirrus emission and emission from late type member galaxies in more distant clusters with smaller angular extent. These problems are alleviated in the case of the two X-ray emitting regions in the Stephan's Quintet compact group of galaxies (SQ, distance=94 Mpc), the so-called "shock region" and the group halo, which are sufficiently resolved to differentiate between galaxies and IGM emission while at the same time being not overly extended from the point of view of detector stability.

The shock region is an X-ray emitting linear feature of length $\approx 40kpc$ (Trinchieri *et al.* 2005), visible also as $H\alpha$ recombination line (Xu *et al.* 1999), radio continuum (Williams *et al.* 2002), warm H2 infrared line emission (Cluver *et al.* 2010). Physically, the unusual multi-phase state of the emitting gas has been interpreted as the result of a high-speed collision between the ISM of the intruder galaxy NGC7318b and a gaseous tidal tail in SQ IGM (Sulentic *et al.* 2001). Since these two colliding regions of gas are, in this interpretation, both of galaxian origin, they were plausibly dusty before the collision heated them to high temperatures. Collisionally–heated dust is therefore expected in the post-shock gas, if dust sputtering has not already completely removed all the dust originally present in the pre-shock gas or if efficient dust injection sources are present.

As already mentioned, the compactness of SQ also allows an attempt to measure the dust emission associated with the group halo. Here the gas is expected to be largely pristine in nature, without having been processed in the ISM of a galaxy. However, dust may still be present if injected into the IGM, for example from halo stars which have been tidally removed from the galaxies, together with ISM material. On sufficiently large time scales these stars are expected to become decoupled from the tidally removed ISM through hydrodynamical interactions between ISM and IGM gas, leaving a population of intergalactic stars in direct contact with the hot IGM.

In this proceedings paper we present some of the results of the analysis performed by Natale *et al.* (2010) (NA10) on the Spitzer data of SQ. A more complete description of the details of this work can be found directly in that paper.

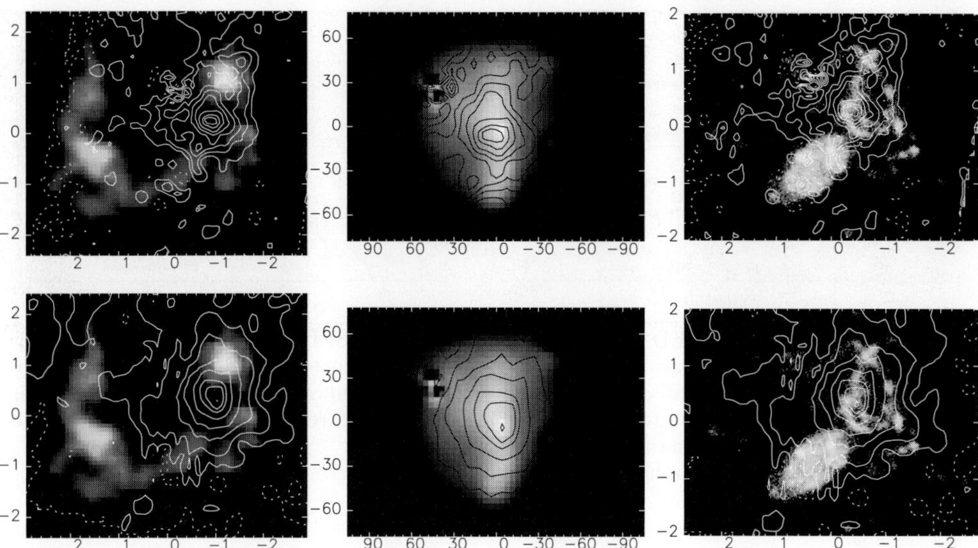

Figure 2. Contours of the FIR residual maps at 70 (upper row) and 160 μm (lower row) overlaid on the HI distribution (VLA, left), the point source subtracted soft X-ray map (XMM-Newton, center) and the FUV map (GALEX, right)

2. The dust emission from X-ray emitting regions in Stephan's Quintet

Fig. 1 shows a multiwavelength view of SQ. In the upper row, one can see the stellar distribution, as traced by the SDSS r-band and GALEX FUV maps (note that NGC7320 is a foreground galaxy), the cold HI distribution (VLA 21cm line) and the soft X-ray emission (XMM-Newton). These maps can be used to compare the stellar and cold HI/hot gas distribution with the PAH/dust emission morphology shown in the lower panels, presenting the Spitzer maps at 8μm (IRAC, stellar light subtracted), 24μm, 70μm and 160μm (MIPS). An accurate description of the Spitzer maps can be found in NA10. Here we simply point out that, by looking at the FIR 70 and 160μm maps, the emission from the shock region becomes more predominant and there is a hint for the presence of an extended FIR emission spatially coincident with the X-ray halo.

To better understand the origin of the dust emission apparently associated with X-ray emitting regions, we developed a simultaneous multi-source fitting technique to remove the sources associated with star formation regions and galaxies in SQ. After the removal of those sources, we obtained FIR residuals maps at 70 and 160μm, which are shown in Fig.2 as contours overlaid on the VLA HI map, the XMM-Newton soft X-ray map (point source subtracted) and the GALEX FUV map. The FIR residual emission is dominated by emission from the shock region, but it also presents an extended FIR emission distribution, apparently associated with the halo of the group. The multiwavelength comparison in Fig.2 shows that the residual FIR emission is not associated with HI, but is very well correlated with X-ray emission instead. This is consistent with the hypothesis of collisionally heated dust emission. However the comparison with the UV map shows that there are many UV sources at the same positions, which can potentially heat the dust as well. These rather contradictory evidences can be clarified by fitting the dust emission SEDs with dust emission templates, as described in the next section.

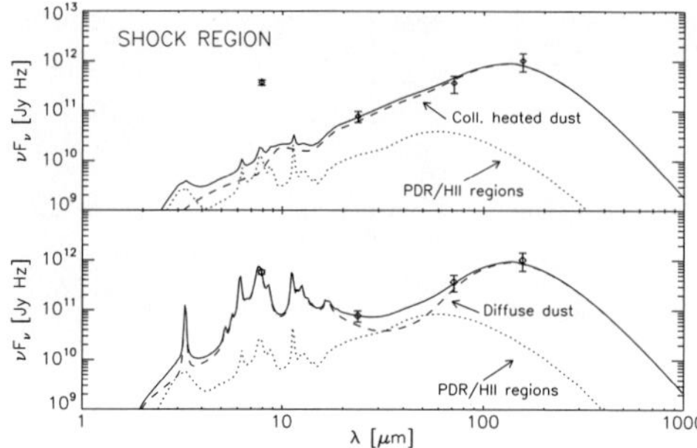

Figure 3. Dust emission SED of the shock region. The upper panel shows the fit using a collisionally heated dust emission template and a star formation region template. The lower panel shows the fit using a diffuse photon-heated dust emission template and a star formation region template.

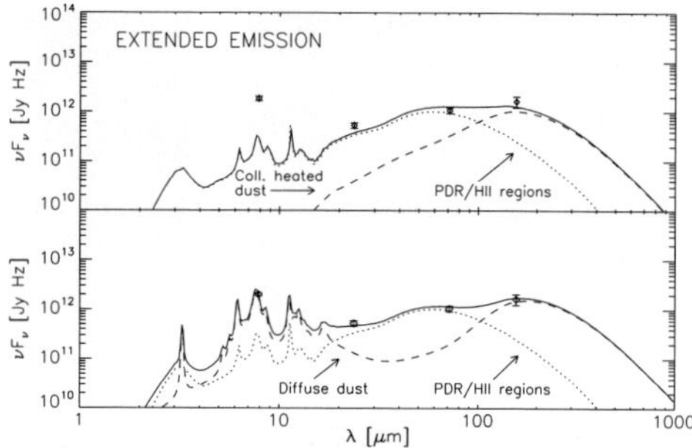

Figure 4. SED of the extended dust emission. The panels show the fits performed with the same kind of templates used for the fitting of the shock region dust emission SED (see Fig. 3)

3. The dust emission SEDs in the shock and halo regions

As shown in NA10, we performed the fit of the dust emission SEDs from the shock and from the halo regions ("the extended FIR emission") in two ways. A first attempt to fit the observed SEDs (see Fig. 3 and 4, upper panels) has been performed by using a combination of a template for collisionally heated dust emission (with plasma parameters fixed to those derived from the X-ray measurements), used to fit mostly the FIR, and a template for warm dust emission for star formation regions, used to fit mostly the MIR emission. We obtained a good fit for the MIR/FIR continuum ($\lambda \geqslant 24\mu$m), but failed to account for the observed 8μm emission. The emission in this band is known to be dominated by PAH line emission. These molecules are extremely fragile when embedded in an environment at temperatures of order of 10^6 K (Micelotta *et al.* 2010). Therefore the relatively high level of observed 8μm emission suggests that a large fraction of the emitting dust is actually embedded in a cold gas reservoir. In the second attempt, shown in the lower panels of Fig.3 and 4, we replaced the template of collisionally heated dust emission with a template for a purely photon heated emission component, calculated as a function of combinations of the ambient optical and UV radiation fields. We found that, using those templates, one is able to fit very well the entire dust emission SEDs. These findings, combined with simple estimations of dust injection rates in the hot gas of the shock and of the halo, support the interpretation that photon heating of dust has a major role in powering the observed MIR and FIR emission. However the few percent of total FIR luminosity that can be powered by collisional heating can still be a dominant cooling path for the hot IGM gas (see NA10 for details).

4. Conclusions and final remarks

Our analysis of the Spitzer data of SQ has shown that most of the dust emission associated with the X-ray emitting shock and halo regions is heated by photons rather than by gas-dust particle collisions, with a major component of the emitting dust being embedded in a cold gas phase (traced by the 8μm PAH dust emission). This implies a multi phase medium in which cold infrared emitting clouds are embedded in the hot X-ray emitting plasma. In the case of the shock region, one possibility is that the emitting dust (whose mass is $M_d \approx 4 \times 10^7$ $M_\odot$) is mainly in the massive amount of molecular gas recently detected by Guillard *et al.* (2012): $M_{H2} \approx 4 \times 10^9\ M_{\odot}$. If these cold molecular clouds, presenting low level of star formation, are optically thin then the dust emission can be powered by the external radiation field produced by the surrounding stellar populations and the shocked gas. However, in an opposite scenario of optically thick clouds, a component of in-situ star formation would be necessary to account for the complete shock region SED.

In the case of the halo, the presence of widespread compact MIR sources far away from the galaxy mainbodies suggests that a substantial part of the FIR extended emission we detected is powered by an extended star formation pattern in SQ. The large scale correspondence between FIR and X-ray emissions might be explained if there is distributed star formation fuelled directly by the cooling of the X-ray halo. This alternative scenario is potentially very important to understand the fuelling of star formation in groups and cluster, and will be further investigated in the context of SQ and other groups.

References

Chelouche, D., Koester, B. P., & Bowen, D. V. 2007, *ApJ*, 671, 97
Cluver, M. E., Appleton, P. N., Boulanger, F., *et al.* 2010, *ApJ*, 710, 248
Draine, B. T. & Salpeter, E. E. 1979, *ApJ*, 231, 438
Giard, M., Montier, L., Pointecouteau, E., & Simmat, E. 2008, *A&A*, 490, 547
Guillard, P. *et al.* 2012, *ApJ*, accepted
Kitayama, T., Ito, Y., Okada, Y., *et al.* 2009, *ApJ*, 695, 1191
Micelotta, E. R., Jones, A. P., & Tielens, A. G. G.. M. 2010, *A&A*, 510, 37
Montier, L. A. & Giard, M. 2004, *A&A*, 417, 401
Natale, G., Tuffs, R. J., Xu, C. K., Popescu, C. C., *et al.* 2010, *ApJ*, 725, 955
Popescu, C. C., Tuffs, R. J., Fischera, J., & Völk, H. 2000, *A&A*, 354, 480
Stickel, M., Lemke, D., Mattila, K., Haikala, L. K., & Haas, M. 1998, *A&A*, 329, 55
Sulentic, J. W., Rosado, M., Dultzin-Hacyan, D., *et al.* 2001, *AJ*, 122, 2993
Trinchieri, G., Sulentic, J., Pietsch, W., & Breitschwerdt, D. 2005, *A&A*, 444, 697
Williams, B. A., Yun, M. S., & Verdes-Montenegro, L. 2002, *AJ*, 123, 2417
Xu, C., Sulentic, J. W., & Tuffs, R. 1999, *ApJ*, 512, 178

Discussion

WANG: Is the dusty matter associated with pre- or after-shock gas?

NATALE: Both the scenarios are possible. We believe that most of the dust emission come from the cold molecular gas embedded in the X-ray emitting diffuse .The origin of this cold gas component will be one of the topics addressed in the talk by Francois Boulanger.

The Spectral Energy Distribution of Galaxies
Proceedings IAU Symposium No. 284, 2011
R.J. Tuffs & C.C. Popescu, eds.

doi:10.1017/S1743921312009386

AKARI observations of the multiphase intergalactic medium of Stephan's Quintet

Toyoaki Suzuki[1], **Hidehiro Kaneda**[2], **Takashi Onaka**[3], **and Tetsu Kitayama**[4]

[1]Institute of Space and Astronautical Science, Japan Aerospace Exploration Agency, 3–1–1 Yoshinodai, Chuo-ku, Sagamihara, Kanagawa 252–5210, Japan

[2]Graduate School of Science, Nagoya University, Furu-cho, Chikusa-ku, Nagoya 464–8602, Japan

[3]Department of Astronomy, Graduate School of Science, The University of Tokyo, 7-3-1 Hongo, Bunkyo-ku, Tokyo 113-0033, Japan

[4]Department of Physics, Toho University, Funabashi, Chiba 274-8510, Japan

email: suzuki@ir.isas.jaxa.jp

Abstract. Stephan's Quintet (SQ, HCG92) is a well studied compact group of galaxies with a disturbed intergalactic medium (IGM). An "intruder" galaxy NGC 7318b is currently colliding with the IGM at a relative velocity of 1000 km s^{-1}, causing a large-scale shock front. We observed SQ with the Far-Infrared Surveyor (FIS) aboard AKARI in four far-infrared (far-IR) bands at 65, 90, 140, and 160μm. The 160μm image clearly shows an additional peak of emission overlying structure extending in the North-South direction along the shock ridge seen in the 140μm band, and in H_2 and X-ray emission. Whereas most of the far-IR emission in the shocked region is from cold dust (20 K), the [CII]158μm emission - whose luminosity is comparable to that of the warm H_2 gas - can significantly contribute to the single peak emission in the 160μm band. We conclude that the [CII] line emission comes from the warm H_2 gas in the shock. Our result represents the first detection of shock-excited [CII] line emission.

Keywords. ISM: structure - galaxies: individual - galaxies: interactions - infrared: ISM

1. Introduction

Stephan's Quintet (SQ, HCG92) is a compact group of galaxies (Hickson 1982) with a disturbed intergalactic medium (IGM). An intruder galaxy, NGC 7318b, is currently colliding with the IGM, causing a large-scale shock front. The shock front has been detected as a ridge in radio and X-ray emission (Allen and Hartsuiker 1972, Pietsch *et al.* 1997). Appleton *et al.* (2006) discovered powerful H_2 rotational line emission from warm molecular gas in the center of the shock ridge with an extremely large equivalent width (EW). This may indicate that far-infrared (far-IR) fine-structure lines such as [CII]158 μm line also show extremely large EWs towards the shock, making a significant contribution to the far-IR luminosity. However, this fact has yet to be revealed observationally.

The Far-Infrared Surveyor (FIS)(Kawada *et al.* 2007) aboard *AKARI*(Murakami *et al.* 2007) has four far-IR bands with central wavelengths of 65, 90, 140, and 160 μm. The well sampled allocation of *AKARI*/FIS 4 bands can provide the spectral information to constrain both the dust temperature and the [CII]158 μm line emission. In this paper we report the observation of SQ with the FIS, and discuss the origin of the far-IR emission in the IGM.

2. Observations and data reduction

SQ was observed with AKARI in the four bands N60 (65 μm), WIDE-S (90 μm), WIDE-L (140 μm), and N160 (160 μm). The four-band images are presented in Fig. 1. At the centers of the shocked region, NGC 7319, and NGC 7320, we derive the flux densities

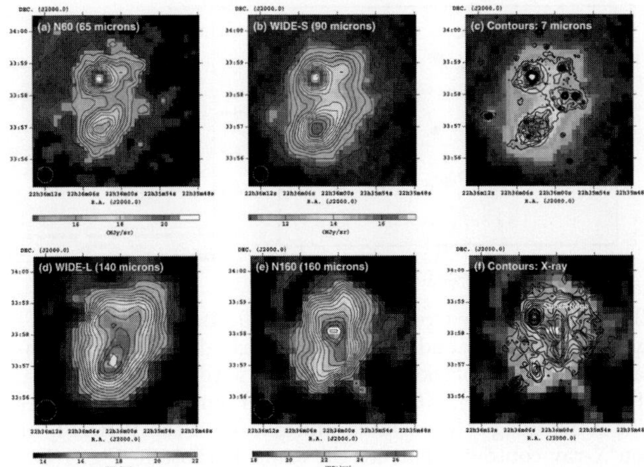

Figure 1. Four-band images of the SQ in the N60 (top-left), WIDE-S (top-middle), WIDE-L (bottom-left), and N160 (bottom-middle) bands. In each image, the FWHM of the PSF is shown in the lower left corner. The upper-right and bottom-right panels show the images of WIDE-S and N160 bands overlaid on *AKARI* 7 μm and XMM/Newton X-ray contours, respectively. The three apertures for photometry of the shocked region ($r = 20''$), NGC 7319 ($r = 30''$), and NGC 7320 ($r = 30''$) are shown in the panels of (c) and (f). As a reference, the shock ridge is shown by the solid white line in the panel (f).

by integrating the surface brightness within circular apertures. Figure 2(a) shows the resulting spectral energy distributions (SEDs) at these positions. Note that the fluxes in Fig. 2(a) are not the fluxes for the whole areas of the shocked region and the galaxies.

3. Results

In Fig. 1, N60 and WIDE-S images show emission from NGC 7319 and NGC 7320. While the WIDE-L image shows a structure extending in the North-South direction along the shock ridge, the N160 image clearly shows an additional single peak of emission overlaid on this structure. There are no corresponding additional peaks associated with the individual galaxies in N160. Figure 1(f) shows that the spatial distribution of the 160 μm emission is quite similar to that of X-ray emission in the shocked region.

The dust temperatures in the shock and surrounding galaxies are about 20-30 K. Because the flux ratio between the N160 and WIDE-L bands is not very sensitive to the dust temperature, the difference in the spatial distribution of far-IR emission between the two bands is hard to explain only by the presence of a cold dust component. An alternative possibility is a contribution from the shock-powered [CII]158 μm line emission to the N160 band. For SQ, the wavelength of [CII]158 μm line is redshifted to 161 μm (Hickson (1992)). Thus, its contribution to the N160 band is larger than to the WIDE-L band. To estimate the contribution from the [CII] line, the WIDE-L flux density $F_{\rm WL}$ is simply subtracted from the N160 flux density $F_{\rm N160}$ by assuming that the dust emission is constant in units of F_ν over the two bands (Fig. 2(b)). By using the same aperture as that in Fig. 1(f), the subtracted flux density $F_{\rm sub}$ is estimated to be 40^{+18}_{-22} mJy. The [CII] luminosity surface density $\Sigma_{L_{\rm [CII]}}$ [erg sec^{-1} kpc^{-2}] is given by $4\pi D^2 (F_{\rm sub} A^{-1})(R_{\rm N160} \Delta\nu^{-1}_{\rm N160} - R_{\rm WL} \Delta\nu^{-1}_{\rm WL})^{-1}$, where D is the distance to the SQ of 94 Mpc, A is the aperture area at the shocked region (260 kpc^2), $R_{\rm N160}$ and $R_{\rm WL}$ are the relative responses of the N160 and WIDE-L bands at 161 μm, respectively, and $\Delta\nu_{\rm N160}$ and $\Delta\nu_{\rm WL}$ are the effective bandwidths of N160 and WIDE-L, respectively. From Kawada *et al.* 2007, $R_{\rm N160}$, $R_{\rm WL}$, $\Delta\nu_{\rm N160}$, and $\Delta\nu_{\rm WL}$ are taken as 0.96, 0.59, 0.4 THz, and 0.8

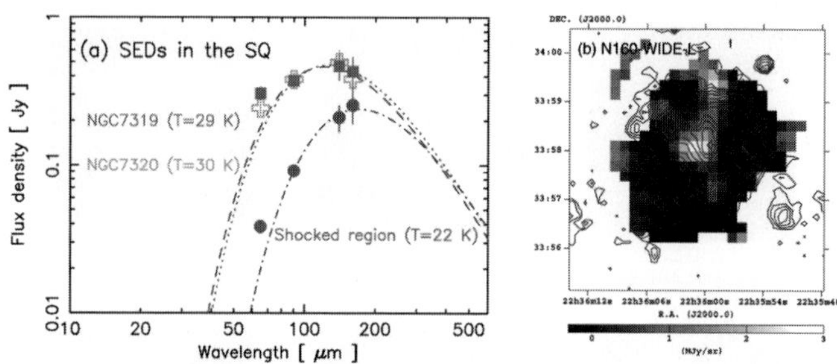

Figure 2. (a) SEDs at the shocked region, NGC 7319, and NGC 7320. The positions of the apertures are shown in Figs 1(c) and (f). The dashed, dotted, and dash-dotted lines show the single temperature modified blackbody spectrum with the emissivity power-law index of unity for NGC 7319 (filled squares), NGC 7320 (croses), and the shocked region (filled circles), respectively. (b) The N160 image shown after subtraction of the WIDE-L image. The contours show XMM/Newton X-ray contours that are the same as those in Fig. 1(f).

THz, respectively. Thus, $\Sigma_{L_{[\mathrm{CII}]}}$ is estimated to be $(1.0^{+0.4}_{-0.5}) \times 10^{39}$ erg sec^{-1} kpc^{-2}, which is comparable to the H_2 line luminosity surface density $\Sigma_{L_{\mathrm{H}_2}}$ of 2×10^{39} erg sec^{-1} kpc^{-2}(Cluver *et al.* 2010). By using the dust emission at 160 μm, the *EW* of the [CII] line is estimated to be $\sim$10 μm, about 10 times larger than that typical for nearby galaxies Boselli *et al.* (2002).

To investigate the possibility of luminous [CII] line emission from the shocked region, the C^+ abundance per hydrogen atom X_{C^+} is estimated with assumptions that the [CII] line emission comes from the warm H_2 gas and is optically thin. Within these assumptions, X_{C^+} is given as

$$X_{\mathrm{C}^+} = 1.1 \times 10^{-4} \left(\frac{\Sigma_{L_{[\mathrm{CII}]}} \; L_\odot \mathrm{kpc}^{-2}}{\Sigma_{\mathrm{M}_{\mathrm{H}_2}} \; M_\odot \mathrm{kpc}^{-2}} \right) \left(\frac{1 + 2\exp(-91 \mathrm{~K}/T) + n_{\mathrm{H}}^{\mathrm{crit}}/n_{\mathrm{H}}}{2\exp(-91 \mathrm{~K}/T)} \right), \tag{3.1}$$

where T is the kinetic temperature of H_2 gas, $n_{\mathrm{H}}^{\mathrm{crit}}$ is the critical density of the [CII] line emission for HI collisions, and n_{H} is the number density of HI. At the shocked region, T and n_{H} are 158 K and $> 10^3$ cm^{-3}, respectively (Cluver *et al.* 2010). $n_{\mathrm{H}}^{\mathrm{crit}}$ is 3.0×10^3 cm^{-3} at 158 K (Langer *et al.* 2010). Thus, X_{C^+} is estimated to be $\sim 1 \times 10^{-4}$. Assuming that the carbon in the [CII] line emitting region is in singly ionized form, the carbon abundance in the shocked region is in agreement with that in an interstellar gas-phase (1.4×10^{-4}, Cardelli *et al.* (1996)). Therefore, the luminous [CII] line emission from the shocked region is physically plausible provided that C^+ is the main carbon form in the warm H_2 gas.

References

Allen, R. J. & Hartsuiker, J. W. 1972, *Nature*, 239, 324

Appleton, P. N. *et al.* 2006, *Ap. Lett.*, 639, L51

Boselli, A., Gavazzi, G., Lequeux, J., & Pierini, D. 2002, *A&A*, 385, 454

Cardelli, J. A., Meyer, D. M., Jura, M., & Savage, B. D. 1996, *ApJ*, 467, 334

Cluver, M. E. *et al.* 2010, *ApJ*, 710, 248

Hickson, P. 1982, *ApJ*, 255, 382

Hickson, P., Mendes de Oliveira, C., Huchra, J. P., & Palumbo, G. G. 1992, *ApJ*, 399, 353

Kawada, M. *et al.* 2007, *PASJ*, 59, 389

Langer, W. D., Velusamy, T., Pineda, J. L., Goldsmith, P. F., Li, D., & Yorke, H. W. 2010, *A&A*, 521L, 17L

Murakami, H. *et al.* 2007, *PASJ*, 59, 369

Pietsch, W., Trinchieri, G., Arp, H., & Sulentic, J. W. 1997, *A&A*, 322, 89

The Spectral Energy Distribution of Galaxies
Proceedings IAU Symposium No. 284, 2011
R.J. Tuffs & C.C. Popescu, eds.

doi:10.1017/S1743921312009398

Radial variations of the SFHs of dwarf irregular galaxies†

Hong-Xin Zhang[1,2,3,4], Deidre A. Hunter[1], Bruce G. Elmegreen[5], Yu Gao[2,3], and Andreas Schruba[6]

[1]Lowell Observatory, 1400 West Mars Hill Road, Flagstaff, Arizona 86001 USA

[2]Purple Mountain Observatory, Chinese Academy of Sciences, 2 West Beijing Road, Nanjing 210008 China

[3]Key Laboratory of Radio Astronomy, Chinese Academy of Sciences, Nanjing 210008 China

[4]Graduate School of the Chinese Academy of Sciences, Beijing 100080 China

[5]IBM T. J. Watson Research Center, PO Box 218, Yorktown Heights, New York 10598 USA

[6]Max-Planck-Institut für Astronomie, Königstuhl 17, 69117 Heidelberg Germany

email: hxzhang@lowell.edu; dah@lowell.edu

Abstract. The LITTLE THINGS project† has compiled multi-wavelength data (including VLA H I-line emission maps, *GALEX* FUV/NUV imagery, *UBV*, narrow-band Hα, and *Spitzer* images) for a representative sample of nearby dwarf irregular (dIrr) galaxies. The broadband data are used to constrain the radial variations of the star formation (SF) rate (SFR) averaged over the past 0.1 Gyr, 1 Gyr and a Hubble time, with a complete library of model SF histories (SFHs). The recent SF of more than $\sim 80\%$ of the dIrrs in our sample has been concentrated in the inner disk, and the SF in the outer disk has been markedly suppressed. This outside-in shrinking of the star-forming disk leaves a down-bending (double exponential) stellar mass surface density ($\Sigma_\star$) distribution. Our findings in dIrrs are in contrast to the inside-out disk growth scenario suggested for luminous spiral galaxies.

Keywords. galaxies: dwarf – galaxies: irregular – galaxies: evolution – galaxies: stellar content

1. Introduction

Dwarf Irregulars (dIrrs) are the most common type of star-forming galaxies in the local universe (e.g. Gallagher & Hunter 1984). Compared to the larger spiral galaxies, gas-rich dIrrs are smaller, bluer, metal-poor, and less luminous. In the classical "inside-out" growth mode, the disk scale length increases with time due to the slower/later accretion of gas at large radii and exhaustion of gas in the centers (e.g. Larson 1976). The negative color gradients and abundance gradients observed in typical spiral galaxies fit well into the inside-out disk growth scenario. However, late-type dIrrs usually show more or less flat gradients of current metallicity (e.g. Croxall *et al.* 2009), implying the inside-out paradigm may not apply to dIrrs.

Lack of large-scale instabilities (Hunter *et al.* 1998), such as spiral arms and bars, in dIrrs means that the present-day stellar population distribution is primarily determined by the *in situ* SFHs, rather than a redistribution process (e.g. resonant scattering with

† Based on data from the LITTLE THINGS Survey (Hunter *et al.*, in preparation), funded in part by the National Science Foundation through grants AST-0707563, AST-0707426, AST-0707468, and AST-0707835 to US-based LITTLE THINGS team members and with generous support from the National Radio Astronomy Observatory.

† Local Irregulars That Trace Luminosity Extremes – The H I Nearby Galaxy Survey, http://www.lowell.edu/users/dah/littlethings/index.html

spiral arms) following *in situ* SF. This makes it possible to probe the stellar disk formation process by studying the present-day stellar population distribution across the disk.

To understand the disk growth process of dIrrs, we derived radial variations of the azimuthal averages of the multi-band spectra energy distributions (SEDs) (i.e. *GALEX* FUV/NUV, *UBV*, narrow-band Hα, and *Spitzer* 3.6 μm) for 34 dIrrs selected from the LITTLE THINGS sample (41 galaxies). The galaxies studied in this work are selected to have the full complement of data available. These surface brightness profiles are fit with stellar population synthesis models. Considering the complex SFHs found in local dIrrs (e.g. Weisz *et al.* 2011), we construct a relatively complete SFH library by dividing the Hubble time into 6 independent age bins, where each bin is assumed to have a constant SFR (see Zhang *et al.* 2011, hereafter Z11).

2. Results

Figure 1 presents the main results of 4 typical galaxies included in our sample. The readers are referred to Z11 for the results of the whole sample. From the *left column* of Figure 1, we can see that all four galaxies exhibit obvious breaks in their surface brightness profiles. The blue compact dwarf (BCD) galaxy Haro 29 shows significant excess starlight in the central regions in all of the bands. This suggests that the central regions have sustained the starburst for an extended period of time. The multi-band surface brightness profiles exhibit different radial variations among different bands interior to the break radius, beyond which the shorter wavelength profiles fall off faster than the longer wavelength profiles. Actually, more than $\sim$ 80% of the whole sample show a steeper radial decline of the surface brightness profiles in the outer disk.

As is shown in the *middle column* of Figure 1, recent SF is concentrated in the inner disk. For all the galaxies shown here except Haro 29, the radial variations of recent SF follow the lifetime average SF $\Sigma_{\rm SFR_{13.7Gyr}}$ much better in the inner disk than in the outer disk. The ratios $\Sigma_{\rm SFR_{0.1Gyr}}/\Sigma_{\rm SFR_{13.7Gyr}}$ and $\Sigma_{\rm SFR_{1Gyr}}/\Sigma_{\rm SFR_{13.7Gyr}}$ exhibit steep declines beyond the break radius. The steeper radial decline of $\Sigma_{\rm SFR_{0.1Gyr}}/\Sigma_{\rm SFR_{13.7Gyr}}$ and $\Sigma_{\rm SFR_{1Gyr}}/\Sigma_{\rm SFR_{13.7Gyr}}$ in the outer disk was found for more than $\sim$ 80% of our whole sample (Figures 6 and 7 in Z11). Therefore, in contrast with the randomly percolating SF scenario suggested for dIrrs (e.g. van Zee 2001), the star-forming disks of dIrrs has been shrinking.

The *right column* of Figure 1 shows the $\Sigma_\star$ profiles derived from our SED modeling. The $\Sigma_\star$ profiles are well fitted with either a single or piece-wise exponential function, which is overplotted on the profiles. Except for Haro 29, which has an up-bending $\Sigma_\star$ profile beyond the break, the other three galaxies have down-bending $\Sigma_\star$ profiles. About 80% of our whole sample of galaxies exhibit obvious down-bending $\Sigma_\star$ profiles (Z11). One important difference between our dIrrs and spiral galaxies is that, a down-bending surface brightness profile usually accompanies a down-bending $\Sigma_\star$ profile, whereas in spiral galaxies the break of a down-bending surface brightness profile is almost gone in the $\Sigma_\star$ profile. For dIrrs, the breaks in the radial profiles of $\Sigma_{\rm SFR_{0.1Gyr}}/\Sigma_{\rm SFR_{13.7Gyr}}$ and $\Sigma_{\rm SFR_{1Gyr}}/\Sigma_{\rm SFR_{13.7Gyr}}$ come with breaks in the $\Sigma_\star$ profiles.

3. Discussion

On average, SF in outer disks has been decreasing more significantly both over time and over radius (Figure 1) compared to inner disks, which would naturally lead to down-bending stellar mass surface density profile, seen in most of our galaxies. This scenario of outside-in depression of SF is different from what may be happening in typical BCD

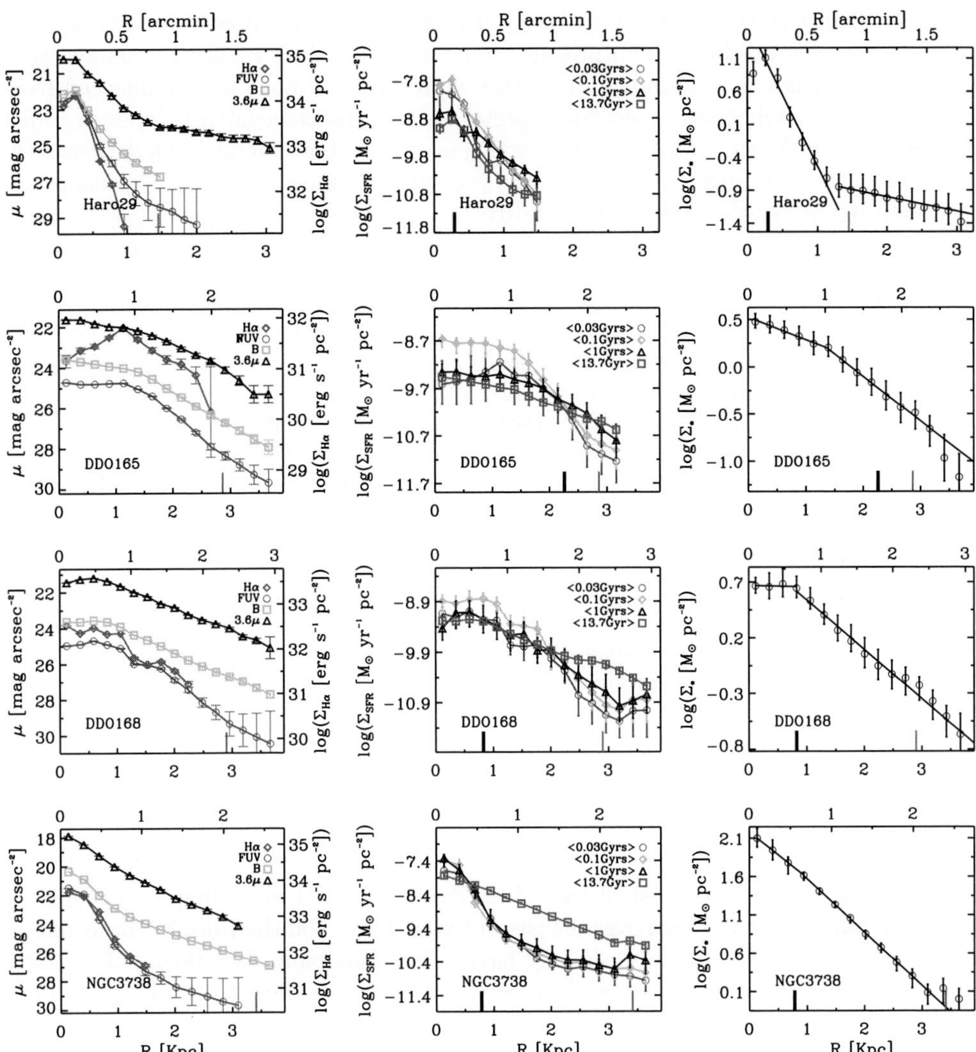

Figure 1. First column: Azimuthally-averaged surface brightness profiles of Hα, FUV, B and 3.6 μm images. The NUV or U-band profiles are plotted instead of FUV if there are no FUV observations. Second column: Radial variations of the SFR averaged over different timescales obtained from our multi-band SED modeling. The averaging timescales are the most recent 0.03 Gyr, the recent 0.1 Gyr, 1 Gyr, and the Hubble time. Third column: Inclination-corrected $\Sigma_\star$ profiles from the SED modeling. The fitted exponential profiles are overplotted as solid lines. The thin vertical line marks the Holmberg radius, and the thick black vertical line marks the V-band scale length.

(e.g. Haro 29) galaxies. The central starburst of BCD galaxies results in faster buildup of the inner stellar disk than the outer part, which could lead to up-bending $\Sigma_\star$ profiles.

In agreement with previous findings in a few nearby dIrrs (e.g. McConnachie *et al.* 2006; Bernard *et al.* 2007; Gallart *et al.* 2008), the star-forming disk of our sample of dIrrs is experiencing outside-in shrinking. The low mass of dIrrs suggests that they are susceptible to both environmental influences (e.g. ram pressure stripping of gas caused by the intragroup or intracluster medium and interactions with neighboring galaxies) and regulation through stellar feedback. The outer disk, especially in relatively low mass systems where the gravitational well and the gas density are relatively low, is more prone to gas removal due to ram pressure stripping. Z11 (their Figure 10) showed that galaxies with a larger tidal index, and hence more likely to have been influenced from the outside, (Karachentsev *et al.* 2004) prefer lower (zero or negative) slopes in radial variations of $\log(\Sigma_{SFR_{0.1Gyr}}/\Sigma_{M_\star})$ and $\log(\Sigma_{SFR_{1Gyr}}/\Sigma_{M_\star})$. Numerical simulations (e.g. Kapferer *et al.* 2008) suggest that both ram pressure stripping and tidal disturbance are needed to reproduce the observations of dIrrs.

References

Bernard E. J., Aparicio A., Gallart C., Padilla-Torres C. P., & Panniello M. 2007, *AJ*, 134, 1124
Croxall, K. V., van Zee, L., Lee, H., *et al.* 2009, *ApJ*, 705, 723
Gallagher, J. S. & Hunter, D. A. 1984, *ARAA*, 22, 37
Gallart, C., Stetson, P. B., Meschin, I. P., Pont, F., & Hardy, E. 2008, *ApJ*, 682, 89
Hunter, D. A., Elmegreen, B. G., & Baker, A. L. 1998, *ApJ*, 493, 595
Kapferer, W., Kronberger, T., Ferrari, C., Riser, T., & Schindler, S. 2008, *MNRAS*, 389, 1405
Karachentsev, I. D., Karachentseva, V. E., Huchtmeier,W. K., & Makarov, D. I. 2004, *AJ*, 127, 2031
Larson, R. B. 1976, *MNRAS*, 176, 31
McConnachie, A. W., Arimoto, N., Irwin, M., & Tolstoy, E. 2006, *MNRAS*, 373, 715
van Zee, L. 2001, *AJ*, 121, 2003
Weisz, D. R., Dalcanton, J. J., Williams, B. F., *et al.* 2011, *ApJ*, 739, 5
Zhang, H.-X., Hunter, D. A., Elmegreen, B. G., Gao, Y., & Schruba, A. 2011, arXiv: 1111.3363 (Z11)

Discussion

AQUAVIVA: What is the star formation history reconstructed in the six bins in the Hubble time?

ZHANG: According to the test of the reliability of our SED modelling, only the star-formation rate (SFR) averaged over the past 0.1 Gyr, 1 Gyr and the Hubble time could be well constrained. This is because we only have a few broadband data points. Therefore, we only focus on the interpretation of the SFR averaged over the 0.1 Gyr, 1 Gyr and the Hubble time.

The Spectral Energy Distribution of Galaxies
Proceedings IAU Symposium No. 284, 2011
R.J. Tuffs & C.C. Popescu, eds.

doi:10.1017/S1743921312009751

Star formation properties in barred galaxies

Zhi-Min Zhou[1], Chen Cao[2] and Hong Wu[1]

[1]National Astronomical Observatories, Chinese Academy of Sciences, Beijing 100012, China
[2]School of Space Science and Physics, Shandong University at Weihai, Weihai, Shandong 264209, China

email: zmzhou@bao.ac.cn

Abstract. Stellar bars are important structures for the internal secular evolution of galaxies. They can drive gas into the central region of galaxies, and result in an enhancement of star formation activity there. Previous studies are limited in the comparisons between barred and unbarred galaxies. Here we try to investigate the connection between star formation activities and different bars, based on multi-wavelength data in a sample of barred spirals. We find that there is no clearly trend of the surface star formation rates in different structures along the bar strength. In addition, there is larger scatter for the properties of star formation activity in the galaxies with middle-strength bars, which may indicate that a variety of star formation stages are more likely associated with these bars.

Keywords. galaxies: evolution — galaxies: photometry — galaxies: structure

1. Introduction

Stellar bars have been found to be common structures in disc galaxies, and they are believed to be the important internal drivers to secular evolution of galaxies (Kormendy & Kennicutt 2004). Bars are able to make the angular momentum of gas and stars redistribution in galactic disks, include large-scale streaming motions in the gas and stars, and result in an increase in the gas mass and enhancement of star formation activity in the inner regions of galaxies (e.g., Ho *et al.* 1997, Athanassoula 2003).

There are strong evidences for the bar-driven secular evolution, both theoretical and numerical. Numerical simulations have established that the gravitational torque of the large-scale bar can make the gas lose angular momentum, and then drive it down to the galactic central regions (Athanassoula 1992, Sellwood & Wilkinson 1993). Many observations also support that barred spirals have higher molecular gas concentrations and enhanced star formation activity in the central kiloparsec than in unbarred systems (e.g., Sakamoto *et al.* 1999, Sheth *et al.* 2005). However, the connection between bars and star formation activity is less clear, and previous studies are limited in the comparisons between barred and unbarred galaxies. To investigate their correlation, we use multi-wavelength data from ultraviolet (UV) to infrared (IR), and try to analyzes a sample of nearby barred galaxies with both weak and strong bars, including from early to late Hubble types spirals.

2. Sample and Data

Our galaxy sample was drawn from three Spitzer local programs: the Spitzer Infrared Nearby Galaxies Survey (SINGS; Kennicutt *et al.* 2003), the Local Volume Legacy (LVL; Dale *et al.* 2009), and the Infrared Hubble Atlas (Pahre *et al.* 2004). From all nearby galaxies in these programs, we took galaxies with bars (Hubble type of SB and SAB),

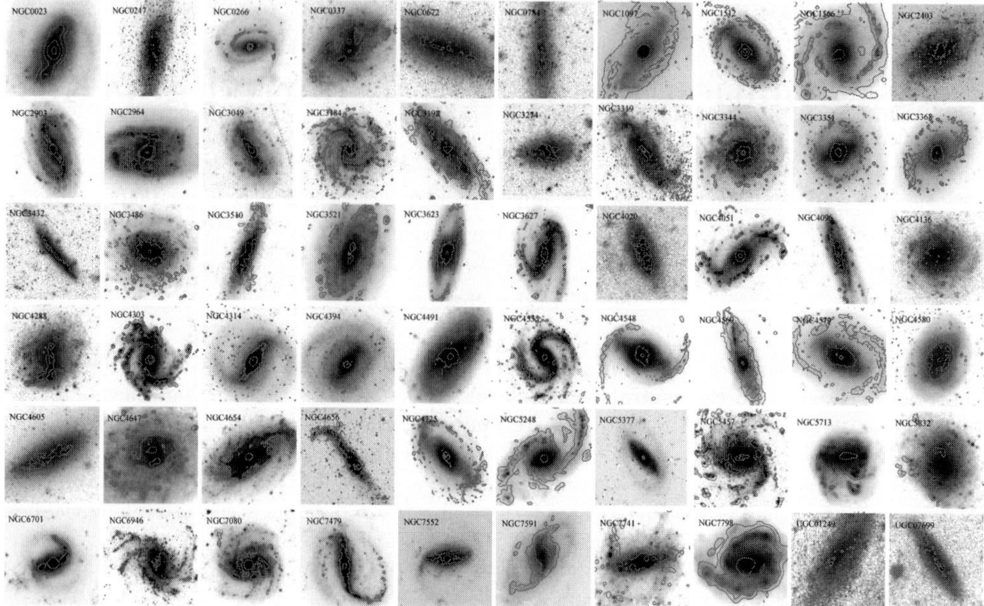

Figure 1. The mass images of the samples. Overplotted are the SFR imaging contours. The images are derived using our data and the corresponding conversion equations. The contours are in arbitrary units, optimized to show and identify the structures of the star formation complexes. The galaxy names are specified in the panels, and north is up and east to the left.

and removed apparently edge-on galaxies and objects with close neighbors. In the final sample, there are 34 SB galaxies and 26 SAB galaxies encompassing the Hubble types SaSm.

In order to investigate the properties of barred galaxies, we archived UV images observed by Galaxy Evolution Explorer (GALEX; Martin *et al.* 2005) and IR images from the Spitzer Space Telescope (Werner *et al.* 2004) data archive. In addition, the narrow-band Hα images from ground-based telescopes were collected to analyze the star formation activities in our sample galaxies. Figure 1 show the images of our sample.

3. Analysis and Results

To further explore possible correlations between the star formation activity in different structural regions and the stellar bars, we performed image decomposition on the 3.6 μm images, which can nicely trace the stellar compositions in nearby galaxies. Based on the decomposition results, we obtained the galaxy structural parameters, estimated the bar strength using ellipticity-based parameter f_{bar}, and derived the star formation activities in different structural regions, i.e., bulges, bars, disks, and the global galaxies.

In Figure 2, we compared the surface SFRs to the bar strength. In general, under visual inspection we see that there is no clearly trend of the surface SFRs in four panels along the bar strength, although there are weak trends such that the surface SFRs in bulges and global galaxies increase with the bar strength along with large scatter. In addition, there is no major discernible difference between early-type and late-type galaxies in their global, bar and disk regions. It is noteworthy that the bulges in our sample have larger scatter around at the middle strength of the bar, which may indicate that a variety of star formation stages in the central region are likely associated with the stellar bars with middle strength, not bars too weak, or too strong. In general, bar-driven secular

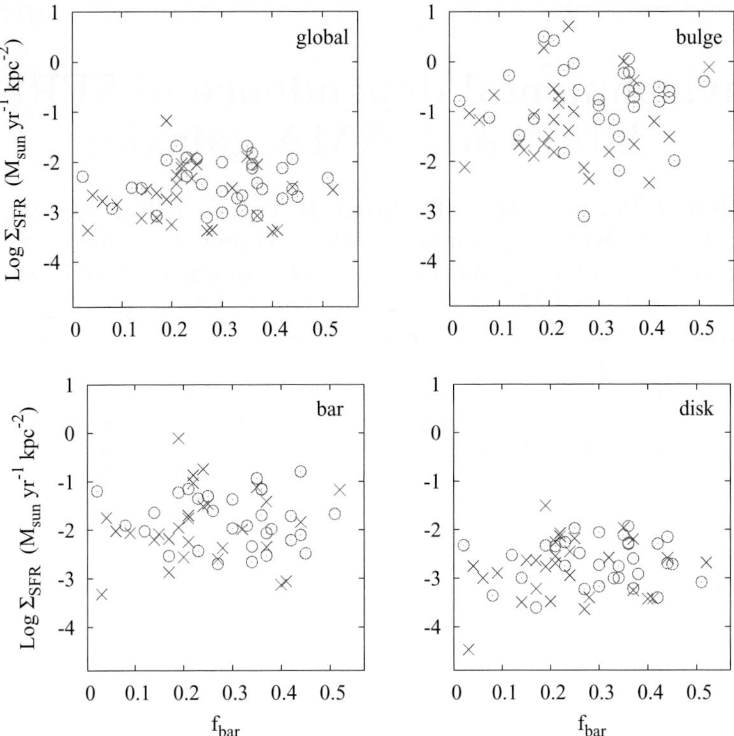

Figure 2. The surface star formation rates as a function of the bar strength f_{bar}. The surface star formation rates in the four panels are form different galactic structures, including bulges, bars and disks, as well as the global galaxies. The four panels use the same signals, circles for early-type galaxies ($1 \leqslant T \leqslant 4$) and crosses for late-type ones ($5 \leqslant T \leqslant 9$).

evolution is a complex process, and depends on many factors, such as time scales and the availability of gas (Coelho & Gadotti 2011). Therefore, in the effect of bars, the star formation activity in the central region of galaxies is likely located in different evolutional stages, even with similar galactic bars.

References

Athanassoula, E. 2003, *MNRAS*, 341, 1179
Athanassoula, E. 1992, *MNRAS*, 259, 345
Coelho, P. & Gadotti, D. A. 2011, *arXiv1111.1736*
Dale, D. A., *et al.* 2009, *ApJ*, 703, 517
Ho, L. C., Filippenko, A. V., & Sargent, W. L. W. 1997, *ApJ*, 487, 591
Kennicutt, R. C. *et al.* 2003, *PASP*, 115, 928
Kormendy, J. & Kennicutt, R. C. 2004, *ARAA*, 42, 603
Martin, D. C., *et al.* 2005, *ApJ*, 619, 1
Pahre, M. A., Ashby, M. L. N., Fazio, G. G., & Willner, S. P. 2004, *ApJS*, 154, 235
Sakamoto, K., Okumura, S. K., Ishizuki, S., & Scoville, N. Z. 1999, *ApJ*, 525, 691
Sellwood, J. A. & Wilkinson, A. 1993, *Rep. Prog. Phys.*, 56, 173
Sheth, K., Vogel, S. N., Regan, M. W., Thornley, M. D., & Teuben, P. J. 2005, *ApJ*, 632, 217
Werner, M. W., *et al.* 2004, *ApJS*, 154, 1

The Spectral Energy Distribution of Galaxies
Proceedings IAU Symposium No. 284, 2011
R.J. Tuffs & C.C. Popescu, eds.

doi:10.1017/S1743921312009404

Environmental dependence of SFRs in late-type GAMA galaxies

Meiert W. Grootes[1], Richard J. Tuffs[1], Ellen Andrae[1], Lee S. Kelvin[2], Jochen Liske[3], Barry F. Madore[4], Cristina C. Popescu[5], Aaron S. G. Robotham[2], Mark Seibert[4], Edward N. Taylor[6], and the GAMA Team

[1]Max-Planck-Institut für Kernphysik, Saupfercheckweg 1, 69117 Heidelberg, Germany ; [2]School of Physics and Astronomy, University of St. Andrews, North Haugh, St. Andrews, KY16 9SS, UK ; [3]European Southern Observatory, Karl-Schwarzshild Str. 2, 85748, Garching, Germany ; [4]Observatories of the Carnegie Institution for Science, 813 Santa Barbara Street, Pasadena, CA 91101, USA ; [5]Jeremiah Horrocks Institute, University of Central Lancashire, Preston PR1 2HE, UK ; [6]Sydney Institute for Astronomy, School of Physics, University of Sydney, NSW 206, Australia

email: meiert.grootes@mpi-hd.mpg.de

Abstract. Present and past gas-fuelling of galaxies is expected to depend upon both the properties of the galaxies themselves, as well as their larger-scale environments. In the case of galaxies in groups the environment, i.e the group mass, can be probed by measuring the velocity dispersion of the group members, as done with the GAMA Galaxy Group catalogue (Robotham *et al.* 2011), probing the halo mass function all the way to small groups. The gas-fuelling rate of normal late-type galaxies can be traced by the SFR under the assumption of a steady state between gas-fuelling and gas-consumption by SF. We present a method to estimate disk opacities from UV/optical photometric characteristics, calibrated using the radiative transfer model of Popescu *et al.* (2011), applied to UV-Opt-FIR GAMA/H-ATLAS photometry for a subset of GAMA galaxies. We use the method to extract attenuation corrected SFRs for a large sample of late-type GAMA galaxies, which we use in an initial application to constrain the dependency of star formation/gas-fuelling in late-type galaxies on mass of parent DMH, and compactness of galaxy group.

Keywords. galaxies: fundamental parameters, galaxies: ISM, galaxies: spiral, dust, extinction

1. Introduction

While the current observed population of stars in the universe is generally accepted as having formed via condensation out of gas in galaxies, the processes governing the introduction of this gas into the galaxies are not yet fully understood. The two main processes under consideration are i) accretion of gas via mergers of previously gas-rich systems, and ii) cooling and direct accretion of gas from the intergalactic medium (IGM). The latter is arguably the more fundamental of the two, in the sense that it is the process by which the first primordial galaxies presumably formed.

Following the current paradigm of structure formation, baryonic structure follows the hierarchical, gravity-driven formation of dark matter halos(DMH). The propensity for gas to cool and accrete, fuelling existing galaxies or creating new ones, is thus expected to depend upon the galaxy's immediate environment, as determined by the DMH, due to virial heating of the gas in the halo. On the other hand, the efficiency of gas-fuelling is also expected to depend on positive/negative feedback effects linked to the properties of the individual galaxies, thus necessitating an approach capable of separating these dependencies. Although considerable theoretical effort has been applied to the subject (Bower *et al.* 2006), comparison of theoretical predictions with observations has hitherto only been possible for the most massive DMHs ($M_{DMH} > 10^{14} M_{\odot}$). Using the Galaxy and Mass Assembly survey (GAMA†, Driver *et al.* 2011), and in particular the GAMA Galaxy Group Catalogue (G^3C, Robotham *et al.* 2011) this range can be extended to DMH with masses on the order of $M_{DMH} \approx 10^{11} M_{\odot}$, probing the mass range of galaxy groups.

Based on the GAMA data, the instantaneous SF rate derived from the NUV can be used as a proxy for the gas-fuelling rate of normal late-type galaxies as appears to be valid for most of these systems (Hopkins & Beacom 2006), creating a large statistical sample of galaxies in a variety of DMHs ranging from group to cluster masses for which gas-fuelling rates can be obtained.

2. Late-type galaxies; opacities and SFRs

The emission of galaxies, especially late-type galaxies, at UV/optical wavelengths is known to be strongly affected by inclination dependent attenuation due to dust concentrated in their disks (Tuffs *et al.* 2004; Pierini *et al.* 2004; Driver *et al.* 2007). As such, any measure of the SFR of galaxies derived from emission in the NUV must be corrected for these effects, in order to obtain an estimate of the sources' intrinsic emission. Estimating the disk opacities of late-type galaxies, allowing attenuation corrections to be applied has, however, proven to be challenging, and various approaches exist. These range from the most powerful using UV-optical-FIR data together with radiative transfer modelling to self-consistently correct for attenuation while taking a source's inclination into account (Popescu *et al.* 2011), to semi-empirical methods based on UV/optical data alone, e.g the UV-spectral slope β (Meurer, Heckmann, & Calzetti 1999) or the Balmer decrement (Calzetti 2001; Wijesinghe *et al.* 2011). Regrettably, full coverage of the UV/optical-FIR spectrum, resp. multiple band UV data and spectroscopy, is often not available, making homogeneous approaches on large statistical samples difficult.

Using the GAMA photometry, stellar masses (Taylor *et al.* 2011), and morphological fits (Kelvin *et al.* 2011) in combination with the *HERSCHEL*-ATLAS (H-ATLAS‡, Eales *et al.* 2010) science demonstration phase sources matched to GAMA sources (Smith *et al.* 2011), we present a correlation between the face-on B-band opacity due to dust τ_B^f and the stellar mass surface density μ_* (Grootes *et al.* 2012a)(cf. Fig. 1)

$$\log(\tau_B^f) = \log\left(\frac{\mu_*}{\mathrm{M_{\odot}kpc}^{-2}}\right) 0.94(\pm 0.06) - 8.0(\pm 0.6) \tag{2.1}$$

for a sample of late-type sources selected in the NUV-r vs. Sersić index n plane (Driver *et al.* (2011); Kelvin *et al.* (2012)), using a separator

$$(\mathrm{NUV} - r) \leqslant -22.35 \log(n) + 11.5 \tag{2.2}$$

† http://www.gama-survey.org ; GAMA is a joint European- Australasian project based around a spectroscopic campaign using the Anglo-Australian Telescope. The GAMA input catalogue is based on data taken from the Sloan Digital Sky Survey and the UKIRT Infrared Deep Sky Survey. Complementary imaging of the GAMA regions is being obtained by a number of independent survey programs including GALEX MIS, VST KIDS, VISTA VIKING, WISE, Herschel-ATLAS, GMRT and ASKAP providing UV to radio coverage. GAMA is funded by the STFC (UK), the ARC (Australia), the AAO, and the participating institutions.

‡ http://www.h-atlas.org ; The Herschel-ATLAS is a project with Herschel, which is an ESA space observatory with science instruments provided by European-led Principal Investigator consortia and with important participation from NASA.

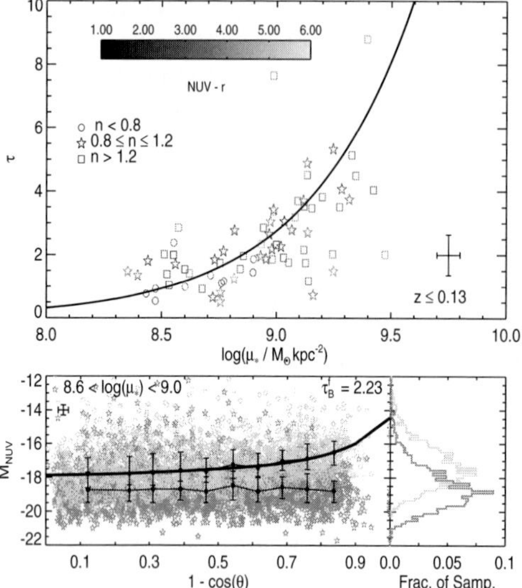

Figure 1. Top: Correlation between τ_B^f and μ_*. The best fit (Eq.2.1) is shown as dash-dotted line. **Bottom:** Distribution of M_{NUV} for a complete sample of GAMA late-type galaxies, characterized by μ_* as a function of inclination θ. The bin wise medians in M_{NUV} and $1-cos(\theta)$ before and after correction are shown by solid lines. The expected uncorrected locus of a fiducial galaxy defined using the model of Popescu *et al.* (2011) with τ_B^f and M_{NUV} corresponding to the medians of the corrected sample is shown as a dash-dotted line. Error bars indicate the interquartile range in each bin.

Figure 2. Top: Median SSFR as a function of M_* in bins of 0.5 dex for 4 bins of DMH mass ranging from $M_{DMH} < 10^{11.75} M_\odot$ to $M_{DMH} > 10^{13.75} M_\odot$ and the comparison field sample. **Bottom:** Median SSFR as a function M_* in bins of 0.5 dex for 4 bins of average link strength σ_{avg} as a probe of local galaxy density and the comparison field sample. Error bars show the interquartile range of the distribution in each bin. The positions of the plotted points have been shifted by ± 0.02 in $\log(M_*/\mathrm{M}_\odot)$ w.r.t. the field sample for visibility.

calibrated on the overlap of GAMA with GALAXY ZOO (Lintott *et al.* 2008)(see Grootes *et al.* 2012b for details). Eq. 2.1 allows τ_B^f to be estimated for statistical samples of late-type galaxies on the basis of optical broad band photometry alone, and is thus homogeneously applicable to large data-sets.

The correlation has been tested using the attenuation-inclination relation for complete samples of late-type galaxies, and successfully corrects the samples for the effects of dust attenuation using the estimated values of τ_B^f, the relation between τ_B^f and intrinsic size for disk galaxies (Möllenhoff, Popescu, & Tuffs 2006), and the radiative transfer model of Popescu *et al.* (2011)(cf. Fig. 1). The result has been found to be consistent with, e.g. the mass-metallicity relation and the M_{gas}/M_* vs. M_* relation and to be driven largely by a rougly linear relation between dust mass and stellar mass (Grootes *et al.* 2012a).

Using this correlation we correct the NUV emission of a complete statistical sample of GAMA late-type galaxies with stellar mass $M_* \geqslant 10^{9.5} M_\odot$ for dust-attenuation and calculate NUV SF rates (SFR) as specified in Wijesinghe *et al.* (2011).

3. Environmental dependence of SFRs

The G^{3}C (Robotham *et al.* 2011) is a catalogue of galaxy groups obtained by applying a friends-of-friends (FoF) algorithm to the GAMA spectroscopic survey which has a redshift completeness of $> 90\%$ to its limiting depth of $r_{pet} < 19.4$, providing a novel environmental metric reliably recovering galaxy groups and providing mass estimates of the host DMHs extending down to $M_{DMH} \approx 10^{11} M_{\odot}$. The specific SFR (SSFR) of galaxies is known to depend on galaxy stellar mass so, to disentangle this effect from environmental effects, we compare the SSFR (corrected as specified above) of late-type GAMA galaxies, selected according to Eq. 2.2, as a function of their stellar mass for different DMH mass bins, and a comparison sample of field galaxies, defined as not being grouped in the G^3C (cf. Fig.2). We find that, while the SSFR of late-type galaxies in halos of $M_{DMH} \leqslant 10^{12.75} M_{\odot}$ is essentially indistinguishable from that of late-type field galaxies at all galaxy stellar masses, the SSFR of galaxies with $M_{*} \leqslant 10^{10.5} M_{\odot}$ in halos with $10^{12.75} M_{\odot} < M_{DMH} \leqslant 10^{13.75} M_{\odot}$ is slightly suppressed w.r.t. the field sample, and strongly suppressed in halos more massive than $10^{13.75} M_{\odot}$. The SSFRs of galaxies with $M_{*} > 10^{10.5} M_{\odot}$ do not display any environmental dependence.

The G^3C also enables the use of grouping parameters as a metric for the group environment. The strength of a link between galaxies in the FoF algorithm can be characterized by the ratio of actual linking length and maximum possible link length, with strongly linked galaxies having small values. Thus, the compactness of a group in redshift space can, to first order, be characterized by the average linking strength σ_{avg} of the group, with small values corresponding to compact groups and larger values to more extended groups. This measure of compactness can be viewed as a proxy for the galaxy density in redshift space, taking only group member galaxies into account. We find that the average SSFR of low mass galaxies is suppressed w.r.t. the field sample in the most compact groups, while the similarity to the field sample increases as the groups become more extended (cf. Fig. 2). Galaxies with $M_{*} > 10^{10.5} M_{\odot}$ don't show environmental dependence, and the SSFR suppression of low mass galaxies is less pronounced as a function of group compactness, than as a function of halo mass.

References

Bower, R. G. & Benson, A. J., Malbon, R., *et al.* 2006, *MNRAS*, 370, 645
Calzetti D. 2001, *PASP*, 113, 1449
Driver, S. P., Popescu, C. C., Tuffs, R. J., *et al.* 2007, *MNRAS*, 379, 1022
Driver, S. P., Hill, D. T., Kelvin, L. S., *et al.* 2011, *MNRAS*, 413, 971
Driver, S. P., *et al.* 2012, in prep.
Eales, S., Dunne, L., Clements, D., *et al.* 2010, *PASP*, 122, 499
Grootes, M. W., *et al.* 2012a, in prep.
Grootes, M. W., *et al.* 2012b in prep.
Hopkins, A. M. & Beacom, J. F. 2006, *ApJ*, 651, 142
Kelvin L. S., Driver, S. P., Robotham, A. S. G., *et al.* 2011, *MNRAS*, submitted
Kelvin L. S., *et al.* 2012, in prep.
Meurer G. R., Heckmann T. M., & Calzetti D. 1999, *ApJ*, 521, 64
Möllenhoff C., Popescu C. C. & Tuffs R. J. 2006, *A&A*, 456, 941
Lintott, C. J., Schawinski, K., Slosar, A., *et al.* 2008, *MNRAS*, 389, 1179
Pierini D., Gordon K. D., Witt A. N., & Madsen G. J. 2004, *ApJ*, 617, 1022
Popescu, C. C. *et al.* 2000, *A&A*, 362, 138
Popescu, C. C., Tuffs, R. J., Dopita, M. A., *et al.* 2011, *A&A*, 527, A109+
Robotham, A. S. G., Norberg, P., Driver, S. P., *et al.* 2011 *MNRAS*,
Smith, D. J. B., Dunne, L., Maddox, S. J., *et al.* 2011, *MNRAS*, 416, 857
Taylor, E. N., Hopkins, A. M., Baldry, I. K., *et al.* 2011, arXiv:1108.0635, accepted by *MNRAS*
Tuffs, R. J., Popescu, C. C., Völk, H. J., Kylafis, N. D., & Dopita, M. A. 2004, *A&A*, 419, 821
Wijesinghe D., Hopkins, A. M., Sharp, R., *et al.* 2011, *MNRAS*, 415, 1002

Discussion

MADORE: Could you remind us what the typical value for the face-on extinction is for spiral galaxies corrected by this method?

GROOTES: This depends both on the wavelength considered, as well as the stellar mass. Additionally there is a spread in the value of opacity for a given stellar mass. Nevertheless a typical spiral galaxy would have a face-on B-band central optical depth of about 4, for which an attenuation of the integrated emission of ~ 0.9 mag in the NUV and ~ 0.3 mag in the g band is found for a median inclination of 60 degrees.

HOLWERDA: How sensitive are you to different scale lengths for dust in the conversion from μ_* to τ_B^f?

GROOTES: Converting μ_* to τ_B^f uses the relative values of the geometrical components of the model to be a correct representation of the galaxy being treated. The model is calibrated on a sample of well-resolved edge-on galaxies with intermediate morphologies along the Hubble sequence of late-type galaxies (see Popescu *et al.* 2000). For these objects the relative scale lengths of the dust and stars is directly determinable through radiation transfer-modelling of the images, and is found to be rather constant. In moving to earlier or later-type disk galaxies there is some degree of flexibility incorporated via the bulge-to-disk ratio, which takes care of the general trend for the overall distribution of older stars in the galaxy (the combined population of old stars in the bulge and disk) to be more centrally concentrated compared to the dust disk when moving towards earlier types on the Hubble sequence. Nevertheless, we expect that our model would start to fail at some level if the ratio of scale lengths of dust in the disk and stars in the disk would markedly change for very early-type or very late type spiral galaxies. As discussed at length in Popescu *et al.* (2011) any such failure would however be recognisable through a shift in the relation between the observed and predicted colour-surface brightness relation of the dust emission of spiral galaxies.

The Spectral Energy Distribution of Galaxies
Proceedings IAU Symposium No. 284, 2011
R.J. Tuffs & C.C. Popescu, eds.

doi:10.1017/S1743921312009416

AKARI/IRC broadband mid-infrared data as an indicator of the Star Formation Rate

Fang-Ting Yuan[1], Tsutomu T. Takeuchi[1], Véronique Buat[2], Sébastien Heinis[2], Elodie Giovannoli[2], Katsuhiro L. Murata[1], Jorge Iglesias-Páramo[3,4] and Denis Burgarella[2]

[1]Division of Particle and Astrophysical Sciences,
Nagoya University, JAPAN
email: yuan.fangting, takeuchi.tsutomu, murata.katsuhiro@g.mbox.nagoya-u.ac.jp

[2]Laboratoire d'Astrophysique de Marseille, OAMP,
Université Aix-Marseille, FRANCE
email: veronique.buat, sebastien.heinis, elodie.giovannoli,
denis.burgarella@oamp.fr

[3]Instituto de Astrofísica de Andalucía (IAA - CSIC),
Glorieta de la Astronomía s.n., SPAIN

[4]Centro Astronómico Hispano Alemán, SPAIN
email: jiglesia@iaa.es

Abstract. With the goal of constructing Star-Formation Rates (SFR) from AKARI Infrared Camera (IRC) data, we analyzed an IR-selected GALEX-SDSS-2MASS-AKARI(IRC & Far-Infrared Surveyor) sample of 153 nearby galaxies. The far-infrared fluxes were obtained from AKARI diffuse maps to correct the underestimation for extended sources raised by PSF photometry. SFRs of these galaxies were derived using the SED fitting program CIGALE. In spite of complicated features contained in these bands, both the *S9W* and *L18W* emissions correlate with the SFR of galaxies. The SFR calibrations using *S9W* and *L18W* are presented for the first time. These calibrations agree well with previous work based on Spitzer data within the scatter, and should be applicable to dust-rich galaxies.

Keywords. infrared: galaxies, ISM: dust, extinction, stars: formation

1. Introduction

Measurements of star formation activity are fundamental to studies of the formation and evolution of galaxies. Numerous efforts have been made to find reliable and convenient star formation rate (SFR) indicators. Apart from the most frequently used indicators, such as FUV continuum, optical recombination lines and FIR continuum (Kennicutt 1998), the mid-infrared (MIR) monochromatic fluxes have also been investigated as SFR indicators. The MIR emission is contributed by several components, including the polycyclic aromatic hydrocarbon (PAH) features, the continuum from stochastically-heated very small grains, silicate absorption at 9.7 and 18 μm, molecular hydrogen lines and fine-structure lines (Leger & Puget 1984; Desert *et al.* 1990; Draine & Li 2007; Treyer *et al.* 2010). the MIR-SFR relation was intensively studied using Spitzer IRAC 8 μm and MIPS 24 μm photometry data and spectral data (e.g., Calzetti *et al.* 2007; Kennicutt *et al.* 2009; Rieke *et al.* 2009). Nevertheless, there is still a debate about the reliability of MIR indicators because of the complicated features it contains.

The recently released AKARI/IRC Point Source Catalogue Version β-1 (hereafter IRCPSC) provides positions and fluxes of sources detected in the all-sky survey in the *S9W* (9 μm) and *L18W* (18 μm) bands (Ishihara *et al.* 2010). Because of the copious amounts

Table 1. Brief summary of the sample.

Survey	Band	Wavelength (μm)	Nb. of sources
GALEX	FUV, NUV	0.153, 0.231	153
SDSS	*u, g, r, i, z*	0.355, 0.469, 0.617, 0.748, 0.893	153
2MASS	*J, H, Ks*	1.244, 1.655, 2.169	153
IRAS*	band-1, 2, 3, 4	12, 25, 60, 100	153
AKARI IRC	*S9W*	9	126
AKARI IRC	*L18W*	18	106
AKARI FIS	*N60, Wide-S, Wide-L, N160**	65, 90, 140, 160	153

* Data were not used for SED fitting.

of MIR data, it would be useful if there is a benchmark of SFR measurement. Compared with the Spitzer 8 and 24 μm bands, the AKARI IRC *S9W* and *L18W* bands cover a wider wavelength range, including silicate absorption features in both bands and emission contributed by large PAH molecules in the *L18W* band. This work is done to investigate whether and to what degree AKARI broadband MIR data could trace SFRs, and how the inclusion of silicate absorption and longer wavelength PAH feature affects the results.

2. Data and Method

We cross-identified IRCPSC with the multi-wavelength data constructed by Takeuchi *et al.* 2010. The original sample is based on a selection using the IRAS-PSCz Saunders *et al.* 2000 and AKARI/FIS Bright Source Catalogue (hereafter FISBSC) data (I. Yamamura *et al.* 2008). These galaxies are then cross-matched to sources detected by GALEX, 2MASS and SDSS. The FISBSC flux density for extended sources would no longer be accurate because of the point source extraction procedure. This is confirmed by a comparison with similar bands in IRAS co-added fluxes, which are specially calculated for extended sources (Saunders *et al.* 2000). Therefore, the fluxes for extended sources are replaced by Äutofluxes derived from AKARI diffuse maps using Source Extractor. The data used in the present study are summarized in Table 1.

The SED fitting program CIGALE (Noll *et al.* 2009) is used to calculate the SFR for our sample. CIGALE was developed to derive highly reliable galaxy properties by fitting the UV/optical SEDs and the related dust emission at the same time, i.e., the stellar population synthesized models are connected with infrared templates by the balance of the energy of dust emission and absorption. A detailed description of CIGALE could be found in Noll *et al.* (2009), Buat *et al.* (2010) and Giovannoli *et al.* (2010) - see also Buat *et al.* (this volume).

3. Result and Conclusion

Regression analysis was conducted to investigate the MIR-SFR relations. SFRs converted from AKARI MIR fluxes were compared with the ones from Spitzer MIR fluxes to test the reliability of AKARI MIR-SFR calibrations. We summarize our results and conclusions as follows:

(*a*) Both 9 μm and 18 μm luminosities correlate with SFRs and thus could be converted into SFRs (Fig. 1). The correlation equations are given by

$$\log \frac{\mathrm{SFR}}{M_\odot/\mathrm{yr}} = (0.99 \pm 0.03) \log \frac{L_9}{L_\odot} - (9.02 \pm 0.32) \qquad (3.1)$$

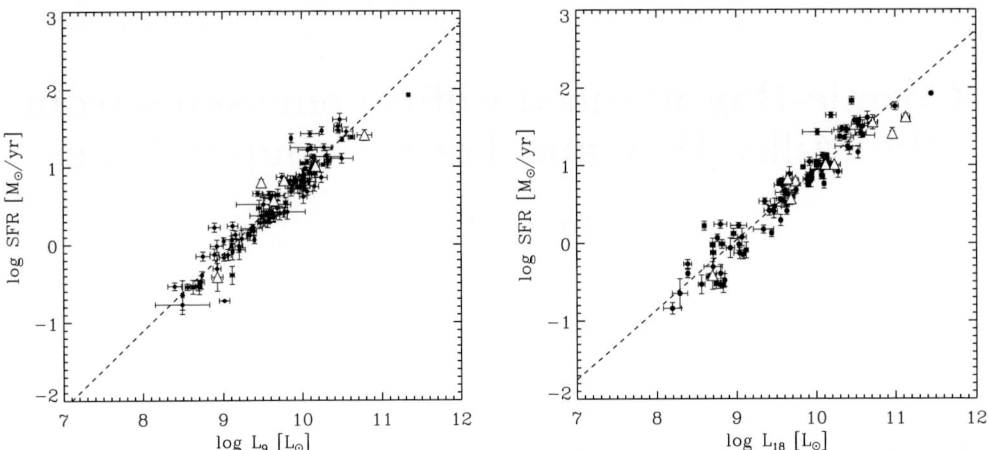

Figure 1. The MIR-SFR relation. The dashed line shows the fitting result. The triangles denote AGNs.

$$\log \frac{\mathrm{SFR}}{M_{\odot}/\mathrm{yr}} = (0.90 \pm 0.03) \log \frac{L_{18}}{L_{\odot}} - (8.03 \pm 0.30). \tag{3.2}$$

The scatters are 0.18 and 0.20 for 9 and 18 μm, respectively.

(*b*) A combination of FUV and MIR luminosities barely reduces the scatters (0.17 and 0.20 dex for 9 and 18 μm, respectively), indicating that unobscured UV photons are not the only reason for variations in the MIR-SFR relation.

(*c*) The comparison of the SFRs derived from Equations 3.1 and 3.2 with the ones derived from Spitzer MIR-SFR relations shows that the silicate absorption included in *S9W* (9 μm) and *L18W* (18 μm) bands little affects the results. The discrepancies, if any, are well within the uncertainties.

(*d*) AGNs in the sample show no discrepancy with normal galaxies in the MIR-SFR diagrams. However, AGNs do show lower MIR values than normal galaxies on average, which probably indicates that small PAH molecules are destructed by the harsh radiation from AGNs.

References

Buat, V., *et al.* 2010, *A&A*, in press

Calzetti, D., *et al.* 2007, *ApJ*, 666, 870

Desert, F.-X., Boulanger, F., & Puget, J. L. 1990, *A&A*, 237, 215

Draine, B. T. & Li, A. 2007, *ApJ*, 657, 810

Giovannoli, E., Buat, V., Noll, S., Burgarella, D., & Magnelli, B. 2011, *A&A*, 525, A150

Ishihara, D., *et al.* 2010, *A&A*, 514, A1

Kennicutt, R. C., Jr. 1998, *ARAA*, 36, 189

Kennicutt, R. C., *et al.* 2009, *ApJ*, 703, 1672

Leger, A. & Puget, J. L. 1984, *A&A*, 137, L5

Noll, S., Burgarella, D., Giovannoli, E., Buat, V., Marcillac, D., & Muñoz-Mateos, J. C. 2009, *A&A*, 507, 1793

Rieke, G. H., Alonso-Herrero, A., Weiner, B. J., Pérez-González, P. G., Blaylock, M., Donley, J. L., & Marcillac, D. 2009, *ApJ*, 692, 556

Saunders, W., *et al.* 2000, *MNRAS*, 317, 55

Takeuchi, T. T., *et al.* 2010, *A&A*, 514, A4

Treyer, M., *et al.* 2010, *ApJ*, 719, 1191

Yamamura, I., Fukuda, Y., & Makiuti, S. 2008, AKARI/FIS all-sky Survey Bright Source Catalogue Version β-1 Release Note (Rev. 1)

The Spectral Energy Distribution of Galaxies
Proceedings IAU Symposium No. 284, 2011
R.J. Tuffs & C.C.Popescu, eds.

doi:10.1017/S1743921312009428

Cosmic-Ray induced diffuse emissions from the Milky Way and Local Group galaxies

Troy A. Porter

Hansen Experimental Physics Laboratory
and
Kavli Institute for Particle Astrophysics and Cosmology
Stanford University, Stanford, USA
email: tporter@stanford.edu

Abstract. Cosmic rays fill up the entire volume of galaxies, providing an important source of heating and ionisation of the interstellar medium, and may play a significant role in the regulation of star formation and galactic evolution. Diffuse emissions from radio to high-energy γ-rays (>100 MeV) arising from various interactions between cosmic rays and the interstellar medium, interstellar radiation field, and magnetic field, are currently the best way to trace the intensities and spectra of cosmic rays in the Milky Way and other galaxies. In this contribution, I describe our recent work to model the full spectral energy distribution of galaxies like the Milky Way from radio to γ-ray energies. The application to other galaxies, in particular the Magellanic Clouds and M31 that are now resolved in high-energy γ-rays by the Fermi-LAT, is also discussed.

Keywords. radiation mechanisms: general, radiation mechanisms: nonthermal, (ISM:) cosmic rays, ISM: magnetic fields, galaxies: ISM, (galaxies:) Local Group, (galaxies:) Magellanic Clouds, gamma rays: observations, infrared: galaxies, radio continuum: galaxies

---

1. Introduction

The luminosity of a star-forming galaxy like the Milky Way (MW) is dominated by the relatively narrow frequency range of the spectral energy distribution (SED) from the ultraviolet (UV) to far infrared (FIR), which is due to stellar emission and dust reprocessing in the interstellar medium (ISM). Related to the birth and death of massive stars, cosmic rays (CRs) are pervasive throughout the ISM (see, e.g., Strong *et al.* 2007 for a recent review). The diffuse emissions arising from various interactions between CRs and the ISM, interstellar radiation field (ISRF – the UV–FIR component of the galactic SED), and magnetic field span radio frequencies to high-energy γ-rays (>100 MeV), but at a lower level of intensity compared to the stellar and dust component. These broadband emissions are currently the best way to trace CR intensities and spectra throughout the MW and other galaxies. Gamma rays are particularly useful in this respect because this energy range gives access to the dominant hadronic component in CRs via the observation of π^0-decay radiation produced by CR nuclei inelastically colliding with the interstellar gas. Understanding the global energy budget of processes related to the injection and propagation of CRs, and how the energy is distributed across the electromagnetic spectrum, is essential to interpret the radio/far-infrared relation (Helou *et al.* 1985, Murphy *et al.* 2006), galactic calorimetry (e.g., Völk 1989), and predictions of extragalactic backgrounds (e.g., Thompson *et al.* 2007, Murphy *et al.* 2008), and for many other studies.

2. Broadband Spectral Energy Distribution for the Milky Way

The MW is the best studied non-AGN dominated star-forming galaxy, and the only galaxy that direct measurements of CR intensities and spectra are available. However, because of our position inside, the derivation of global properties is not straightforward and requires detailed models of the spatial distribution of the emission. In recent work by Strong *et al.* (2010) using the GALPROP (e.g., Strong *et al.* 2000, Moskalenko *et al.* 2002; also see http://galprop.stanford.edu) and FRaNKIE (Fast Radiative Numerical Kode for Interstellar Emission – Porter *et al.* 2008) codes, we have calculated the broadband SED of the MW for the first time. Emission by stars and dust is included, together with the diffuse emissions for different CR propagation models consistent with local CR data. Figure 1 (left) shows the broadband luminosity spectrum of the Galaxy, including the input luminosity for CRs for a 4 kpc halo using a diffusive-reacceleration CR propagation model. Strong *et al.* (2010) considered a range of halo sizes (2–10 kpc). For this range, the relative decrease in the injected CR proton and helium luminosities is $\sim 10\%$. For smaller halo sizes, the CRs escape quicker requiring more injected power to maintain the local CR spectrum. In addition, for larger halo sizes CR sources located at further distances can contribute to the local spectrum, which is the normalisation condition, hence less power is required. Contrasting with the CR nuclei, the injected primary CR electron luminosity *increases* with z_h, which is required to counter the increased inverse-Compton (IC) energy losses in the halo from the longer escape time.

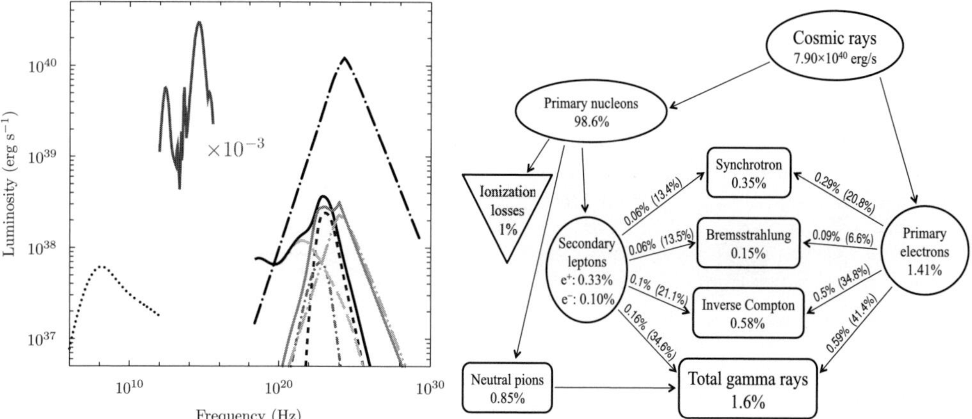

Figure 1. *Left:* Global CR-induced luminosity spectra of the MW for a diffusive-reacceleration CR propagation model with $z_h = 10$ kpc. Line styles: ISRF, including optical and infrared *scaled by factor* 10^{-3} (solid, left side of figure) and components for propagation model – Cosmic rays, protons (long dash dot), helium and heavier nuclei (long dash double dot), primary electrons and secondary electrons and positrons (long dash triple dot); CR-induced diffuse emissions, IC (long dash), bremsstrahlung (medium dash dot), π^0-decay (small dash), total γ-rays (solid, right side of figure), synchrotron (dotted). *Right:* Luminosity budget of the MW for the CR propagation model shown in the left panel. The percentage figures are shown with respect to the total injected luminosity in CRs, 7.9×10^{40} erg s^{-1}. The percentages in brackets show the values relative to the luminosity of their respective lepton populations (primary electrons, secondary electrons/positrons).

Figure 1 (right) illustrates the detailed energy budget for the propagation model shown in Fig. 1 (left). The energy channelled into γ-ray and other secondary production from the CR nuclei component is only a very small fraction of the total power injected in these particles. For the model shown, the total of synchrotron, IC, and bremsstrahlung luminosities for the primary electrons accounts for over half of the total luminosity injected in these particles. Note that it is the IC emission at γ-ray energies that is responsible for the majority of the energy losses, not synchrotron radiation as usually assumed. Including the contribution by secondary electron/positron production, approximately half of the total output in γ-rays is provided by these particles, even though they count for only $\sim 2\%$ of the injected power. Non-AGN dominated galaxies like the MW turn out to be very good lepton calorimeters if *all* the energy-loss processes are taken into account.

3. Gamma Rays from Local Group Galaxies

Until recently, the MW was the only galaxy that was resolved in high-energy γ-rays. However, observations by the *Fermi* Large Area Telescope (LAT) have extended the sample of resolved star-forming galaxies to include the Magellanic Clouds (Abdo *et al.* 2010, Abdo *et al.* 2010) and M31 (Abdo *et al.* 2010).

Figure 2 (left) shows the background subtracted counts map for a $20° \times 20°$ region of interest (ROI) surrounding the LMC. The remaining feature is extended emission that is spatially confined to within the LMC boundaries, which are traced by the iso column density contour $N_{\rm H} = 1 \times 10^{21}$ H cm^{-2} of neutral hydrogen in the LMC (Kim *et al.* 2005). The extended γ-ray emission from the LMC can be resolved into several components. The brightest emission feature is located near $(\alpha_{J2000}, \delta_{J2000}) \approx (05^{\rm h}40^{\rm m}, -69°15')$, which is close to the massive star-forming region 30 Doradus (30 Dor) that houses the two Crab-like pulsars PSR J0540−6919 and PSR J0537−6910. Excess γ-ray emission is also seen toward the north and the west of 30 Dor. These bright regions are embedded within a more extended and diffuse glow that covers an area of approximately $5° \times 5°$.

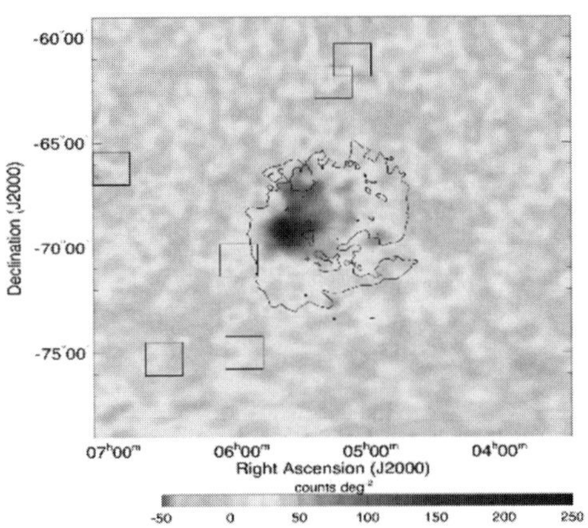

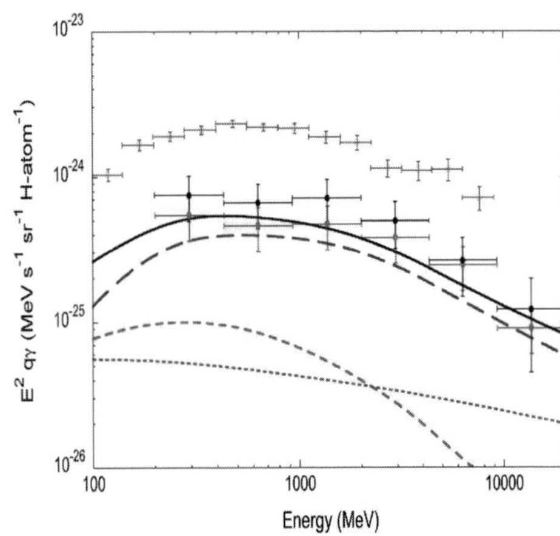

Figure 2. *Left:* Gaussian kernel ($\sigma = 0.2°$) smoothed count map for the ROI about the LMC after subtraction of the background model for the energy range 200 MeV – 20 GeV and for a pixel size of $0.1° \times 0.1°$. Overlaid is the N(H I) contour of 1×10^{21} H cm^{-2} of the LMC to indicate the extent and shape of the galaxy. The boxes show the locations of the 6 point sources (background blazars) that were included in the background model. *Right:* Differential average γ-ray emissivity spectrum for the LMC. Data for models $\mathcal{H}_1$ (upper filled circles) and $\mathcal{H}_2$ (lower filled circles) (see text) and the local ISM emissivity (Abdo *et al.* 2009 – upper grey data points) are shown. Model lines: solid, predicted total γ-ray emissivity computed in the framework of a one-zone model for $\mathcal{H}_2$; long dashed, π^0-decay; short dashed, bremsstrahlung; dotted, IC. See Abdo *et al.* 2010 for details.

Most of the gas in the LMC is found in the form of neutral hydrogen and helium with only about 1% of the total mass ionised. However, the distribution of H II provides the best fit to the observed γ-ray emission. It is characterised by a strong emission peak near 30 Dor which is attributed to the intense ionising radiation of the massive stars in this highly active region. Even accounting for the 30 Dor emission, the H II distribution is still a significantly better tracer than the neutral gas of the residual emission. This contrasts with the MW, where the majority of the diffuse γ-ray emission appears to trace the neutral gas.

Abdo *et al.* 2010 tested two hypotheses for the origin of the γ-ray emission from the LMC: ($\mathcal{H}_1$) all γ-ray emission from the LMC is attributed to diffuse emission from CR interactions, and ($\mathcal{H}_2$) only emission from a disk-like component arises from CR interactions, while the γ-rays from 30 Dor originate from other sources. The emissivity spectrum derived using these two hypotheses for the LMC is shown in Fig. 2 (right). The integrated emissivity >100 MeV for the LMC is $\sim 2-4$ times lower than the locally derived emissivity (also shown in the figure). Similar analysis for the SMC (Abdo *et al.* 2010) and M31 (Abdo *et al.* 2010) derive emissivities >100 MeV that are $\sim 6-7$ and ~ 2 times lower, respectively. Assuming that the proton-to-electron ratio in these galaxies above a few tens of GeV is similar to that in the MW and that all of the emission is from diffuse processes, these values point to a general picture where the intensity of the emission is controlled by the power injected by CR sources scaling approximately with the relative star-formation rates in the different systems. Future work combining the γ-ray, radio, and other data will result in broadband SEDs of these galaxies spanning more than 20 decades in frequency, which can be used to investigate a variety of phenomena related to how energy is injected and cycled within the ISM of galaxies other than the MW.

4. Acknowledgements

The *Fermi*-LAT Collaboration acknowledges support from a number of agencies and institutes for both development and the operation of the LAT as well as scientific data analysis. These include NASA and DOE in the United States, CEA/Irfu and IN2P3/CNRS in France, ASI and INFN in Italy, MEXT, KEK, and JAXA in Japan, and the K. A. Wallenberg Foundation, the Swedish Research Council and the National Space Board in Sweden. Additional support from INAF in Italy and CNES in France for science analysis during the operations phase is also gratefully acknowledged. GALPROP development is supported via NASA Grant Nos. NNX10AE78G and NNX09AC15G.

References

Abdo, A. A., Ackermann, M., Ajello, M., *et al.* 2009, *ApJ*, 703, 1249
Abdo, A. A., Ackermann, M., Ajello, M., *et al.* 2010, *A&A*, 512, 7
Abdo, A. A., Ackermann, M., Ajello, M., *et al.* 2010, *A&A*, 523, L2
Abdo, A. A., Ackermann, M., Ajello, M., *et al.* 2010, *A&A*, 523, 46
Helou, G., Soifer, B. T., & Rowan-Robinson, M. 1985, *ApJL*, 298, L7
Kim, S., Staveley-Smith, L., Dopita, M. A., *et al.* 2005, *ApJS*, 143, 487
Moskalenko, I. V., Strong, A. W., Ormes, J. F., *et al.* 2002, *ApJ*, 565, 280
Murphy, E. J., Helou, G., Braun, R., *et al.* 2006, *ApJ*, 638, 157
Murphy, E. J., Helou, G., Kenney, J. D. P., *et al.* 2008, *ApJ*, 678, 828
Porter, T. A., Moskalenko, I. V., Strong, A. W., *et al.* 2008, *ApJ*, 682, 400
Strong, A. W., Moskalenko, I. V., & Reimer, O. 2000, *ApJ*, 537, 763
Strong, A. W., Moskalenko, I. V., & Ptuskin, V. S. 2007, *Ann. Rev. Nuc. Part. Sci.* , 57, 285
Strong, A. W., Porter, T. A., Digel, S. W., *et al.* 2010, *ApJL*, 722, 58
Thompson, T. A., Quataert, E., & Waxman, E. 2007, *ApJ*, 654, 219
Völk, H. J. 1989, *A&A*, 218, 67

Discussion

TABATABAEI: Have you tried to derive the XCO conversion factor based on your gamma-ray data?

PORTER: Concerning the question of whether XCO can be constrained using gamma-rays, I would say, yes, provided all details of cosmic ray propagation and origin are known. In practice, it is problematic to obtain XCO throughout the Galaxy using gamma-ray data.

The Spectral Energy Distribution of Galaxies
Proceedings IAU Symposium No. 284, 2011
R.J. Tuffs & C.C. Popescu, eds.

doi:10.1017/S174392131200943X

The H.E.S.S. view of the Milky Way in TeV light

Christoph Deil[1] **for the H. E. S. S. collaboration**
[1]Max-Planck-Institute for Nuclear Physics, P.O. Box 103980, 69029 Heidelberg, Germany
email: Christoph.Deil@mpi-hd.mpg.de

Abstract. Since 2003 the H.E.S.S. collaboration has been operating an array of four imaging Cherenkov telescopes in the Khomas Highlands of Namibia. H.E.S.S. can detect gamma rays in the energy range 100 GeV to 100 TeV, has a large field of view (5 degree), good angular resolution (0.1 degree), energy resolution (15%) and sensitivity (a 1% Crab flux point source is detected at 5 sigma significance in 25 h). About half of the available observing time has been spent on the Milky Way, either in scan mode or on individual sources, resulting in the detection of more than 60 Galactic TeV sources. In this talk the two most numerous source classes will be discussed, pulsar wind nebulae and supernova remnants. For the identification and understanding of the TeV emission seen by H.E.S.S. additional measurements of non-thermal emission, mainly in the radio, X-ray and lower-energy gamma-ray bands, are critical. Since August 2008 the Fermi Large Area Telescope has been scanning the whole sky in the energy range from 20 MeV to more than 300 GeV and has detected about 200 Galactic sources as well as diffuse emission from the Milky Way. This talk will give an overview of Galactic H.E.S.S. observations in the multi-wavelength context, with a focus on Fermi.

Keywords. surveys, gamma rays: observations, cosmic rays, Galaxy: structure

1. Introduction

Most of the contributions to this conference on 'The spectral energy distribution (SED) of Galaxies' study the thermal emission by the stars and dust in the Milky Way and other Galaxies. Yet all of this emission lies in the rather narrow energy band from 0.01 to 10 eV and represents only a small part of the total SED. Figure 1 illustrates the full SED, including the nonthermal emission components which range from low-frequency radio at 10^{-8} eV to very-high-energy gamma rays at 10^{14} eV. Here we give a short introduction to the progress made in the last years by the surveys of the gamma-ray emission of the Milky Way by the H.E.S.S. (`http://www.mpi-hd.mpg.de/hfm/HESS/`) and Fermi (`http://fermi.gsfc.nasa.gov/`) telescopes.

The Milky Way contains populations of cosmic particle accelerators that produce high-energy cosmic ray electrons, positrons and nuclei, typically with power-law spectra, reaching in some cases 10^{15} eV or more. The electrons interact with ambient magnetic fields and emit synchrotron radiation from radio to X-ray energies. They also produce inverse Compton gamma-ray emission in interactions with ambient low-energy photon fields such as the cosmic microwave background or stellar, dust and synchrotron radiation. Nuclei interact with surrounding gas and produce gamma-ray emission via neutral pion decay. Figure 1 illustrates the resulting SED components. The luminosity of the emission in each case is proportional to the product of cosmic ray and target density, both of which can be a few orders of magnitude higher inside and near the cosmic accelerators compared to the average densities in the Milky Way of $B \sim 1\,\mu$G and $n \sim 1\,\mathrm{cm}^{-3}$. The lifetime of most cosmic accelerators is 10^3 to 10^6 years, yet the average time it takes a cosmic

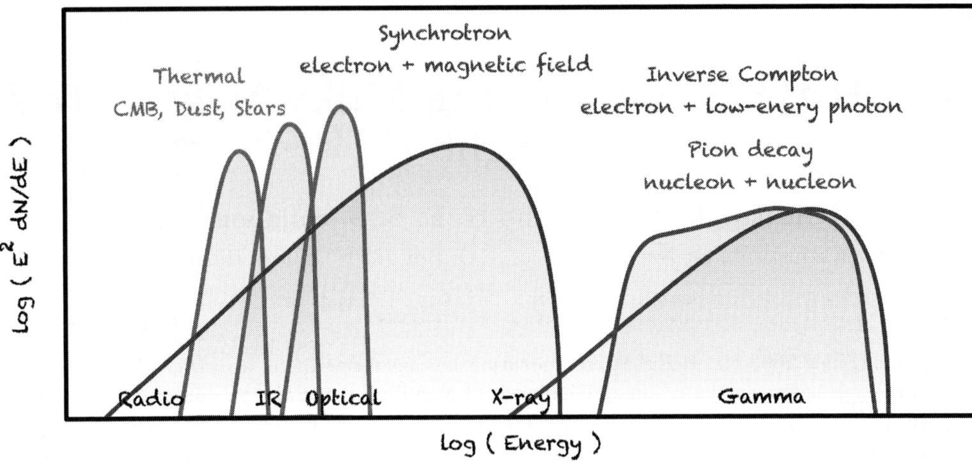

Figure 1. Illustration of the non-thermal SED components produced by cosmic rays as well as the thermal components (cosmic microwave background, emission by dust and stars) which can represent target photon fields for interaction with high-energy electrons, producing inverse Compton emission gamma rays.

ray particle to diffuse (due to gyromotion in the turbulent Galactic magnetic field) out of the Milky way is 10^7 years, so in addition to the sources there is a diffuse emission component arising from a 'sea of cosmic rays' interacting with passive targets.

The largest Galactic source class for the Fermi LAT survey at GeV energies are pulsars. For the H.E.S.S. survey at TeV energies the largest identified source classes are pulsar wind nebulae (PWNe) and supernova remnants (SNRs), sometimes interacting with nearby molecular clouds (MCs). Pulsars and PWNe are thought to be mostly electron (and positron) accelerators (Gaensler & Slane 2006), which originate from pair production in the pulsar magnetosphere (Daugherty & Harding 1982). It is possible that electrons and nuclei are 'ripped out' of the pulsar and thus PWNe have a more complicated composition than an electron-positron plasma (Horns *et al.* 2007). SNRs accelerate both electrons and nuclei from the thermal gas present in the shock wave acceleration zone. For most SNRs observed in gamma rays it is unclear which emission mechanism—leptonic or hadronic—dominates, the observed SEDs can be explained using either or a mixed model. For example for RX J1713.7-3946 a leptonic model is favoured (Abdo and Fermi LAT Collaboration 2011), for Tycho's SNR a hadronic model (Morlino & Caprioli 2011).

2. H.E.S.S. Galactic plane survey

H.E.S.S. is an array of imaging atmospheric Cherenkov telescopes that performs pointed observations with a field of view of ~ 5 deg. It detects extended air showers in the Earth's atmosphere induced by gamma rays and a background of cosmic rays and reconstructs the energy and direction of each event. Since the start of its operations in 2003 it has spent about half of the available observing time (~ 1000 h/yr) on the Galactic plane survey (GPS) region $l = -100$ to $+60$ deg and $b = -3.5$ to $+3.5$ deg, resulting in ~ 2300 h of good-quality observations. The GPS observations are again split 50/50 on systematic scan-mode observations and observations on specific targets that have been identified at other wavelengths or discovered as 'hotspots' in the scan-mode observations—see Gast *et al.* (2011) for the latest update. The resulting sensitivity (expressed in units of the

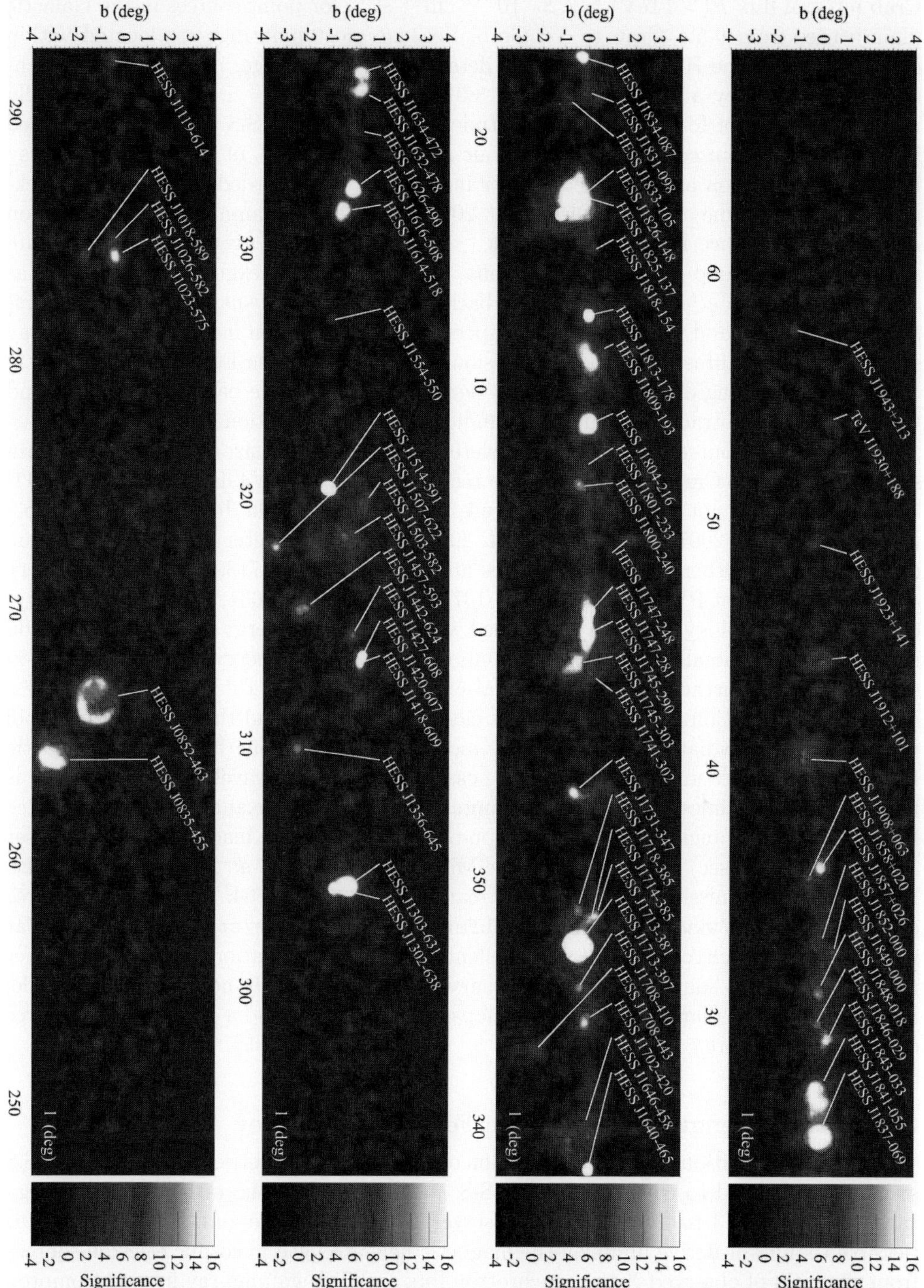

Figure 2. Latest significance map for the H.E.S.S. Galactic plane survey. This image shows for each position the statistical significance of the observed excess over the background within a correlation radius of 0.22 deg. The color transition to red occurs at approx. 7 sigma, which is typically chosen as the requirement for detection. Figure reproduced from Gast *et al.* (2011).

Crab integral flux $F(>1\text{TeV}) = 2.26 \cdot 10^{-11}$ cm^{-2} s^{-1}) for point sources in the Galactic plane varies from 0.5% Crab to 3% Crab. Because most pointings are located within $b = \pm 1$ deg and the H.E.S.S. sensitivity deteriorates at the edges of the FOV, the sensitivity increases by a factor of 3 at $b \sim \pm 3$ deg. Latitudes $b = \pm 4$ deg are practically not surveyed except for a few small longitude regions. The H.E.S.S. sensitivity is limited by background from cosmic-ray-induced air showers. Over 99% of the events the array initially triggered on and recorded shower images can be discarded as cosmic-ray background during offline analysis (Ohm *et al.* 2009). Still for the gamma-hadron separation cuts used to produce Figure 2, only 2.2k ($\sim$ 6%) of the 3.6M events used to compute that significance map are actually photons. The remaining background is modelled as described in Berge *et al.* (2007). For the background methods usually used to make images (field of view and ring method) and spectra (reflected region method) of gamma-ray sources, possible diffuse gamma-ray emission on the scale of the FOV (5 deg diameter) or even smaller (ring diameter of 0.5 to 4 deg) is included in the background model and thus effectively subtracted in flux and significance map production.

More than 60 sources have been discovered. They range in size from spatially unresolved faint (0.5% Crab) sources like the recent discovery of W49B ((Brun *et al.* 2011, Brun *et al.* 2011), not yet labeled in Figure 2) to the SNR Vela Jnr (HESS J0852-463, (Aharonian *et al.* 2007, Aharonian *et al.* 2007)) with a diameter of 2 deg and a flux of $\sim$ 70% Crab. Other prominent sources are the SNR RX J1713.7-3946 and two very extended and bright PWNe, HESS J1825-137 and HESS J1303-631, for which an energy-dependent morphology could be established, with the higher-energy ($\sim$ 10 TeV) emission concentrated in a small area around the pulsar and the lower-energy ($\sim$ 1 TeV) emission extending much further out. The physical explanation is that the acceleration process and thus the injection of electrons occurs close to the pulsar and then the particles cool via adiabatic or radiation losses as they move away. About one third of the Galactic H.E.S.S. sources are unidentified. In some cases no good multiwavelength counterpart is known, and sometimes several viable counterparts exist. For example the point source HESS J1745-290 coincides with the position of the supermassive black hole Sgr A* (offset $8 \pm 9_{stat} \pm 9_{sys}$ arcsec) in the centre of the Milky Way (Acero *et al.* 2010), but could also be explained as emission from the PWN candidate G359.95-0.04. The H.E.S.S. source catalog at `http://www.mpi-hd.mpg.de/hfm/HESS/pages/home/sources/` and TeVCat `http://tevcat.uchicago.edu/` are excellent sources of information and contain further references for each source. The H.E.S.S. survey data is currently not publicly available, although there is a plan to publish GPS maps in FITS format and a corresponding source catalog in the future.

3. Multiwavelength context: The Fermi LAT survey

For the identification and characterization of the physical properties of Galactic H.E.S.S. sources it is crucial to combine the H.E.S.S. measurement in the rather narrow energy range (typically 0.3 to 30 TeV) with observations at lower energies. For example it is possible to infer or derive limits on the magnetic field strength in cosmic electron sources from the ratio of observed X-ray synchrotron flux and TeV gamma-ray inverse Compton flux. At optical and infrared wavelengths the thermal starlight and dust emission typically dominates by several orders of magnitude over the nonthermal emission, which is the reason why radio and X-ray observations are often used to search for counterparts and to understand the physical conditions in TeV sources.

In 2008 the Fermi LAT pair production telescope started to perform an all-sky survey, the data is publicly available. Figure 3 shows the count maps obtained in four decades of

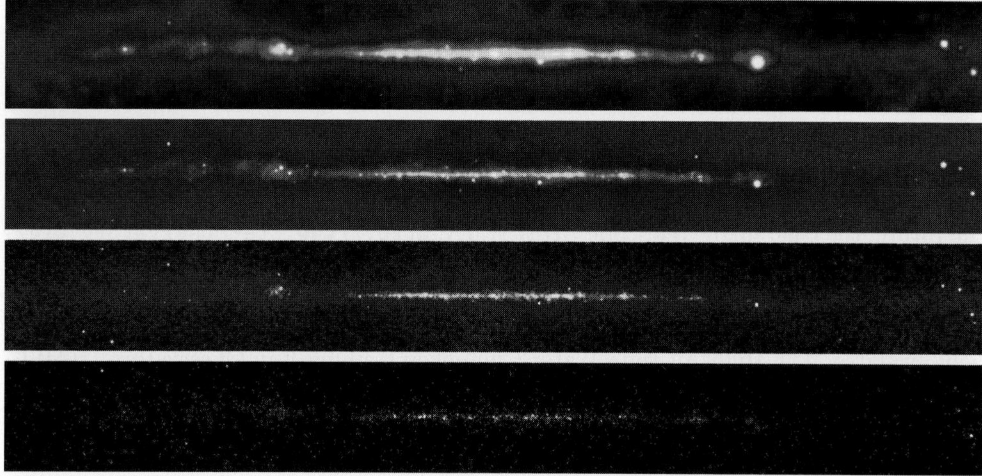

Figure 3. Fermi LAT count maps of the Galactic plane (CAR projection, GLON = -180 to +180 deg, GLAT = -10 to 10 deg, 0.1 deg binning, smoothed with 0.5 deg Gaussian for better visibility of the single events in the high-energy band). Top to bottom 0.1 to 1 GeV (20M events), 1 to 10 GeV (3.5M events), 10 to 100 GeV (90k events), 100 to 1000 GeV (3.5k events). 31 months exposure, P6_V3_DATACLEAN cuts.

energy (0.1 – 1– 10 – 100 – 1000 GeV) using 31 months of data. The diffuse emission from the Milky Way as well as individual sources are clearly visible. At low energies the Fermi measurements in the Galactic plane are limited by the low angular resolution ($\sim$ 10 deg at 100 MeV, improving by almost 2 orders of magnitude at the highest energies), due to a high level of diffuse background emission and high source density. At high energies the sensitivity is limited by the accumulated exposure, above 100 GeV a cluster of only a handful of photons represents a highly significant source. The average exposure at high energies for 31 months of data in the inner Galaxy is $\sim 8 \cdot 10^{10}$ cm^2 s. Even for one of the brightest sources in the sky, the Crab nebula, only $\sim$ 50 photons will be detected above 100 GeV given its integral flux of $F(> 100\text{GeV}) \sim 6 \cdot 10^{-10}$ cm^{-2} s^{-1}(Meyer *et al.* 2010).

The Fermi collaboration has published the 2FGL catalog (The Fermi-LAT Collaboration 2011), where they have modelled the total emission as the sum of the isotropic and Galactic diffuse background components plus 1861 point and 12 extended sources. This works extremely well for the extragalactic sky. For the Galactic plane they find $\sim$200 sources, although one has to note that given the high source density and the fact that many sources are actually extended, there are cases where big and bright sources are represented as multiple entries in the catalog (e.g. three 2FGL sources for Vela Jnr). The latitude distribution is very narrow, even though Fermi has a better sensitivity for sources at higher latitudes, because the Galactic diffuse emission there is weaker. It seems that even though the H.E.S.S. GPS region only covers 3% of the sky, most Galactic sources above the H.E.S.S. sensitivity limit have already been found.

4. Outlook

Currently the H.E.S.S. collaboration is constructing a fifth, much larger H.E.S.S. II telescope (600 m^2 reflector) in the centre of the array of the four H.E.S.S. I telescopes (100 m^2 reflector each). H.E.S.S. II will allow measurements in the gap in the measured SEDs around 100 GeV that is currently present for most gamma-ray sources seen by Fermi and H.E.S.S. as well as provide better sensitivity over the full energy range (Becherini *et al.* 2010). It is worth noting that due to its smaller FOV compared to H.E.S.S. I (3.5 instead of 5 deg), a survey of large parts of the Milky Way is not possible. Instead H.E.S.S. II (and H.E.S.S. I - the full array will most likely not be split up to observe different targets at the same time) will focus on exciting astrophysical questions of individual sources. For example we hope that with H.E.S.S. II we will be able to observe emission from gamma-ray bursts and the pulsed emission component from pulsars. Somewhat further in the future, the Cherenkov Telescope Array (CTA, `http://www.cta-observatory.org/`) will give an order of magnitude more sensitive (milli-Crab), higher-resolution view of the Galaxy in the energy range 10 GeV to 100 TeV. It will be operated as an observatory where everyone can apply for observation time and after some time all data becomes publicly available. The High-Altitude Water Cherenkov Gamma-Ray Observatory (HAWC, `http://hawc.umd.edu/`) is currently being constructed in Mexico and will survey the whole southern hemisphere in the energy range 100 GeV to 1000 TeV. CTA and HAWC will be complementary instruments. HAWC has a large field of view and high duty cycle, whereas CTA has a low energy threshold, high sensitivity and good angular resolution. At least for the next five years, though, the H.E.S.S. Galactic plane survey in combination with the Fermi all-sky and other lower-energy surveys will be the best dataset to study the populations of Galactic cosmic accelerators.

References

Abdo, A. A. and Fermi LAT Collaboration 2011, *ArXiv e-prints*, 1103.5727
Acero, F. *et al.* 2010, *MNRAS*, 402, 1877
Aharonian, F. *et al.* 2007, *ApJ*, 661, 236
Becherini, Y. *et al.* 2010, *American Institute of Physics Conference Series*, 1085, 738
Berge, D., Funk, S., & Hinton, J. 2007, *A&A*, 466, 1219
Brun, F. *et al.* 2011, *ArXiv e-prints*, 1104.5003
Daugherty, J. K. & Harding, A. K. 1982, *ApJ*, 252, 337
The Fermi-LAT Collaboration 2011, *ArXiv e-prints*, 1108.1435
Gaensler, B. M. & Slane, P. O. 2006, *ARA&A*, 44, 17
Gast, H. *et al.* 2011, *Proc. 32nd ICRC, Beijing*, in press
Horns, D. *et al.* 2007, *Ap&SS*, 309, 189
Meyer, M., Horns, D., & Zechlin, H. 2010, *MNRAS*, 523, A2
Morlino, G. & Caprioli, D. 2011, *ArXiv e-prints*, 1105.6342
Ohm, S., van Eldik, C., & Egberts, K. 2009, *Astroparticle Physics*, 31, 383

Discussion

ROMERO: You mention that the γ-ray emission from RX J1713-39 is likely to be leptonic. This is in contradiction with one of the claims of the previous speakers. Can you justify your claim please?

DEIL: The claim that the emission is likely to be leptonic was made by the Fermi Collaboration based on their measured spectrum (combined with the HESS spectrum). I'm not sure which previous speaker or leptonic model you are referring to.

PORTER: I'd like to comment on the question whether pair creation is important in SNRs: The turnover at an energy of a few tens of TeV in leptonic spectra are due to Inverse Compton in the Klein-Nishina regime on the infrared component of the interstellar radiation field. Pair creation is not important below 50 TeV on the galactic interstellar radiation field.

The Spectral Energy Distribution of Galaxies
Proceedings IAU Symposium No. 284, 2011
R.J. Tuffs & C.C. Popescu, eds.

doi:10.1017/S1743921312009441

The Galactic Centre - a laboratory for starburst galaxies (?)

Roland M. Crocker

Max-Planck-Institut für Kernphsik, P.O. Box 103980
Heidelberg, Germany

email: Roland.Crocker@mpi-hd.mpg.de

Abstract. The Galactic centre – as the closest galactic nucleus – holds both intrinsic interest and possibly represents a useful analogue to starburst nuclei which we can observe with orders of magnitude finer detail than these external systems. The environmental conditions in the GC – here taken to mean the inner 200 pc in diameter of the Milky Way – are extreme with respect to those typically encountered in the Galactic disk. The energy densities of the various GC ISM components are typically ∼two orders of magnitude larger than those found locally and the star-formation rate density ∼three orders of magnitude larger. Unusually within the Galaxy, the Galactic centre exhibits hard-spectrum, diffuse TeV ($=10^{12}$ eV) gamma-ray emission spatially coincident with the region's molecular gas. Recently the nuclei of local starburst galaxies NGC 253 and M82 have also been detected in gamma-rays of such energies. We have embarked on an extended campaign of modelling the broadband (radio continuum to TeV gamma-ray), non-thermal signals received from the inner 200 pc of the Galaxy. On the basis of this modelling we find that star-formation and associated supernova activity is the ultimate driver of the region's non-thermal activity. This activity drives a large-scale wind of hot plasma and cosmic rays out of the GC. The wind advects the locally-accelerated cosmic rays quickly, before they can lose much energy in situ or penetrate into the densest molecular gas cores where star-formation occurs. The cosmic rays can, however, heat/ionize the lower density/warm H_2 phase enveloping the cores. On very large scales (∼10 kpc) the non-thermal signature of the escaping GC cosmic rays has probably been detected recently as the spectacular 'Fermi bubbles' and corresponding 'WMAP haze'.

Keywords. Galaxy: center, cosmic rays, stars: formation, gamma rays: theory

---

1. Introduction

In part, this talk constitutes a general advertisement to the wider astronomical community concerning the utility of γ-ray data: analysis of GeV and TeV data – especially combined with data at other wavelengths – sometimes allows one to constrain the processes occurring or conditions prevailing in astrophysical regions with much greater precision than when recourse is made only to data in 'traditional' wavebands. I will illustrate this general philosophy through an analysis of the inner ∼200 pc in diameter of our own Galaxy (hereafter the 'Galactic centre' or GC). To be sure, this region is important in its own right: as, by definition, the closest galactic nucleus our 'up-close' view of the GC is of great interest for what it tells us – by analogy – about the activity of galactic nuclei in a cosmological context. But an understanding of the GC – responsible for ∼10% of the Galaxy's *massive* star-formation (Figer *et al.* 2004) and, of course, the host of its supermassive black hole – is a prerequisite for understanding the overall ecology of the Galaxy as I hope to make clear below.

The ISM conditions in the GC are certainly extreme in comparison with those encountered in the Galactic disk: the energy densities/pressures of the various ISM phases are two orders of magnitude higher than in the Galaxy at large, typically 100's of eV per cc. The magnetic field amplitude, for instance, through the region is at least 50 μG (Crocker *et al.* 2010a). Equally, the plasma pressures and temperatures, light field, and gas turbulence are all much higher than locally (and all roughly in equipartition with each other). All this is driven by star-formation which peaks sharply over the same inner $\sim$200 pc region of the Galaxy to a star-formation-rate area density *three* orders of magnitude larger than the average value in the disk (e.g., Crocker *et al.* 2011b).

Extending this general picture that the conditions in the GC may more closely resemble those encountered in a starburst environment, my recent work with collaborators (Crocker *et al.* 2011a, Crocker & Aharonian 2011, Crocker *et al.* 2011b) has revealed a number of interesting facts about the GC: i) the star-formation (and resultant supernova activity) in the GC drives a 'super'-wind out of the region (more correctly, an outflow rather than a wind as the ejected material does not reach escape velocity); ii) the GC outflow advects co-mingled plasma and non-thermal particle populations to large distance from the plane and – I claim – the non-thermal γ-ray and microwave signatures of this process have recently been found; and iii) despite the broad similarity to a starburst alluded to above, the GC is basically in steady state: the current star-formation rate and consequent activity is typical for the time-averaged state of the GC over timescales approaching 10 Gyr.

2. Multi-Wavelength Indications of a GC Outflow

Unusually in the Galaxy, the GC is a source of extended, *diffuse* TeV emission as revealed by the HESS telescope (Aharonian *et al.* 2006). This emission, extending over $\sim$1.5° in Galactic longitude, is spatially correlated with the column of molecular gas, showing peaks corresponding to the positions of the densest parts of the giant molecular complexes – Sgr. B, C, etc – inhabiting the region. Such a correlation is expected in the case that the observed γ-rays originate from *hadronic* interactions, i.e., the collisions between non-thermal protons (and heavier ions in general) and ambient gas, exactly the collision processes typically investigated at the LHC.

On even wider scales than for the TeV emission ($\sim$ 6° in Galactic longitude: LaRosa *et al.* 2005, Crocker *et al.* 2010a), radio continuum observations show that the GC is a distinct source of diffuse, $\sim$GHz, non-thermal emission. Such emission must be due to the synchrotron losses experienced by a wide-spread population of cosmic ray electrons inhabiting the GC.

Despite the fact of this wide-spread non-thermal emission, the GC is actually significantly *under*luminous in both radio continuum and $\sim$TeV (and $\sim$GeV) γ-ray wavebands – given the amount of star-formation currently going on there – as we now explain. Firstly, placing the GC on a plot of its 60 μm vs. 1.4 GHz luminosity, one determines that radio continuum emission from this system falls one order of magnitude (i.e., $\sim 4\sigma$) short with respect to the expectation afforded by the FIR-radio continuum correlation (e.g., Condon 1992).

Why might this be? Although many details about the FIR-RC correlation remain unclear – including why the correlation is, in general, so tight over so many orders of magnitude in luminosity (see contribution by T. Thompson to this volume) – it must be that the correlation is ultimately related to the phenomenon of *massive* star-formation in galaxies. Massive stars and their products dominate both the thermal and non-thermal radiative output from systems. On the one hand, they emit most of the optical and UV photons that are – to a greater or lesser degree – reprocessed by ambient dust into infrared wavelengths; FIR can, therefore, generally be treated as a direct tracer of the current star-formation rate in the system. On the other, the lives of massive stars end in supernova explosions whose shock waves, expanding into the local ISM, are the sites where non-thermal particle populations (i.e., cosmic rays) are accelerated. The subsequent synchrotron emission emitted by cosmic ray electrons as they gyrate around the local magnetic field dominates, in general, the ∼GHz radio continuum output of star-forming galaxies. Given, then, that both sources of radiation are tied back to the rate at which massive stars are being formed, in the case that a system is calorimetric Völk (1989) to the electron population it accelerates (i.e., such electrons lose their energy radiatively in situ rather than carrying it away), there should be a correlation between FIR and RC emission as observed.

Equally, given that cosmic ray hadrons are accelerated by the same supernova shocks, were galaxies or systems to be calorimetric to these accelerated cosmic ray ion populations, then one would expect a linear scaling (Thompson *et al.* 2007) between their (non-thermal) γ-ray and their FIR emission. Comparison between the GC system and expectation from this theoretical scaling places the GC's TeV emission at only ∼1% of expectation (with the system's GeV emission as detected by the *Fermi* satellite by Chernyakova *et al.* 2010 at about 10% of expectation, but substantially polluted by point sources in the field)

What explains this phenomenology – why is the GC so under-luminous in non-thermal emission? Generically, three explanations present themselves:

(*a*) The system suffered a significant a starburst event more recently than the 10^7 year lifetimes of the massive stars which dominate its radiative output. Of course, there is some stochasticity in the GC's star-formation rate: the GC has apparently undergone individual bursts of star-formation that have led to the creation of its super-stellar clusters (GC, Quintuplet, Arches) on these sort of timescales. The evidence, however, from stellar luminosity function studies is that the rate of star-formation inferred from the creation timescale of these clusters is not atypical of (in fact somewhat less than) the GC's long-term averaged star-formation rate (Figer *et al.* 2004). In fact, the evidence that the GC's star-formation rate has been stable for a timescale approaching 10 Gyr (Figer *et al.* 2004, Maness *et al.* 2007) seems increasingly firm. So this explanation does not fly.

(*b*) A second possibility is that conditions in the GC render its supernova remnant population inefficient cosmic ray accelerators. *Prima facie*, this is not unreasonable: the very large *volumetric-average* gas density of the region would be expected to imply strong ionisation cooling of low-energy cosmic rays, perhaps preventing their acceleration to higher energies. Our detailed modelling of the region (Crocker *et al.* 2011b), however, shows that this explanation also cannot work: GC SNRs are *at least* typically 'efficient' as cosmic ray accelerators, losing $\gtrsim$10% of their total 10^{51} erg mechanical energy into freshly-accelerated non-thermal particles.

(*c*) Finally, it may be that the GC is not a calorimeter to either cosmic ray electrons or protons.

Given the exclusion of the first two possibilities, it seems that the third explanation given above must hold. But here an immediate challenge is encountered: the diffuse, non-thermal radiation described above is very hard, consistent with $dN/dE \propto E^{-2.3}$ type population of both emitting electrons and protons. This is quite different to the situation generally encountered in the Galactic disk (and directly measured at Earth), where the cosmic ray spectrum is $\sim$E$^{-2.7}$, significantly steeper than the expectation for the distribution of cosmic rays injected into the ISM following diffusive (first order) Fermi acceleration at astrophysical shocks. So some process acts to steepen the Galaxy's steady state cosmic ray population and – as has been known for decades – a natural explanation of this steepening is that it is due to the energy-dependence of the confining effect of the

Galaxy's magnetic field. Cosmic rays scatter on magnetic field inhomogeneities, effectively diffusing through the Galaxy's volume but with a diffusion coefficient that (given the growth of a particle's gyroradius with momentum) grows with energy: high-energy particles are kept for a shorter time than low energy ones. This cannot be happening in the GC: the hard spectrum of the in situ particle population is completely consistent with the expectation for the injection distribution. Thus, if some process is acting to transport particles away – as apparently required on the basis of the evidence described above – this process must act without prejudice as to particle energy.

This requirement is naturally met by a large-scale outflow or wind of a few hundred km/s. In fact, there is ample observational evidence for such an outflow (reviewed in Crocker *et al.* 2011b) and observations of the star-forming external galaxies would tell us to expect one (e.g., Strickland & Heckman 2009). Indeed, scaling the results of Martin (2005) according to the GC's estimated SFR areal density, we find that the expectation afforded by observations of external galaxies is that the GC should drive an outflow with a speed of $\sim$400 km/s.

Particularly compelling evidence comes from the radio continuum observations of Law (2010) who traces emission along two radio continuum spurs north of Sgr A and Sgr C, showing that these spurs form part of a larger, coherent structure (the GC Lobe) but, more relevantly, that the emission from these structures exhibits a non-thermal spectrum that steepens with Galactic latitude. This is a clear signal of the synchrotron ageing of an electron population as it carried away from its injection site in the plane.

Another interesting piece of evidence is the (apparent) existence of a 'very-hot', diffuse plasma through the region as traced by X-ray emission (Koyama *et al.* 1989, Muno *et al.* 2004). This plasma – with a temperature of $\sim$8 keV – would be substantially hotter than anything generally encountered in Galactic plane SNRs. Whether it exists or is, in fact, an illusion created by the effect of many of faint, point-like sources (Revnivtsev *et al.* 2009), remains controversial, however. But it is certainly true that such hot, diffuse plasmas are detected in the star-forming nuclei of star-forming galaxies and are, in fact, required to provide (a large fraction of) the mid-plane pressure required to launch the outflows from these systems (e.g., Strickland & Heckman 2009).

So it seems that there is a star-formation driven outflow out the GC. With the understanding that this wind exists, at the back-of-the-envelope level we can employ the contrast between the observed and (calorimetrically) predicted luminosities of the GC in RC and γ-ray wavebands to tell us something about the system's environmental parameters. The basic considerations here are that the relevant radiative energy loss timescale must be slow enough with respect to the outflow transport time that we only detect (radiatively) a small fraction of the power injected into both cosmic ray protons and electrons (Crocker *et al.* 2011a). This means that if one of the parameters governing radiative losses – the magnetic field for synchrotron losses of the electrons, the gas density for hadronic losses of the protons – is 'dialed-up', there must be commensurate rise in the outflow speed.

But the outflow speed cannot rise without bound: the kinetic and thermal power of the outflow can do no more than saturate that total mechanical power delivered by the region's supernovae (and, sub-dominantly, its stellar winds). On the other hand – and speaking more loosely – we have an expectation from observation of external starburst nuclei that the thermalization efficiency of the outflow is not likely to be much below $\sim$10% (i.e., at least 10% of the mechanical power delivered by supernovae ends up heating or moving the ISM: e.g., Strickland & Heckman 2009).

These considerations imply that the mean magnetic field and gas density the non-thermal particles encounter as they escape from the GC system fall in the rough ranges

100-300 μG and 5-20 cm^{-3}, respectively. This first determination confirms that the ISM magnetic field is very strong as previously indicated. The second is very interesting as it implies the cosmic rays 'see' a gas density less than the volumetric average in the region which is ~ 100 cm^{-3}, dominated by the very high density ($\gtrsim 10^4$ cm^{-3}) but small filling factor cores of the region's giant molecular cloud complexes. Thus cosmic rays seems to be somehow excluded from the densest parts of the molecular clouds. In fact, this is likely not so mysterious: the existence of the general outflow means that the particles only remain in the system for a limited time, a time presumably too short for them to either diffuse or be convected into the hearts of the highly turbulent molecular gas distributions. This would make an interesting contrast between the GC and the situation recently claimed (Papadopoulos (2010)) for starburst systems where cosmic rays have been alleged to have a crucial role in changing the chemistry in the densest molecular core regions, thus crucially altering the conditions for star formation.

3. Connection to the Fermi Bubbles

Given that most cosmic rays leave the system before they lose much energy, we can infer that the GC launches about $\sim 10^{39}$ erg/s into non-thermal particles into the Galactic bulge on an outflow. Independent considerations suggest that it has been sustaining this activity for at least a few Gyr. What is the implication of this?

The 'Fermi Bubbles' constitute one of the most interesting recent discoveries (Su *et al.* (2010)) in high-energy astronomy: these are enormous structures, discovered in ~GeV γ-ray data from the *Fermi*-LAT, that extend ~10 kpc north and south from the Galactic plane above the Galactic centre. They are characterised by an unusually-hard spectrum, $dF_\gamma/dE_\gamma \propto E_\gamma^{-2.1}$ and have a total luminosity 4×10^{37} erg/s. Most researchers have considered the general idea that the γ-ray emission from these structures arises from the inverse-Compton (IC) emission from a (mysterious) population of cosmic ray electrons. Given that the spectrum of the Bubbles displays no obvious variation with Galactic latitude, however, it is necessary that the photon background being up-scattered by this putative electron population is the CMB. This, in turn, implies that the electrons have an energy scale ~TeV and consequently short IC loss times, $\sim 10^6$ years. So, given the vast extension of the Bubbles, these electrons either have to be delivered very quickly – presumably on an AGN-outflow originating at Sgr A* (Guo & Mathews 2011) – or accelerated in-situ by first- (Cheng *et al.* 2011) or second-order (Mertsch & Sarkar 2011) Fermi acceleration processes.

We have recently considered an alternative explanation that – because of the long loss times on the low-density plasma of the Bubbles – escapes the difficulties facing any leptonic mechanism, namely, that the Bubbles' γ-ray emission arises from the hadronic collisions of a population of cosmic ray *protons* (and heavier ions) populating their interiors (Crocker & Aharonian 2011). This explanation requires i) (given adiabatic and ionisation energy losses) a total cosmic ray hadron power $\sim 10^{39}$ erg/s that ii) (essentially because of the same long loss time referred to above) has been injected quasi-continuously into the Bubbles for a timescale of $\gtrsim 8$ Gyr. Note that exactly these requirements are matched by the GC CR outflow that we identified above *on the basis of completely independent considerations* to do with observations at radio continuum and TeV γ-ray wavebands of the inner ~ 200 pc of the Galaxy. This putative solution fits nicely from a number of other perspectives:

(*a*) The hard-spectrum of the emission is also explained: by construction, the cosmic rays injected into the Bubbles are trapped so there is no energy-dependent loss process acting to modify the in-situ, steady state distribution away from the injection spectrum and the daughter γ-rays will trace this hard, parent proton distribution.

(*b*) On the other hand, π^0-decay kinematics enforces a down-turn below ~GeV on a spectral energy distribution plot of the emitted γ-radiation; such a downturn in robustly detected, at least qualitatively, in the Bubbles' spectra (Su *et al.* 2010).

(*c*) The total enthalpy of the Bubbles can be calculated to be $\sim 10^{57}$ erg – this can be supplied by the GC outflow over the same long, multi-Gyr timescales required for the hadronic γ-ray scenario.

(*d*) The total plasma mass of the Bubbles is $\sim 10^8\ M_\odot$ (Su *et al.* (2010)) – this mass can also be explained given the rate of mass flux in the GC outflow and assuming the same long timescales.

(*e*) Dynamically, the Bubbles end up being slightly over-pressured but slightly under-dense with respect to the surrounding halo plasma, with internal energy density supplied approximately equally by cosmic rays and their interior hot plasma. They can be expected to rise slowly under buoyancy.

(*f*) The hadronic scenario naturally predicts concomitant secondary electron production within the Bubbles; these secondaries would synchrotron-radiate on the Bubble's magnetic field thereby explaining the coincident (at lower Galactic latitude) 'WMAP haze' detected (Finkbeiner 2004, Dobler & Finkbeiner 2008) at microwave frequencies.

Of course, all this requires that the Fermi Bubbles are very old structures – almost as old as the Galaxy – and that they can trap TeV cosmic rays for multi-Gyr timescales. In fact, our scenario implies that they would be calorimeters for GC activity over the history of the Milky Way. This is an interesting prospect indeed.

4. Conclusions

(*a*) Our modelling shows that the GC environment is an analogue to that found in starbursts to the extent that we can show that the energy-density in both thermal and non-thermal ISM components is ~2 orders of magnitude higher than typically encountered in the Galactic disk.

(*b*) Another similarity to starburst nuclei is the driving of a powerful outflow from the system; in the GC case, at least, this outflow does not reach escape velocity, however.

(*c*) On the other hand, star-formation in the region seems to have been sustained at more-or-less the current rate for many billions of years: most star-formation that has taken place in the system is not really 'bursty', though there are, of course, stochastic variations in the star-formation rate when one looks with sufficiently fine grain.

(*d*) We find that cosmic rays do not penetrate into the cores of the giant molecular clouds. This is presumably because their short residence time in the region does not permit this. A contrasting situation where cosmic rays crucially modify ISM chemistry in star-forming regions has recently been claimed for starbursts (Papadopoulos 2010).

(*e*) On the other hand, ionising collisions experienced by the cosmic rays may explaining the anomalously hot and high ionisation state inferred for the (low density) envelope molecular phase identified in H_3^+ absorption studies (Goto *et al.* 2008).

Finally we remark that all the high-energy activity identified above can be explained as a result, ultimately, of the power injected through the process of (massive) star-formation; the super-massive black hole is not *required* to have much influx beyond the inner few pc. A final speculation, in fact, is that a significant consequence of the sustained, star-forming activity of the GC is to *prevent* significant activity of the black hole.

5. Acknowledgements

I gratefully acknowledge the contribution of my collaborators – David Jones, Felix Aharonian, Casey Law, Fulvio Melia, Juergen Ott, and Tomo Oka – to the research presented here and thank the organisers for the invitation to speak at SED2011.

References

Aharonian, F. A., *et al.* 2006, *Nature*, 439, 695
Chernyakova, M., Malyshev, D., Aharonian, F. A., Crocker, R. M., & Jones, D. I. 2011, ApJ, 726, 60
Condon, J., 1992, AAP, 30, 575
Crocker, R. M., Jones, D. I., Melia, F., Ott, J., & Protheroe, R. J. 2010, Nature, 463, 65
Crocker, R. M., Jones, D. I., Aharonian, F., *et al.* 2011a, MNRAS., 411, L11
Crocker, R. M. & Aharonian, F. A. 2011, *Physical Review Letters*, 106, 101102
Crocker, R. M., Jones, D. I., Aharonian, F., Law, C. J., Melia, F., Oka, T., & Ott, J. 2011, MNRAS., 413, 763
Cheng, K.-S., Chernyshov, D. O., Dogiel, V. A., Ko, C.-M., & Ip, W.-H. 2011, ApJ, 731, L17
Dobler, G. & Finkbeiner, D. P. 2008, ApJ, 680, 1222
Figer, D., *et al.* 2004, ApJ, 601, 319
Finkbeiner, D. P. 2004, arXiv:astro-ph/0409027
Goto, M., *et al.* 2008, ApJ, 688, 306
Guo, F. & Mathews, W. G. 2011, arXiv:1103.0055
Koyama, K., Awaki, H., Kunieda, H., Takano, S., & Tawara, Y. 1989, Nature, 339, 603
LaRosa, T. N., Brogan, C. L., Shore, S. N., Lazio, T. J., Kassim, N. E., & Nord, M. E. 2005, ApJ, 626, L23
Law, C. J. 2010, ApJ, 708, 474
Maness, H., *et al.* 2007, ApJ, 669, 1024
Martin, C. L. 2005, ApJ, 621, 227
Mertsch, P. & Sarkar, S. 2011, *Physical Review Letters*, 107, 091101
Muno, M. P. *et al.* 2004, ApJ, 613, 326
Papadopoulos, P. P. 2010, ApJ, 720, 226
Revnivtsev, M., Sazonov, S., Churazov, E., Forman, W., Vikhlinin, A., & Sunyaev, R. 2009, Nature, 458, 1142
Strickland, D. K. & Heckman, T. M. 2009, ApJ, 697, 2030
Su, M., Slatyer, T. R., & Finkbeiner, D. P. 2010, ApJ, 724, 1044
Thompson, T. A., Quataert, E., & Waxman, E. 2007, ApJ, 654, 219
Völk, H. J. 1989, AAP, 218, 67

Discussion

GALLAGHER: You began the talk by asking if the galactic Center is a starburst analogue. Have you perhaps shown that this is not the case? For example, a starburst nucleus such as NGC253 lies on the FIR-radio relationship, has 10^3 x the local ISM energy density, but not such a strong B-field etc. Is it possible that the SMBH is in fact playing some role in the Galactic case?

CROCKER: Yes - I think the GC is not a complete analogue to real starbursts. Though the energy density in the ISM components is certainly much higher than is typical for the disk, I have argued that it is undergoing steady-state star formation. Moreover it does not seem to be an electron or proton calorimeter. My own opinion is that the SMBH does not have an important role on distances of a few hundreds of pc.

BUREAU: In relation to the lack of γ-rays compared to that expected from the extension of the FIR-RC correlation, the problem may not be with the cosmic rays and γ-rays, but rather with the FIR and the whole chain of star formation-related events which presumably give rise to the FIR-radio-continuum-γ-ray relation. Just like the CO in the early-type galaxies I talked about, the Galactic Centre is located in the rapidly varying part of the potential well where the rotation curve is rapidly rising. This will yield significantly different shear and epicyclic frequencies than the outer parts of disks where the relations are normally studied, which is bound to affect the star formation processes and thus the related observables. So the dynamics rather than the physical conditions may be the key difference here.

In relation to Jay Gallaghers' comment, NGC253 is essentially a bulgeless disk galaxy, contrary to the Galaxy (and early-type galaxies), which may explain why it follows the usual correlations.

CROCKER: OK. However, our detailed modelling of the region's non-thermal emission independently established the power going into the acceleration of these particles (around 10^{39}erg/s). This power is consistent with the expectation were one to estimate a supernova rate from the FIR emission and assume 10% of supernova total energy (10^{51}erg) goes into cosmic rays.

ROMERO: What is the timescale of adiabatic cooling for protons in your model? Do adiabatic losses play any role at all in you model?

CROCKER: The proton outflow launched from the Galactic centre carries a nonthermal power of 1039 erg/s. Around half of this power is lost into adiabatic losses. Note that in our model the adiabatic loss time is long because the Bubbles have to be rising very slowly over timescales ¿few Gyr. (This is comparable to the pp loss time by construction).

The Spectral Energy Distribution of Galaxies
Proceedings IAU Symposium No. 284, 2011
R.J. Tuffs & C.C. Popescu, eds.

doi:10.1017/S1743921312009453

The star formation rate in the Milky Way: Results from stars and planetary nebulae

Walter J. Maciel[1], Helio J. Rocha-Pinto[2], and Roberto D. D. Costa[1]

[1]Departmento de Astronomia, Universidade de São Paulo
Rua do Matão 1226, 05508-090 São Paulo SP, Brazil
email: maciel@astro.iag.usp.br, roberto@astro.iag.usp.br

[2]Observatório do Valongo, Universidade Federal do Rio de Janeiro
Ladeira Pedro Antonio 43, 20080-090 Rio de Janeiro RJ, Brazil
email: helio@ov.ufrj.br

Abstract. The star formation rate (SFR) is still a poorly known characteristic of the Milky Way, especially concerning the possibility of an irregular SFR, compared to a constant rate. Some recent results based on the distribution of dwarf stars with chromospheric ages suggest at least two major bursts in the past 10 Gyr, while other investigations are consistent with an approximately constant SFR. The SFR also shows important spatial variations, particularly concerning the radial variations along the galactic disc. In this work, we investigate two different problems relative to the galactic SFR: (i) We estimate the star formation rate in the galactic disc based on the age distribution of the planetary nebula central stars (CSPN), and compare these results with previous investigations based on dwarf stars. The CSPN ages were derived on the basis of five different methods, involving the observed nebular metallicities and kinematical properties; (ii) We derive radial abundance gradients from several elements in planetary nebulae, and compare these results with recent determinations based on younger objects, such as HII regions and cepheid variables. Since the gradients are linked to the formation process of the galactic disc, we can estimate the spatial variation of the SFR. Preliminary results indicate that at least one major star formation burst is obtained, as well as a relatively smooth variation of the SFR along the galactocentric radius.

Keywords. Galaxy: formation, ISM: planetary nebuale

1. Introduction

Radial abundance gradients have long been considered as a key ingredient of the chemical evolution of spiral galaxies, especially when their magnitude, space and time variations are taken into account. These gradients are caused essentially by differences in the star formation rate in different regions of the galactic disc, so that they can give valuable information on the star formation process. We presently have detailed information on the gradients not only for the Milky Way, but also for several galaxies in the Local Group. Such information comes basically from young objects (HII regions, OB stars and associations, and cepheids), as well as from evolved objects (open clusters and planetary nebulae, PN). In this work we estimate the star formation rate (SFR) on the basis of (i) some recent determinations of the abundance gradients based on young as well as aged objects, and (ii) the determination of individual ages for a large sample of central stars of planetary nebulae (CSPN).

2. The CSPN age distribution

We have made an effort to derive the age distrbution of the CSPN, in view of the fact that this information is essential in the interpretation of the abundance gradients and their time variations. Several methods have been developed based both on age-metallicity relations and on their kinematic properties (Maciel *et al.* 2010, 2011). Maciel *et al.* (2010) suggested that most objets studied have ages under 6 Gyr, and the average distribution peaks somewhere between 1 and 3 Gyr, depending on the assumptions made for the relation between the stellar masses and the corresponding ages. More recently, Maciel *et al.* (2011), based on kinematic ages of the CSPN, confirmed the previous results, but suggested generally lower ages for the CSPN.

3. Average gradients in the galactic disc

Gradients from young objects are observed in cepheids, HII regions, OB stars and associations. There are presently some controversies concerning the behviour of the gradients along the galactocentric radius, but an average gradient of -0.03 to -0.06 dex/kpc is obtained for the Milky Way, with similar values for some of the galaxies in the Local Group. Concerning evolved objects, the most recent results apply to open clusters and planetary nebulae, and the gradients are typically in the range -0.03 to -0.10 dex/kpc for oxygen and sulphur (Maciel *et al.* 2003, 2005, Henry *et al.* 2010, Andreuzzi *et al.* 2011).

A comparison of the oxygen abundances with Fe abundances, which are not well determined in PN, indicates a tight correlation with a slope in the range 1.0 to 1.2 (Maciel 2002), so that O/H and S/H gradients from PN can be converted into [Fe/H] gradients. The PN gradients are probably slightly larger than the gradients derived from the younger objects, but the difference seems now marginal compared to our previous investigations (Maciel *et al.* 2005), which may explain the essentially similar gradients measured in PN and HII regions in some Local Group galaxies (Stanghellini *et al.* 2010, Magrini *et al.* 2009) or in cepheids (Luck *et al.* 2011). The similarity of the gradients can be confirmed, for example, from the compilation by Spitoni & Matteucci (2011). Taking into account the uncertainties in the determinations, it can be concluded that the abundance gradients have had a small change in the last 3-4 Gyr, slower than usually assumed. Earlier epochs are more difficult to assess, but results from older PN and open clusters (ages greater than 5 Gyr) generally suggest steeper gradients.

4. Effects on the star formation rate

(i) Recent investigations based on chromospheric ages of dwarf stars in the Milky Way disc suggest that the star formaton rate may have suffered time variations, especially in the last 5 or 6 Gyr, so that our Galaxy would have some characteristics resembling starburst galaxies. From the age distribution of the CSPN one can obtain some evidence of galactic bursts, assuming that the birth rate of the planetary nebulae is representative of the SFR in these systems. A first examination of the CSPN age distribution would suggest an enhanced rate a few Gyr ago, depending on the adopted hypotheses on the mass-age relationship. Figure 1a shows a comparison of the derived uncorrected SFR from Rocha-Pinto *et al.* (2000a,b) and the results from the present investigation. The dotted line refers to the original results by Rocha-Pinto *et al.* (2000b), while the solid and dashed lines show the SFR based on the PN age determinations from Maciel *et al.* (2011) and Maciel *et al.* (2010), respectively. It can be concluded that the age distribution of the CSPN supports the idea that at least one era of enhanced star formation activity

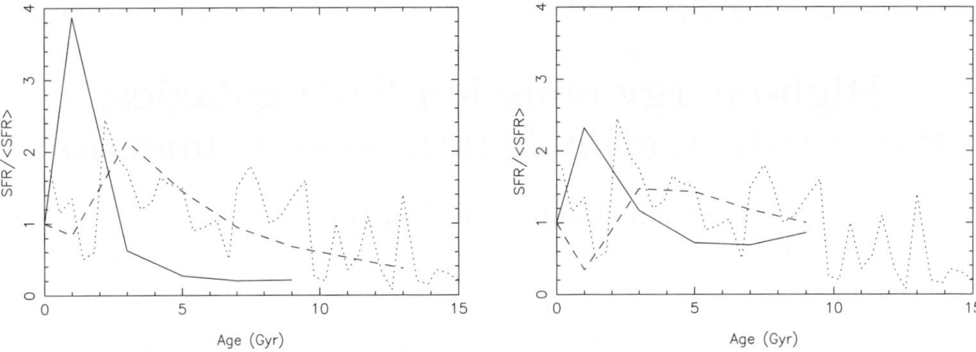

Figure 1. Time variation of the star formation rate. (a) uncorrected, (b) corrected. Dotted lines are the SFR by Rocha-Pinto *et al.* (2000), solid lines reflect results from Maciel *et al.* (2011), and the dashed lines are from the earlier work by Maciel *et al.* (2010).

occurred in the Milky Way in the last few Gyr. Figure 1b shows the SFR after applying a correction factor obtained using the initial mass function for the dying stars at each age interval. The peaks are somewhat decreased, but an important enhancement is still observed, especially for the kinematic ages.

(ii) Considering the similarity of the abundance gradients for young and evolved objects, it can be concluded that the average gradients have not changed much during the last 3 to 4 Gyr, approximately, depending on the adopted ages of the CSPN. Therefore, the SFR has probably not changed much as a function of the galactocentric radius during this period. Adopting an average gradient of -0.05 dex/kpc, and assuming that the abundance gradient is directly proportional to the SFR, the difference in the SFR along the galactic disc is about a factor of 4 in the region where the gradients are best measured, between the galactocentric distances R = 4 kpc and R = 14 kpc, approximately. This implies a smooth variation of the SFR along the galactocentric radius, in very good agreement with the radial distribution of the SFR as summarized for example by Portinari & Chiosi (1999).

Acknowledgements

This work was partially supported by FAPESP and CNPq.

References

Andreuzzi, G., Bragaglia, A., Tosi, M., & Marconi, G. 2011, *MNRAS* 412, 1265
Henry, R. B. C., Kwitter, K. B., Jaskot, A. E., *et al.* 2010, *ApJ* 724, 748
Luck, R. E., Andrievsky, S. M., Kovtyukh, V. V., *et al.* 2011, *AJ* 142, 51
Maciel, W. J. 2002, *Rev. Mex. A&A SC* 12, 207
Maciel, W. J., Costa, R. D. D., & Idiart, T. E. P. 2010, *A&A* 512, A19
Maciel, W. J., Costa, R. D. D., & Uchida, M. M. M. 2003, *A&A* 397, 667
Maciel, W. J., Lago, L. J., & Costa, R. D. D. 2005, *A&A* 433, 127
Maciel, W. J., Rodrigues, T. S., & Costa, R. D. D. 2011, *Rev. Mex. A&A*, 47, 401
Magrini, L., Stanghellini, L., & Villaver, E. 2009, *ApJ* 696, 729
Portinari, L. & Chiosi, C. 1999, *A&A* 350, 827
Rocha-Pinto, H. J., Scalo, J., Maciel, W. J., & Flynn, C. 2000a, *ApJ* 531, L115
Rocha-Pinto, H. J., Scalo, J., Maciel, W. J., & Flynn, C. 2000b, *A&A* 358, 869
Spitoni, E. & Matteucci, F. 2011, *A&A* 531, A72
Stanghellini, L., Magrini, L., Villaver, E., & Galli, D. 2010, *A&A* 521, A3

The Spectral Energy Distribution of Galaxies
Proceedings IAU Symposium No. 284, 2011
R.J. Tuffs & C.C. Popescu, eds.

doi:10.1017/S1743921312009465

High-energy emission from galaxies: the star-formation/gamma-ray connection

Stefan Ohm[1,2]† and Jim Hinton[1]

[1]X-ray and Observational Astronomy Group, Department of Physics and Astronomy, University of Leicester, LE1 7RH, UK
email: jim.hinton@le.ac.uk

[2]School of Physics and Astronomy, University of Leeds, LS2 9JP, UK
email: stefan.ohm@le.ac.uk

Abstract. The impact of non-thermal processes on the spectral energy distributions of galaxies can be dramatic, but such processes are often neglected in considerations of their structure and evolution. Particle acceleration associated with high mass star formation and AGN activity not only leads to very broad band (radio-γ-ray) emission, but may also produce very significant feedback effects on galaxies and their environment. The recent detections of starburst galaxies at GeV and TeV energies suggest that γ-ray instruments have now reached the critical level of sensitivity to probe the connection between particle acceleration and star-formation in galaxies. In this paper we will try to summarise this recent progress, put it into a multi-wavelength context and also discuss the prospects for more precise and sensitive γ-ray measurements with the upcoming CTA observatory.

Keywords. acceleration of particles, radiation mechanisms: non-thermal, (stars:) supernovae: general, (ISM:) cosmic rays, Galaxy: center, galaxies: starburst, gamma rays: observations, gamma rays: theory

---

1. Introduction

The most dramatic examples of high-energy emission from galaxies are found in Active Galactic Nuclei (AGN), where particle acceleration occurs in jets powered by supermassive black holes. In such systems non-thermal emission is often seen across the whole electromagnetic spectrum (in systems where the jets are aligned with the line-of-sight to the observer – commonly known as blazars). In the more common case that jets are not aligned to the observer, non-thermal dominance occurs only in the radio, X-ray and γ-ray domains. The high-energy emission of these objects has been discussed by many authors (see e.g. Wagner 2008 for a compilation of VHE γ-ray blazars), here we focus on the less dramatic, but potentially very importent non-thermal emission of "inactive" galaxies, where the spectral energy distribution (SED) is dominated by stellar processes.

In the Milky Way (MW), we know that the population of ultra-relativistic particles, known as cosmic rays (CRs), plays an important dynamical role; being close to pressure equilibrium with thermal gas and magnetic fields in the Interstellar Medium (ISM). Our knowledge of this population has in the past come largely from direct (local) measurements and from radio-synchrotron emission from the energetically less important relativistic electrons, rather than the nuclei that make up 99% of the CRs.

The process of star formation (SF), particularly of massive stars, is now known to lead to astrophysical particle acceleration, and hence γ-ray emission, via a number of different

† SO acknowledges the support of the Humboldt foundation by a Feodor-Lynen research fellowship.

objects and phenomena. The well-established accelerators are associated with the end-products of the massive-stellar lifecycle: supernova remnants (SNR) and pulsars and their associated pulsar wind nebulae (PWNe). Over the last two decades, measurements in the X-ray and γ-ray domains have shown that SNRs are efficient particle accelerators (see e.g. Vink 2011), with increasing evidence for the acceleration of nuclei as well as electrons (e.g. Aharonian *et al.* 2008, Abdo *et al.* 2010b). Less secure are the apparent association of γ-ray emission with colliding stellar winds in binary systems (Farnier *et al.* 2011) and the collective effect of winds in clusters (see e.g. Abramowski *et al.* 2012).

Independent of the dominant contributor to the CR population, the expectation from our own Galaxy is that all massive star-forming regions should be associated with particle acceleration and hence γ-ray emission, at some level. Indeed, recent progress in the field of high-energy (HE) and very-high-energy (VHE) γ-ray astronomy, driven by the satellite-borne instrument *Fermi*-LAT and ground-based VHE telescopes such as H.E.S.S. and VERITAS, has led to the identification of a growing population of high-energy-emitting star-forming galaxies (including the nearby starburst galaxies M 82 and NGC 253). We focus on these new developments below, discussing in particular the case of starburst galaxies, and finally discuss the prospects for the next-generation γ-ray mission CTA.

2. Particle energy losses and γ-ray production

Ultra-relativistic electrons and nuclei suffer energy losses by a number of different mechanisms, which lead to the production of photons from the radio to VHE γ-ray regime. Here we consider the primary energy-loss mechanisms in turn, giving example loss-timescales appropriate for the environment in starburst regions. CR nuclei predominantly lose energy in strong nuclear interactions with ambient matter, producing π^0-decay γ-ray emission primarily above a few hundred MeV. For a typical starburst region the (fairly energy-independent) interaction timescale is $t_{pp} \approx 10^5\,(n/250\,\mathrm{cm}^{-3})^{-1}$ years, where n is the average ambient density in hydrogen atoms per cm^3, with an average energy loss of ~50% per collision. For electrons a number of different energy-loss mechanisms play a significant role. For the highest energy electrons synchrotron and inverse Compton (IC) emission dominate with energy loss timescales ($E/dE/dt$) of $t_{\mathrm{sync}} \approx 200\,(E/\mathrm{TeV})^{-1}$ years and $t_{\mathrm{IC}} \approx 300\,(E/\mathrm{TeV})^{-1}$ years, for a starburst-like environment with an average magnetic field strength of 250μG, and a radiation field energy density of $1000\,\mathrm{eV\,cm}^{-3}$. At electron energies below a few GeV Bremsstrahlung and (at the lowest energies) Coulomb losses become increasingly important, impacting on the shape of the equilibrium (i.e. cooled) electron spectrum and hence the spectrum of the non-thermal emission (see Figure 1 and Section 4 for more details). In such an environment a 1 TeV electron will produce IC and synchrotron photons with typical energies of $\approx$150 GeV (for black-body target photons with $T = 50\,$K) and $\approx$20 keV, respectively.

These timescales are so short that it is often very difficult for high energy electrons to escape from star-forming regions. Nuclei, in contrast may be removed by diffusion or bulk motion before significant energy losses occur. Starburst galaxies often drive winds which carry away particles from the nuclear region into the inter-galactic medium. This advection of particles is determined by the wind speed and its scale height, and can be characterised by an average residence time of the particles in the nucleus – which is around 2×10^5 years for the $\sim 500\,\mathrm{km\,s}^{-1}$ starburst wind of NGC 253. More details on energy loss and γ-ray production mechanisms in general can be found in Aharonian (2004), Longair (2011), Hinton & Hofmann (2009). For more details on the environment and high-energy emission of starbursts see for example Paglione *et al.* (1996), Lacki *et al.* (2011) and references therein.

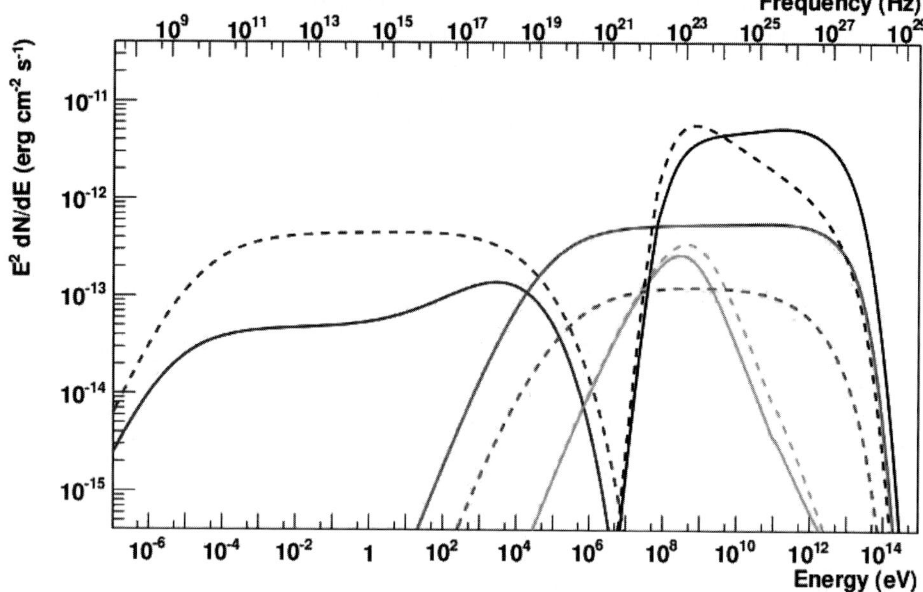

Figure 1. Simple model SED for a representative starburst region at a distance of 3.5 Mpc. Contributions of IC (10^2-10^{14}eV), synchrotron (10^{-7}-10^6eV), Bremsstrahlung (10^4-10^{12}eV) and π^0-decay (10^7-10^{14}eV) are shown for a single-zone, time-dependent model for the continuous injection of electrons and protons over 2×10^5 yr. For the solid lines (model 1), a magnetic field strength of $B = 100\,\mu$G, a radiation field energy density of $U_{\rm rad} = 2500$ eV of a black-body with temperature 50 K, and an average density $n_{\rm H}$ of 250 particles per cm^3 are assumed. The energy input is 3×10^{40} erg s^{-1} for electrons and an order of magnitude higher for protons. An injection spectrum index of $\alpha = 2.0$ and maximum accelerated particle energy of $E_{\max} = 100$ TeV are used for this model. Dashed lines (model 2) illustrate the effect of reducing $U_{\rm rad}$ by a factor of 10 and doubling the magnetic field strength. A proton spectral index of 2.3 is used in this second case. A colour version of this figure is available online (Ohm & Hinton 2012).

3. Non-thermal emission from Starburst galaxies

The presence of high-energy electrons in the archetypal starburst galaxies NGC 253 and M 82 was identified through low-frequency radio-synchrotron emission measurements in the late 1970's (e.g. Shimmins & Wall 1973) and early 1980's (e.g. Laing & Peacock 1980), respectively. VHE γ-ray emission from these two objects has only recently been reported by Acero *et al.* (2009) and Acciari *et al.* (2009), respectively. Both galaxies have also been detected in the HE domain using data obtained in observations with the LAT instrument onboard the *Fermi* satellite (Abdo *et al.* 2010a). Both starbursts are right at the sensitivity limit of *Fermi*-LAT and ground-based VHE γ-ray instruments – their detection required very long exposure times and they comprise the weakest γ-ray source class detected so far. Within large errors, the HE and VHE γ-ray spectrum of M 82 can be described by a single power-law in energy with photon index $\Gamma \approx 2.3$. Similarly, the extrapolated HE γ-ray spectrum of NGC 253 ($\Gamma \approx 2.3$) is consistent with the integrated VHE γ-ray flux reported by H.E.S.S. In the following, we will outline the importance of different energy-loss processes of nuclei and electrons in a typical starburst environment as discussed in Section 2 and illustrate the effect on the SED of changing key parameters of the system with a very simple model.

Fig. 1 shows a illustrative model SED for the particle injection and cooling in an environment similar to that found in the starburst nuclei of NGC 253 and M 82. Electrons and protons are injected over the assumed particle residence time of 2×10^5 years. As this timescale is longer than the cooling timescales for electrons in the system, this SED is representative of the system at equilibrium (i.e. the present day SED – given the typical starburst lifetime of $\sim 10^7$ years), provided that the escape probability is energy-independent. The IC, synchrotron and bremsstrahlung emission for electrons is shown for "typical" and "extreme" values of the radiation field energy density and magnetic field strength. For model 1 a comparatively low B-field strength of $100\,\mu$G and rather high radiation field density of $2500\,\mathrm{eV\,cm^{-3}}$ are assumed. In this model a high-energy upturn in the synchrotron spectrum can be seen – a consequence of IC-dominated cooling in the Klein-Nishina regime. In model 2 the magnetic field strength is assumed to be $B = 200\,\mu$G and $U_{\mathrm{rad}} = 250\,\mathrm{eV\,cm^{-3}}$, and synchrotron losses always dominate over IC. The energy input of $3 \times 10^{41}\,(3 \times 10^{40})\,\mathrm{erg\,s^{-1}}$ for protons (electrons) represents an electron-to-proton ratio of 1/10 and has been derived assuming one supernova explosion every decade and a particle acceleration efficiency of 10%. As the electron-to-proton ratio at injection in our own galaxy is likely $\sim$1/100, this SED illustrates that π^0-decay γ-rays are very likely to dominate the SED at GeV energies. For electrons with energies much smaller than $\approx$1 GeV, Coulomb cooling starts to become significant and results in a hardening of the radio spectrum in the 100 MHz to 1 GHz frequency range.

These qualitative results should be compared to the more sophisticated modelling of non-thermal emission of starbursts in (for example) Paglione *et al.* (1996), Domingo-Santamaría & Torres (2005), Persic *et al.* (2008) or Lacki *et al.* (2011).

4. A comparison of NGC 253 and the Milky Way

Now that γ-ray emission from external star-forming galaxies has been established, we have the opportunity for the first time to compare the high-energy emission of the MW to other systems and test our ideas on relativistic particle production and transport. As a very nearby starburst with well-established radio, GeV and TeV emission, NGC 253 is well suited to such a comparison.

The synchrotron emission from starburst galaxies is in general correlated to their far-infrared (FIR) luminosity and hence star-formation rate (van Buren & Greenhouse 1994). The radio emission of NGC 253 is coincident with the $\sim$0.5 kpc central molecular zone (CMZ), which has a star-formation rate exceeding that of the entire MW. Fig. 2 provides a multi-frequency view of NGC 253 on the scale of the galaxy as a whole and close to the nucleus. The nucleus exhibits bright infra-red and radio emission, and an outflow is apparent in thermal X-rays. In a recent CO study Sakamoto *et al.* (2011) found striking morphological similarities between the CMZ of NGC 253 and that of the MW (see Fig. 2). The physical size of the CMZ as a whole, and that of the prominent CO peaks (20$-$50 pc), are remarkably similar in the two systems. Although the molecular mass estimate is afflicted by a considerable error in the CO-to-H_2 conversion ratio, the Giant Molecular Cloud (GMC) complexes in NGC 253 have significantly higher CO($2-1$) intensities, suggesting order of magnitude higher densities in NGC 253 and a factor 10 higher total mass.

Table 1 summarises SF and γ-ray emission related quantities for the NGC 253 starburst nucleus, for the MW CMZ and for the MW as a whole. The SF rate of the MW and the CMZ are inferred from number counts of young stellar objects and the application of a stellar evolution model. The supernova rate of the CMZ has been estimated by a number of different methods including the FIR-supernova rate relation, stellar composition models, and pulsar population studies (see Crocker *et al.* 2011 for a detailed discussion). Since the distance to NGC 253 is much larger than to the centre of our Galaxy, young

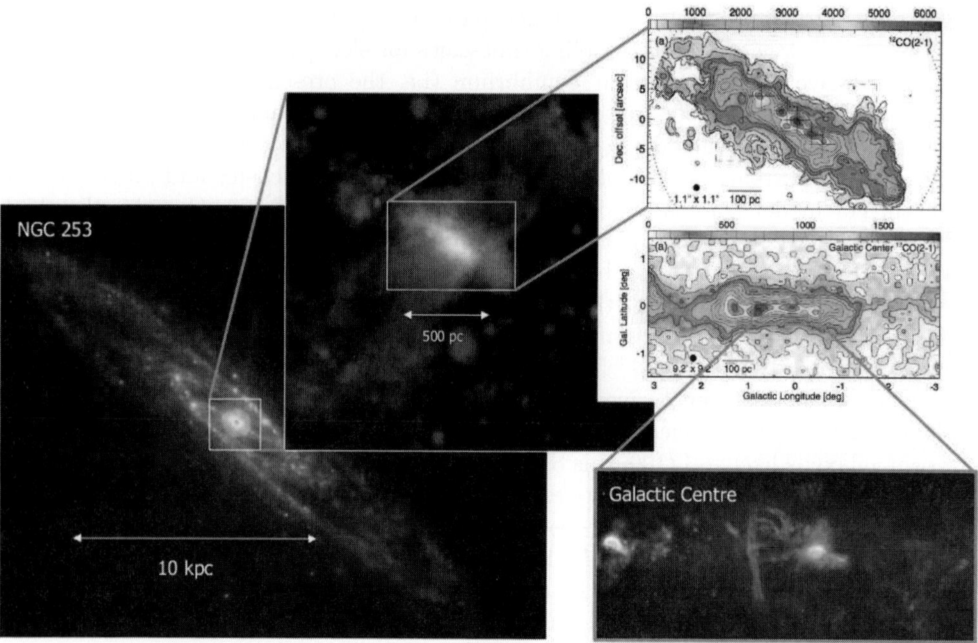

Figure 2. Composite Spitzer and GALEX image of NGC 253 on the bottom left (24μm, 8μm, far-UV) and zoom into its central region (Hα, 20cm, $0.2-1.5$ keV, reproduced from Heesen *et al.* (2011)). Also shown is the CMZ of our Galaxy as seen by Bolocam, the VLA and Spitzer in the bottom right (Credit: A. Ginsburg, John Bally, F. Yusef-Zadeh, Bolocam GPS team; GLIMPSE II team). The CMZ of NGC 253 and the Milky Way show striking similarities in ^{12}CO$(2-1)$ line emission (see text for more details). Image reproduced from Sakamoto *et al.* (2011). A colour version of this figure is available online (Ohm & Hinton 2012).

stellar objects are too faint to be detected by current instruments, hence the SFR has to be inferred by indirect approaches such as the scaling relation between SFR and IR flux (Kennicutt 1998). The supernova rate also has to be indirectly inferred, from e.g. iron-line measurements. The inferred SFR and supernova rate in the starburst nucleus of NGC 253 are orders of magnitude larger than in the CMZ.

Under the assumption that particle acceleration predominantly occurs in SNR shells, or is simply driven by massive star formation, and the particle acceleration efficiency ($\epsilon_{\rm CR}$) is fixed, the CR injection power should correlate with the supernova rate. For a fixed efficiency of γ-ray production (ϵ_γ) from accelerated particles, the γ-ray flux should therefore be proportional to the supernova rate. Table 1 gives the ratio of the γ-ray luminosity in the GeV and TeV bands to the supernova rate in the three systems considered. Comparing the entire Milky Way with the NGC 253 starburst, a factor of a few difference apparently exists in the product $\epsilon_{\rm CR}\epsilon_\gamma$. A much more dramatic difference exists between the apparently much more similar regions of the NGC 253 starburst and the CMZ, with $\epsilon_{\rm CR}\epsilon_\gamma$ smaller by 1-2 orders of magnitude with respect to NGC 253. This observation however, can be understood in the following picture: The convective escape time, i.e. the residence time of particles in both systems, is comparable if the CMZ in the MW drives a nuclear wind, similar in terms of wind speed and size, to the one observed in NGC 253 (see e.g. Crocker *et al.* 2011). The order of magnitude higher density in the NGC 253 starburst region on the other hand, results in an order of magnitude higher fraction of energy lost by an average proton in p-p interactions. Hence, the efficiency of converting

Table 1. Comparison of NGC 253 and both the CMZ of the Milky Way and our galaxy as a whole. GeV and TeV luminosities are quoted for 0.1-100 GeV and 0.1-100 TeV respectively.

	CMZ NGC 253	CMZ Milky Way	Milky Way	References
size (FWHM, kpc^2)	$\sim 0.5 \times 0.1$	$\sim 0.5 \times 0.1$	$\sim 25 \times 0.1$	(1,2)
$L_{\rm IR}\,(L_\odot)$	$\sim 3 \times 10^{10}$	$\sim 4 \times 10^8$	$(1-3) \times 10^{10}$	(1,3,4)
$M_{\rm gas}\,(M_\odot)$	$\sim 5 \times 10^8$	$\sim 5 \times 10^7$	$\sim (5-10) \times 10^9$	(1,5)
SFR ($M_\odot\,{\rm yr}^{-1}$)	3.5*	0.08	$0.7 - 1.5$	(6,7,8)
$\nu_{\rm SN}$ (yr^{-1})	0.03-0.08	$\sim (2-8) \times 10^{-4}$	0.02 ± 0.01	(9,10,11,12)
$L_{\rm GeV}\,(L_\odot)$	$\sim 3 \times 10^6$	$\sim 400\ (2100)^\dagger$	$\sim 2 \times 10^5$	(13,14)
$L_{\rm TeV}\,(L_\odot)$	$\sim 4 \times 10^5$	$\sim 100\ (160)^\ddagger$	$\sim 8 \times 10^3$	(14,15)
$L_{\rm GeV}/\nu_{\rm SN}$ ($L_\odot$ Myr)	40-100	0.5-10	6-20	
$L_{\rm TeV}/\nu_{\rm SN}$ ($L_\odot$ Myr)	5-15	0.1-1	0.3-0.8	

References:
(1) Sakamoto *et al.* (2011), (2) Combes (1991), (3) Sodroski *et al.* (1995), (4) Paladini *et al.* (2007), (5) Pierce-Price *et al.* (2000), (6) Melo *et al.* (2002), (7) Immer *et al.* (2011), (8) Robitaille & Whitney (2010), (9) van Buren & Greenhouse (1994), (10) Engelbracht *et al.* (1998), (11) Crocker *et al.* (2011), (12) Diehl *et al.* (2006), (13) Abdo *et al.* (2010a), (14) Strong *et al.* (2010) (15) Aharonian *et al.* (2006)
Notes:
* The star-formation rate has been inferred from the correlation with the *IR* luminosity.
† GeV luminosity of the CMZ as predicted by the *Fermi*-LAT galactic diffuse model (*gal_2yearp7v6_v0.fits* from `http://fermi.gsfc.nasa.gov/ssc/data/access/lat/BackgroundModels.html`). Numbers in brackets include point-like sources from the Fermi 2-year catalogue (Abdo *et al.* 2011) in the region-of-interest. These numbers represent upper limits since they include contributions of diffuse emission and unrelated point-like sources along the line-of-sight.
‡ TeV luminosity of the CMZ as measured by H.E.S.S. (Aharonian *et al.* 2006). Numbers in brackets include the H.E.S.S. GC point-like source HESSJ1745–290 (Aharonian *et al.* 2009) and G 0.9+0.1 (Aharonian *et al.* 2005).

CR energy into γ rays (ϵ_γ) is expected to be an order of magnitude increased relative to the CMZ in the MW.

5. Prospects for CTA

The upcoming Cherenkov Telescope Array project (Actis *et al.* 2011) represents a major step forward in both the sensitivity and precision of γ-ray astronomy. Compared to current instruments it will provide a broader energy coverage and overlap in energy with the *Fermi* satellite, a much better angular resolution, and a factor of 10 better sensitivity. This will allow us to measure the spectra of M 82 and NGC 253 in great detail and to search for spectral features such as the emergence and subsequent disappearance of an IC component. If the γ-ray emission from the known starbursts originates from the GMC complex seen e.g. in CO, it might be possible to detect a significant extension in the γ-ray emission, given the arc-minute-scale angular resolution of CTA. Finally, the greatly improved sensitivity might allow us to detect other nearby galaxies such as Andromeda (M 31) and/or to establish new source classes such as ultra-luminous infrared galaxies, helping us to gain a deeper insight into the relationship between star formation and particle acceleration.

References

Abdo, A. A., *et al.* (*Fermi*-LAT Collaboration) 2010, *ApJL*, 709, L152
Abdo, A. A., *et al.* (*Fermi*-LAT Collaboration) 2010, *ApJ*, 718, 348
Abdo, A. A., *et al.* (*Fermi*-LAT Collaboration) 2011, *arXiv*, 1108.1435
Abramowski, A., *et al.* (H.E.S.S. Collaboration) 2012, *A&A*, 537, A114
Acciari, V. A., *et al.* (VERITAS Collaboration) 2009, *Nature*, 462, 770
Acero, F., *et al.* (H.E.S.S. Collaboration) 2009, *Science*, 326, 1080

Actis, M., *et al.* 2011, *Experimental Astronomy*, 32, 193
Aharonian, F. A. & Very high energy cosmic gamma radiation: a crucial window on the extreme Universe, 2004, *World Scientific Publishing Company; 1st edition*
Aharonian, F., *et al.* (H.E.S.S. Collaboration) 2005, *A&A*, 432, L25
Aharonian, F., *et al.* (H.E.S.S. Collaboration) 2006, *Nature*, 439, 695
Aharonian, F., *et al.* (H.E.S.S. Collaboration) 2008, *A&A*, 481, 401
Aharonian, F., *et al.* (H.E.S.S. Collaboration) 2009, *A&A*, 503, 817
Combes, F. 1991, *ARA&A*, 29, 195
Crocker, R. M., *et al.* 2010, *MNRAS*, 413, 763
Diehl, R., *et al.* 2006, *Nature*, 439, 45
Domingo-Santamaría, E. & Torres, D. F., 2005, *A&A*, 444, 403
Engelbracht, C. W., *et al.* 1998, *ApJ*, 505, 639
Farnier, C., Walter, R., & Leyder, J.C., 2011 *A&A*, 526A, 57
Heesen, V., Beck, R., Krause, M., & Dettmar, R. J. 2011, *A&A*, 535, A79
Hinton, J. A. & Hofmann, W. 2009, *ARA&A*, 47, 523
Immer, K., Schuller, F., Omont, A., & Menten K. M. 2011, *A&A*, in press
Kennicutt, R. C., Jr. 1998, ARA&A, 36, 189
Melo, V. P., *et al.* 2002, *ApJ*, 574, 709
Lacki, B. C., *et al.* 2011, *ApJ*, 734, 107
Laing, R. A. & Peacock, J. A. 1980, *MNRAS*, 190, 903
Longair, M. S. 2011, High Energy Astrophysics *Cambridge University Press; 3rd edition*
Ohm, S. & Hinton, J. A. 2012, *arXiv*, 1202.0260
Paladini, R., *et al.* 2007, *A&A*, 465, 839
Paglione, T. A. D., Marscher, A. P., Jackson, J. M., & Bertsch, D. L. 1996, *ApJ*, 460, 295
Persic, M., Rephaeli, Y., & Arieli, Y. 2008, *A&A*, 486, 143
Pierce-Price, D., *et al.* 2000, *ApJ*, 545, L121
Robitaille, T. P. & Whitney, B. A. 2010, *ApJ*, 710, L11
Sakamoto, K., *et al.* 2011, *ApJ*, 735, 19
Shimmins, A. & Wall, J. 1973, *Australian J. Phys.*, 26, 93
Sodroski, T. J., *et al.* 1995, *ApJ*, 452, 262
Strong A. W., *et al.* 2010, *ApJL*, 722, L58
Van Buren, D. & Greenhouse, M. A. 1994, *ApJ*, 431, 640
Vink, J. 2011, *arXiv*, 1112.0576
Wagner, R. M. 2008, *MNRAS*, 385, 11935

The Spectral Energy Distribution of Galaxies
Proceedings IAU Symposium No. 284, 2011
R.J. Tuffs & C.C. Popescu, eds.

doi:10.1017/S1743921312009477

Dark gas: a new possible link between low and high-energy phenomena

Kazufumi Torii[1], Yasuo Fukui[1], Hidetoshi Sano[1], Junki Sato[1], Takeshi Okuda[1], Hiroaki Yamamoto[1], Akiko Kawamura[2], Norikazu Mizuno[2], Toshikazu Onishi[3] and Hideo Ogawa[3]

[1]Department of Physics and Astrophysics, Nagoya University, Furo-cho, Chikusa-ku, Nagoya, Aichi 464-8601, Japan

[2]National Astronomical Observatory of Japan, Mitaka, Tokyo 181-8588, Japan

[3]Department of Astrophysics, Graduate School of Science, Osaka Prefecture University, 1-1 Gakuen-cho, Naka-ku, Sakai, Osaka 599-8531, Japan

email: `torii@a.phys.nagoya-u.ac.jp`

Abstract. Galactic-scale studies of γ-rays and sub-mm radiation suggest that a significant amount of neutral interstellar medium is not detectable either in CO or HI (Grenier *et al.* 2005; Ade *et al.* 2011). This component is called "dark gas". Here we argue that cool and dense atomic gas without molecules is responsible for the dark gas. This interpretation is supported by a recent finding of cool HI gas corresponding to the TeV γ-ray shell in the SNR RX J1713.7-3946 (Fukui *et al.* 2011). Such HI gas is not recognized under a usual assumption of optically thin HI emission but is identified by a careful analysis considering optically thick HI. The typical column density of such HI gas is a few times 10^{21} cm^{-2} and is also identified as visual extinction.

Keywords. ISM: clouds; cosmic rays; supernova remnants, gamma rays: observations, dark gas

---

Interstellar medium (ISM) consists of gas and dust and the gas is either in molecular or atomic form if neutral. The molecular gas typically has temperatures of $\sim$10 K and densities of $\sim$100–1000 cm^{-3}, and is detected in the rotational transitions of CO and other molecules. On the other hand, the atomic gas has higher temperatures, $\sim$100 K, and lower densities, $\sim$1 cm^{-3}, and is detected in the 21 cm transition of HI. It is important to understand physical conditions of the neutral ISM quantitatively in order to elucidate the formation of stars and the interaction between cosmic rays (CR) and the ISM, since the ISM plays a major role in the evolution of the Galaxy.

It has been argued that there is yet an additional component in the ISM, so called "dark gas", which is not detectable either by CO or HI. The dark gas is suggested from γ-ray observations of EGRET as well as sub-mm observations by Planck (Grenier *et al.* 2005; Ade *et al.* 2011). CR protons interact with the ISM protons to produce neutral pions which decay into two γ-rays. The dark gas may provide a possible link between CR protons and the ISM. We here discuss the origin and implications of the dark gas in the light of a new study of the ISM protons in a TeV γ-ray SNR.

1. Dark gas

Grenier *et al.* (2005) compared the EGRET γ-ray data with CO and HI in order to test if the γ-rays show spatial correspondence with the neutral gas. The distribution of the γ-rays should correspond to the ISM proton distribution if the CR density distribution is uniform. Grenier *et al.* (2005) found that the γ-ray distribution shows generally a good correspondence with CO and HI whereas there is some residual γ-rays that has no

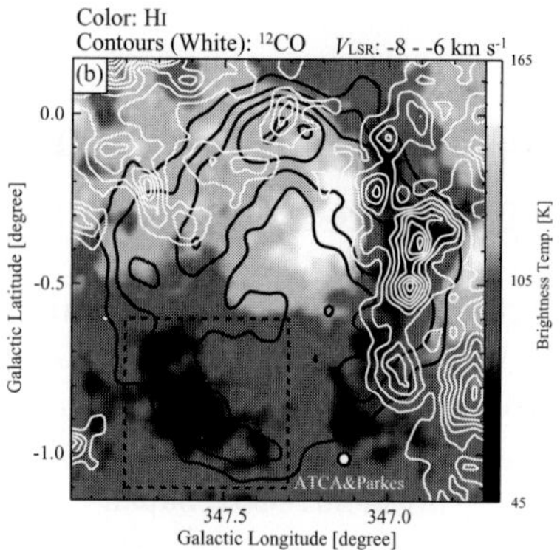

Figure 1. HI distribution towards RX J1713.7-3946 (grey scale). White and black contours show ^{12}CO J=1–0 and TeV γ-rays, respectively. Dotted box is the region shown in Figure 2(a).

correspondence either with CO or HI. The authors called this residual component "dark gas" and estimated that the dark gas has proton column densities $N_{\rm p}$ of $\sim 3{\times}10^{20}$–3×10^{21} cm^{-2} and masses of 20–100 % of those estimated from CO and HI.

More recently, observations with the Planck satellite showed that the mm and sub-mm dust emission also has a residual component not explicable either by CO or HI (Ade *et al.* 2011). The dark gas identified in the Planck data has mass ~30 % larger than those estimated from CO and HI in a column density range of $N_{\rm p}{\sim}1$–$5{\times}10^{21}$ cm^{-2}. These properties appear consistent with those of the dark gas derived from the EGRET data.

Some of the previous works discussed that lower density molecular gas without CO may correspond to the dark gas. Wolfire *et al.* (2010) studied a steady state chemistry model of a GMC and argued that GMCs may have molecular envelopes where CO is photo-dissociated. Such envelopes may explain the dark gas. It is however to be noted that the time scale of H_2 formation on the dust grain surface is fairly long as $t = 10^7$ year at a HI density of 100 cm^{-2} (e.g., Koyama & Inutsuka 2000) and the evolutionary time scale of a GMC, ~10 Myrs (e.g., Fukui and Kawamura 2010), may be only marginally long for H_2 formation from HI (c.f., Wolfire *et al.* 2010). The dark gas is so far identified only for the local clouds at galactic latitudes higher than 10° where only low-mass clouds are distributed (Grenier *et al.* 2005; Ade *et al.* 2011). Such molecular clouds have smaller evolutionary timescales less than 10^6 yrs, too short to form H_2 envelopes from HI, and it may be unlikely that the molecular gas without CO corresponds to the dark gas.

2. TeV γ-ray SNR RX J1713.7-3946

We shall here discuss cool and dense atomic gas as an alternative candidate of the dark gas. A recent study of cool HI gas seen as self-absorption in the TeV γ-ray SNR RX J1713.7-3946 provides a hint to the alternative (Fukui *et al.* 2011). We shall summarize the main points of the work below from Fukui *et al.* (2011).

RX J1713.7-3946 is the brightest TeV γ-ray supernova remnant (SNR) (Aharonian *et al.* 2006). CO clouds interacting with the SNR were discovered in the 2.6 mm CO emission (Fukui *et al.* 2003). Recently, Fukui *et al.* (2011) carried out a combined analysis of both the ^{12}CO J=1–0 and HI and showed a good spatial correspondence with the TeV γ-rays. Figure 1 indicates that the HI distribution shows two dark regions; one of them in the west coincides with CO whereas the other in the southeast, the dark HI SE cloud, shows no CO emission. The dark HI SE cloud corresponds well with the TeV γ-ray shell and the authors interpreted that the HI has a broad self-absorption as shown in

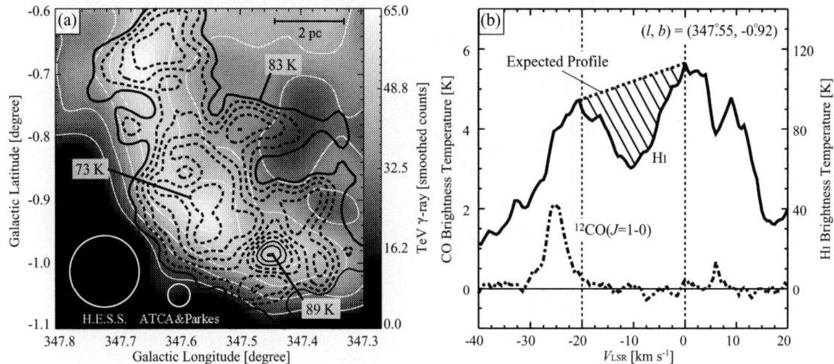

Figure 2. (a) The TeV γ-rays and HI distributions toward the SE cloud. (b) Example spectra of HI and ^{12}CO J=1–0 in the SE cloud. The shaded area shows an expected HI profile.

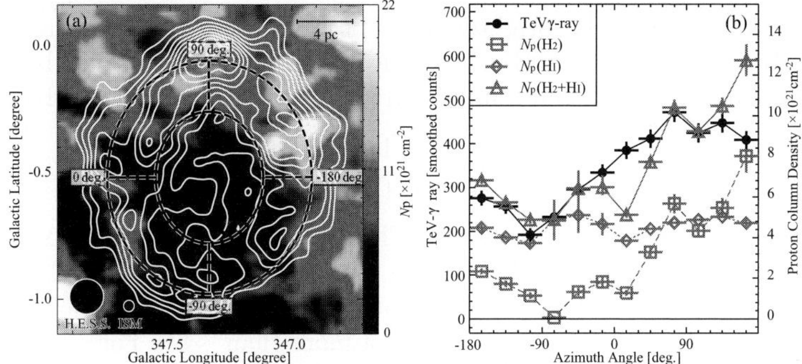

Figure 3. (a) Distributions of column density of the total ISM protons $N_{\rm p}(\mathrm{H_2{+}HI})$. Contours show the TeV γ-rays. (b) Azimuthal distributions of $N_{\rm p}(\mathrm{H_2})$, $N_{\rm p}(\mathrm{HI})$, $N_{\rm p}(\mathrm{H_2{+}HI})$, and TeV γ-ray-smoothed counts per beam in the ring shown in the figure (a).

Figure 2. The visual extinction is also enhanced along the shell in the SE cloud, lending another support for the self-absorption interpretation.

Figures 3(a) shows the distribution of the total ISM protons. We find it corresponds well to the γ-rays spatially. The azimuthal distribution of the γ-rays and the total ISM protons shown in Figure 3(b) also shows the correspondence; in particular, the dark HI SE cloud makes the correspondence significantly better than in the case of H_2 alone.

The observed dark gas properties are summarized as follows; spin temperature $T_{\rm s}$, column density $N_{\rm p}(\mathrm{HI})$ and optical depth τ are $\sim$40 K, $\sim 5\times10^{21}$ cm^{-2} and 0.7–1.4, respectively. These characteristics are consistent with those of Planck and EGRET. If we assume the optically thin HI in the dark HI SE cloud, we have an estimate of HI column density $\sim$30–50 % less than the self-absorption case. We therefore suggest that such cool and dense HI gas is ubiquitous and may be responsible for the dark gas.

3. Cool and dense HI gas as the dark gas candidate

We shall discuss general properties of the cool and dense HI gas as the candidate of the dark gas. The formation of such HI gas is naturally explained by compression due to stellar winds of high mass stars in a typical time scale of 1 Myr (e.g., Yamamoto *et al.* 2006). The equation 3.1 is used to estimate the column density of HI, $N_{\rm p}(\mathrm{HI})$.

$$N_{\rm p}(\mathrm{HI}) = \frac{32\pi\nu k_{\rm b}}{3c^3 h A_{\rm ul}} \int_{\nu_1}^{\nu_2} \Delta T_{\rm b} d\nu = 1.823 \times 10^{18} \times W(\mathrm{HI}) \ (\mathrm{cm}^{-2}), \qquad (3.1)$$

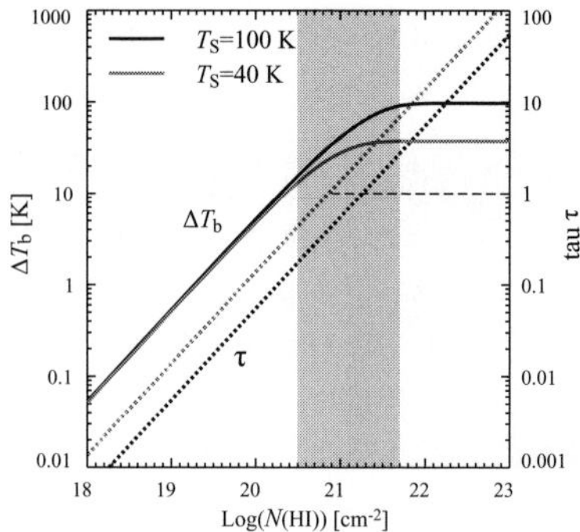

Figure 4. $\Delta T_{\rm b}$ and τ variations depending on $N_{\rm p}$(HI). Here Δv of 10 km s^{-1} is assumed. The filled area shows a $N_{\rm p}$ range of the dark gas (Grenier *et al.* 2005; Ade *et al.* 2011).

where $\Delta T_{\rm b}$ and W(HI) is the brightness temperature and integrated intensity of HI, respectively. Here optically thin HI is assumed. It usually works well, but we have to consider optical depth τ and spin temperature $T_{\rm s}$ in the cool and dense HI as follows,

$$\tau = \frac{1}{1.823 \times 10^{18}} \frac{1}{T_{\rm s}\,({\rm K})} \frac{N_{\rm p}({\rm HI})\,({\rm cm}^{-2})}{\Delta v\,({\rm km\,s}^{-1})}, \tag{3.2}$$

where Δv is the velocity width of HI profile. The brightness temperature $\Delta T_{\rm b}$ is described as $\Delta T_{\rm b}=(T_{\rm s}-T_{\rm bg})(1-\exp(-\tau))$ where $T_{\rm bg}$ is temperature of background radiation field.

Figure 4 shows variations of $\Delta T_{\rm b}$ and τ depending on $N_{\rm p}$(HI). $\Delta T_{\rm b}$ becomes saturated at $N_{\rm p}$(HI) of more than 10^{21} cm^{-2} and at τ of more than $\sim$1.0. For such a moderately optically thick condition, we underestimate the column density $N_{\rm p}$(HI) in the optically thin assumption (see Eq. 3.1). The filled area in the figure shows the estimated column densities $N_{\rm p}$ of the dark gas obtained with the Planck and EGRET results, which correspond well to the regime where $\Delta T_{\rm b}$ becomes saturated.

The spin temperature $T_{\rm s}$ of HI is $\sim$100 K or higher in warm neutral medium and decreases with density from 100 K to 10 K in a range of $N_{\rm p} \sim 10^{20}$–10^{21} cm^{-2} because of higher shielding of stellar photons and enhanced cooling with density increase. (e.g., Figure 2 in Goldsmith *et al.* 2007). Because the cooling timescale is much shorter than the H_2 formation timescale (e.g., Koyama & Inutsuka 2000), the cool and dense atomic gas of 100 cm^{-3} can stay atomic without forming molecules for $\sim 10^7$ years.

We summarize that the present work has shown cool and dense HI gas is a viable candidate for the dark gas which is not explicable either by CO or by the optically thin HI analysis. Further detailed properties of the cool and dense HI remains to be explored.

References

Ade, P. A. R. *et al.* 2011, arXiv: 1101.2029v1
Aharonian, F. A., *et al.* 2006, *A&A*, 449, 223
Goldsmith, P. F., Li, D., & Krčo, M. 2007, *ApJ*, 654, 273
Grenier, I. A., Casandjian, J.-M., & Terrier, R. 2005, *Science*, 307, 1292
Fukui, Y., *et al.* 2003, *PASJ*, 55, 61
Fukui, Y., *et al.* 2011, *ApJ*, in press
Fukui, Y. & Kawamura, A. 2010, *ARAA*, 48, 547
Koyama, H. & Inutsuka, S. 2000, *ApJ*, 532, 980
McClure-Griffiths, N. M. *et al.* 2005, *ApJS*, 158, 178
Sato, F. & Fukui, Y. 1978, *ApJ*, 83, 1607
Yamamoto *et al.* 2006, *ApJ*, 642, 307

The Spectral Energy Distribution of Galaxies
Proceedings IAU Symposium No. 284, 2011
R.J. Tuffs & C.C. Popescu, eds.

doi:10.1017/S1743921312009489

Cosmic rays and high energy emission from starburst galaxies

Brian C. Lacki[1] **and Todd A. Thompson**[2]

[1]Jansky Fellow & Institute for Advanced Study; Einstein Drive, Princeton, NJ 08540, USA
[2]Department of Astronomy and Center for Cosmology and AstroParticle Physics, Ohio State University; 140 W. 18th Ave, Columbus, OH 43210, USA

email: brianlacki@ias.edu

Abstract. The nearby starburst galaxies M82 and NGC 253 are now detected in GeV and TeV γ-rays, allowing us to directly study cosmic rays (CRs) in starburst galaxies. Combined with radio observations, the detections constrain the propagation and density of CRs in these starbursts. We discuss the implications for "proton calorimetry", whether CR protons cool through pion losses before escaping these galaxies. The ratio of γ-ray and radio luminosities constrains how much of the CR electron cooling is due to synchrotron losses. As for leptonic emission, we predict that synchrotron and Inverse Compton emission make up $\sim$1-10% of the unresolved hard X-ray emission from M82, and a few percent or less of the total X-ray emission from starbursts. A detection of these components would inform us of the magnetic field strength and 10 - 100 TeV electron spectrum. We conclude by discussing the prospects for detecting leptonic MeV γ-rays from starbursts and the cosmic γ-ray background.

Keywords. galaxies: starburst, gamma rays: theory, gamma rays: observations, X-rays: galaxies

---

As the conference theme says, galaxies are detected over more than 18 orders of magnitude in energy. This is not just true for the Milky Way and active galactic nuclei anymore, but now for starburst galaxies as well: M82 is detected from 38 MHz (0.2 μeV; Kellermann, Pauliny-Toth, & Williams 1969) to 4 TeV (970 YHz; Acciari *et al.* 2009). Thermal emission lies in the middle of the spectrum (millimeter to X-rays), but the rest is nonthermal. Nonthermal emission comes mainly from cosmic rays (CRs), high energy particles accelerated by supernova remnant shocks or other star formation processes. Protons (and other nuclei) make up most of the energy density in CRs above a GeV, but CR electrons and positrons ($e^{\pm}$) are also present. Starbursts have much star formation in small volumes, resulting in large CR energy densities.

CRs radiate in a variety of ways. These processes have different spectra, peaking at different energies. Radiation can be leptonic, from CR $e^{\pm}$, or hadronic, from CR protons. Synchrotron radiation is emitted by $e^{\pm}$ gyrating in magnetic fields and is a very broad continuum (with a characteristic frequency $\nu_{\rm synch} \propto BE_e^2$, where E_e is the electron energy and B is the magnetic field strength) spanning from radio to X-rays. Inverse Compton (IC) is emitted by $e^{\pm}$ upscattering individual photons; it too is a very broad continuum (characteristic IC photon energy $E_{\rm IC} \propto E_e^2 \varepsilon$, where ε is the energy of a typical low energy photon), stretching from X-rays to γ-rays. Bremsstrahlung is emitted by $e^{\pm}$ being deflected by the electric fields of ambient atoms and therefore correlates with gas density; it is a narrow continuum feature ($E_{\rm brems} \propto E_e$) that peaks at a few hundred MeV. Finally, CR protons can inelastically collide with ambient gas atoms to produce pions, uncharged (π^0) 1/3 of the time or charged (π^+ or π^-) 2/3 of the time. Neutral pions quickly decay into pionic γ-rays; because of the kinematics of the pion production process, the pionic γ-rays form a plateau ($E_{\rm pionic} \propto E_p$) reaching up to $\gtrsim$ 100 TeV, but dropping off quickly below ~ 70 MeV. If we had the full broadband spectra of starburst galaxies we would learn both about their CRs and the environments they travel in.

The GeV–TeV Band: CR Protons at Last. Although protons make up the bulk of the CR energy density in star-forming galaxies, only the recent detections of starbursts by Fermi (Abdo *et al.* 2010), HESS (Acero *et al.* 2009), and VERITAS (Acciari *et al.* 2009) gave us direct evidence for them. Previously, their presence was merely inferred from equiparition arguments on radio observations (e.g., Akyuz, Brouillet, & Ozel 1991). The γ-ray detections indicate $U_{\rm CR} \approx 100-300$ eV cm^{-3} within M82 and NGC 253's starbursts (Acero *et al.* 2009; Lacki *et al.* 2011).

Unlike the Milky Way, starbursts may be "proton calorimeters", in which most of the energy in CR protons goes into making pions, since they have high gas densities. In this limit, 1/3 of the injected CR proton power ($L_{\rm CR}$) is converted into pionic γ-rays; the rest goes into secondary $e^{\pm}$ and neutrinos. The γ-ray detections allow us to estimate the fraction $F_{\rm cal}$ of CR power that ends up in pionic products. We have some idea of the power being pumped into CRs in M82 and NGC 253: it should scale with the supernova (or massive star formation) rate. From their IR luminosities, we estimate supernova rates of 0.06 yr^{-1} in M82 and 0.02 yr^{-1} in NGC 253's core (though with large uncertainties – 0.1 to 0.3 yr^{-1} are sometimes quoted). Assuming that each supernova injects 10^{50} ergs of CR protons, we find that $F_{\rm cal} = 3L_{\gamma}/L_{\rm CR} \approx 40\%$ of the CR power in M82 and NGC 253 goes into pionic losses (Lacki *et al.* 2011). By contrast, in the Milky Way only a few percent of the CR proton power ends up in pions (e.g., Strong *et al.* 2010).

Pionic γ-rays necessarily come with pionic $e^{\pm}$: if these cooled only by synchrotron emission, the *Fermi* detections imply that M82 and NGC 253 should be very bright in GHz radio. Since half as much power in γ-rays goes into secondary $e^{\pm}$ as pionic γ-rays, and synchrotron radiation spreads each dex in $e^{\pm}$ energy into two dex of synchrotron frequency, we expect that $\nu L_{\nu}(\text{GeV}) f_{\rm synch} \beta = 4\nu L_{\nu}(\text{GHz})$, where $\beta \approx 1$ corrects for differences in observed $e^{\pm}$ energy and $f_{\rm synch}$ is the fraction of $e^{\pm}$ power going into synchrotron. In fact, M82 and NGC 253 are much less luminous in radio than in GeV γ-rays, implying that $f_{\rm synch} \approx 0.1$. This indicates the importance of non-synchrotron energy losses – IC, bremsstrahlung, ionization, or escape – for these $e^{\pm}$ (Lacki *et al.* 2011).

The keV Band: Crucial Information Buried by Foregrounds. Several decades ago, it was thought that starbursts would be bright sources of IC X-rays, because they have intense infrared radiation fields and a supply of CR $e^{\pm}$ evident from their radio emission (e.g., Hargrave 1974). Since $e^{\pm}$ of similar energies emit GHz radio synchrotron and keV IC, a ratio of the power in these two bands could tell us directly about the magnetic field strength in starbursts. However, a variety of arguments indicate starburst galaxies have strong magnetic fields (e.g., Völk 1989; Thompson *et al.* 2006; Robishaw, Quataert, & Heiles 2008; Persic *et al.* 2010; Crocker *et al.* 2010), meaning that the ratio of radio and IC emission should be large. Furthermore, the amount of GeV emission observed limits the IC emission since the IC continuum should stretch from X-rays to TeV energies.

But another source of nonthermal X-ray emission in starbursts is synchrotron emission from 10 - 100 TeV $e^{\pm}$. The strong magnetic fields in starbursts means that lower energy $e^{\pm}$ can contribute to synchrotron than in normal galaxies (c.f. Protheroe & Wolfendale 1980, Aharonian & Atoyan 2000). These $e^{\pm}$ can include primaries, pionic secondaries, or pair $e^{\pm}$ produced by far-infrared photons annihilating 10 - 100 TeV photons through the $\gamma + \gamma \rightarrow e^{+} + e^{-}$ process (e.g., Aharonian, Atoyan & Nagapetyan 1983; Torres 2004). The synchrotron and IC contributions have different spectra: IC emission should be hard ($\Gamma \approx 1 - 1.5$) whereas synchrotron emission should be soft ($\Gamma \approx 2$).

Using one-zone models to solve the leaky box equation for CR populations (c.f. Torres 2004; Lacki, Thompson, & Quataert 2010), we have calculated the nonthermal leptonic

X-ray emission from several nearby starbursts (Lacki & Thompson 2010b). After applying constraints from the observed radio and γ-ray emission, we find that only a few percent of the diffuse, unresolved X-ray emission from these starbursts is synchrotron or IC. If the gas density is low, the synchrotron X-ray contribution depends strongly on where the primary electron spectrum (assumed to be a single power law) cuts off, since there are few secondaries or pair $e^{\pm}$ at 10 – 100 TeV. In M82, synchrotron and IC account for only $\lesssim 10\%$ of the diffuse, unresolved X-ray emission, most of which probably comes from hot X-ray gas (Strickland & Heckman 2007). The diffuse X-ray emission in turn is only a minority of the total X-ray emission, which largely is expected to come from X-ray binaries (e.g., Persic *et al.* 2004). In Arp 220, synchrotron may be about a quarter of the observed 2 – 10 keV X-rays if it is not absorbed by hydrogen, but hydrogen absorption from the huge gas column densities likely reduces that fraction. In general we find that synchrotron and IC are no more than $\sim 2-3\%$ of the total 2 – 10 keV luminosity of starbursts.

The MeV Band: The Hole in our Rainbow. GeV and TeV observations tell us about protons but only indirectly connect with GHz-emitting $e^{\pm}$, while keV observations must contend with large foregrounds. In between those energies, in the MeV band, leptonic emission should be the primary source of luminosity (Fig. 1).

Detailed models of CR populations in M82 and NGC 253 predict IC and bremsstrahlung emission at MeV energies (e.g., de Cea del Pozo *et al.* 2009). Unfortunately, it is dimmer than the pionic emission, since there are more $\geqslant$ GeV protons than $e^{\pm}$. An even bigger problem is that current instruments' sensitivity to MeV γ-rays are far behind that of *Fermi*-LAT at 30 MeV and above. These sensitivities must improve by a factor of *a thousand* to detect M82 and NGC 253, the brightest starbursts in the sky.

Could we instead detect all the starbursts in the Universe in the form of the diffuse γ-ray background (c.f. Pavlidou & Fields 2002, Thompson, Quataert, & Waxman 2007)?

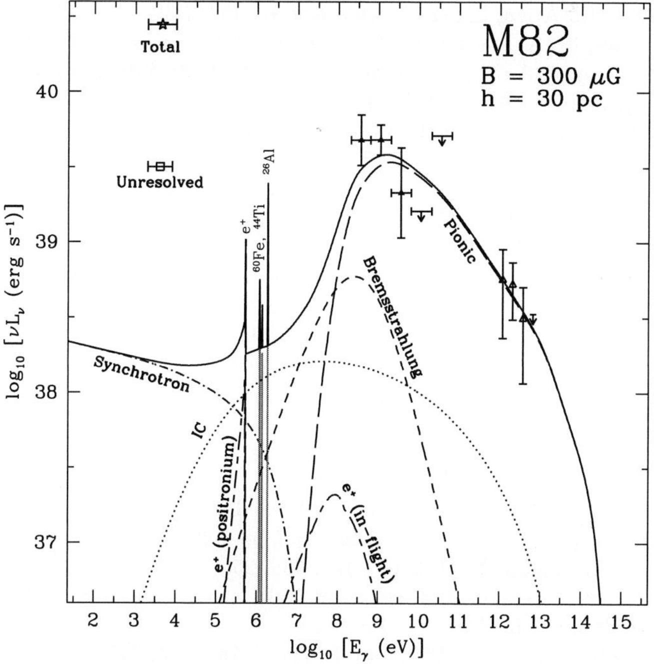

Figure 1. Predicted high energy spectrum of M82 (ignoring X-ray absorption). Data points are Cappi *et al.* 1999, Strickland & Heckman 2007, Abdo *et al.* 2010, and Acciari *et al.* 2009.

By using the radiation spectra from one-zone models of the Milky Way, M82, and NGC 253 as templates for star-formation throughout the Universe, we recently calculated the expected γ-ray background at MeV and GeV energies. We find that starbursts have much absolute power at GeV energies, but normal galaxies have higher MeV to GeV ratios. In our fiducial model, starbursts are more important above a few GeV, normal galaxies below that, together making up $\sim 1/3$ of the GeV γ-ray background and a few percent of the MeV background (Lacki *et al.* 2011, in prep.).

In high-z normal galaxies, IC losses off the CMB rapidly dominate the lifetimes of CR $e^{\pm}$ because $t_{\rm IC}^{\rm CMB} \propto U_{\rm CMB}^{-1} \propto (1+z)^{-4}$. Thus power that would normally go into other energy loss processes (including radio emission) or escape instead gets channeled into the MeV-scale IC emission. Since protons escape easily in normal galaxies, the leptonic contribution can fill in the "pion bump". In starburst galaxies, the IC losses from the CMB are insignificant compared to all the other strong losses in the extreme environments (the "buffering" of Lacki & Thompson 2010a). Furthermore, in the more proton calorimetric starbursts, the pionic emission is more dominant.

Acknowledgements

BCL is supported by a Jansky Fellowship from the National Radio Astronomy Observatory. NRAO is operated by Associated Universities, Inc., under cooperative agreement with the National Science Foundation. BCL also acknowledges discussions and collobration with John Beacom and Shunsaku Horiuchi on the forthcoming paper on the MeV background. TAT is supported in part by an Alfred P. Sloan Foundation Fellowship and by NSF grant AST-0908816.

References

Abdo, A. A., Ackermann, M., Ajello, M., *et al.* 2010, *ApJL*, 709, L152
Acciari, V. A., Aliu, E., Arlen, T., *et al.* 2009, *Nature*, 462, 770
Acero, F., Aharonian, F., Akhperjanian, A. G., *et al.* 2009, *Science*, 326, 1080
Aharonian, F. A., Atoyan, A. M., & Nagapetyan, A. M. 1983, *Astrophysics*, 19, 187
Aharonian, F. A. & Atoyan, A. M. 2000, *A&A*, 362, 937
Akyuz, A., Brouillet, N., & Ozel, M. E. 1991, *A&A*, 248, 419
Cappi, M., Persic, M., Bassani, L., *et al.* 1999, *A&A*, 350, 777
Crocker, R. M., Jones, D. I., Melia, F., Ott, J., & Protheroe, R. J. 2010, *Nature*, 463, 65
de Cea del Pozo, E., Torres, D. F., Rodriguez, A. Y., & Reimer, O. 2009, arXiv:0912.3497
Hargrave, P. J. 1974, *MNRAS*, 168, 491
Kellermann, K. I., Pauliny-Toth, I. I. K., & Williams, P. J. S. 1969, *ApJ*, 157, 1
Lacki, B. C., Thompson, T. A., & Quataert, E. 2010, *ApJ*, 717, 1
Lacki, B. C. & Thompson, T. A. 2010a, *ApJ*, 717, 196
Lacki, B. C. & Thompson, T. A. 2010b, arXiv:1010.3030
Lacki, B. C., Thompson, T. A., Quataert, E., Loeb, A., & Waxman, E. 2011, *ApJ*, 734, 107
Pavlidou, V. & Fields, B. D. 2002, *ApJL*, 575, L5
Persic, M., Rephaeli, Y., Braito, V., *et al.* 2004, *A&A*, 419, 849
Persic, M. & Rephaeli, Y. 2010, *MNRAS*, 403, 1569
Protheroe, R. J. & Wolfendale, A. W. 1980, *A&A*, 92, 175
Robishaw, T., Quataert, E., & Heiles, C. 2008, *ApJ*, 680, 981
Strickland, D. K. & Heckman, T. M. 2007, *ApJ*, 658, 258
Strong, A. W., Porter, T. A., Digel, S. W., *et al.* 2010, *ApJL*, 722, L58
Thompson, T. A., *et al.* 2006, *ApJ*, 645, 186
Thompson, T. A., Quataert, E., & Waxman, E. 2007, *ApJ*, 654, 219
Torres, D. F. 2004, *ApJ*, 617, 966
Völk, H. J. 1989, *A&A*, 218, 67

The Spectral Energy Distribution of Galaxies
Proceedings IAU Symposium No. 284, 2011
R.J. Tuffs & C.C. Popescu, eds.

doi:10.1017/S1743921312009490

Cosmic ray production and emission in M82

Tova Yoast-Hull[1], John Everett[1,2], J. S. Gallagher III[2] and Ellen Zweibel[1,2]

[1]Department of Physics, University of Wisconsin-Madison, Madison, WI 53706, USA
[2]Department of Astronomy, University of Wisconsin-Madison, Madison, WI 53706, USA

email: yoasthull@wisc.edu

Abstract. Starting from first principles, we construct a simple model for the evolution of energetic particles produced by supernovae in the starburst galaxy M82. The supernova rate, geometry, and properties of the interstellar medium are all well observed in this nearby galaxy. Assuming a uniform interstellar medium and constant cosmic-ray injection rate, we estimate the cosmic-ray proton and primary & secondary electron/positron populations. From these particle spectra, we predict the gamma ray flux and the radio synchrotron spectrum. The model is then compared to the observed radio and gamma-ray spectra of M82 as well as previous models by Torres (2004), Persic *et al.* (2008), and de Cea del Pozo *et al.* (2009). Through this project, we aim to build a better understanding of the calorimeter model, in which energetic particle fluxes reflect supernova rates, and a better understanding of the radio-FIR correlation in galaxies.

Keywords. galaxies: individual (M82), galaxies: starburst, cosmic rays, gamma rays: theory, radio continuum: galaxies

1. Introduction

M82 is a nearby starburst galaxy. The supernova rate, geometry, and density and extent of the interstellar medium are all well observed. The galaxy is modeled as a calorimeter with a supernova-powered cosmic ray model. The calorimeter model for galaxies predicts that all the energy input from supernovae is expended within the galaxy and that both the far-infrared and radio synchrotron emission from cosmic rays are proportional to the supernova rate (Völk 1989). Thus, by starting with the supernova rate, the cosmic ray particle production can be determined.

Our objective is to build a simple model of cosmic ray interactions that reproduces the gamma ray flux and synchrotron emission of M82 and is readily scalable to other systems. Here we present preliminary results from this investigation.

2. Procedure

Our single zone model is based on the following assumptions: uniform density (180 cm^{-3}), magnetic field (120 μG), radius (300 pc), supernova rate (0.3/yr); equilibrium for particle injection and energy losses (no diffusion); a power-law injection spectrum (p$\sim$2.1); and a particle acceleration efficiency from supernovae (10% of SN energy, $E_{51} = E_{SN}/10^{51} erg$, to protons and 2% of the proton energy to electrons).

Assuming an equilibrium situation, we know that the particle spectrum is given by $N(E) = Q(E) \cdot \tau(E)$ where $\tau(E)$ is the loss lifetime. The source function is given by $Q(E) = AE^{-P}$ where the proportionality constant, A, is related to the supernova rate ν, the particle acceleration efficiency η, and the starburst region volume V, such that

$\int_{E_{min}}^{E_{max}} Q(E)EdE = \frac{\eta\nu E_{51}}{V}$. This provides the source function for the protons which can be related to the source function for the electrons, $Q_e = Q_p \cdot N_e/N_p$. From the proton spectrum, we calculate the pion source function by weighting the differential cross sections for pion production by the proton spectrum and integrating over the proton energy (Torres 2004).

$$q_\pi(E_\pi) = cn_H \int_{E_{th}}^{\infty} N_p(E_p) \frac{d\sigma(E_p, E_\pi)}{dE_p} dE_p \tag{2.1}$$

From the π^- and π^+ source functions, we can calculate the secondary electron and positron source functions by integrating over γ_π and γ_e (Schlickeiser 2002).

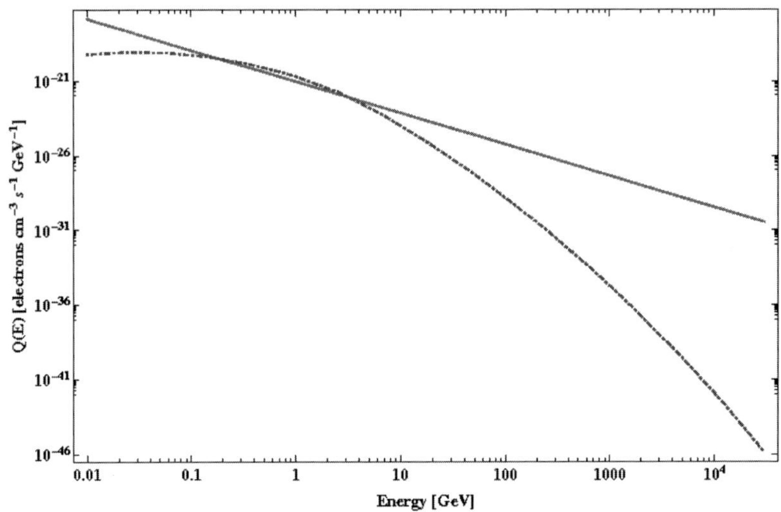

Figure 1. Electron source function: primary (solid) and secondary (dashed) electrons.

3. Results

Then, from the π^0 source function, we can calculate the gamma ray emissivity as follows (Schlickeiser 2002).

$$q_\gamma(E_\gamma) = \int_{E_{min}}^{\infty} \frac{q_{\pi^0}(E_{\pi^0})}{2\sqrt{E_{\pi^0}^2 - m_{\pi^0}^2 c^4}} dE_{\pi^0} \tag{3.1}$$

From the source function,we can calculate the predicted flux as $F = q_\gamma V/(4\pi d^2)$, where d is the distance to M82. Thus, we see that a simple one zone model can provide a reasonable fit to the gamma ray flux from M82, see Fig. 2a.

To calculate the synchrotron spectrum, we know that $J(\nu)d\nu = -\frac{dE}{dt} N(E)dE$, where N(E) includes the electron spectrum (both primaries and secondaries from pion decays) and the secondary positron spectrum (Longair 2011). The energy loss due to synchrotron emission is given by

$$-\frac{dE}{dt} = \frac{4}{3}\sigma_T c \left(\frac{E}{m_e c^2}\right)^2 \frac{B}{2\mu} \tag{3.2}$$

In the synchrotron spectrum, we also include an estimate of free-free absorption for a completely ionized medium. In recent work, we have found that the addition of free-free

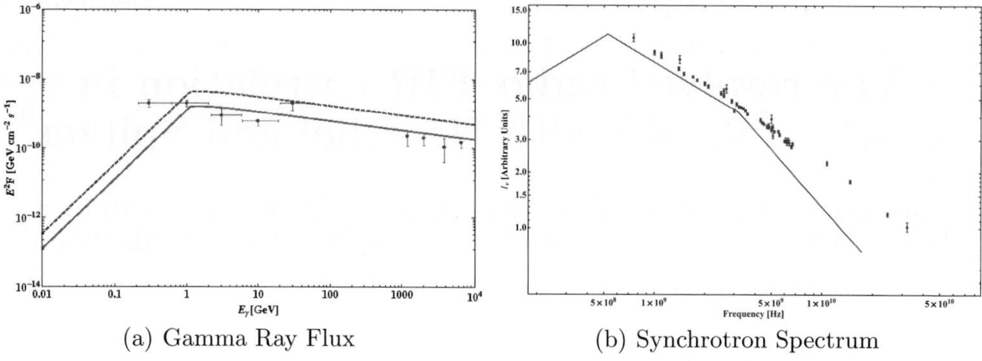

(a) Gamma Ray Flux (b) Synchrotron Spectrum

Figure 2. Left: Gamma ray flux from M82. Observational data points between 0.1 GeV and 100 GeV were observed with Fermi and data points above 1 TeV were observed with VERITAS. Right: Synchrotron spectrum for M82. Observation data points taken from Klein *et al.* (1988) and Williams & Bower (2009).

emission required us to develop a multi-zone density model. It is assumed that there is a constant source function and no self-absorption for the preliminary discussion.

4. Conclusions

Because of the dense molecular interstellar medium, intense radiation field, and high magnetic field in M82, the particle energy loss lifetimes are very short, supporting the calorimeter model. Due to the short lifetimes most protons interact producing pions which decay making $e^{+/-}$, and so we find that the secondary electron densities are comparable to those for primary electrons. Our predicted gamma ray flux matches well with data from Fermi and VERITAS. We also find that the predicted synchrotron spectrum begins to turn over at lower frequencies due to free-free absorption. The results from our simple and intuitive model compare well with the results of more complex models.

Acknowledgments

This work was supported in part by NSF grants AST-0708967, AST-0903900, PHY-0821899 (to CMSO), PHY-09069061 (to the IceCube Collaboration).

References

de Cea del Pozo, E., Torres, D. F., & Rodríguez, A. Y. 2009, *2009 Fermi Symposium*
Klein, U., Wielebinski, R., & Morsi, H. W. 1988, *A&A*, 190, 41K
Longair, M. S. 2011, *High Energy Astrophysics* (Cambridge: University Press), p. 212
Schlickeiser, R. 2002 *Cosmic Ray Astrophysics* (Berlin: Springer), p. 115
Torres, D. F. 2004, *ApJ*, 617, 966
Williams, P. K. G. & Bower, G. C. 2009, *ApJ*, 710, 1462
Völk, H. J. 1989, *A&A*, 218, 67V

The Spectral Energy Distribution of Galaxies
Proceedings IAU Symposium No. 284, 2011
R.J. Tuffs & C.C.Popescu, eds.

doi:10.1017/S1743921312009507

The resolved radio–FIR correlation in nearby galaxies with Herschel and Spitzer

Fatemeh S. Tabatabaei[1], Eva Schinnerer[1], Eric Murphy[2], Rainer Beck[3], Annie Hughes[1], Brent Groves[1] and The KINGFISH Team

[1]Max-Planck-Institut für Astronomie, Königstuhl 17, 69117-Heidelberg, Germany

[2]The Observatories of the Carnegie Institution for Science, CA 911101, USA

[3]Max-Planck-Institut für Radioastronomie, Auf dem Hügel 69, 53121-Bonn, Germany

email: taba@mpia.de

Abstract. We investigate the correlation between the far-infrared (FIR) and radio continuum emission from NGC6946 on spatial scales between 0.9 and 17 kpc. We use the Herschel PACS (70, 100, 160μm) and SPIRE (250μm) data from the KINGFISH project. Separating the free-free and synchrotron components of the radio continuum emission, we find that FIR is better correlated with the free-free than the synchrotron emission. Compared to a similar study in M33 and M31, we find that the scale dependence of the synchrotron–FIR correlation in NGC6946 is more similar to M31 than M33. The scale dependence of the synchrotron–FIR correlation can be explained by the turbulent-to-ordered magnetic field ratio or, equivalently, the diffusion length of the cosmic ray electrons in these galaxies.

Keywords. Star formation, ISM, Magnetic field, Cosmic ray electrons.

---

1. Introduction

The correlation between the radio and far-infrared (FIR) luminosities of galaxies has been shown to be invariant over more than 4 orders of magnitude over a broad range of redshift of (e.g. Sargent *et al.* 2010). It is only recently that variations in the radio-FIR correlation have become apparent when studying correlations on different scales within galaxies. Detailed multi-scale analysis of the radio and FIR data showed that the smallest scale on which the radio-FIR correlation holds is not the same in the LMC (Hughes *et al.* 2006), M51 (Dumas *et al.* 2010), M31 and M33 (Tabatabaei *et al.* 2007a, submitted); the reason is still unclear. Could massive star formation alone cause such variations? What is the influence of the dust heating sources? Are differences in the magnetic field structure important? And what is the role of the propagation of cosmic ray electrons (CREs)?

One fact that makes the link between star formation and FIR less obvious is that FIR emission consists of at least two emission components: cold dust and warm dust. The cold dust emission may not be directly linked to the young stellar population, but is rather powered by the interstellar radiation field (ISRF, Xu 1990). Furthermore, it is necessary to distinguish between the two main radio continuum (RC) components, free-free and synchrotron emission. The free-free emission from electrons in HII regions around young, massive stars is expected to be closely connected to the warm dust emission that is heated by those same stars. Although the synchrotron-emitting CREs also originate from star forming regions (i.e. supernovae remnants; the final episodes of massive stars), the synchrotron–FIR correlation may not be as tight as the free-free–FIR correlation *locally*, as a result of diffusion of the cosmic ray electrons from their production sites.

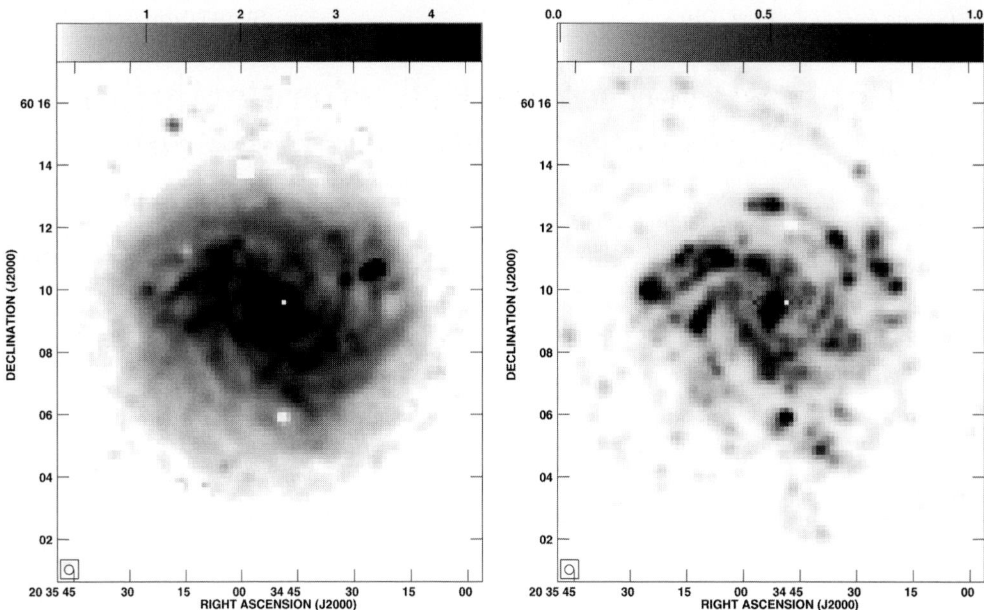

Figure 1. 20cm synchrotron (left) and free-free emission (right) from NGC6946 at 18" resolution. The units are in mJy/beam. Here external radio sources have been blanked.

2. Free-free and synchrotron emission from NGC 6946

Tabatabaei *et al.* (2007b) presented a method to separate the free-free and synchrotron components of the RC emission from M 33, in which a de-reddened Hα map was used as a template for the free-free emission. Using the same method, we derive the free-free emission from NGC6946 at 20 cm. Subtraction of the free-free emission from the observed 20 cm (VLA + Effelsberg, Beck 1991) emission then yields the synchrotron map (Fig. 1). Diffusion/propagation of the CREs is evident from the synchrotron map, which shows a diffuse extended emission as well as strong emission from the galaxy center, giant star forming regions, and spiral arms. The thermal free-free emission has a more clumpy distribution following star forming regions. The thermal fraction is about 8% at 20 cm.

3. Wavelet correlation

Using a 2D continuous wavelet transformation ('Pet-hat', Frick *et al.* 2001), we decompose the Herschel PACS/SPIRE FIR and the 20cm maps into 8 spatial scales (a) from 0.9 ($\sim$ twice our linear resolution of 18") to 17 kpc. For each scale of decomposition, we derive cross-correlation coefficients (r_w) between the FIR and RC maps. Figure 2 shows $r_w(a)$ between the 70μm and the free-free/synchrotron emission in NGC 6946, compared with similar correlations in M 33 and M 31 (Tabatabaei *et al.* submitted).

4. Results

The free-free–FIR correlation is stronger than the synchrotron–FIR correlation in NGC 6946 as seen in the other two galaxies. Interestingly, the free-free–FIR correlation exhibits a more or less similar trend in these galaxies, unlike the synchrotron–FIR correlation which shows more variation with scale a. In NGC 6946, the synchrotron–FIR correlation decreases towards smaller scales. Such a trend is also seen in M31, M51 (Dumas *et al.*

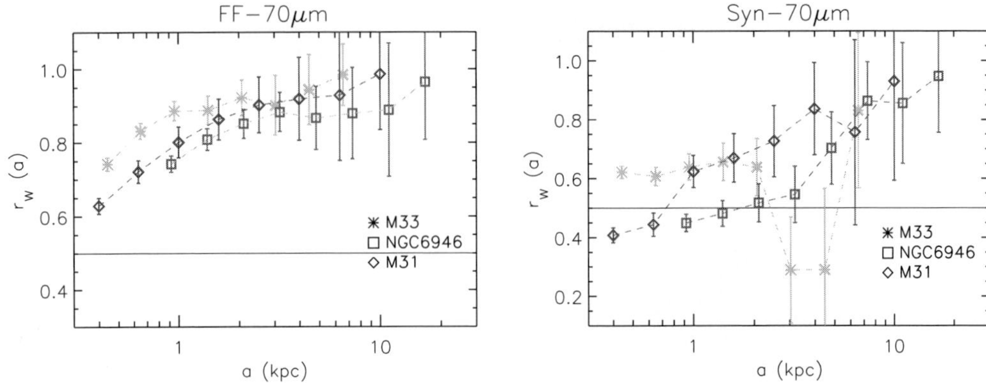

Figure 2. Multi-scale correlations between 70μm FIR and free-free (left), synchrotron (right) emission from NGC6946, compared to those in M31 and M33.

2011), and the LMC (Hughes *et al.* 2006) but not in M33 which shows a stronger correlation on small scales and weaker on larger scales (3-5 kpc). What causes these variations? Assuming that massive stars are the common source of the free-free, FIR (via heating the dust), and synchrotron emission, the synchrotron–FIR variations must be linked to variations in magnetic fields (structure and strength) and diffusion/propagation of CREs. One would expect a better correlation on small scales ($\sim$ 100 pc) in galaxies with higher star formation rate (SFR) due to the stronger turbulent magnetic field and/or high concentration of young CREs near their star forming regions (e.g. Murphy *et al.* 2006). This is evident for M33 which has a 5 times higher SFR surface density, Σ(SFR), than M31 (Tabatabaei & Berkhuijsen 2010). This cannot be investigated in NGC6946, since there is no information for scales $<$ 1 kpc. On the other hand, including the LMC and M51 suggests that the smallest scale on which the radio–FIR correlation holds ($r_w > 0.5$), is better correlated with the ratio of turbulent-to-ordered magnetic field, which in turn is related to the diffusion length of CREs (Yan & Lazarian 2004), than with Σ(SFR) (Tabatabaei *et al.* submitted).

On large scales ($a > 1$ kpc), the synchrotron emission is due to an ordered (uniform) magnetic field and diffused, older CREs. The ordered magnetic field is controlled by various dynamical, environmental effects caused by, e.g., galactic rotation, density waves, shear and anisotropic compressions in spiral arms (e.g. Fletcher *et al.* 2011). A good synchrotron–FIR correlation on large scales is expected when enough diffuse CREs exist and that they have opportunity to explore the gaseous/dusty components of a galaxy (Hughes *et al.* 2006). This would require a confinement of the magnetic field to the gas. We plot a schematic view of this condition in Fig. 3, showing sources of the FIR and synchrotron emission and factors which control their correlation on small and large scales.

5. Summary

We derived radio–FIR correlations as a function of scale for NGC6946, separately for free-free and synchrotron emission, and for various Herschel FIR bands (70, 100, 160, and 250μm). Our main results are summarized as follows:

- We find strong synchrotron emission from the central part and around star-forming regions. This is explained by the direct proportionality between SFR and turbulent magnetic field/ young CREs. In addition a smooth diffuse emission is present.

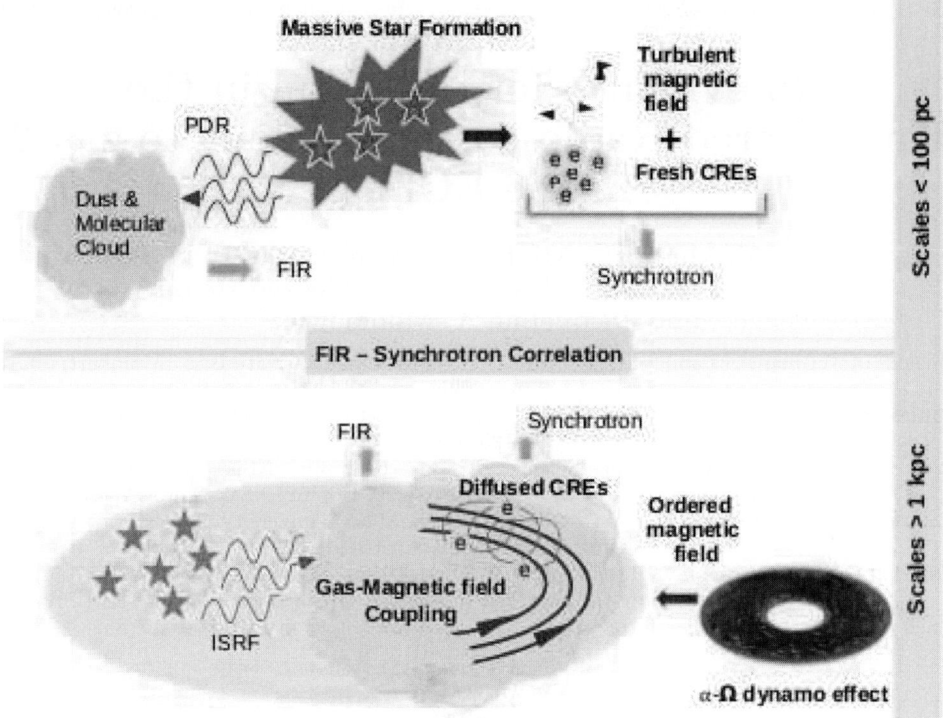

Figure 3. A sketch of various sources of the synchrotron and FIR emission and important factors controlling the synchrotron-FIR correlation on small and large scales in a galaxy.

- The wavelet power spectrum of the synchrotron emission is stronger on scales > 1 kpc than < 1 kpc, possibly due to fast diffusion of CREs in the strong, ordered magnetic field.
- The FIR and free-free emissions are correlated on all scales; a better correlation is found with warmer dust.
- The FIR and synchrotron emission may not be correlated on some scales, due to diffusion of CREs and/or magnetic field structures.

Last but not least, our results show that the radio-FIR correlation depends not only on star formation but also on properties of the magnetic field in galaxies.

References

Beck, R. 1991, *A&A*, 251, 15
Dumas, G., Schinnerer, E., Tabatabaei, F. S., *et al.* 2011, *AJ*, 141, 41
Fletcher, A., Beck, R., Shukurov, A., Berkhuijsen, E. M., & Horellou, C. 2011, *MNRAS*, 412, 2396
Frick, P., Beck, R., Berkhuijsen, E. M., & Patrickeyev, I. 2001, *MNRAS*, 327, 1145
Hughes, A., Wong, T., Ekers, R., *et al.* 2006, *MNRAS*, 370, 363
Murphy, E. J., Helou, G., Braun, R., *et al.* 2006, *ApJL*, 651, L111
Tabatabaei, F. S., Beck, R., Krause, M., *et al.* 2007b, *A&A*, 466, 509
Tabatabaei, F. S., Beck, R., Krügel, E., *et al.* 2007a, *A&A*, 475, 133
Tabatabaei, F. S. & Berkhuijsen, E. M. 2010, *A&A*, 517, A77
Xu, C. 1990, *ApJ*, 365, 47
Yan, H. & Lazarian, A. 2004, *ApJ*, 614, 757

The Spectral Energy Distribution of Galaxies
Proceedings IAU Symposium No. 284, 2011
R.J. Tuffs & C.C.Popescu, eds.

doi:10.1017/S1743921312009519

No evidence for evolution in the Far-Infrared-Radio correlation out to z $\sim$ 2 in the ECDFS

Minnie Y. Mao[1,2,3,4], Minh T. Huynh[2,5], Ray P. Norris[3], Mark Dickinson[6], Dave Frayer[7], George Helou[2] and Jacqueline A. Monkiewicz[8]

[1]School of Mathematics and Physics, University of Tasmania, Private Bag 37, Hobart, 7001, Australia
[2]Infrared Processing and Analysis Center, California Institute of Technology, Pasadena CA 91125, USA
[3]CSIRO Astronomy and Space Science, PO Box 76, Epping, NSW, 1710, Australia
[4]Australian Astronomical Observatory, PO Box 296, Epping, NSW, 1710, Australia
[5]International Centre for Radio Astronomy Research, M468, University of Western Australia, Crawley, WA 6009, Australia
[6]National Optical Astronomy Observatory, 950 North Cherry Avenue, Tucson, AZ 85719, USA
[7]National Radio Astronomy Observatory, PO Box 2, Green Bank, WV 24944, USA
[8]School of Earth and Space Exploration, Arizona State University, Tempe, AZ 85287, USA

email: minnie.mao@csiro.au

Abstract. The Far-Infrared Radio Correlation (FRC) is the tightest and most universal correlation known among global parameters of galaxies. Here we present the results of our investigation of the 70 μm FRC of starforming galaxies in the Extended Chandra Deep Field South (ECDFS) out to z > 2. In order to quantify the evolution of the FRC we used both survival analysis and stacking techniques, which gave similar results. We also calculated the FRC using total infrared luminosity and rest-frame radio luminosity, qTIR, and find that qTIR is constant (within 0.22) over the redshift range 0 - 2. We see no evidence for evolution in the FRC at 70 μm, which is surprising given the many factors that are expected to change this ratio at high redshifts.

Keywords. galaxies: evolution, galaxies: formation, infrared: galaxies, radio continuum: galaxies

1. Introduction

The correlation between the far-infrared (FIR) and radio emission for star-forming galaxies in the local Universe was first observed by van der Kruit (1973). The correlation is linear, spans five orders of magnitude of bolometric luminosity and has been shown to hold for a wide-range of Hubble types (e.g. Helou, Soifer & Rowan-Robinson 1985).

The far-reaching nature of the FRC has made it a valuable diagnostic and astronomers have used it to identify radio-loud AGN, define the radio luminosity/SFR relation and, at higher redshifts, estimate distances of sub-mm galaxies with no optical counterparts. Consequently it is of great importance to determine whether the FRC holds at high redshifts.

The FRC may fail at higher redshifts for a number of reasons. Electrons are expected to lose energy by inverse Compton (IC) interactions with the cosmic microwave background (CMB), whose energy density scales as $(1+z)^4$, implying a lower level of radio emission at higher redshifts. Moreover, synchrotron emission is proportional to the magnetic field strength squared, so evolution of magnetic field strength should affect the FRC at higher

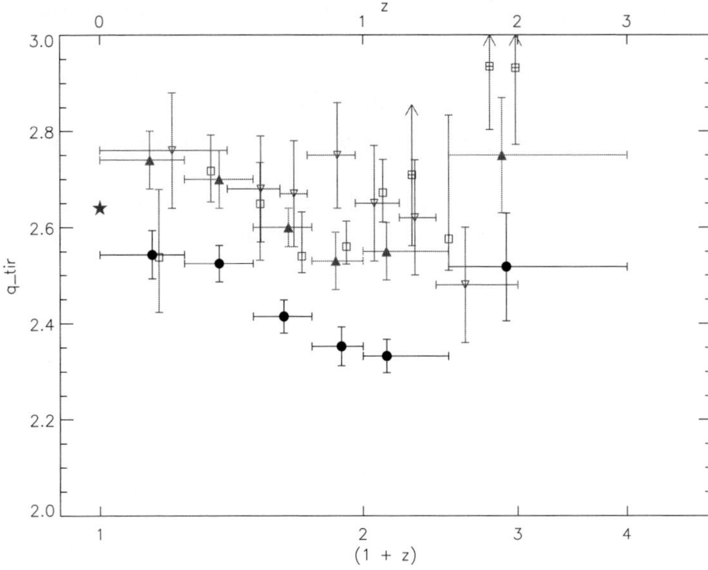

Figure 1. Median q_{TIR} values for different redshift bins. The black filled circles show the median q_{TIR} values for sources with both 70 μm and 1.4 GHz detections and the filled triangles show the median q_{TIR} value for all sources taking into account the lower limits using survival analysis. The grey upside-down open triangles show the median q_{TIR} values derived by Bourne *et al.* (2010) and the grey open squares show the median q_{TIR} values derived by Sargent *et al.* (2010b) for star-forming galaxies. Vertical error bars are standard errors for both our dataset and Bourne's data, but the vertical error bars for Sargent's data are upper and lower 95% confidence levels. The star at $(1+z) = 1$ represents the median $q_{TIR} = 2.64 \pm 0.02$ from Bell (2003).

redshifts. Changes in the spectral energy distributions may also be expected due to evolution in dust properties and metallicity. However, current studies show no firm evidence for evolution in the FRC (e.g., Garrett 2002; Appleton *et al.* 2004; Seymour *et al.* 2009; Bourne *et al.* 2010; Ivison *et al.* 2010a,b; Sargent *et al.* 2010a,b; Huynh *et al.* 2010).

2. Data

Our project studies the FRC's dependence on redshift using deep 70 μm data from the Far-Infrared Deep Extragalactic Survey (FIDEL, PI: Dickinson), which uses the Spitzer Space Telescope, and 1.4 GHz data from the Very Large Array (VLA) in the Extended Chandra Deep Field South (ECDFS). This work is the first to use such deep 70 μm data to study the evolution of the FRC based on individual sources. We focus on 70 μm data because it is an excellent tracer of star-formation as it probes closer to the 100 μm star-formation peak in the IR SED.

Our final catalogue comprises 617 70 μm sources, 91% of which have redshift information, over a third of which are spectroscopically determined. 353 of these sources have radio counterparts from Miller *et al.* (2008). In order to account for the radio non-detections we used both a stacking analysis and Survival Analysis. Our sample detects LIRGs to z $\sim$ 1.25 and ULIRGs to z $\sim$ 3 (Mao *et al.* 2011).

3. The FRC Shows No Evidence for Evolution with z

We computed q_{TIR} for all sources that had redshift information using:

$$q_{TIR} = \log(L_{IR}/L_{1.4\,GHz}), \tag{3.1}$$

where L_{IR} is the total infrared luminosity, and $L_{1.4GHz}$ is the rest-frame 1.4 GHz luminosity.

Figure 1 shows the median q_{TIR} for only the detected sources (black circles), and for all sources by using survival analysis (red triangles).

Our q_{TIR} values are all within ~0.22 of each other. Our values agree, within the errors, with the work of Sargent *et al.* (2010b), and also agree with the work of Bourne *et al.* (2010), with the exception of the redshift bins $0.75 < z < 1$ and $z > 1.5$, where our values differ by $< 2\sigma$ compared to the values given by Bourne *et al.* (2010) for similar redshift ranges.

4. Summary and Conclusions

We have used the deepest 70 μm data to date to study the FRC out to $z > 2$ of ULIRGs in ECDFS. To quantify the evolution of the FRC we binned our data in redshift and calculate the FRC using luminosities and find that evolution in q_{TIR} is constrained within 0.22. A more detailed discussion of these results and their implications may be found in Mao *et al.* (2011)

The fact that we see no evidence for evolution is very surprising. Perhaps IC cooling and other effects such as evolution of the magnetic field strength, or evolution of dust properties are insignificant to $z \sim 2$, or perhaps there is a complex interplay between these factors conspiring in the preservation of the FRC at higher redshifts.

In the near future *Herschel* will measure the FIR properties of normal galaxies to $z \sim 1$, and ULIRGs to $z \sim 4$. This will allow us to study the FRC to higher redshifts and hence gain a better understanding of the evolution of star-forming galaxies.

References

Appleton P. N., *et al.* 2004, *ApJS*, 154, 147

Bell E. F. 2003, *ApJ*, 586, 794

Bourne N., Dunne L., Ivison R. J., Maddox S. J., Dickinson M., & Frayer D. T. 2010, *MNRAS*, 1524

Garrett M. A. 2002, *A&A*, 384, L19

Helou G., Soifer B. T., & Rowan-Robinson M. 1985, *ApJ*, 298, L7

Huynh M. T., Gawiser E., Marchesini D., Brammer G., & Guaita L. 2010, *ApJ*, 723, 1110

Ivison R. J., *et al.* 2010, *MNRAS*, 402, 245

Ivison R. J., *et al.* 2010, *A&A*, 518, L31

Mao M. Y., Huynh M. T., Norris R. P., Dickinson M., Frayer D., Helou G., & Monkiewicz J. A. 2011, *ApJ*, 731, 79

Miller N. A., Fomalont E. B., Kellermann K. I., Mainieri V., Norman C., Padovani P., Rosati P., & Tozzi P. 2008, *ApJS*, 179, 114

Sargent M. T., *et al.* 2010, *ApJS*, 186, 341

Sargent M. T., *et al.* 2010, *ApJ*, 714, L190

Seymour N., Huynh M., Dwelly T., Symeonidis M., Hopkins A., McHardy I. M., Page M. J., & Rieke G. 2009, *MNRAS*, 398, 1573

van der Kruit P. C. 1973, *A&A*, 29, 263

The Spectral Energy Distribution of Galaxies
Proceedings IAU Symposium No. 284, 2011
R.J. Tuffs & C.C. Popescu, eds.

doi:10.1017/S1743921312009520

The non-thermal broadband spectral energy distribution of radio galaxies

Gustavo E. Romero[1,2,†]

[1]Instituto Argentino de Radioastronomía (IAR, CCT La Plata, CONICET), C.C.5, (1894) Villa Elisa, Buenos Aires, Argentina

[2]Facultad de Ciencias Astronómicas y Geofísicas, Universidad Nacional de La Plata, Paseo del Bosque s/n, 1900 La Plata, Argentina

email: romero@fcaglp.unlp.edu.ar romero@iar-conicet.gov.ar

Abstract. I present a model for the non-thermal production of electromagnetic radiation in the jets of radio galaxies. The model goes far beyond the simple one-zone models usually applied to these sources. The transport equation is solved in the co-moving frame of the jet, taken into account the inhomogeneous structure of the outflow. Energy distributions for all types of particles are then obtained in a self-consistent way, including protons, electrons, and secondaries. The spectral energy distribution resulting from all relevant radiative processes is computed, including synchrotron radiation, relativistic Bremsstrahlung, proton-proton collisions and subsequent decays, photo-meson production, radiation from pairs formed by photon absorption and injection from decays, as well as direct pair production. Absorbing fields in the host galaxy are considered when computing the final SED. The model is applied to Centaurus A and compared with the available multi-wavelength data.

Keywords. Galaxies: active, galaxies: jets, gamma rays: theory, radiation mechanisms: non-thermal.

---

1. Introduction

Radio galaxies can display electromagnetic emission across the entire spectrum, reaching in some cases muti-TeV energies. The most active radio galaxies are classified as Faranoff-Riley type I (FR I) and Faranoff-Riley type II (FR II). The former are intrinsically weaker sources, with extended radio lobes and plumes. The latter are powerful sources, with higher accretion rates and more luminous jets, that terminate in bright hot spots. Nearby FR I radio galaxies such as Centaurus A (Cen A) and M87 have been recently detected at very high-energies. These galaxies show significant activity in the central region, with strong variability reported at different wavelengths. This variability indicates that a significant part of the non-thermal emission is produced in the inner part of the relativistic jets, close to the central source. In what follows I present the outlines of a model that can explain the broadband nuclear emission of active radio galaxies and I show an application of such a model to the specific case of Cen A. For additional details the reader is referred to Reynoso *et al.* (2011).

2. Basics of the model

I assume that the jet is launched from the black hole's ergosphere, being completely formed at a distance $z_0 = 50R_{\rm g}$, and with an initial bulk Lorentz factor Γ_0. Equipartition

† Member of CONICET.

holds between the magnetic and kinetic energy at the base of the jet. This is required if the plasma is set in motion by magneto-centrifugal effects. The kinetic power is given by $L_{\rm j}^{\rm (kin)} = \frac{q_{\rm j}}{2} L_{\rm Edd}$, where $L_{\rm Edd}$ is the Eddington luminosity and $0 < q_{\rm j} < 1$ is a numerical factor. The jet is modeled as a cone with an half-opening angle ξ. Most of its content is thermal plasma. The value of the magnetic field at the position z_0 in jet is

$$B_0 = \sqrt{\frac{8 L_{\rm j}^{\rm (kin)}}{[r_{\rm j}(z_0)]^2 \, v_{\rm b}}}, \tag{2.1}$$

where $r_{\rm j}(z_0) = z_0 \tan\xi$ is the jet radius at z_0.

The jet magnetic energy is gradually converted into bulk kinetic energy and the field decreases as the flow accelerates. The later is parametrized by

$$B(z) = B_0 \left(\frac{z_0}{z}\right)^m, \tag{2.2}$$

with $m \in \{1, 2\}$. If all the magnetic energy is transformed into bulk kinetic energy, the Lorentz factor of the jet increases with z approximately as

$$\Gamma_{\rm b}(z) \approx 1 + \frac{B_0^2 z_0^2 \tan^2\xi}{8 \dot{m}_{\rm j} c} \left[2 - \left(\frac{z_0}{z}\right)^{2m-2}\right].$$

In turn, a fraction $q_{\rm rel}$ of the bulk kinetic power is transformed into highly relativistic particles in a region that starts at $z_{\rm acc}$. In this region the bulk kinetic energy is well below the magnetic one (sub-partition condition), and shocks can develop. Both primary electrons and protons are assumed to be injected at $z_{\rm acc}$, over a length Δz along the jet, with a power law distribution in the frame of the jet,

$$Q'_{e,p}(z', E') = \frac{K_{e,p}}{4\pi} \left(\frac{z'_{\rm acc}}{z'}\right) \left(\frac{E'}{m_i c^2}\right)^{-s} e^{-E'/E'_{\rm max}}, \tag{2.3}$$

in units of $({\rm GeV}^{-1}{\rm s}^{-1}{\rm cm}^{-3}{\rm sr}^{-1})$.

The constants $K_{e,p}$ are fixed by normalizing the injected power as

$$L'_{e,p} = \int_{V'} dV' \int_{E'_{\rm min}} dE' E' Q'(z', E'). \tag{2.4}$$

The acceleration mechanism might be diffusive shock acceleration, but other possibilities cannot be ruled out.

The maximum energies of the injected particles are determined from the balance between the acceleration rate and the cooling rates. The proton to electron ratio at the injection point is given by the parameter $a = L'_p/L'_e$, where $L'_e + L'_p = q_{\rm rel} L_{\rm j}^{\rm (kin)}$.

The main cooling processes are synchrotron emission for electrons and protons, and pp, $p\gamma$ interactions and adiabatic cooling for protons. Inverse Compton (IC) interactions with synchrotron photons in the jet are also considered. The expression for the acceleration rate is: $t_{\rm acc}^{-1} = \eta c e B'(z)/E'$. For the other rates the reader can see Romero & Vila (2008) and Vila & Romero (2010).

The steady-state particle distributions $N_i(E, z)$ in the jet can be found by solving a 1-dimensional transport equation with cooling and convection along the jet, with a bulk velocity $v \approx c$, as seen in the observer frame:

$$v \frac{\partial N_i}{\partial z} + \frac{\partial \left[b_i(z, E) N_i\right]}{\partial E} + \frac{N_i}{T_{\rm dec}(E)} = Q_i(z, E). \tag{2.5}$$

This equation is solved first for electrons, assuming no IC cooling. This first approximation to $N_e(z,E)$ is then iteratively improved taking into account also the IC cooling rate t_{IC}^{-1} to solve the transport equation. The distribution of protons is obtained considering also the cooling through $p\gamma$ interactions with the low energy photons in the jet. These inelastic interactions, together with the pp collisions, give rise to pions and their decay products, which include muons and neutrinos. The injection of pions $Q_\pi(z,E)$ can be worked out using the proton distribution $N_p(z,E)$. The distribution of pions $N_\pi(z,E)$ found with Eq. (2.5) is then considered to obtain the injection of muons $Q_\mu(z,E)$, which in turn is used to obtain the muon distribution $N_\mu(z,E)$. The different particle species are responsible for the radiative processes described in the next section.

3. Radiative processes

In the jet frame, where the particle distributions are isotropic, the synchrotron emissivity is

$$Q'_{\gamma,\mathrm{syn}}(E'_\gamma,z') = \frac{\epsilon'_{\mathrm{syn}}(E'_{\mathrm{ph}},z')}{4\pi E'_\gamma}, \tag{3.1}$$

where $\varepsilon'_{\mathrm{syn}}$ is the power per unit energy per unit volume of the synchrotron photons,

$$\varepsilon'_{\mathrm{syn}}(E'_{\mathrm{ph}},z') = \left(\frac{1-e^{-\tau_{\mathrm{SSA}}(E'_{\mathrm{ph}},z')}}{\tau_{\mathrm{SSA}}(E'_{\mathrm{ph}},z')}\right)\int_{E'_e}^{\infty} dE' 4\pi P_{\mathrm{syn}} N'_e(E',z'). \tag{3.2}$$

Here, P_{syn} is the usual synchrotron power per unit energy emitted by the electrons.

The effect of synchrotron self-absorption (SSA) within the jet is determined by an optical depth

$$\tau_{\mathrm{SSA}}(E'_{\mathrm{ph}},z') = \int_0^{\frac{r_{\mathrm{j}}(z')}{\sin\theta'}} dl' \alpha_{\mathrm{SSA}}(E'_{\mathrm{ph}},z',l').$$

θ' is the viewing angle in the jet frame, and the SSA coefficient α_{SSA} is evaluated with its standard expression along the path in the jet. The IC emissivity is

$$Q'_{\gamma,\mathrm{IC}}(E'_\gamma,z') = \frac{r_0^2 c}{2}\int_{E'^{(\mathrm{min})}_{\mathrm{ph}}}^{E'_\gamma} dE'_{\mathrm{ph}} \frac{n'_{\mathrm{ph}}(E'_{\mathrm{ph}},z')}{E_{\mathrm{ph}}} \int_{E'_{\mathrm{min}}}^{E'_{\mathrm{max}}} dE' \frac{N'_e(E',z')}{\gamma_e^2} F(q). \tag{3.3}$$

The integration limits are:

$$E'_{\mathrm{min}} = \frac{E'_\gamma}{2} + \frac{m_e c^2}{2}\sqrt{\frac{E'_\gamma}{2E'_{\mathrm{ph}}} + \frac{E'^2_\gamma}{2m_e^2 c^4}} \quad \text{and } E'_{\mathrm{max}} = \frac{E'_\gamma}{1-\frac{E'_\gamma}{E'_{\mathrm{ph}}}}. \tag{3.4}$$

As for the emission of protons, the relevant processes are pp and $p\gamma$ interactions. The gamma-ray emissivity is

$$Q'_{\gamma,pp}(E'_\gamma,z') = n'_{\mathrm{c}}(z')c\int_0^1 \frac{dx}{x} N'_p\left(\frac{E'_\gamma}{x},z'\right) F_\gamma\left(x,\frac{E'_\gamma}{x}\right)\sigma_{pp}^{(\mathrm{inel})}\left(\frac{E'_\gamma}{x}\right), \tag{3.5}$$

where the function $F_\gamma(x,E')$ is the same as defined by Kelner *et al.* (2006), for a proton energy $E' = E'_\gamma/x$. The target-proton density in the jet frame is $n'_{\mathrm{c}}(z') = n_{\mathrm{c}}(z'\Gamma_{\mathrm{b}})/\Gamma_{\mathrm{b}}$. Its value in the observer frame is

$$n_{\mathrm{c}}(z) = \frac{(1-q_{\mathrm{rel}})\dot{m}_{\mathrm{j}}}{m_p\pi z^2\tan^2\xi_{\mathrm{j}} v_{\mathrm{b}}(z)}. \tag{3.6}$$

4. Spectral energy distributions

The total luminosity L_γ is calculated as

$$L_\gamma = 4\pi \int_V dV E_\gamma Q_\gamma(E_\gamma, z).$$

where the photon emissivity $Q_\gamma(E_\gamma, z)$ corresponds to the observer frame.

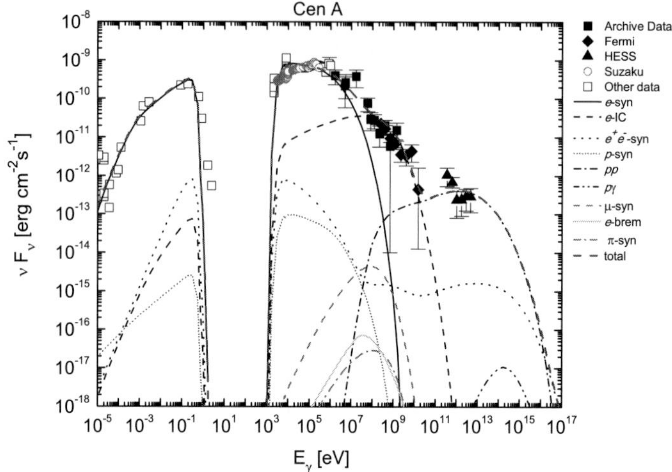

Figure 1. Lepto-hadronic model for the broadband spectral energy distribution of Centaurus A inner region.

The parameter $a = L'_p/L'_e$ takes the values >> 1. The main contributions arise from synchrotron emission of electrons, inelastic *pp* collisions and synchrotron radiation of secondary electron-positron pairs. The role of internal absorption is discussed by Reynoso *et al.* (2011). In Fig. 1 we show the application of the model to the available data for the inner region of Cen A.

The particle distributions described here correspond to a steady-state jet. Temporal variability can be introduced in different forms according to the timescales involved: 1) For long timescales (days or more), with a time-dependent injection $Q_i(z, E, t)$ and a time-derivative term $\partial N/\partial t$ in the transport equation. 2) For short timescales (hours or less), through the entrainment of magnetized clouds or stars in the jet as proposed by Araudo *et al.* (2010), or by bends or turbulence produced through instabilities (Romero 1995).

Acknowledgments

My work on this topic has been done in collaboration with M.M. Reynoso and M.C. Medina. I thank them both for fruitful discussions. This research was supported by CONICET (Argentina) through grant PIP 112-200901-00078.

References

Araudo, A. T., Bosch-Ramon, V., & Romero, G. E. 2010, *A&A*, 522, A97
Kelner, S. R., Aharonian, F. A., & Bugayov, V. V. 2006, *Phys. Rev. D*, 74, 034018
Reynoso, M. M., Medina, M. C., & Romero, G. E. 2011, *A&A*, 531, A30
Romero, G. E. & Vila, G. S. 2008, *A&A*, 485, 623
Romero, G. E. 1995, *Astrophys. Sp. Sc.*, 234, 49
Vila, G. S. & Romero, G. E. 2010, *MNRAS*, 403, 1457

The Spectral Energy Distribution of Galaxies
Proceedings IAU Symposium No. 284, 2011
R.J. Tuffs & C.C. Popescu, eds.

doi:10.1017/S1743921312009532

The challenging SED of AP Librae

David Sanchez[1], Berrie Giebels[2] and Pascal Fortin[2]
on behalf of the H.E.S.S and *Fermi*-LAT collaborations

[1]Max Planck Institut fr Kernphysik, Heidelberg
[2]LLR/Ecole Polytechnique/CNRS/IN2P3

email: `david.sanchez@mpi-hd.mpg.de, berrie@in2p3.fr, fortin@llr.in2p3.fr`

Abstract. Matching the broad-band emission of active galaxies with the predictions of theoretical models can be used to derive constraints on the properties of the emitting region and to probe the physical processes involved. AP Librae is the third low frequency peaked BL Lac (LBL) detected at very high energy (VHE, E>100GeV) by an Atmospheric Cherenkov Telescope; most VHE BL Lacs (34 out of 39) belong to the high-frequency and intermediate-frequency BL Lac classes (HBL and IBL). LBL objects tend to have a higher luminosity with lower peak frequencies than HBLs or IBLs. The characterization of their time-averaged spectral energy distribution is challenging for emission models such as synchrotron self-Compton (SSC) models.

Keywords. gamma rays: observations, galaxies: active, BL Lacertae objects: individual

1. Introduction

Of the 125 sources detected at TeV energies, 41 are blazers. The vast majority (34) are HBLs. Only four are LBLs: S5 0716+714, 1ES 1215+303, BL Lacertae, and AP Librae.

LBLs tend to have a lower synchrotron emission peak than HBLs, but a higher synchrotron peak luminosity. Synchrotron Self-Compton (SSC) models successfully reproduce the time-average Spectral Energy Distribution (SED) of HBLs but have difficulties for LBLs. The characterization of the SED of LBLs from radio to TeV energy is of primary importance to constrain emission models.

AP Librae ($z = 0.049$) was recently detected by the Fermi Large Area Telescope (LAT) Atwood *et al.* (2009) as a bright source (0FGL J1517.9-2423) Abdo *et al.* (2009) and subsequently detected at VHE by the High Energetic Stereoscopic System (H.E.S.S.).

2. Observations and Analysis

AP Librae has been observed by H.E.S.S. during 10.9 hours of live time after the detection. Data were analyzed with the *Model* analysis method (de Naurois *et al.* 2009) leading to a detection with a significance of 7σ for an excess of 81 γ-rays. The spectrum is compatible with a power law of index $\Gamma = 2.45 \pm 0.20$ with a differential flux at the decorrelation energy $E_0 = 0.664$TeV of $I = (1.63 \pm 0.23) \times 10^{-12}\text{cm}^{-2}\text{s}^{-1}\text{TeV}^{-1}$.

The analysis includes 3 years of data (from August 4, 2008 to August 4, 2001) taken in the normal all sky survey mode. A binned likelihood method implemented in the `gtlike` tool, part of the *Fermi* analysis software (ScienceTools v9r24p0), was used to derive the spectral parameters of the source. Events from the P7 SOURCE class with an energy between 300 MeV and 300 GeV were considered. The instrument was described by the Instrumental Response Functions (IRFs) P7SOURCE_V6.

The source spectrum is well described by a simple power law of index $\Gamma = 2.09 \pm 0.04$ for a total flux above 300 MeV of $(1.98 \pm 0.10) \times 10^{-8}\text{cm}^{-2}\text{s}^{-1}$. A fit with a log-parabola function does not improve the fit significantly.

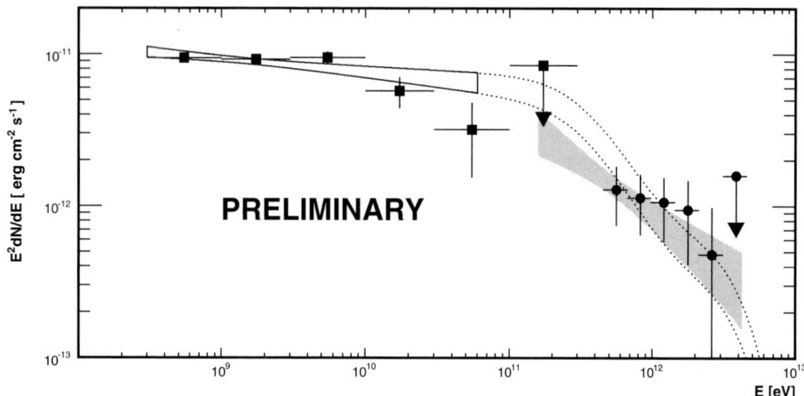

Figure 1. *Fermi*-LAT (square/black) and H.E.S.S. (circle/grey) SEDs. The contours correspond to the 1σ error. The dotted line is the *Fermi* error extrapolated and corrected for EBL using Franceschini *et al.* (2008).

The H.E.S.S. and *Fermi* spectra are shown in Figure 1. The *Fermi* spectrum has been extrapolated towards the H.E.S.S energy range and corrected for Extragalactic Background Light (EBL) absorption using the model of Franceschini *et al.* (2008). This extrapolation is in good agreement with the data points with a χ^2 of 5.7 for 5 degrees of freedom. The corresponding χ^2 probability is 34%.

3. Spectral Energy Distribution

The Spectral Energy Distribution of AP Librae is presented on Figure 2. Contemporaneous data are from UVOT, *Swift*-XRT, *RXTE*, *Fermi* and H.E.S.S. Archival data from NED have also been added. The most important feature of this SED is the broadness of the high energy component, usually assumed to come from inverse Compton (IC) scattering. The low energy cutoff (≈ 0.1 keV) of the synchrotron emission and the spectral index ($\Gamma = 1.60 \pm 0.06$) of the *Swift/RXTE* spectrum clearly show that the X-rays are produced by IC emission. This is quite remarkable since for the HBLs, the X-ray emission is synchrotron emission. This makes AP Librae an interesting source, with a fairly narrow synchrotron component but a broad high-energy component.

The energy of the photons produced by synchrotron is $E_s \propto \gamma^2$. In a SSC framework, assuming IC scattering in the Thomson regime, the photon energy is $E_c \propto \gamma^4$. A ratio between the IC width divided by the synchrotron width of 2, in logarithmic scale, can then easily be explained by SSC models. This is somehow optimistic while Klein-Nishina effects tend to narrow the IC component since the photon energy in this regime is $E_c \propto \gamma^2$. Then, one can expect to find a ratio between 1 and 2.

A fair estimation of the width of the components is obtained by fitting with a 3rd order polynomial and computing the size at half maximum. Following this procedure, the IC component is found to have a FWHM more than a factor of 4 greater than the synchrotron component, larger than a ratio of 2 found in SSC framework.

The result of a simple one zone SSC calculation is also shown in Fig. 2 as well as the model from Tavecchio *et al.* (2010). The electron distribution is described by a broken power law of index $S_1 = 2$ below an energy break $\gamma_b = 1.4 \times 10^4$ and $S_2 = 4.9$ above. A magnetic field B of 0.1G and $\delta = 29$ were chosen for the model, so as to reproduce the low energy and X-ray components, but the model cannot reproduce the *Fermi* and H.E.S.S. data. Adding another spectral component to reproduce the H.E.S.S data does not help

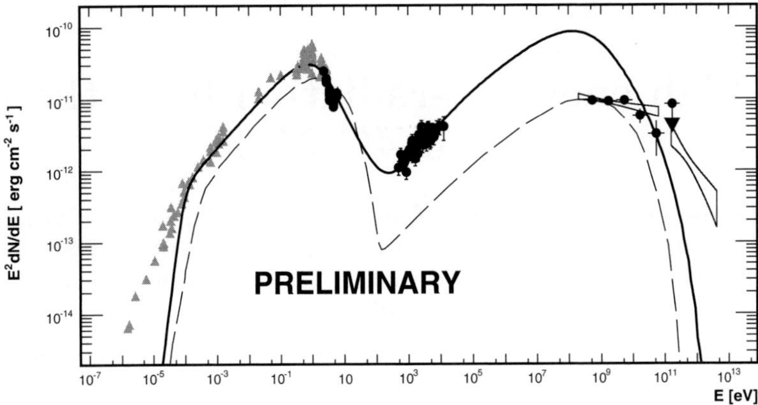

Figure 2. The time-averaged SED of AP Librae. The solid line is our SSC model; dashed line is the model from Tavecchio *et al.* (2010).

since the *Fermi* spectrum is already overestimated by over a factor of 10. For other LBLs, similar difficulties arose - the SED can only be reproduced by adding a component or using multiple emission zones (Abdo *et al.* 2011, Anderhub *et al.* 2009). Thus, TeV LBLs are a challenging and constraining class of objects for the emission models.

Acknowledgement

The *Fermi* LAT Collaboration acknowledges support from a number of agencies and institutes for both development and the operation of the LAT as well as scientific data analysis. These include NASA and DOE in the United States, CEA/Irfu and IN2P3/CNRS in France, ASI and INFN in Italy, MEXT, KEK, and JAXA in Japan, and the K. A. Wallenberg Foundation, the Swedish Research Council and the National Space Board in Sweden. Additional support from INAF in Italy and CNES in France for science analysis during the operations phase is also gratefully acknowledged.

The support of the Namibian authorities and of the University of Namibia in facilitating the construction and operation of H.E.S.S. is gratefully acknowledged, as is the support by the German Ministry for Education and Research (BMBF), the Max Planck Society, the French Ministry for Research, the CNRS-IN2P3 and the Astroparticle Interdisciplinary Programme of the CNRS, the U.K. Science and Technology Facilities Council (STFC), the IPNP of the Charles University, the Polish Ministry of Science and Higher Education, the South African Department of Science and Technology and National Research Foundation, and by the University of Namibia. We appreciate the excellent work of the technical support staff in Berlin, Durham, Hamburg, Heidelberg, Palaiseau, Paris, Saclay, and in Namibia during the construction and operation of the equipment.

References

Anderhub, H *et al.* *ApJL*, 704, L129

Abdo, A. A. *et al.* 2009, *ApJS*, 183, 46

Abdo, A. A. *et al.* 2001, *ApJ*, 730, 101

Abdo, A. A. *et al.* 2011, *ApJS*, Submitted

Atwood, W. B. *et al.*, 2009, *ApJ*, 697, 1071

Franceschini, A. *et al.* 2008, *A&A*, 487, 837

de Naurois, M. & Rolland, L. 2009, *Astroparticle Physics*, 32, 231

Tavecchio, F. *et al.* 2010, *MNRAS*, 401, 1570

Tavecchio, F., Maraschi, L., & Ghisellini, G. 1998, *ApJ*, 509, 608

The Spectral Energy Distribution of Galaxies
Proceedings IAU Symposium No. 284, 2011
R.J. Tuffs & C.C. Popescu, eds.

doi:10.1017/S1743921312009544

Very high energy γ-radiation from the radio quasar 4C 21.35

Josefa Becerra González[1,2], Laura Maraschi[3], Daniel Mazin[4,5], Elisa Prandini[6], Koji Saito[5], Julian Sitarek[7], Antonio Stamerra[8], Fabrizio Tavecchio[3], Tomislav Terzić[9], Aldo Treves[10], MAGIC Collaboration[11]

[1]Inst. de Astrofsica de Canarias, E-38200 La Laguna, Tenerife, Spain; [2]Depto. de Astrofsica, Universidad de La Laguna, E-38206 La Laguna, Spain; [3]INAF National Institute for Astrophysics, I-00136 Rome, Italy; [4]IFAE, Edifici Cn., Campus UAB, E-08193 Bellaterra, Spain; [5]Max-Planck-Institut fr Physik, D-80805 Mnchen, Germany; [6]Universit di Padova and INFN, I-35131 Padova, Italy; [7]University of d, PL-90236 Lodz, Poland; [8]Universit di Siena, and INFN Pisa, I-53100 Siena, Italy; [9]University of Rijeka, HR-51000 Rijeka, Croatia, email: tterzic@phy.uniri.hr; [10]Universit dellnsubria, Como, I-22100 Como, Italy; [11]The full list of collaborators can be found at: wwwmagic.mppmu.mpg.de

Abstract. A very high energy (VHE) γ-radiation was detected from a flat spectrum radio quasar (FSRQ) 4C 21.35 (PKS1222+21) by MAGIC (Major Atmospheric Gamma Imaging Cherenkov) telescopes on June 17th 2010. 4C 21.35 is only the 3rd FSRQ detected in VHE γ-rays. With its hard spectrum ($\Gamma = 2.72 \pm 0.34$) with no apparent cut-off at energies below 130 GeV and an extremely fast variation of flux (doubling in $8.6^{+1.1}_{-0.9}$ minutes), this detection poses a challenge to existing models of VHE γ-radiation from FSRQs. The most important results of observations performed by MAGIC telescopes are presented here, as well as some possible explanations of those results.

Keywords. 4C 21.35, PKS 1222+216, MAGIC, gamma-rays

---

1. MAGIC Telescopes

The Major Atmospheric Gamma Imaging Cherenkov (MAGIC) experiment is a system of two 17 m ground-based IACT (Imaging Atmospheric Cherenkov Telescopes). The telescopes record Cherenkov radiation from particle showers created when a γ-ray enters the atmosphere. These are currently the world's largest telescopes of their kind, and are sensitive to γ-rays of energies above 50–60 GeV. The experiment is situated in the Observatorio del Roque de los Muchachos on the Canary Island La Palma (28°N, 18°W), 2200 m a.s.l. The performance of the MAGIC telescope stereo system is discussed in Aleksić *et al.* (2011a).

2. MAGIC Observations, Data Analysis and Results

4C 21.35 is a Flat Spectrum Radio Quasar (FSRQ). An alternative designation (and the one usually used in VHE (very high energy) astronomy) is PKS 1222+216. It is only the 3rd FSRQ detected in the VHE γ-ray band so far, and it is the 2nd most distant VHE γ-ray source (z = 0.432 Osterbrock & Pogge (1987)) with a well-determined redshift.

MAGIC observed 4C 21.35 between May 1st and June 19th 2010. During that period a total of 16.5 hours of data was collected. A first hint of VHE γ-ray signal was spotted on May 3rd. However, a positive detection (with significance above 5σ) came on June

17th. The discovery was reported on in Aleksić *et al.* (2011b). The details of MAGIC data analysis can be found in Moralejo *et al.* (2009) and Aleksić *et al.* (2010).

4C 21.35 was observed on June 17th for 30 minutes under moderate moonlight conditions. The energy threshold for this observation was 70 GeV. An excess of 190 γ-like events was collected (with a γ-ray rate of 6 min^{-1}) which translates to 10.2 σ. The light curve is shown in Fig.1. The constant flux hypothesis is rejected with high confidence ($P < 1.1 \times 10^{-5}$). The light curve was fitted by a linear and exponential fit. Both fits are acceptable, but the time scale can be determined from exponential fit only, which implies the flux doubling time of $8.6^{+1.1}_{-0.9}$ minutes. This is the fastest time variation ever observed in FSRQ, and among the shortest time scales measured on TeV emitters. The spectrum is shown in Fig.2. The observed spectrum has a slope of $\Gamma = 3.75 \pm 0.27$. VHE γ-rays interact with EBL (extragalactic background radiation) producing $e^- - e^+$ pairs. In order to obtain the intrinsic spectrum of the source, γ-rays that were absorbed have to be added to the measured spectrum. The expected absorption is calculated using EBL models. The intrinsic spectrum with the slope $\Gamma = 2.72 \pm 0.34$ is obtained using an EBL model from Dominguez *et al.* (2011). We see no evidence of a strong intrinsic cut-off below 130 GeV.

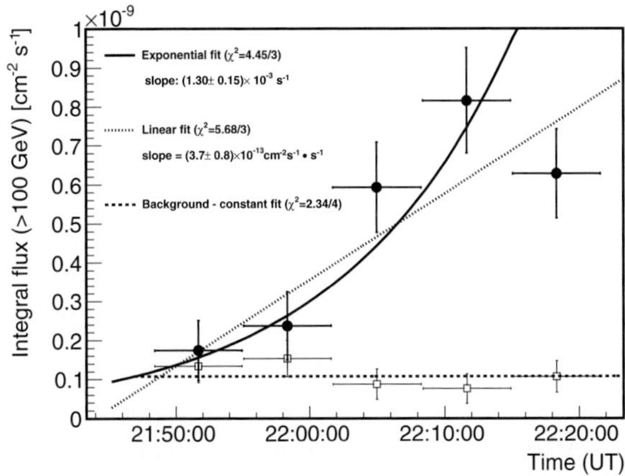

Figure 1. The light curve of 4C 21.35 on June 17th 2010. Black full circles represent the signal, and grey empty squares the background. The full black line is an exponential fit of the signal, while the linear fit is represented by the black dotted line. The background was fitted by a constant shown by the grey dashed line.

3. Discussion

Since we see no evidence of a cut-off in the spectrum at energies below 130 GeV, we conclude that the site for production of VHE γ-rays must be somewhere outside the BLR (broad line region) (see for example Ghisellini & Tavecchio (2009), Reimer (2007), Tavecchio & Mazin (2009), Liu & Bai (2006)). The size of the BLR is estimated to be of the order of 10^{17} cm. At the same time the fast variability of the flux implies an extremely compact emission region with size of the order of 10^{14} cm. These two results taken together exclude the "canonical" emission scenario. One possible explanation of these features is a presence of a compact emission region within large scale jet as already suggested in Ghisellini & Tavecchio (2008), Giannios, Uzdensky & Begelman (2009) and Marscher & Jorstad (2010). Another possibility is a strong jet recollimation at certain point. See for example Nalewajko & Sikora (2009), Bromberg & Levinson (2009), or Stawarz *et al.* (2006). A two-zone scenario was proposed in Tavecchio *et al.* (2011).

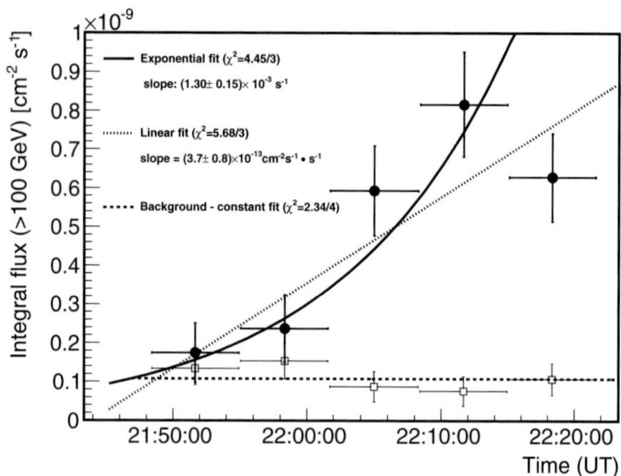

Figure 2. The Differential energy spectrum of 4C 21.35 above 70 GeV. The measured spectrum is represented by black circles, and the black line is the power law fit for observed spectrum. The blue squares represent the intrinsic spectrum, and the blue dashed line is the best fit to a power law of the intrinsic spectrum. The upper limits are given at 95% C.L. and indicated as arrows both for the measured and the intrinsic spectrum. The grey shaded area represents the systematic uncertainties of the analysis.

According to this model the rapidly varying VHE γ-ray emission originates in a small portion of the outflow blob outside BLR, while the emission at lower frequencies originates from the standard emission region. The standard jet can be either inside or outside of the BLR.

A report on a full multi wavelength campaign is in preparation, including a cross correlation study between radio, optical, x-ray and γ-ray data. We hope and expect that the results of that work will shed more light on this intriguing problem.

References

Aleksić, J., *et al.* (the MAGIC Collaboration) 2010, *ApJL*, 726, 58

Aleksić, J., *et al.* (the MAGIC Collaboration) 2011a, *accepted to Astroparticle Physics, arXiv:1108.1477*

Aleksić, J., *et al.* (the MAGIC Collaboration) 2011b, *ApJL*, 730, L8

Bromberg, O. & Levinson, A. 2009, *ApJ*, 699, 1274

Dominguez, A. *et al.* 2011, *MNRAS*, 410, 2556

Ghisellini, G. & Tavecchio, F. 2008, *MNRAS*, 386, 28

Ghisellini, G. & Tavecchio, F. 2009, *MNRAS*, 397, 985

Giannios, D., Uzdensky, D. A., & Begelman, M. C. 2009, *MNRAS*, 395, 29

Liu, H. T. & Bai, J. M. 2006, *ApJ*, 653, 1089

Marscher, A. P. & Jorstad, S. G. 2010, in: T. Savolainen, E. Ros, R. W. Porcas,& J. A. Zensus (eds.), *Rapid Variability of Gamma-ray Emission from Sites near the 43 GHz Cores of Blazar Jets*, Proc. Fermi meets Jansky: AGN in Radio and Gamma-Rays (Bonn: MPI), p. 171

Moralejo, A., *et al.* (the MAGIC Collaboration) 2009, *Proceedings of 31st ICRC, Łódź, Poland, arXiv:0907.0943*

Nalewajko, K. & Sikora, M. 2009, *MNRAS*, 392, 1205

Osterbrock, D. E. & Pogge, R. W. 1987, *ApJ*, 323, 108

Reimer, A. 2007, *ApJ*, 665, 1023

Stawarz, L., Aharonian, F., Kataoka, J., *et al.* 2006, *MNRAS*, 370, 981

Tavecchio, F. & Mazin, D. 2009, *MNRAS*, 392, L40

Tavecchio, F., Becerra-Gonzalez, J., Ghisellini, G., *et al.* 2011, *A&A*, 534, A86

The Spectral Energy Distribution of Galaxies
Proceedings IAU Symposium No. 284, 2011
R.J. Tuffs & C.C. Popescu, eds.

doi:10.1017/S1743921312009556

Synchrotron radiation from giant $e^{\pm}$ pair halos

A. Eungwanichayapant[1,3], W. Maithong[1] and D. Ruffolo[2,3]

[1]School of Science, Mae Fah Luang University, Chiang Rai, Thailand, 57100
[2]Department of Physics, Faculty of Science, Mahidol University, Bangkok, Thailand, 10400
[3]Thailand Center of Excellence in Physics (ThEP), Chiang Mai, Thailand, 50202

email: anant.e@gmail.com

Abstract. A giant $e^{\pm}$ pair halo is formed by electromagnetic cascades developing around an AGN under the intergalactic magnetic field (1nG - 1μG). Many studies have been focussed on the pair halos in the gamma band because it has been predicted that the $e^{\pm}$s in the pair halos up-scatter the Cosmic Microwave Background (CMB) to be gamma-rays. However, the pair halos do not emit only gamma photons but also X-ray photons via synchrotron radiation. In this paper, the Spectral Energy Distributions (SEDs) and the angular distributions of the synchrotron radiation of the pair halos from the Monte Carlo simulations will be discussed.

Keywords. electromagnetic cascades, giant electron/positron pair halos, intergalactic magnetic field

---

1. Introduction

In 1994, the model of giant $e^{\pm}$ pair halo was proposed (Aharonian *et al.* 1994). It shows the calculated spectral energy distributions of the gamma-rays from pair halos. Later on, the angular distributions of the gamma-ray from the pair halos were calculated (Eungwanichayapant & Aharonian 2009). Most of the works on the giant $e^{\pm}$ pair halos have been focused on radiation in the gamma band. Observing of the pair halos have also been done only in the gamma-rays. The first attempt was done by the HEGRA group (Aharonian *et al.* 1999) and recently by Ando and Kusenko (2010).

X-rays from synchrotron radiation is another band that might have the potential for searching the pair halos. We started in this direction by calculating the SEDs and angular distributions of the X-ray from pair halos in various situations.

Giant $e^{\pm}$ Pair Halos Model

The absorption of intrinsic gamma-rays via $\gamma\gamma$ pair production produces $e^{\pm}$ pairs. If the intergalactic magnetic field is strong enough (1nG - 1μG), the $e^{\pm}$ will gyrate before scattering the CMB photons because the gyro radius of the $e^{\pm}$s is smaller than the inverse Compton scattering mean free path (Fig. 2). As a result, the electromagnetic cascades developing around the central source produce a spherical cloud of $e^{\pm}$ pairs. During the gyration of the pairs, they radiate potentially observable synchrotron photons.

2. Methodology

Monte Carlo simulations were used to follow the cascades developing around an AGN. The emission time distributions, $T(E)$, of every cascading $e^{\pm}$ were computed for the calculation of the synchrotron energy distribution:

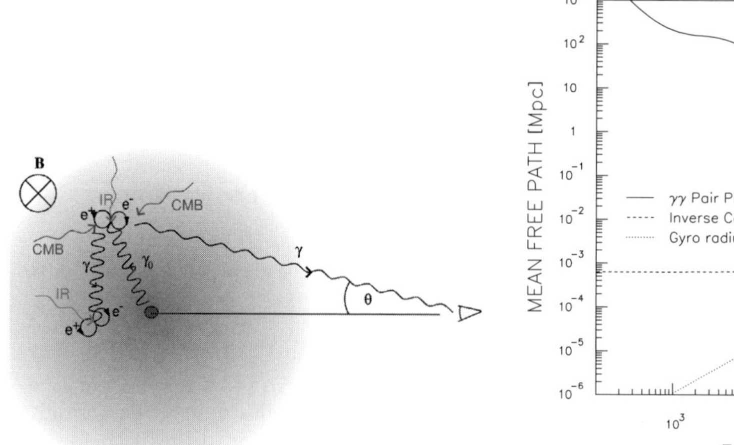

Figure 1. Physical model of a giant $e^{\pm}$ pair halo.

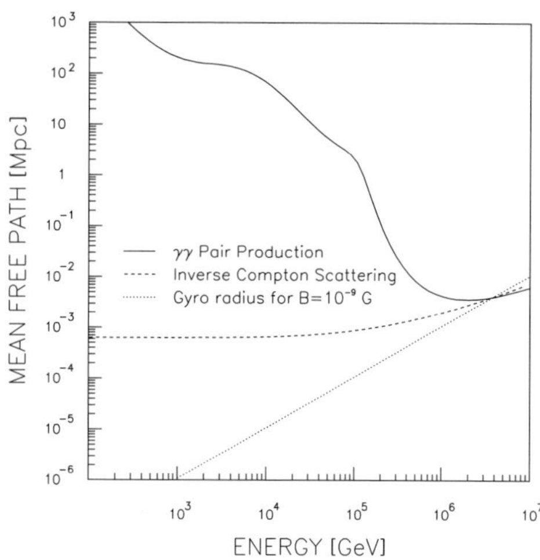

Figure 2. Comparison of the important lengths

$$\frac{dE}{d\nu} = \int \frac{\sqrt{3}e^3 B \sin\alpha}{2\pi mc^2} G(\frac{\nu}{\nu_c}) T(E) dE, \tag{2.1}$$

and converted to the pair halo flux by

$$F = \frac{\mathrm{L}_0}{\mathrm{E}_{\gamma_0} 4\pi d_s^2} E, \tag{2.2}$$

where L_0 is gamma luminosity of the AGN, E_{γ_0} is the total energy of the intrinsic gamma photons emitted by the AGN, and d_s is the distance from the AGN.

The technique of the observer sphere as described in Eungwanichayapant & Aharonian (2009) was applied to find the angle distance. Every time an $e^{\pm}$ emits synchrotron photons, a synchrotron ray was generated randomly. The ray would cross the sphere of radius d_s somewhere and make an angle, θ, to the normal of the sphere at the crossing point. The angle θ is interpreted as the angular distance from the source of the $e^{\pm}$.

3. Results

The spectral energy distributions and angular distributions of the Synchrotron photons from giant $e^{\pm}$ pair halos shown here are the pair halos from monoenergetics sources located at $z = 0.129$ †.

The results in Fig. 3 and Fig. 4 show that when the intrinsic gamma energy, E_0, increased, the SED extended to higher energy, more energy flux contained in higher energy band, but remain the same level in lower energy band. Interestingly, the angular distributions were sensitive to the intrinsic gamma energy between 100 - 500 TeV.

The results in Fig. 5 and Fig. 6 show that when intergalactic magnetic field, B, is increased, the SED extends to higher energy, the energy flux increasing in all bands. The synchrotron pair halos were more extended when the magnetic field got close to 1 μG. However, when the magnetic field was stronger than $\approx 3\mu$G‡ the pair halos became more

† the redshift of 1ES1426+428, one of the promising candidate for detecting a pair halo
‡ This is the level that energy density of the magnetic field and background photon field are equal

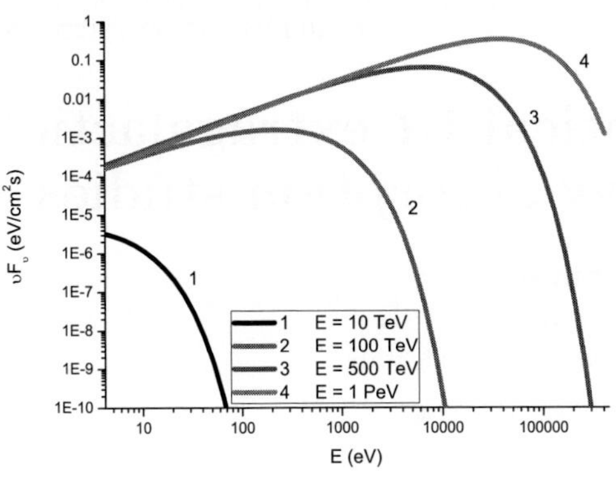

Figure 3. Pair halo SEDs in difference monoenergetic spectra.

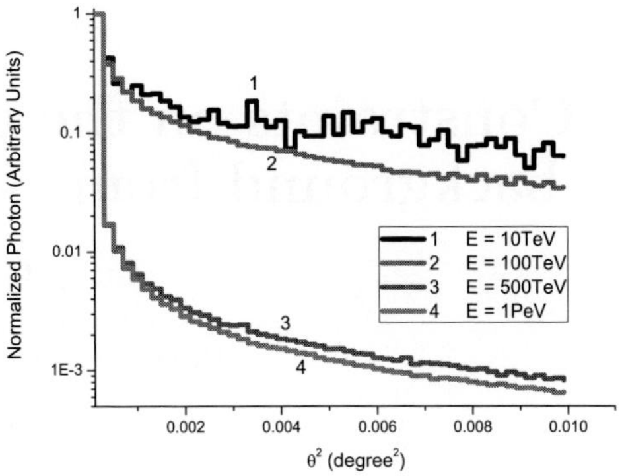

Figure 4. Pair halo angular distributions in difference monoenergetic spectra.

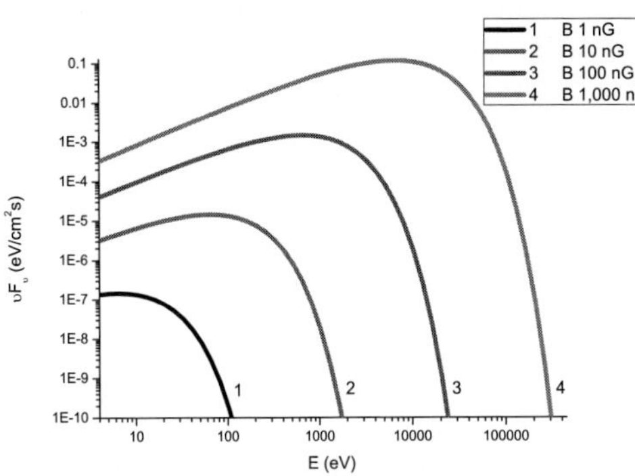

Figure 5. Pair halo SEDs in difference magnetic field strengths.

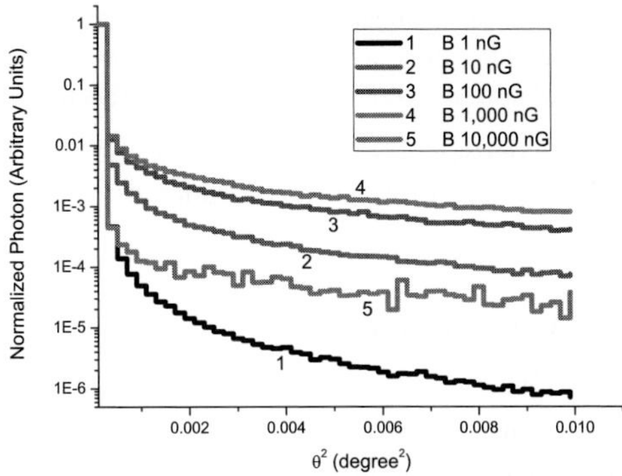

Figure 6. Pair halo angular distributions in difference magnetic field strengths.

centrally peaked since $e^{\pm}$s lose most of their energy via synchrotron radiation and could not generate next generation cascades.

4. Conclusion

The giant $e^{\pm}$ pair halos not only radiate gamma-rays but also X-rays via synchrotron radiation. Our results show that we can gain information about the intergalactic magnetic field at the central source redshift from the SEDs and angular distributions of the pair halos. However, observing the pair halos can be challenging for present X-ray observatories because their energy flux sensitivities do not go down to the predictions for the pair halo levels.

References

Aharonian, F. A., Coppi, P. S., & Voelk, H. J. 1994, *ApJ*, 423, L5

Aharonian F. A. *et al.* 1999, *A&A*, 349, 11

Ando, S. & Kusenko, A. 2010, *ApJ*, 722, L39

Eungwanichayapant, A. & Aharonian, F. A. 2009, *IJMPD*, 18(06), 911

The Spectral Energy Distribution of Galaxies
Proceedings IAU Symposium No. 284, 2011
R.J. Tuffs & C.C. Popescu, eds.

doi:10.1017/S1743921312009568

Constraints on the optical-IR extragalactic background from γ-ray absorption studies

Luigi Costamante[1,2]

[1]Stanford University, Stanford, USA
[2]INAF, Brera Observatory, Milano, Italy

email: luigic2011@gmail.com

Abstract. Very high energy (VHE $\gtrsim$ 0.1 TeV) gamma-rays from extragalactic sources, interacting by $\gamma - \gamma$ collisions with diffuse intergalactic radiation fields, provide an alternative way to constrain the diffuse background light, completely independent of direct measurements. The limits depend however on our knowledge of the physics of the gamma-ray sources. After clarifying the interplay between background light and VHE spectra, I summarize the extent and validity of the obtainable limits, and where future improvements can be expected.

Keywords. galaxies: active, BL Lacertae objects: general, cosmology: diffuse radiation, gamma rays: observations, infrared: galaxies.

---

1. Introduction

VHE γ-rays† from extragalactic sources provide an alternative and completely independent way, with respect to direct measurements, for probing the diffuse Extragalactic Background Light (EBL, meaning from UV to far infrared wavelengths; e.g. Hauser & Dwek 2001). This approach is based on the study of "absorption" features imprinted on the GeV-TeV spectra by the interaction with EBL photons through the pair-creation process ($\gamma\gamma \rightarrow e^+e^-$, see e.g. Aharonian 2001 and references therein). Blazars represent a very useful class of γ-ray beamers, being numerous over a wide range of redshifts, very luminous (enhanced by relativistic beaming) and long-lasting sources. However, they are far from being standard candles, and therefore the constraints on the EBL that can be derived are *always* heavily dependent on our understanding of the blazar intrinsic emission and physical properties.

Nevertheless, it is possible to derive meaningful constraints if the intrinsic blazar properties implied by certain EBL spectra are very far from the observed or expected range, and/or become inconsistent with other features in a blazar's spectral energy distribution (SED). The observations of the last 5 years have allowed substantial progress, and provided the strongest constraints to date. Here both the main results and the limits of their validity are discussed.

2. Diagnostics: how absorption deforms TeV spectra

The convolution of the $\gamma - \gamma$ cross section with the EBL spectrum yields an energy-dependent attenuation. Because the optical depth τ mostly increases with γ-ray energy, the emerging γ-ray spectrum is observed steeper than the initial, intrinsic spectrum

† In the following I will use 'VHE' and 'TeV'as interchangeable terms, meaning ±1 decade around 1 TeV. The same for 'HE' (high-energy, >0.1 GeV) and 'GeV'. Considering where detectors have the best sensitivity and resolution for typical astronomical spectra, Cherenkov telescopes are in fact TeV instruments and the Fermi-LAT detector is a GeV instrument.

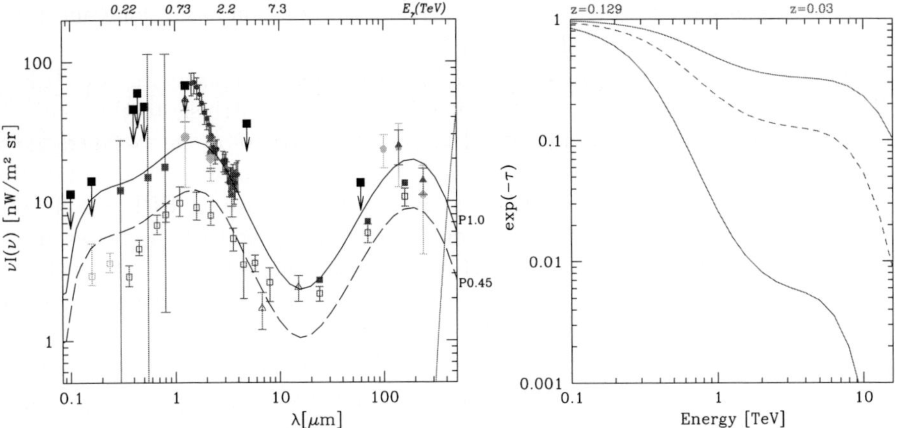

Figure 1. Left: SED of the EBL, as was discussed in the year 2005 (from Aharonian *et al.* 2006). Open symbols show lower limits from galaxy counts, while filled symbols correspond to direct estimates. The annotations on the upper axis denote the VHE energy corresponding to the peak of the $\gamma - \gamma$ cross section. The two curves (identical except for normalization) are plotted as simple shapes reproducing the main EBL features. Right: the attenuation curves $e^{-\tau}$ corresponding to the EBL curves in the left panel (lower solid and dashed lines), and at two different redshifts for the upper EBL curve (lower and upper solid lines). The attenuation curves represent directly the shape of an observed VHE spectrum, if the initial spectrum were a power-law parallel to the upper axis. All calculations performed assuming $H_0 = 70$ km/s/Mpc, $\Omega_m = 0.3$, $\Omega_\Lambda = 0.7$.

emitted by the source (namely, the photon index $\Gamma_{obs} \geqslant \Gamma_{int}$). This can be seen in Fig. 1, where the attenuation factor $e^{-\tau}$ is plotted as a function of energy. Limits on the EBL can be thus derived if the intrinsic spectrum required by a particular EBL intensity or shape is too anomalous (typically too hard) with respect to the known physics of the source (Costamante *et al.* 2004). Several points can be noted from Fig. 1:

1) different EBL wavebands (having different EBL spectra) affect each part of the γ-ray spectrum in a different way, yielding a complex deformation shape. However, over some ranges (namely 0.2-1 TeV and 2-8 TeV), it approximates a power-law shape. That is, if the intrinsic spectrum is a power-law, the observed spectrum can be well fitted by a power-law of steeper index.

2) The amount of steepening $\Delta\Gamma$ increases both with distance and EBL intensity. Thus redshift gives leverage: the same difference in EBL flux $\Delta F_{\rm EBL}$ causes a higher $\Delta\Gamma$ at larger distances. More distant sources provide therefore more sensitivity for EBL constraints, but at the same time they suffer stronger attenuation, yielding much lower statistics. As a result, there is an optimal range of redshifts that provides the best compromise between sensitivity and statistical uncertainty, and thus gives the most stringent EBL limits. With the present detectors, it is around $z \sim 0.15 - 0.25$.

3) If the EBL number density $n(\epsilon) \propto \epsilon^{-\beta}$ (i.e. the EBL SED $\propto \lambda^{\beta-2}$), the optical depth $\tau(E_\gamma)$ becomes $\propto E_\gamma^{\beta-1}$. Where $\beta \approx 1$, the optical depth thus becomes independent of energy (Aharonian 2001). In such case *there is no steepening*: the absorbed spectrum reproduces the original shape, simply attenuated by a constant factor. This is what partly happens in the EBL waveband 3-10 μm, according to all most recent EBL calculations (e.g. Franceschini *et al.* 2008, Dominguez *et al.* 2011, Gilmore *et al.* 2011, Kneiske & Dole 2010). It yields a flattening feature in the attenuation curve between ~1-2 and 8-9

TeV. There, the observed γ-ray spectrum partly recovers its intrinsic slope and thus can appear harder than in the 0.2-1 TeV band †

4) Therefore, *NO cutoff is produced by EBL absorption between 0.2 and 8 TeV*! Any cutoff or steepening seen in this energy range is most likely intrinsic to the source spectrum. Fig. 1 also clarifies why almost all "data points" reported on the so-called "gamma-ray horizon" plots in the literature (showing the E_γ at which $\tau = 1$ as a function of redshift) are basically meaningless: there is no way to measure an EBL cutoff between 0.1 and 10 TeV using only VHE data. The only two cutoffs from EBL are between the HE and VHE bands, i.e. between 20 and 200 GeV (thus requiring a Fermi-LAT spectrum), or above 8-10 TeV (for which only two sources have been detected with sufficient statistics so far).

5) The γ-ray steepening is determined by the difference in optical depth between the two ends of an observed VHE band. Therefore any spectral steepening (and consequently any EBL upper limit) can always be counteracted and cancelled by an appropriate increase of the optical depth (i.e. of the EBL flux) at the lower γ-ray energies, so as to equalize the optical depths. In the 0.1-1 TeV range, this is obtained by increasing the UV flux with respect to the NIR flux (see e.g Aharonian *et al.* 2006). Therefore, *any limit on the EBL depends always on an assumption on the UV flux*, by definition. All claims of "new" methods for "model-independent" or "shape-independent" EBL limits (e.g. Mazin & Raue 2007, Hinton & Hofmann 2009) are therefore essentially incorrect, even if their (hidden) assumptions are reasonable and in fact commonly used (e.g. assuming that the UV flux is always lower than the optical-NIR flux).

3. Optical-NIR constraints

Until 2005, the large uncertainty especially around 0.8-3 μm (a factor 5-10$\times$ between lower limits and direct estimates) caused a fundamental ambiguity in the interpretation of the gamma-ray spectra of blazars. An observed TeV spectrum could be the result either of a hard intrinsic spectrum attenuated by a high-density EBL, or of a soft intrinsic spectrum less absorbed by a low-density EBL. In both cases, the required intrinsic VHE spectrum was well within the typical range shown so far by blazars, so no useful constraint could be derived (with the exception of the 2-10 μm range, see Aharonian *et al.* 2003). This ambiguity essentially undermined our capability to study blazars, since it corresponded to large differences both in the energy and luminosity of the γ-ray peak in the SED.

In 2005, however, the HESS observations of the BL Lac objects 1ES 1101-232 (z=0.186) and H 2356-309 (z=0.165) provided a fundamental breakthrough (Aharonian *et al.* 2006). The observed γ-ray spectra (detected between 0.2 and 1-3 TeV) were much harder than expected for their redshift, implying extremely hard intrinsic spectra ($\Gamma_{int} \lesssim 0$) for high EBL densities (Fig. 2). Such hard spectra were never seen in the closer, less absorbed objects and were at odds with all the known blazar physics and phenomenology. They were also not supported by same-epoch multiwavelength observations of their SED. A low EBL intensity could instead accommodate all the new data within the typical range of blazar properties. Assuming that the intrinsic spectra were not harder than Γ=1.5 an upper limit could be derived, which resulted in limts on the EBL very close to the lower limits given by the integrated light of known galaxies (details in Aharonian *et al.* 2006).

† In fact, $\gamma - \gamma$ absorption can make a gamma-ray spectrum even *harder* than originally emitted, when $n(\epsilon) \propto \epsilon^{>-1}$ (e.g. for a Planckian distribution). This is indeed an effective way to explain very hard γ-ray spectra (Aharonian *et al.* 2008).

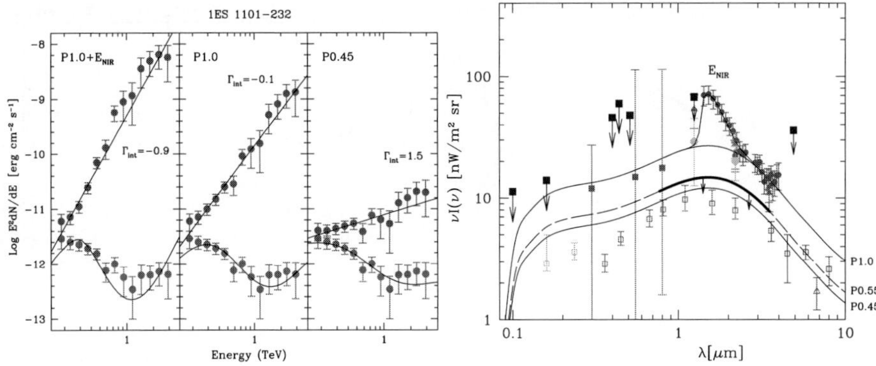

Figure 2. Left: HESS spectra of 1ES 1101-232, corrected for absorption with three different EBL SEDs, as labelled (from Aharonian *et al.* 2006). Lower points: observed data. Upper points: absorption-corrected data. The lines show the best fit power-laws to the reconstructed spectrum, and the corresponding shape after absorption. Right: upper limit for the EBL (black shaded region), derived assuming a blazar spectrum with $\Gamma \geqslant 1.5$.

This result is now further corroborated by several other sources and observations (e.g. Aharonian *et al.* 2007a , Acciari *et al.* 2009, Albert *et al.* 2009).

This result has three main consequences: 1) it pins down the origin of the EBL, showing that at these wavelengths it is strongly dominated by the direct starlight from galaxies, excluding a strong contribution from other sources like Pop-III stars (Santos *et al.* 2002); 2) it means that the intergalactic space is more transparent to γ-rays than previously thought, thus enlarging the γ-ray horizon; 3) it strongly reduces the ambiguity on TeV spectra of blazars.

3.1. *The (in)famous limit of* Γ=*1.5.*

There is a lot of confusion in literature on the validity of the Γ=1.5 hardness limit and on the reasons for its adoption. It is important to understand that *this is NOT the hardest possible theoretical spectrum* in blazars, and was never introduced as such (see Aharonian *et al.* 2006). Indeed there are many possible mechanisms to produce extremely hard gamma-ray spectra in blazars, such as bulk-motion comptonization (Aharonian 2001), internal absoption on narrow-band photon fields (Aharonian *et al.* 2008), uncooled particle acceleration spectra or fine-tuned shock acceleration (yielding however $\Gamma \sim 1.2$, e.g. Stecker *et al.* 2007), a low-energy cutoff in the particle distribution at very high energy (e.g. Katarzynski *et al.* 2006, Lefa *et al.* 2011), or relativistic Maxwellian distributions (e.g. Saugé & Henri 2004, Lefa *et al.* 2011).

The Γ=1.5 limit is a reference value, a benchmark that so far marks *the borderline between reality and speculation.* Observations do show that blazars can have photon spectra with $\Gamma \gtrsim 1.5$ (directly seen in synchrotron or inverse Compton emissions), and these can be produced with standard shock acceleration and cooling mechanisms without invoking special conditions or fine-tuning. On the other hand, much harder spectra have never been significantly detected in blazars so far at high particle energies, neither directly or by synchrotron emission. The alternative scenarios, though not excluded by observations, are not suggested by them either, and in some cases require extreme fine-tuning to avoid conflicts with multiwavelength SED data.

Furthermore, the Γ=1.5 value should not be considered a sharp, hard limit. At any given redshift, EBL absorption establishes a one-to-one relation between the observed and intrinsic γ-ray spectral slope. Thus a different assumption shifts the EBL limit accordingly, but note that only a large change (e.g. $\Gamma_{int} << 1$) can effectively impact the

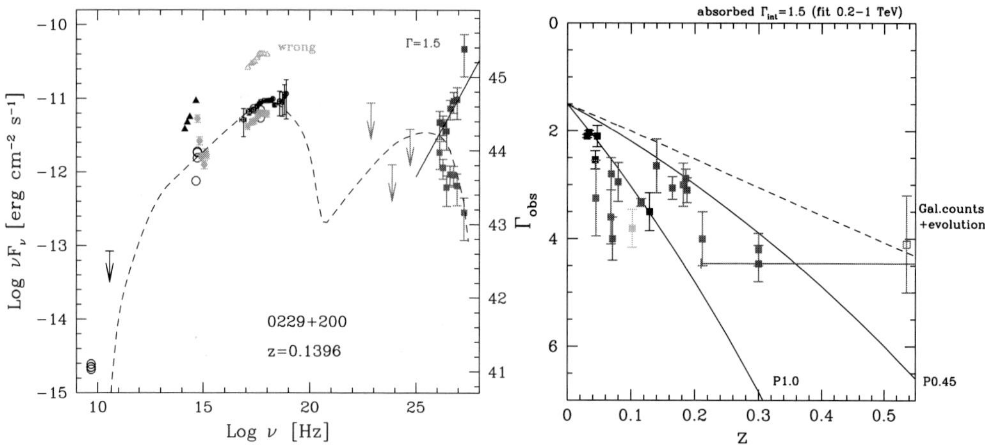

Figure 3. Left: the SED of 1ES 0229+200, showing the correct X-ray data from Swift (filled circles), at much lower flux than the wrong X-ray data (upper open triangles) used in Tavecchio *et al.* 2009. Right: observed VHE photon indices of all detected TeV blazars as a function of redshift. The lines represent the expected observed (i.e. absorbed) photon index of a source with an intrinsic spectrum $\Gamma = 1.5$, when fitted in the range 0.2-1 TeV, for different EBL levels (as labelled). The separation of the data from the lines at each redshift shows how much the intrinsic spectrum is harder or softer than 1.5. Adapted from Costamante 2006.

conclusion of a low EBL. For example, in the case of 1ES 1101-232, a shift of $\Delta\Gamma = 0.3$ (e.g. by assuming $\Gamma_{int} \sim 1.2$) affects the EBL limit by no more than 10% (Aharonian *et al.* 2006). The optical-NIR upper limit is therefore robust to such changes in blazar assumptions, which are of the same order of the statistical and systematic uncertainties in the γ-ray spectra.

The only relevant issue left, therefore, is to assess - theoretically and observationally - if blazars do have emission components characterized by *extremely* hard spectra (e.g. $\Gamma = 0.7$ or less) or not, and how they behave.

4. Why a low EBL density is still the most likely solution.

Despite the several possible scenarios for producing very hard spectra in blazars, there are two main reasons why a low EBL density seems still the correct solution.

The first concerns the spectrum of the emitting particles. Most of these scenarios invoke very hard particle spectra, either as a pile-up/Maxwellian distribution or as a low-energy cutoff, and suppressing radiative cooling. These mechanisms look feasable even in a one-zone synchrotron self-Compton scenario for blazars (see Lefa *et al.* 2011 for a recent comprehensive discussion). However, since the synchrotron emission (which in these objects corresponds to optical–X-ray energies) traces directly the particle spectrum, such hard features should appear in the synchrotron spectrum as well, at some energies and at least in some epochs. Instead, they have never been observed in more than 30 years of X-rays and optical/IR observations of blazars, requiring a "cosmic conspiracy" to always hide the hard emission below a more normal component or in seldom-observed bands (mm). Intriguingly, there was a recent claim of possible observational evidence for a low-energy cutoff in the Swift data of 1ES 0229+200, as an unusually high X-ray-to-UV flux ratio Tavecchio *et al.* 2009. However, a bug was found in the calculation of the effective area of the X-ray data, and the correct result (see Fig. 3) show a spectrum in line with all other objects and observations. Future observations with ALMA (mm) and

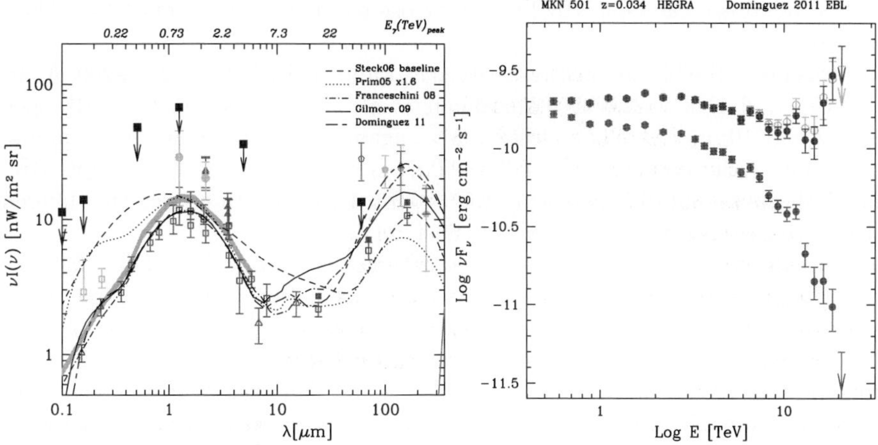

Figure 4. Left: SED of the EBL with the most recent models. The grey band shows the EBL upper limits in the Opt-NIR range (with the $+1\sigma$ statistical uncertainty from the VHE data) re-calculated using the EBL shape by Franceschini *et al.* (2008). Right: HEGRA spectrum of Mkn 501, as observed (red, lower points) and absorption-corrected with two recent EBL-model calculations (Dominguez *et al.* 2011 and Gilmore *et al.* 2011, upper black-blue filled and grey-green open points, respectively). Both cases show an upturn.

NuSTAR (hard X-ray) may provide further insights. Without invoking a hard particle spectrum, internal absorption on a narrow-banded photon field can make the gamma-ray spectrum extremely hard (Aharonian *et al.* 2008) However, large fluxes in the GeV range are required, and these seem now excluded by the low fluxes or non-detection of the hard-TeV sources with Fermi-LAT (Abdo *et al.* 2010).

The second reason for a low EBL density (and thus for $\Gamma_{int} \gtrsim 1.5$) is that a high EBL would require a dramatic change of blazar properties in a very narrow range of redshifts (see Fig. 3). A high EBL would create a sharp dicotomy in the spectra of blazars around $z = 0.15$: all sources below this redshift would have $\Gamma \geqslant 1.5$, while all blazars above this redshift would always have $\Gamma < 1.5$, all other properties being equal or very similar. There is no known reason, observational bias or cosmic physical parameter that can explain such an abrupt change of properties over such a small redshift range. Instead, a low EBL intensity makes the range of TeV blazar spectra consistent among each other for all redshifts sampled so far.

Though not yet “bullet-proof”, until proven otherwise a low EBL seems the preferable solution according to Occam’s razor. (see also Madau & Silk 2005 for the cosmological problems to explain a high EBL flux).

5. The Chain of Constraints

Since the EBL photon field is the same for all sources, and evolving in redshift, the hardness limits on the γ-ray spectrum from different blazars and in different energy bands can be combined to constrain the EBL over a wide range of wavelengths, and in a more stringent way than allowed by each single object (Costamante *et al.* 2004, Dwek & Krennrich 2005, Mazin & Raue 2007). Each new spectrum can take advantage of the previous limits, forming a chain of constraints that starts with (and depends on) the assumption on the UV flux. With the reasonable hypothesis that the UV background is lower than the Opt-NIR one (e.g. around 2-6 nW/m^{2}sr), the blazar spectra between 0.1

and 1 TeV pin down the Opt-NIR flux very close to the galaxy counts limit (Aharonian *et al.* 2006).

With this limit, the blazar spectra between $\sim$ 1 and 10 TeV (1ES 0229+200; Aharonian *et al.* 2007b and 1ES 1426+428; Aharonian *et al.* 2003) constrain the EBL spectrum between 2 and 10μm to a slope close to λ^{-1}, again very close to the lower limits from galaxy counts (Aharonian *et al.* 2007b). This is possible because the Opt-NIR limit precludes the possibility of increasing the $\sim$1 μm EBL flux to reduce the difference in optical depth between 1 and 10 TeV (see Sect. 2).

The consequent upper limit at 10 μm constrains the rising of the EBL SED towards the far-infrared hump, which is determined by warm and cold dust emission. This band is probed by γ-rays between $\sim$ 8 and 100 TeV. So far, only two close-by objects have been measured up to 15-20 TeV, thanks to HEGRA observations: Mkn 421 (z=0.031; Aharonian *et al.* 1999) and Mkn 501 (z=0.034; Aharonian *et al.* 2001). Interestingly, the HEGRA spectrum of Mkn 501 seems to have problems with the most recent EBL calculations, which cause an up-turn or pile-up at the highest energies (see Fig. 4). This is the modern version of the so called "TeV-FIR background crisis" (see e.g. Aharonian 2001), caused by a first very high estimate of the 50 μm EBL flux. However the information on warm dust and the γ-ray statistic are still insufficient to draw any sensible conclusion. More data in the $>$ 10 TeV range are needed (hopefully from CTA).

6. Future and Conclusions

VHE data sample mainly the local EBL ($z = 0 - 0.5$). To probe the EBL evolution over cosmic time and in the UV region, sources up to $z = 3 - 4$ and data in the 10-100 GeV range are needed. The first results from Fermi-LAT (Abdo *et al.* 2010b) are in agreement with the VHE limits and with the most recent EBL evolution models. In this respect, however, GRBs might turn out to be more useful γ-ray sources, given that EBL and blazar-intrinsic evolutionary effects might be very difficult to disentangle (see e.g. Reimer 2007).

At VHE, on te other hand, CTA will NOT improve the Opt-NIR EBL per se, contrary to various claims. The reason is twofold. Firstly, there is not much room for improvement! The present upper limits already match the lower limits by galaxy counts quite closely (see Fig. 4). Secondly, even with infinite γ-ray statistics, there remains the unavoidable systematic uncertainty in blazar modelling. The small $\Delta\Gamma \sim 0.1 - 0.3$ induced by the residual EBL uncertainty between lower and upper limits can be typically accommodated with very small changes in blazar parameters, for a given set of multiwavelength data.

The main question to be addressed is whether our blazar assumptions are correct (in which case the EBL flux is already pinned down) or are not (and by how much). In the latter case, the entire construction must be revisited, since our understanding of the gamma-ray sources would have changed completely. It is on this aspect, on blazar physics, that CTA is expected to provide the most significant improvements and to test our assumptions by finding counter examples. This can be done by monitoring more low-redshift objects in the TeV range (to detect directly $\Gamma < 1$) and by measuring more high-redshift spectra (to find those for which $\Gamma = 1.5$ is incompatible with galaxy counts; Costamante 2007).

In conclusion, a low EBL close to galaxy counts seems the "convergent solution", despite some uncertainties in blazar physics. However, there are still some fundamental aspects of the acceleration and emission mechanisms in blazars that are not yet understood, which present/future observations are/will test. The EBL limits obtained so far are robust against small changes in assumptions about blazars, but they all depend, by construction, on the assumption of a low diffuse UV flux.

Ackowledgements:
I would like to thank the Conference Organizers for the invitation and financial support, and the Max-Planck-Institut für Kernphysik for hospitality and support during my visit.

References

Abdo, A. A., *et al.* (Fermi Coll.) 2010, *ApJ*, 715, 429
Abdo, A. A., *et al.* (Fermi Coll.) 2010b, *ApJ*, 723, 1082
Acciari, V. A., *et al.* (VERITAS Coll.) 2009, *ApJ*, 695, 1370
Aharonian, F. A., *et al.* (HEGRA Coll.), 1999, *A&A*, 350, 757
Aharonian, F. 2001, *Proceedings 27th ICRC (Hamburg)*, Invited, Rapporteur, and Highlight Papers, 250
Aharonian, F. A., *et al.* (HEGRA Coll.), 2001, *A&A*, 366, 62
Aharonian, F. A., *et al.* (HEGRA Coll.), 2003, *A&A*, 403, 523
Aharonian, F. A., *et al.* (H. E. S. S. Coll.) 2006, *Nature*, 440, 1018
Aharonian, F. A., *et al.* (H. E. S. S. Coll.) 2007, *A&A*, 473, 25
Aharonian, F. A., *et al.* (H. E. S. S. Coll.) 2007b, *A&A*, 475, L9
Aharonian, F. A., Khangulyan, D., & Costamante, L. 2008, *MNRAS*, 387, 1206
Albert, J., *et al.* (MAGIC Coll.) 2008, *Science*, 320, 1752
Costamante, L., *et al.* 2004, *NewAR*, 48, 469
Costamante, L., 2007, *Ap&SS*, 309, 487
Dominguez, A. *et al.* 2011, *MNRAS*, 410, 2556
Dwek, E. & Krennrich, F., 2005, *ApJ*, 618, 657
Franceschini, A., *et al.* 2008, *A&A*, 487, 837
Gilmore, R. C., *et al.* 2011, *MNRAS*, submitted (arXiv:1104.0671)
Hauser, M. G. & Dwek, E. 2001, *ARAA*, 39, 249
Hinton, J. A. & Hofmann, W., 2009, *ARAA*, 47, 523
Katarzynski, K., *et al.* 2006, *MNRAS*, 368, L52
Kneiske, T. M. & Dole, H. 2010, *A&A*, 515, A19
Lefa, E., Rieger, F. M., & Aharonian, F., 2011, *ApJ*, 740, 64
Madau, P. & Silk, J. 2005, *MNRAS*, 359, L37
Mazin, D. & Raue, M. 2007, *A&A* 471, 439
Reimer, A., 2007, *ApJ*, 665, 1023
Santos, M. R. *et al.*, 2002, *MNRAS*, 336, 1082
Saugé, L. & Henri, G. 2004, *ApJ* 616, 136
Stecker, F. W., Baring, M. G., & Summerlin, E. J. 2007, *ApJ*, 667, L29
Tavecchio, F., *et al.* 2009, *MNRAS*, 399L, 59

Discussion

MATTILA: Does MAGIC give additional (or better) constraints to the EBL at optical wavelengths, where it should be more sensitive than HESS because of its lower γ-ray energy range ? (I am referring to the 2008 Science paper on 3C279).

COSTAMANTE: In this case (3C 279), not really. The problem is given by the much higher uncertainty in the spectrum. It is also caused by 1) the smaller detected band (less than half a decade, with respect to a full decade in energy for 1ES 1101-232 and the other BL Lacs), and 2) the much lower S/N. The lower γ-ray energy range probes slightly shorter wavelengths, but it also reduces the sensitivity of the γ-ray spectrum to EBL changes, with respect to the 0.2-2 TeV range, despite the higher redshift. In the blue-UV range ("max EBL" curve in their Fig. S2), however, the values are basically arbitrary, and should not be considered limits at all (since a higher UV flux would also make the MAGIC spectrum softer, see Sect 2).

In summary, that observation does not provide more stringent constraints on the optical background, though the mere detection itself corroborates the idea of a low overall EBL (adopting historical fluxes for 3C 279). It is also important to bear in mind that, for this source, there is a further systematic uncertainty due to internal absorption of UV photons from the broad line region (this object has broad lines in the optical spectrum).

The Spectral Energy Distribution of Galaxies
Proceedings IAU Symposium No. 284, 2011
R.J. Tuffs & C.C.Popescu, eds.

doi:10.1017/S174392131200957X

Spectrophotometric measurement of the Extragalacic Background Light

Kalevi Mattila[1], Kimmo Lehtinen[1], Petri Väisänen[2], Gerhard von Appen-Schnur[3], and Christoph Leinert[4]

[1]University of Helsinki, FI-00014 Helsinki, Finland
[2]South African Astronomical Observatory and SALT, Cape Town, South Africa
[3]Astronomisches Instititut, Ruhr-Universität-Bochum, D-44801 Bochum, Germany
[4]Max-Planck-Institut für Astronomie, D-69117 Heidelberg

email: mattila@cc.helsinki.fi

Abstract. The Extragalactic Background Light (EBL) at UV, optical and NIR wavelengths consists of the integrated light of all unresolved galaxies along the line of sight plus any contributions by intergalactic matter including hypothetical decaying relic particles. The measurement of the EBL has turned out to be a tedious problem. This is because of the foreground components of the night sky brightness, much larger than the EBL itself: the Zodiacal Light (ZL), Integrated Starlight (ISL), Diffuse Galactic Light (DGL) and, for ground-based observations, the Airglow (AGL) and the tropospheric scattered light. We have been developing a method for the EBL measurement which utilises the screening effect of a dark nebula on the EBL. A differential measurement in the direction of a high-latitude dark nebula and its surrounding area provides a signal that is due to two components only, i.e. the EBL and the diffusely scattered ISL from the cloud. We present a progress report of this method where we are now utilising intermediate resolution spectroscopy with ESO's VLT telescope. We detect and remove the scattered ISL component by using its characteristic Fraunhofer line spectral signature. In contrast to the ISL, in the EBL spectrum all spectral lines are washed out. We present a high quality spectrum representing the difference between an opaque position within our target cloud and several clear OFF positions around the cloud. We derive a preliminary EBL value at 400 nm and an upper limit to the EBL at 520 nm. These values are in the same range as the EBL lower limits derived from galaxy counts.
Unit: We will use in this paper the abbreviation 1 cgs = 10^{-9}erg s^{-1}cm^{-2}sr^{-1}Å^{-1}

Keywords. Cosmology: diffuse radiation, Galaxy: solar neighborhood

1. Introduction

The importance of the Extragalactic Background Light (EBL) for cosmology has long been recognized (e.g. Partridge & Peebles 1967). The EBL at UV, optical and near infrared wavelengths consists of the integrated light of all unresolved galaxies along the line of sight plus any contributions by intergalactic gas and dust and by hypothetical decaying relic particles. A large fraction of the energy released in the Universe since the recombination epoch is expected to be contained in the EBL. An important aspect is the balance between the UV-NIR and the far infrared EBL: what is lost by dust obscuration in the UV-NIR will re-appear in the FIR. Some central, but still largely open astrophysical problems to be addressed through EBL measurements include the formation and early evolution of galaxies and the star formation history of the Universe. Because of the foreground components, much larger than the EBL itself, the measurement of the EBL has turned out to be a tedious problem. As a consequence we lack a generally accepted measured value of the optical EBL. For a review see Leinert *et al.* (1998).

Recently, Bernstein *et al.* (2002a,b) have announced "the first detection" of the EBL at 300, 550, and 800 nm. They used a combination of space borne (HST) and ground based (Las Campanas) measurements in their method. However, Mattila (2003) has argued that they neglected important effects of the atmospheric scattered light and had a problem with the inter-calibration of the two telescopes used. Therefore, the claim for a detection of the EBL appeared premature. After reanalysis of their systematic errors Bernstein *et al.* (2005) and Bernstein (2007) came to the same conclusions which they formulated as follows: " ... the complexity of the corrections required to do absolute surface (spectro)photometry from the ground make it impossible to achieve 1% accuracy in the calibration of the ZL." and "...the only promising strategy ... is to perform all measurements with the same instrument, so that the majority of corrections and foreground subtractions can be done in a relative sense, before the absolute calibration is applied."

In this situation it is highly desirable to obtain a measurement of the EBL with an independent method which utilises one and the same instrument for all sky components and is virtually free of the large foreground components, the Airglow (AGL), Zodiacal Light (ZL) and tropospheric scattered light.

2. The dark cloud shadow method.

We have been developing over several years a method for the measurement of the EBL which utilises the screening effect of a dark nebula on the background light (see Mattila 1990 for a review and previous results). A differential measurement of the night-sky brightness in the direction of a high galactic latitude dark nebula and its surrounding area, which is (almost) free of obscuring dust, provides a signal that is due to two components only: (1) the EBL and (2) the diffusely scattered starlight from interstellar dust in the cloud (and to smaller extent also in its surroundings). All the large foreground components, i.e. the ZL, the AGL, and the tropospheric scattered light, are completely eliminated (see Fig. 1a). The direct starlight down to $\sim$21-23 mag can be eliminated by selecting the measuring areas with the help of deep images. At high galactic latitudes ($|b| > 30$ deg), the star density is sufficiently low to allow blank areas of sufficient size to be easily found. Models of the Galaxy show that the contribution from unresolved stars beyond this limiting magnitude is of minor or no importance (Mattila 1976). If the scattered light from the interstellar dust were zero (i.e. if the grain albedo $a = 0$), then the difference in surface brightness between a transparent comparison area and the dark nebula would be due to the EBL only, and an opaque nebula would be darker by the amount of the EBL intensity (dashed line in Fig. 1c). The scattered light is not zero, however. A dark nebula in the interstellar space is always exposed to the radiation field of the integrated Galactic starlight, which gives rise to a diffuse scattered light (shaded area in Fig. 1c). Because the intensity of this scattered starlight in the dark nebula is equal to or larger than the EBL, its separation is the main task in our method. The key issue in the dark cloud method is to get a reliable estimate for the scattered light. This can be achieved by means of spectroscopy. With intermediate-resolution spectroscopy ($R \approx 800$) it is possible to determine the strengths of the stronger Fraunhofer lines in the spectrum of the excess surface brightness of the dark cloud. Spectroscopic measurement of a faint surface brightness signal, only $1 - 10\%$ above the dark sky level has been possible with the VLT/FORS long slit spectroscopy.

Target cloud and positions. The high galactic latitude dark nebula Lynds 1642 ($l = 210.9\,\text{deg}, b = -36.7\,\text{deg}$) has been chosen as our primary target. It has a high obscuration ($A_V > 15$ mag) in the centre and areas of good transparency ($A_V \approx 0.15$ mag) in its immediate surroundings within $\sim 1-2$ deg. Its declination of $-14.5\,\text{deg}$ allows observations at low airmasses from Paranal. Our measuring positions ($2'\phi$) for intermediate band photometry (Mattila and Schnur 1990) are shown in Fig. 2 as circles and the positions where photometry and VLT/FORS long slit spectra (2"x6.8') were taken, as squares. Differential measurements with repeated ON/OFF switching cycle of $\sim$ 0.5 h were used to eliminate airglow variations. The spectral range was 360 – 600 nm and the resolution $\Delta\lambda =$ 1.1 nm. Stamps of 10'x10' size centered on the VLT/FORS positions, adopted from DSS blue plates, are shown in the margins: the dark central position #8 ($A_V > 15$ mag) in upper left, followed clockwise by the two intermediate opacity positions #9 and 42, and then the transparent OFF positions. Two observed spectra "dark nebula minus surrounding sky" are shown in Fig. 3: One is for the opaque central position #8 with $A_V > 15$ mag, the other for the average of two intermediate opacity positions, #9 and #42 with $A_V \approx 1$ mag. The crosses show the results of previous intermediate band photometry by Mattila and Schnur (Mattila 1990) with error bars and filter band widths indicated. The spectra have been re-scaled by a multiplicative factor to adjust them to the more reliably calibrated photometry. The intermediate opacity positions are used to validate the model of Galactic starlight spectrum, needed to accurately separate the Galactic contribution from the observed spectrum "dark position minus clear sky". In combination with the dark central position they enable us to disentangle the effects of extinction and reddening by dust from the spectral shape of the Galactic starlight.

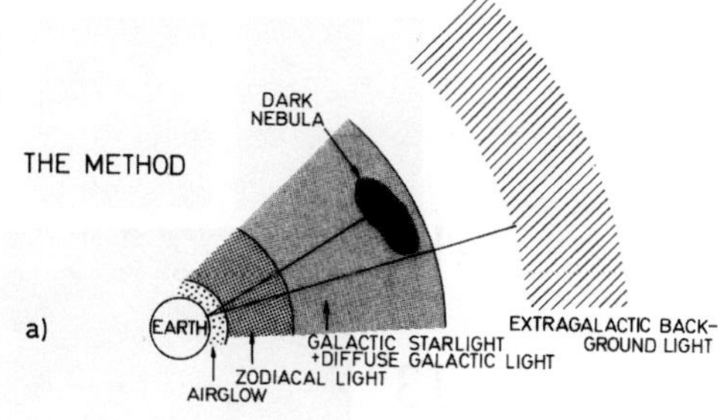

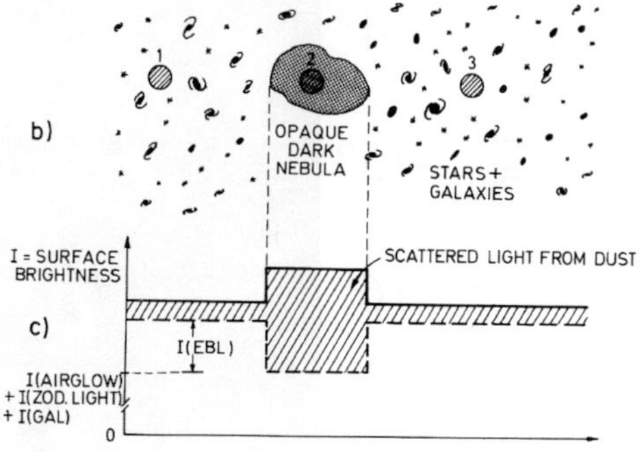

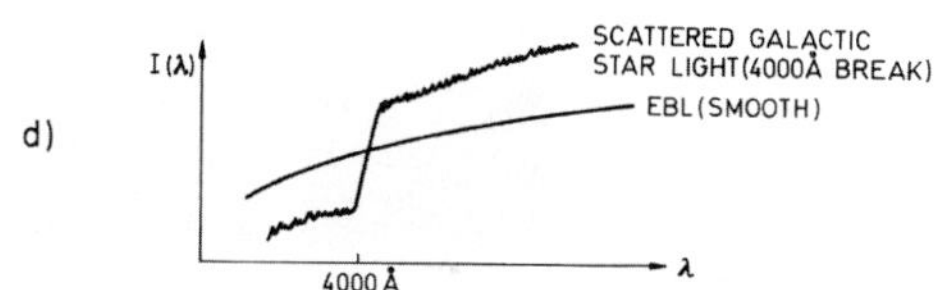

Figure 1. EBL measurement with the dark cloud shadow method

Component separation method. Our separation method utilises the difference in the spectra of the EBL and the integrated Galactic starlight (see Fig 1d). While the scattered Galactic starlight spectrum has the characteristic stellar Fraunhofer lines and the discontinuity at 400 nm the EBL spectrum is a smooth one without these features. This can be understood because the radiation from galaxies and other luminous matter over a vast redshift range contributes to the EBL, thus washing out any spectral lines or discontinuities. The spectrum of the integrated Galactic starlight can be synthesised by

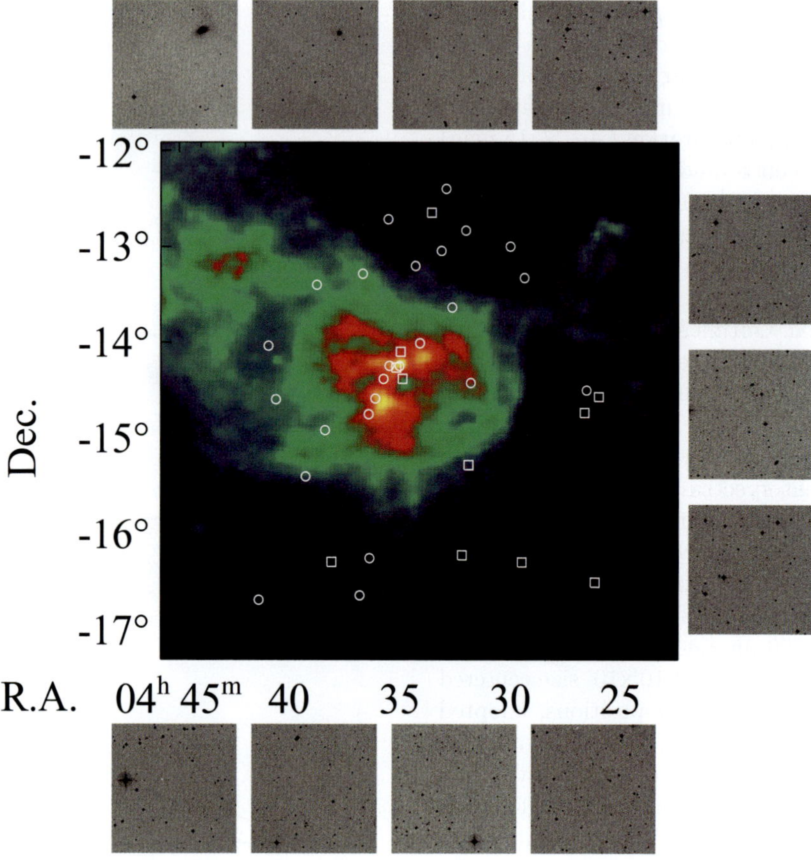

Figure 2. The observed positions in the L 1642 cloud area. superimposed on an IRAS 100 μm map. For details see text.

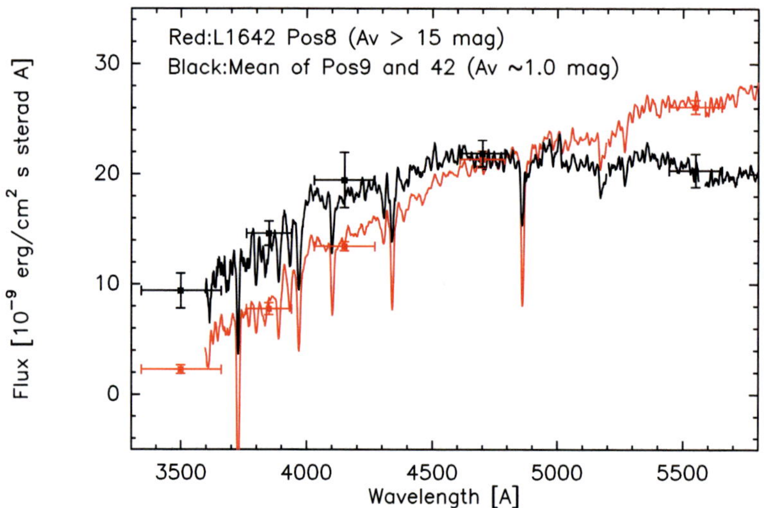

Figure 3. Observed spectrum "dark nebula minus surrounding sky" for the central opaque position #8 and the mean of two translucent positions, #9 and #42.

using the known spectra of representative stars of different spectral classes, as well as the distribution of stars and dust in the Solar neighbourhood. Such synthetic spectra have been calculated using the model of Mattila (1980a,b) and the recent STELIB library of stellar spectra (Le Borgne *et al.* 2003). The resulting ISL spectrum, mean over sky, is shown in Fig. 4 (the uppermost blue line). This spectrum, smoothed to our FORS resolution of 1.1 nm, scaled and reddened to correspond to the dark cloud scattered light spectrum, shows several strong Fraunhofer lines and the discontinuity at 400 nm. For the separation of the scattered light we cannot, however, rely on the integrated starlight *model* alone. We have to test and validate this model by observations at intermediate opacity positions of the dark cloud where most of the EBL is transmitted through the cloud and contributes only little to the surface brightness difference "dark cloud minus surroundings".

3. Separation of the scattered light

The observed surface brightness difference, ΔI, "dark nebula minus surrounding sky" consists of the following main components (followed by + or - depending on whether it makes a positive or negative contribution to ΔI): **(1)** Starlight scattered by dust in the nebula (+); **(2)** Starlight scattered by dust in the comparison fields of the surrounding sky (-); **(3)** diffuse line emission by ionised gas in the comparison fields beyond the distance of the dark cloud (-); **(4)** the Extragalactic Background Light (present in the comparison fields) (-).

In order to derive an estimate for the EBL it is necessary to separate the contribution of scattered light (components 1 and 2). When applying the dark cloud method to intermediate band photometry, we utilised the 400 nm discontinuity which is strong and well defined in the integrated starlight spectrum and is present also in the scattered light spectra as shown in Fig. 3. However, its value as determined using intermediate bands is strongly influenced by the wavelength-dependent extinction and multiple scattering in the cloud which need to be modeled.

We use our model ISL spectra and the observed VLT/FORS spectrum for the opaque central position #8 to separate the scattered light component. The scattered light spectrum is assumed to be a copy of the ISL spectrum but to be reddened by a factor linearly proportional to the wavelength. We show in Fig. 4 an overall model fit for 360 - 550 nm assuming that $I(EBL) = 2$ cgs at OFF and $= 0$ at the #8 ON position. The "ON-position" model spectrum (blue line) is the mean ISL spectrum over the whole sky, suitably scaled (by factor 0.18 at 400 nm) and linearly reddened (by 0.075 per 100 nm). The "OFF-position" model spectrum (green line) is the ISL spectrum for $|b| = 37$ deg, scaled according to the waveleghth dependent extinction (determined from 2MASS JHK stellar and ISOPHOT 200 μm surface photometry). The ON minus OFF model spectrum is shown as red line. The observed ON minus OFF spectrum for Pos #8 (as in Fig. 3) is shown as black line. As can be seen the fit is almost perfect except for the Balmer lines. These lines are contaminated by the excess emission of diffuse ionised gas in the OFF positions. At this stage we do not attempt to model this component (shown as magenta line in Fig. 4 on arbirtrary scale) and exclude the Balmer lines from our fitting procedure.

4. EBL estimates from the dark cloud method

Although a good fit of the observed spectrum by scattered ISL only can be achieved there are small differences in the depths of the Fraunhofer lines (other than the Balmer Lines) indicating that $I(EBL) \neq 0$. We have made fits over suitably selected narrow

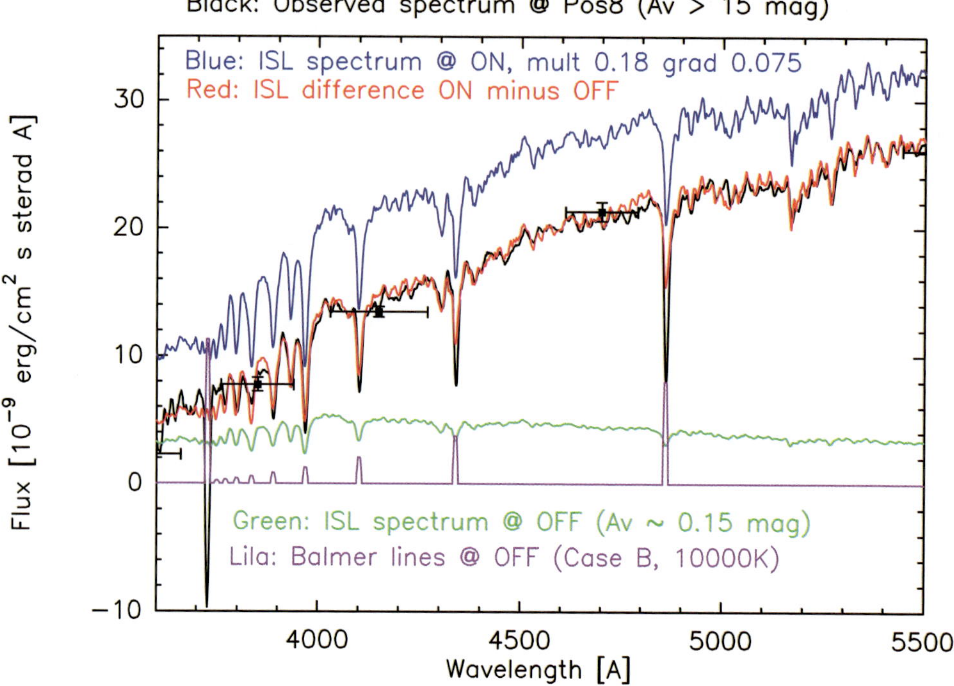

Figure 4. Fitting the central opaque position #8 spectrum with the ISL spectrum. I(EBL) = 2 cgs has been assumed for OFF positions. See text for details.

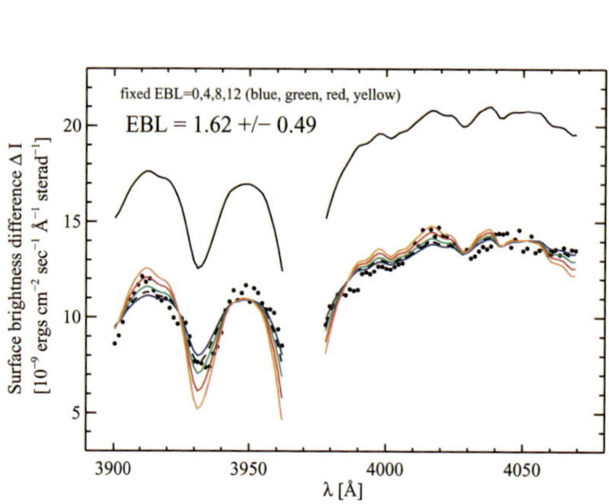

Figure 5. Fitting of the 390 - 407 nm spectral range of the opaque position 8 spectrum with the Integrated Starlight spectrum and different assumed values of the EBL.

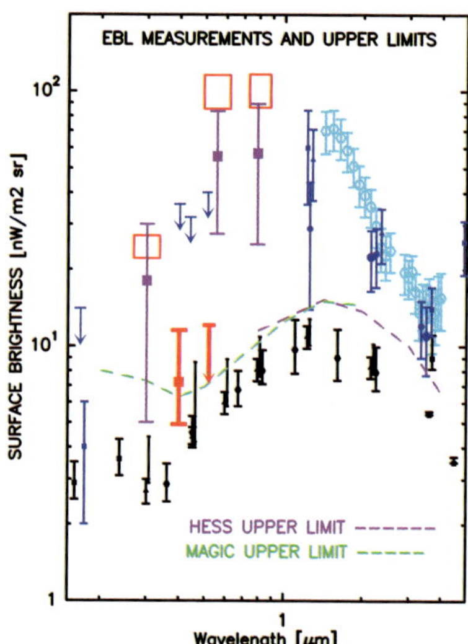

Figure 6. Compilation of current EBL measurements, lower and upper limits. See text for symbols.

wavelength intervals. The EBL is assumed to be constant over the wavelength slot in question. The ISL scaling factor and wavelegth dependent gradient are free parameters. The fit to the range 390 - 407 nm is shown in Fig. 5. The CaII H line at 396.9 nm was not included because it is contaminated by the Hϵ line of ionised gas. Our choice for another suitable slot with strong Fraunhofer lines was 510 - 535 nm. By fitting the observed

spectrum with an ISL spectrum and EBL contributions of different amounts we obtain the following EBL values for these two slots:
$I(EBL)(400 \text{ nm}) = 1.6 \pm 0.5$ cgs (1σ) or 6.4 ± 2.0 nW m^{-2} sterad^{-1}
$I(EBL)(520 \text{ nm}) \leqslant 1.6$ cgs (2σ) or $\leqslant 8.3$ nW m^{-2} sterad^{-1}

The statistical error (1σ) is small enough for putting to the EBL significant limits comparable with the EBL minimum value of $\sim 0.8 - 1.0$ cgs for $\lambda = 360 - 810$ nm derived by summing up deep galaxy counts (Madau and Pozzetti 2000).

Systematic errors. (1) Blocking by the dark cloud is not complete for incident isotropic diffuse radiation like the EBL. For $A_V = 16$ mag the blocking factor is 0.86 (see Mattila 1976, Table A2). (2) The systematic error caused by our ISL model is tested by observing the semi-transparent positions with $A_V \approx 1$ mag for which the scattered starlight has the same spectrum but the EBL contribution is only $\sim 30\%$ of that for $A_V = 15$ mag. They give $I(EBL) \approx 0$ (within 1σ statistical errors) and may thus indicate a need to increase the EBL upper boundary error bar by $\sim 0.3 \times I(EBL)$ or ~ 0.5 cgs.

With a 14% blocking factor correction and increasing the upper bound of the errors (systematic) by 0.5 cgs we end up with the following preliminary EBL estimates:
$I(EBL)(400 \text{ nm}) = 1.8^{+1.0}_{-0.5}$ cgs (1σ) or 7.2^{+4}_{-2} nW m^{-2} sterad^{-1}
$I(EBL)(520 \text{ nm}) \leqslant 2.3$ cgs (2σ) or $\leqslant 12.0$ nW m^{-2} sterad^{-1}

We show in Fig. 6 a compilation from UV to NIR (0.1 - 5 μm) of recent EBL measurements and upper limits (colour symbols) as well as lower limits from galaxy counts (black symbols). Upper limits from the TeV gamma-ray absorption method are shown as dashed lines. The resulting values from the present paper are shown as red solid square with error bars at 400 nm and upper limit at 520 nm. The reanalysis by Mattila (2003) of the Bernstein *et al.* (2002) EBL values are shown as large red rectangles, indicating upper limits. In agreement with these, the reanalysis by Bernstein (2007) resulted in values that were substantially increased from their original estimates. They are shown as the solid magenta squares with error bars. For references to the three upper limits at 400 - 520 nm see Leinert *et al.* (1998), and for the other points see Mattila (2006) and Dominguez *et al.* (2011). For a new EBL estimate (not shown in Fig. 6) based on reanalysis of Pioneer10/11 data see Matsuoka *et al.* (2011) and these Proceedings.

References

Bernstein, R. A. (2007), *ApJ*, 666, 663
Bernstein, R. A. *et al.* 2002a, *ApJ*, 571, 56
Bernstein, R. A. *et al.* 2002b, *ApJ*, 571, 85
Bernstein, R. A. *et al.* 2005, *ApJ*, 632, 713
Dominguez *et al.* 2011, *MNRAS*, 410, 2556
Le Borgne, J.-F. *et al.* 2003, *A&A*, 402, 433
Leinert, Ch. *et al.* 1998, *A&AS*, 127, 1
Madau, P. & Pozzetti, L. 1967, *MNRAS*, 312, L9
Matsuoka, Y. *et al.* 2011, *ApJ*, 736, 119
Mattila, K. 1976, *A&A*, 47, 77
Mattila, K. 1980a, *A&A*, 82, 373
Mattila, K. 1980b, *A&AS*, 39, 53
Mattila, K. 1990, *IAUS*, 139, 257
Mattila, K. 2003, *ApJ*, 591, 119
Mattila, K. 2006, *MNRAS*, 372, 1253
Partridge, B., & Pebbles, P. J. E. 1967, *ApJ*, 148, 377

Discussion

CLEMENTS: Any prospect of extending techniques to the near-IR?

MATTILA: Yes, it might be feasible up to 0.8-1 μm. But the increasing contribution by the - even more rapidly varying - airglow features would make it mandatory to use a rapid ON-OFF duty cyle, perhaps with the "nod & shuffle" technique. The other (better) alternative would be to do the measurements from outside the airglow layer.

TUFFS: Would there be any improvement in accuracy with which the scattered light component can be determined by measuring the FIR/submm SED of the cloud?

MATTILA: We have made absolute photometry with the ISOPHOT instrument on board ISO at 120 and 200 μm for all our photometric/spectroscopic positions. These FIR data are very helpful (even essential) to get a reliable estimate of the remaining small scattered starlight contribution in the transparent positions around the cloud.

The Spectral Energy Distribution of Galaxies
Proceedings IAU Symposium No. 284, 2011
R.J. Tuffs & C.C. Popescu, eds.

doi:10.1017/S1743921312009581

Cosmic Optical Background: the view from Pioneer 10/11

Y. Matsuoka[1], N. Ienaka[2], K. Kawara[2], and S. Oyabu[1]

[1]Graduate School of Science, Nagoya University, Furo-cho, Chikusa-ku, Nagoya, Japan

[2]Institute of Astronomy, The University of Tokyo, Osawa 2-21-1, Mitaka, Tokyo, Japan

email: matsuoka@a.phys.nagoya-u.ac.jp

Abstract. We present the new constraints on the cosmic optical background (COB) obtained from an analysis of the *Pioneer 10/11* Imaging Photopolarimeter (IPP) data. After careful examination of data quality, the usable measurements free from the zodiacal light are integrated into sky maps at the blue ($\sim$0.44 μm) and red ($\sim$0.64 μm) bands. Accurate starlight subtraction is achieved by referring to all-sky star catalogs and a Galactic stellar population synthesis model down to 32.0 mag. We find that the residual light is separated into two components: one component shows a clear correlation with thermal 100 μm brightness, while another betrays a constant level in the lowest 100 μm brightness region. Presence of the second component is significant after all the uncertainties and possible residual light in the Galaxy are taken into account, thus it most likely has the extragalactic origin (i.e., the COB). The derived COB brightness is $(1.8 \pm 0.9) \times 10^{-9}$ and $(1.2 \pm 0.9) \times 10^{-9}$ erg s^{-1} cm^{-2} sr^{-1} Å^{-1} at the blue and red band, respectively, or 7.9 ± 4.0 and 7.7 ± 5.8 nW m^{-2} sr^{-1}. Based on a comparison with the integrated brightness of galaxies, we conclude that the bulk of the COB is comprised of normal galaxies which have already been resolved by the current deepest observations. There seems to be little room for contributions of other populations including "first stars" at these wavelengths. On the other hand, the first component of the IPP residual light represents the diffuse Galactic light (DGL)—scattered starlight by the interstellar dust. We derive the mean DGL-to-100 μm brightness ratios of 2.1×10^{-3} and 4.6×10^{-3} at the two bands, which are roughly consistent with the previous observations toward denser dust regions. Extended red emission in the diffuse interstellar medium is also confirmed.

Keywords. cosmology: observations — dark matter — diffuse radiation — dust, extinction — galaxies: evolution — infrared: ISM — methods: data analysis — space vehicles

---

1. Introduction

The cosmic optical background (COB) is the optical component of the extragalactic background light, which is the integrated radiation from all light sources outside the Galaxy. Dominant contribution to the COB comes from stellar nucleosynthesis in galaxies at redshifts $z < 10$, while other mechanisms such as mass accretion to super massive black holes in active galactic nuclei, gravitational collapse of stars, and particle decay can contribute to the COB brightness. As a fossil record of light production activity in the Universe, the COB conveys information on the cosmic star formation history including birth and death of Population III stars. However, robust detection of the COB has long been hampered by the extremely bright foreground emissions. While expected brightness of the COB is around 1 bgu $\equiv 1 \times 10^{-9}$ erg s^{-1} cm^{-2} sr^{-1} Å^{-1}, the terrestrial airglow and the zodiacal light (ZL) are a few orders of magnitude brighter than this level at optical wavelengths (Leinert *et al.* 1998). The diffuse Galactic light (DGL), which refers to scattered starlight by the interstellar dust, is another but much fainter component of the diffuse light of the night sky. In order to overcome this long-standing problem of foreground removal, we made use of the imaging data obtained by the Imaging Photopolarimeters on board the *Pioneer 10* and *11* spacecrafts (Matsuoka *et al.* 2011). The method and obtained results are presented here.

2. Observations and Reductions

The detailed description of the observations and the primary data processing can be found in, e.g., Weinberg *et al.* (1974) and Gordon *et al.* (1998). Using the *Pioneer 10* IPP data collected at various heliocentric distances R, Hanner *et al.* (1974) find that the ZL brightness is below the detectable level of the instrument when the spacecraft is at beyond $R = 3.26$ AU. Hence the data obtained at the larger distances are most suitable for an analyses of diffuse radiations outside the ZL clouds.

We performed thorough cleaning of the data by removing those with negative flux record, possible contamination of the scattered sunlight, abnormal values of fluxes and/or colors, and so on. Then all the good-quality data were integrated into a single sky map at each of the IPP blue ($B_{\rm IPP}$) and red ($R_{\rm IPP}$) bands, using a similar algorithm to 1st iteration of the "maximum correlation method" (Aumann *et al.* 1990). The final maps have the angular resolution of $\sim 0^\circ.7$. The possible systematic uncertainties were estimated by dividing the data into various subgroups with several criteria such as spacecrafts and heliocentric distances and comparing the sky maps created from each of them. The full description of the data reduction process is given in Matsuoka *et al.* (2011)

3. Galactic Light

Outside the detectable ZL clouds, the dominant brightness component incident on the IPPs is Galactic starlight. Since the contribution of brightest stars has already been subtracted when the archival data were created, we had to subtract the contribution of fainter stars from the IPP measurements. Integrated brightness of relatively bright ($V < 11$ mag) and faint ($V > 11$ mag) stars were calculated from the Tycho-2 Catalog (Høg *et al.* 2000) and the *Hubble Space Telescope* (*HST*) Guide Star Catalog II (GSC-II) version 2.3 (Lasker *et al.* 2008). Furthermore, the contributions of stars even fainter than the GSC-II detection limits were estimated, down to 32.0 mag, using a star count model provided by a stellar population synthesis code TRILEGAL (Girardi *et al.* 2005). Transformations of stellar magnitudes in the different passband systems were derived using the Bruzual-Persson-Gunn-Stryker (BPGS) stellar spectral atlas. We integrated the starlight-subtracted IPP brightness into sky maps with the same algorithm as described above, which gave us the "diffuse emission maps" comprised of diffuse Galactic and extragalactic emission components.

The DGL is attributed to dust and gas particles in the interstellar medium (ISM). Dominant contribution to the optical DGL comes from scattering of the interstellar radiation field (ISRF) by the dust. We estimate the DGL contribution to IPP diffuse emission brightness by separating out the component which correlates with the diffuse Galactic far-IR emission. The good correlation between the DGL and far-infrared (IR) emission brightness is a natural result from the fact that both emissions are caused by the interstellar dust exposed to the ISRF, and is actually observed in several regions (compiled in Bernstein *et al.* 2002). With the DGL-to-100 μm brightness ratio $a_{\rm d}$, the IPP diffuse emission brightness $S^{\rm diffuse}$ can be written as:

$$\begin{aligned} S^{\rm diffuse} &= S^{\rm DGL} + S^{\rm COB} \\ &= a_{\rm d}(S^{\rm diffuse}_{100\mu{\rm m}} - S^{\rm CIB}_{100\mu{\rm m}}) + S^{\rm COB} \end{aligned}$$

where $S^{\rm DGL}$ and $S^{\rm COB}$ are the DGL and COB brightness at the IPP bands, and $S^{\rm diffuse}_{100\mu{\rm m}}$ is diffuse 100 μm brightness observed outside the ZL clouds. The cosmic infrared background (CIB) component $S^{\rm CIB}_{100\mu{\rm m}}$ is subtracted from $S^{\rm diffuse}_{100\mu{\rm m}}$ in the above equation, leaving only Galactic component.

4. Results and Discussion

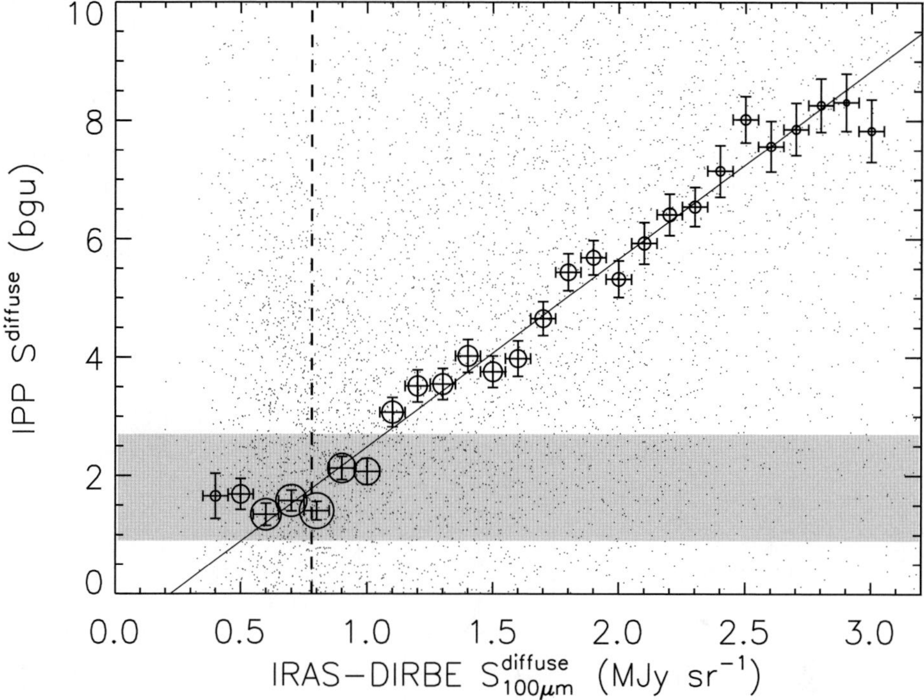

Figure 1. Observed IPP diffuse emission brightness $S^{\rm diffuse}$ at $B_{\rm IPP}$ versus the diffuse 100 μm brightness $S^{\rm diffuse}_{100\mu m}$ (*dots*). The open circles and error bars represent mean values of $S^{\rm diffuse}$ and their errors in the $S^{\rm diffuse}_{100\mu m}$ bins. The sizes of the circles are proportional to numbers of the data points in the bins. The solid line shows the regression line at $S^{\rm diffuse}_{100\mu m} > 1.0$ MJr sr^{-1}, while the dashed line shows the CIB brightness $S^{\rm CIB}_{100\mu m}$ reported by Lagache *et al.* (2000). The shaded area shows 1σ confidence interval of our final COB estimate.

We show the measured IPP diffuse emission brightness $S^{\rm diffuse}$ versus the diffuse 100 μm brightness $S^{\rm diffuse}_{100\mu{\rm m}}$ outside the ZL clouds in Figure 1, at $B_{\rm IPP}$ band as an example. Our analysis focuses on the lowest brightness region with $S^{\rm diffuse}_{100\mu{\rm m}} < 3.0$ MJy sr^{-1} and with the IPP map coverage at the Galactic latitudes $|b| > 35°$, which corresponds to about a quarter of the whole sky. The diffuse 100 μm brightness is taken from Schlegel *et al.* (1998). We clearly detect the linear correlations between $S^{\rm diffuse}$ and $S^{\rm diffuse}_{100\mu{\rm m}}$ at the large $S^{\rm diffuse}_{100\mu{\rm m}}$, and more importantly, the flattening of these relations at $S^{\rm diffuse}_{100\mu{\rm m}} < 0.8$ MJy sr^{-1} in the both IPP bands. The inflection points are in very good agreement with the CIB brightness reported by Lagache *et al.* (2000), which had been predicted in a Monte-Carlo simulation (Matsuoka *et al.* 2011). Our estimates of the COB brightness are derived as the $S^{\rm diffuse}$ values on the observed $S^{\rm diffuse} - S^{\rm diffuse}_{100\mu{\rm m}}$ linear correlations made by the DGL, at the point where 100 μm brightness equals to the CIB. Taking all the uncertainties in the measurement and data reduction processes into account, the obtained results are

$S^{\rm COB} = 1.8 \pm 0.9$ and 1.2 ± 0.9 bgu, or 7.9 ± 4.0 and 7.7 ± 5.8 nW m^{-2} sr^{-1}, at $B_{\rm IPP}$ and $R_{\rm IPP}$ bands, respectively.

We compile the current measurements of the cosmic background and the integrated brightness of galaxies at ultraviolet, optical, and near-IR wavelengths in Figure 2. In contrast to Bernstein (2007), our new results of the COB brightness are mildly larger than and consistent within 1σ uncertainty with the integrated brightness of galaxies measured in the *Hubble* deep field (HDF; Madau & Pozzetti 2000). They are on the smooth extension of the upper limits found by Aharonian *et al.* (2006). With our best estimates, approximately 60 % and 90 % of the COB have already been resolved into discrete galaxies in the HDF at 0.44 μm ($B_{\rm IPP}$) and 0.64 μm ($R_{\rm IPP}$), respectively. On the other hand, Totani *et al.* (2001) demonstrate that 60 – 90 % and 80 – 100 % of the total light from galaxies have been resolved at 0.45 and 0.61 μm. The above facts indicate that bulk of the COB are comprised of normal galaxies, and there are little room for contributions of other populations at these wavelengths.

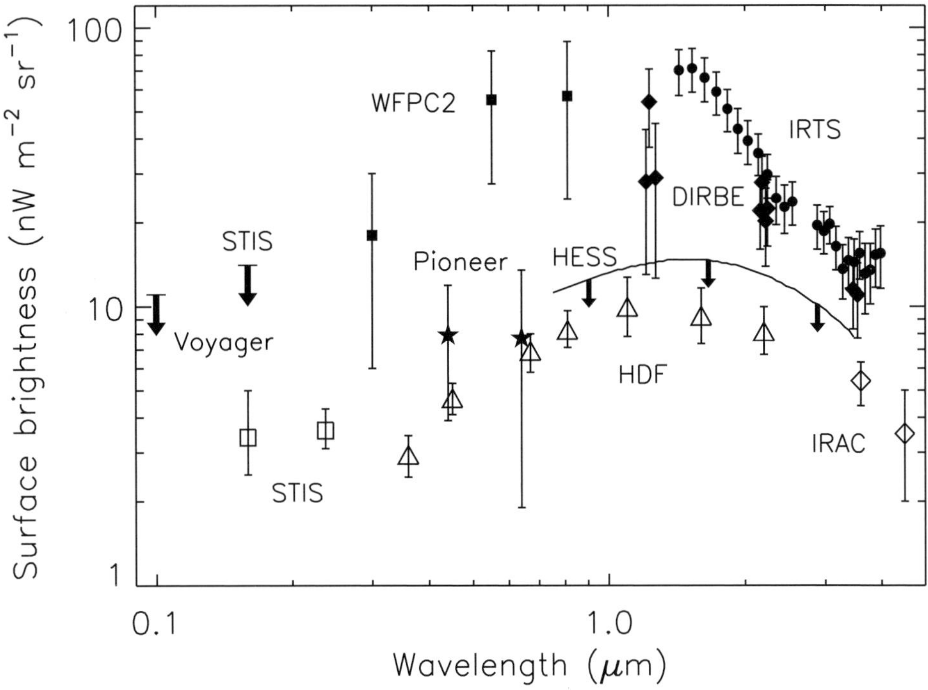

Figure 2. Current measurements of the cosmic background (filled symbols) and the integrated brightness of galaxies (open symbols). At optical wavelengths, squares and stars represent the results by Bernstein (2007) and Matsuoka *et al.* (2011). The solid line with arrows between 0.8 and 4 μm represent the upper limits reported by Aharonian *et al.* (2006). The integrated brightness of galaxies at the corresponding wavelengths are given by Madau & Pozzetti (2000) (triangles). Further details of this figure are described in Matsuoka *et al.* (2011).

References

Aharonian, F., Akhperjanian, A. G., Bazer-Bachi, A. R., *et al.* 2006, *Nature*, 440, 1018
Aumann, H. H., Fowler, J. W., & Melnyk, M. 1990, *AJ*, 99, 1674
Bernstein, R. A. 2007, *ApJ*, 666, 663
Bernstein, R. A., Freedman, W. L., & Madore, B. F. 2002, *ApJ*, 571, 56

Girardi, L., Groenewegen, M. A. T., Hatziminaoglou, E., & da Costa, L. 2005, *A&A*, 436, 895
Gordon, K. D., Witt, A. N., & Friedmann, B. C. 1998, *ApJ*, 498, 522
Hanner, M. S., Weinberg, J. L., Deshields, L. M., II, Green, B. A. & Toller, G. N. 1974, *J. Geophys. Res.*, 79, 3671
Høg, E., Fabricius, C., Makarov, V. V., *et al.* 2000, *A&A*, 355, L27
Lagache, G., Haffner, L. M., Reynolds, R. J., & Tufte, S. L. 2000, *A&A*, 354, 247
Lasker, B. M., Lattanzi, M. G., McLean, B. J., *et al.* 2008, *AJ*, 136, 735
Leinert, C., Bowyer, S., Haikala, L. K., *et al.* 1998, *A&AS*, 127, 1
Madau, P. & Pozzetti, L. 2000, *MNRAS*, 312, L9
Matsuoka, Y., Ienaka, N., Kawara, K., & Oyabu, S. 2011, *ApJ*, 736, 119
Schlegel, D. J., Finkbeiner, D. P., & Davis, M. 1998, *ApJ*, 500, 525
Totani, T., Yoshii, Y., Iwamuro, F., Maihara, T., & Motohara, K. 2001, *ApJ*, 550, L137
Weinberg, J. L., Hanner, M. S., Beeson, D. E., Deshields, L. M., II & Green, B. A. 1974, *J. Geophys. Res.*, 79, 3665

Discussion

MATILLA: This has been a very careful re-analysis of the valuable Pioneer 10/11 background starlight data base. Still, I have two questions on the systematic errors which are crucial for the final result:

Firstly, the IPP gives systematic errors of 8% (Blue) and 13% (Red), but never less than 4 S_{10} $(V)_{G2V}$. You announce and use, instead of these, ten times smaller errors of 1% (or 0.4 background). Can you defend these small systematic errors?

Secondly, the IRAS 100 μm surface brightness values, even after rescaling with COBE/DIRBE, do not have an accurate zero point better than $\sim$ 1 MJy/sterad (perhaps). How does this uncertainty influence your EBL value derived from the I $(\text{optical})_{diffuse}$ vs IRAS 100 μm diagram?

MATSUOKA: With regard to the first question, we have performed the direct comparison between the brightness measured by the IPPs and the independent star catalogues, and found systematic differences of a few percent. I wonder whether the level of $\sim$ 10% systematic errors you mentioned apply to the lowest brightness regions on which our analysis is based.

With regard to the second point, indeed the CIB brightness of Lagache *et al.* (2000), based on the same COBE-DIRBE calibration as the Schlegel *et al.* (1998) 100 μm map we used, is reported to be affected by the residual zodiacal light (ZL). But our results suggest that the similar amount of residual ZL is present in the Schlegel map, and this kind of uncertainty cancels out in our method (see our paper). In any case, we expect to have more reliable results when future far-IR maps, e.g. from the AKARI survey, are available.

The Spectral Energy Distribution of Galaxies
Proceedings IAU Symposium No. 284, 2011
R.J. Tuffs & C.C. Popescu, eds.

doi:10.1017/S1743921312009593

An empirical approach to the extragalactic background light from AEGIS galaxy SED-type fractions

Alberto Domínguez[1,2,3]

[1]UCO/Lick Observatory, Dept. of Astronomy & Astrophysics, University of California, Santa Cruz, CA 95064, USA

[2]Dept. of Physics, University of California, Santa Cruz, CA 95064, USA

[3]Now at: Dept. of Physics & Astronomy, University of Caliornia, Riverside, CA 92521, USA

email: `alberto.dominguez@ucr.edu`

Abstract. The extragalactic background light (EBL) is of fundamental importance both for understanding the entire process of galaxy evolution and for γ-ray astronomy. However, the overall spectrum of the EBL between 0.1 and 1000 μm has never been determined directly, neither from observed luminosity functions (LFs), over a wide redshift range, nor from any multiwavelength observation of galaxy spectral energy distributions (SEDs). The evolving overall spectrum of the EBL is derived here utilizing a novel method based on observations only. It is emphasized that the local EBL seems already well constrained from the UV up to the mid-IR. Different independent methodologies such as direct measurement, galaxy counts, γ-ray attenuation and realistic EBL modelings point towards the same EBL intensity level. Therefore, a relevant contribution from Pop III stars to the local EBL seems unlikely.

Keywords. galaxies: formation, galaxies: evolution, cosmology: observations – diffuse radiation

1. Introduction

The extragalactic background light (EBL) is the accumulated radiation in the Universe from the star formation process, plus a contribution from active galactic nuclei (AGNs). These photons mostly lie in the range ~0.1-1000 μm. The direct measurement of the EBL is a very difficult task subject to high uncertainties. This is mainly due to the contribution of zodiacal light, some orders of magnitude larger than the EBL (e.g., Hauser & Dwek 2001; Chary & Pope 2010). Interestingly, Matsuoka *et al.* (2011) have recently claimed a detection of the EBL free of zodiacal light. Other observational approaches set reliable lower limits on the EBL, such as measuring the integrated light from discrete extragalactic sources (e.g., Madau & Pozzetti 2000; Fazio *et al.* 2004; Keenan *et al.* 2010). On the other hand, there are phenomenological approaches in the literature that predict an overall EBL model (i.e., between 0.1 and 1000 μm and for any redshift). These are basically of the four kinds described in Domínguez *et al.* (2011a) and enumerated here in Table 1. Generally, any EBL modeling is built from two main quantities: one describing the galaxy density evolution over time and another one describing the overall galaxy emission (stellar component plus absorption/re-emission by dust). Table 1 briefly summarizes how these two main quantities are treated in the most relevant EBL modelings in the bibliography. We stress that the previous four-type classification is based upon how the different methodologies describe the galaxy density evolution.

We consider the theoretical approach taken in Somerville *et al.* (2011) and Gilmore *et al.* (2011) as complementary to our observationally motivated one to eventually reach

Type of modeling & refs.	Galaxy density evolution	Galaxy emission
(i) Forward evolution, e.g., Somerville *et al.* (2011), Gilmore *et al.* (2011)	**Semi-analytical models**	**Modeled**. Stellar emission: Bruzual & Charlot (2003) (BC03); Dust absorption: Charlot & Fall (2000); Dust re-emission: templates by Rieke *et al.* (2009)
(ii) Backward evolution, e.g., Franceschini *et al.* (2008)	**Observed** local-optical galaxy luminosity function (LF, for starburst population) and near-IR galaxy LF observed up to $z = 1.4$ (for elliptical and spiral populations)	**Modeled**. A few galaxy types morphologically classified based on optical images.
(iii) Inferred evolution, e.g., Finke *et al.* (2010), Kneiske & Dole (2010)	**Parameterization** of the history of the star formation rate density of the Universe	**Modeled**. Stellar emission: Single bursts of solar metallicity from Bruzual & Charlot (1993) (in Kneiske & Dole 2010)/BC03 (in Finke *et al.* 2010); Dust absorption: General extinction law; Dust re-emission: Modified black bodies. AGN galaxies are not considered.
(iv) Observed evolution, Domínguez *et al.* (2011a)	**Observed** near-IR galaxy LF up to $z = 4$	**Observed**. Based on multiwavelength photometry from the UV up to MIPS 24 for $\sim$ 6000 galaxies up to z=1. Consider 25 different galaxy types including AGN galaxies.

Table 1. Classification and comparison of the main characteristics of recent EBL modelings.

a complete understanding of galaxy evolution. Approaches of type (ii) are potentially problematic because they imply extrapolations backwards in time of local or low-redshift luminosity functions (LFs). Intrinsically different galaxy populations exist at high redshifts, which cannot be accounted for by these extrapolations. In particular, Franceschini *et al.* (2008) use observed LFs in the near-IR from the local Universe to $z = 1.4$ to describe the elliptical and spiral populations, and only local information to describe irregular/starbursting galaxies. They distinguish between galaxy morphologies using images from different instruments. Different local LFs and data sets in the IR are used to constrain the mid- and far-IR background. Their modelling is complex and not reproducible. Despite these problems, this methodology is based upon LFs, which are directly observed and well understood unlike type (iii) models based on parameterizations of the history of the SFR density of the Universe, a quantity with large uncertainties and biases.

One important application of the EBL for γ-ray astronomy is to recover the unattenuated spectra of extragalactic sources. This will not be discussed in this proceedings paper but we refer the interested reader to Domínguez *et al.* 2011a,b for a discussion about this.

2. Methodology

Our model is based on the rest-frame K-band galaxy LF from Cirasuolo *et al.* (2010) and on multiwavelength galaxy data from the All-wavelength Extended Groth Strip International Survey (AEGIS†, Davis *et al.* 2007) of about 6000 galaxies in the redshift

† http://aegis.ucolick.org/

range $0.2 - 1$. The Cirasuolo *et al.* (2010) LF is used to count galaxies (and therefore to normalize the total EBL spectral intensity) at each redshift. The LF as well as our galaxy sample is divided into three magnitude bins according to the absolute rest-frame K-band magnitude, i.e., faint, middle and bright. Within every magnitude bin, an SED type is statistically attached to each galaxy in the LF assuming SED-type fractions that are a function of redshift within those magnitude bins. This is estimated by fitting our AEGIS galaxy sample to the 25 galaxy-SED templates from the SWIRE‡ library (Polletta *et al.* 2007). Then, luminosity densities are calculated from these magnitude bins from every galaxy population at all wavelengths, and finally all the light at all redshifts is added up to get the overall EBL spectrum.

3. Results and conclusions

Fig. 1 shows the local EBL, with its uncertainties, compared with direct and indirect observational data, and other EBL models. Other quantities such as EBL evolution† are discussed in Domínguez *et al.* (2011a). Fig. 1 suggests that the EBL coming from galaxies is already well constrained in the region from the UV up to the mid-IR but not in the far-IR. The EBL measurements free of zodiacal light in two optical bands by Matsuoka *et al.* (2011) agree with our EBL estimations. Furthermore, galaxy counts from very deep surveys taken with very sensitive instruments (Madau & Pozzetti 2000; Fazio *et al.* 2004; Keenan *et al.* 2010) should be considered as a good estimation of the true EBL from galaxies. On the other hand, different fully independent modelings based on different methodologies and galaxy data sets such as Franceschini *et al.* (2008), Gilmore *et al.* (2011), Domínguez *et al.* (2011a) agree in the specific intensity level of the EBL. In particular, galaxy count data are in excellent agreement with our EBL estimations. From these results, a relevant contribution from Pop III stars to the local EBL seems unlikely.

Summarizing, the best available data sets are used in our modeling (Cirasuolo *et al.* 2010's LF and the AEGIS galaxy catalogue) observed over a wide redshift range. This model has the following main advantages over other existing EBL models: transparent methodology, reproducibility, and -very important- utilizing direct galaxy data. The galaxy evolution is directly observed in the rest-frame K band up to $z = 4$. Observed galaxies up to $z = 1$ from the UV up to 24 μm with SEDs of 25 different types (from quiescent to rapidly star-forming galaxies and including AGN galaxies) are taken into account in the same observational framework. A study of the uncertainties in the model (such as uncertainties in the Schechter parameters of the Cirasuolo *et al.* (2010) LF and the errors in the photometric catalogue) is made directly from the data.

It is concluded that the EBL from galaxies seems already well constrained from UV to mid-IR wavelengths, even though uncertainties are still large in the far-IR. Furthermore, discoveries of γ-rays from distant blazars (e.g., Aleksić *et al.* 2011a,b,c) support the EBL specific intensity level derived from galaxy count and recent EBL models such as Gilmore *et al.* (2011), Franceschini *et al.* (2008), and Domínguez *et al.* (2011a). We highlight that the EBL specific intensity calculated with our method matches lower limits from galaxy counts, which implies the highest transparency of the Universe to γ-ray allowed by standard physics (see Domínguez *et al.* 2011c). This predicts a promising future for the new generation of imaging atmospheric Cherenkov telescopes, namely CTA.

‡ http://www.iasf-milano.inaf.it/~polletta/templates/swire_templates.html
† EBL specific intensities are publicly available at http://side.iaa.es/EBL

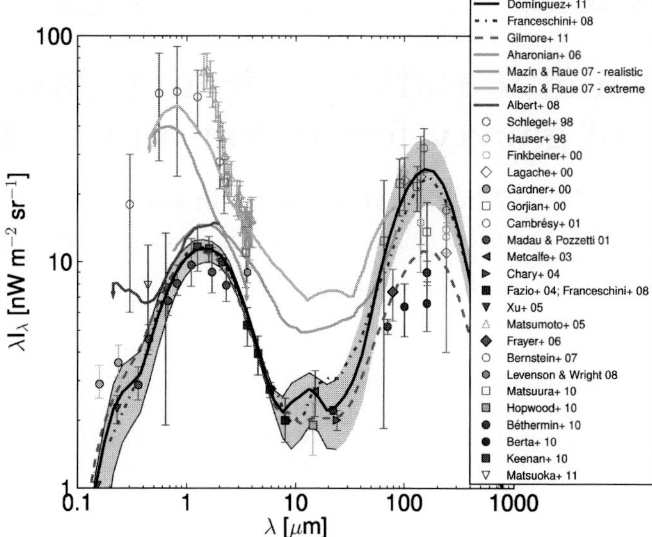

Figure 1. The solid-black line is the extragalactic background light calculated from our methodology. Direct data, data from galaxy count, upper limits from γ-ray astronomy and other recent EBL modelings are shown as well (see Domínguez *et al.* 2011a for details).

Acknowledgements

I am grateful for financial support from a Fermi grant to participate in the IAU 284 Symposium on The Spectral Energy Distribution of Galaxies.

References

Aleksić, J., *et al.* 2011a, *ApJ*, 726, 58
Aleksić, J., *et al.* 2011b, *ApJ*, 730, L8
Aleksić, J., *et al.* 2011c, *A&A*, 530, A4
Bruzual, A. G. & Charlot, S. 1993, *ApJ*, 405, 538
Bruzual, A. G. & Charlot, S. 2003, *MNRAS*, 344, 1000
Charlot, S. & Fall, S. M. 2000, *ApJ*, 539, 718
Chary, R.-R. & Pope, A. 2010, *arXiv*:1003.1731
Cirasuolo, M., *et al.* 2010, *MNRAS*, 401, 1166
Davis, M., *et al.* 2007, *ApJ*, 660, L1
Domínguez A., *et al.* 2011a, *MNRAS*, 410, 2556
Domínguez A., *et al.* 2011b, Proceeding of the 25th Texas Symposium, arXiv:1103.4534
Domínguez, A., Sánchez-Conde, M. A., & Prada, F. 2011c, *JCAP* (in press), *arXiv*:1106.1860
Fazio, G. G., *et al.* 2004, *ApJS*, 154, 39
Finke, J. D., Razzaque, S., & Dermer, C. D. 2010, *ApJ*, 712, 238
Franceschini, A., Rodighiero, G., & Vaccari, M. 2008, *A&A*, 487, 837
Gilmore, R. C., Somerville, R. S., Primack, J. R., & Domínguez, A. 2011, *arXiv*:1104.0671
Hauser, M. G. & Dwek, E. 2001, *ARA&A*, 39, 249
Keenan, R. C., Barger, A. J., Cowie, L. L., & Wang, W.-H. 2010, *ApJ*, 723, 40
Kneiske, T. M. & Dole, H. 2010, *A&A*, 515, A19
Matsuoka, Y., Ienaka, N., Kawara, K., & Oyabu, S. 2011, *ApJ*, 736, 119
Madau, P. & Pozzetti, L. 2000, *MNRAS*, 312, L9
Polletta, M., *et al.* 2007, *ApJ*, 663, 81
Rieke, G. H., *et al.* 2009, *ApJ*, 692, 556
Somerville, R. S., Gilmore, R. C., Primack, J. R., & Domínguez, A. 2011, *arXiv*:1104.0669

The Spectral Energy Distribution of Galaxies
Proceedings IAU Symposium No. 284, 2011
R.J. Tuffs & C.C. Popescu, eds.

doi:10.1017/S174392131200960X

Panchromatic radiation from galaxies as a probe of galaxy formation and evolution

Michael Rowan-Robinson

Astrophysics Group, Blackett Laboratory, Imperial College, London SW7 2AZ

email: m.rrobinson@imperial.ac.uk

Abstract. I review work on modelling the infrared and submillimetre SEDs of galaxies. The underlying physical assumptions are discussed and spherically symmetric, axisymmetric, and 3-dimensional radiative transfer codes are reviewed. Models for galaxies with Spitzer IRS data and for galaxies in the Herschel-Hermes survey are discussed. Searches for high redshift infrared and submillimetre galaxies, the star formation history, the evolution of dust extinction, and constraints from source-counts, are briefly discussed.

Keywords. infrared galaxies, radiative transfer, star formation

---

1. Early work on radiative transfer models for infrared sources

During the 1960s and early 1970s very simplistic models began to be developed for infrared sources associated with circumstellar dust shells, HII regions and star-forming regions. These models assumed that the dust grain density n(r) and temperature T(r) have a power-law dependence on radius r and that the dust is optically thin. Scattering was neglected and so the emission was simply calculated as a sum of $n(r)Q_\nu B_\nu(T(r))$, integrated over r.

The first complete solution of the spherically symmetric radiative transfer problem for an optically thick dust cloud was by Leung (1975). He used detailed grain properties and carried out a full solution of the moment equations, including anisotropic scattering. The grain temperature was solved for by assuming radiative balance. Yorke & Krugel (1977) modelled the dynamical evolution of an HII region, including the interaction between gas and dust. Radiative transfer was approximated by diffusion. These two papers marked a tremendous step forward in the quality of models for infrared sources. To take advantage of numerical techniques which had been used in modelling stellar atmospheres, Leung essentially assumed that the radiation field is isotropic through most of the cloud (the Eddington approximation).

Rowan-Robinson (1980) carried out a numerical solution of the full radiative transfer equation. The intensity, $I_\nu(\theta, r)$, was calculated throughout the cloud using a grid of values in ν, r, θ. This showed that the Eddington approximation is a poor assumption throughout a dust cloud and revealed the phenomenon of sideways beaming of the radiation field at the inner edge of cloud. Viewed from a distance a dust cloud surrounding a star should show a ring of bright emission in the mid infrared, coming from the inner edge of the dust cloud. A second interesting phenomenon revealed by a full solution of the radiative transfer equation is that of back radiation onto the star. The dust cloud radiates at the star and so the net spectrum of emission emerging from the star is no longer a blackbody, but has absorption bands at the wavelengths of dust emission features (Rowan-Robinson 1982). Rowan-Robinson & Harris (1982, 1983ab) applied this code to all the bright circumstellar dust shells in the AFGL survey.

2. Physical assumptions

In modelling the infrared spectrum of star-forming regions and of galaxies, the key issues are the nature of the dust grains and their spatial distribution. A range of grain species and grain radii are needed to account for the main observed features of the interstellar dust extinction profile with wavelength: carbonaceous grains, silicates, and PAHs. While the larger grains tend to be amorphous in most astrophysical environments, the smaller grains may be crystalline, and the very smallest grains, the PAHs, are essentially large molecules. The possibility that the different grain species may be aggregated into larger fractal-like structures has not been much explored.

The earliest radiative transfer models were targeted at either circumstellar dust shells or HII regions associated with massive star formation. For the latter the assumption of spherical symmetry is valid for the early, compact phase of HII region evolution. But many of the more evolved HII regions show a blister-like geometry, as the HII region breaks out of the parent molecular cloud. A starburst in a galaxy will involve stars in both stages of development as well as in the later supernova phase. Not many published models attempt to capture this variety of geometries. In ultraluminous infrared galaxies we are almost certainly dealing with multiple star-forming clouds, so the heating of a cloud by its neighbours can probably not be neglected and a full 3-D treatment is needed.

When we look at the wonderful images of the Milky Way produced by the Herschel mission, we can not fail to be struck by the chiaroscuro distribution of the dust. We see a complex alternation of optically thick and optically thin regions, illuminated from many directions. At the very least a clear distinction between optically thick dust clouds and the optically thin intervening medium needs to be maintained in the models.

3. Spherically symmetric radiative transfer models for starbursts

Despite their rather limited physical validity, it is not surprising that spherically symmetric radiative transfer models have been widely deployed in the literature for massive star-forming regions (e.g. Krugel & Siebenmorgen 1994, Silva *et al.* 1998). Efstathiou, Rowan-Robinson & Siebenmorgen (2000) made such models slightly more realistic by following the evolution of a spherically symmetric HII region within a dense molecular cloud. They assume an embedded phase, lasting for 10^7 years. Following the star's explosion as a supernova, the evolution continues as an expanding neutral shell, lasting from 10^7 to 10^8 years. At 10^8 years the emergent spectrum is virtually indistinguishable from that of the general optically thin interstellar medium (the 'cirrus'). The time-sequence of SEDs calculated by Efstathiou *et al.* (2000) poses the interesting question: can we identify starbursts of different ages observationally (see section 8 below)?

Subsequent work on radiative transfer models for starbursts includes the models of Takagi *et al.* (2003), which includes distributed stars and dust satisfying a King law; the programme of Dopita *et al.* (2005, 2006ab), Groves *et al.* (2008) which incorporate detailed HII region physics; and the large library of parameterised models generated by Siebenmorgen & Krugel (2007).

4. 2-D, axially symmetric, radiative transfer models for infrared sources

The problem of radiative transfer in an axially symmetric dust cloud was first solved by Eftstathiou & Rowan-Robinson (1991) and applied by them to the problem of models for protostars. More recent axially symmetric models for protostars have been developed by Whitney *et al.* (2003, 2004).

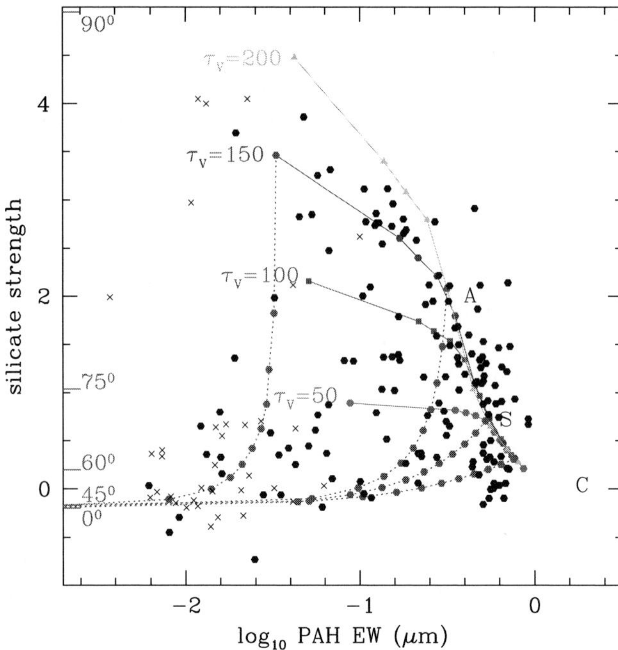

Figure 1. 9.7 μm silicate depth versus 6.2 μm PAH equivalent width for starbursts and AGN (Spoon *et al.* 2007). The observed points are well represented by the grid of models (Rowan-Robinson & Efstathiou 2009).

Axially symmetric radiative transfer models for AGN dust tori were developed by Pier & Krolik (1992), Granato & Danese (1994), and Efstathiou & Rowan-Robinson (1995). The main issue addressed by these papers was why we do not see the 10 μm silicate feature strongly in emission from Type I QSOs. Granato and Danese proposed to solve this by reducing the abundance of silicon, while Efstathiou and Rowan-Robinson suggest that a tapered disk geometry shields the hot dust from our direct line of sight.

High quality mid infrared spectra from the Spitzer IRS for a large sample of starbursts and AGN allow us to test these ideas. Rowan-Robinson & Efstathiou (2009) have shown that the suite of starburst models of Efstathiou *et al.* (2000) and AGN dust tori models of Efstathiou & Rowan-Robinson (1995) successfully explain the distribution of starbursts and AGN in the diagnostic diagram proposed by Spoon *et al.* (2007), 9.7 μm silicate depth versus 6.2 μm PAH equivalent width (Fig 1). While there is aliasing of the models in this diagram, since different combinations of starburst and AGN dust torus models can predict the same point in the diagram, the model sequences do populate the distribution of observed points well. In particular the range of AGN dust tori silicate optical depths, from very deep absorption to weak emission is well matched. It will be interesting to model the individual spectra to gain a more precise determination of the model parameters.

5. 3-D radiative transfer models for clumpy AGN dust tori

A first attempt to approximate a fully 3-dimensional radiative transfer was made by Rowan-Robinson (1995), who argued that the dust torus around AGN might comprise a large number of small but optically thick clouds. The emission from the clouds was approximated by a slab model, and because these clouds are found to be approximately isothermal, the absence of a 10 μm emission feature could be explained. Nenkova *et al.*

(2002, 2008), advocated a 5-10 clump model for NGC1068, also using a slab treatment of the radiative transfer. Fully 3-dimensional radiative transfer codes based on Monte Carlo methods have been developed by Dullemond & van Bemmel (2005), Shartmann *et al.* (2005, 2008), and Hoenig *et al.* (2006). These codes have the capacity to fully track the observed chiaroscuro of the dust distribution in galaxies and AGN.

6. Radiative transfer models for quiescent disk galaxies ('cirrus')

For optically thin dust in the interstellar medium of galaxies we only need to ensure thermal balance between the grains and the interstellar radiation field. The earliest radiative transfer model proposed for the 'cirrus' component in galaxies found by IRAS was given by Rowan-Robinson & Crawford (1989), who modelled all IRAS galaxies detected in all four IRAS wavebands in terms of three components: an M82-like starburst, an AGN dust torus and a cirrus component. This cirrus model was extended by Rowan-Robinson (1992), who found that the key parameter in characterising the cirrus spectrum was the ratio of the intensity of the radiation field to that of the solar neighbourhood, ψ. The shape of the spectrum of the interstellar radiation field was of less importance.

Silva *et al.* (1998) substantially improved this treatment by modelling the interstellar dust in galaxies as a spheroidal distribution. Regions of massive star formation were modelled with a spherically symmetric code to generate a combined cirrus and starburst SED. A chemical evolution code followed the star formation rate, the gas fraction and metallicity, with the gas being divided between the star-forming molecular clouds and the diffuse medium. Further detailed models of spiral galaxies have been developed by Popescu *et al.* (2000, 2011), Misiriotis *et al.* (2001), Dale *et al.* (2001), Tuffs *et al.* (2004), Piovan *et al.* (2006).

Efstathiou & Rowan-Robinson (2003) developed a simple suite of cirrus models for local galaxies, assuming an optically thin interstellar medium, characterised by the extinction A_V (not $>> 1$). The radiation field was determined by Bruzual & Charlot (1993) starburst models, parameterised by the age t_*, and an exponential decay time τ (for local galaxies, $t_* = 0.25$ Gyr, $\tau = 5$-11 Gyr). Galaxies are then characterised by a single mean intensity, ψ = bolometric intensity/solar neighbourhood intensity. They also found that cirrus models, with slightly higher A_V and ψ (~2-3 times higher), can also fit high-z galaxies from the SCUBA blank-field surveys. Restricting analysis to SCUBA sources: (a) which have been confirmed by submm interferometry, or (b) sources from 8 mJy survey which have radio associations, they found that 70% of submillimetre galaxies (16/23) were successfully modelled by cirrus model. More recently Efstathiou & Siebenmorgen (2009) have confirmed this result by modelling 12 submillimetre galaxies which have IRS spectroscopy. They found that 3 were fitted with a pure cirrus model, the rest with a combined cirrus+starburst model.

7. Infrared templates

Several groups have generated suites of models for starbursts, AGN dust tori, and for quiescent (cirrus) galaxies and these have been applied to modelling the infrared and submillimetre SEDs of hundreds of thousands of galaxies. Typically each component has two or three key parameters (for starbursts, the initial optical depth and the age; for AGN dust tori, the inclination and torus geometry; and for cirrus, ψ and perhaps the SED of radiation field). It is beyond the scope of this short review to try to review all these SED modelling efforts. Instead I pick out a couple of issues that have emerged from modelling galaxies detected by Herschel.

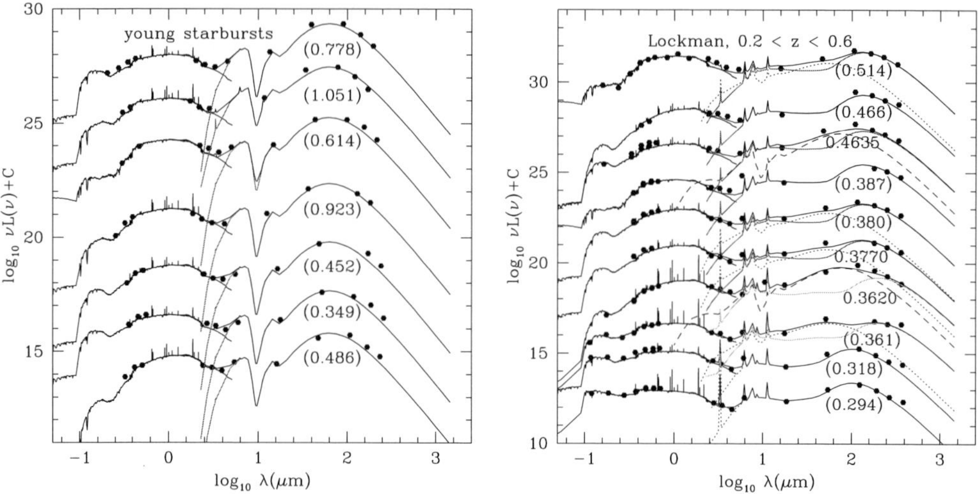

Figure 2. L: Young starbursts in the Hermes survey. R: Galaxies in the Hermes-Lockman area fitted with infrared templates, including cirrus models with $\psi = 1$(objects 5,6,8,10 from bottom) and 0.1 (objects 3, 4 from bottom). Galaxies are labelled with their photometric redshift. From Rowan-Robinson *et al.* (2010).

The first question is: are we beginning to be able to recognise young starbursts ? Rowan-Robinson *et al.* (2010) find a number of galaxies in the Herschel-Hermes survey which seem to require a young starburst model (Fig 2L). Secondly, how can we understand the very cold dust that seems to be seen in some Herschel (and Planck) galaxies ? Fig 2R shows a number of relatively local galaxies which seem to require cirrus components with $\psi = 1$. This is not too startling since such galaxies were noted by Rowan-Robinson (1992) in the IRAS data. What is surprising is that there also seem to be some galaxies with $\psi = 0.1$. If this really is optically thin dust, then it must come from extended dust distributions on scales of tens of kiloparsecs. Bendo (2012, this volume) argues that the cool dust seen in the outer regions of nearby galaxies is heated by evolved stars not star formation. Such cool dust may be seen in dwarf galaxies (Madden 2012, this volume): an example is the dwarf galaxy NGC1705 studied by O'Halloran *et al.* (2010), for which they estimate a dust temperature of 5.8 K.

8. Evolution of galaxies - the infrared view: when was the first infrared light ?

Turning to the evolution of galaxies over cosmological times, the search for high redshift ($z > 5$) galaxies is highly dependent on spectral synthesis models matched to near infrared photometry (Schaerer 2012, this volume). Leitherer (2012, this volume) has given a very nice review of issues in stellar spectral synthesis.

One further issue is that dust is a potential source of aliasing in $z > 5$ photometric redshifts, since extinction of radiation shortward of the Balmer break can mimic the Lyman break. Young stars forming today are emerging from $A_V \sim 100$ clouds. What is the likely dust optical depth in star-forming clouds at high redshift ? We know that quasars at $z > 6$ already have carbon, so these regions have already formed stars, and some have submillimetre emission so dust has formed.

When was the first (rest-frame) infrared light ? Current ideas suggest that reionization took place between $z = 11$ and $z = 6$, the first stars formed at $z \sim 30$, and the first galaxies at z ~ 10-20. How long did it take to get $A_V > 1$ in star-forming clouds and hence most starlight being reprocessed into the infrared ?

If supernovae make enough dust to provide this opacity, then we need only 10^6 years from the formation of the first normal massive stars. However if we need AGB stars to make the dust, then we need at least 10^9 years. It is interesting to note that 1 Gyr from z=10 would be z=4, so redshift 4 is not an unreasonable estimate for the moment when star formation starts to be universally a far infrared phenomenon.

In principle we can learn a lot about this era from dwarf galaxies, with metallicity extending down to 1/30th solar. The halo of our Galaxy also has stars with even lower metallicity, which could be from the z$>$10 era.

9. Submillimetre galaxies

The first z $>$ 2 IRAS galaxy (IRAS F10214+4723) turned out to be bright at 850 μm (Rowan-Robinson *et al.* 1991, 1993). Franceschini *et al.* (1994) pointed out that the negative K-correction at submillimetre wavelengths makes these a window to high redshift. However, the observed redshift-distribution of submillimetre galaxies peaks at z ~ 2-3 (Chapman *et al.* 2005). Source-count models seem to need a redshift cutoff at ~ 4 to explain 850 and 1100 μm counts (see section 12). The Herschel-Hermes survey detects plenty of z $>$ 3 galaxies (eg Rowan-Robinson *et al.* 2010) and Herschel-ATLAS has detected a z = 4.2 starburst galaxy (Cox *et al.* 2011). But it still seems surprising that we have not detected more $z > 4$ submillimetre galaxies.

Blain (1996) emphasised that gravitational lensing could be a big factor in explaining the high luminosities of submillimetre galaxies. It is true that a significant fraction of IRAS hyperluminous infrared galaxies are lensed (Rowan-Robinson & Wang 2010 estimate the lensed percentage to be 10-30%). Negrello *et al.* (2010) show 5 lensed galaxies from the Herschel-Atlas survey. They estimate that a significant fraction (50%?) of bright ($>$100mJy) 500 μm sources are lenses and predict that more than 100 lenses will be found. The combination of the negative K-correction and gravitational lensing should make it relatively easy to search for z = 4-6 infrared galaxies, if they are common.

10. Star formation history

Madau (1996) used Lilley *et al.* survey and HDF to estimate star-formation rate as a function of redshift. Rowan-Robinson *et al.* (1997) used ISO data to derive much higher star-formation rates in the range z = 0-1 ie star-formation rates derived from ultraviolet and optical data need a strong dust correction. In using the far infrared luminosity to estimate star-formation rates it is important to correct for heating by evolved stars (Bell 2003, Rowan-Robinson 2003a). With this correction, the far infrared luminosity is a good estimator for the obscured star formation rate. A recent summary of estimates using different methods has been given by Hopkins (2007). The mean star-formation rate increases steeply from z = 0-1, peaks at z = 1 - 2.5 and then declines towards higher redshift. This star-formation history can then be used to predict how the mean extinction in galaxies should vary with redshift (Rowan-Robinson 2003b, Rowan-Robinson *et al.* 2008). Fig 3 shows observed points are from a study of photometric redshifts in the Hubble Deep Field. The curves are closed box galaxy models with constant yield, instantaneous recycling, and a star-formation history consistent with the observed history, and in which all heavy elements are assumed to be in dust.

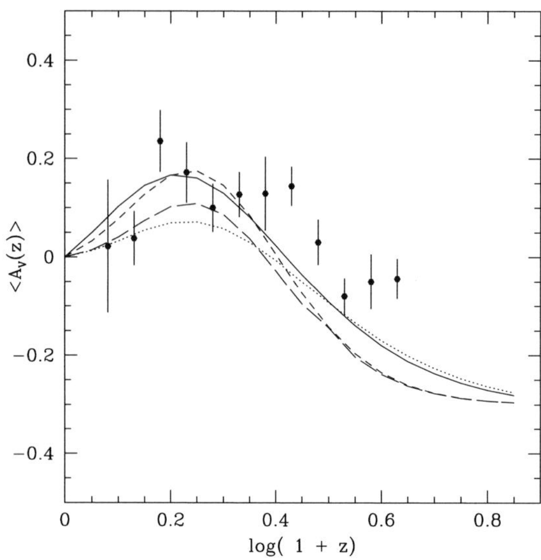

Figure 3. Mean A_V versus z for galaxies in HDFN (Rowan-Robinson 2003). Curves are from closed box, instantaneous recycling, models for the evolution of gas and dust in galaxies.

11. Stellar mass, star formation rate, dust mass

The importance of modelling infrared and submillimetre SEDs with radiative transfer models is that key secondary parameters like star-formation rate and dust mass can be extracted. From stellar spectral synthesis models for optical and near infrared SEDS, the stellar mass can be calculated. For example the SWIRE Photometric Redshift Catalogue provides over 1 million redshifts (Rowan-Robinson *et al.* 2008). For most of these, stellar masses are estimated. For those detected at 8 or 24 μm (25 % of the catalogue), the star-formation rate and dust mass are estimated, as well as the luminosities in each contributing component to the SED fit. Recently a new version, based on improved and extended photometry, has been made available
(http://astro.ic.ac.uk/~mrr/swirephotzcat/zcatrev12ff2.dat.gz).

From stellar synthesis models, we can determine $M_*/L(3.6)$, and then use $L(3.6)$ to get M_*. The evolution of $M_*/L(3.6)$ with time is approximated as $\sim (a + bt^{-0.6})$, which is found to be accurate to 5%.

For starbursts we can estimate the star-formation rate from $\phi_*(M_\odot/yr) = 2.2$ x $10^{-10}(L60/L_\odot)$ (Rowan-Robinson 2001). The specific star formation rate ϕ_*/M_* is then a measure of t_{sf}^{-1}, the inverse of the time to generate the observed stellar mass, forming stars at the current observed rate.

From radiative transfer models for the infrared templates, we can estimate mass of dust, M_{dust} in each component, and sum this. Figure 4 shows the dust mass versus stellar mass, in two different redshift ranges. Infrared galaxies at $z \sim 2$ have both higher specific star-formation rates and dust masses compared to local galaxies.

12. Far infrared and submillimetre source-counts

An important way to constrain the star-formation history and AGN evolution history of galaxies is through models for infrared and submillimetre source-counts and the integrated background radiation. Recent source-count models which try to fit source-counts across the infrared and submillimetre band include those by Rowan-Robinson (2009), Valiante *et al.* (2009) and Bethermin *et al.* (2011). My models appear to require

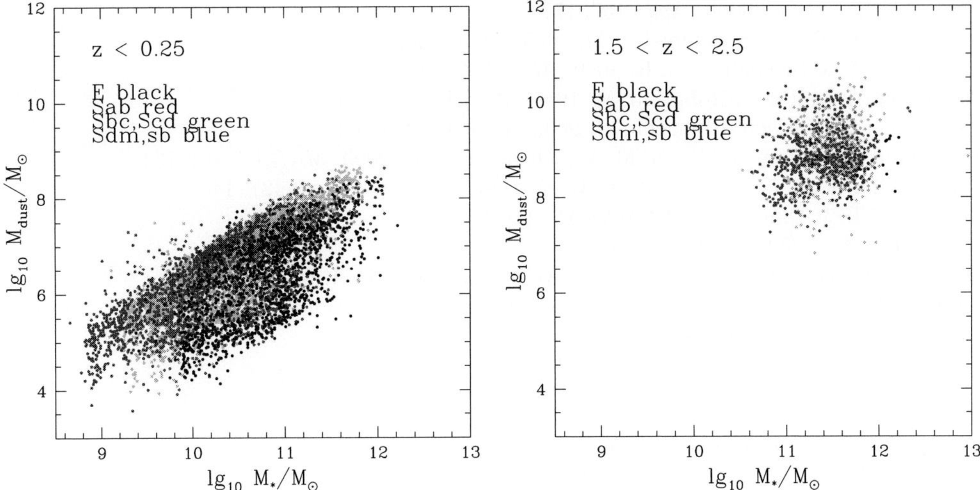

Figure 4. Dust mass versus stellar mass at z < 0.25 and 1.5 < 2.5 for galaxies in the SWIRE survey (Rowan-Robinson *et al.* 2008). Galaxies detected at $z \sim 2$ in infrared surveys are much dustier, and have much higher specific star formation rate, than their local counterparts.

a redshift cutoff at around $z \sim 4$ to be consistent with submillimetre counts and the background radiation.

Oliver *et al.* (2010) have presented counts from the Herschel-Hermes survey at 250, 350, 500 μm and find that the counts, together with a P(D) analysis, account for a large fraction of infrared background. Amblard *et al.* (2011) have analysed clustering in the submillimetre background and conclude that submillimetre galaxies reside in dark matter halos with mass > $3\text{x}10^{11} M_{\odot}$. This minimum mass scale for sub-mm galaxies corresponds to the most efficient dark matter mass scale for star-formation in the Universe.

13. Conclusions

(1) Modelling of ultraviolet to submillimetre SEDs has reached new levels of mathematical and physical sophistication. The use of the simplistic models of the 1960s can not be justified. At the very least the large published libraries of models calculated from radiative transfer codes should be used.

(2) Interesting cosmological insights can be expected from new missions (eg Herschel) and facilities (eg ALMA).

(3) It will be important to understand cold dust, and phenomena at low abundances.(4) It may be hard work finding infrared galaxies at z > 4 because only unusual environments have enough dust.

References

Amblard, A., *et al.* 2011, *Nature*, 470, 510
Bell, E. F. 2003, *ApJ*, 586, 794
Bethermin, M., *et al.* 2011, *AA*, 258, 4
Blain, A. W. 1996, *MNRAS*, 283, 1340
Bruzual, A. G. & Charlot, S. 1993, *ApJ*, 405, 538
Chapman, S., *et al.* 2005, *ApJ*, 622, 772
Cox, P., *et al.* 2011, *ApJ*, 740, 63
Dale, D. A., Helou, G., Contursi, A., Silbermann, N. A., & Kolhatkar, S. 2001, *ApJ*, 549, 215.
Dopita, M. A., *et al.* 2005, *ApJ*, 619, 755
Dopita, M. A., *et al.* 2006a, *ApJ* 647, 244

Dopita, M. A., *et al.* 2006b, *ApJS* 167, 177
Dullemond, C. P. & van Bemmel, I. M. 2005, *A&A*, 436, 47
Efstathiou, A. & Siebenmorgen R. 2009, *A&A*, 502, 541
Efstathiou, A. & Rowan-Robinson, M. 1995, *MNRAS*, 273, 649
Efstathiou, A. & Rowan-Robinson, M. 2003, *MNRAS*, 343, 322
Efstathiou, A., Rowan-Robinson, M. & Siebenmorgen R. 2000, *MNRAS*, 313, 734
Franceschini, A., Mazzei, P., de Zotti, G., & Danese, L. 1994, *ApJ*, 427, 140
Granato, G.L. & Danese L. 1994,*MNRAS*, 268, 235
Groves, B., *et al.* 2008, *ApJS*, 176, 438
Hoenig, S. F., Beckert T., Ohnaka K. & Weigelt G. 2006, *A&A*, 452, 459
Hopkins, A. M. 2007, in 'At the Edge of the Universe', eds. J. Afonso *et al.*, ASP Conf. Ser. 380, 423
Kruegel, E. & Siebenmorgen, R. 1994, *A&A*, 282, 407
Leung, C. M. 1975, *ApJ*, 199, 340
Misiriotis, A., *et al.* 2001, *A&A*, 372, 775
Negrello, M., *et al.* 2010, *Science*, 330, 800
Nenkova, M., Ivezic, Z. & Elitzur M. 2002, *ApJ*, 570, L9
Nenkova, M., Ivezic, Z. & Elitzur M. 2008, *ApJ*, 685, 160
O'Halloran, B., *et al.* 2010, *A&A*, 518, L58
Oliver, S., *et al.* 2010, *A&A*, 518, L21
Pier, G. L. & Krolik J. 1992, *ApJ*, 401, 99
Piovan, L., Tantalo, R., & Chiosi, C. 2006, *MNRAS*, 366, 923.
Popescu, C. C., *et al.* 2000, *A&A*, 362, 138
Popescu, C. C., *et al.* 2011, *A&A*, 527, 109
Rowan-Robinson, M. 1980, *ApJS* 44, 403
Rowan-Robinson, M. 1982, in 'Submillimetre Wave Astronomy', eds. J.E. Beckman and J.P. Phillips (CUP), p.47
Rowan-Robinson, M. & Harris S. 1982, *MNRAS*, 200, 197
Rowan-Robinson, M. & Harris S. 1983, *MNRAS*, 202, 767, 797
Rowan-Robinson, M. & Crawford, J. 1989, *MNRAS*, 238, 523
Rowan-Robinson, M., *et al.* 1991, *Nature*, 351, 719
Rowan-Robinson, M. *et al.* 1993, *MNRAS*, 261, 513
Rowan-Robinson, M. 1992, *MNRAS* 258, 787
Rowan-Robinson, M. 1995, *MNRAS*, 272, 737
Rowan-Robinson, M. *et al.* 1997, *MNRAS*, 289, 490
Rowan-Robinson, M. 2001, *ApJ*, 549, 745
Rowan-Robinson, M. 2003a, *MNRAS*, 344, 13
Rowan-Robinson, M. 2003b, *MNRAS*, 345, 819
Rowan-Robinson, M., *et al.* 2008, *MNRAS*, 386, 697
Rowan-Robinson, M. 2009, *MNRAS*, 394, 117
Rowan-Robinson, M. & Efstathiou, A. 2009, *MNRAS*, 399, 615
Rowan-Robinson, M. & Wang, L. 2010, *MNRAS*, 406, 720
Rowan-Robinson, M. *et al.* 2010, *MNRAS*, 409, 2
Schartmann, M., *et al.* 2005, *A&A*, 437, 861
Schartmann, M., *et al.* 2008, *A&A*, 482, 67
Siebenmorgen, R. & Krugel E. 2007, *A&A*, 461, 445
Silva, L., Granato, G. L., Bressan, A., & Danese, L. 1998, *ApJ*, 509, 103
Spoon, H., *et al.* 2007, *ApJL*, 654, L49
Takagi, T., Arimoto, N., & Hanami, H. 2003, *MNRAS*, 340, 813
Tuffs, R. J., *et al.* 2004, *A&A*, 419, 821
Valiante, E., *et al.* 2009, *ApJ*, 701, 1814
Whitney, B. A., *et al.* 2003, *ApJ*, 591, 1049
Whitney, B. A., *et al.* 2004, *ApJ*, 617, 1177
Yorke, H. W. & Kruegel, E. 1977, *A&A* 54, 183

Discussion

AQUAVIVA: You mentioned that Planck has mainly found low-redshift submm galaxies so far. Can you comment on the implications of this finding?

ROWAN-ROBINSON: The issue for point-source detection by Planck is confusion with cirrus, as the Planet beam is comparable to the IRAS 100 μm beam. There do not appear to be many high galaxies in the ERCSC. I can't predict what the situation will be for the final point-source catalogue.

The Spectral Energy Distribution of Galaxies
Proceedings IAU Symposium No. 284, 2011
R.J. Tuffs & C.C. Popescu, eds.

doi:10.1017/S1743921312009611

Photometric studies of PAH emission from distant infrared galaxies

Toshinobu Takagi[1], Hideo Matsuhara[1], Takehiko Wada[1], Youichi Ohyama[2] and the AKARI extragalactic team

[1]Institute of Space and Astronautical Science, JAXA, Sagamihara, Kanagawa 229-8510, Japan

[2]Academia Sinica, Institute of Astronomy and Astrophysics, Taiwan

email: takagi@ir.isas.jaxa.jp

Abstract. Using extensive mid-IR datasets from AKARI, i.e. 9 band photometry covering the wavelength range from $2\,\mu$m to $24\,\mu$m and the unbiased spectroscopic survey for sources with $S_\nu(9\mu\mathrm{m}) > 0.3$ mJy, we investigated the PAHs emission features in distant starburst galaxies. PAH-selected galaxies, selected with an extremely red mid-IR colour due to PAHs, are found to have a peculiar rest-frame 11-to-8 μm flux ratio, which is systematically smaller than nearby starburst/AGN spectral templates. This may indicate a systematic difference in the physical condition of the ISM between nearby and distant starburst galaxies.

Keywords. infrared: galaxies, galaxies: starburst, ISM: evolution, dust

1. AKARI extragalactic surveys around the NEP

Along with the all-sky survey, AKARI performed 5088 pointed observations in selected areas of sky during its cold phase with liquid helium. Using 13 per cent of these pointed-observation opportunities, we conducted an extragalactic survey around the North Ecliptic Pole (NEP). A prominent characteristic of this survey is its comprehensive mid-IR wavelength coverage, using 9 photometric bands to span the wavelength range from 2 to $24\,\mu$m. Furthermore, we utilized the slitless spectroscopic capability of the IRC onboard AKARI for an unbiased spectroscopic survey at mid-IR wavelengths.

Photometric survey (NEP-Deep): The NEP survey is two-tiered, consisting of the NEP-Deep and NEP-Wide surveys with circular areas of $0.6\,\mathrm{deg}^2$ and $5.8\,\mathrm{deg}^2$, respectively. The field configuration of the NEP survey is shown in Figure 1. Here we mainly use the data from the NEP-Deep survey, which has follow-up observations over the whole electromagnetic spectrum from X-ray to radio. Detailed descriptions of the NEP-Deep survey and the mid-IR source catalogue are given in Wada *et al.* (2008) and Takagi *et al.* (2012, in press), respectively.

Slitless spectroscoPIC surveY of galaxies (SPICY): In the SPICY program, we obtained $5-15\,\mu$m spectra of a flux-limited sample of mid-IR sources with $S_\nu(9\,\mu\mathrm{m}) >$ 0.3 mJy. The SPICY fields are patchily distributed around the NEP-Deep field, covering $\sim$1000 arcmin^2 in total. Most of the SPICY fields lie inside the NEP-Deep field. The spectral resolution of the IRC slitless spectroscopy is $R \simeq 50$ and therefore any narrow fine-structure lines would be smoothed out. The PAH 6.2, 7,7, and $8.6\,\mu$m peaks were simultaneously fitted along with the continuum. From these PAH features, we measured the PAH luminosity as well as the redshift for more than 50 galaxies at $z < 0.5$. The PAH

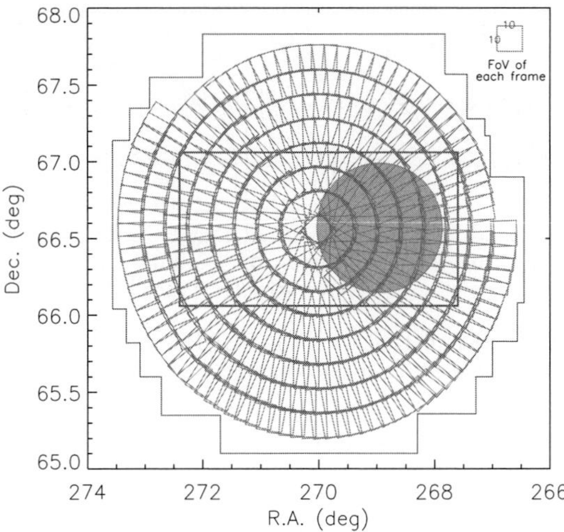

Figure 1. The field-of-view configuration of the NEP survey. The circular area depicted with boxes indicate the NEP-Wide field. The smaller shaded circular area is for the NEP-Deep coverage. The other squarish fields indicate the field of optical follow-up observations.

luminosity [$\nu L_\nu(7.7\,\mu\mathrm{m})$] is found to be 10^9–$10^{11} L_\odot$. The SPICY project and spectral analyses will be presented elsewhere (Ohyama *et al.* in prep).

2. PAH-selected galaxies

The excellent mid-IR wavelength coverage of the NEP survey enables us to photometrically identify galaxies whose mid-IR emission is clearly dominated by PAHs. Takagi *et al.* (2007) demonstrated that the AKARI/IRC all-band photometry is capable of identifying the approximate spectral shape of the PAH emission, specifically the steep rise in flux at the blue side of the PAH 6.2 μm feature. Based on this fact, we selected PAH-dominated galaxies at $z \sim 0.5$ using a single colour, i.e. the 11-to-7 μm flux ratio (see Takagi *et al.* 2010). This flux ratio corresponds to the PAH-to-stellar luminosity ratio for galaxies at $z \sim 0.5$. We call galaxies with this flux ratio greater than 8 "PAH-selected galaxies". These galaxies are expected to have high specific star formation rates (SFR).

For local galaxies, such a large PAH-to-stellar luminosity ratio is found only in intensive star-forming regions (Ohyama *et al.* in prep). For the PAH-selected galaxies, we found a greater PAH-to-stellar luminosity ratio in global measurements, indicating that intensive star formation spreads across the whole body of the galaxies.

3. SED fitting

We analysed the SED of PAH-selected galaxies using an evolutionary SED model of starburst galaxies (Takagi *et al.* 2003 – SBURT), and also with spectral templates of local galaxies. The SBURT model is a radiative transfer model of the spherical starburst region, which has centrally concentrated stars and uniformly distributed dust clouds. Fitting parameters are the starburst age, total mass of the system, the compactness of the starburst region (controlling the optical depth), and the type of the extinction curve (MW, LMC, or SMC). In Fig. 2, we show the example of the resulting best-fitting SED models for optical-NIR and AKARI/IRC all-band photometries. PAH-selected galaxies

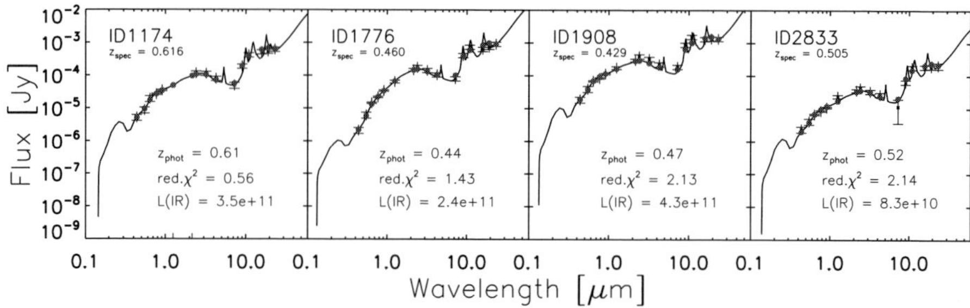

Figure 2. Example of the SED fitting with the SBURT model. Filled circles and data with error bars indicate the model (with filter convolution) and observed fluxes, respectively.

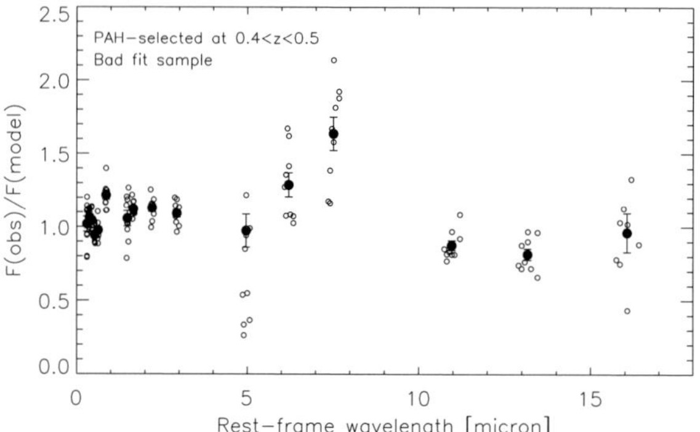

Figure 3. The ratio of the observed to model fluxes of the PAH-selected galaxies at $0.4 < z < 0.5$ versus rest-frame wavelength. The circles with error bars indicate the average flux ratios.

are all fitted with the dust model of the MW, which has the highest PAH fraction in dust, i.e. 5 per cent.

We found that the SBURT model can provide acceptable models for the half of the PAH-selected galaxies based on the χ^2 values. For the remaining half of the sample, we show the ratio of the observed-to-model fluxes as a function of the rest-frame wavelength in Fig. 3. These flux ratios indicate that the fit is rather good, except for the 5–7 μm wavelength range where the most prominent PAH features are found. A similar trend can be seen for the good fit sample as well.

4. PAH interband flux ratio

The SED fitting analysis indicate a systematic variation of the PAH features, compared to the local template of the PAH features. Fig. 3 shows that the model reproduces the rest-frame 11–15 μm emission well, but indicates a large anomaly in the rest-frame 11/8 μm flux ratio, i.e. the flux ratio of C-H to C-C mode emission of PAHs. A possible origin of this deviation is the ionization state of PAHs, which affect the absorption cross section of PAHs specifically around 8 μm (e.g. Li & Draine 2001).

From the SPICY dataset, we measured the inter-band flux ratio of PAH 6.2-to-7.7 μm and found that this inter-band ratio is similar to the values of nearby starburst galaxies. Unfortunately, the wavelength coverage of the SPICY is not large enough to cover the PAH 11.3 μm. However, the excellent wavelength coverage of photometric data allows us to estimate the PAH luminosity without a significant k-correction. With the SPICY data,

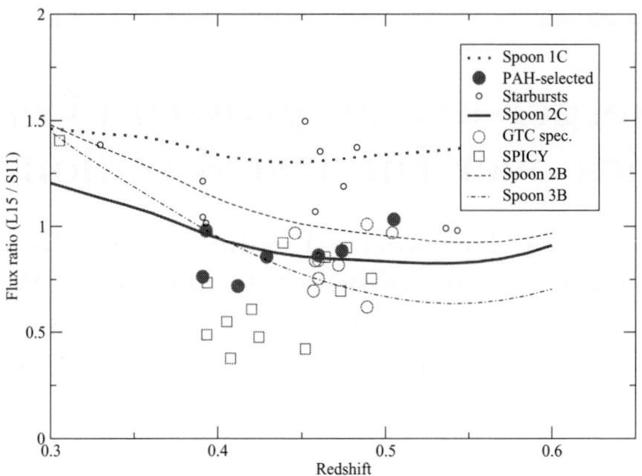

Figure 4. The 15-to-11 μm flux ratio of PAH-selected galaxies at $z_{\rm spec} \sim 0.4$. This flux ratio corresponds to the 11-to-8 μm flux ratio at $z = 0.4$. Different lines indicate the flux ratios expected from nearby starburst galaxies/AGNs taken from Spoon *et al.* (2007).

we confirmed that photometrically measured PAH 7.7 μm luminosities linearly correlate with the spectroscopic measurements.

We investigated the rest-frame 11/8 μm flux ratio of the PAH-selected galaxies at $z \sim 0.4$, for which a spectroscopic redshift is available. At this redshift, the rest-frame 11/8 μm flux ratio can be measured from the AKARI 15-to-11 μm flux ratio. In Fig. 4 we compare these flux ratios with those expected from a spectral template of Spoon *et al.* (2007) which has spectra with a wide variety of the PAH equivalent widths and silicate absorption. We found that the observed flux ratios of the PAH-selected galaxies are too small to be reproduced with the nearby spectral template. This trend is consistent with the scenario that PAHs are mostly ionized in the PAH-selected galaxies.

In a resolved image of M82 (see Arimatsu *et al.*; these proceedings), similar anomalies of the inter-band flux ratio can be found only in small spots around the centre. In PAH-selected galaxies, the anomaly is found for the global measurements, indicating a systematic difference in the physical condition of the ISM between nearby and distant starburst galaxies.

References

Li, A. & Draine, B. T. 2001, *ApJ*, 554, 778
Spoon, H. W. W., Marshall, J. A., Houck, J. R., *et al.* 2007, *ApJ*, 654, L49
Takagi, T., Matsuhara, H., Wada, T., *et al.* 2007, *PASJ*, 59, 557
Takagi, T., Arimoto, N., & Hanami, H. 2003, *MNRAS*, 340, 813
Takagi, T., Ohyama, Y., Goto, T., *et al.* 2010, *A&A*, 514, A5+
Wada, T., Matsuhara, H., Oyabu, S., *et al.* 2008, *PASJ*, 60, 517

Discussion

CLEMENTS: How large is the sample of objects where you see this change in PAH SEDs?

TAKAGI: There are approximately 100 sources with S > 0.3 mJy at 9 μm for which we obtained MIR spectra. We detected PAH emission for half of them.

The Spectral Energy Distribution of Galaxies
Proceedings IAU Symposium No. 284, 2011
R.J. Tuffs & C.C. Popescu, eds.

doi:10.1017/S1743921312009623

Minor-merger-driven growth of early-type galaxies over the last 8 billion years

S. Kaviraj[1], R. M. Crockett[2], J. Silk[2], R. S Ellis[3], S. K. Yi[4], R. W. O'Connell[5], R. Windhorst[6] and B. C. Whitmore[7]

[1]Blackett Laboratory, Imperial College London, London SW7 2AZ

[2]Dept. of Physics, Denys Wilkinson Building, Keble Road, OX1 3RH

[3]Department of Astronomy, California Institute of Technology, Pasadena, CA 91125, USA

[4]Department of Astronomy, Yonsei University, Seoul 120-749, Republic of Korea

[5]Department of Astronomy, University of Virginia, Charlottesville, VA 22904-4325, USA

[6]School of Earth and Space Exploration, Arizona State University, Tempe, AZ 85287-1404, USA

[7]Space Telescope Science Institute, 3700 San Martin Drive, Baltimore, MD 21218, USA

email: s.kaviraj@imperial.ac.uk

Abstract. We summarise recent progress in understanding the star formation activity in early-type galaxies (ETGs), using recent studies that leverage photometry in the rest-frame ultraviolet (UV) wavelengths. While classically thought to be old, passively-evolving systems, recent UV studies have revealed widespread star formation in ETGs, with ~20% of the stellar mass in today's ETGs forming at late epochs ($z < 1$). A strong correlation is found between the presence of morphological disturbances and blue UV colours, suggesting that the star formation is merger-driven. However, the major merger rate at late epochs is far too low to satisfy the number of disturbed ETGs, indicating that *minor* mergers drive the star formation in these galaxies over the latter half of cosmic time. Together with the recent literature which suggests that minor mergers may drive the size evolution of massive ETGs, these results highlight the significant role of minor mergers in driving the evolution of massive galaxies in the low and intermediate-redshift Universe.

Keywords. galaxies: elliptical and lenticular, cD, galaxies: formation, galaxies: evolution, galaxies: interactions, galaxies: high-redshift, ultraviolet: galaxies

1. Introduction

Massive early-type galaxies (ETGs) dominate the stellar mass density in the local Universe. Hosting over half the stars at the present day, they hold a unique record of the evolution of the visible Universe over cosmic time. In recent decades, observational astronomy has focussed largely on the optical wavelengths, and our understanding of ETG evolution has naturally been shaped by such studies. The optical properties of ETGs (e.g. red optical colours and high ratios of alpha elements to iron), motivated a long-standing classical model (termed 'monolithic collapse'), which postulated their formation in short, efficient bursts of star formation at high redshift ($z \gg 1$) followed by purely passive evolution thereafter (e.g. Bower *et al.* 1992; Renzini 2006). However, monolithic collapse is difficult to reconcile with the standard hierarchical paradigm of galaxy evolution, in which merger-driven and quiescent star formation is expected to take place in galaxies over the lifetime of the Universe.

Indeed, it can be shown that the *optical* predictions of semi-analytical models based on the standard paradigm are virtually indistinguishable from those of monolithic collapse

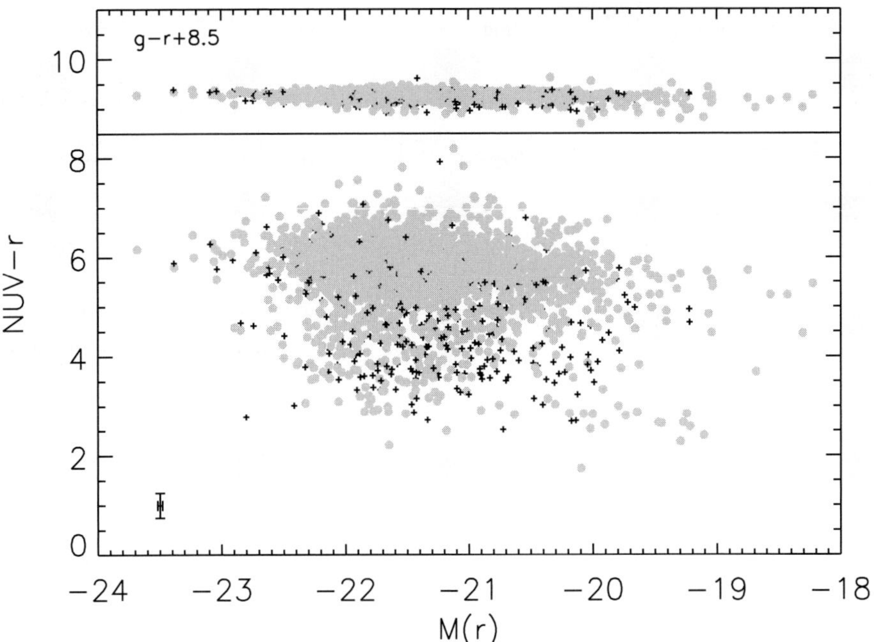

Figure 1. The optical (top panel) and the $NUV-r$ (bottom panel) colour-magnitude relations (CMRs) of our early-type galaxies. We intentionally show the optical CMR on the same scale as its UV counterpart, to highlight the significant difference in their respective scatters. The small grey crosses represent galaxies which have Type II AGN. The maximum contamination in the UV flux due to Type II AGN is less than 15% (0.15 mag), which is much smaller than the observed scatter in the UV colours. The NUV filter is centred at 2300 Å. This figure has been adapted from Kaviraj *et al.* 2007, ApJS, 173, 619.

(Kaviraj *et al.* 2005). This is because star formation in ETGs at late epochs ($z < 1$) is too weak to significantly impact the optical spectrum, implying that the optical wavelengths are not a good discriminant between the monolithic and hierarchical models. Accurately quantifying the star formation history of ETGs at late epochs requires a sensitive indicator of weak star formation, such as the rest-frame ultraviolet wavelengths (UV; 1200-3000 Å), which is more than an order of magnitude more sensitive to star formation than the optical wavelengths.

2. Persistent star formation in early-type galaxies since $z \sim 1$

A series of recent studies has exploited rest-frame UV photometry of ETGs, to accurately measure the stellar mass growth in these galaxies since $z \sim 1$. Using data from NASA's *GALEX* UV space telescope, combined with optical data from the Sloan Digital Sky Survey (SDSS), Kaviraj *et al.* (2007) showed that, while the *optical* colours of local ($0 < z < 0.1$) ETGs are indeed red, they exhibit almost 5 mags of spread in their rest-frame UV colours (Figure 1). Furthermore, the UV fluxes in the bluest 30% of ETGs are larger than the maximum UV flux that can be expected from old, extreme-horizontal-branch stars, which suggests that *at least* 30% of local ETGs contain unambiguous signatures of recent star formation. The mass fraction in young stars is of the order of a few percent and the luminosity-weighted average age of the young stellar populations is ~0.5 Gyrs (Kaviraj *et al.* 2007).

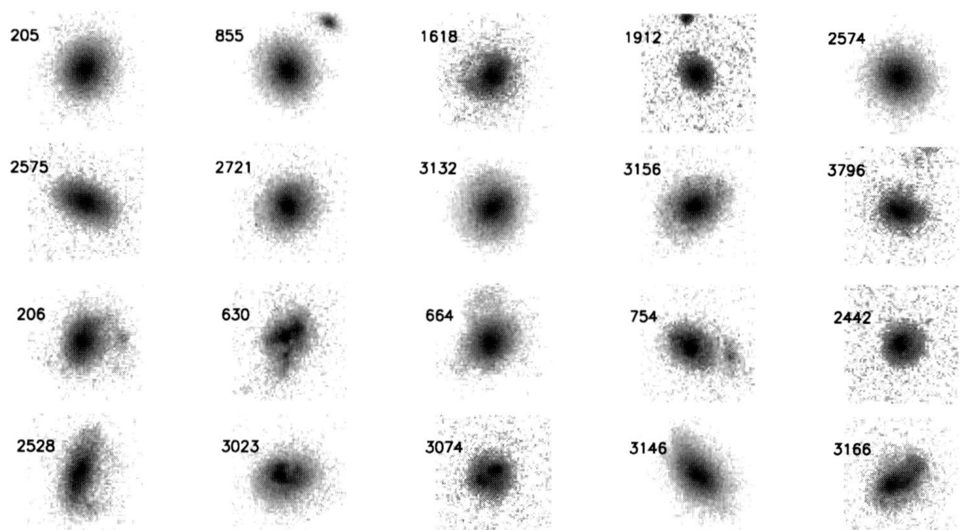

Figure 2. Examples of ETGs that are relaxed i.e. do not show any signs of morphological disturbances (rows 1 and 2) and ETGs that are morphologically disturbed (rows 3 and 4). This figure has been reproduced from Kaviraj *et al.* 2011, MNRAS, 411, 2148.

This result was subsequently consolidated by studying ETGs at intermediate redshift ($0.5 < z < 1$). At these epochs the Universe is too young for the horizontal branch to be in place, making the UV a rather 'clean' indicator of recent star formation. Combining data from three intermediate-redshift surveys (MUSYC, COMBO-17 and GEMS) Kaviraj *et al.* (2008) demonstrated that the scatter in the UV colours of ETGs remained unchanged across the entire redshift range $0 < z < 1$. Together with the previous work using *GALEX*, this study concluded that *ETGs of all luminosities form stars over the last 8 billion years, with massive (more luminous than L_*) ETGs forming ~20% of their stellar mass after $z \sim 1$.*

3. Minor mergers as the principal driver of star formation in early-type galaxies

While the studies described above established the presence of star formation in ETGs, ruling out the classical monolithic collapse model, they offered little insight into the sources of gas that drives this star formation. There are several potential sources of gas in ETGs, such as internal stellar mass loss, accretion from hot gas or ambient HI reservoirs and mergers. Stellar mass loss does not provide enough gas at the present day, ruling it out as the principal driver of star formation in local ETGs (Kaviraj *et al.* 2007). Indeed, the wider literature indicates that the interstellar medium of local ETGs may be external in origin. For example, results from the recent SAURON survey, which has performed integral-field-spectroscopy of local ETGs, indicate that the gas is often kinematically-decoupled from the stars (Sarzi *et al.* 2006). Measurements of dust masses in ETGs typically yield values that are in excess of the maximum value expected from stellar mass loss (e.g. Merluzzi 1998). Furthermore, more than 70% of ETGs appear morphologically disturbed in deep optical imaging (van Dokkum 2005).

To establish the role of mergers in driving the star formation in ETGs, Kaviraj *et al.* explored whether a correlation exists between the presence of morphological disturbances and blue UV colours. This is a difficult exercise to perform at low redshift, where imaging

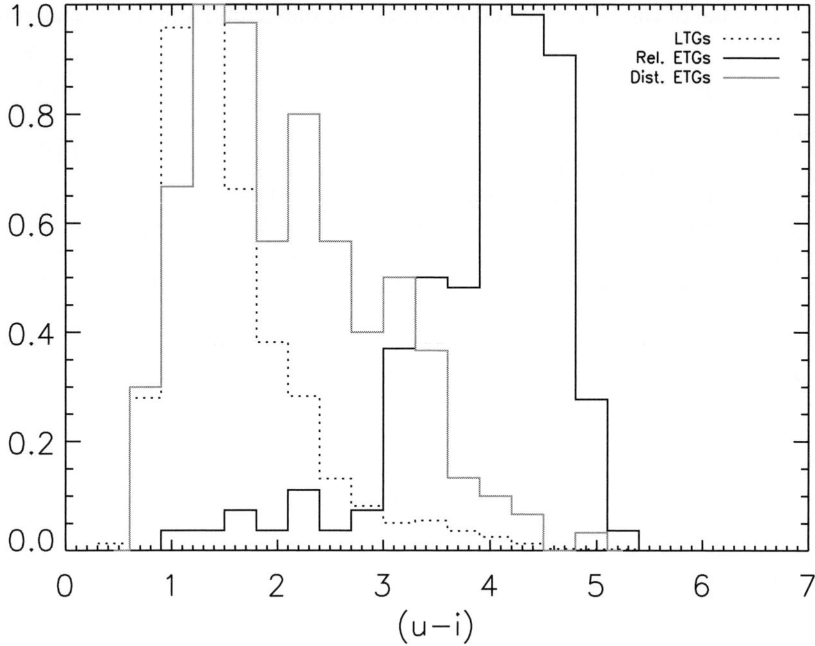

Figure 3. Rest-frame UV colour histograms for the relaxed ETG, disturbed ETG and late-type populations. Note that all histograms are normalised to 1. This figure has been adapted from Kaviraj et al. 2011, MNRAS, 411, 2148.

from surveys such as the SDSS lack the resolution and depth to reveal faint morphological disturbances from e.g. minor mergers. However, this is possible using *Hubble Space Telescope* (HST) imaging at intermediate redshift. using surveys such as COSMOS. Kaviraj *et al.* found that ETGs at $z \sim 0.6$ that had blue UV colours were indeed highly likely to exhibit morphological disturbances (see Figures 2 and 3). However, the major merger rate at late epochs is several factors too low to satisfy the number of disturbed ETGs, *indicating that minor mergers (with mass ratios between 1:4 and 1:10) drive the star formation in ETGs over the last 8 billion years.* It is worth noting that the strong correlation between disturbed morphology and star formation also indicates that processes such as accretion from hot gas and ambient HI reservoirs are not responsible for triggering the star formation in these galaxies.

Although significant progress has been made over the past five years in deciphering the evolution of ETGs, only the broad outline of their star formation histories has been firmly established. While ~80% of the stars in these galaxies are old, the remaining minority formed over the last 8 billion years via minor mergers between ETGs and small, gas-rich companions. However, several outstanding questions remain. *What are the detailed properties (mass ratios, satellite gas fractions, etc.) of the minor mergers that drive star formation in ETGs at late epochs? At what redshifts and through what processes (e.g. major mergers, cold streams) did the old, dominant stellar populations in today's ETGs form?* The former can be answered using detailed spatially-resolved UV-optical analyses of local ETGs using new instruments such as HST's Wide Field Camera 3 (WFC3; e.g. Crockett *et al.* 2011), while the latter can be explored by studying newborn ETGs at the epoch of peak cosmic star formation, using new surveys of the high-redshift Universe that leverage the WFC3's near-infrared capabilities (e.g. CANDELS). The next five years offer us the tantalising possibility of accurately tracing the formation history of ETGs over

90% of the lifetime of the Universe and revolutionise our understanding of this important class of galaxy.

References

Bower R. G., Lucey J. R., & Ellis R. 1992, *MNRAS*, 254, 589
Crockett R. M., Kaviraj S., Silk J. I., *et al.* 2011, *ApJ*, 727, 115
Kaviraj S., Devriendt J. E. G., Ferreras I., & Yi S. K. 2005, *MNRAS*, 360, 60
Kaviraj S., Khochfar S., Schawinski K., *et al.* 2008, *MNRAS*, 388, 67
Kaviraj S., Schawinski K., Devriendt J. E. G., *et al.* 2007, *ApJS*, 173, 619
Merluzzi P. 1998, *A&A*, 338, 807
Renzini A. 2006, *ARAA*, 44, 141
Sarzi M., Falcón-Barroso J., Davies R. L., *et al.* 2006, *MNRAS*, 366, 1151
van Dokkum P. G. 2005, *AJ*, 130, 2647

Discussion

LEITHERER: Is there evidence for higher supernova rates in early-type galaxies with blue colours?

KAVIRAJ: Yes there is. Higher type Ic rates are found in ETGs with blue colours. Recent work also shows most type II rates are enhanced but sample sizes are quite small at the moment.

The Spectral Energy Distribution of Galaxies
Proceedings IAU Symposium No. 284, 2011
R.J. Tuffs & C.C. Popescu, eds.

doi:10.1017/S1743921312009635

Charting the evolution of the ages and metallicities of massive galaxies since $z = 0.7$

Anna Gallazzi[1], Eric F. Bell[2], Stefano Zibetti[1], Daniel Kelson[3] and Jarle Brinchmann[4]

[1]Dark Cosmology Center, University of Copenhagen, Niels Bohr Institute, Juliane Maries Vej 30, 2100 Copenhagen, Denmark

[2]Department of Astronomy, University of Michigan, Ann Arbor, MI 48109, USA
[3]Observatories of the Carnegie Institution of Washington, Pasadena, CA 91101, USA
[4]Leiden Observatory, Leiden University, 2300RA, Leiden, the Netherlands

email: gallazzi@dark-cosmology.dk

Abstract. Detailed studies of the stellar populations of intermediate-redshift galaxies can shed light onto the processes responsible for the significant evolution of the massive galaxy population since $z < 1$. We have undertaken such a study by means of deep rest-frame optical spectroscopy with IMACS on Magellan on a sample of ~80 galaxies selected from CDFS to have stellar masses $> 10^{10} M_{\odot}$ and redshift $0.65 < z < 0.75$. We analyse stellar absorption line strengths and interpret them with a Monte Carlo library of star formation histories to derive constraints on mean stellar ages, metallicities and stellar masses. We present here the first characterization of the stellar mass–metallicity and stellar mass–age relations at z~0.7 and their evolution to the present-day.

Keywords. galaxies: evolution, galaxies: stellar content, galaxies: high-redshift

1. Introduction

In the local Universe bulge-dominated red-sequence galaxies contribute more than half of the total stellar mass and metal density (Bell *et al.* 2003, Baldry *et al.* 2004, Gallazzi *et al.* 2008). Half of the stellar mass density of this population has come into place in the past 8 billion years (Bell *et al.* 2004, Faber *et al.* 2007, Brown *et al.* 2007). Over the same epoch, the total star formation rate density, dominated by blue galaxies with masses between 10^{10} and $10^{11} M_{\odot}$, has declined by almost a factor of 10 (Hopkins & Beacom 2006). Detailed studies of the stellar population properties of $z < 1$ galaxies (i.e. the progenitors), in combination with star formation diagnostics and morphological characterization, can shed light on the relative importance of various physical processes potentially responsible for the star formation quenching at $z < 1$ and the ensuing build-up of the red-sequence. So far only a few studies have addressed the evolution in stellar population properties in the epoch $0 < z < 1$ and they are generally based on co-added spectra (e.g. Schiavon *et al.* 2006, Cool *et al.* 2008, Sánchez-Blázquez *et al.* 2009). Therefore, they cannot distinguish between the evolution of individual galaxies already on the red-sequence at intermediate redshift and the evolution of the population through addition of quenched blue galaxies.

2. Sample and physical parameters estimates

In order to tackle the issues above, we have obtained deep multi-object rest-frame optical spectroscopy with IMACS on Magellan for a representative sample of 77 massive galaxies at redshift z~0.7, allowing us to measure velocity dispersions and stellar absorption features and estimate the stellar population properties for individual objects.

The sample has been selected from the CDFS field of the COMBO-17 survey (Wolf *et al.* 2004) to have redshift in the range $0.65< z <0.75$ and an apparent R-band magnitude brighter than 22.7. This magnitude cut allows us to map the red-sequence down to its completeness limit at z=0.7 of $M_* \sim 3 \times 10^{10} M_\odot$ (Borch *et al.* 2006) and to sample the massive end of the blue-cloud. These galaxies have been observed with a single mask for a total exposure time of 10 hours, reaching an average S/N of 15 per Å for the most massive galaxies and of 8 for lower mass ones. The spectra have a resolution of 6.25Å FWHM (~3.4Å rest-frame) and span the wavelength range 6500 – 9500Å, covering the most significant absorption features from the 4000Å-break up to Fe5335. Moreover, this sample has multi-wavelength coverage from the UV to 24μm, including HST imaging.

We measure velocity dispersions down to ~100 km/s for 70% of the sample, with an accuracy of $\lesssim$15% for 45% of the sample, using the pPXF code by Cappellari & Emsellem (2004). Stellar absorption features are measured off the spectra corrected for nebular emission lines using PLATEFIT (Tremonti *et al.* 2004). We derive constraints on stellar mass, light-weighted age and stellar metallicity following the Bayesian approach developed by Gallazzi *et al.* (2005) to analyse SDSS spectra of local galaxies. We interpret an optimal set of absorption indices ($D4000_n$, $H\delta_A + H\gamma_A$, $H\beta$, $[Mg_2Fe]$, $[MgFe]'$,the same used at low redshift and affected the least by α/Fe variations) with a large spectral library based on Bruzual & Charlot (2003) models and Monte Carlo star formation histories and metallicities, and derive probability density functions of galaxy physical parameters. Stellar ages and masses are constrained with a typical accuracy of 0.1 dex, while stellar metallicity is constrained on average within 0.25 dex, with a stronger dependence on spectral S/N.

3. Mass–metallicity and mass–age relations at $z = 0.7$

Locally, galaxies show clear average trends of increasing stellar age and metallicity with increasing mass. The scatter in these relations is lowest above the 'transition mass' of $3 \times 10^{10} M_\odot$, where red-sequence galaxies dominate (Gallazzi *et al.* 2005, Kauffmann *et al.* 2003). We observe a similar trend in the mean luminosity-weighted age of $z = 0.7$ galaxies increasing with stellar mass from $\sim$ 2Gyr at $3\times10^{10} M_\odot$ to $\sim$ 5Gyr at $3\times10^{11} M_\odot$. The distribution in age is narrow at masses $> 10^{11} M_\odot$, while the scatter increases by 0.1 dex below this mass, qualitatively consistent with an evolution with redshift of the characteristic transition mass (c.f. Pannella *et al.* 2009). This is shown in the upper panel of Fig.1 where the $z = 0.7$ galaxies are compared to the local relation from Gallazzi *et al.* (2005). As expected the overall distribution in stellar age of $z = 0.7$ galaxies is shifted to younger ages than present-day galaxies. However, under the assumption of passive evolution, $z = 0.7$ galaxies could only contribute to the oldest fraction of present-day massive galaxies. This means that passive evolution is not viable already at masses $> 10^{11} M_\odot$ or that lower-mass $z = 0.7$ galaxies must evolve to contribute to the younger end of present-day massive galaxies.

In the lower panel of Fig.1 we show the distribution in stellar metallicity as a function of stellar mass at $z = 0.7$, compared to the local relation. In addition to stellar age, the data suggest a mild evolution also in stellar metallicity since $z = 0.7$. Somewhat surprisingly, this is particularly the case for the most massive galaxies. Their specific star formation rate, estimated from UV and IR luminosities as in Bell *et al.* (2005), is typically lower than $3 \times 10^{-11} yr^{-1}$, hence we do not expect these massive galaxies to enrich their stellar populations significantly. On the contrary, around and below $10^{11} M_\odot$, galaxies have higher specific SFR ($\gtrsim 10^{-10} yr^{-1}$) and a large scatter in stellar metallicity, with a fraction of them having metallicities comparable to local galaxies. These intermediate-mass $z = 0.7$ star-forming galaxies could contribute to the more metal-rich and younger

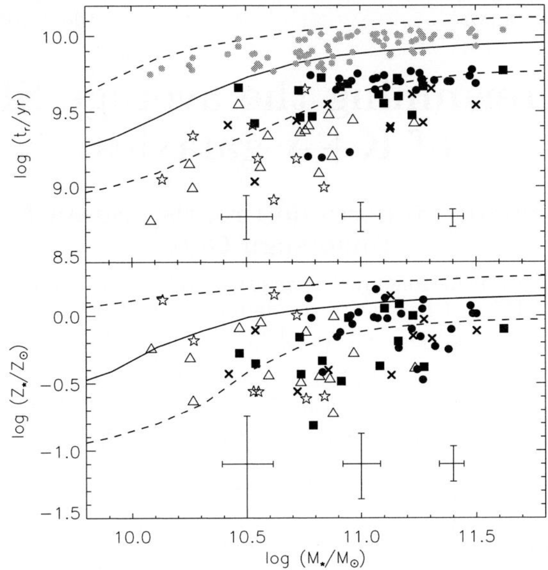

Figure 1. Luminosity-weighted age (*upper panel*) and stellar metallicity (*lower panel*) as a function of stellar mass for our sample galaxies at $z = 0.7$. Average uncertainties are indicated for three mass bins. Different symbols reflect different values of specific star formation rates (circles: $\log(SFR/M_*) < -11$; squares: $-11 < \log(SFR/M_*) < -10.5$; cross: $-10.5 < \log(SFR/M_*) < -10$; traingles: $-10 < \log(SFR/M_*) < -9.5$; stars: $\log(SFR/M_*) > -9.5$). Solid and dashed lines show the median and scatter of the local relations from Gallazzi *et al.* (2005). Grey points in the upper panel indicate the expected location of $z = 0.7$ galaxies if they evolved passively until today.

fraction of present-day massive galaxies. Interestingly, this would be consistent with the age-metallicity anti-correlation observed at fixed mass, or velocity dispersion, in local quiescent galaxies (e.g. Graves *et al.* 2010, Gallazzi *et al.* 2006).

References

Baldry, I. K., Glazebrook, K., Brinkmann, J., *et al.* 2004, *ApJ*, 600, 681
Bell, E. F., McIntosh, D. H., Katz, N., *et al.* 2003, *ApJS*, 149, 289
Bell, E. F., Wolf, C., Meisenheimer, K., *et al.* 2004, *ApJ*, 608, 752
Bell, E. F., Papovich, C., Wolf, C., *et al.* 2005, *ApJ*, 625, 23
Borch, A., Meisenheimer, K., Bell, E. F., *et al.* 2006, *A&A*, 453, 869
Brown, M. J. I., Dey, A., Jannuzi, B. T., *et al.* 2007, *ApJ*, 654, 858
Bruzual, G. & Charlot, S. 2003, *MNRAS*, 344, 1000
Cappellari, M. & Emsellem, E. 2004, *PASP*, 116, 138
Cool, R. J., Eisenstein, D. J., Fan, X., *et al.* 2008, *ApJ*, 682, 919
Faber, S. M., Willmer, C. N. A., Wolf, C., *et al.* 2007, *ApJ*, 665, 265
Gallazzi, A., Charlot, S., Brinchmann, J., *et al.* 2005, *MNRAS*, 362, 41
Gallazzi, A., Charlot, S., Brinchmann, J., *et al.* 2006, *MNRAS*, 370, 1106
Gallazzi, A., Brinchmann, J., Charlot, S., *et al.* 2008, *MNRAS*, 383, 1439
Graves, G. J., Faber, S. M., & Schiavon, R. P. 2010, *ApJ*, 721, 278
Hopkins, A. M. & Beacom, J. F. 2006, *ApJ*, 651, 142
Kauffmann, G., Heckman, T. M., White, S. D. M., *et al.* 2003, *MNRAS*, 341, 54
Pannella, M., Gabasch, A., Goranova, Y., *et al.* 2009, *ApJ*, 701, 787
Sánchez-Blázquez, P., Jablonka, P., Noll, S., *et al.* 2009, *A&A*, 499, 47
Schiavon, R. P., Faber, S. M., Konidaris, N., *et al.* 2006, *ApJ*, 651, L93
Tremonti, C. A., Heckman, T. M., Kauffmann, G., *et al.* 2004, *ApJ*, 613, 898
Wolf, C., Meisenheimer, K., Kleinheinrich, M., *et al.* 2004, *A&A*, 421, 913

The Spectral Energy Distribution of Galaxies
Proceedings IAU Symposium No. 284, 2011
R.J. Tuffs & C.C. Popescu, eds.

doi:10.1017/S1743921312009647

Determining the average SFR of K+A galaxies

Danielle M. Nielsen[1], Roberto de Propris[2], Susan E. Ridgway[3] and Tomotsugu Goto[4]

[1]Dept. of Astronomy, University of Wisconsin – Madison
475 N. Charter Street, Madison, WI 53706, USA

[2]Cerro Tololo Inter-American Observatory, La Serena, Chile

[3]National Optical Astronomy Observatory,
950 N. Cherry Avenue, Tucson, AZ 85719, USA

[4]Subaru Telescope, Hilo, HI, USA

email: nielsen@astro.wisc.edu

Abstract. We stack FIRST survey cutout images of 811 K+A galaxies to derive a mean 1.4 GHz radio image of our sample from which we measure a mean K+A flux density of 56 μJy. We carry out Monte Carlo simulations by randomly selecting radio-quiet white dwarfs to create 10,000 stacks equivalent to our K+A stack. From the measured fluxes of these stacks, we establish a 5σ detection limit of 43 μJy for stacked images. For the average redshift of our sample, we find an mean star formation rate of $\sim$1.7 $M_{\odot}$ yr^{-1}. We split the sample by age and find a mean radio flux of 60 μJy, which corresponds to a star formation rate of 1.6 $M_{\odot}$ yr^{-1}, for galaxies with starburst ages less than 250 Myr.

Keywords. galaxies: starburst, radio continuum

---

1. Introduction

Named for their spectra when discovered, K+A galaxies looked like a K-giant with strong Balmer absorption lines. These galaxies have been classified as post-starburst galaxies as the presence of Balmer absorption lines, and the lack of [OII] and Hα emission lines imply an abrupt truncation of a recent episode of star formation (Dressler & Gunn 1983). K+A galaxies were first discovered in intermediate redshift clusters where many galaxies are found to have star formation highly obscured by dust (Dressler *et al.* 2007). We determine the current star formation rate (SFR) of K+A galaxies to test this scenario.

2. K+A sample & data

We draw a sample of 811 K+A galaxies from the updated catalog of Goto (2007) assembled from DR7 of the SDSS. The redshifts of our sample range from $0.02 < z < 0.4$. Our data are drawn from the Faint Images of the Radio Sky at Twenty Centimeters (FIRST) survey (Becker *et al.* 1995); for each galaxy, we extract a $1'$ cutout from the FIRST database.

3. Radio stacking & Monte Carlo simulation

Many of our sources are too faint to be detected by the FIRST survey (detection limit $\sim$1 mJy), therefore we stack FIRST cutouts of K+A galaxies to create an average radio image of our sample (see White *et al.* 2007). This average K+A galaxy image is shown

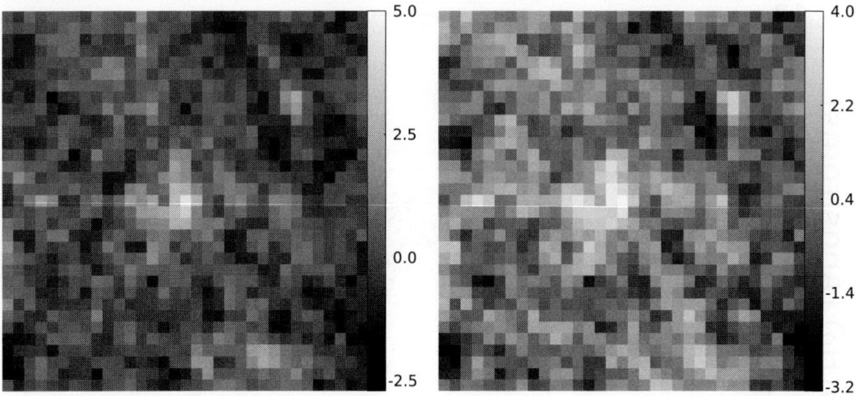

Figure 1. Stacked images of K+A galaxies, greyscale units are in μJy. *Left:* full sample of 811 galaxies. *Right:* 456 K+A galaxies with burst ages less than 250 Myr.

on the left in Fig. 1. We measure the flux in an aperture equivalent to the FIRST beam and find the mean radio flux of K+A galaxies to be 56 μJy.

We estimate the noise of stacked FIRST images with a Monte Carlo simulation. From a catalog of 8,495 white dwarfs – known radio-quiet sources – we create 10,000 stacks of 811 randomly selected white dwarfs (equivalent to the stack of K+A galaxies). We then measure the fluxes of these stacks as before, from which we derive a 5σ detection limit of 43 μJy for our stacked image, thus our average K+A image yields a significant detection.

4. Star formation rate

From the mean 1.4 GHz flux, we calculate the absolute luminosity for the average redshift of our sample, $z = 0.143$. The absolute luminosity can then be converted to a SFR (Yun *et al.* 2001).

$$L_{1.4GHz} = 4\pi D_L^2 S_{1.4GHz}(1+z)^{\alpha}/(1+z), \tag{4.1}$$

$$\text{SFR } (\text{M}_\odot \text{ yr}^{-1}) = 5.9 \times 10^{-22} L_{1.4GHz}, \tag{4.2}$$

For our average K+A galaxy, we find a SFR of 1.7 $\text{M}_\odot$ yr^{-1}.

5. Selecting "young" sources

Although the full sample shows a low average SFR, we note that there are galaxies with significant detections for a single FIRST image. We find 82 galaxies with fluxes in excess of a 3σ detection (435 μJy); 31 galaxies have fluxes in excess of a 5σ detection (725 μJy). The corresponding SFR calculation for these galaxies ranges from 0.45 to 2,000 $\text{M}_\odot$ yr^{-1} indicating another source of 1.4 GHz emission for some galaxies.

To determine if K+A galaxies that have experienced a starburst more recently have greater 1.4 GHz fluxes, we select a subsample of 456 "young" K+A galaxies with burst ages less than 250 Myr and perform the same analysis. Ages are determined from Fig. 2 in which we plot Hδ and D4000 against models from GALAXEV (Bruzual & Charlot 2003). In Fig. 2, galaxies with fluxes in excess of 435 μJy are plotted as filled squares.

Again, we stack this subsample of 456 galaxies (shown on the right in Fig. 1) and find a mean flux of 61 μJy. The Monte Carlo simulation gives a 5σ detection of 58 μJy. Therefore, our subsample is found to be at the detection limit. This flux yields an average SFR of 1.6 $\text{M}_\odot$ yr^{-1}.

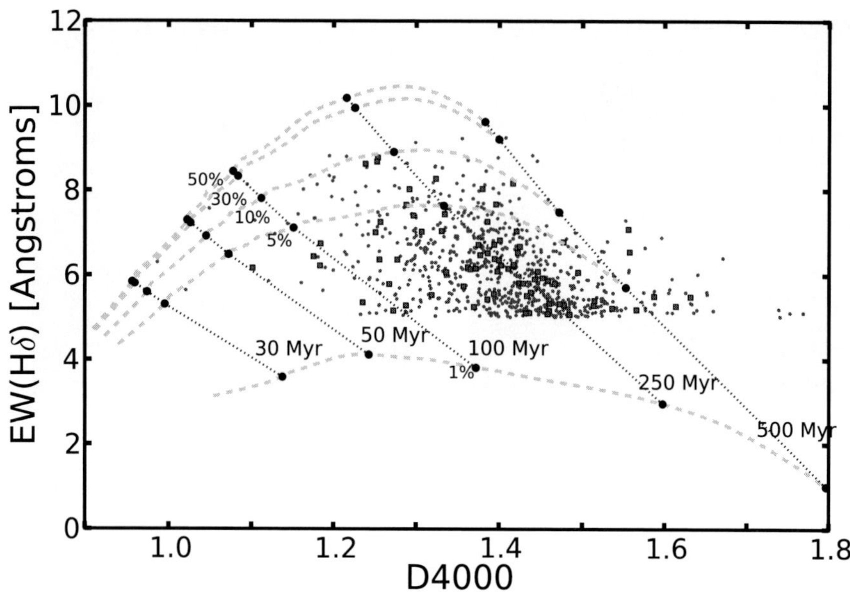

Figure 2. We plot Hδ and D4000 and overplot models to find starburst ages. Dashed lines indicate model galaxies evolved over 10 Gyr with an instantaneous starburst of mass 1, 5, 10, 30 and 50%. The dotted lines indicate 30, 50, 100, 250 and 500 Myr after the burst. Galaxies with 1.4 GHz fluxes greater than 435 μJy are plotted as filled squares and galaxies with less plotted as grey dots.

6. Discussion

We find an average SFR of 1.7 $M_{\odot}$ yr^{-1} for K+A galaxies. However we find there are individual galaxies with significant (>435μJy) 1.4 GHz detections. The population of galaxies with detections does not correspond to the subsample of "young galaxies which have experience starbursts in the past 250 Myr. Instead this population displays a range of burst ages comparable to that of the full sample. We found the average SFR of galaxies with recent starbursts to be 1.7 $M_{\odot}$ yr^{-1}.

Our average SFR agrees well with previous work of Goto (2004): for 15 of the nearest K+A galaxies, he found an upper limit of <15 $M_{\odot}$ yr^{-1}.

References

Becker, R. H., White, R. L., & Helfand, D. J. 1995, *ApJ*, 450, 559

Bruzual, G. & Charlot, S. 2003, *MNRAS*, 344, 1000

Dressler, A. & Gunn, J. E. 1983, *ApJ*, 270, 7

Dressler, A., *et al.* 2009, *ApJ*, 693, 140

Goto, T. 2004, *A&A*, 427, 125

Goto, T. 2007, *MNRAS*, 381, 187

White, R. L., Helfand D. J., Becker R. H., Glikman E., & de Vries W. 2007, *ApJ*, 654, 99

Yun, M. S., Reddy, N. A., & Condon, J. J. 2001, *ApJ*, 554, 803

The Spectral Energy Distribution of Galaxies
Proceedings IAU Symposium No. 284, 2011
R.J. Tuffs & C.C. Popescu, eds.

doi:10.1017/S1743921312009659

High resolution SMA imaging of (ultra)-luminous infrared galaxies

Qinghua Tan[1,2,3], Yu Gao[1,2], Zhong Wang[4], and Vivian U[4,5]

[1]Purple Mountain Observatory, Chinese Academy of Sciences, Nanjing 210008, China
[2]Key Laboratory of Radio Astronomy, Chinese Academy of Sciences, Nanjing 210008, China
[3]Graduate School of CAS, Beijing 100039, China
[4]Harvard-Smithsonian Center for Astrophysics, Cambridge, MA02138, USA
[5]Institute for Astronomy, University of Hawaii, 2680 Woodlawn Drive, Honolulu, HI 96822, USA

Abstract. We present preliminary results on Submillimeter Array (SMA) observations of three Luminous Infrared Galaxies (LIRGs) and one infrared Quasi Stellar Object (IR QSO). The galaxies were observed at sub-kpc spatial resolution in lines of CO and ^{13}CO, as well as in the continuum at 1.3mm. The moment maps show that the molecular gas in these galaxies is distributed in rotating disks with velocity gradients ranging from ~30 to 120 km s^{-1} kpc^{-1}. Combined with archival CO J=3-2 data, the spatial distributions of the CO J=3-2/J=2-1 ratios shows clear variations across the galaxies. The brightness temperature of the overlap region in VV 114 is found to be lower than that in the nuclear region of VV 114E, suggesting that that the bulk of molecular gas in this region is sub-thermalized.

Keywords. galaxies: individual (VV114, NGC 1614, IRAS 17578-0400, IRAS F22454-1744) - galaxies: interactions - galaxies: starburst

1. Introduction

Strong interactions or mergers between gas-rich disk galaxies is considered to be one of the primary mechanisms to trigger the intense star formation in luminous infrared galaxies (Sanders & Mirabel 1996). Tidal interaction between progenitor galaxies will transport the gas to the nucleus, condense the gas and trigger a widespread starburst. Since the proposal that ULIRGs hosting an AGN are transition objects that eventually evolve into optical QSOs (Sanders *et al.* 1988), more recent studies have focused on the evolutionary connection (Hopkins *et al.* 2006). However, for the galaxies at high redshifts, two mechanisms, major merger and cold mode accretion, have been suggested as drivers of active star formation (Narayanan *et al.* 2010; Keres *et al.* 2005). Thus, systematic studies of local luminous infrared galaxies are essential for the understanding of galaxy evolution.

Because molecular gas provides fuel for star formation and can carry the shocks, transporting the angular momentum during the galaxy interaction, the physical properties, distribution and kinematics of molecular gas are important for understanding the physical processes in galaxy mergers. Massive molecular gas reservoirs in (U)LIRGs have been revealed in previous studies and high resolution observations show that a substantial fraction of molecular gas lies within their central kpc regions (Downes & Solomon 1998; Wilson *et al.* 2008). Therefore, together with multiwavelength imaging studies, high resolution mapping of molecular gas in a large sample of (U)LIRGs will help provide a comprehensive picture of star formation activity during the merging process.

Table 1. Summary of basic properties

Galaxy	cz (km s^{-1})	logL_{FIR} ($L_{\odot}$)	Array Config.	Freq. (GHz)	Beam Size (arcsec)	Spatial Scale (kpc/1″)	Vel. Gradient[1] (km s^{-1} kpc^{-1})
VV 114	6016	11.71	Extended	230	1.52×1.36	0.40	100[1]
NGC 1614	4778	11.65	Extended	230	1.22×1.06	0.32	120
IRAS 17578-0400	4210	11.48	Compact	230	3.09×2.50	0.28	50
IRAS F22454-1744	35166	12.13	Extended	230	1.81×1.19	2.10	30

[1] *Velocity gradient along the north-south direction.*

In this paper, we present new data obtained with the SMA for the first four galaxies in the CO J=2-1 line. In addition, we present preliminary results on the CO J=3-2/J=2-1 line ratio distribution by combining the new data with archival CO J=3-2 data.

2. Observations and Data reduction

Our data were taken with the SMA between 2009 August and 2010 May. Each galaxy was observed in CO and ^{13}CO lines and in the 1.3mm continuum. The HPBW of the primary beam is ~52″ at these observing frequencies. The array configuration and synthesized beam size for each galaxy are listed in Table 1.

The initial data calibration was carried out using the MIR software package. Observations of planets or moons such as Uranus and Callisto were taken for determining the flux of the gain calibrator, which is a nearby quasar observed every ~20 mins to calibrate phase gain variations with time. Bandpass calibration was determined using a strong quasar. The further editing and imaging were done using the MIRIAD package.

3. Results

The high resolution CO J=2-1 integrated intensity, velocity field and dispersion maps for the four galaxies observed with the SMA are shown in Figure 1. The CO emission was found to be distributed over the central 2-6 kpc region of the rotating disk, with peak velocity dispersion ranging from 150 to 250 km s^{-1}. On a linear scale, the observed velocity gradient over the CO emission region is between 30 and 120 km s^{-1} kpc^{-1}. The global average line ratio of CO/^{13}CO J=2-1 for the three galaxies with ^{13}CO detection is ~25, which is comparable to the line ratio of CO/^{13}CO J=1-0 found in LIRGs (Aalto *et al.* 1991).

For VV 114, the significantly improved high resolution CO J=2-1 image shows a barlike structure with two peaks and a molecular tail, similar to that seen in CO J=3-2 (Iono *et al.* 2004). The irregular velocity field appears to reveal two distinct kinematical components with major axes along the north-south and east-west directions respectively. The CO map of IRAS 17578-0400 exhibits two emission peaks. In the extension to the northwest, the velocity contours are more irregular, perhaps caused by an additional kinematical component. Both the position-velocity diagram along the major axis and the channel map show that the secondary peak in the southeast is a single component slightly offset in the position, and based on the optical image which also exhibits two peaks, we consider the secondary CO peak might come from a separate physical component. For the luminous infrared QSO IRAS F22454-1744, the gaussian fit to the image implies a slightly resolved source, with a deconvolved source size of FWHM = 3.1″ × 1.3″, corresponding to an average radius of ~2.4 kpc. Both the extended gas distribution and the ordered velocity gradient found in moment maps seems to suggest a rotating

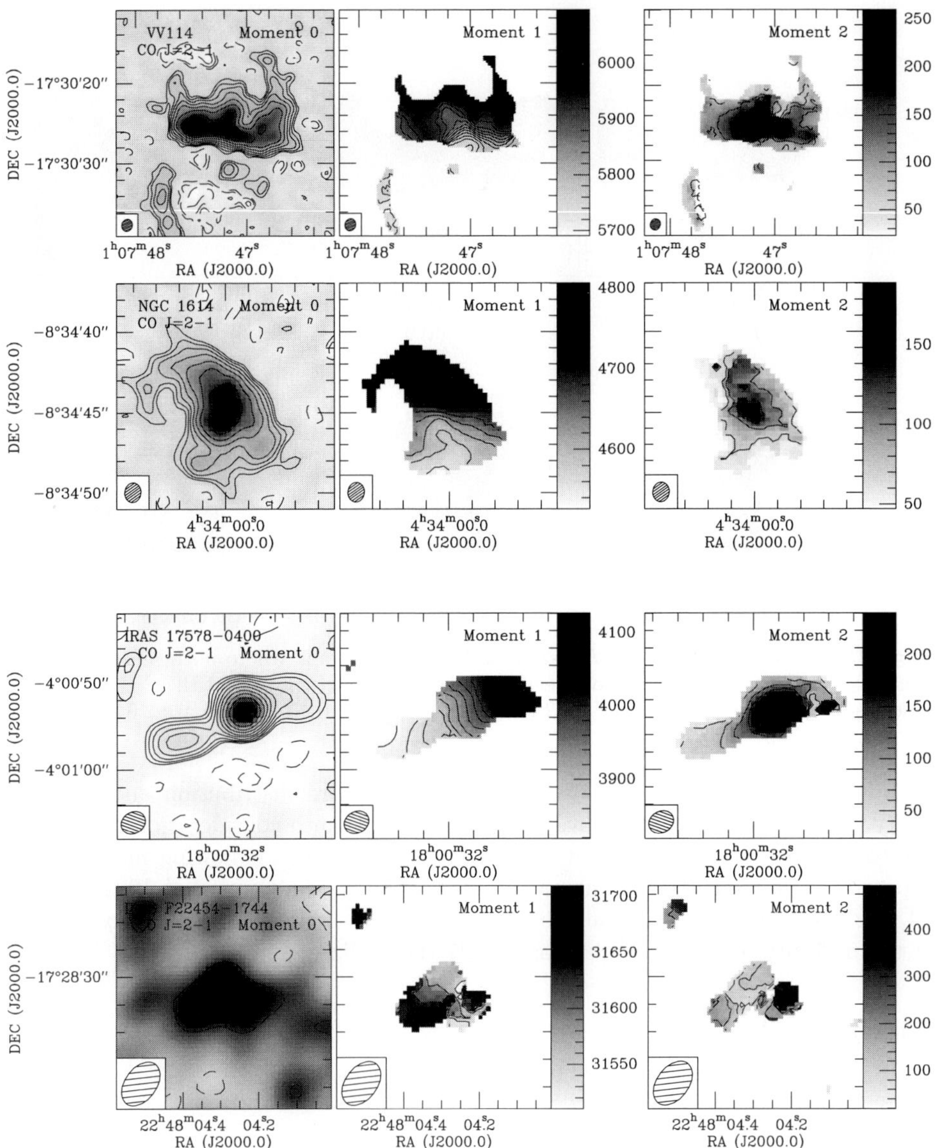

Figure 1. SMA CO J=2-1 moment maps. The lowest contour of moment 0 map is $\pm 2\sigma$ and contours increase by factors of 1.5 (except IRAS F22454 with contour increase in steps of 1σ). The moment 1 and 2 maps were made using the positive signal $> 3\sigma$ and blanked where the integrated intensity is $<2\sigma$. The synthesized beam size for each map is listed in Table 1. The unit of the colorbar is km s^{-1}. For VV 114 and NGC 1614, the moment maps were derived from the combination of Extended and Compact data.

disk structure. However, observations with higher resolution and sensitivity are urgently needed for further confirmation.

Combined with the archival CO J=3-2 data (Wilson *et al.* 2008), we convolved the CO J=2-1 emission to the resolution of the J=3-2 image (Fig. 2). The line ratio map of CO J=3-2/J=2-1 clearly shows that the gas temperature varies across the galaxy. The overlap region of VV 114 exhibits lower line ratios than those measured towards the nuclear region in VV 114E, suggesting that cold and extensive molecular gas are

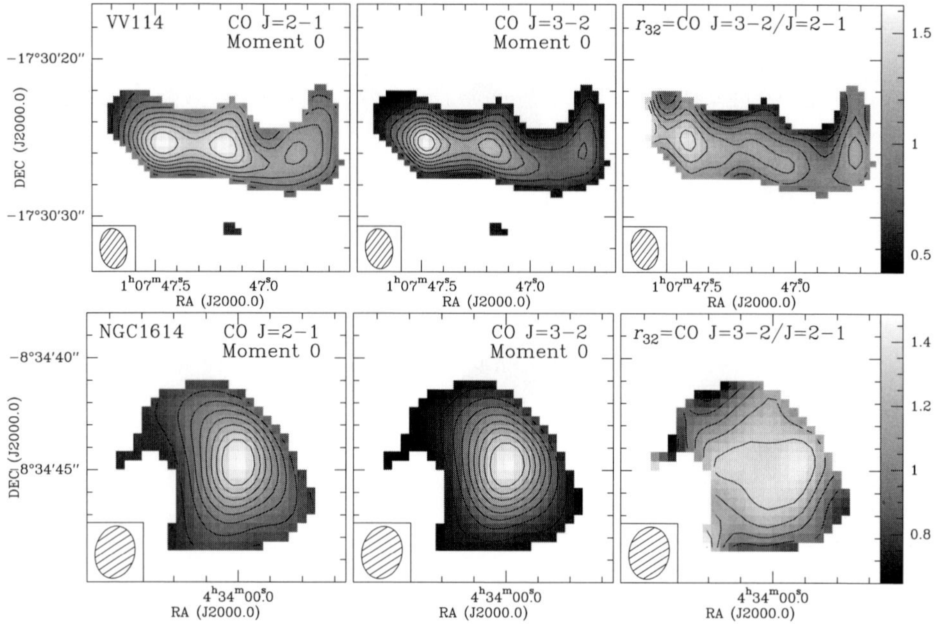

Figure 2. CO J=2-1,J=3-2 and J=3-2/J=2-1 line ratio maps where both measurements are above 3σ. The CO J=2-1 maps have been convolved to the resolution of the J=3-2 maps. The synthesized beam sizes of the CO J=3-2 maps for VV 114 and NGC 1614 are $2.54'' \times 1.79''$ and $2.36'' \times 1.85''$, respectively

predominant in this region. Taking the gas temperature distribution and the largest velocity dispersion found in the overlap region (Fig. 1) into consideration, this result is likely to indicate that a significant amount of molecular gas is flowing into this region and possibly forming the new dynamical center of the system (Yun *et al.* 1994). For NGC 1614, the line ratio has a value above unity, and varies little across the galaxy, implying that the majority of the molecular gas is fully thermalized.

References

Aalto, S., Black, J., Johansson, L., & Booth, R. 1991, *A&A*, 249, 323
Downes, D. & Solomon, P. 1998, *ApJ*, 507, 615
Keres, D., Katz, N., Weinberg, D., *et al.* 2005, *MNRAS*, 363, 2
Hopkins, P., Somerville, R., Hernquist, L., *et al.* 2006, *ApJ*, 652, 864
Iono, D., Ho, P., Yun, M., *et al.* 2004, *ApJ*(Letters), 616, L63
Narayanan, D., Hayward, C., Cox, T., *et al.* 2010, *MNRAS*, 401, 1613
Sanders, D., Soifer, B., Elias, J., Neugebauer, G., & Matthews, K. 1988, *ApJ*(Letters), 328, L35
Sanders, D. & Mirabel, I. 1996, *ARAA*, 34, 749
Wilson, C., Petitpas, G., Iono, D., *et al.* 2008, *ApJ*, 178, 189
Yun, M., Scoville, N., & Knop, R. 1994, *ApJ*(Letters), 430, L109

The Spectral Energy Distribution of Galaxies
Proceedings IAU Symposium No. 284, 2011
R.J. Tuffs & C.C. Popescu, eds.

doi:10.1017/S1743921312009660

Multi-wavelength probes of distant lensed galaxies

Stephen Serjeant

Dept. of Physical Sciences, The Open University, Milton Keynes, MK7 6AA, UK
email: s.serjeant@open.ac.uk

Abstract. I summarise recent results on multi-wavelength properties of distant lensed galaxies, with a particular focus on *Herschel*. Submm surveys have already resulted in a breakthrough discovery of an extremely efficient selection technique for strong gravitational lenses. Benefitting from the gravitational magnification boost, blind mm-wave redshifts have been demonstrated on IRAM, SMA and GBT, and follow-up emission line detections have been made of water, [OIII], [CII] and other species, revealing the PDR/XDR/CRDR conditions. I also discuss HST imaging of submm lenses, lensed galaxy reconstruction, the prospects for ALMA and e-Merlin and the effects of differential magnification. Many emission line diagnostics are relatively unaffected by differential magnification, but SED-based estimates of bolometric fractions in lensed infrared galaxies are so unreliable as to be useless, unless a lens mass model is available to correct for differential amplification.

Keywords. gravitational lensing, surveys, galaxies: active, galaxies: high-redshift, galaxies: ISM, galaxies: starburst, cosmology: dark matter, infrared: galaxies, radio continuum: galaxies, submillimeter

---

1. Introduction

Gravitational lens magnification is a fast route to faint object and faint population studies. This technique has been exploited to reveal or constrain the populations responsible for integrated far-IR and submm extragalactic backgrounds (e.g. Smail *et al.* 1997, Knudsen *et al.* 2008). Lensing is also one of the few available means of inferring the distribution of dark matter in galaxies (e.g. Gavazzi *et al.* 2007). Rare lens configurations even have the promise of yielding $\Omega_{\rm M}$ and w to 10%; this requires 50 double Einstein rings, requiring in turn a parent sample of thousands of lenses (Gavazzi *et al.* 2008). Indeed many current and future applications of lensing are limited by sample size (Treu 2010). Furthermore, most lens surveys have selected strongly for foreground lenses at relatively low redshifts, e.g. $z < 0.2$, restricting studies of the evolving dark matter distribution in galaxies. These are strong drivers for both larger and higher redshift samples of lenses.

The past year has seen an explosion of interest in submm-selected strong gravitational lenses, mainly driven by *Herschel* open and guaranteed time surveys and their follow-ups. New techniques pioneered with *Herschel* are capable of detecting lenses in much higher numbers and to much higher redshifts than previous optical/near-IR surveys such as SLACS or SLQS (Bolton *et al.* 2006, Oguri *et al.* 2006). This paper will briefly review some recent key infrared lensing results with a particular focus on the multiwavelength spectral energy distributions and on new data from *Herschel*.

2. Galaxy cluster lenses

Some of the first images made public from *Herschel* were the SPIRE guaranteed time $250 - 350 - 500\,\mu$m images of the galaxy cluster lens Abell 2218. This well-known system is one of several to have been used to constrain the 850 μm background population (e.g. Knudsen *et al.* 2008). Recently, Hopwood *et al.* (2010) had also used the Japanese *AKARI* facility to make its deepest 15 μm galaxy population constraints in this field, finding they could account for 87% of the predicted extragalactic background at this wavelength. At the time of writing, the SPIRE data in this cluster has not been published despite the data release, but Hopwood *et al.* (in prep.) have stacked the *AKARI* population in the submm maps and found that at least 41% (27%) of the background at 500 μm (250 μm) is attributable directly to the 15 μm population. At far-IR wavelengths, Altieri *et al.* (2010) presented *Herschel* guaranteed time PACS data of this cluster at 100 − 160 μm, demonstrating a constraint on the source counts comparable with that of the ultra-deep GOODS-North maps and accounting for (55 ± 24)% and (77 ± 31)% of the DIRBE direct measurements at 100 μm and 160 μm respectively. The prospects are clearly excellent for more extensive *Herschel* cluster lens work (e.g. Egami *et al.* 2010).

3. Field lenses: the promise of Herschel

3.1. *Proofs of concepts*

Gravitational lenses were originally hard to discover. For example, the Cosmic Lens All-Sky Survey (CLASS) observed nearly 12,000 flat-spectrum radio sources, finding 16 lenses (Myers *et al.* 2003). The large public SDSS database led to more rapid lens discovery (e.g. Bolton *et al.* 2006), though at low lens redshifts by virtue of the selection criteria.

The steep source counts in the submm imply a strong magnification bias. Prior to *Herschel*, the expectation was that about 50% of galaxies with 500 μm flux densities above $S_{500} > 100$ mJy would be lenses, with the remainder easily identifiable as local galaxies from optical imaging or as radio-loud AGN from radio data (Negrello *et al.* 2007). This prediction was supported by the mm-wave SPT counts. Spectacular confirmation came with the first *Herschel* candidates from H-ATLAS: in its first five lens candidates, Negrello *et al.* (2010) showed clear evidence of 2-image and 4-image lensing configurations from SMA data while Keck imaging showed only a foreground elliptical. There were therefore 5 lenses from first 5 candidates! The technique is much more sensitive to higher-redshift foreground lenses than SDSS-based selection and holds the promise of quickly and efficiently selecting large numbers of lenses.†

An associated breakthrough has been "blind" submm/mm-wave redshifts (e.g. Lupu *et al.* 2010, Frayer *et al.* 2011, Scott *et al.* 2011) with Z-spec (CSO) and Zpectrometer (GBT), with IRAM confirmations. This finally circumvents the torturous process of multi-stage multi-wavelength cross-IDs and exhaustive 8/10-m-class spectroscopic campaigns. For lensing systems, it provides rapid and unequivocal confirmation of a submm emitter background to the optical galaxy.

3.2. *Spectroscopic diagnostics*

The redshift determination of blind submm/mm-wave spectroscopy also makes follow-up far-IR and submm emission line diagnostics possible. SPIRE FTS observations of redshift $z = 3.0$ H-ATLAS lens ID 81 by Valtchanov *et al.* 2011 found first detection of 88 μm [OIII] line at $z > 0.05$. The high [OIII]/far-IR ratio and the limit on [OI]/[CII] suggests an AGN contributes ionizing radiation, as also suggested by the high radio/far-IR ratio. Cox *et al.* (2011) detect multiple CO transitions and [CII] in the $z = 4.3$ H-ATLAS galaxy SDP ID 141; their interpretation is of a PDR with a warm ($\simeq$40 K) dense (10^4 cm^{-3})

† The *Herschel* lensing was also covered by the BBC series Bang Goes The Theory in September 2011, in which the Open University are co-producers and partners.

gas, and the low line/far-IR ratio suggests again a high ionization parameter. Progress in detecting these diagnostics is fast: Lupu *et al.* 2011 present early results from high-density gas tracers in lensed submm galaxies, including HNC, HCN, HCO^+ and ^{13}CO.

Omont *et al.* 2011 detected an H_2O transition in the $z = 2.3$ H-ATLAS galaxy SDP ID 17, arguing against a PDR on the grounds that the luminosity ratio $L(H_2O\ 2_{02} - 1_{01})/L(CO\ (8-7)$ is comparable to that of Mrk 231. This, plus the fact that $L(H_2O\ 2_{02} - 1_{01})/L_{far-IR}$ is greater than that of Mrk 231, suggests the presence of an AGN. Van der Werf *et al.* (2011) detected four rotational H_2O transitions in the $z = 3.9$ lensed quasar APM 08279+5255, inferring the presence of warm gas (105 ± 21 K). An immediate corollary of both detections is that ALMA will detect H_2O in many more submm lenses.

3.3. *Imaging diagnostics*

Early work on the SEDs of the submm lenses found the background sources are very difficult or impossible to detect in optical imaging, while *Spitzer* $3.6 - 4.5\,\mu$m data can be used to detect the background sources (Hopwood *et al.* 2011). The immediate corollary is that *there is a class of lenses that are missed entirely in optically-selected samples.* The first HST data on these lenses (Negrello *et al.* in prep.) finds the first submm-selected Einstein rings, and with the benefit of the prior from the HST imaging, faint features can now be discerned in the ground-based imaging. Submm galaxies are known to have a wide variety of colours and obscurations so one might expect many to be detectable by *Euclid* at least in its near-infrared channel. The *Herschel* samples will be a crucial training set for strong lens discovery in *Euclid.*

3.4. *The luminosity function and refined lens selection*

Lapi *et al.* 2011 derived the submm luminosity function (LF) of H-ATLAS galaxies at $z > 1$ using a far-IR/submm colour-based photometric redshift estimator, calibrated against redshift determinations from CO spectroscopy. Local spiral galaxies and lensed sources were eliminating by rejecting galaxies with optical IFs. The derived evolving LF agrees with the predictions of an updated version of the Granato *et al.* 2001 model.

The steepness of the bright-end slope of the submm LF, together with the submm K-corrections, are the underlying causes of the steepness of the submm source counts. However the steepness of the LF suggests a means to improve the efficiency of the lens selection, and to extend it to fainter submm fluxes: if one can select galaxies at the bright end of the *luminosity function* then they should be prone to magnification bias in a similar way to the bright submm source counts. To do this requires relaxing the constraint in the Lapi *et al.* 2011 analysis of rejecting galaxies with optical IDs. Instead, González-Nuevo *et al.* (2011) use a high-redshift selection of submm galaxies based on their far-IR/submm colours, then use VIKING near-IR data and search for optical/submm photometric inconsistencies. This approach has led to the discoveries of lens candidates fainter than the 100 mJy 500 μm flux threshold of Negrello *et al.* (2010). The source counts of their lens candidates agree with magnification bias predictions of the Lapi *et al.* 2011 luminosity functions. Extrapolating to the full H-ATLAS survey, the authors estimate an astonishing $> 10^3$ strong lenses can be reliably detected in H-ATLAS alone, several times more than from using a simple monochromatic submm flux cut.

4. Differential magnification: curse or blessing?

Gravitational lensing is purely geometrical and therefore wavelength-independent. None-theless, in any astrophysical lens the magnification varies with position in the source plane, so if the background source is extended with intrinsic colour gradients, then the lensed source may have colours different to the unlensed case. This differential magnification could have a very significant effect on broad-band SEDs as well as on emission line diagnostics. Many authors assume, implicitly or explicitly, that these differential effects can be neglected, but this assumption is often wholly unjustified. Serjeant (2011) makes a statistical assessment of differential magnification for various photometric and spectroscopic diagnostics, a range of source redshifts and a constant comoving density of lenses. Here, I present the special case of a lens redshift ($z_l = 0.9$) and source redshift ($z_s = 2.286$) identical to those in the lens system IRAS F10214+4724.

The model background source is based on the Cosmic Eyelash (Swinbank *et al.* 2010), QSO 1148+5251 (Walter *et al.* 2009) and Mrk 231 (Van der Werf *et al.* 2010). There are three continuum components: an AGN, a starburst and cirrus, with relative bolometric fractions of 0.3:0.5:0.2. The AGN has a torus SED from Nenkova *et al.* 2008 peaking in the mid-IR, modelled as a circular region with 0.1 kpc radius. The starburst region is concentrated in four identical 50 pc-radius GMCs as per the Cosmic Eyelash, with a starburst SED from Efstathiou *et al.* (2000). The cirrus is an extended ellipse with 2.5 kpc major axis, ellipticity of 0.4 and a cirrus SED from Efstathiou *et al.* (2000). There are also emission line regions: a cool CO region tracing the cirrus with clumps modelled in the LVG approximation with a kinetic temperature of 20 K, density 100 cm^{-3}, CO column 10^{14} cm^{-2} and an external background temperature of $2.73(1 + z_s)$ K; warm CO regions with 0.4 kpc radii centred on the GMCs, with LVG model temperatures of 45 K, densities 10^4 cm^{-3} and CO column 10^{18} cm^{-2}; a circular warm H_2O region with 120 pc radius and a cool H_2O region with major axis 1 kpc (both H_2O regions follow the precedent of Mrk 231); a circular [CII] emission region offset by 0.6 kpc from the AGN, with a radius of 0.75 kpc, following the precedent of QSO J1148+5251. The foreground lens is modelled as the sum of a de Vaucouleurs' profile and a Navarro-Frenk-White profile, both typical of the SLACS population. The simulation is conducted at a resolution of 0.001″.

A clear result of these simulations is that the apparent bolometric fractions of strong lenses (magnification $\mu > 10$) depend very strongly on the selection wavelength. Fig. 1 shows the bolometric fractions for a simulated lensed galaxy selected at 60 μm and at 500 μm, while Fig. 2 (left) shows the AGN bolometric fraction for the lensed galaxies selected at these wavelengths. At this source redshift, 60 μm (observed-frame) is close to the peak of the AGN bolometric output, while observations at 500 μm predominantly sample the star-forming and cirrus components. Clearly the probability distribution of the AGN bolometric fraction depends very strongly on the selection wavelength, though some lens configurations are common to both wavelengths. The effect is much weaker for moderate-magnification systems ($2 < \mu < 5$); in these systems, the observed bolometric fractions *are* unbiased estimators of the underlying values.

Another striking result is that the warm and cool H_2O regions are also subject to differential magnification, at least in strongly-lensed systems (magnifications $\mu > 10$). Fig. 2 shows the relative magnification factors of the warm and cool H_2O emission line regions. These components contribute a similar flux in Mrk 231 for upper energy levels under 200 K, while the warm component dominates at higher temperatures. There is clearly a long tail in Fig. 2 in which the warm component is boosted relative to the cooler, more extended component. A similar tail is present in the H_2O *vs.* 500 μm magnifications, and in H_2O *vs.* GMCs. The presence of this tail provides an alternative explanation for the detection of H_2O ($2_{02} - 1_{11}$) in an H-ATLAS galaxy (Omont *et al.* 2011) because $\simeq 2/3$ of the flux of this line in Mrk 231 comes from the compact warm component. Having said that, at moderate magnifications ($2 < \mu < 5$) there are no appreciable differential magnification effects in the H_2O lines.

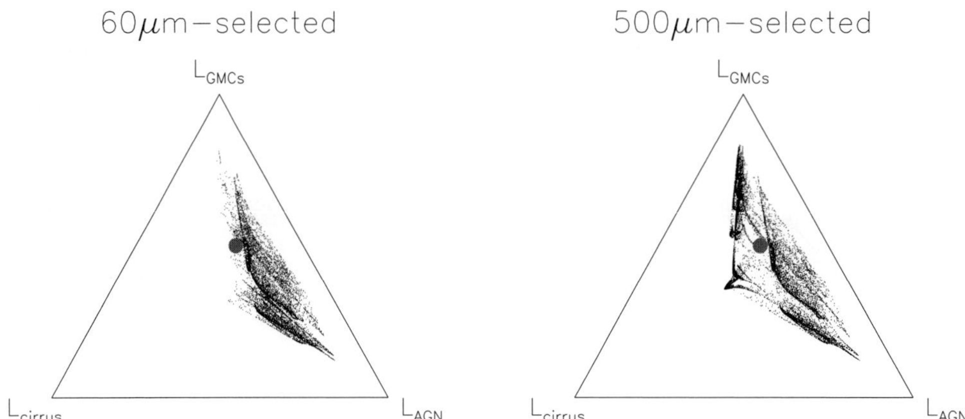

Figure 1. Bolometric fractions of strongly-lensed galaxies (magnifications $\mu > 10$, small dots, sparse-sampled for clarity) compared to the underlying background source fractions (large filled circle). These are ternary diagrams, i.e. the bolometric fraction is proportional to the perpendicular distance from an edge. At each apex, the bolometric fraction at that apex for that component is 100%, while the fraction is 0% at the opposite edge. Note the different distributions in 500 μm-selected and 60 μm-selected lenses.

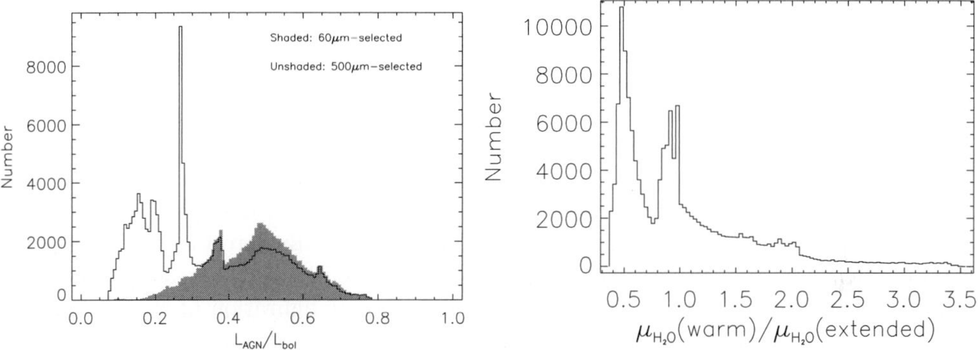

Figure 2. Left: AGN bolometric fractions in the simulated lensed galaxy, with monochromatic magnifications $\mu > 10$. Note that the probability distribution of bolometric fractions depends strongly on the selection wavelength. Right: relative magnification factors of the warm and cool H_2O lines discussed in the text, for a 500 μm-selected lenses with magnification $\mu > 10$.

Differential magnification also affects the observed bolometric fraction of the [CII] line (e.g. Valtchanov *et al.* 2011). However, in this case the bolometric fractions vary by $\simeq 0.1 - 0.2$ dex, comparable to the measurement uncertainties. Simulating the [CII] and CO line diagnostic diagram in Valtchanov *et al.* 2011 (not shown), it appears that the dynamic range of the diagnostic is much larger than that of the differential magnification effects, so this diagnostic is only weakly affected by differential lensing.

Finally, there is a serious distortion of the CO ladder. Fig 3 shows the predicted lensed CO Spectral Line Energy Distributions (SLEDs), compared to that of the underlying background source. This time, even moderate-magnification systems are affected. The warm CO is located very close to the GMCs in the model galaxy, but this does not necessarily guarantee that the warm CO is boosted relative to the cool component, even if selecting at 500 μm. Unlike the case of the [CII] and CO line diagnostic diagram in

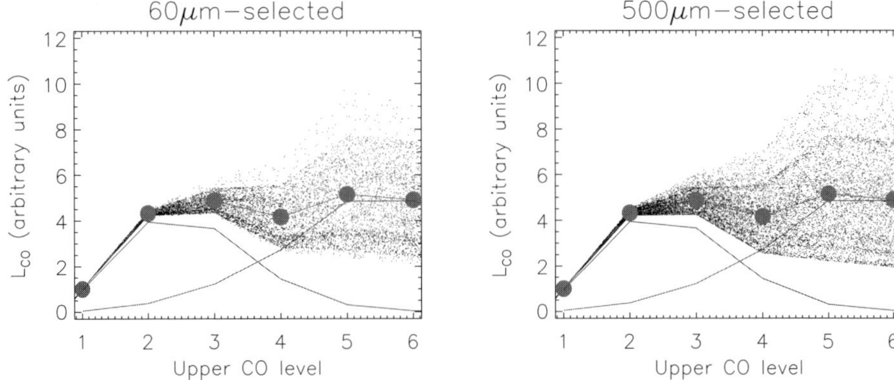

Figure 3. CO Spectral Line Energy Distribution (SLED) for the underlying source (large points), comprised of the warm and cool components (filled lines). Also shown as a shaded region is the range spanned by the differentially magnified simulated galaxies. Clearly, irrespective of selection wavelength, the CO SLED is strongly affected. This effect is more or less equally present for magnification $\mu > 10$ lenses (shown) and $2 < \mu < 5$ lenses.

Valtchanov *et al.* 2011, the useful dynamic range of the CO SLED is much smaller, making it much more sensitive to differential magnification effects.

Differential magnification is therefore a serious problem if its effects are not accounted for. Both the SED and emission line ratios are significantly distorted in galaxies with magnifications $\mu > 10$, and the CO SLED is very strongly distorted in any system with $\mu > 2$. However if a foreground mass model is available and the source is resolved, it can nevertheless work to our advantage. Lensing conserves surface brightness, so regions of high magnification are subject to angular magnification. This may, for example, provide uniquely high angular resolution views of active nuclei or star-forming regions in infrared-luminous galaxies. Observations by ALMA or eMerlin will easily detect lens morphologies, and an eMerlin legacy survey follow-up of submm-selected lenses is already underway.

These results are robust to changes in the assumed configuration of the background source. Only if the background source is broadly homogeneous could strong differential magnification effects be avoided. One might argue that the number of GMC regions could be much larger, e.g. to resemble those in $z = 0$ spiral discs (and at variance with observations of high-redshift infrared-luminous galaxies), but an active nucleus will inevitably create a colour gradient in the background source. Serjeant (2011) shows that the same effects are present for a constant comoving population of lenses, generalising the fixed lens redshift of $z_l = 0.9$ in this paper, and for different source redshifts.

5. Conclusions

Galaxy cluster lenses are a rapid route to probing the populations dominating the cosmic infrared backgrounds, both through direct detections and stacking analyses. The bulk of the observed or predicted cosmic infrared backgrounds at 15 μm, 100 μm, 160 μm and 850 μm have now been resolved into individual sources through ultra-deep imaging in the fields of foreground gravitationally-lensing galaxy clusters. Meanwhile, there is a clear need for larger numbers of strong galaxy-galaxy lens systems, particularly at higher lens redshifts. Following the spectacular confirmation of the ∼ 100% gravitational lens selection efficiency from submm-selected lenses, there is now a tremendous scope for such lens discovery in the submm with *Herschel.* A further breakthrough has been "blind" submm/mm-wave redshift determination, i.e. without recourse to the torturous multi-stage multi-wavelength cross-identification and long-term 8m/10m-class optical/near-IR spectroscopy. Emission line diagnostics already yielding constraints on the physical conditions in these lensed galaxies. However, differential magnification effects cannot be neglected. In particular, the observed CO SLED in any lensed system can easily be un-representative of the SLED in the underlying source. Furthermore, the uncorrected bolometric fractions of strongly-lensed galaixes (magnifications $\mu > 10$) inferred from broad-band SEDs are so unreliable as to be useless.

SS thanks friends and colleagues the H-ATLAS (Eales *et al.* 2010) and HerMES consortia, whose results are quoted in this paper, STFC (grant ST/G002533/1) for financial support, and the conference organisers for the kind invitation.

References

Altieri, B., *et al.* 2010, *A&A*, 518, L17
Bolton, A. S., *et al.* 2006, *ApJ*, 638, 703
Cox, P., *et al.* 2011, *ApJ* in press (arXiv:1107.2924)
Eales, S. A., *et al.* 2010, *PASP*, 122, 499
Efstathiou, A., Rowan-Robinson, M., & Siebenmorgen, R. 2000, *MNRAS*, 313, 734
Egami, E., *et al.* 2010, *A&A*, 518, L12
Frayer, D. T., *et al.* 2011, *ApJL*, 726, 22
Gavazzi, R., *et al.* 2007, *ApJ*, 667, 176
Gavazzi, R., *et al.* 2008, *ApJ*, 677, 1046
González-Nuevo, J., *et al.* 2011, *ApJ*, submitted
Granato, G. L., *et al.* 2001, *MNRAS*, 324, 757
Hopwood, R., *et al.* 2010, *ApJL*, 728, 4
Hopwood, R., *et al.* 2011, *ApJL*, 728, 4
Knudsen, K. K., *et al.* 2008, *MNRAS*, 384, 1161
Lapi, A., *et al.* 2011, *ApJ*, in press (arXiv:1108.3911)
Lupu, R., *et al.* 2010, ApJ submitted (arXiv:1009.5983)
Lupu, R., *et al.* 2011, in The Molecular Universe, Proceedings of the 280th Symposium of the International Astronomical Union, Toledo, Spain, May 30-June 3, 2011
Myers, S. T., *et al.* 2003, *MNRAS*, 341, 1
Negrello, M., *et al.* 2007, *MNRAS*, 377, 1557
Negrello, M., *et al.* 2010, *Science*, 330, 800
Nenkova, M., Sirocky, M. M., Ivesić, Ž., & Elitzur, M. 2008, ApJ, 685, 147
Oguri, M., *et al.* 2006, *AJ*, 132, 999
Omont, A., *et al.* 2011, *A&A*, 530, L3O
Scott, K. S., *et al.* 2011, *ApJ*, 733, 29
Serjeant, S. 2011, *MNRAS* submitted
Smail, I., *et al.* 1997, *ApJL*, 490, L5
Swinbank, A. M., *et al.* 2010, *Nature*, 464, 733
Treu, T. 2010, *ARA&A*, 48, 87
Walter, F., *et al.* 2009, *Nature*, 457, 699
Valtchanov, I., *et al.* 2011, *MNRAS*, in press (arXiv:1105.3929)
Van der Werf, P. P., *et al.* 2010, *A&A*, 518, L42
Van der Werf, P. P., *et al.* 2011, *ApJ*, 741, L38

The Spectral Energy Distribution of Galaxies
Proceedings IAU Symposium No. 284, 2011
R.J. Tuffs & C.C. Popescu, eds.

doi:10.1017/S1743921312009672

Cosmic Infrared Background ExpeRiment (CIBER): A probe of Extragalactic Background Light from reionization

Asantha Cooray[1], Jamie Bock[2], Mitsunobu Kawada[3], Brian Keating[4], Andrew Lange[2], Dae-Hee Lee[5], Louis Levenson[2], Toshio Matsumoto[6], Shuji Matsuura[6], Tom Renbarger[4], Ian Sullivan[2], Kohji Tsumura[6], Takehiko Wada[6], and Michael Zemcov[2]

[1]Center for Cosmology, University of California, Irvine, USA
[2]Department of Physics, Caltech, Pasadena USA
[3]Department of Physics, Nagoya University, Japan
[4]Department of Physics, University of California, La Jolla, USA
[5]Korea Astronomy and Space Science Institute, Daejeon, Korea
[6]Institute of Space and Astronautical Sciences, JAXA, Japan

email: acooray@uci.edu

Abstract. The Cosmic Infrared Background ExpeRiment (CIBER) is a rocket-borne absolute photometry imaging and spectroscopy experiment optimized to detect signatures of first-light galaxies present during reionization in the unresolved IR background. CIBER-I consists of a wide-field two-color camera for fluctuation measurements, a low-resolution absolute spectrometer for absolute EBL measurements, and a narrow-band imaging spectrometer to measure and correct scattered emission from the foreground zodiacal cloud. CIBER-I was successfully flown in February 2009 and July 2010 and four more flights are planned by 2014, including an upgrade (CIBER-II). We propose, after several additional flights of CIBER-I, an improved CIBER-II camera consisting of a wide-field 30 cm imager operating in 4 bands between 0.5 and 2.1 microns. It is designed for a high significance detection of unresolved IR background fluctuations at the minimum level necessary for reionization. With a FOV 50 to 2000 times larger than existing IR instruments on satellites, CIBER-II will carry out the definitive study to establish the surface density of sources responsible for reionization.

Keywords. diffuse radiation, large-scale structure of universe, infrared: general

1. Introduction

The optical and UV radiation from sources present during reionization is now present in the near-infrared with a small, but non-negligible, contribution to the Extragalactic Background Light (EBL). Searches for this radiation based on absolute photometry have proven problematic due to confusion with the Zodiacal foreground. Instead of the absolute background, in Cooray *et al.* (2004), we proposed to develop a near-infrared sounding rocket experiment, CIBER, to conduct a deep search for extragalactic background fluctuations from the epoch of reionization associated with first-light galaxies.

The EBL spectrum contains all radiative information from the reionization epoch (Santos *et al.* 2003; Kashlinsky *et al.* 2004; Fernandez & Komatsu 2006; Cooray *et al.* 2009). We expect that the EBL contains diffuse signatures of reionization, such as Ly-α background radiation redshifted to near-IR wavelengths today. Remnants of these first stars, black holes in the form of miniquasars, will also contribute to the EBL (Cooray & Yoshida 2004).

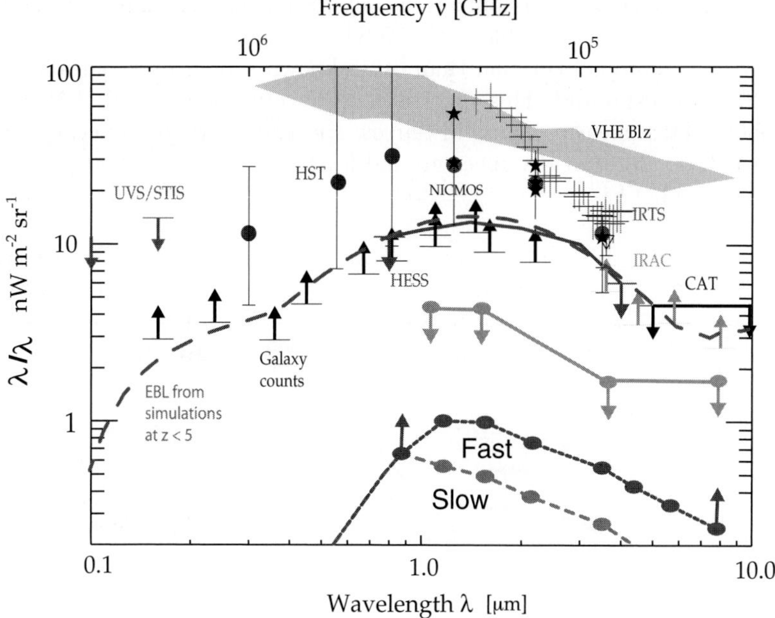

Figure 1. Summary of EBL observations at infrared and optical wavelengths, showing upper limits, reported residuals after subtraction of local foregrounds, reported detections with absolute photometry (DIRBE and IRTS), and integrated galaxy counts (lower limits). The large experimental scatter at 1.2 μm is notable, with Zodiacal removal suspected as a prime source of systematic error (Dwek et al. 2005). A recent upper limit based on TeV absorption using HESS (Aharonian et al. 2006) contradicts the excess reported by several authors at 1-2 μm. The indirect EBL measurements with TeV spectra, however, do not provide consistent results since a previous independent estimate suggests a higher background (the region labeled VHE Blz; Schroedter 2005). The long-dashed line shows the integrated galaxy light associated with galaxies formed at $z < 5$ based on a semi-analytical model (Primack *et al.* 2008). The dotted and dashed lower lines show the estimated EBL from $z > 6$ for fast and slow reionization histories, lower limits since the calculation is based on the minimum UV luminosity density needed to reionize and maintain the ionized state of the intergalactic medium (Chary & Cooray 2011). The light grey upper limits show constraints on the first-light EBL from reported fluctuations in deep Spitzer and NICMOS images.

Interestingly, integrated individual galaxy counts appear to fall short of the EBL measured with absolute photometry at near-infrared wavelengths (Fig. 1). For instance in the 1-3 μm band, galaxies contribute an intensity of $\sim$10 nW/m^2 sr (Madau & Pozzetti 2000). In contrast, measurements of the extragalactic background light (EBL) by DIRBE and the IRTS at the same wavelengths range from 10 nW/m^2 sr at 3.6 μm (Levenson & Wright 2008) to up to 60 nW/m^{2}sr at 1.2 μm (Cambresy *et al.* 2001; Matsumoto *et al.* 2005). It is highly unlikely that this whole difference is associated with sources during reionization, since such a large intensity requires unphysical requirements on star-formation (Madau & Silk 2005).

Though the total luminosity produced by sources responsible for reionization is uncertain, a lower limit can be placed assuming a minimal number of photons to produce and maintain reionization, given existing information on reionization, rest-frame UV luminosity functions of galaxies at $z > 6$, stellar mass estimates for galaxies at $z \sim 6$, among others (Chary & Cooray 2011). Such minimal reionization scenarios produce an EBL $\sim$ 1 nW/m^2 sr, a level undetectable by current absolute photometry measurements which, after dedicated space-borne measurements, show large discrepancies. As the spectral signature in EBL from reionization contains integrated emission from all sources, including fainter ones undetectable with JWST, it captures the exact reionization history and provides more information than other known probes of reionization, including CMB and 21-cm background. This spectral feature could be resolved in the future with absolute photometric measurements in narrow spectral bands between 1-2 μm with an out-of-Zodi EBL explorer at distances around 5 AU (Cooray *et al.* 2009).

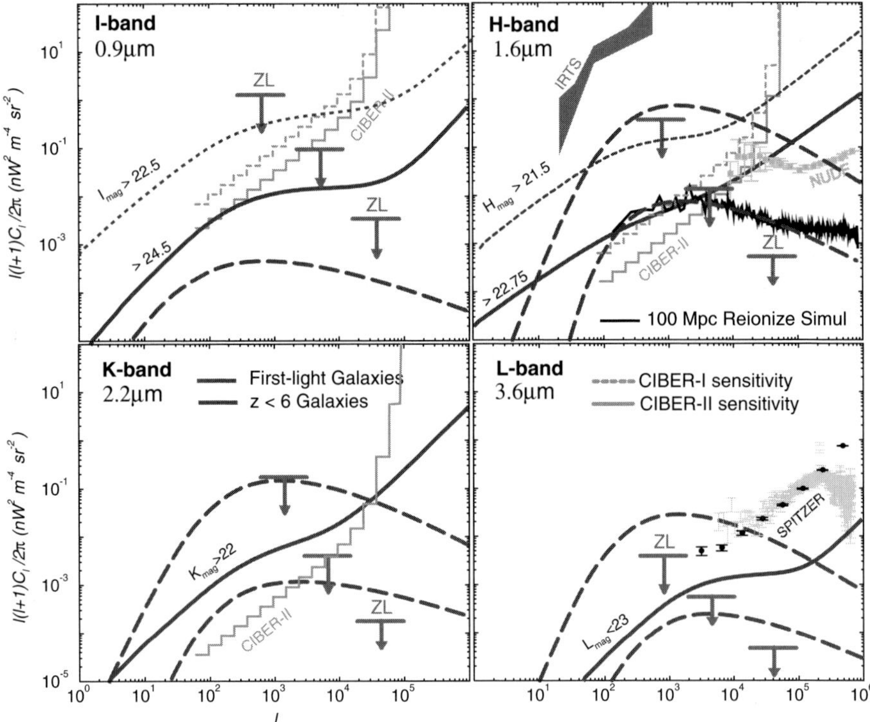

Figure 2. Spatial power spectrum of EBL fluctuations in standard IR bands. The long-dashed curves show the power spectra of fluctuations from first-light galaxies forming over the redshift interval $8 < z < 15$. The top long-dashed line shows the case where fluctuations are normalized to *Spitzer* measurements (Kashlinsky *et al.* 2008). The bottom long-dashed line shows the minimal signal necessary to produce reionization. Light grey points show *Spitzer* measurements by Kashlinsky *et al.* while black points show residuals from Cooray *et al.* after subtracting faint, blue dwarfs detected in deep HST ACS images. The thick-solid curves give estimated fluctuations from known galaxies, as a function of magnitude cutoff, based on a galaxy distribution model based on the halo appoach (Cooray & Sheth 2002; Cooray 2006) matched to existing clustering data (e.g., Sullivan *et al.* 2007). The galaxy cutoff taken for CIBER is 25% pixel removal using deep ancillary source catalogs. CIBER-II (thick-solid line) has a lower residual local galaxy foreground due to its smaller pixel size compared with CIBER-I (small-dashed line). The stair-steps in three panels shows the binned statistical sensitivity of CIBER-I and CIBER-II instruments in a single 50s observation. Zodiacal light is known to be spatially uniform, shown by the upper limits. The fluctuations in the upper-right panel are the EBL fluctuations resulting from a 2048^3 particle, large volume (100 Mpc) numerical simulation of reionization completed by the Princeton group (Trac & Cen 2007) with a combination of both Pop II and Pop III stars in first-light galaxies, as analyzed by the UCI team. The shape of fluctuations is model independent and shows the overall bump at $\ell \sim 1000$ in agreement with the analytical model.

Even in the relatively recent times since reionization it appears there may be an under accounting of galaxies. The current star-formation in Lyman-break galaxies (LBGs) at z $\sim$ 6 (Bouwens *et al.* 2006) falls about a factor of 6 to 9 below the minimum required to maintain an ionized IGM given canonical estimates for the clumping factor of the gas and the escape fraction of ionizing photons from galaxies. The upper limits on the ultraviolet luminosity function suggests a negative evolution with increasing redshift, in the sense that galaxies at the bright end of the UV luminosity function at z $\sim$ 6 were fainter in the ultraviolet. The implication is that the contribution from star-formation in faint galaxies, below the detection threshold of current surveys, is higher than has been estimated. The *Spitzer* stellar mass estimates suggest a top-heavy IMF for high redshift galaxies (Chary 2008), suggesting a different stellar population in those galaxies than in the local Universe. If current galaxy surveys under account for galaxy evolution or miss populations, the deficit may be uncovered in a careful study of the EBL.

A first-light galaxy EBL component from reionization and any contribution from faint sources at $z \sim 6$ unresolved by existing deep, pencil surveys can be uncovered using a careful study of background fluctuations (Cooray *et al.* 2004). The technique involves the measurement of the angular power spectrum via an anisotropy study, similar in method to well-established techniques for studying CMB anisotropy. First-light galaxies have predictable clustering, determined by the growth of dark matter perturbations in ΛCDM cosmology, and will produce EBL fluctuations with a characteristic spatial power spectrum. Zodiacal light is known to be spatially uniform on arcminute scales. What is uncertain is the amplitude of fluctuations since it depends on the astrophysics of reionization and the importance of, say, fainter sources at $z \sim 6$ to 8 that are responsible for maintaining reionization. A fluctuation measurement has the potential to detect a first-light galaxy component to the EBL at much fainter levels than can be probed with absolute photometry.

Since our work on CIBER began (Cooray *et al.* 2004; Bock *et al.* 2006), *Spitzer* IRAC has carried out a study of EBL clustering anisotropy. After a deep removal of both point sources, and deconvolving emission from the extended PSF, Kashlinsky *et al.* (2005; 2007) claim detection of first-light galaxy fluctuations based on a deviation from a shot-noise power spectrum on the largest angular scales.

Some doubt may be cast on the interpretation of first sources due to the angular scales and wavelengths covered by IRAC. Due to IRAC's small field of view, the power spectrum from high-z objects must be carefully separated from fainter, unresolved foreground galaxies (Cooray *et al.* 2007; Chary *et al.* 2008). In addition, fluctuation measurements have been made at shorter near-IR wavelengths in narrow, but very deep, NICMOS images (Thompson *et al.* 2007). The images again show a fluctuation signal (Fig 2). The spectrum of fluctuations measured between *Spitzer* IRAC and HST NICMOS appears to be approximately flat, and the authors state it does not correspond to the shape expected for $z > 8$ sources (Thompson *et al.* 2007), but rather to a low redshift population unresolved in both *Spitzer* and HST.

Considering the difficulty of the measurement, we believe the amplitude between NICMOS and *Spitzer* is quite uncertain. We also note that narrow fields such as GOODS and NICMOS UDF are affected by cosmic variance, especially at $z > 6$, where the correlation length of halos responsible for reionization is greater than the field size. These caveats do not rule out the possibility that the *Spitzer* fluctuations, at least at some fractional level, reveal a first-light component. The initial results are clearly exciting, and have important implications on galaxy formation, but IR background fluctuations must be further studied at shorter wavelengths, where the signatures from reionization are expected to dominate.

2. CIBER-I

The CIBER instrument currently consists of two wide-field imagers to measure EBL fluctuations, a Low-Resolution Spectrometer (LRS) to probe the EBL via absolute photometry, and a Narrow-Band Spectrometer (NBS), to measure the brightness of the Zodiacal light by the Ca-II 854.2 Fraunhofer line (Bock *et al.* 2006). Two imaging cameras operate at 0.9 and 1.6 μm and probe EBL fluctuations over a wide 4 sqr. degree field of view, allowing measurements over the distinctive peak in the power spectrum at $\ell \sim 2000$ (0.1 degrees).

With the NBS, CIBER will also test the large EBL intensity ($\sim$50 nW/m^2 sr) reported by DIRBE and IRTS. The measurement is not limited by the sensitivity of the instruments, but primarily by the Zodiacal foreground and to a lesser extent by the astrophysical systematic errors in removing stars and scattered starlight, expected to be $\sim$ 2% of the Zodiacal brightness.

As the residual EBL spectrum closely resembles that of Zodiacal light (Dwek *et al.* 2005), there is the possibility of a large $\sim$25% error in the Zodi model used by both DIRBE and IRTS teams. The NBS has sufficient sensitivity and systematic error control to precisely measure the Zodiacal amplitude at 854.2 nm. This measurement can be extrapolated to the DIRBE 1.2 and 2.2 μm bands based on the Zodiacal spectrum measured by the LRS. We also propose several additional flights to observe multiple lines of sight through the Zodiacal cloud, measured between a span of 6 months, to vary the solar elongation angle. The observation windows are carefully chosen so these fields are within the DIRBE observations at the same time of year. If the DIRBE models are incorrect at the 20% level, this should become readily apparent, and the multiple observations are necessary to help us understand how the Zodi models should be corrected.

CIBER-I was launched from White Sands Missile Range in New Mexico on a Terrier-Black Brant sounding rocket on February 2009 and in July 2010. The vehicle performed as expected, and the payload achieved an apogee of 320 km in both flights. All flight events went according to plan and we hope to release our first science results within a year (a result based on CIBER-I involving the zodiacal light spectrum and the silicate absorption appears in the literature; Tsumura *et al.* 2010). With the second flight, we now have a first measurement of the EBL out to angular scales grater than 30 arcminutes at 1 and 1.6 μm with CIBER. This fluctuation analysis will test the claims of Kashlinsky *et al.* (2007) at shorter near-IR wavelengths. The results will be published over the coming year. CIBER-I has two more flights scheduled, with the next panned for February 2012.

3. CIBER-II

After the planned fourth flight in 2013, we plan a more capable camera designed to probe fluctuations down to the low level of minimal reionization fluctuations, with a factor of $\sim$ 10 improvement in sensitivity compared to that of CIBER-I (see Fig. 2). The CIBER-II camera consists of a 30 cm telescope operating simultaneously in four bands between 0.5 and 2.1 μm with each imaging the same 2 sqr. degree field of view. The four bands are matched to identify the spectral dependence of the reionization contribution from foreground and Zodiacal fluctuations. As is the case with CIBER-I, CIBER-II will image wide fields with existing deep coverage in optical and near-IR with *Spitzer* and ground-based instruments.

The cameras are designed for high sensitivity to surface brightness in the short amount of time available during a sounding rocket flight. In comparison with space-borne telescopes, CIBER-II measures fluctuations on large angular scales, on both sides of the expected peak in the power spectrum. These are also the angular scales best suited for distinguishing this signal from local galaxies, and systematic effects due to its distinctive power spectrum. By measuring large angular scales, potential problems with source removal are less serious than in a small deep measurement. CIBER-II is designed to have a very large instrumental FOV, the figure of merit relevant for measuring surface brightness, reaching the required raw sensitivity even in a short sounding rocket flight compared with the long integrations possible on a satellite. It is expected that CIBER-II will carry out the definitive study to establish the surface density of sources responsible for reionization and its results will complement number counts at $z > 6$ from JWST, especially at the bright end, to put a complete picture of reionization together. NASA has already approved two flights with CIBER-II in 2013 and 2014.

Acknowledgements

CIBER is funded by NASA APRA NNG05WC18G (at Caltech) and NNX07AG43G (at UCI). AC thanks funding from NSF CAREER AST-0645427, Award 1310310 from Spitzer, and HST-AR-11241/11242 from STScI. AC also thanks the organizers for the invitation to present this work and for the hospitality.

References

Aharonian, F., *et al.*, 2006, *Nature*, 440, 1018
Bock, J., *et al.* 2006, *New Astronomy Reviews*, 50, 215
Bouwens, R., *et al.* 2006, *ApJ*, 653, 53
Cambresy, L., *et al.* 2001, *ApJ*, 555, 563
Chary, R.-R. 2008, *ApJ*, 680, 32
Chary, R.-R., Cooray, A., & Sullivan, I. 2008, *ApJ*, 681, 53
Chary, R.-R. & Cooray, A. 2011, in preparation
Cooray, A. & Yoshida, N. 2004, *MNRAS*, 351, L71
Cooray, A. & Sheth, R. 2002, *Physics Report*, 372, 1 arXiv:astro-ph/0206508
Cooray, A. 2006, *MNRAS*, 365, 842
Cooray, A., Bock, J., Keating, B., Lange, A., & Matsumoto, T. 2004, *ApJ*, 606, 611
Cooray, A., *et al.* 2007, *ApJ*, 659, L91
Cooray, A., *et al.* 2009, arXiv.org:0902.2372
Dwek, E., Arendt, R., & Krennrich, F. 2005, *ApJ*, 635, 784
Fernandez, E. & Komatsu, E. 2006, *ApJ*, 646, 703
Kashlinsky, A., *et al.* 2004, *ApJ*, 608, 1
Kashlinsky, A., *et al.* 2005, *Nature*, 438, 45
Kashlinksy, A., *et al.* 2007, *ApJ*, 654, L5
Levenson, L. & Wright, E., 2008, *ApJ*, 683, 585
Madau, P. & Silk, K. 2005, *MNRAS*, 359, L37
Madau, P. & Pozzetti, L. 2000, *MNRAS*, 312, L9
Matsumoto, T., *et al.* 2005, *ApJ*, 626, 31
Primack, J., Gilmore, R. & Somerville, R. arXiv.org:0811.3230
Santos, M. R., Bromm, V., & Kamionkowski, M. 2002, *MNRAS*, 336, 1082
Schroedter, M. 2005, *ApJ*, 628, 617
Sullivan, I., *et al.* 2007, *ApJ*, 657, 37
Thompson, R. *et al.* 2008, arXiv:0706.0547
Trac, H. & Cen, R. 2007, *ApJ*, 671, 1
Tsumura, K., *et al.* 2010, *ApJ*, 719, 394

Discussion

MADORE: To reach 0.1 % accuracy on the EBL how large an area do you have to survey to beat cosmic variance?

COORAY: ZEBRA will cover 10 fields as a function of galactic latitude with each covering about 10 sq. degrees. 10 fields are adequate to beat the cosmic variance.

The Spectral Energy Distribution of Galaxies
Proceedings IAU Symposium No. 284, 2011
R.J. Tuffs & C.C. Popescu, eds.

doi:10.1017/S1743921312009684

Evolutionary Map of the Universe

Ray P. Norris

CSIRO Astronomy and Space Science, PO Box 76, Epping, NSW, 1710, Australia

email: Ray.Norris@csiro.au

Abstract. EMU is a wide-field radio continuum survey planned for the new Australian Square Kilometre Array Pathfinder (ASKAP) telescope, due to be completed in 2012. The primary goal of EMU is to make a deep ($\sim 10\mu$Jy/bm rms) radio continuum survey of the entire Southern Sky at 1.4 GHz, extending as far North as $+30^\circ$ declination, with a 10 arcsec resolution. EMU is expected to detect and catalog about 70 million galaxies, including typical star-forming galaxies up to $z = 1$, powerful starbursts to even greater redshifts, and AGNs to the edge of the Universe. EMU will undoubtedly discover new classes of object. Here I present the science goals and survey parameters.

Keywords. surveys, radio continuum: galaxies, cosmological parameters

1. Introduction

Until recently, most large radio surveys could only detect radio-loud Active Galactic Nuclei (AGN) and very nearby star-forming galaxies. As a result, radio data were of little interest to those modelling the Spectral Energy Distribution (SED) of populations of galaxies. However, we are now going through a period of massive change in radio-astronomy. New technology being developed for the pathfinder instruments of the Square Kilometre Array (SKA) are enabling radio surveys far deeper than before, so that radio data will start to become increasingly important for studies of multiwavelength SEDs. Here I describe the largest such survey, EMU (Evolutionary Map of the Universe), which will reach a similar sensitivity (~ 10 μJy/beam) to the deepest current small-area surveys, but over the entire visible sky. At that sensitivity, EMU will be able to trace the evolution of galaxies over most of the lifetime of the Universe.

Fig. 1 shows the major 20-cm continuum radio surveys. The largest existing radio survey, shown in the top right, is the wide but shallow NRAO VLA Sky Survey (NVSS) (Condon *et al.* 1998). The most sensitive existing radio survey is the deep but narrow Lockman Hole observation (Owen & Morison 2008) in the lower left. All current surveys are bounded by a diagonal line that roughly marks the limit of available telescope time of current-generation radio telescopes. The region to the left of this line is currently unexplored, and this area of observational phase space presumably contains as many potential new discoveries as the region to the right.

2. ASKAP: the Australian SKA Pathfinder

The Australian SKA Pathfinder (ASKAP) is a new radio telescope being built both to test and develop aspects of potential SKA science and technology, and to demonstrate the capabilities of the Australian SKA candidate site. However, ASKAP is a major telescope in its own right, likely to generate significant new astronomical discoveries. ASKAP (Johnston *et al.* 2007, 2008; Deboer *et al.* 2009) will consist of 36 12-metre

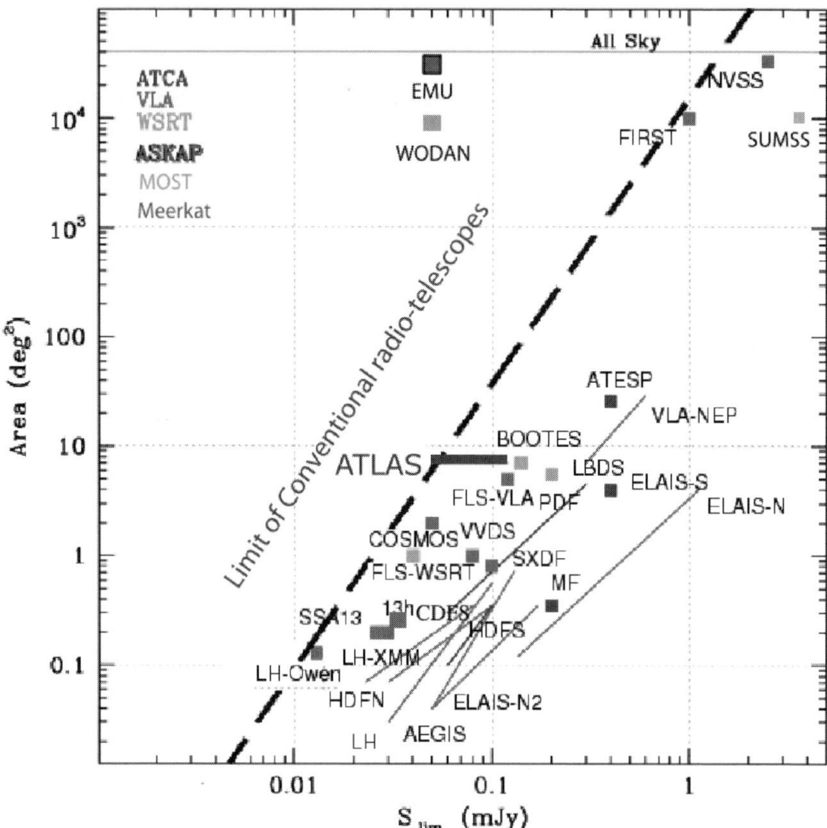

Figure 1. Comparison of EMU with existing deep 20 cm radio surveys. Horizontal axis is 5-σ sensitivity, and vertical axis shows the sky coverage. The diagonal dashed line shows the approximate envelope of existing surveys, which is largely determined by the availability of telescope time. The squares in the top-left represent the EMU survey, discussed in this paper, and the complementary WODAN (Röttgering *et al.* 2010b) survey which has been proposed to cover the sky North of $+30°$.

antennas spread over a region 6 km in diameter. Although the array of antennas is no larger than many existing radio telescopes, the feed array at the focus of each antenna is revolutionary, with a phased-array feed (PAF) of 96 dual-polarisation pixels, designed to work in a frequency band of 700–1800 MHz, with an instantaneous bandwidth of 300 MHz. This will replace the single-pixel feeds that are almost universal in current-generation synthesis radio telescopes. As a result, ASKAP will have a field of view up to 30 deg^2, enabling it to survey the sky thirty times faster than existing synthesis arrays, and allowing surveys of a scope that cannot be contemplated with current-generation telescopes. To ensure good calibration, the antennas are a novel 3-axis design, with the feed and reflector rotating to mimic the effect of an equatorial mount, ensuring a constant position angle of the PAF and sidelobes on the sky.

As the result of a competitive selection process, ASKAP design is being driven by ten survey science projects, with priority being given to two key projects: WALLABY (an all-sky HI survey) and EMU (an all-sky radio continuum survey).

3. EMU: Evolutionary Map of the Universe

The primary goal of EMU (Norris *et al.* 2011a) is to make a deep (10 μJy/beam rms) radio continuum survey of the entire Southern Sky, extending as far North as +30°. EMU will cover roughly the same fraction (75%) of the sky as the benchmark NVSS survey (Condon *et al.* 1998), but will be 45 times more sensitive, and will have an angular resolution (10 arcsec) 4.5 times better. Because of the excellent short-spacing *uv* coverage of ASKAP, EMU will also have higher sensitivity to extended structures. Like most radio surveys, EMU will adopt a 5-σ cutoff, leading to a source detection threshold of 50 μJy/beam. EMU is expected to generate a catalogue of about 70 million galaxies, and all radio data from the EMU survey will be placed in the public domain as soon as the data quality has been assured.

EMU differs from many previous surveys in that the survey includes cross-identification with major surveys at other wavelengths, and will produce public-domain VO-accessible catalogues as "value-added" data products. This is facilitated by the growth in the number of large southern hemisphere telescopes and associated planned major surveys spanning all wavelengths. In addition, EMU is collaborating closely with complementary radio surveys such as WODAN (Röttgering *et al.* 2010b), LOFAR (Röttgering *et al.* 2010a), and Meerkat-MIGHTEE. To help plan EMU, a pilot survey, ATLAS, is being conducted over 7 sq. deg. of the CDFS-SWIRE and ELAIS-SWIRE fields (Norris *et al.* 2006; Middelberg *et al.* 2008).

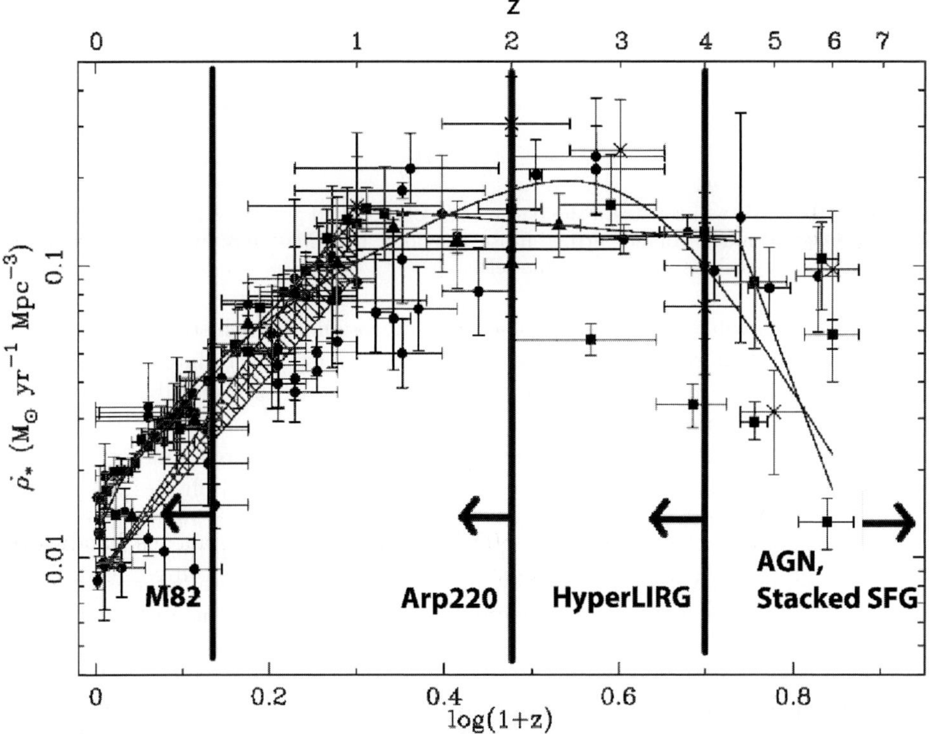

Figure 2. The cosmic star formation rate as a function of cosmic time, adapted from Hopkins & Beacom (2006), showing the EMU 5σ detection limits for an M82, Arp220, and a HyperLIRG. AGNs and stacked star-forming galaxies can be detected to arbitrarily high redshifts. The plot is very uncertain beyond z=2, and may be level to a higher redshift, because of uncertain extinction corrections. EMU should be able to distinguish betwen alternative models at high redshift.

4. Science

Broadly, the key science goals for EMU, detailed in Norris *et al.* (2011a), are:

• To trace the evolution of star-forming galaxies from $z = 2$ to the present day, and stack to even higher redshift. Radio data are unaffected by dust, enabling us to use radio flux as a sensitive and accurate measure of star formation rate (Garn *et al.* 2009).

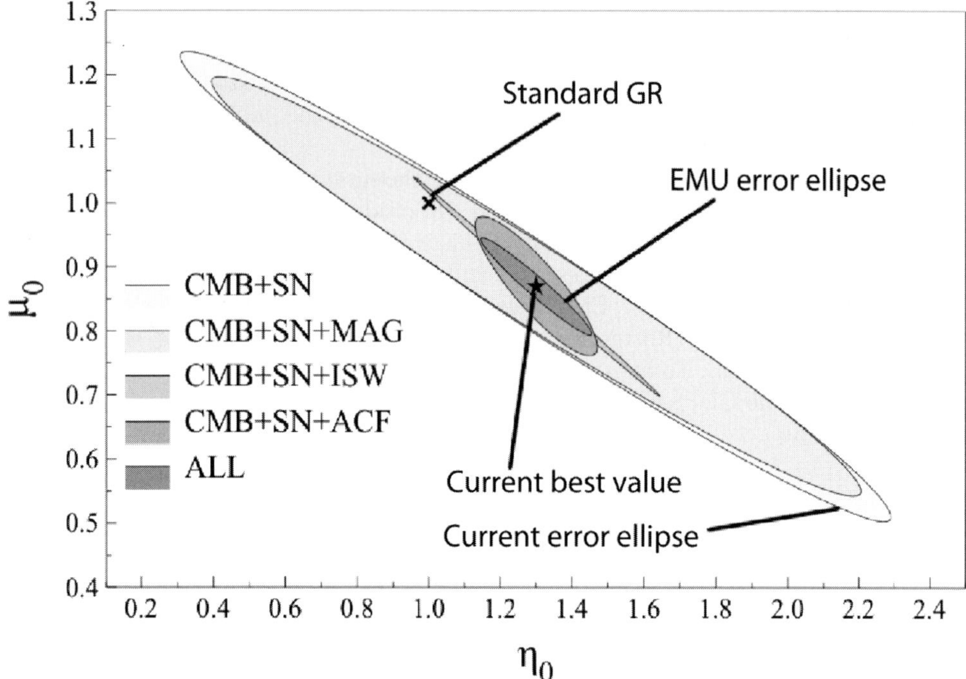

Figure 3. How EMU will constrain the parameters of modified gravity, adapted from Raccanelli *et al.* (2011). The much smaller error ellipse provided by EMU may exclude alternatives to General Relativity such as braneworld models.

• To trace the evolution of massive black holes throughout the history of the Universe, and understand their relationship to star formation. Particularly interesting are those AGN buried beneath many magnitudes of extinction, which are invisible at optical/IR wavelengths (e.g. Norris *et al.* 2011b,c).

• To use the distribution of radio sources to explore the large-scale structure and cosmological parameters of the Universe, and to test fundamental physics. For example, by using brute-force statistics to measure the effect of weak gravitational lensing and the Integrated Sachs-Wolfe effect, Raccanelli *et al.* (2011) have shown that EMU will make the best measurement yet of dark energy and modified gravity parameters (see Fig. 3).

• To determine how radio sources populate dark matter halos, as a step towards understanding the underlying astrophysics of clusters and halos. For example, we expect to detect thousands, and possibly hundreds of thousands, of clusters by using wide-angle-tail galaxies as cluster probes (e.g. Mao *et al.* 2010).

• To create the most sensitive wide-field atlas of Galactic continuum emission yet made in the Southern Hemisphere, addressing areas such as star formation, supernovae, and Galactic structure. For example, we expect to detect thousands of new radio stars, and have started the SCORPIO pilot survey (Umana *et al.*, in preparation) of 4 sq. deg. of the Galactic Plane, to guide our survey strategy.

• To explore an uncharted region of observational parameter space, with a high likelihood of finding new classes of object. In particular, we will systematically mine the database of EMU data to search for sources whose properties lie outside the parameters of known objects or phenomena.

References

Condon, J. J., *et al.* 1998, *AJ*, 115, 1693
Deboer, D. R., *et al.* 2009, *IEEE Proceedings*, 97, 1507
Garn, T., Green, D. A., Riley, J. M., & Alexander, P. 2009, *MNRAS*, 397, 1101
Hopkins, A. M. & Beacom, J. F. 2006, *ApJ*, 651, 142
Johnston, S., *et al.* 2007, *PASA*, 24, 174
Johnston, S., *et al.* 2008, *Experimental Astronomy*, 22, 151
Mao, M., *et al.* 2010, *MNRAS*, 406, 2578
Middelberg, E., *et al.* 2008, *AJ*, 135, 1276
Norris, R. P., *et al.* 2006, *AJ*, 132, 2409
Norris, R. P., *et al.* 2011a, *PASA*, 28, 215
Norris, R. P., *et al.* 2011b, *ApJ*, 736, 55
Norris, R. P., Lenc, E., Roy, A. L., & Spoon, H. 2011c, submitted to *MNRAS* (arXiv:1107.3895)
Owen, F. N. & Morison, G. E. 2008, *AJ*, 136, 1889
Raccanelli, A., *et al.* 2011, submitted to *MNRAS* (arXiv:1108.0930)
Röttgering, H., *et al.* 2010a, PoS (ISKAF2010)050
Röttgering, H., *et al.* 2010b, http://www.astron.nl/radio-observatory/apertif-eoi-abstracts-and-contact-information

Discussion

CLEMENTS: The lurking AGN in ULIRGs were very interesting, but won't they ruin your radio-FIR photo-Zs?

NORRIS: Yes it's very important that we are able to remove them and it is a challenge to do so. Fortunately we are getting good at developing techniques for achieving this. For example, we will have radio polarisation data for all our sources, and virtually every polarised source is an AGN. In addition we will use fields like ATCA-ATLAS and COSMOS to estimate any remaining contamination from AGN, which we can then use to make any necessary correction for them.

The Spectral Energy Distribution of Galaxies
Proceedings IAU Symposium No. 284, 2011
R.J. Tuffs & C.C. Popescu, eds.

doi:10.1017/S1743921312009696

Extragalactic science with ALMA

George J. Bendo[1] and the UK ALMA Regional Centre Node

[1]UK ALMA Regional Centre Node, Jodrell Bank Centre for Astrophysics
School of Physics and Astronomy, University of Manchester
Oxford Road, Manchester M13 9PL, United Kingdom
email: george.bendo@manchester.ac.uk

Abstract. The Atacama Large Millimeter/submillimeter Array (ALMA) is a telescope comprising 66 antennas that is located in the Atacama Desert in Chile, one of the driest locations on Earth. When the telescope is fully operational, it will perform observations over ten receiver bands at wavelengths from 9.5-0.32 mm (31-950 GHz) with unprecedented sensitivities to continuum emission from cold (<20 K) dust, Bremsstrahlung, and synchrotron emission as well as submillimetre and millimetre molecular lines. With baselines out to 16km and dynamic reconfiguration, ALMA will achieve spatial resolutions ranging from 3″ to 0.010″, allowing for detailed imaging of continuum or molecular line emission from 0.1-1 kpc scale gas and dust discs in high-redshift sources or 10-100 pc scale molecular clouds and substructures within nearby galaxies. Science observations started on 30 September 2011 with 16 antennas and four receiver bands on baselines up to 400 m. The telescope's capabilities will steadily improve until full operations begin in 2013.

Keywords. telescopes, submillimeter, instrumentation: interferometers

1. Introduction

Science observations by the Atacama Large Millimetre/submillimeter Array (ALMA) started on 30 September 2011. At the time of this writing, ALMA still has limited science capabilities. The telescope has only 16 antennas operational and bands 3,6,7, and 9 available (Mathys 2011). Nonetheless, ALMA is still able to produce both spectral line and continuum maps with sensitivities that are either comparable to or better than other millimetre and submillimetre telescopes currently in operation. When fully operational, ALMA will be the most sensitive millimetre and submillimetre telescope for both continuum and spectral line observations. Specifications of the bands available for use on ALMA are given in Table 1.

The first call for proposals was for Cycle 0, which runs from 30 September 2011 to 30 June 2012. A total of 919 proposals were submitted, with 112 projects being listed as the highest priority observations. The earliest extragalactic targets will include observations of nearby active galactic nuclei including Centaurus A, NGC 1068, and NGC 1097; nearby starburst galaxies such as M83 and NGC 253; infrared luminous galaxies such as Arp 220, NGC 1614, and NGC 6240; gravitational lenses; and high-redshift submillimetre galaxies. In nearby galaxies, the observations will be used to map dust and gas on 10-100 pc scales so as to examine the dynamics and excitation of gas within the centres of the galaxies. For high redshift sources, the observations will be used to identify the optical counterparts of sources and to resolve submillimetre and millimetre gas and dust emission on kpc scales. A complete list of accepted observing proposals is given at http://www.eso.org/public/archives/releases/sciencepapers/eso1137/eso1137.pdf .

Table 1. ALMA band specifications

Band	Frequency (GHz)	Wavelength (μm)	Primary Beam ($''$)	Largest Angular Scales ($''$)	Angular Resolution ($''$)	
					Compact	Extended
1	31.3-45	6.7-9.5	145-135	93	13-9	0.14-0.1
2	67-90	3.3-4.5	91-68	53	6-4.5	0.07-0.05
3[2]	84-116	2.6-3.6	72-52	37	4.9-3.6	0.05-0.038
4	125-163	1.8-2.4	49-37	32	3.3-2.5	0.035-0.027
5	163-211	1.4-1.8	37-29	23		
6[2]	211-275	1.1-1.4	29-22	18	2.0-1.5	0.021-0.016
7[2]	275-373	0.80-1.1	22-16	12	1.5-1.1	0.016-0.012
8	385-500	0.60-0.78	16-12	9	1.07-0.82	0.011-0.009
9[2]	602-720	0.42-0.50	10-8.5	6	0.68-0.57	0.007-0.006
10	787-950	0.32-0.38	7.7-6.4	3	0.52-0.43	0.006-0.005

Notes:
[1] These data are for when ALMA is fully operational. The data are from *Observing with ALMA: A Primer for Early Science* (Schieven 2010).
[2] These are bands available in Cycle 0 onwards. Other bands will be developed and made available later.

2. ALMA support

ALMA is a partnership between North America, Europe, and East Asia in cooperation with the Republic of Chile. Each region has its own ALMA Regional Centre (ARC), with telescope activity coordinated by the Joint ALMA Observatory (http://www.almaobservatory.org/). Additional details on the organisational structure are given by the *ALMA Cycle 0 Proposer's Guide* (Mathys 2011) and *Observing with ALMA: A Primer for Early Science* (Schieven 2010). The ARCs provide a wide range of support services for users. First of all, the ARCs provide user support with proposal preparation and submission, including tutorials on using the ALMA Observing Tool, which is used to prepare and submit proposals. For proposals that are accepted for observation, the ARCs will create and evaluate the scheduling blocks, which are used as instructions on performing the observations. Following the completion of observations, the ARCs will provide assistance with data reduction. The ARCs not only provide periodic workshops on using CASA (the data reduction software for ALMA) but also provide one-to-one user support for data processing. The ARCs also work on general software development and engage in public outreach work. The three ARCs are:

• The European Southern Observatory (http://almascience.eso.org/) for Europe. (The European ARC also has nodes in the Czech Republic, France, Germany, Italy, the Netherlands, Sweden, and the United Kingdom that provide local support for ALMA users.)
• The National Radio Astronomy Observatory (https://science.nrao.edu/facilities/alma) for North America.
• The National Astronomical Observatory of Japan (http://almascience.nao.ac.jp) for East Asia.

References

Mathys, G. 2011, *ALMA Cycle 0 Proposer's Guide* (Garching: ESO), Version 1.1

Schieven, G., 2011, *Observing with ALMA: A Primer for Early Science*, (Victoria: NRC Herzberg Institute of Astrophysics), ALMA Doc. 0.1, ver. 2.2

The Spectral Energy Distribution of Galaxies
Proceedings IAU Symposium No. 284, 2011
R.J. Tuffs & C.C. Popescu, eds.

doi:10.1017/S1743921312009702

Looking at the distant universe with the MeerKAT Array (LADUMA)

B. W. Holwerda[1], S.–L. Blyth[2], A. J. Baker[3] and the LADUMA team

[1]European Space Agency (ESTEC), Keplerlaan 1, 2200 AV Noordwijk, The Netherlands
[2]ACGC, Department of Astronomy, University of Cape Town, South Africa
[3]Rutgers University, USA

email: benne.holwerda@esa.int

Abstract. The MeerKAT (64 x 13.5m dish radio interferometer) is South Africa's precursor instrument for the Square Kilometre Array (SKA), exploring dish design, instrumentation, and the characteristics of a Karoo desert site and is projected to be on sky in 2016. One of two top-priority, Key Projects is a single deep field, integrating for 5000 hours total with the aim to detect neutral atomic hydrogen through its 21 cm line emission out to redshift unity and beyond. This first truly deep HI survey will help constrain fueling models for galaxy assembly and evolution. It will measure the evolution of the cosmic neutral gas density and its distribution over galaxies over cosmic time, explore evolution of the gas in galaxies, measure the Tully-Fisher relation, measure OH maser counts, and address many more topics. Here we present the observing strategy and envisaged science case for this unique deep field, which encompasses the Chandra Deep Field-South (and the footprints of GOODS, GEMS and several other surveys) to produce a singular legacy multi-wavelength data-set.

Keywords. surveys, galaxies: evolution galaxies: high-redshift galaxies: ISM galaxies: luminosity function, mass function galaxies: statistics cosmology: observations radio lines: galaxies

---

1. Introduction

Over the past 8 Gyr, galaxies have evolved dramatically, with the number density of luminous, blue, star-forming galaxies dropping by an order of magnitude (e.g., Bell *et al.* 2005) and the number of luminous red galaxies increasing by a similar amount (Driver *et al.* 1998; Bell *et al.* 2005). Similarly since $z \sim 1$, the star formation rate density in the Universe has plummeted by an order of magnitude (e.g., Madau *et al.* 1998; Hopkins *et al.* 2008), yet the density of neutral hydrogen (HI), the fuel reservoir for star-formation, may have remained constant (Fig. 1). Given that, locally, the overall HI mass and star-formation rate appear to be linked for galaxies (Kennicutt 1998), this points to significant redistribution of the gas reservoir over this epoch.

To complement our view of high redshift galaxy populations (the stellar, ionised, and molecular gas), a key science goal of the Square Kilometre Array (SKA Carilli & Rawlings 2004) is to explore the atomic hydrogen (HI) out to $z \sim 1$ and to explore how galaxies form and evolve. While SKA will study HI in large numbers of galaxies at $z \sim 1$, MeerKAT (Booth *et al.* 2009; de Blok *et al.* 2009), with its high sensitivity and wide instantaneous bandwidth, is the choice pathfinder instrument with which to begin the study of HI in galaxies out to redshift unity. We therefore are undertaking the LADUMA† (Looking At the Distant Universe with the MeerKAT Array) survey for a 5000 hour observation of a single field (Holwerda & Blyth 2010a,b; Holwerda *et al.* 2011a). This survey in combination with existing and planned complementary multi-wavelength

† Literally, It thunders. in the Zulu language: http://www.ast.uct.ac.za/laduma/Home.html

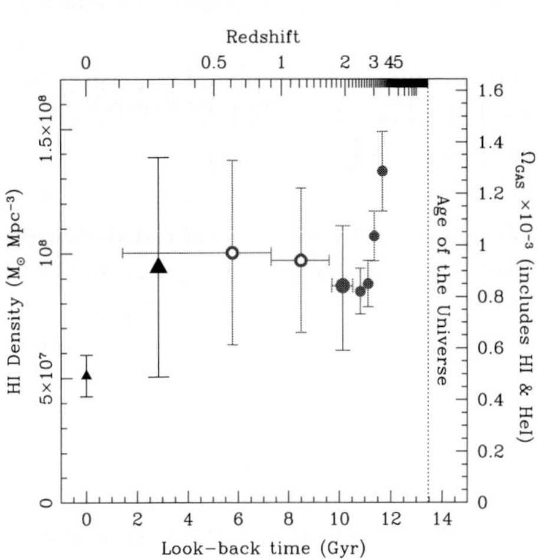

Figure 1. The cosmic HI density from a variety of methods; direct HI line detections (black tringles Zwaan *et al.* 2005; Lah *et al.* 2007), and Ly-α absorption estimates (open and filled circles Rao *et al.* 2006; Prochaska *et al.* 2005; Noterdaeme *et al.* 2009, respectively).

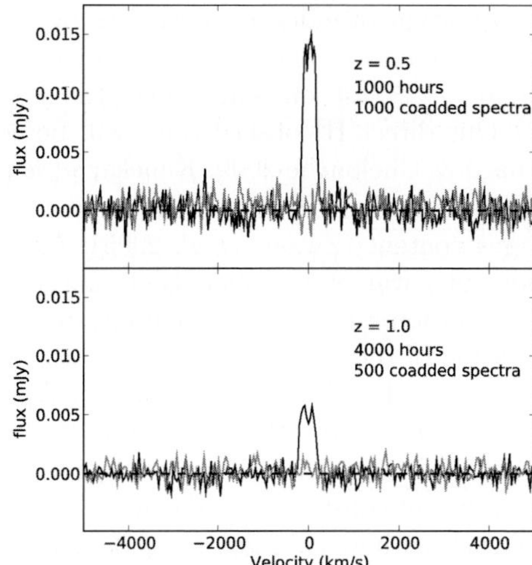

Figure 2. Simulated stacking results (black lines) for z = 0.5 (top panel) and z = 1.0 (bottom panel), (Δz = 0.1). Galaxies were simulated according to the Oxford S^3 database (Obreschkow *et al.* 2009).

observations will provide the first complete picture of galaxy evolution from $0 < z \lesssim 1.4$ and serve as a key benchmark for future studies with the SKA.

2. Observing Strategy

Target Field The Extended Chandra Deep Field-South was chosen for its wealth of multi-wavelength data and existing spectroscopy (e.g., Cardamone *et al.* 2010; Balestra *et al.* 2010) as well as continuous visibility from the MeerKAT site.
Data The final 5000 hour integration results in a trumpet-shaped data-cube†, as the primary beam widens with redshift, covering an ever larger volume with redshift.
Direct Detections & Line Stacking Given the shape of the survey volume and the gradually decreasing sensitivity with distance, our science will be done with a mix of direct detections (for low redshifts and/or high HI masses) and the stacking of multiple objects with low signal-to-noise (see also Meyer *et al. this volume*). Fig. 2 illustrates the line stacking technique. Using the known positions and redshifts of a sample of sources, their radio spectra are aligned to a common (rest) frequency frame and combined. One can determine the mean HI mass and possibly the HI profile of a sub-class of galaxies. This technique will require an order of magnitude more spectroscopic redshifts than currently available (Cardamone *et al.* 2010; Balestra *et al.* 2010).

3. LADUMA Science

The LADUMA data will address a multitude of science questions;
Cosmic Hydrogen Density (Ω_{HI}) Summing over all objects, one obtains an estimate of the *total* atomic hydrogen volume density of the Universe at a given epoch. Given that the

† The vuvuzela is a uniquely South African trumpet primarily used for encouragement during soccer matches, as most of the world knows after the 2010 World Cup.

star-formation rate density decreases by a factor ten (Hopkins *et al.* 2008, *this volume)* and specific star-formation rate by a factor three (Noeske *et al.* 2007), over the redshift range $0 < z \lesssim 1.4$, one can expect the star-formation fuel to be depleted, but at what rate? Our direct HI observations will be compared to those Lyα and Mg II absorbers (Ménard & Chelouche 2009; Kanekar *et al.* 2009).

HI Mass Function (HIMF) is the number density of galaxies as a function of their neutral gas content (Zwaan *et al.* 2005). An accurate measurement of this quantity is an important parameter in models of galaxy evolution where processes like gas inflow and star formation are sought to be understood. Evolution in the breakpoint, slope, and normalization of the HIMF can be used to test hierarchical galaxy formation and cold flow models (van der Heyden *et al.* 2009). And finally, we can also explore environmental effects on the HIMF (Bouchard *et al.* 2009).

Galaxy Evolution The HI content of different populations as a function of redshift will place a useful constraint on our understanding of their evolution. We will address questions such as how cold gas mass depends on halo mass? what is the relation between stellar and gas mass over time(e.g., Kannappan 2004), and what is the relation between the cold gas mass and specific star-formation rate?

The Tully-Fisher Relation; Rotationally supported disks lie on the Tully-Fisher relation (Tully & Fisher 1977). For small high-redshift samples, the stellar T-F relation has been determined (e.g. Kassin *et al.* 2007; Puech *et al.* 2010), but the LADUMA data will provide HI data for the first time at higher redshift with the advantage of probing the whole disk (and consequently halo), and with sufficient statistics to determine the normalization and slope of the T-F relation at different redshifts.

Bonus Science includes the number densities and strengths of OH masers over cosmic time (see Briggs 1998), the merger rate from close companions (e.g., Holwerda *et al.* 2011b), and the possible detection of faint 21 cm absorption (e.g., the "Cosmic Web").

Concluding Remarks: The LADUMA survey promises to facilitate a range of exciting science and provide the deepest HI observations available until the SKA is commissioned.

References

Balestra, I. *et al.* 2010, *A&A*, 512, A12+

Bell, E. F. *et al.* 2005, *ApJ*, 625, 23

Booth, R. S. *et al.* 2009, ArXiv e-prints/0910.2935

Bouchard, A. *et al.* 2009, in Panoramic Radio Astronomy: Wide-field 1-2 GHz Research on Galaxy Evolution

Briggs, F. H. 1998, *A&A*, 336, 815

Cardamone, C. N. *et al.* 2010, *ApJS*, 189, 270

Carilli, C. L. & Rawlings, S. 2004, *New A Rev.*, 48, 979

de Blok, W. J. G. *et al.* 2009, in Panoramic Radio Astronomy: Wide-field 1-2 GHz Research on Galaxy Evolution

Driver, S. P. *et al.* 1998, *ApJ*, 496, L93+

Holwerda, B. & Blyth, S. 2010a, in ISKAF2010 Science Meeting

Holwerda, B. W. & Blyth, S. 2010b, ArXiv e-prints/1007.4101

Holwerda, B.W. *et al.* 2011a, in *Bulletin of the AAS*, Vol. 43, Abstract #217, #433.17

Holwerda, B. W. *et al.* 2011b, ArXiv e-prints

Hopkins, P. F. *et al.* 2008, *ApJ*, 679, 156

Kanekar, N. *et al.* 2009, *MNRAS*, 396, 385

Kannappan, S. J. 2004, *ApJ*, 611, L89

Kassin, S. A. *et al.* 2007, *ApJ*, 660, L35

Kennicutt, Jr., R. C. 1998, *ApJ*, 498, 541

Lah, P. *et al.* 2007, *MNRAS*, 376, 1357
Madau, P. *et al.* 1998, *ApJ*, 498, 106
Ménard, B. & Chelouche, D. 2009, *MNRAS*, 393, 808
Noeske, K. G. *et al.* 2007, *ApJ*, 660, L47
Noterdaeme, P. *et al.* 2009, *A&A*, 505, 1087
Obreschkow, D. *et al.* 2009, *ApJ*, 703, 1890
Prochaska, J. X. *et al.* 2005, *ApJ*, 635, 123
Puech, M. *et al.* 2010, *A&A*, 510, A68+
Rao, S. M. *et al.* 2006, *ApJ*, 636, 610
Tully, R. B. & Fisher, J. R. 1977, *A&A*, 54, 661
van der Heyden, K. *et al.* 2009, in Panoramic Radio Astronomy: Wide-field 1-2 GHz Research on Galaxy Evolution
Zwaan, M. A. *et al.* 2005, *MNRAS*, 359, L30

The Spectral Energy Distribution of Galaxies
Proceedings IAU Symposium No. 284, 2012
R.J. Tuffs & C.C. Popescu, eds.

doi:10.1017/S1743921312009714

Conference Summary

Jay Gallagher[1], **Carol Lonsdale**[2], **and Gustavo Bruzual**[3,4]

[1]Univ. of Wisconsin, Dept. of Astronomy,
5534 Sterling 475 N. Charter St., Madison, WI 53706-1582, USA
email:jsg@astro.wisc.edu

[2]NRAO, 520 Edgemont Road, Charlottesville, VA, 22903, USA
email:clonsdal@nrao.edu

[3]Centro de Radioastronomía y Astrofísica, UNAM, Campus Morelia
Apartado Postal 3-72, Morelia, Michoacán, 58090 México.
email:g.bruzual@crya.unam.mx

[4]On sabbatical leave from Centro de Investigaciones de Astronomía (CIDA)
Apdo. Postal 264, Mérida 5101, Venezuela

1. ISM & the Emergent Light from Galaxies

If galaxies consisted only of stars, and some early-type systems in general and dwarf spheroidal galaxies in particular fit this prescription, then the calculation of the SED in principle is straightforward. The emergent luminosity at any wavelength simply is the sum over all the luminosities of all the stars in the system. This can be calculated, of course provided that one has a complete understanding of stellar populations, which remains a non-trivial issue. Most galaxies, however, also contain an interstellar medium (ISM). The ISM absorbs, scatters and reprocesses the radiation and relativistic particles from sources within galaxies, primarily stars and AGN. That the ISM is neither isotropic nor homogeneous adds to the challenge of how to properly account for its influence on the luminosity emerging from galaxies.

Light interactions within the ISM naturally vary with wavelength. For example in the Lyman continuum, absorption via bound-free transitions of H and He are extremely important and give rise to strong emission lines in the ultraviolet and especially the optical to near-infrared regions. The presence of especially H-recombination lines offers a key window into the numbers of short-lived O-stars and thus galactic star formation rates. They also affect colours of galaxies and therefore need to be incorporated in stellar population models. While we see that this is done in modern models, the issue remains as to what is the best prescription for the inclusion of emission line features and how this should vary with host galaxy properties.

Interstellar dust is a critical factor in spectra from the mid-UV to near-IR, where it absorbs and scatters light, with absorbed energy being re-radiated into the thermal infrared-millimeter regime. The most basic approach is to deal with the balance between absorbed and reradiated luminosity components with a correction for scattering. Dust scattering is problematic because it involves a phase function that can perturb the angular distribution of the radiated light. For example, forward scattering by dust can enhance the intensity of light scattering out in the polar direction from a disk starburst with a bipolar wind, as in an M82-type of situation. Polarization thus remains an important diagnostic for assessing the importance of scattered light in galaxies.

The effects of dust naturally are nonlinear, $I = I_0 e^{-\tau}$. As each sightline through a galaxy has its unique τ_i, even the case of a galaxy where the ISM is assumed to have uniform properties is complicated. The choice of the mean obscuration to each point on the projected image of a galaxy, $\overline{\tau(x, y)}$, is not readily obtained from observations. The exceptions are cases where discreet objects can be studied, e.g., stars or star clusters, in which dust obscuration maps in principle can be obtained. Comparisons of these types of maps with results based on the more traditional integrated light methods is offering important tests of our current models.

Of course in real galaxies dust properties, e.g. column densities and compositions, also can substantially vary with location. This behaviour occurs in the Milky Way, as well as being seen between different types of galaxies. We therefore have seen the considerable effort devoted to empirically assessing variations in dust properties through measurements of the influence of dust on emergent spectra, as, for example, in the "Calzetti (*et al.*) law" approach to estimating ultraviolet dust obscuration.

While we have seen progress in this area, dealing with the varying properties of dust and thus its range of interactions with light within galaxies continues to be a key issue. As galaxies evolve their chemical compositions, radiation fields, gas density distributions, magnetic fields structures and cosmic ray properties all change. These factors in turn are expected to have some influence over the composition, sizes and optical properties of dust grains. One basic question concerns the degree to which the dusty ISM of galaxies experience convergent evolution. Does the dust in giant elliptical galaxies have properties that are close to that in highly evolved spirals such as the Milky Way?

From the perspective of this meeting, we see that the effects of the ISM are no longer the wild card of unknowns that they once were thought to present. Bulk properties of nearby galaxies are seen to behave in reasonably regular ways that indicate dust properties are only mildly variable. The situation becomes a bit less clear as one goes to extreme objects. In these cases the combination of access to the absorbed and especially re-emitted light from galaxies offered by the *Spitzer* and *Herschel* space observatories, as well as ground-based millimeter and submillimeter telescopes along with improved simulations, offer a pathway to further improvements.

2. The Milky Way

Many stores in cities in the Anglophonic world display a "shop locally" sign, indicating their view of the advantages of keeping resources and often obtaining goods and services from one's own region. A similar claim can be made from the discussion of the Milky Way system in this meeting. Our home Galaxy and its companions offer tremendous advantages for studies of the SEDs of other galaxies:

- *Location:* The Sun is in the disk of the Milky Way which offers us a uniquely deep and high resolution view of a presumably fairly typical Galactic disk, as well as high quality access to its other stellar structures: bulge, halo, and nucleus.
- *Satellites:* We are even more fortunate that the Milky Way offers such a range of satellite galaxies. In particular, systems like the Magellanic Clouds are only rarely found in close proximity to giant spirals, and these allow us to study the individual stars and details of ISM structures in low metallicity, low mass star-forming galaxies.
- *6-D:* The Milky Way is the *only* galaxy where the three-dimensional spatial array of individual stars along with their space motions can be measured with any precision. Less accurate but still reasonable distances also can be obtained for interstellar clouds, thereby allowing us to build models of the true distribution of stars and gas in physical and velocity spaces.

2.1. *Stellar Populations*

When dealing with stellar populations we normally invoke what might be thought of as a kind of soft generalized cosmological principle: stellar populations throughout the universe are similar to those found in the Milky Way and its satellites. This applies particularly to low mass stars, which can only be detected in and around the Galaxy, as well as other key features of the stellar population such as binary and multiple stars along with individual intrinsic properties, such as ages, chemical abundances, and rotation. Nearby stars then offer the best tests for our models of stellar structure and evolution that form the foundations of stellar population studies in external galaxies. Our home system also is the place where the idealized concept of the existence of a stellar initial mass function (IMF) can best be directly tested by counting stellar numbers as functions of mass in a range of systems.

As we have heard, the universal IMF assumption can then be tested through studies of other, more distant systems, but without the local standard the IMF situation would be even more difficult than it already is. The good and perhaps somewhat surprising result is that thus far variations in the IMF in the key $\sim$0.5-1$M_\odot$ range, where much of the stellar mass resides, appear to be relatively small over a wide range of chemical abundance and age. Whether this pleasant congruence in the lower IMF extends to galaxies with very different evolutionary histories, such as giant ellipticals, remains uncertain.

It is less clear how routinely the low mass and upper IMFs connect in the manner seen near the Sun, and whether the slope and cutoff of the upper IMF depend on the scale of each star forming event. Uncertainties associated with the concept of a nearly universal IMF thus remain, and require quantitative measurements along with investments in theory to understand if and when significant IMF variations occur. For now, however, the standard assumption that the IMF determined in the Galaxy is a valid approximation appears to hold at a minimum in systems that are similar to our own, and does not seem to show signs of possible problems until galaxies with extremes (high or low) in star forming properties are encountered.

2.2. *The ISM*

Our location within the Milky Way's disk offers a close-up view of the structure of the ISM which reveals its high level of complexity. The densest regions, of course, are molecular and sites of star formation while the diffuse ISM actually contains a spectrum of gas densities and temperatures. This mix is pervaded by magnetic fields and cosmic rays giving rise to the synchrotron component of the Milky Way's SED. The current generation of ISM models feature the role of turbulence in transporting energy between scales and in helping to create density variations that then evolve under the effects of cooling or heating, gravity, magnetic pressures, etc. As we have seen at this meeting, the Milky Way's key position as the natural laboratory for research into the micro-physics of the ISM remains secure.

This is also an area where we are aided by an observational information explosion. The ISM of the Milky Way has been observed from the γ-rays to radio. In each spectral region we are gaining the resolution to map structures that lead to models of interactions between sources and their surroundings that also modify the emergent SEDs. An issue in this case is how best to combine the growing understanding of small scale processes into models for the interpretation of integrated properties of external galaxies?

One example of a step in this direction is offered by studies of spiral structure within the Milky Way, where we see the details of cloud formation and evolution into star forming sites, with measurements of external galaxies that offer overviews of spirals arm structures and kinematics. This investment thus leads to an improved ability to model the SED from the spiral arm components of galaxies, which in grand-design spirals are major sites for the formation of massive stars. We then saw how this approach can be expanded to a more general exploration of the role of filamentary gas structures in star formation and thus also in predicting the SED from young regions in a variety of extragalactic systems.

An even more local example of foundations supplied by the Milky Way comes from cosmic ray studies. As we have learned that cosmic ray energy densities are likely to be significant in a range of galaxies, the only direct measurements of the cosmic ray energy spectrum comes from observations made in and around the solar system. These data also provide empirical limits on the critical ratio of electron to proton cosmic ray numbers as well as indications of the chemical composition of the cosmic ray nuclei.

2.3. *The Nucleus*

Like other galaxies, the Milky Way contains a dense stellar nucleus within which lurks our supermassive black hole (SMBH). Depending on one's perspective, the low level of activity from our SMBH can be seen as unfortunate. But in any case, given the distance of $\sim$8 kpc to the Galactic center, we can observe our nucleus with superb precision and resolution across the electromagnetic spectrum. Many key phenomena are found to take place on small spatial scales that are impossible to directly observe in external galaxies.

How does this information then connect to the more general problem of SEDs from extragalactic systems? The most direct approach could be to consider how the various components contribute to the radiated nuclear power of the Milky Way, which then offers a template for interpreting the SEDs of other normal galaxies. The knowledge of the masses of potentially star-forming molecular clouds, their locations, and physical states gives us some idea of the range of possible states for our nucleus without and even with a significant power contribution from our SMBH.

Dealing with the nucleus is perhaps where the disparity between Galactic and extragalactic spatial scales is most prominently seen. While scales of $\sim$500-1000 au can be accessed in several spectral regions in the Milky Way's nucleus, our limit at M31 naturally is 100 times worse, implying a loss of a factor of $\geqslant 10^4$ in the information contained in a spatial map due to both resolution and dynamic range limits in dealing with a more distant object. And the situation is even worse for the nearest AGN, e.g. in NGC 5128 at 3.7 Mpc. Bridging this gap, and the associated chasm between power production in quiescent and highly active nuclei, is essential for improving our ability to assess the impact of nuclear activity and associated starbursts on the evolution of galaxies from measurements of their multi-wavelength SEDs.

This meeting provides a concrete demonstration of how progress in understanding the SEDs is derived from a combination of improved modelling techniques, better understanding and representations of the small scale physics, and a vast array of multi-wavelength observations. We have been presented with a grand opportunity to combine multi-wavelength measurements of galaxies with varying degrees of spatial resolution, with results from simulations to derive intrinsic properties of their stars and AGN. This process is also one that does not simply fit in to one of the classic subfields of astronomy. Instead it requires a melding of multiple fields, e.g., from stellar physics to properties of interstellar dust and from stellar population to radiative transfer models as well as observations across the electromagnetic spectrum. This meeting stands as an example of the benefits that can be derived from discussions amongst experts across many fields who have related interests in common.

While the proceedings of this IAU Symposium demonstrate the excellent progress that has been achieved in the past several years, it also underscores the degree to which we are dealing with work in progress. We then can take away a few overarching points from this meeting regarding the future of inquires into the origins and evolution of the spectra of galaxies and their subsystems:

• We are passing an initial era of qualitative comparisons and moving into a time when quantitative efforts are increasingly central to further progress. On the theoretical side advances are coming from the combination of more sophisticated stellar models–interiors and atmospheres–along with improved simulations of radiative transfer through disk galaxies and around AGN. Observationally the range of multi-wavelength data on consistent flux scales, higher spectral and spatial resolution, and better laboratory data combine to support empirical improvements. To this end it might be useful to consider building a library of standardized SEDs for the best observed galaxies as a challenge for modelers.

• We have seen that conditions in the Milky Way are frequently the foundations on which models are build for interpreting other systems. It is immediately clear that this approach cannot be universal, e.g. it does not offer much insight into the SEDs of blazers. A better understanding of the circumstances under which our existing models begin to falter–e.g. how much AGN contribution, or how much of a starburst, or how low of a metallicity–could be useful in helping to focus future efforts.

• As always in astronomy, we need to be alert to identify and deal with the unexpected. In an era of surveys there can be a tendency to put aside outliers and to focus on means trends in samples. Yet a full model should be able to deal with both aspects of any sample. We need to bear in mind that some of the astrophysical "signal" from key physical processes can show up in the form of the cosmic "noise" contributed by objects with unique or unexpected properties.

3. The Infrared Universe

Our understanding of the infrared-millimeter Universe has changed dramatically in recent years thanks to ISO, Spitzer, AKARI, BLAST, many ground-based submillimeter/millime-ter observatories, and most recently Herschel and PLANCK and the future is bright and exciting as the WISE data are released and ALMA comes on line. We heard of several interesting developments from Herschel observations, and we were treated to one of the first results from ALMA data.

Herschel has opened up the 200-500μm range to major investigation, and a result that is emerging from several studies is the widespread existence of excesses at long wavelengths, compared to greybodies with dust emissivity β=2 commonly fit to far-infrared SEDs, that may be attributed to large masses of cold dust in large disk systems. L. Dunne presented a summary of dust temperatures from redshifts up to 4 from H-ATLAS, BLAST and SCUBA, which illustrated nicely the selection bias towards warm dust emission of earlier $\lambda < 200\mu$ missions, notably IRAS, and the colder temperatures we can now access with the later missions. The newer SED templates of the H-ATLAS sample, presented in D. Smith's poster, are considerably colder than current libraries. L. Dunne finds strong evolution of the dust mass function out to z$=$0.5. M. Rowan-Robinson has pointed out the importance of massive cool dust disks for some time for SCUBA and Spitzer-selected samples, and presented fits to Herschel HerMES moderate redshift galaxies which require cold dust (10-13K) that must be extended on scales of tens of kpc if optically thin.

A significant cold dust component is evident in well studied nearby galaxies. In a Herschel imaging study of M31 B. Groves demonstrated strong dust temperature gradients across the disk, with temperatures dropping below 15K at a radius of 15kpc. E. Mentuch finds the dust-to-star mass profile ratio increases with radius in M51 and that dust temperature correlates with Hα intensity, with highest temperatures in the spiral arms. G. Bendo presented Spitzer and Herschel colour maps of M81, M83 and NGC 2403 and by comparing with Hα and 1.6μm concludes that the 250-500μm emission originates primarily in a diffuse medium heated by evolved stars. A poster by M. Lam comes to a similar conclusion based on a worsening correlation between progressively longer wavelength Herschel-ATLAS bands and Hα for a sample of SDSS galaxies. C. Popescu described her most recent modelling of such a diffuse cool/cold disk. A similar long wavelength excess is also seen in the Sombrero, described by I. DeLooze, however she finds that cold clumps with no embedded sources provide more of the excess emission than a diffuse medium. In a poster B. Holwerda presented first results of a Herschel study of nearby edge-on disks which has the potential to determine the vertical extent of dust disks, finding that NGC 891 has a dust disk of similar size as the stellar disk plus an extended diffuse dusty component associated with the HI disk. Some ellipticals also show cold dust, as illustrated in an H-ATLAS sample with mean dust temperature 20.4K and dust mass 3.7 x 10^8 $M_\odot$ in a poster by N. Agius.

Low metallicity dwarfs have also been found to have submillimeter excesses indicating the possible existence of a cold dust component, as described by U. Lisenfeld and S. Madden. A poster by A. Remy illustrated the Herschel colours of the Herschel Dwarf Galaxies Survey, showing excess at 350μm for some systems; in some cases an excess is seen only beyond 500μm. In low metallicity systems there is the issue of how large amounts of cool dust can be effectively shielded from the strong UV radiation fields. In a study of the LMC presented in a poster, F. Galliano concludes that the 500μm excess is not likely to originate with very cold dust and may indicate grains with larger submillimeter opacity than Galactic grains. All these studies serve to emphasize that far-infrared and submillimeter emission is often not a very good tracer of the star formation rate.

An intriguing discovery of an HI disk 130 kpc in diameter in the zone-of-avoidance was presented by R. Kraan-Kortweg. This system, unlike other known giant HI galaxies, is actively forming stars at a LIRG-like rate of 35 $M_\odot$/yr. Since it's optically undetectable behind $A_v \sim$7.5mag, it was imaged with Spitzer revealing a 50kpc star forming disk rich in PAHs. It appears to be a disk starburst building around an old bulge, perhaps a local analog of z$\sim$1 systems with large SFRs due to large gas reservoirs in large disks.

4. Turbulence and Shocks

The longstanding question of how star formation is triggered and controlled was addressed with new data that probes turbulent processes and shocks. B. Elmegreen gave an overview of a wide range of star formation triggers – gravitational collapse in spiral-compressed gas; large rings with triggered or lingering star formation along the edges; filament collapse – based on some of the fantastically detailed images, including the Antennae, M51 and Galactic regions, available from HST and other instruments. He concludes that the overall star formation rate is governed by the total mass of cold gas and the details are dominated by complex shock dynamics. F. Boulanger presented one of the first ALMA results, which illustrates shock effects on a small scale: a compact clump in the ALMA CO(3-2) Science Verification map of the Antennae which coincides with a compact H_2 source that is a massive turbulent cloud at the interface of two complexes separated by 150 km/s. Boulanger proposed that turbulence is regulating the efficiency of star formation, quenching it when turbulence governs the gas dynamics, and that star formation proceeds when turbulent energy dissipates, based on H_2 tracing of shocked and turbulent ISM in radio galaxies, cooling flows, and other systems including Stefan's Quintet (SQ). A detailed study of SQ, which displays a massive inter-galactic shock in

H_2 containing 4 x 10^9 $M_\odot$ of cold gas with suppressed star formation, was presented by G. Natale, and the poster describing AKARI observations of SQ by T. Suzuki presented the first detection of shock-excited C[II] 158μm line emission.

5. Is There a Universal Star Formation Law?

Several talks in this meeting addressed the problem of the determination of the Star Formation Law. We should ask ourselves if there is any evidence for a universal Star Formation Law, i.e. a single slope, a single proportionality constant between SFR and mass density. Evidence does not seem very convincing, or this law depends strongly on environment. A few numbers taken from Y. Gao's talk illustrate the situation: According to Schmidt (1959), SFR $\propto \rho(HI)^n$, with $n = 1-3$, mostly $2-3$ in the ISM of our Galaxy. Kennicutt (1989) concludes that for disks the average SFR $\propto \rho(HI + H_2)^n$, but n is not well constrained, $1-3$, with a wide spread. Kennicutt (1998) compares the star formation law for total $(HI + H_2)$ gas with that derived in dense gas regions. The results are not conclusive, in dense gas regions SFR $\propto M(H_2)$ with a non-unique slope: $n = 1, 1.4, 1.7$. This key issue should be solved in the near future. One possible conclusion is that there is not a universal star formation law.

6. SPS Models and Spectral Fits

In this meeting we have seen new developments and many applications of stellar population synthesis models. Including stochastic effects in the number of stars drawn from an otherwise continuous and analytical IMF, as well as following the evolution of multiple systems (binary stars), and using evolutionary tracks computed for rotating stars, C. Leitherer has obtained a significant decrease in the M/L ratio of the model galaxies, arguing that this requires a revision of the IMF and star formation rates derived from models which ignore these effects. D. Schaerer has shown us convincingly that the effects of nebular emission should be included in the spectral models of distant star-forming galaxies if we want to determine reliable physical parameters for these galaxies. P. Prugniel has presented an interesting approach to include variable amounts of the $[\alpha/\mathrm{Fe}]$ ratio in observed stellar spectra using theoretical model atmospheres to compute the differential effect in selected spectral features. He has been successful in fitting full spectra to galaxies and star clusters and measuring $[\alpha/\mathrm{Fe}]$ in these systems. Several excellent talks and posters considered deriving the physical properties of galaxies by fitting optical SEDs, making some assumptions about absorption and extinction of stellar light by dust grains, and computing the energy radiated in the IR. E. Da Cunha showed how these models have been used to derive star formation histories, stellar mass, and dust content of large sample of galaxies. V. Acquaviva reminded us how the MCMC technique can be used in SED fitting to derive the best estimates for various physical parameters together with their error.

Our understanding of the relevance of TP-AGB stars in the NIR spectrum of stellar populations is not yet final. S. Meidt and S. Zibetti, as well as some already published papers, report that they do not see evidence for an increased contribution of TP-AGB stars in the NIR (w.r.t. BC03). We expect that this long lasting problem will be solved hopefully soon. Tracks should predict the right number of stars in the TP-AGB phase, and the NIR stellar SEDs used to describe these stars, including the reddening by the dusty envelope, should be as close as possible to real spectra. Physical properties like the mass of galaxies, derived from NIR-observed fluxes, depend critically on an adequate treatment of this stellar phase in population synthesis models.

It is important to understand what the results of these fits are telling us. Are these models just purely mathematical exercises, or do they resemble the physics that takes place in real galaxies? Most of the parameters that go into this modelling are quite uncertain themselves: IMF, M_{UP}, M_{LOW}, stellar metallicity distribution, number of ionizing photons, the reddening law, the amount of dust, the kind of dust, etc. What happens if we vary these quantities? How robust are the derived physical properties of the galaxies to these changes? Can we constrain these quantities from the ratio of IR to UV-optical flux or are we getting out what we put in? Tools are available to do a more detailed study that treats properly radiative transfer of radiation in a 2D or 3D stellar/dust distribution (e.g. Popescu *et al.*).

Progress in modelling stellar populations has been slow but continuous. Key issues remaining in stellar population synthesis models are being solved, as new ingredients (empirical and theoretical) become available. The advice is to use with caution the results from SPS models and fitting algorithms, and to add error bars to the values of the physical parameters that you derive.

7. Acknowledgements

G. Bruzual acknowledges support by the National Autonomous University of México (UNAM), through grant IA102311.

Author Index